Y0-AGT-570

The Enzymes

VOLUME VI

CARBOXYLATION AND DECARBOXYLATION (NONOXIDATIVE)

ISOMERIZATION

Third Edition

CONTRIBUTORS

ELIJAH ADAMS
ALFRED W. ALBERTS
H. A. BARKER
ANN M. BENSON
ELIZABETH A. BOEKER
BOB B. BUCHANAN
IRWIN FRIDOVICH
LUIS GLASER
JORAM HELLER
P. W. HOLLOWAY
HAROLD M. KOLENBRANDER
M. DANIEL LANE
ERNST A. NOLTMANN
E. J. PECK, JR.
W. J. RAY, JR.
MICHAEL C. SCRUTTON
STANLEY SELTZER
MARVIN I. SIEGEL
ESMOND E. SNELL
THRESSA C. STADTMAN
PAUL TALALAY
MERTON F. UTTER
P. ROY VAGELOS
MARCIA WISHNICK
HARLAND G. WOOD
MURRAY R. YOUNG

ADVISORY BOARD

THE ENZYMES

Edited by PAUL D. BOYER

Molecular Biology Institute and
Department of Chemistry
University of California
Los Angeles, California

Volume VI

CARBOXYLATION AND DECARBOXYLATION (NONOXIDATIVE)

ISOMERIZATION

THIRD EDITION

ACADEMIC PRESS New York and London 1972

ACADEMIC PRESS, INC.
111 Fifth Avenue, New York, New York 10003

United Kingdom Edition published by
ACADEMIC PRESS, INC. (LONDON) LTD.
24/28 Oval Road, London NW1 7DD

LIBRARY OF CONGRESS CATALOG CARD NUMBER: 75-117107

PRINTED IN THE UNITED STATES OF AMERICA

Contents

List of Contributors

Numbers in parentheses indicate the pages on which the authors' contributions begin.

ELIJAH ADAMS (479), Department of Biological Chemistry, University of Maryland School of Medicine, Baltimore, Maryland

ALFRED W. ALBERTS (37), Department of Biological Chemistry, Washington University School of Medicine, St. Louis, Missouri

H. A. BARKER (509), Biochemistry Department, University of California, Berkeley, California

ANN M. BENSON (591), Department of Pharmacology and Experimental Therapeutics, The Johns Hopkins University School of Medicine, Baltimore, Maryland

ELIZABETH A. BOEKER (217), Laboratory of General and Comparative Biochemistry, National Institute of Mental Health, National Institutes of Health, Bethesda, Maryland

BOB B. BUCHANAN (193), Department of Cell Physiology, University of California, Berkeley, California

IRWIN FRIDOVICH (255), Department of Biochemistry, Duke University Medical Center, Durham, North Carolina

LUIS GLASER (355), Department of Biological Chemistry, Washington University School of Medicine, St. Louis, Missouri

JORAM HELLER (573), Jules Stein Eye Institute, UCLA School of Medicine, Los Angeles, California

P. W. HOLLOWAY (565), Department of Biochemistry, University of Virginia School of Medicine, Charlottesville, Virginia

HAROLD M. KOLENBRANDER (117), Department of Chemistry, Central College, Pella, Iowa

M. DANIEL LANE (169), Department of Physiological Chemistry, The Johns Hopkins School of Medicine, Baltimore, Maryland

ERNST A. NOLTMANN (271), Department of Biochemistry, University of California, Riverside, California

E. J. PECK, JR. (407), Department of Biological Sciences, Purdue University, Lafayette, Indiana

W. J. RAY, JR. (407), Department of Biological Sciences, Purdue University, Lafayette, Indiana

MICHAEL C. SCRUTTON (1), Department of Biochemistry, Temple University School of Medicine, Philadelphia, Pennsylvania

STANLEY SELTZER (381), Department of Chemistry, Brookhaven National Laboratory, Upton, New York

MARVIN I. SIEGEL (169), Department of Physiological Chemistry, The Johns Hopkins School of Medicine, Baltimore, Maryland

ESMOND E. SNELL (217), Department of Biochemistry, University of California, Berkeley, California

THRESSA C. STADTMAN (539), Laboratory of Biochemistry, National Heart and Lung Institute, National Institutes of Health, Bethesda, Maryland

PAUL TALALAY (591), Department of Pharmacology and Experimental Therapeutics, The Johns Hopkins University School of Medicine, Baltimore, Maryland

MERTON F. UTTER (117), Department of Biochemistry, Case Western Reserve University School of Medicine, Cleveland, Ohio

P. ROY VAGELOS (37), Department of Biological Chemistry, Washington University School of Medicine, St. Louis, Missouri

MARCIA WISHNICK* (169), Department of Biochemistry, The New York University School of Medicine, New York, New York

HARLAND G. WOOD (83), Department of Biochemistry, Case Western Reserve University School of Medicine, Cleveland, Ohio

MURRAY R. YOUNG (1), Department of Biochemistry, Northwestern University School of Medicine, Chicago, Illinois

* Present address: Department of Biochemistry, Public Health Research Institute of the City of New York, New York, New York.

Preface

With the publication of this volume, covering the nonoxidative carboxylases and decarboxylases, and the isomerases, the Third Edition of "The Enzymes" is well past its midpoint. Volume VII, now in preparation, will cover elimination, aldol cleavage and condensation, other C—C cleavage, and phosphorolysis. Subsequent volumes will cover group transfer, ATP-linked syntheses and related processes, and oxidation–reduction.

From any viewpoint, the scope of present information about enzymes is impressive. But I am much more impressed with the breadth and quality of our knowledge at this stage than I was when the edition was launched. In one sense, as I write the Preface for each volume, I am somewhat in the position of a loving father of a large family: he cannot tell which of his children he loves best, but readily praises the qualities of any child being discussed. Thus I feel that the present volume is a worthy companion of other volumes in the treatise.

Much of this volume concerns enzymic reactions of carbon dioxide. Far from being a metabolic end product, carbon dioxide is a vital, versatile metabolite. Its enzymology is of central importance to such processes as photosynthesis and fatty acid synthesis. Similarly, the isomerases are also to be admired as they execute change with great facility. Students of these fields will find rich reading in the contributions contained in this volume.

Advisory Board members who contributed to the planning of the present volume are Robert H. Abeles, B. L. Horecker, Edwin G. Krebs, and P. Roy Vagelos. We are indebted to them for much of the quality of the contents of the volume. Also, the technical competence and fine cooperation of the staff of Academic Press and of my own office assistants merit continued recognition.

PAUL D. BOYER

Contents of Other Volumes

Volume I: Structure and Control

Volume II: Kinetics and Mechanism

Volume III: Hydrolysis: Peptide Bonds

Volume IV: Hydrolysis: Other C–N Bonds, Phosphate Esters

Volume V: Hydrolysis (Sulfate Esters, Carboxyl Esters, Glycosides), Hydration

Volume VII: Elimination and Addition, Aldol Cleavage and Condensation, Other C–C Cleavage, Phosphorolysis, Hydrolysis (Fats, Glycosides)

1

Pyruvate Carboxylase

MICHAEL C. SCRUTTON • MURRAY R. YOUNG

I. Introduction

In 1960 Utter and Keech (*1*) reported the presence in chicken liver mitochondria of an enzyme which catalyzed the direct carboxylation of pyruvate to yield oxalacetate. This enzyme, which catalyzes reaction (1) (*2*) was given the trivial name, pyruvate carboxylase, and is desig-

1. M. F. Utter and D. B. Keech, *JBC* **235,** PC17 (1960).

$$\text{Pyruvate} + \text{ATP} + \text{HCO}_3^- \xrightleftharpoons{\text{Me}^{2+},\ \text{acetyl-CoA}} \text{oxalacetate} + \text{ADP} + \text{P}_i \quad (1)$$

nated as pyruvate: CO_2 ligase (ADP); EC 6.4.1.1. Subsequent studies have indicated that this enzyme has a wide tissue and species distribution. Pyruvate carboxylase activity has been detected in most mammalian tissues examined although high maximal catalytic capacities for this enzyme (expressed on the basis of milligram protein) are confined to liver, kidney cortex, and adipose tissue (*2–4*). The enzyme is also present in avian and amphibian liver (*2–5*), and in yeasts and other fungi (*6–9*) and bacteria (*10–12*). All pyruvate carboxylases detected thus far in vertebrate tissues require the presence of an acyl-CoA, e.g., acetyl-CoA, in order to catalyze a significant rate of pyruvate carboxylation (*2–5, 13*) as shown in reaction (1). However, microbial pyruvate carboxylases exhibit a wide diversity in their requirement for activation by this cofactor (cf. Section IV,A), although in other respects the reaction catalyzed is identical to reaction (1). The wide tissue and species distribution observed for pyruvate carboxylase probably reflects the multiple roles proposed for this enzyme in metabolism. These roles include involvement in (a) gluconeogenesis from three-carbon precursors, e.g., lactate and alanine (*1, 2*); (b) the anaplerotic synthesis of oxalacetate (*14*); and (c) glyceroneogenesis and lipogenesis (*15*).

II. Properties and Mechanism

A. General Properties and Minimal Mechanism

Studies on pyruvate carboxylases purified from both vertebrate and microbial sources indicate that certain general properties related to the

2. M. F. Utter and D. B. Keech, *JBC* **238,** 2603 (1963).
3. F. J. Ballard, R. W. Hanson, and L. Reshef, *BJ* **119,** 735 (1970).
4. J. W. Anderson, *BBA* **208,** 165 (1970).
5. M. C. Scrutton, *Metab. Clin. Exp.* **20,** 168 (1971).
6. M. Losada, J. L. Canovas, and M. Ruiz-Amil, *Biochem. Z.* **340,** 60 (1964).
7. S. J. Bloom and M. J. Johnson, *JBC* **237,** 2718 (1962).
8. H. J. Stan and J. Schormuller, *BBRC* **32,** 289 (1968).
9. S. A. Overman and A. H. Romano, *BBRC* **37,** 457 (1969).
10. W. Seubert and U. Remberger, *Biochem. Z.* **334,** 401 (1961).
11. E. S. Bridgeland and K. M. Jones, *BJ* **104,** 9P (1967).
12. T. K. Sundaram, J. J. Cazzulo, and H. L. Kornberg, *BBA* **192,** 355 (1969).
13. H. V. Henning and W. Seubert, *Biochem. Z.* **340,** 160 (1964).
14. H. L. Kornberg, *Essays Biochem.* **2,** 1 (1966).
15. R. W. Hanson and F. J. Ballard, *BJ* **105,** 529 (1967).

catalytic mechanism of this enzyme may be characteristic of all pyruvate carboxylases regardless of the species from which the enzyme is obtained.

1. *The Presence of Bound Biotin*

All pyruvate carboxylases examined thus far are rapidly inactivated by incubation in the presence of avidin, a protein which has a high affinity for biotin (*16*). This inactivation is irreversible under most conditions (but cf. *17*) and is prevented by preincubation of the avidin preparation with excess biotin. Identification of pyruvate carboxylases as biotin-proteins is confirmed by detection of stoichiometric concentrations of this coenzyme in purified preparations of the enzymes from avian liver (*18, 19*), mammalian liver (*19, 20*), *Saccharomyces cerevisiae* (*21, 22*), and *Pseudomonas citronellolis* (*23*). The biotin-protein stoichiometry approaches 4 moles/mole for all these pyruvate carboxylases except the enzyme from *P. citronellolis* which contains 2 moles/mole (*18–23*). In addition, incorporation of radioactivity into purified preparations of the pyruvate carboxylases from *S. cerevisiae* and *P. citronellolis* occurs at a constant ratio with enzymic activity when these microorganisms are grown in media containing ^{3}H-biotin (*23, 24*). *In vitro* activation of the apopyruvate carboxylase from *Bacillus stearothermophilus* requires the presence of biotin in addition to ATP and Me^{2+} (*25, 26*). *In vivo* administration of biotin to chickens raised under conditions of biotin deficiency results in a rapid activation of pyruvate carboxylase in the livers of these birds, which is not prevented by concurrent administration of inhibitors of *de novo* protein synthesis (*26a*).

2. *Partial Reactions and the Role of Bound Biotin*

Inactivation of pyruvate carboxylases by incubation with avidin indicates that the biotinyl residues of these enzymes have an important

16. N. M. Green, *BJ* **89,** 599 (1963).
17. M. C. Scrutton and A. S. Mildvan, *Biochemistry* **7,** 1490 (1968).
18. M. C. Scrutton and M. F. Utter, *JBC* **240,** 1 (1965).
19. J. C. Wallace and M. C. Scrutton, unpublished observations (1969).
20. W. R. McClure, H. A. Lardy, and H. P. Kneifel, *JBC* **246,** 3569 (1971).
21. M. R. Young, B. Tolbert, R. C. Valentine, J. C. Wallace, and M. F. Utter, *Federation Proc.* **27,** 522 (1968).
22. M. C. Scrutton, M. R. Young, and M. F. Utter, *JBC* **245,** 6220 (1970).
23. B. L. Taylor and M. F. Utter, personal communication (1971).
24. M. R. Young, B. Tolbert, and M. F. Utter, "Methods in Enzymology," Vol. 13, p. 250, 1969.
25. J. J. Cazzulo, T. K. Sundaram, and H. L. Kornberg, *Nature* **223,** 1137 (1970).
26. J. J. Cazzulo, T. K. Sundaram, and H. L. Kornberg, *Nature* **227,** 1103 (1970).
26a. A. D. Deodhar and S. P. Mistry, *BBRC* **34,** 755 (1969).

role in the catalytic mechanism. This role has been defined by examination of the properties of the exchanges of $^{32}P_i$ with ATP and of ^{14}C-pyruvate with oxalacetate which are catalyzed by all pyruvate carboxylases examined thus far. In all instances demonstration of exchange of $^{32}P_i$ with ATP requires the addition of ADP, HCO_3^-, and Me^{2+} (*10, 27–29*) while exchange of ^{14}C-pyruvate with oxalacetate occurs in the absence of other reaction components (*10, 27–30*). A general minimal mechanism for pyruvate carboxylases may be formulated on the basis of these data [reactions (2) and (3)] (*27*) which is analogous to that proposed for other biotin carboxylases (*31*):

$$\text{E-biotin} + \text{ATP} + HCO_3^- \underset{}{\overset{Me^{2+}}{\rightleftharpoons}} \text{E-biotin} \sim CO_2 + \text{ADP} + P_i \tag{2}$$

$$\text{E-biotin} \sim CO_2 + \text{pyruvate} \rightleftharpoons \text{E-biotin} + \text{oxalacetate} \tag{3}$$

The exchange of $^{32}P_i$ with ATP is activated by addition of an acyl-CoA, e.g., acetyl-CoA, for the pyruvate carboxylases from mammalian and avian liver (*27, 29*) and from *S. cerevisiae* (*28*). Although addition of acetyl-CoA also stimulates the rate of exchange of ^{14}C-pyruvate with oxalacetate catalyzed by pyruvate carboxylase from rat liver (*29*), the presence of this cofactor has no effect on the rate of pyruvate–oxalacetate exchange by the other pyruvate carboxylases examined (*27, 28, 30*). Hence reaction (2) provides the primary locus for activation of pyruvate carboxylases by acetyl-CoA in accord with the suggestion that the rate of pyruvate carboxylation is limited by a step in the formation of E-biotin $\sim CO_2$ from ATP + HCO_3^- (*27*).

In the case of pyruvate carboxylase from avian liver, further support for the proposed minimal mechanism [reactions (2) and (3)] is provided by isolation of the E-biotin $\sim CO_2$ intermediate. Formation of this intermediate from enzyme + $H^{14}CO_3^-$ requires the presence of ATP, Me^{2+}, and acetyl-CoA and is prevented by preincubation of the enzyme with avidin. However, transfer of $^{14}CO_2$ from the isolated intermediate to pyruvate occurs in the absence of other reaction components. The E-biotin $\sim CO_2$ intermediate formed by pyruvate carboxylase is acid labile but is stabilized by treatment with diazomethane (*27*). These studies suggest identification of the intermediate as 1′-*N*-carboxybiotinyl enzyme by analogy with more definitive analyses performed for other

27. M. C. Scrutton, D. B. Keech, and M. F. Utter, *JBC* **240,** 574 (1965).
28. J. J. Cazzulo and A. O. M. Stoppani, *ABB* **121,** 596 (1967).
29. W. R. McClure, H. A. Lardy, and W. W. Cleland, *JBC* **246,** 3584 (1971).
30. J. Gailiusis, R. W. Rinne, and C. R. Benedict, *BBA* **92,** 595 (1964).
31. J. Knappe, *Ann. Rev. Biochem.* **39,** 736 (1970).

biotin carboxylases (*31–33*). However, the alternate postulate (*34*) that *O*-carboxybiotinyl enzyme is the structure of the intermediate which participates in catalysis cannot be excluded for this, or any other, biotin carboxylase.

Initial rate studies of the overall reaction for the pyruvate carboxylases from chicken liver (*35*), rat liver (*36*), and *Aspergillus niger* (*37*) which indicate that these enzymes utilize a Ping-Pong Bi Bi Uni Uni mechanism (*38*) are also consistent with the minimal mechanism proposed in reactions (2) and (3) (cf. Section II,B).

3. *Requirement for* HCO_3^-, $MeATP^{2-}$, *and Free* Me^{2+}

Several further generalizations have emerged from studies of the substrate and activator species which participate in catalysis by various pyruvate carboxylases. First, the enzymes from chicken liver (*39*) and *S. cerevisiae* (*40*) have been shown to utilize HCO_3^- rather than CO_2. Since a similar requirement is observed in studies with other biotin carboxylases (*41*), participation of HCO_3^- may be characteristic of this class of enzymes.

Second, the concentration of Me^{2+} required for maximal activation of all pyruvate carboxylases examined thus far is severalfold greater than the optimal concentration of ATP (*12, 20, 42–44*). This relationship indicates that the presence of both $MeATP^{2-}$ and also Me^{2+} may be required for catalysis. A requirement for $MeATP^{2-}$ has been rigorously established for the pyruvate carboxylases from chicken liver (*45*), rat liver (*20*), sheep kidney cortex (*43, 46*), and *S. cerevisiae* (*44*). ATP^{4-}

32. J. Knappe, B. Wenger, and U. Wiegand, *Biochem. Z.* **337,** 232 (1963).
33. M. D. Lane and F. Lynen, *Proc. Natl. Acad. Sci. U. S.* **49,** 379 (1964).
34. T. C. Bruice and A. F. Hegarty, *Proc. Natl. Acad. Sci. U. S.* **65,** 805 (1970).
35. R. E. Barden, C. H. Fung, M. F. Utter, and M. C. Scrutton, *JBC* **246** (in press) (1972).
36. W. R. McClure, H. A. Lardy, M. Wagner, and W. W. Cleland, *JBC* **246,** 3579 (1971).
37. H. M. Feir and I. Suzuki, *Can. J. Biochem.* **47,** 697 (1969).
38. W. W. Cleland, *BBA* **67,** 104 (1963).
39. T. G. Cooper, T. T. Tchen, H. G. Wood, and C. R. Benedict, *JBC* **243,** 3857 (1968).
40. T. G. Cooper, M. R. Young, and M. F. Utter, unpublished observations (1969).
41. Y. Kaziro, L. F. Hass, P. D. Boyer, and S. Ochoa, *JBC* **237,** 1460 (1962).
42. D. B. Keech and M. F. Utter, *JBC* **238,** 2609 (1963).
43. D. B. Keech and G. J. Barritt, *JBC* **242,** 1983 (1967).
44. J. J. Cazzulo and A. O. M. Stoppani, *BJ* **112,** 747 (1969).
45. M. C. Scrutton, unpublished observations (1967).
46. J. McD. Blair, *FEBS Letters* **2,** 245 (1969).

is found to be a competitive inhibitor of several of these enzymes with respect to $MeATP^{2-}$ (*43, 44, 47*). Although the requirement for free Me^{2+} also appears characteristic of all pyruvate carboxylases, differing specificities for Me^{2+} activation are observed. Thus Mg^{2+}, Mn^{2+}, and Co^{2+} activate the pyruvate carboxylases from chicken (*42*) and rat (*20*) liver, but only Mn^{2+} is effective as an activator for the enzyme from *B. stearothermophilus* (*12*), and only Mn^{2+} and Mg^{2+} for the enzyme from *S. cerevisiae* (*6*). Other divalent metal ions, e.g., Ca^{2+}, Zn^{2+}, Cu^{2+}, and Ni^{2+}, typically act as competitive inhibitors with respect to free Mg^{2+} (*20, 45, 48*).

4. *Effects of Monovalent Cations*

A requirement for activation by a monovalent cation, e.g., K^+, is observed for all pyruvate carboxylases examined thus far including the enzymes from chicken liver (*45*), rat liver (*20*), *S. cerevisiae* (*49*), and *A. niger* (*37*). Studies with the rat liver enzyme implicate the first partial reaction [reaction (2)] as the site of monovalent cation activation since K^+ is required for exchange of $^{32}P_i$ into ATP but has no effect on exchange of ^{14}C-pyruvate into oxalacetate (*29*).

The specificity of Me^+ activation differs markedly for the various pyruvate carboxylases. Thus, K^+, Rb^+, Cs^+, NH_4^+, Tl^+, and $tris^+$ are effective activators of the enzyme from chicken liver; Na^+ is a weak activator, and Li^+ is an inhibitor. Most of the effective activators exhibit activator constants (at pH 7.8) in the range 10–30 m*M* (*45*). Similar effects of many of these cations are also observed for the enzyme from *S. cerevisiae* (*28*). However, for pyruvate carboxylase from rat liver (*20*), K^+, Rb^+, Cs^+, and NH_4^+ are effective activators which exhibit K_A values in the range of 2 m*M*; Na^+ and Li^+ are weak activators; and $tris^+$ is neither an activator nor an inhibitor. The activation of pyruvate carboxylase from chicken liver by $tris^+$ is of special interest since this cation acts as an inhibitor of most enzymes which are activated by K^+ (cf. *50*).

5. *A Generalized Minimal Mechanism for Pyruvate Carboxylase*

The properties which may be typical of all pyruvate carboxylases are summarized below in the form of a generalized minimal mechanism for this enzyme [reactions (4) and (5)]:

47. J. M. Wimhurst and K. L. Manchester, *BJ* **120,** 79 (1970).
48. M. C. Scrutton, M. R. Olmsted, and M. F. Utter, "Methods in Enzymology," Vol. 13, p. 235, 1969.
49. B. Tolbert and M. F. Utter, personal communication (1969).
50. G. F. Betts and H. J. Evans, *BBA* **167,** 193 (1968).

$$\text{E-biotin} + \text{MeATP}^{2-} + \text{HCO}_3^- \underset{}{\overset{\text{Me}^{2+},\ \text{Me}^+}{\rightleftharpoons}} \text{E-biotin} \sim \text{CO}_2 + \text{MeADP}^- + \text{P}_i \quad (4)$$

$$\text{E-biotin} \sim \text{CO}_2 + \text{pyruvate} \rightleftharpoons \text{E-biotin} + \text{oxalacetate} \quad (5)$$

Several additional properties which also appear characteristic of all pyruvate carboxylases, e.g., the presence of bound metal ion, the "two-site" mechanism, are described in subsequent sections (Sections II,B and II,D). This uniformity in catalytic properties contrasts markedly with the wide diversity in regulatory properties which is observed for pyruvate carboxylases purified from different sources (cf. Section IV,A).

B. Product Inhibition Studies—The Two-Site Mechanism

Further insight into the mechanism of catalysis by the pyruvate carboxylases from chicken liver (*35*), rat liver (*36*), and *A. niger* (*37*) has been provided by product inhibition analyses of the overall reaction catalyzed by these enzymes. The most unusual feature of these studies is the observation, in most instances, of competitive relationships when the concentrations of a substrate and a product of the same partial reaction are varied, since for all the simple Ping-Pong Bi Bi Uni Uni mechanisms noncompetitive relationships are predicted when these substrates and products are varied as a pair (*51*). For example, for all these pyruvate carboxylases oxalacetate acts as a competitive inhibitor of HCO_3^- fixation when pyruvate is the varied substrate, and $MgATP^{2-}$ acts as a competitive inhibitor of oxalacetate decarboxylation when either $MgADP^-$ or P_i is the varied substrate (*35–37*). The product inhibition data obtained are described by a two-site Ping-Pong Bi Bi Uni Uni mechanism in which the two partial reactions [reactions (4) and (5)] occur at separate catalytic sites as shown in Fig. 1. Thus, in contrast to the classic assumption in studies on enzyme kinetics that catalytic and active sites are equivalent (*38*), the mechanism proposed for these pyruvate carboxylases requires the assumption that at least two catalytic sites are present in each active site. In the case of chicken liver pyruvate carboxylase the kinetic studies provide no evidence for interaction between the two catalytic sites; hence, spatial separation of these sites on the protein appears likely (*35*). However, limited site–site interaction may occur for the rat liver enzyme since the K_m for HCO_3^- is a function of the pyruvate concentration at higher levels of this substrate (*36*). If the catalytic sites are spatially separated on the protein, the biotin residues may provide the means of communication between

51. W. W. Cleland, *BBA* **67**, 188 (1963).

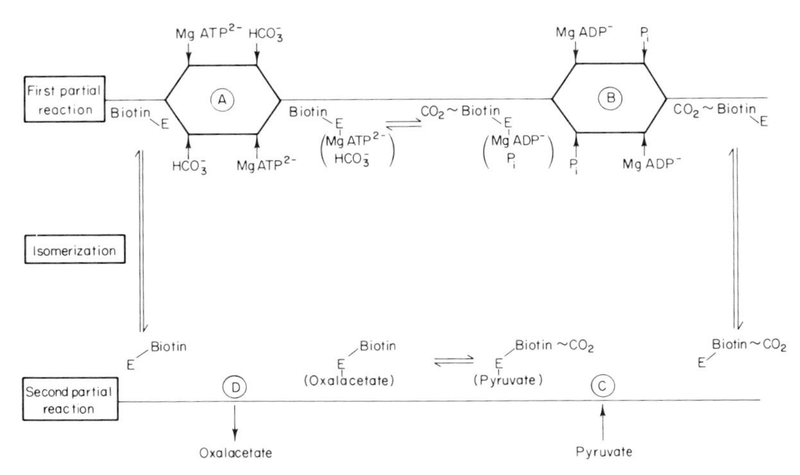

FIG. 1. A kinetic scheme describing the reaction catalyzed by pyruvate carboxylase from chicken liver. This scheme is based on the unpublished data of R. E. Barden, C. H. Fung, M. F. Utter, and M. C. Scrutton and does not include the abortive complexes mentioned in the text.

the sites since these residues are mounted on a 14-Å flexible arm if they are attached to the ϵ-NH_2 of a lysine residue, as is the case in other biotin carboxylases (*32, 33*). Movement of the biotin residue between the catalytic sites may be facilitated by the difference in net charge which characterizes the carboxylated and noncarboxylated states.

The two-site mechanism proposed for pyruvate carboxylase (Fig. 1) is analogous to the mechanism proposed for methylmalonyl-CoA-oxalacetate transcarboxylase by Northrop and Wood (*52, 53*) as the result of a similar kinetic analysis. Furthermore, recent studies on acetyl-CoA carboxylase from *Escherichia coli* (*54–56*) have demonstrated that two separable subunits carry the sites which are responsible for catalysis of partial reactions analogous to those shown in reactions (2) and (3) while a third small subunit, the biotin carrier protein, bears the biotinyl residue of this enzyme. A two-site mechanism analogous to that shown for pyruvate carboxylase in Fig. 1 may therefore be characteristic of all biotin enzymes.

52. D. B. Northrop and H. G. Wood, *JBC* **244,** 5820 (1969).

53. D. B. Northrop, *JBC* **244,** 5808 (1969).

54. A. W. Alberts, A. M. Nervi, and P. R. Vagelos, *Proc. Natl. Acad. Sci. U. S.* **63,** 1319 (1969).

55. P. Dimroth, R. B. Guchhait, E. Stoll, and M. D. Lane, *Proc. Natl. Acad. Sci. U. S.* **67,** 1353 (1970).

56. R. B. Guchhait, J. Moss, W. Sokolski, and M. D. Lane, *Proc. Natl. Acad. Sci. U. S.* **68,** 653 (1971).

C. The First Partial Reaction

If the rapid equilibrium assumption which is required in order to derive the rate equation for the mechanism of Fig. 1 by the procedure of Cha (*35*, *57*) is applied to substrate binding and product release, the kinetic analysis for pyruvate carboxylase from chicken liver provides evidence for a random order mechanism in the first partial reaction [reaction (4)] (Fig. 1). The analysis also indicates that kinetically significant ternary (E-$MgADP^-$—HCO_3^-; E—P_i—HCO_3^-) and quaternary (E—$MgADP^-$—P_i—HCO_3^-) abortive complexes are formed since the product inhibition relationships observed for components of reaction (4) differ in some respects from those predicted for the simplest rapid equilibrium random Bi Bi mechanism (*35*). The ternary and quaternary abortive complexes which are proposed on the basis of the kinetic analysis are those whose formation would be predicted on the basis of structural similarity to the catalytically reactive complexes (Fig. 1).

For pyruvate carboxylase from rat liver formation of E-biotin $\sim CO_2$ from ATP + HCO_3^- [reaction (4)] has been suggested to occur in a concerted process with participation of a histidine residue (*20*) since (a) the requirements for observation of exchange of ^{14}C-ADP with ATP (P_i, HCO_3^-, Me^{2+}, and acetyl-CoA) are complementary with those for observation of exchange of $^{32}P_i$ with ATP (ADP, HCO_3^-, Me^{2+}, and acetyl-CoA (*29*); and (b) the variation of V_{max} with pH indicates that a residue with $pK = 7.0$ participates in the rate-limiting step (*20*). However, for pyruvate carboxylase from chicken liver the properties of the exchange of ^{14}C-ADP with ATP do not resemble those observed for exchange of $^{32}P_i$ with ATP, since (a) addition of only Me^{2+} is required for observation of exchange of ^{14}C-ADP with ATP; and (b) inactivation of the overall reaction caused by incubation at 2°, or in the presence of avidin or sulfhydryl reagents, does not result in concomitant inactivation of this exchange reaction (*58*). Although participation of an E-P_i intermediate would provide the simplest explanation for the differing properties observed for the ATP–P_i and ATP–ADP exchange reactions, this mechanism is excluded by (a) the initial rate and product inhibition analysis of the overall reactions (*35*) (cf. Fig. 1), and (b) the failure to observe exchange of $H^{14}CO_3^-$ with oxalacetate in the absence of added ADP (*58*). The anomalous properties of the ATP–ADP exchange reaction may, therefore, result from formation of an E—P_i intermediate on an abortive pathway which is not kinetically significant in the context of catalysis of pyruvate carboxylation (*58*).

57. S. Cha, *JBC* **243,** 820 (1968).
58. M. C. Scrutton and M. F. Utter, *JBC* **240,** 3714 (1965).

It is therefore apparent that the chemical events which are responsible for formation of the carboxylated enzyme in reaction (4) are still largely obscure.

D. The Second Partial Reaction—The Presence and Role of Bound Metal Ion

1. *The Nature of the Bound Metal Ion in Various Pyruvate Carboxylases*

Since the initial discovery that pyruvate carboxylase from chicken liver is a Mn metalloenzyme (*17, 60*) this metal or Mg, Zn, or Co, has been demonstrated to be a constituent of various pyruvate carboxylases (Table I). All pyruvate carboxylases which have been examined have been found to contain bound metal and in all cases, except the enzyme from rat liver which has not been subjected to a comprehensive survey analysis (*20*), bound metal and biotin are present at approximately equimolar concentration suggesting a catalytic role for the metal ion

TABLE I
Bound Metal Content of Various Pyruvate Carboxylases

Source of enzyme	Bound metal(s) present	Approx. stoichiometry g-atoms/mole biotin	Ref.
Chicken liver	Mn	1	(*17, 59*)
Chicken liver (Mn-deficient chickens)[a]	Mg	1	(*60*)
Turkey liver	Mn	1	(*19*)
Calf liver	Mn	0.5	(*19*)
	Mg	0.5	
Rat liver	Mn	(0.5)[b]	(*20*)
S. cerevisiae	Zn	1	(*22*)
S. cerevisiae from medium containing 2 m*M* Co^{2+}	Zn	0.5	(*22*)
	Co	0.5	

[a] Inclusion of suboptimal concentrations of Mn in the diet results in production of a preparation containing both Mg and Mn. The ratio of Mg:Mn in the purified enzyme is a function of the dietary Mn content in the suboptimal range.

[b] For a sample of 60% maximal specific activity. This enzyme has not been assayed for the presence of other bound metals, e.g., Mg.

59. M. C. Scrutton, M. F. Utter, and A. S. Mildvan, *JBC* **241**, 3480 (1966).
60. P. Griminger and M. C. Scrutton, *Federation Proc.* **29**, 765 (1970).

been suggested that the electrophilic character of the metal ion might promote carboxyl transfer from the E-biotin ~ CO_2 intermediate to pyruvate [reaction (5)] by facilitating the departure of a proton from the methyl group of pyruvate (Fig. 2) (*60, 70*). Operation of a mechanism of this type is supported by observations of (a) a primary isotope rate effect when the extent of CO_2 fixation by pyruvate carboxylase is compared with the extent of release of radioactivity from ^{3}H-pyruvate (*76*), and (b) carboxylation with retention of configuration in studies when pyruvates of differing chiralities are used as substrates (*76*). The data do not, however, discriminate between mechanisms in which either a 1′-*N*-carboxybiotinyl residue (Fig. 2A) (*60, 70*) or an *O*-carboxybiotinyl residue (Fig. 2B) (*34*) serves as the carboxyl donor. These mechanisms differ from previous proposals (*60, 70*) in which the carboxyl group of the carboxybiotinyl residue (or the β-carboxyl group of oxalacetate) was shown as providing a ligand to the bound metal in addition to that provided by the carbonyl group of the keto acid. The data presently available appear more consistent with monodentate coordination of both pyruvate *and* oxalacetate and provide no evidence for interaction of the carboxybiotinyl residue with the bound metal. In addition, it is questionable whether transcarboxylation would be facilitated if the carboxyl group, which is to be transferred, is coordinated to the bound metal.

The mechanisms of Fig. 2 are based on data obtained in studies on pyruvate carboxylase from chicken liver. Although the similarities in the catalytic properties (cf. Sections II,A and II,B) suggest that one of these mechanisms may also describe the role of the bound metal ion (Table I) in catalysis by other pyruvate carboxylases, the only support for this postulate is provided by the general observation of inhibition by oxalate (*19, 36, 62, 77*). Additional evidence supporting the proposed general role for bound metal ion in biotin-dependent transcarboxylation involving α-keto acids (*78*) is, however, provided by data obtained for other enzymes which catalyze a partial reaction analogous to reaction (5), e.g., methylmalonyl-CoA-oxalacetate transcarboxylase (*79, 80*), a biotin-containing oxalacetate decarboxylase from *Aerobacter aerogenes* (*81*).

76. I. A. Rose, *JBC* **245,** 6025 (1970).
77. E. Palacian, G. deTorrontegui, and M. Losada, *BBRC* **24,** 644 (1966).
78. M. C. Scrutton, *in* "Inorganic Biochemistry" (G. L. Eichhorn, ed.). Elsevier, Amsterdam, 1972 (in press).
79. D. B. Northrop and H. G. Wood, *JBC* **244,** 5801 (1969).
80. H. G. Wood, "The Enzymes," 3rd ed., Vol. VI, p. 83, 1972.
81. J. R. Stern, *Biochemistry* **6,** 3545 (1967).

III. Structure

Structural studies on the pyruvate carboxylase molecule are of particular interest since this enzyme possesses several features which are unusual among regulatory enzymes, including multiple levels of quaternary structure and cooperative interactions which are restricted to the effector sites. In addition, a wide diversity of responses to the metabolites (acetyl-CoA, acetoacetyl-CoA, and L-aspartate) which regulate the activity of pyruvate carboxylase is observed for the various forms of this enzyme obtained from different species.

A. Molecular Parameters and Quaternary Structure

Highly purified preparations of pyruvate carboxylase have been obtained from a variety of sources including mammalian liver (*20, 83*), mammalian kidney cortex (*84*), avian liver (*18, 83*), *S. cerevisiae* (*85*), and *P. citronellolis* (*23*). The molecular parameters which characterize some of these enzymes are summarized in Table II. Although in all cases these enzymes are large proteins, the $s^{\circ}_{20,w}$ determined in sedimentation velocity studies on the purified enzymes from chicken liver, *S. cerevisiae,* and *P. citronellolis* is in reasonable agreement with the sedimentation coefficient of the catalytically functional species determined as described by Cohen *et al.* (*82, 86*). In this latter method the sedimentation velocity is determined from the change in absorbance which results as the enzyme molecules are centrifuged through the reaction mixture. The observed agreement between these sedimentation coefficients indicates that catalysis of CO_2 fixation on pyruvate is not accompanied by association or dissociation of the enzyme protein.

The first level of quaternary structure in the pyruvate carboxylases purified from chicken liver and *S. cerevisiae* has been defined by examination of negatively stained preparations of these two enzymes in the electron microscope (*85, 87*). Although in both cases the fundamental

82. B. L. Taylor, R. E. Barden, and M. F. Utter, *Federation Proc.* **30**, 1057 (1971).
83. J. C. Wallace and M. F. Utter, personal communication (1969).
84. R. Bais and D. B. Keech, personal communication (1971).
85. M. R. Young, B. Tolbert, M. F. Utter, and R. C. Valentine, unpublished observations (1970) (manuscript in preparation).
86. R. Cohen, B. Giraud, and A. Messiah, *Biopolymers* **5**, 203 (1967).
87. R. C. Valentine, N. G. Wrigley, M. C. Scrutton, J. J. Irias, and M. F. Utter, *Biochemistry* **5**, 3111 (1966).

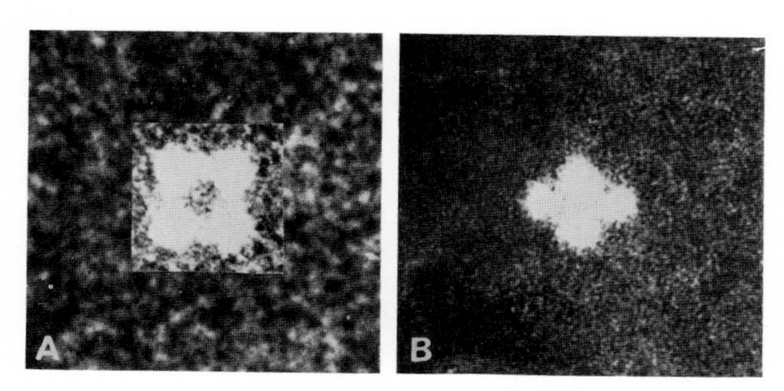

FIG. 3. Electron micrographs of pyruvate carboxylases from (A) chicken liver and (B) *S. cerevisiae*. The micrographs shown were obtained by photographic superimposition of multiple images. Overall magnification factors are 1.25×10^6 (chicken liver) and 1.0×10^6 (*S. cerevisiae*).

organization is tetrameric at this level, these two enzymes differ in that pyruvate carboxylase from chicken liver appears as a square tetramer and pyruvate carboxylase from *S. cerevisiae* as a rhomboid tetramer (Fig. 3). Similar studies for pyruvate carboxylases purified from turkey and calf liver (*88*) indicate that at the first level the quaternary structure of these enzymes resembles that observed for the enzyme from chicken liver. In view of (a) the similar appearance in the electron microscopy of the protomers of the vertebrate liver enzymes (cf. Fig. 3) and (b) the content of bound metal (Mn and/or Mg) (Table I) and biotin which approaches 4 g-atoms (moles)/mole (*17, 19*), the catalytic centers of these enzymes may be symmetrically distributed in the tetramer, although experimental proof for this postulate is lacking at present. The rhomboidal symmetry exhibited by pyruvate carboxylase from *S. cerevisiae* (Fig. 1) and the detection of dimers when preparations of this enzyme are fixed with glutaraldehyde and examined in the electron microscope (*85*) appear most simply consistent with the proposal that the protomers are nonidentical. However, these data are also satisfied if identical protomers are arranged in a tetramer such that the two dimers are joined by heterologous bonds (*85*). This latter postulate is in better accord with the bound metal (zinc) and biotin content of pyruvate carboxylase from *S. cerevisiae* which also approaches 4 g-atoms (moles)/mole (*22*) (Table I).

In contrast to these pyruvate carboxylases, all of which exhibit a tetrameric organization of the protomers, preliminary studies on pyruvate carboxylase from *P. citronellolis* indicate a dimeric organization of

88. J. C. Wallace, M. F. Utter, and R. C. Valentine, personal communication (1969).

TABLE II

Molecular Properties of Some Pyruvate Carboxylases

Source of enzyme	$\bar{M}_w$	$s^{\circ}_{20,w}$ (Protein)	$s_{20,w}$ (Cohen's method)	Ref.
Chicken liver	480,000–550,000[a]	14.8–17.0[b]	16.7	*(18, 83, 89)*
Rat liver	500,000–600,000[c]	15.1[d]	—	*(20)*
Sheep kidney cortex	430,000–465,000	15.1[d]	—	*(84)*
S. cerevisiae	600,000	15.8–17.0[b]	17.0	*(85, 89)*
P. citronellolis	270,000	12.8	13.4	*(82, 89)*

[a] The value of 660,000 reported previously *(18)* appears to be too high. This discrepancy may result from aggregate formation in earlier studies.

[b] The higher values for $s^{\circ}_{20,w}$ were obtained in the presence of reaction components and in a buffer containing 50% D_2O. Although the presence of these components may effect $s^{\circ}_{20,w}$, it is considered more probable that the discrepancy results from the use of inadequate density and viscosity corrections in the earlier studies *(18, 89)*.

[c] Estimated from the biotin content and subunit molecular weight.

[d] Since the status of the viscosity and density corrections employed in these studies s uncertain, these values represent a lower limit.

protomers since (a) the biotin content approaches 2 moles/mole, and (b) the molecular weight of this enzyme is approximately one-half that observed for all other pyruvate carboxylases examined thus far (Table II) *(89)*. This difference in protomer organization is of special interest since the catalytic activity of pyruvate carboxylase from *P. citronellolis* is not affected by any of the metabolites, e.g., acetyl-CoA and L-aspartate, which regulate catalysis by other microbial pyruvate carboxylases *(23)* (cf. Section IV,A).

Since the molecular weights of the protomers of pyruvate carboxylase from avian liver as observed by electron microscopy may be estimated as approximately 150,000 *(87)*, it is possible that these protomers may themselves be composed of smaller subunits which are not resolved in the electron microscope under the conditions used. However, SDS-electrophoresis studies on the pyruvate carboxylases from rat liver and sheep kidney cortex indicate that under these conditions dissociation occurs to yield a single species of molecular weight 125,000–130,000 *(20, 84)*, which can presumably be identified as the protomer. Evidence for smaller subunits of differing molecular weights is obtained only if the rat liver enzyme is subjected to SDS-electrophoresis after denaturation with guanidine hydrochloride and subsequent carboxymethylation *(20)*.

89. B. L. Taylor, R. E. Barden, and M. F. Utter, personal communication (1970).

B. Effect of Mild Denaturation and of Chemical Modification on the Structure and Catalytic Activity of Pyruvate Carboxylase

1. *Incubation at 2°*

In contrast to most enzymes which become more stable as the temperature is decreased, pyruvate carboxylase from avian liver is most stable in the region 20°–30°C and is rapidly inactivated when incubated in dilute buffer at temperatures below 10°C (*18, 90*). This "cold lability" has been observed for several other enzymes besides this pyruvate carboxylase (cf. *90*) but is still a relatively unusual phenomenon in enzymology. Sedimentation velocity (*18, 90*) and electron microscopic (*87*) analysis indicates that the inactivation of pyruvate carboxylase which results from incubation at 2° is accompanied respectively by (a) conversion of the native tetramer (15 S) to a smaller molecule which exhibits a sedimentation coefficient of approximately 7 S, and (b) disappearance of the tetrameric molecules which are observed in negatively stained preparations of the native enzyme. Regain of the catalytic activity which results from rewarming of cold inactivated preparations to 23°C in the presence of ATP is accompanied by restoration of the original sedimentation pattern and reappearance of tetrameric molecules in the electron microscope (*87, 90*). Although these data suggest the existence of a simple equilibrium between the native 15 S tetramer, which is the catalytically functional species, and inactive 7 S protomers, further studies have revealed a more complex situation (*90*). First, at least two processes, only one of which is reversible on rewarming, contribute to the inactivation which results from incubation at 2°. Neither of these processes appears to obey a simple rate law, and their relative contribution to the overall rate of inactivation is affected by parameters such as pH, protein concentration, and salt concentration. Second, under most conditions cold inactivation is accompanied by dissociation, and both effects are prevented by addition of acetyl-CoA. However, in the presence of ATP reversible inactivation occurs without concomitant dissociation. Reactivation induced by rewarming to 23°C is enhanced in the presence of ATP and obeys a first-order rate law. These two observations indicate the existence of an inactive tetramer (B_4 in Fig. 4). Third, conversion to high molecular weight aggregates (C_n) is observed after either prolonged incubation at 2°C or on rewarming cold-inactivated preparations in the absence of ATP.

90. J. J. Irias, M. R. Olmsted, and M. F. Utter, *Biochemistry* **8**, 5136 (1969).

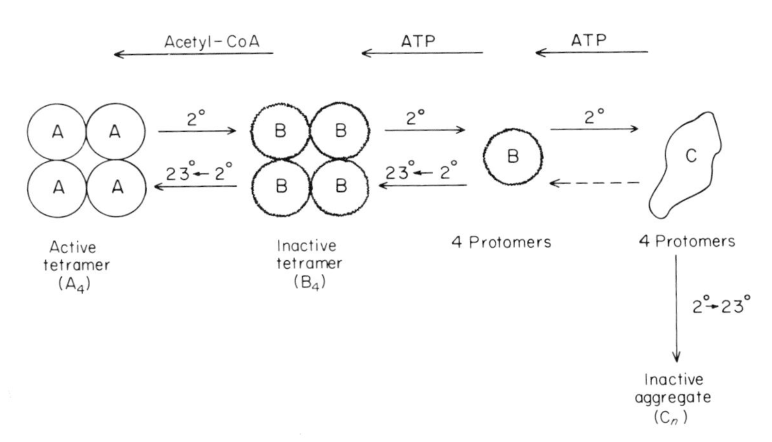

FIG. 4. A scheme describing the effects of mild denaturation on the structure of pyruvate carboxylase and the influence of various reaction components on the denaturation process. This scheme is a simplified form of that proposed by Irias *et al.* (*90*), which provides a more complete description of the system.

All these observations can be explained by a scheme which requires the existence of two protomer (B and C) and two tetramer (A_4 and B_4) conformations (Fig. 4) (*90*). The two tetramers (A_4 and B_4) are proposed to be in reversible equilibrium with the protomer (B). This protomer (B) can also be converted in a slow (and possibly irreversible) reaction to a second protomer (C) which has a conformation predisposing it to nonspecific aggregation. In addition to the effects of reaction components (Fig. 4), the rate of both reversible and irreversible inactivation at 2° is decreased in the presence of D_2O (*45*) or, in H_2O, by addition of high concentrations of mono- or polyhydric alcohols, e.g., methanol and sucrose (*18*, *90*). Such effects, which suggest that protein–water interaction is involved in inactivation at 2°, provide support for the postulate that hydrophobic bonds play an important role in maintenance of the tertiary and quaternary structure of pyruvate carboxylase from avian liver (*18*, *90*). Although cold lability is also observed for pyruvate carboxylases purified from *A. niger* (*37*) and *A. globiformis* (*11*), it is not a property of this enzyme in all species since apparently cold-stable pyruvate carboxylases have been obtained from calf liver (*83*), rat liver (*20*), *S. cerevisiae* (*24*), and *P. citronellolis* (*23*).

2. *Urea and pH*

Incubation of pyruvate carboxylase from avian liver with low concentrations of urea (0.4 M) at 23°C causes an inactivation and dissociation of pyruvate carboxylase which can be prevented by addition of acetyl-CoA and which, in these respects, resembles the effects of incubation at 2°. Reversibility of the effects of urea has not, however, been demonstrated (*90*).

Inactivation which is prevented by addition of acetyl-CoA is also observed when pyruvate carboxylase is incubated at 23° under either acid (pH 6.2) or alkaline (pH 7.95) conditions. The inactivation resulting from incubation at pH 7.95 may also resemble that caused by incubation at 2° since the effect is accompanied by disappearance of the 15 S species and is reversible on readjustment to neutral pH (*90*). However, formation of the 7 S (protomer) species under these conditions is not clearly documented. This latter observation could be explained if exposure to alkaline pH increases the rate of conversion of protomer B to protomer C (Fig. 4) with an increase in the extent of aggregate (C_n) formation. Increased aggregate formation is observed under these conditions (*90*).

3. *Sulfhydryl Reagents*

Inactivation of pyruvate carboxylase from avian liver also results from incubation of the enzyme with various sulfhydryl reagents (*42*). When the reagent employed introduces an uncharged residue on the reacting sulfhydryl groups (Type I), e.g., *N*-ethylmaleimide, a biphasic inactivation is observed (*91*). In the first phase of the reaction inactivation occurs without significant alteration in the sedimentation pattern and tetrameric molecules are produced which exhibit a reduced level of catalytic activity as compared with the native enzyme. These partially active tetramers are of particular interest since they possess properties of activation by acetyl-CoA which differ from those observed for native pyruvate carboxylase (*92*). Since other catalytic parameters are unaffected by this partial modification which is associated with the reaction of approximately 70% of the sulfhydryl groups of the enzyme (*91*), it is apparent that the effect is specifically directed to the regulatory properties. During the second phase of inactivation by *N*-ethylmaleimide, aggregation to high molecular weight species occurs without apparent involvement of a stable protomer (*91*).

In contrast, if the sulfhydryl reagent used introduces a charged residue (Type II), e.g., *p*-chloromercuribenzoate, 5,5′-dithiobis(2-nitrobenzoate) and is added in 10- to 20-fold molar excess over the tetramer concentration, the inactivation observed is accompanied by dissociation of the 15 S tetramer to yield a 7 S protomer. Both inactivation and dissociation are reversible on addition of a thiol, e.g., dithioerythritol (*91*).

91. E. Palacian and K. E. Neet, *BBA* **212**, 158 (1970).
92. E. Palacian and K. E. Neet, personal communication (1970).

When the time of incubation and/or the molar excess of sulfhydryl reagent is increased, a process which is not reversible on addition of thiol makes an increasing contribution to the overall inactivation. This irreversible phase of inactivation may be associated with formation of high molecular weight aggregates (*91*).

Comparison with the properties of inactivation caused by incubation at 2° (Section III,B,1) indicates that the effect of Type II sulfhydryl reagents on the structure of pyruvate carboxylase from avian liver may be adequately described by a scheme similar to that of Fig. 4. Inactivation by Type I sulfhydryl reagents may also be described by Fig. 4 if a labile protomer is assumed to be an intermediate in the formation of aggregates which occurs during the second phase of inactivation. The differing effects of Types I and II sulfhydryl reagents on the structure of pyruvate carboxylase could therefore result from the relative stabilities of the protomers produced although an alternate postulate which does not depend on the participation of protomers as intermediates in the aggregation process cannot be excluded. A relationship between the effects of these two types of sulfhydryl reagent is indicated by the observation that prior treatment of pyruvate carboxylase with *N*-ethylmaleimide to yield the partially active tetramer protects the enzyme against dissociation induced by subsequent addition of 5,5′-dithiobis(2-nitrobenzoate) (*91*).

4. *Maleic Anhydride*

Treatment of the pyruvate carboxylases purified from avian liver and *S. cerevisiae* with maleic anhydride, which reacts with the amino and sulhydryl groups of proteins (*93*), causes rapid inactivation of both these enzymes, but differing patterns of dissociation are observed. Under these conditions modification of pyruvate carboxylase from *S. cerevisiae* results initially in the formation of an inactive tetramer (*94*). This inactive tetramer then dissociates more slowly to yield first 9–10 S species, which may be tentatively identified as the dimer observed in electron microscopic analysis of glutaraldehyde-fixed preparations, and, subsequently, the 6 S protomer (*85*). Similar treatment of pyruvate carboxylase from avian liver yields the 6–7 S protomers directly without the participation of a stable dimeric intermediate in a reaction which appears similar to that induced by incubation at 2° or treatment with Type II sulfhydryl reagents (*88*). In neither case have studies been

93. B. S. Hartley, *BJ* **109,** 805 (1970).

94. M. R. Young, M. F. Utter, and R. C. Valentine, unpublished observations (1969).

performed to define the nature of the residues modified by maleic anhydride under the conditions employed.

5. *Other Reagents*

Keech and Farrant (*95*) have reported that 1-fluoro-2,4-dinitrobenzene reacts with the ε-amino group of a single lysyl residue in pyruvate carboxylase purified from sheep kidney cortex causing inactivation of this enzyme. Since only acetyl-CoA among the reaction components protects against this inactivation, the lysyl residue attacked appears to be implicated either in the binding of acetyl-CoA or in mediation of the effect(s) of this activator on the catalytic site (*95*). Similar effects are not however observed when the ε-amino groups of pyruvate carboxylase from avian liver are modified using the specific reagent, 2-methoxy-5-nitrotropone (*45, 96*).

Investigation of the role of other amino acid residues (e.g., tyrosine, tryptophan, and histidine) in the structure or catalytic function of pyruvate carboxylase from avian liver has been frustrated by the presence of the many highly reactive sulfhydryl groups of this enzyme.

C. Immunochemical Studies

Antibody prepared to pyruvate carboxylase from chicken liver interacts equally strongly with the enzyme obtained from chicken or turkey liver, but only weak interaction is observed with the enzyme purified from calf liver and no interaction with the enzyme purified from *S. cerevisiae*. Conversely, antibody prepared to pyruvate carboxylase from *S. cerevisiae* fails to interact with any of the enzymes obtained from mammalian or avian liver which have been examined thus far (*83*). These immunological relationships are in accord with the structural differences between the avian liver and yeast enzymes which are observed in the electron microscope (Fig. 3). The structural differences between the mammalian and avian liver enzymes which are detected in the immunochemical studies do not, however, appear to result from differences in the organization of the protomers in the tetramer.

In addition, recent studies have shown that antibody prepared to pyruvate carboxylase purified from rat liver mitochondria cross-reacts with the enzyme in other rat tissues, e.g., kidney cortex, brain, and adipose tissue (*3*).

95. D. B. Keech and R. K. Farrant, *BBA* **151**, 493 (1968).
96. H. Tamaoki, Y. Murase, S. Minato, and K. Nakanishi. *J. Biochem.* (*Tokyo*) **62,** 7 (1967).

IV. Regulatory Properties

Since as noted above (Section I) pyruvate carboxylase has been implicated as a key enzyme in several metabolic pathways (*14, 15, 97*), the existence of a variety of regulatory mechanisms for this enzyme in different tissues and species might be anticipated. Potential regulatory mechanisms described thus far for pyruvate carboxylases in the different species examined include (a) activation by acyl derivatives of coenzyme A or, less effectively, by coenzyme A itself; (b) inhibition by acetoacetyl-CoA; (c) inhibition by L-aspartate; and (d) effects of acyl derivatives of coenzyme A and of L-aspartate on the rate of synthesis of holopyruvate carboxylase from apopyruvate carboxylase plus biotin.

A. Acyl Derivatives of Coenzyme A and Related Compounds

1. *General Properties*

Pyruvate carboxylases have been described which (a) are fully active in the absence of added acyl-CoA (Types IV and V); (b) are stimulated by addition of an acyl-CoA (Type III) (*98*); and (c) are essentially inactive in the absence of an added acyl-CoA, e.g., acetyl-CoA (Types I and II). The sources of these various pyruvate carboxylases and a further explanation of their activation status is given in Table III. The pyruvate carboxylases which are essentially inactive in the absence of added acyl-CoA (Types I and II) are of special interest since this situation is rarely observed for activators other than metal ions (but cf. *99, 100*). The pyruvate carboxylases obtained from rat liver (*20, 29*) and *B. stearothermophilus* (*101*) appear to exhibit a limited degree of catalytic activity in the absence of added acyl-CoA, although the significance of this acyl-CoA-independent activity remains to be established. However, for pyruvate carboxylase from chicken liver the absolute re-

97. M. F. Utter, D. B. Keech, and M. C. Scrutton, *Advan. Enzyme Regulation* **2**, 49 (1964).

98. M. F. Utter, M. C. Scrutton, M. R. Young, B. Tolbert, J. C. Wallace, J. J. Irias, and R. C. Valentine, *Abstr. 7th Intern. Congr. Biochem. Tokyo, 1967* Vol. 2, p. 247. Sci. Council Japan, Tokyo, 1968.

99. R. L. Metzenberg, M. Marshall, P. P. Cohen, and W. G. Miller, *JBC* **234,** 1534 (1959).

100. M. Marshall, R. L. Metzenberg, and P. P. Cohen, *JBC* **236,** 2229 (1961).

101. J. J. Cazzulo, T. K. Sundaram, and H. L. Kornberg, *Proc. Roy. Soc.* **B176,** 1 (1970).

TABLE III
REGULATORY PROPERTIES OF VARIOUS PYRUVATE CARBOXYLASES

Type	Source	Acyl-CoA activation			Inhibition by L-aspartate	Ref.
		Effect	n[a] (pH 7.8)	Most effective activator[b]		
IA	Avian liver	Required[c]	3	Acetyl-CoA	No	(*1, 102*)
IB	Mammalian liver and kidney cortex	Required[d]	2	Acetyl-CoA	No	(*5, 20, 103, 104*)
II	*Bacillus*	Required[d]	3	Acetyl-CoA	Yes	(*101*)
III	*Saccharomyces*	Stimulatory[e]	1	Palmityl-CoA	Yes	(*6, 77, 98*)
IV	*Aspergillus*	None	—	—	Yes	(*37*)
V	*Pseudomonas*	None	—	—	No	(*10, 23*)

[a] Obtained from Eq. (6) (*105*):

$$\log_{10} \frac{V}{V_{max} - V} = n \log_{10} [\text{acyl-CoA}] - \log_{10} K \qquad (6)$$

[b] On the basis of K_A, the concentration of acyl-CoA required to achieve 50% maximal activation.

[c] Initial rate of CO_2 fixation is increased more than 1000-fold in the presence of a saturating concentration of acetyl-CoA (*102*).

[d] Initial rate of CO_2 fixation is increased approximately 50-fold in the presence of a saturating concentration of acetyl-CoA (*20, 101*).

[e] Initial rate of CO_2 fixation is increased approximately 3-fold in the presence of a saturating concentration of acyl-CoA if the assay system contains 5 mM $KHCO_3$. The extent of activation is increased if the K^+ and/or HCO_3^- concentration is decreased (*106*).

quirement for activation by an acyl-CoA is well established since it is observed over a wide range of pH (6.5–10.0) and temperature (0°–44°), at concentrations of the substrates ranging from one order of magnitude below to two orders of magnitude above their Michaelis constants and over a wide range of concentrations and at different levels of purification of the enzyme (*45, 102*). Although conditions may exist under which this enzyme exhibits a significant rate of catalysis in the absence of an acyl-CoA, the data presently available are not indicative of such a possibility. It is, therefore, necessary to consider that acyl-CoA's might activate *this* pyruvate carboxylase as a result of direct participation in the catalytic mechanism. Several observations appear, however, to be inconsistent with such a possibility. First, although all pyruvate carboxylases examined thus far utilize a similar catalytic mechanism (Section II), these enzymes exhibit wide diversity in response to addition of an acyl-CoA (Table III) (*1, 5, 20, 101–104*). Second, no carbon interchange between the substrates and acetyl-CoA is observed during catalysis by pyruvate carboxylase from avian liver (*2*). And, finally, carboxyl or acyl transfer reaction sequences, which would explain the cofactor role for acetyl-CoA in the pyruvate carboxylase reaction (*97*), are excluded for the enzyme from avian liver by the failure to observe the exchange reactions which are predicted by these mechanisms (*42, 97, 102*). The carboxyl transfer mechanism is also inconsistent with the effective activation of pyruvate carboxylase from avian liver by high concentrations of coenzyme A (Table IV).

The relationship between the initial rate of CO_2 fixation, or oxalacetate decarboxylation, and the acyl-CoA concentration exhibits cooperative character for Types I and II pyruvate carboxylases (Table III). When the data obtained at pH 7.4– 7.8 are analyzed according to the empirical equation of Hill (*105*), the Hill coefficient (n) obtained which provides an indication of the extent of cooperativity, approaches 3.0 for the pyruvate carboxylases from chicken liver (*102*) and *B. stearothermophilus* (*101*) and 2.0 for pyruvate carboxylases from mammalian liver (*5*) and kidney cortex (*104*). Both the Hill coefficient and the activator constant observed for activation of pyruvate carboxylase from avian and mammalian liver by acetyl-CoA are unaffected by variation of the concentrations of other reaction components but are sensitive functions of pH in the range 7.0–9.0 (*20, 35, 102*) thus permitting variations

102. M. C. Scrutton and M. F. Utter, *JBC* **242,** 1723 (1967).
103. W. R. McClure and H. A. Lardy, *JBC* **246,** 3591 (1971).
104. G. J. Barritt, D. B. Keech, and A. M. Ling, *BBRC* **24,** 476 (1966).
105. J. Monod, J. P. Changeux, and F. Jacob, *JMB* **6,** 306 (1963).

in $[H^+]$ to modulate the response of pyruvate carboxylase to this activator.

In contrast, pyruvate carboxylase from *S. cerevisiae* exhibits a hyperbolic relationship between initial rate of CO_2 fixation and acyl-CoA concentration under most conditions (*106*) (but cf. Section IV,B).

2. *Specificity of Activation*

a. The Acyl Group. Some data illustrating differences in the specificity of the activation of the pyruvate carboxylases from chicken liver and *S. cerevisiae* by various acyl-CoA's are summarized in Table IV. The use of highly purified preparations of CoASH has revealed that this compound is a weak activator of both these enzymes. Therefore, in contrast to previous suggestions (*107*), the major difference in activation

TABLE IV
COMPARISON OF SPECIFICITY OF ACYL-CoA ACTIVATION OF PYRUVATE CARBOXYLASES FROM CHICKEN LIVER AND *S. cerevisiae*[a]

Compound	Pyruvate carboxylase from chicken liver		Pyruvate carboxylase from *S. cerevisiae*
	K_A (μM)[b]	K_i (μM)[b]	K_A (μM)[b]
CoASH	1000	—	900[c]
Acetyl-CoA	2	—	92
n-Butyryl-CoA	75	—	34
n-Valeryl-CoA	—	90	—
n-Hexanoyl-CoA	—	—	7
n-Octanoyl-CoA	—	—	0.6
Palmityl-CoA	—	—	0.2
Benzoyl-CoA	—	No inhibition[d]	7
Malonyl-CoA	—	470	No effect[e]

[a] Data compiled from the unpublished observations of C. H. Fung, B. Tolbert, M. R. Young, and M. F. Utter. The data for pyruvate carboxylase from chicken liver differ in several respects from previous reports (*102, 107*) and reflect the availability of more highly purified preparations of coenzyme A and the various acyl derivatives.

[b] Determination at pH 7.8 in the presence of either 5 mM (pyruvate carboxylase from *S. cerevisiae*) or 15 mM (pyruvate carboxylase from chicken liver) $KHCO_3$. Serum albumin was omitted from the assay system.

[c] The value reported previously (*107*) was in error by an order of magnitude because of typographical error.

[d] At 100 μM.

[e] At 100 μM in the presence or absence of acetyl-CoA.

106. B. Tolbert and M. F. Utter, personal communication (1970).

107. M. F. Utter and M. C. Scrutton, *Current Topics Cellular Regulation* **1**, 253 (1969).

specificity between these two pyruvate carboxylases is observed in the relationship between the structure of the acyl group and the effectiveness of activation expressed as K_A (Table IV).

For the chicken liver enzyme activation is observed only if the acyl group contains no more than four carbon atoms and carries no negatively charged (carboxyl) group (Table IV). All activating acyl-CoA's exhibit similar V_{max} values but differences are observed in K_A, the activator concentration required to obtain 50% V_{max}. Use of this latter criterion identifies acetyl-CoA as the most effective activator of pyruvate carboxylase from chicken liver (Table IV). If, however, the acyl group contains five carbon atoms, e.g., *n*-valeryl-CoA, or bears a carboxyl group, e.g., malonyl-CoA, the acyl-CoA becomes an inhibitor of this pyruvate carboxylase which exhibits competitive behavior with respect to acetyl-CoA (Table IV). A marked increase in the chain length of the acyl group, e.g., palmityl-CoA, further modifies the acyl-CoA to a species which has no detectable effect on the catalytic properties of pyruvate carboxylase from avian liver. Less extensive studies for pyruvate carboxylases from mammalian liver suggest that these enzymes exhibit a similar acyl group specificity to that observed for the enzyme from chicken liver although differences are apparent in the apparent K_A observed for a given acyl-CoA (*5*, *108*). In addition, pyruvate carboxylases from mammalian liver may be less sensitive to inhibition by carboxylated acyl-CoA's, e.g., malonyl-CoA (*103*).

A very different acyl group specificity is observed for pyruvate carboxylase from *S. cerevisiae*. The apparent K_A becomes progressively more favorable as the chain length and/or hydrophobicity of the acyl group is increased (Table IV). Thus, palmityl-CoA and *n*-octanoyl-CoA, which are the most effective activators described thus far, exhibit activator constants two orders of magnitude more favorable than that observed for acetyl-CoA (Table IV). Introduction of a carboxyl group into the acyl residue, e.g., malonyl-CoA, renders the acyl-CoA ineffective as either an activator or an inhibitor of this enzyme (Table IV). Activation of pyruvate carboxylase from *S. cerevisiae* by acyl-CoA's appears therefore to result from occupancy of a site which is extremely hydrophobic but yet relatively free of steric restrictions.

b. The Coenzyme A Residue. Analogs of acetyl-CoA have also been tested in which the coenzyme A portion of the molecule has been modified by (a) removal of the adenosine phosphate moiety (acetyl-pantetheine and acetyl-4′-phosphopantetheine), (b) removal of the 3′-phosphate (acetyl-3′-dephospho-CoA), or (c) replacement of $-NH_2$ by

108. J. C. Wallace, C. H. Fung, and M. F. Utter, personal communication (1969).

–OH at C-6 of the adenine ring (acetyl-deamino-CoA). All of these analogs act as competitive inhibitors of pyruvate carboxylase from avian liver with respect to acetyl-CoA but are inert as activators indicating the importance of an intact coenzyme A residue to the activation process. The inhibitor constants for these analogs lie in the range 0.2–0.4 m*M* except for acetyl-3′-dephospho-CoA ($K_i = 1.2$ m*M*) (*109*). This indication of an important role for the 3′-phosphate in the binding of acyl-CoA is consistent with the observation that certain divalent anions, e.g., SO_4^{2-}, are potent competitive inhibitors with respect to acetyl-CoA and also mimic some of the effects of this activator on the conformation of pyruvate carboxylase (cf. Section IV,A,3) (*110*).

Activation of pyruvate carboxylase from *S. cerevisiae* also requires the presence of an intact coenzyme A residue. Both acetyl- and palmitylpantetheine and acetyl-4′-phosphopantetheine are effective inhibitors, whereas acetyl-3′-dephospho-CoA appears ineffective either as an activator or as an inhibitor of this enzyme (*106*). The observation of inhibition by palmitylpantetheine excludes the possibility that the activating effect of palmityl-CoA could result from a nonspecific interaction arising from the hydrophobic character of this molecule (cf. *111*). In addition, since the acylpantetheines are inhibitory in the presence *and* absence of an activating acyl-CoA (*106*), the acyl-CoA independent activity exhibited by this pyruvate carboxylase is probably not attributable to the presence of bound activator. This suggestion is supported by (a) the effects of [K^+] and [HCO_3^-] on the extent of the acyl-CoA-independent catalytic activity for a given preparation and (b) the failure to observe incorporation of ^{14}C-pantothenate into the enzyme purified from a pantothenate-requiring strain of *S. cerevisiae* (*106*).

3. *Effects on Parameters Reflecting Enzyme Conformation*

a. Structural Parameters. Addition of acetyl-CoA protects pyruvate carboxylase from avian liver against inactivation and dissociation induced by incubation at 2° or in the presence of 0.4 *M* urea (cf. Section III) and against inactivation and aggregation induced by incubation at pH 6 or in the presence of 0.5 m*M* SDS (*90, 102*). However, the presence of this activator has no effect on (a) the extent of reactivation and of tetramer formation which occurs when cold-inactivated preparations are rewarmed to 23° (*90*) and (b) the dissociation to small (2.6 S) subunits resulting from exposure to 0.4 *M* SDS or 2 *M* guanidine hydrochloride

109. C. H. Fung and M. F. Utter, personal communication (1970).
110. C. H. Fung and M. C. Scrutton, unpublished observations (1970).
111. K. Taketa and B. M. Pogell, *JBC* **241, 720** (1966).

(*102*). The effect of acetyl-CoA, therefore, appears to be limited to stabilization of the active tetramer. A gross change in enzyme structure is not, however, involved since neither the sedimentation behavior nor the appearance in the electron microscope is significantly altered by addition of acetyl-CoA (*102, 112*).

Several lines of evidence support the proposal that the effects of acyl-CoA's on the structure and activity of pyruvate carboxylase from avian liver are derived from the same primary effect on the protein conformation. First, a sigmoid relationship with acetyl-CoA concentration is observed for the effects of this cofactor on both the catalytic activity and the extent of protection against inactivation at 2° (*90, 102*). Second, the K_d for acetyl-CoA estimated from the inactivation studies approximates the K_A obtained in initial rate studies at a given pH (*5, 90*). And, third, the specificity of the effect of acyl-CoA's and related compounds on inactivation at 2° resembles that observed for the activation of CO_2 fixation (*90*) (Section IV,A,2).

b. Environment of the Biotin Residues. A more precise indication of the nature of the conformational change which accompanies activation by an acyl-CoA is provided by examination of the effect of the activator on the rate of inactivation of pyruvate carboxylase by avidin. Inactivation resulting from reaction of avidin with the biotinyl residues in the catalytic site of this enzyme is essentially irreversible and is first order in enzyme sites (*58*). The effects of various reaction components on this rate of inactivation therefore provides a probe for conformational changes in the region of the active site since the role of the biotinyl residues in the catalytic mechanism is clearly established (cf. Section II,A). At low concentration ($<150\ \mu M$) addition of acetyl-CoA increases the rate of inactivation by avidin indicating that the biotinyl residues which form a key region of the active site of pyruvate carboxylase have become more accessible to the bulky avidin molecule. The maximal increase in the inactivation rate (approximately 5-fold) is observed in the presence of 150 μM acetyl-CoA (*102*). However, no precise correlation is possible with the effect of this and other acyl-CoA's on the structure and/or activity of pyruvate carboxylase from chicken liver. At higher acetyl-CoA concentrations the inactivation rate is observed to decrease again, possibly because of the presence of impurities in the acetyl-CoA preparation used for these studies.

It should be noted that the rate of inactivation of pyruvate carboxylase by avidin is also affected by addition of other reaction components, e.g., ATP, Me^{2+}, pyruvate, and oxalacetate (*45, 58, 68*).

112. R. C. Valentine and M. C. Scrutton, unpublished observations (1965).

c. Other Effects on the Conformation of the Protein. Studies on the specific modification of the environment or reactivity of amino acid residues of pyruvate carboxylases in the presence of an activating acyl-CoA have been less fruitful. The environments of the tyrosyl and tryptophyl residues are often sensitive to changes in enzyme conformation (*113*). In the case of pyruvate carboxylase from avian liver, however, these residues appear to be insensitive since no significant alterations in either the ultraviolet absorption or the protein fluorescence spectrum of this enzyme are observed on addition of acetyl-CoA (*114*). Inactivation of pyruvate carboxylase from sheep kidney cortex by 1-fluoro-2,4-dinitro-benzene resulting from reaction with a single lysyl residue is prevented by addition of acetyl-CoA, apparently implicating this residue in the activation process (*95*). Similar effects are not observed for the avian liver enzyme (*107*).

d. Acetoacetyl-CoA. Most acyl-CoA's which fail to activate pyruvate carboxylase from chicken liver are either inert, e.g., palmityl-CoA, or are competitive inhibitors with respect to acetyl-CoA, e.g., malonyl-CoA. However, acetoacetyl-CoA is unique among all acyl CoA's tested thus far since (a) it acts as a noncompetitive inhibitor with respect to all reaction components including acetyl-CoA, and (b) it is the only inhibitor which modifies the cooperative relationship between the initial rate and the acetyl-CoA concentration. Thus, in the presence of saturating concentrations of acetoacetyl-CoA a hyperbolic saturation relationship is observed for activation by acetyl-CoA (*115, 116*). These properties suggest that pyruvate carboxylase carries a specific site(s) for acetoacetyl-CoA although binding studies are necessary to substantiate this proposal. Furthermore, since β-hydroxybutyryl-CoA, the other partner of the redox couple, is a very weak activator of pyruvate carboxylase from chicken liver, the effects of these two acyl-CoA's provide a mechanism for indirect regulation of the rate of pyruvate carboxylation by the NAD^+–NADH ratio (*116*).

B. L-ASPARTATE

L-Aspartate has no effect on the catalytic properties of pyruvate carboxylases obtained from mammalian (*103*) or avian (*48*) liver, but is a potent inhibitor of some microbial pyruvate carboxylases (Table III).

113. D. B. Wetlaufer, *Advan. Protein Chem.* **17**, 304 (1962).
114. R. A. Harvey and M. C. Scrutton, unpublished observations (1969).
115. M. F. Utter and C. H. Fung, *Z. Physiol. Chem.* **351**, 284 (1970).
116. C. H. Fung and M. F. Utter, *Federation Proc.* **29**, 542 (1970).

TABLE V
INTERACTIONS BETWEEN THE EFFECTS OF ACYL-COA'S AND OF L-ASPARTATE ON THE CATALYTIC PROPERTIES OF CERTAIN MICROBIAL PYRUVATE CARBOXYLASES

Type[a]	Source	Effects of Acyl-CoA	Characteristics of inhibition by L-aspartate	Ref.
IV	*A. niger*	None	Linear, noncompetitive inhibitor with respect to substrates, e.g., pyruvate	(*37*)
III	*S. cerevisiae*	Stimulatory	Cooperative: $n^b \approx 1.4$ in the absence of acyl-CoA. In the presence of acyl-CoA increases are observed in (a) the concentration of L-aspartate required to achieve 50% inhibition and (b) in n which approaches 3.0 at saturating levels of acyl-CoA	(*77, 106, 118*)
II	*B. stearothermophilus*	Required	Cooperative: $n^b \approx 3.0$ in the presence or absence of acetyl-CoA. The concentration of L-aspartate required to achieve 50% inhibition increases as the acetyl-CoA concentration is increased	(*101*)

[a] Compare Table III.
[b] Determined as described by Taketa and Pogell (*117*).

Several interesting relationships emerge from a comparison of the effects of acyl-CoA's and of L-aspartate on the catalytic properties of these pyruvate carboxylases (Table V). First, L-aspartate is either a linear noncompetitive inhibitor with respect to the substrates (Type IV) (*37*) or has no inhibitory effect (Type V) (*23*) for those pyruvate carboxylases which are fully active in the absence of an acyl-CoA. In contrast, L-aspartate is a cooperative inhibitor of those microbial pyruvate carboxylases which are activated by addition of an acyl-CoA (Types II and III). For these enzymes the extent of cooperativity in the relationship between initial rate and L-aspartate concentration reflects that observed for the relationship between initial rate and acyl-CoA concentration (Tables III and V).

Second, the properties of interaction between the effects of acyl CoA's and L-aspartate on those pyruvate carboxylases which are influenced by both these effectors (Types II and III) appear to be related to the nature of the relationship between initial rate and effector concentration which is observed in the absence of the second effector. If this relation-

117. K. Taketa and B. M. Pogell, *JBC* **240**, 651 (1965).
118. J. J. Cazzulo and A. O. M. Stoppani, *ABB* **127**, 563 (1968).

ship is markedly cooperative ($n \approx 3$), under such conditions (Type II), interaction between these effectors appears to be confined to modification of the concentration required to achieve 50% maximal activation (acetyl-CoA) or inhibition (L-aspartate) (*101*) (Table V). However, if a minimal extent of cooperativity is observed in the absence of the second effector ($n \approx 1.0$–1.4) (Type III), both the concentration required to achieve 50% maximal response *and* the extent of cooperativity in the relationship between initial rate and effector concentration are increased by addition of, or increases in the concentration of, the second effector (Table V). Thus, for example, in the presence of saturating levels of L-aspartate, the relationship between initial rate and acyl-CoA concentration is markedly sigmoid in contrast to the essentially hyperbolic relationship which is observed for the Type III pyruvate carboxylases in the absence of this inhibitory effector (*118*). A similar response is observed when either acetyl-CoA or palmityl-CoA is used as the activator (*106, 118*).

The interaction between the effects of acyl-CoA's and L-aspartate on the catalytic properties of pyruvate carboxylase from *S. cerevisiae* presents an interesting contrast to the interrelationships between the effects of acetyl-CoA (activator) and acetoacetyl-CoA (inhibitor) on catalysis by pyruvate carboxylase from avian liver (*116*) (cf. Section IV,A,4). For this latter enzyme the presence of the inhibitory effector reduces the extent of cooperativity in the relationship between initial rate and activator concentration, which is markedly cooperative in absence of the inhibitor.

The effects of acetyl-CoA and L-aspartate on the catalytic activity of pyruvate carboxylase from *S. cerevisiae* are qualitatively similar to the effects of these metabolites on the properties of PEP carboxylase purified from the Enterobacteriaceae (*119, 120*). Although such similarities appear indicative of a role for acetyl-CoA and L-aspartate in regulation of the rate of the anaplerotic synthesis of oxalacetate from PEP or pyruvate in microorganisms (*14*), this postulate does not appear consistent with the absence of effects of these metabolites on the catalytic properties of pyruvate carboxylase from *P. citronellolis* (*10, 23*).

C. NADH

NADH inhibition of pyruvate carboxylase from *S. cerevisiae* has been reported by Cazzulo and Stoppani (*44*). The inhibition is com-

119. J. L. Canovas and H. L. Kornberg, *Proc. Roy. Soc.* **B165,** 189 (1966).
120. P. Maeba and B. D. Sanwal, *BBRC* **21,** 503 (1965).

petitive with respect to acetyl-CoA and appears specific for this nucleotide since NAD^+, and NADPH, and $NADP^+$ are ineffective (*44*).

D. Phosphoenolpyruvate

Phosphoenolpyruvate has been reported as an inhibitor of pyruvate carboxylases obtained from chicken (*116*) and rat (*47*) liver. However, recent studies for the enzyme from chicken liver indicate that the inhibition results from the presence of an unidentified contaminant in certain phosphoenolpyruvate preparations (*115*).

E. Regulation of Holopyruvate Carboxylase Synthesis

Synthesis of the holopyruvate carboxylase of *B. stearothermophilus* from the apoenzyme and biotinyl-AMP is subject to regulation by the same effectors which control the catalytic activity of the holoenzyme (*25*, *26*). Thus, synthesis of the holoenzyme from the apoenzyme and biotinyl-AMP (or biotin, ATP, and Mg^{2+}) requires addition of acetyl-CoA as a cofactor and is inhibited by addition of L-aspartate. The rate of synthesis of holoenzyme is a cooperative function of both acetyl-CoA and L-aspartate concentration (*26*). Although the site of action of these effectors remains to be determined, it seems most likely that they influence the rate of holoenzyme synthesis as the result of alterations induced in the conformation of the apoenzyme. However, direct effects on the catalytic properties of holopyruvate carboxylase synthetase cannot be excluded.

Acetyl-CoA does not appear to be required for the synthesis of holopyruvate carboxylase in extracts prepared from the livers of biotin-deficient chickens (*121*).

V. Conclusion

In this chapter we have attempted to summarize the present status of knowledge of the properties and structure of the pyruvate carboxylases which have been obtained from various organisms. Although these studies are still far from complete, sufficient data have been accumulated to permit certain tentative conclusions. It appears that all pyruvate carboxylases studied thus far exhibit similar catalytic properties (Section

121. M. Madapally and S. P. Mistry, *BBA* **215**, 316 (1970).

II) but that marked differences are observed when the regulatory properties of these enzymes are examined (Section IV). In addition, although all pyruvate carboxylases appear to be large proteins which exhibit protomer and possibly subunit levels of organization, marked differences are observed both in the arrangement of the protomers in the oligomer and also in the nature of the interactions between the protomers. Since the data on the pyruvate carboxylases are among the most extensive reported thus far for any regulatory enzyme, it seems possible that constancy in catalytic properties as contrasted with variability in regulatory properties may be characteristic of the evolution of such enzymes. It is however apparent from the studies performed thus far that the pattern of development of the various regulatory properties is not a simple one (cf. Table III) and that a much wider sampling of pyruvate carboxylases will be required in order to clarify the nature of the evolutionary relationships.

Discussion of the physiological significance of the catalytic and regulatory properties of the pyruvate carboxylases has been omitted from this chapter since this aspect has been considered previously (*5, 107, 122*).

ACKNOWLEDGMENTS

The authors are grateful to Drs. M. F. Utter, B. L. Tolbert, R. E. Barden, H. A. Lardy, W. R. McClure, K. E. Neet, and E. Palacian and to Messrs. C. H. Fung and B. L. Taylor for permission to quote from their unpublished data.

122. M. C. Scrutton and M. F. Utter, *Ann. Biochem.* **37**, 249 (1968).

2

Acyl-CoA Carboxylases

ALFRED W. ALBERTS • P. ROY VAGELOS

I. Introduction

Acyl-CoA carboxylases [acyl-CoA:carbon dioxide ligases (ADP)] are enzymes containing covalently bound biotin that catalyze an ATP-dependent carboxylation of an acyl-CoA. Three enzymes of this type have been intensively studied, β-methylcrotonyl-CoA carboxylase, propionyl-CoA carboxylase, and acetyl-CoA carboxylase. One or all of these enzymes play central roles in the metabolism of most if not all forms of life.

A number of excellent reviews have appeared describing these enzymes (*1–10*). In addition there are several reviews on the chemistry of biotin and its role in nutrition (*9–17*). This report will attempt to summarize the studies that have been carried out on the structure and metabolic role of β-methylcrotonyl-CoA carboxylase, propionyl-CoA carboxylase, and acetyl-CoA carboxylase. Acetyl-CoA carboxylase is discussed in the greatest detail since this enzyme has been recently dissociated into subunits (*18*), and studies of these subunits have allowed a new level of understanding of biotin enzymes.

II. β-Methylcrotonyl-CoA Carboxylase

A. Reaction Catalyzed

β-Methylcrotonyl-CoA carboxylase catalyzes the biotin-dependent carboxylation of β-methylcrotonyl-CoA to form β-methylglutaconyl-

1. S. Ochoa and Y. Kaziro, *Federation Proc.* **20,** 982 (1961).
2. S. J. Wakil, *Ann. Rev. Biochem.* **33,** 369 (1962).
3. J. Knappe and F. Lynen, *Colloq. Ges. Physiol. Chem.* **14,** 265 (1963).
4. Y. Kaziro and S. Ochoa, *Advan. Enzymol.* **26,** 312 (1964).
5. P. R. Vagelos, *Ann. Rev. Biochem.* **33,** 139 (1964).
6. S. Ochoa and Y. Kaziro, *Comp. Biochem.* **16,** 210 (1965).
7. H. G. Wood and M. F. Utter, *Essays Biochem.* **1,** 1 (1965).
8. F. Lynen, *BJ* **102,** 381 (1967).
9. J. Knappe, *Ann. Rev. Biochem.* **39,** 757 (1970).
10. J. Moss and M. D. Lane, *Advan. Enzymol.* **35,** 321 (1971).
11. S. P. Mistry and K. Dakshinamurti, *Vitamins Hormones* **22,** 1 (1964).
12. F. A. Robinson, "The Vitamin Co-factors of Enzyme Systems," p. 497. Pergamon, Oxford, 1966.
13. S. A. Koser, "Vitamin Requirements of Bacteria and Yeasts," p. 37. Thomas, Springfield, Illinois, 1968.
14. W. H. Sebrell, Jr. and R. S. Harris, *in* "The Vitamins," Vol. 2, p. 262. Academic Press, New York, 1968.
15. T. W. Goodwin, "The Biosynthesis of Vitamins and Related Compounds," p. 145. Academic Press, New York, 1963.
16. H. H. Mitchell, "Comparative Nutrition of Man and Domestic Animals," Vol. 2, p. 150. Academic Press, New York, 1964.
17. S. F. Dyke, "The Chemistry of the Vitamins," p. 161. Wiley (Interscience), New York, 1965.
18. A. W. Alberts and P. R. Vagelos, *Proc. Natl. Acad. Sci. U. S.* **59,** 561 (1968).

CoA in a two-step reaction (*19*). Reaction (1) represents the ATP-de-

$$\text{ATP} + \text{HCO}_3^- + \text{Biotin-E} \overset{\text{Mg}^{2+}}{\rightleftharpoons} \text{CO}_2^-\text{-Biotin-E} + \text{ADP} + \text{P}_i \quad (1)$$

$$\text{CO}_2^-\text{-Biotin-E} + \text{CH}_3\text{—}\overset{\overset{\text{CH}_3}{|}}{\text{C}}\text{=CH—CO—SCoA} \rightleftharpoons \text{CH}_3\text{—}\overset{\overset{\text{CH}_2\text{—COO}^-}{|}}{\text{C}}\text{=CH—CO—SCoA} + \text{Biotin-E} \quad (2)$$

$$\text{ATP} + \text{HCO}_3^- + \text{CH}_3\text{—}\overset{\overset{\text{CH}_3}{|}}{\text{C}}\text{=CH—CO—SCoA} \xrightleftharpoons{\text{Mg}^{2+}, \text{Biotin-E}} \text{CH}_3\text{—}\overset{\overset{\text{CH}_2\text{—COO}^-}{|}}{\text{C}}\text{=CH—CO—SCoA} \quad (3)$$

pendent carboxylation of the biotin-enzyme forming carboxybiotin-enzyme, and reaction (2) represents the transfer of the carboxyl group from the carboxybiotin-enzyme to the acceptor molecule, β-methylcrotonyl-CoA. The product of the reaction is β-methylglutaconyl-CoA. This reaction sequence is the model for all of the ATP-dependent biotin carboxylases, and it will therefore be discussed in detail below.

B. Historical Background and Metabolic Significance

In a series of investigations by Plaut and Lardy (*20, 21*) and by Coon (*22*) it was shown that rat liver preparations converted leucine to acetoacetate and acetyl-CoA. ^{14}C from $H^{14}CO_3^-$ was stoichiometrically incorporated into the carboxyl carbon of acetoacetate, and this incorporation was significantly reduced in preparations from animals maintained on a biotin deficient diet. Further studies showed that liver mitochondria from biotin deficient rats were unable to convert isovaleric acid or β-methylcrotonic acid to acetoacetic acid (*23*). Subsequent evidence from Coon's laboratory (*24–26*) indicated that leucine initially underwent transamination to form α-ketoisocaproic acid; oxidative decarboxylation of the latter gave rise to isovaleryl-CoA. Isovaleryl-CoA was shown to undergo dehydrogenation to form β-methylcrotonyl-CoA which, in turn, was hydrated by extracts of rat liver to yield β-hydroxyisovale-

19. F. Lynen, J. Knappe, E. Lorch, G. Jutting, and E. Ringelmann, *Angew. Chem.* **71,** 481 (1959).
20. G. W. E. Plaut and H. A. Lardy, *JBC* **186,** 705 (1950).
21. G. W. E. Plaut and H. A. Lardy, *JBC* **192,** 435 (1951).
22. M. J. Coon, *JBC* **187,** 71 (1950).
23. J. E. Fisher, *Proc. Soc. Exptl. Biol. Med.* **88,** 227 (1955).
24. B. K. Bachhawat, W. G. Robinson, and M. J. Coon, *JACS* **76,** 3098 (1954).
25. B. K. Bachhawat, W. G. Robinson, and M. J. Coon, *JBC* **216,** 727 (1955).
26. B. K. Bachhawat, W. G. Robinson, and M. J. Coon, *JBC* **219,** 539 (1956).

ryl-CoA. Although preliminary experiments suggested that β-hydroxy-isovaleryl-CoA was carboxylated to form β-hydroxy-β-methylglutaryl-CoA directly (*24–26*), later experiments (*27, 28*) indicated that β-hydroxyisovaleryl-CoA is not on the main pathway of leucine degradation; it is formed only in an abortive side reaction. Experiments with a microbial enzyme system reported from Lynen's laboratory (*19, 27, 28*) established that β-methylcrotonyl-CoA is the intermediate which undergoes biotin-dependent carboxylation, and the product of this carboxylation reaction was identified as β-methylglutaconyl-CoA. β-Methylglutaconyl-CoA is subsequently hydrated to form β-hydroxy-β-methylglutaryl-CoA which can be either degraded to acetoacetate and acetyl-CoA or it can serve as a precursor in steroid biosynthesis. These reactions have been confirmed in a number of biological systems (*29–31*).

The discovery that biotin is involved in leucine metabolism and that it participates directly in the β-methylcrotonyl-CoA carboxylase reaction suggested that β-methylcrotonyl-CoA or some earlier metabolites or metabolite derivatives of this pathway might accumulate in biotin deficiency states. In line with this Tanaka and Isselbacher (*32*) have reported that when biotin deficient rats were given large amounts of leucine there occurred a marked increase in the amount of β-hydroxy-isovaleric acid in the urine as well as small but definite increases in isovaleric and β-methylcrotonic acids in the plasma. An inborn error of leucine metabolism in humans causes β-hydroxyisovaleric aciduria. In one patient large amounts of β-hydroxyisovaleric acid and β-methylcrotonic acid conjugated to glycine were excreted in the urine (*33*). These compounds, which were found in significant quantities in both healthy parents and two siblings, who were probably heterozygotes, were not detected in a large series of controls. The genetic defect in this family is probably localized to the β-methylcrotonyl-CoA carboxylase.

27. F. Lynen, *Proc. Intern. Symp. Enzyme Chem., Tokyo Kyoto, 1957* p. 57. Maruzen, Tokyo, 1958.
28. F. Lynen, *J. Cellular Comp. Physiol.* **54**, Suppl. 1, 33 (1959).
29. A. del Campello-Campbell, E. E. Dekker, and M. J. Coon, *BBA* **31**, 290 (1959).
30. H. C. Rilling and M. J. Coon, *JBC* **235**, 3087 (1960).
31. M. J. Coon and W. G. Robinson, "Methods in Enzymology," Vol. 5, p. 451, 1962.
32. K. Tanaka and K. J. Isselbacher, *Lancet* **II**, 930 (1970).
33. L. Eldjarn, E. Jellum, O. Stokke, H. Pande, and P. E. Waaler, *Lancet* **II**, 521 (1970).

C. Distribution

This enzyme has not been studied in many sources. As noted above, it is found in the mitochondria of rat liver, and it has been partially purified from chicken and ox (*29–31*). In addition, it was partially purified from extracts of a *Mycobacterium* and purified to homogeneity and crystallized from extracts of a species of *Achromobacter* (*19, 34–37*) isolated by the enrichment culture technique with isovaleric acid as carbon source. The specific activity of the *Achromobacter* enzyme is related to the state of growth, reaching its highest value during logarithmic growth, and then rapidly decreasing as the cells approach stationary phase (*37*). The enzyme has also been partially purified from a species of *Pseudomonas* grown on hexane (*31*).

D. Substrate Specificity

β-Methylcrotonyl-CoA carboxylase catalyzes the carboxylation of β-methylcrotonyl-CoA to form β-methylglutaconyl-CoA. The apparent substrate affinity constants for the enzyme purified from *Achromobacter* are 8.3×10^{-5} *M* for ATP, 3×10^{-3} *M* for HCO_3^-, and 1.2×10^{-5} *M* for β-methylcrotonyl-CoA (*36*). For the reaction Mg^{2+} is required. However, neither the specificity for the metal requirement nor the optimal concentration of Mg^{2+} has been discussed in the literature. It should be noted that high levels of $MgCl_2$, 7.5 m*M*, compared to ATP concentrations of 0.5 m*M*, are used in the assay. The significance of this is not known. Two other thioesters, acetoacetyl-CoA and crotonyl-CoA, are carboxylated at a rate 14 and 20%, respectively, of the rate of carboxylation of β-methylcrotonyl-CoA, whereas propionyl-CoA and acetyl-CoA are carboxylated at much slower rates (2.6 and 2%, respectively). The relatively high rates of carboxylation of crotonyl-CoA and acetoacetyl-CoA have been attributed to the similarity in structure between the natural substrate, β-methylcrotonyl-CoA, crotonyl-CoA, and the enol form of acetoacetyl-CoA (*36*). The pH optimum of the reaction is 7.9–8.3.

34. J. Knappe, H. G. Schlegel, and F. Lynen, *Biochem. Z.* **335,** 101 (1961).

35. F. Lynen, J. Knappe, E. Lorch, G. Jutting, E. Ringelmann, and J. P. Lachance, *Biochem. Z.* **335,** 123 (1961).

36. R. H. Himes, D. L. Young, E. Ringelmann, and F. Lynen, *Biochem. Z.* **337,** 48 (1963).

37. R. Apitz-Castro, K. Rehn, and F. Lynen, *European J. Biochem.* **16,** 71 (1970).

The carboxylation reaction involves sulfhydryl groups as indicated by the effects of thiol reagents on the enzymic activity (*36*). Thus iodoacetamide, *N*-ethylmaleimide, and *p*-mercuribenzoate are potent inhibitors. The inhibition by *p*-mercuribenzoate is reversed by 2-mercaptoethanol. Protection against inhibition by *N*-ethylmaleimide is afforded by prior incubation of the enzyme with a combination of ATP, $MgCl_2$, and HCO_3^-, or a combination of ADP, $MgCl_2$, and P_i. These experiments suggested that a thiol group was involved in the binding of ATP to the enzyme or that ATP binding and/or carboxybiotin formation caused structural changes which led to burying of the sulfhydryl site. No protection was afforded by incubation with β-methylcrotonyl-CoA.

E. Molecular Characteristics

The crystalline *Achromobacter* β-methylcrotonyl-CoA carboxylase has a molecular weight of 760,000 daltons as determined by sedimentation velocity (*37*). It has a $s_{20,w}$ of 20.7 S and a $D_{20,w}$ of 2.8×10^{-7} cm^2/sec. There are 1.27 μmoles of biotin per gram of protein or 4 moles/mole of enzyme, indicating a tetrameric structure typical of several other biotin enzymes (*9, 36, 37*). Further evidence for a tetrameric structure was obtained by electron microscopy where only particles of about one-fourth the expected size were seen although ultracentrifugation showed no sign of dissociation (*37*). The enzyme has an isoelectric point in the vicinity of pH 3.5, as expected from the high percentage (23%) of glutamic and aspartic acid residues (*37*).

F. Mechanism of Action

Much of our present knowledge of the mechanism of action of biotin carboxylases is derived from the pioneering work of Lynen and his colleagues on β-methylcrotonyl-CoA carboxylase (*19*). The involvement of covalently bound biotin in this enzyme was first suggested by the finding that extracts of liver from biotin deficient rats were devoid of carboxylase activity (*38*). Direct evidence for the presence of biotin was obtained by Lynen *et al.* (*19*) who showed that the biotin content of enzyme preparations purified from *Mycobacterium* sp. was directly proportional to the enzyme activity. In addition, avidin, a protein found in egg white which forms a complex with biotin, specifically inhibited carboxylase activity. A great aid in the study of this enzyme was the important discovery that β-methylcrotonyl-CoA carboxylase catalyzes

38. J. F. Woessner, Jr., B. K. Bachhawat, and M. J. Coon, *JBC* **233**, 520 (1958).

compared to mitochondria of normal rats, thereby implicating biotin in propionate carboxylation. Methylmalonyl-CoA was identified as the product of propionyl-CoA carboxylation by Flavin *et al.* (*50*) and by Katz and Chaikoff (*51*). When propionyl-CoA carboxylase was obtained pure, it was discovered that the carboxylation product is *S*-methylmalonyl-CoA which is racemized to its optical enantiomorph, *R*-methylmalonyl-CoA, by a specific racemase (*52–56*). *R*-Methylmalonyl-CoA is converted by a vitamin B_{12} coenzyme-dependent mutase to succinyl-CoA (*57–59*) in a reaction which involves an intramolecular shift of the thioester carboxyl unit to the methyl carbon (*54, 60, 61*).

Since the main pathway of propionate oxidation in animal tissues involves the carboxylation of propionyl-CoA to methylmalonyl-CoA, the hereditary absence of propionyl-CoA carboxylase leads to severe metabolic disorders and early death (*62, 63*). High concentrations of propionate are found in the plasma and urine of these patients (*62*). The enzyme defect in this inborn error of metabolism was discovered by Hsia *et al.* (*64, 65*) who found that fibroblasts cultured from a child with this condition were unable to oxidize ^{14}C-propionate to $^{14}CO_2$ but were able to oxidize ^{14}C-methylmalonate to $^{14}CO_2$. Further experiments showed that the patient's fibroblasts contained no propionyl-CoA carboxylase while fibroblasts from both parents contained decreased levels of this enzyme.

An alternate route for methylmalonyl-CoA metabolism was suggested by Lynen and Tada (*66*) and by Wawszkiewicz and Lynen (*67*) who proposed that the macrocyclic lactone, erythronolide, the ground

50. M. Flavin, P. J. Ortiz, and S. Ochoa, *Nature* (*London*) **176**, 823 (1955).
51. J. Katz and I. L. Chaikoff, *JACS* **77**, 2659 (1955).
52. R. Mazumder, T. Sasakawa, Y. Kaziro, and S. Ochoa, *JBC* **236**, PC53 (1961).
53. R. Mazumder, T. Sasakawa, Y. Kaziro, and S. Ochoa, *JBC* **237**, 3065 (1962).
54. C. S. Hegre, S. J. Miller, and M. D. Lane, *BBA* **56**, 538 (1962).
55. M. Sprecher, M. J. Clark, and D. B. Sprinson, *BBRC* **15**, 581 (1964).
56. J. Retey and F. Lynen, *BBRC* **16**, 358 (1964).
57. M. Flavin and S. Ochoa, *JBC* **229**, 965 (1957).
58. S. Gurani, P. Mistry, and B. C. Johnson, *BBA* **38**, 187 (1960).
59. R. Mazumder, T. Sasakawa, and S. Ochoa, *JBC* **238**, 50 (1963).
60. H. Eggerer, P. Overath, F. Lynen, and E. R. Stadtman, *JACS* **82**, 2643 (1960).
61. R. W. Kellermeyer and H. G. Wood, *Biochemistry* **1**, 1124 (1962).
62. D. Gompertz, C. N. Storrs, D. C. K. Bau, T. J. Peters, and E. A. Hughes, *Lancet* **I**, 1140 (1970).
63. N. D. Barnes, D. Hull, L. Balgobin, and D. Gompertz, *Lancet* **II**, 244 (1970).
64. Y. E. Hsia, K. J. Scully, and L. E. Rosenberg, *Lancet* **I, 757** (1969).
65. Y. E. Hsia, K. J. Scully, and L. E. Rosenberg, *J. Clin. Invest.* **50,** 127 (1971).
66. F. Lynen and M. Tada, *Angew. Chem.* **73**, 513 (1961).
67. E. J. Wawszkiewicz and F. Lynen, *Biochem. Z.* **340**, 213 (1964).

structure of erythromycin A, an antibiotic produced by *Streptomyces erythraeus,* is synthesized in a condensation decarboxylation reaction between units of methylmalonyl-CoA with propionyl-CoA acting as primer. Experimental support for this hypothesis has been obtained by Kanada and Corcoran (*68*) and Grisebach *et al.* (*69*).

C. Distribution

Propionyl-CoA carboxylase has been purified and crystallized from mitochondria obtained from pig heart (*57, 70–75*) and purified from bovine liver (*76–80*) and rat liver mitochondria (*81*). In a study of the distribution of propionyl-CoA carboxylase in rat liver cells, Scholte (*82*) found that 75% of the enzymic activity was localized in the mitochondria. The activity in the mitochondria was stimulated 8-fold by sonication. Similar results were obtained with two other mitochondrial enzymes, glutamic dehydrogenase and malonyl-CoA decarboxylase. Within the mitochondria the enzyme is either very loosely bound to the inner membrane or it is free, as judged by comparison with marker enzymes. In microorganisms the enzyme has been purified from extracts of *Mycobacterium smegmatis* (*45*) and from *Rhodosporillum rubrum* grown photoheterotrophically on a wide variety of carbon sources (*48, 83*). In addition, the enzyme has been identified in extracts of *Nocardia corallina* (*84*), *Streptomyces erythraeus* (*67*), *Pseudomonas citronellolis* (*85*), *Mycobacterium phlei* (*85*), and *Bacillus cereus* (*85*).

68. T. Kanada and J. W. Corcoran, *Federation Proc.* **20,** 273 (1961).
69. H. Grisebach, H. Achenbach, and W. Hofheinz, *Z. Naturforsch.* **15b,** 560 (1960).
70. M. Flavin, H. Castro-Mendoza, and S. Ochoa, *JBC* **229,** 981 (1957).
71. A. Tietz and S. Ochoa, *JBC* **234,** 1394 (1959).
72. Y. Kaziro, E. Leone, and S. Ochoa, *Proc. Natl. Acad. Sci.* U. S. **46,** 1319 (1960).
73. Y. Kaziro, S. Ochoa, R. C. Warner, and J. Y. Chen, *JBC* **236,** 1917 (1961).
74. Y. Kaziro, A. Grossman, and S. Ochoa, *JBC* **240,** 64 (1965).
75. Y. Kaziro, "Methods in Enzymology," Vol. 13, p. 181, 1969.
76. D. R. Halenz and M. D. Lane, *JBC* **235,** 878 (1960).
77. M. D. Lane, D. R. Halenz, D. P. Kosow, and C. S. Hegre, *JBC* **235,** 3082 (1960).
78. M. D. Lane, "Methods in Enzymology," Vol. 5, p. 576, 1962.
79. D. R. Halenz, J. Feng, C. S. Hegre, and M. D. Lane, *JBC* **237,** 2140 (1962).
80. C. S. Hegre and M. D. Lane, *BBA* **128,** 172 (1966).
81. H. Y. Neujahr and S. P. Mistry, *Acta Chem. Scand.* **17,** 1140 (1963).
82. H. R. Scholte, *BBA* **178,** 137 (1969).
83. M. Knight, *BJ* **84,** 170 (1962).
84. C. L. Bangh, D. S. Bates, G. W. Claus, and C. H. Werkman, *Enzymologia* **23,** 15 (1961).
85. A. W. Alberts, unpublished observations (1970).

D. Substrate Specificity

Propionyl-CoA carboxylase catalyzes the carboxylation of propionyl-CoA to form *S*-methylmalonyl-CoA. The apparent substrate affinity constants for the enzymes isolated from both pig heart and bovine liver mitochondria are essentially identical (*71, 73, 79, 80*). Thus, the apparent K_m for propionyl-CoA ranged from $2.0 \times 10^{-4}\ M$ to $2.6 \times 10^{-4}\ M$, for ATP it was $8.0 \times 10^{-5}\ M$ to $55.0 \times 10^{-5}\ M$, and for bicarbonate it was from $1.9 \times 10^{-3}\ M$ to $2.3 \times 10^{-3}\ M$. Propionyl-CoA carboxylase catalyzes the carboxylation of a wide variety of acyl-CoA esters (*71, 73, 79, 80*) including acetyl-CoA, butyryl-CoA, and valeryl-CoA at rates greatly reduced from that found with propionyl-CoA; however, all these thioesters had K_m values similar to that for propionyl-CoA. The enzyme purified from *M. smegmatis* (*45*) has a K_m for propionyl-CoA of $5.2 \times 10^{-5}\ M$ while that from *R. rubrum* (*48*) has a K_m of $1.3 \times 10^{-4}\ M$. The *R. rubrum* enzyme also carboxylates acetyl-CoA and butyryl-CoA at reduced rates.

A striking difference between the mammalian and bacterial enzymes is the pH optimum of the reaction (*45, 73*). Thus in the mammalian system there is a sharp optimum between pH 8.0 and 8.5 (*73*) while the *M. smegmatis* enzyme has a rather broad optimum ranging from 6.8 to 8.7 (*45*).

Bovine liver and pig heart propionyl-CoA carboxylase are activated by certain monovalent cations (*81, 86–88*). Thus the bovine liver enzyme is stimulated 7- to 9-fold by K^+, Rb^+, Cs^+, or NH_4^+ ions, while Na^+, Li^+, or tetramethylammonium ions showed little or no activation (*86*). No effect of these cations was noted on the K_m values for the different substrates. The rate of carboxylation catalyzed by the pig heart enzyme was only increased 2- to 2.5-fold by monovalent cations (*81, 87, 88*). However, the K_m value for bicarbonate was decreased from 8×10^{-3} to $3 \times 10^{-3}\ M$ while the K_m values for the other substrates were unchanged by the monovalent cations.

Mammalian and bacterial propionyl-CoA carboxylases are sulfhydryl enzymes (*73, 79, 80*). They show marked sensitivity to sulfhydryl poisons such as *p*-mercuribenzoate [100% inhibition of the pig heart enzyme (*73*) at $5 \times 10^{-6}\ M$] and *N*-ethylmaleimide and require sulfhydryl compounds for maximal activity. Bovine liver propionyl-CoA carboxylase is protected by propionyl-CoA from *p*-mercuribenzoate inhibition (*80*).

86. A. J. Georgio and G. W. E. Plaut, *BBA* **139,** 487 (1967).
87. H. Y. Neujahr, *Acta Chem. Scand.* **17,** 1777 (1963).
88. J. B. Edwards and D. B. Keech, *BBA* **159,** 167 (1968).

E. Molecular Characteristics

Propionyl-CoA carboxylase has been crystallized and extensively characterized from mitochondrial extracts of pig heart (*1, 4, 6, 73–75*). This enzyme has a molecular weight of 700,000, and it contains one mole of bound biotin per 175,000 g of enzyme, indicating that the enzyme might consist of four subunits of molecular weight 175,000. The enzyme has a sedimentation coefficient of 19.7 S. In the presence of 7 *M* urea the enzyme dissociates into inactive subunits of 2.5 S (*73*). The molecular weight of these subunits is unknown. The bovine liver mitochondrial enzyme appears to have very similar molecular properties (*76–79*) although is has not been as well characterized as the pig heart enzyme. The isoelectric point of the crystalline pig heart enzyme is 6.1 (*73*).

Biotin Binding Site

Propionyl-CoA carboxylase is a biotin enzyme, and its activity is greatly depressed in biotin deficient rats (*49, 89, 90*). Kosow and Lane (*90*) showed that incubation of liver slices from biotin deficient rats with (+)-biotin led to a rapid restoration of enzymic activity. Incubation of partially purified preparations obtained from livers of biotin deficient rats with ATP and (+)-^{14}C-biotin yielded propionyl-CoA carboxylase labeled with ^{14}C-biotin (*91, 92*). This preparation was hydrolyzed with Pronase, a bacterial proteolytic enzyme (*93*), and ^{14}C-labeled biocytin was isolated from the hydrolyzate and identified by paper and ion exchange chromatography (*92*). This experiment indicated that the biotin of propionyl-CoA carboxylase is covalently linked to an ϵ-amino group of lysine, as was shown with β-methylcrotonyl-CoA carboxylase. Biocytin has also been identified as a component of propionyl-CoA carboxylase of pig heart (*74*), ox liver (*44*), and *M. smegmatis* (*45*).

An enzyme has been purified from livers of biotin deficient rats that catalyzes a Mg^{2+}-ATP-dependent synthesis of propionyl-CoA carboxylase holoenzyme from ^{14}C-biotin and the apoenzyme (*94–98*). No other co-

89. H. A. Lardy and J. Adler, *JBC* **219,** 933 (1956).
90. D. P. Kosow and M. D. Lane, *BBRC* **4,** 92 (1961).
91. D. P. Kosow and M. D. Lane, *BBRC* **5,** 191 (1961).
92. D. P. Kosow and M. D. Lane, *BBRC* **7,** 439 (1962).
93. M. Nomoto, Y. Narahashi, and M. Murakami, *J. Biochem.* (*Tokyo*) **45,** 593 (1960).
94. D. P. Kosow, S. C. Huang, and M. D. Lane, *JBC* **237,** 3633 (1962).
95. J. L. Foote, J. E. Christner, and M. J. Coon, *Federation Proc.* **21,** 239 (1962).
96. J. L. Foote, J. E. Christner, and M. J. Coon, *BBA* **67,** 676 (1963).
97. L. Siegel, J. L. Foote, and M. J. Coon, *JBC* **240,** 1025 (1965).
98. H. C. McAllister and M. J. Coon, *JBC* **241,** 2855 (1966).

factors are required for this reaction and biotinyl adenylate was identified as an intermediate (*97, 99*). A similar enzymic activity has been obtained from extracts of *Propionibacterium shermanii* (*94, 100*); this enzyme catalyzes the synthesis of both pig heart propionyl-CoA carboxylase holoenzyme and bacterial methylmalonyl-CoA transcarboxylase holoenzyme. These holoenzyme synthetases apparently specifically select one of many lysine residues of the apoenzyme in forming the holoenzyme. The fact that the bacterial enzyme catalyzes the synthesis of an animal holocarboxylase and a bacterial holotranscarboxylase suggests that the site of biotin binding must be very similar in these two different biotin enzymes (*94, 98, 100*).

F. Mechanism of Action

Propionyl-CoA carboxylase catalyzes the two-step carboxylation of propionyl-CoA shown in reactions (8)–(10). The first direct evidence that CO_2^--enzyme participates in biotin-dependent carboxylation reactions was obtained by Kaziro and Ochoa (*41*) with pig heart propionyl-CoA carboxylase. Incubation of the enzyme with either Mg^{2+}-ATP and $H^{14}CO_3^-$ for the forward reaction or with ^{14}C-carboxyl-labeled methylmalonyl-CoA for the backward reaction led to the isolation of enzyme containing approximately one mole of $^{14}CO_2^-$ per mole of biotin. Incubation of the $^{14}CO_2^-$-enzyme with propionyl-CoA gave rise to ^{14}C-methylmalonyl-CoA and incubation with Mg^{2+}, ADP, and $^{32}P_i$ gave rise to ^{32}P-ATP. Both of these reactions were inhibited by avidin.

Further information on the mechanism of action of propionyl-CoA carboxylase was obtained from the study of exchange reactions which measure the partial reactions catalyzed by this enzyme (*72, 101, 102*), reactions (11)–(13).

$$\text{ATP} + {}^{32}\text{P}_i \overset{Mg^{2+}}{\rightleftharpoons} {}^{32}\text{P-ATP} + \text{P}_i \qquad (11)$$

$$\text{ATP} + {}^{14}\text{C-ADP} \overset{Mg^{2+}}{\rightleftharpoons} {}^{14}\text{C-ATP} + \text{ADP} \qquad (12)$$

$$^{14}\text{C-Propionyl-CoA} + \text{methylmalonyl-CoA} \rightleftharpoons \text{propionyl-CoA} + {}^{14}\text{C-methylmalonyl-CoA} \qquad (13)$$

The results of experiments with propionyl-CoA carboxylase were very similar to those described with β-methylcrotonyl-CoA carboxylase (see

99. L. Siegel, J. L. Foote, J. E. Christner, and M. J. Coon, *BBRC* **13,** 307 (1963).
100. M. D. Lane, D. L. Young, and F. Lynen, *JBC* **239,** 2858 (1964).
101. D. R. Halenz and M. D. Lane, *BBA* **48,** 425 (1961).
102. Y. Kaziro, L. F. Hass, P. D. Boyer, and S. Ochoa, *JBC* **237,** 1460 (1962).

Section I,F); i.e., both ADP and HCO_3^- were required for the ATP-$^{32}P_i$ exchange, and the three reactions were inhibited by avidin (*19*).

The exchange reactions have been used to study the function of monovalent cations in the carboxylase reaction (*86*). It was found that the three exchange reactions as well as the overall reaction were stimulated by K^+. This is in contrast to the requirement for divalent cation which is only required for the ATP-dependent carboxylation of biotin-enzyme [reaction (8)] and the exchange reactions involving ATP [reactions (11) and (12)]. Divalent cation is not required for the transcarboxylation partial reaction [reaction (9)] which is measured most conveniently as the exchange reaction between ^{14}C-propionyl-CoA and methylmalonyl-CoA [reaction (13)]. Of interest in this study was the finding that the overall carboxylase activity in the presence or absence of monovalent cations showed a pH optimum of pH 8–8.5 with either Mn^{2+} or Mg^{2+} ions. However, the ATP-P_i and ADP-ATP exchanges had pH optima of 6.8–7.0 and showed a marked preference for Mn^{2+} ions. Again in these exchange reactions monovalent cations stimulated but had no effect on the divalent cation requirement. Since the exchanges reflect the partial reactions of the propionyl-CoA carboxylase reaction, these results presented an enigma for which no explanation was offered.

The apparent equilibrium constant of the reaction catalyzed by pig heart propionyl-CoA carboxylase has been determined by Kaziro *et al.*, reaction (14) (*74*).

$$K' = \frac{[\mathrm{ADP}][\mathrm{P_i}]\,[\text{methylmalonyl-CoA}]}{[\mathrm{ATP}][\mathrm{HCO_3^-}]\,[\text{propionyl-CoA}]} \tag{14}$$

At pH 8.1 and 28° it averaged 5.7. The free-energy change, $\Delta F'_{301} = -1078$ cal/mole, indicates that the reaction is readily reversible, the carboxylation of propionyl-CoA being slightly favored.

The mechanism of transfer of the carboxyl group of carboxybiotin-enzyme to propionyl-CoA [reaction (9)] has been studied by Retey and Lynen (*103*), Arigoni *et al.* (*104*), and Prescott and Rabinowitz (*105*). The product of the reaction is methylmalonyl-CoA in the *S* configuration. 2-^{3}H-Propionyl-CoA was chemically synthesized with the isotope in either the 2-*S* or 2-*R* position. With these compounds it was shown that the 2-*R* hydrogen enters the solvent while the 2-*S* hydrogen is retained in the methylmalonyl-CoA formed. Thus, during the transcarboxylation step of propionyl-CoA carboxylation, configuration is retained at the C-2 position. The rate at which the 2-*R* hydrogen enters

103. J. Retey and F. Lynen, *Biochem. Z.* **342,** 256 (1965).
104. D. Arigoni, F. Lynen, and J. Retey, *Helv. Chim. Acta* **49,** 311 (1966).
105. D. J. Prescott and J. L. Rabinowitz, *JBC* **243,** 1551 (1968).

the solvent is the same as the rate of methylmalonyl-CoA formation, indicating an absence of an isotope effect. During this process there was no evidence for activation of the C–H bond by the enzyme (*105*). A mechanism proposed for the reaction involves simultaneous proton removal and proton transfer (*103*).

IV. Acetyl-CoA Carboxylase

A. Reaction Catalyzed

Acetyl-CoA carboxylase catalyzes the biotin-dependent carboxylation of acetyl-CoA to form malonyl-CoA in a two-step reaction (*1–10*), reactions (15)–(17).

$$\text{ATP} + \text{HCO}_3^- + \text{biotin-E} \rightleftharpoons \text{CO}_2^-\text{-biotin-E} + \text{ADP} + \text{P}_i \quad (15)$$

$$\text{CO}_2^-\text{-biotin-E} + \text{CH}_3\text{CO—SCoA} \rightleftharpoons {}^-\text{OOCCH}_2\text{CO—SCoA} + \text{biotin-E} \quad (16)$$

$$\text{Sum: ATP} + \text{HCO}_3^- + \text{CH}_3\text{CO—SCoA} \xrightleftharpoons{\text{Me}^{2+},\ \text{biotin-E}} {}^-\text{OOCCH}_2\text{CO—SCoA} + \text{ADP} + \text{P}_i \quad (17)$$

B. Historical Background and Metabolic Significance

The idea that biotin was in some way involved in fatty acid biosynthesis was derived from the finding that oleic acid had a sparing effect on the biotin requirement of lactic acid bacteria (*14, 106–109*). Stimulation of fatty acid synthesis by bicarbonate was demonstrated by Brady and Gurin (*110*) and by Lyon *et al.* (*111*) in rat liver sections. The conclusive evidence for the participation of bicarbonate in the synthesis of long-chain fatty acids from acetate was shown by Klein (*112*) using a cell-free fatty acid synthesizing system from yeast, by Squires *et al.* (*113*) with a system from avocado, and by Gibson *et al.* (*114*) with a system from avian liver. The elucidation of the catalytic role of bicarbonate

106. V. R. Williams and E. A. Freger, *JBC* **166**, 355 (1946).
107. R. L. Potter and C. A. Elvehjem, *JBC* **169**, 63 (1947).
108. A. E. Axelrod, K. Hoffman, and B. F. Daubert, *JBC* **169**, 761 (1947).
109. W. L. Williams, H. P. Broquist, and E. E. Snell, *JBC* **170**, 619 (1947).
110. R. O. Brady and S. Gurin, *JBC* **199**, 421 (1952).
111. I. Lyon, R. P. Geyer, and L. D. Marshall, *JBC* **217**, 757 (1957).
112. H. P. Klein, *J. Bacteriol.* **73**, 530 (1957).
113. C. Squires, P. K. Stumpf, and C. Schmid, *Plant Physiol.* **33**, 365 (1958).
114. D. M. Gibson, E. B. Titchener, and S. J. Wakil, *BBA* **30**, 376 (1958).

in fatty acid synthesis was obtained in Wakil's laboratory (*114–119*). The fatty acid synthesizing system of pigeon liver was separated into two protein fractions, and one of these fractions contained a significant amount of biotin. Wakil (*116*) demonstrated that this fraction catalyzed the first step in fatty acid synthesis, the Me^{2+}-ATP dependent carboxylation of acetyl-CoA to form malonyl-CoA. Avidin inhibited this reaction, thus showing that biotin is involved in this carboxylation. This important finding, that the first step in fatty acid biosynthesis is the formation of malonyl-CoA, has since been confirmed in numerous laboratories and the enzyme system catalyzing this reaction, acetyl-CoA carboxylase, has been the subject of intensive investigations. The role of malonyl-CoA as the condensing unit in fatty acid biosynthesis has been well documented (*5, 8, 120*). In addition to being involved in fatty acid biosynthesis, malonyl-CoA has been reported to be an intermediate in the biosynthesis of biotin (*121*), thus showing that a biotin enzyme, acetyl-CoA carboxylase, is required in the biosynthesis of biotin. Malonyl-CoA is also directly involved in aromatic acid biosynthesis (*66, 120*). Thus, Lynen has shown that 1 mole of acetyl-CoA condenses with 3 moles of malonyl-CoA to form 6-methylsalicylic acid. Other compounds that are probably produced by a related mechanism include griseofulvin, terramycin, and penicillic acid (*122, 123*).

C. Distribution

Acetyl-CoA carboxylase is widely distributed in nature; this enzyme probably accounts for the majority of malonyl-CoA formation in most classes of organisms. Originally shown to be present in extracts of avian liver by Wakil (*116*) and subsequently purified from this source (*119, 124–129*), it has also been purified from rat epididimal adipose tissue

115. S. J. Wakil, E. B. Titchener, and D. M. Gibson, *BBA* **29,** 225 (1958).
116. S. J. Wakil, *JACS* **80,** 6465 (1958).
117. S. J. Wakil, E. B. Titchener, and D. M. Gibson, *BBA* **34,** 227 (1959).
118. S. J. Wakil and D. M. Gibson, *BBA* **41,** 122 (1960).
119. M. Waite and S. J. Wakil, *JBC* **237,** 2750 (1962).
120. F. Lynen, *Federation Proc.* **20,** 941 (1961).
121. A. Lezius, E. Ringelmann, and F. Lynen, *Biochem. Z.* **336,** 510 (1963).
122. R. J. Light and C. D. Hager, *ABB* **125,** 326 (1968).
123. P. Dimroth, H. Walter, and F. Lynen, *European J. Biochem.* **13,** 109 (1970).
124. S. Numa, E. Ringelmann, and B. Reidel, *BBRC* **24,** 750 (1966).
125. T. Goto, E. Ringelmann, B. Reidel, and S. Numa, *Life Sci.* **6,** 785 (1967).
126. C. Gregolin, E. Ryder, A. K. Kleinschmidt, R. C. Warner, and M. D. Lane, *Proc. Natl. Acad. Sci. U. S.* **56,** 148 (1966).

(*130–132*), bovine adipose tissue (*133*), rat liver (*134, 135*), rat mammary gland (*136*), brewer's yeast (*137*), wheat germ (*138*), and *Escherichia coli* (*18, 139–146*). In addition it has been identified in a number of the microorganisms including *Lactobacillus plantarum* (*147, 148*), *Mycobacterium avium* (*149*), *Mycobacterium phlei* (*85*), *Bacillus cereus* (*85*), *Pseudomonas citronellolis* (*85*), and *Brevibacterium thiogenitalis* (*150*).

Most of the present evidence indicates that in animal sources acetyl-CoA carboxylase is located in the soluble supernatant fraction of homogenized cells. However, Margolis and Baum (*151*) reported that the car-

---

127. C. Gregolin, E. Ryder, R. C. Warner, A. K. Kleinschmidt, and M. D. Lane, *Proc. Natl. Acad. Sci. U. S.* **56,** 1751 (1966).
128. C. Gregolin, E. Ryder, and M. D. Lane, *JBC* **243,** 4227 (1968).
129. C. Gregolin, E. Ryder, R. C. Warner, A. K. Kleinschmidt, H. Chang, and M. D. Lane, *JBC* **243,** 4236 (1968).
130. P. R. Vagelos, A. W. Alberts, and D. B. Martin, *BBRC* **8,** 4 (1962).
131. P. R. Vagelos, A. W. Alberts, and D. B. Martin, *JBC* **238,** 533 (1963).
132. P. R. Desjardins and K. Dakshenamurti, *Federation Proc.* **28,** 557 (1969).
133. A. K. Kleinschmidt, J. Moss, and M. D. Lane, *Science* **166,** 1276 (1969).
134. M. Matsuhashi, S. Matsuhashi, and F. Lynen, *Biochem. Z.* **340,** 263 (1964).
135. S. Nakanishi and S. Numa, *European J. Biochem.* **16,** 161 (1970).
136. A. L. Miller and H. R. Levy, *JBC* **244,** 2334 (1969).
137. M. Matsuhashi, S. Matsuhashi, and F. Lynen, *Biochem. Z.* **340,** 243 (1964).
138. M. D. Hatch and P. K. Stumpf, *JBC* **236,** 2879 (1961).
139. P. R. Vagelos, A. W. Alberts, J. Elovson, and G. L. Powell, *in* "Physiology of Adipose Tissue" (J. Vague, ed.), p. 2. Excerpta Med. Found., Amsterdam, 1969.
140. A. W. Alberts, A. M. Nervi, and P. R. Vaglos, *Proc. Natl. Acad. Sci. U. S.* **63,** 1319 (1969).
141. A. M. Nervi, A. W. Alberts, and P. R. Vagelos, *ABB* **143,** 401 (1971).
142. A. W. Alberts, S. Gordon, and P. R. Vagelos, *Proc. Natl. Acad. Sci. U. S.* **68,** 1259 (1971).
143. R. R. Fall, A. M. Nervi, A. W. Alberts, and P. R. Vagelos, *Proc. Natl. Acad. Sci. U. S.* **68,** 1512 (1971).
144. P. Dimroth, R. B. Guchhait, E. Stoll, and M. D. Lane, *Proc. Natl. Acad. Sci. U. S.* **67,** 1353 (1970).
145. R. B. Guchhait, J. Moss, W. Sokolski, and M. D. Lane, *Proc. Natl. Acad. Sci. U. S.* **68,** 653 (1971).
146. P. Dimroth, R. B. Guchhait, and M. D. Lane, *Z. Physiol. Chem.* **352,** 351 (1971).
147. J. Birnbaum, *ABB* **132,** 436 (1969).
148. J. Birnbaum, *J. Bacteriol.* **104,** 171 (1970).
149. M. Kusunose, E. Kusunose, and Y. Yamamura, *J. Biochem.* (*Tokyo*) **46,** 525 (1959).
150. T. Kanzaki, K. Isobe, H. Okazaki, and H. Fukuda, *Agr. Biol. Chem.* (*Tokyo*) **33,** 771 (1969).
151. S. A. Margolis and H. Baum, *ABB* **114,** 445 (1966).

boxylase is associated with the microsomal fraction of pigeon liver. Similar results were found for the carboxylase of lactating rabbit mammary gland (*152*). In addition Yates *et al.* (*153*) have apparently localized acetyl-CoA carboxylase to the membranes of the granular cytoplasmic reticulum in hepatocytes of glutaraldehyde-fixed rat liver. In these studies the authors noted the formation of lead phosphate precipitate when fixed liver sections were incubated with ATP, $MnCl_2$, bicarbonate, acetyl-CoA, and biotin. Omission of any of these components or the inclusion of avidin prevented the formation of precipitate. The precipitate appeared to be associated with the membranes of the granular reticulum; none was seen over the mitochondrial matrix, membranes of the Golgi complex, microbodies, or plasma membranes with the exception of the bile canaliculi. Since there is some discrepancy in the results of the various approaches that have attempted to localize this enzyme within animal cells, further studies are required. Of interest, of course, is the possibility that acetyl-CoA carboxylase and the fatty acid synthetase complex might be closely associated in the cell.

In green plants, acetyl-CoA carboxylase appears to be localized within the chloroplast, although thus far it has not been possible to isolate it directly from this source (*154*). Acetyl-CoA carboxylase has not been localized in microbial systems.

D. Substrate Specificity

Acetyl-CoA carboxylase catalyzes the carboxylation of acetyl-CoA to form malonyl-CoA. The rat liver enzyme has undergone extensive kinetic analysis (*155*). The apparent K_m values were found to be $2.5 \times 10^{-5}\,M$ for acetyl-CoA and $2.5 \times 10^{-3}\,M$ for bicarbonate. Two values for the apparent K_m for ATP could be derived from Lineweaver-Burk plots at low ($1.5 \times 10^{-5}\,M$) and high ($4 \times 10^{-5}\,M$) ATP concentrations, suggesting that at high concentrations, ATP might act as an allosteric effector. Similar nonlinear plots for ATP were reported with acetyl-CoA carboxylases from avian liver (*128*) and rat mammary gland (*136*). With the exception of the enzyme from *E. coli* (*18*), acetyl-CoA carboxylase from other sources carboxylates propionyl-CoA almost as well as acetyl-CoA (*128, 134, 136, 137*). Either Mg^{2+} or Mn^{2+} is utilized by acetyl-CoA car-

152. D. J. Easter and R. Dils, *BBA* **152,** 653 (1968).
153. R. D. Yates, J. A. Higgins, and R. J. Barrnett, *J. Histochem. Cytochem.* **17,** 379 (1969).
154. D. Burton and P. K. Stumpf, *ABB* **117,** 604 (1966).
155. T. Hashimoto and S. Numa, *European J. Biochem.* **18,** 319 (1971).

boxylase with somewhat different specificities depending on the enzyme source and the pH of the assay (*18, 85, 128, 136, 156*).

Brady and Gurin in 1952 (*110*) reported that fatty acid synthesis in cell-free extracts of pigeon liver was greatly stimulated by citrate. A similar effect was subsequently observed in extracts from several sources (*157–161*), and a number of tricarboxylic acid cycle intermediates were found to be stimulatory in these systems although citrate and isocitrate were most effective. An early suggestion that citrate acted as a source of TPNH or CO_2 (*162*), both of which are required for fatty acid synthesis, was discarded when Porter *et al.* (*163*) and Numa *et al.* (*164*) showed that citrate could not be replaced by either a TPNH-generating system or by CO_2. The citrate effect on fatty acid synthesis was localized to the first enzyme in the pathway, acetyl-CoA carboxylase, simultaneously in three laboratories (*119, 165, 166*). Martin and Vagelos (*165*) showed that citrate did not act as a carboxyl donor during the carboxylation of acetyl-CoA, and it did not exchange with the HCO_3^- added to the medium. Vagelos *et al.* (*130, 131*) established that citrate is an allosteric activator of acetyl-CoA carboxylase. These investigators showed that activation of the enzyme by citrate was associated with a conversion of the inactive protomeric form to an active aggregated form of the enzyme; the activation and aggregation caused by citrate were freely reversible and parallel effects. These observations were confirmed by studies of acetyl-CoA carboxylase in a number of animal systems including rat liver (*134, 167*), avian liver (*124–127, 168*), bovine adipose tissue (*133*), and rat mammary gland (*136*). Citrate has no apparent effect on the purified acetyl-CoA carboxylase from *E. coli* or brewer's yeast (*18, 137*). A stimulatory effect of citrate and other compounds on

156. R. M. Scorpio and E. J. Masoro, *BBRC* **31**, 950 (1968).
157. N. L. R. Bucker, *JACS* **75**, 498 (1953).
158. G. Popjak and B. Tietz, *BJ* **56**, 46 (1954).
159. R. G. Langdon, *JBC* **226**, 615 (1957).
160. M. D. Siperstein and V. M. Fagan, *J. Clin. Invest.* **37**, 1185 (1958).
161. S. Abraham, K. J. Mathes, and I. L. Chaikoff, *JBC* **235**, 2551 (1960).
162. R. O. Brady, A. M. Mamoon, and E. R. Stadtman, *JBC* **222**, 795 (1956).
163. J. W. Porter, S. J. Wakil, A. Tietz, M. I. Jacob, and D. M. Gibson, *BBA* **25**, 35 (1957).
164. S. Numa, M. Matsuhashi, and F. Lynen, *Biochem. Z.* **334**, 203 (1961).
165. D. B. Martin and P. R. Vagelos, *JBC* **237**, 1787 (1962).
166. M. Matsuhashi, S. Matsuhashi, S. Numa, and F. Lynen, *Federation Proc.* **21**, 288 (1962).
167. S. Numa, W. M. Bortz, and F. Lynen, *Advan. Enzyme Regulation* **3**, 407 (1965).
168. M. D. Lane, J. Edwards, E. Stoll, and J. Moss, *Vitamins Hormones* **28**, 345 (1970).

the acetyl-CoA carboxylase from baker's yeast has not been fully characterized (*169*).

E. Molecular Characteristics

Extensive studies have been carried out on the avian liver (*124–129*, *168*, *170*), bovine adipose tissue (*133*), and on the rat liver enzyme (*134*, *135*, *155*, *171*, *172*). A summary of some of the molecular properties of the enzymes from these sources is shown in Table I. The avian liver enzyme has been shown to exist in two forms, an active polymeric form with a molecular weight ranging from 4 to 8 million and an inactive protomeric form with a weight of 410,000 (*129*, *168*). An electron microscopic investigation of the two forms revealed that the inactive protomeric form appeared as particles with a diameter of 80–140 Å (*133*). The active polymeric form is composed of 10–20 protomers, and it appears in the electron microscope as filaments with dimensions of 0.2–0.5 μ in length and 80–100 Å in width.

A number of factors have been investigated that effect the interconversion between the protomeric and polymeric forms of the enzyme (*124*, *129*, *168*). Aggregation of the enzyme is favored by high protein concentration, anions (such as citrate, isocitrate, malonate, tricarballylate, sulfate, and phosphate), acetyl-CoA, and pH 6.5–7. Dissociation to the protomeric form is favored by alkaline pH, Cl^-, low protein concentration, and carboxylation of the enzyme; Mg^{2+}-ATP and malonyl-CoA also favor dissociation, and this may result from carboxylation of the enzyme which occurs in the presence of these compounds (*129*, *168*). However, direct binding of these compounds, in the absence of carboxylation, might also promote dissociation as indicated by the finding that either ATP or Mg^{2+} alone converts the large form into an intermediate-size form (*170*). As can be seen from Table I, carboxylases from other sources behave in a manner similar to avian liver acetyl-CoA carboxylase. Attempts have been made to dissociate only the avian liver enzyme into its component subunits (*129*, *168*). As shown in the table, treatment with 0.1–1.0% sodium dodecyl sulfate leads to the formation of subunits of about 110,000 molecular weight. Such denaturation by sodium dodecyl sulfate is irreversible, and the subunits are apparently inactive.

169. R. K. Rasmussen and H. P. Klein, *BBRC* **28**, 415 (1967).
170. S. G. Numa, T. Goto, E. Ringelmann, and B. Riedel, *European J. Biochem.* **3**, 124 (1967).
171. S. Numa, E. Ringelmann, and F. Lynen, *Biochem. Z.* **343**, 243 (1965).
172. S. Numa and E. Ringelmann, *Biochem. Z.* **343**, 258 (1965).

TABLE I
MOLECULAR PROPERTIES OF ACETYL-CoA CARBOXYLASE

Source	Conditions	Form	Sedimentation velocity: Density gradient (S)	Sedimentation velocity: Anal. ultracentrifugation (S)	Molecular weight (sedimentation equilibrium)	Ref.
Avian liver	Citrate, isocitrate or phosphate	Polymeric (filamentous)	47–50	55–59	4,000,000–8,000,000	*126, 127, 129, 168, 170*
	0.5 *M* NaCl, pH 8–9	Protomeric	13–15	13.1	410,000	
	0.1–1.0% SDS	Subunit		4.0	110,000	
Bovine adipose tissue	Citrate	Polymeric (filamentous)	47–50	68	Several million	*133*
	0.5 *M* NaCl, pH 8–9	Protomeric	13–15	14.4	550,000	
Rat liver	Citrate	Polymeric	51	57		*135*

F. Subunit Structure and Function

1. *Demonstration of Active Subunits in E. coli*

As noted above attempts to dissociate the animal acetyl-CoA carboxylase into active subunits have been unsuccessful. Earlier experience with the fatty acid synthetase indicated that this enzyme system in animals exists as a tight multienzyme complex that cannot be dissociated into active components, whereas the fatty acid synthetase of *E. coli* readily dissociates into active component enzymes (*5*, *139*). *Escherichia coli* acetyl-CoA carboxylase was therefore examined to discover whether it might be dissociable into active subunits. Initial experiments with the *E. coli* enzyme indicated that it could be partially purified by ammonium sulfate precipitation (*18*, *139*). Attempts at further purification by fractionation with alumina $C\gamma$ led to complete loss of enzymic activity which could, however, be reconstituted when two fraction (E_a and E_b) from the alumina $C\gamma$ step were mixed. Table II shows the requirements for acetyl-CoA carboxylase activity as measured by the formation of ^{14}C-malonyl-CoA from acetyl-CoA, ATP, $MnCl_2$, and $H^{14}CO_3^-$.

An initial indication of the roles of the two protein fractions in the carboxylation reaction was obtained with the use of avidin, the specific biotin binding protein. Incubation of E_a with avidin, followed by the addition of excess biotin to inactivate the avidin prior to the addition to E_b, led to complete loss of acetyl-CoA carboxylase activity. On the other hand, when E_b was initially incubated with avidin in a similar manner, there was no loss of carboxylase activity. These experiments indicated that only E_a contained covalently bound biotin. When E_a and E_b were

TABLE II
Requirements for Malonyl-CoA Formation

Reaction system	Malonyl-CoA (nmoles)
Complete[a]	3.7
Complete minus E_a	0
Complete minus E_b	0
Complete minus $MnCl_2$	0
Complete minus ATP	0.28
Complete minus Acetyl-CoA	0

[a] The complete reaction mixtures contained 55 mM imidazole-HCl buffer, pH 6.7; 0.44 mM $MnCl_2$; 0.44 mM ATP; 0.3 mM acetyl-CoA; 14 mM $KH^{14}CO_3$ (0.2 μCi/μmole); and 0.005 unit each of E_a and E_b in a total volume of 0.09 ml.

partially purified, they were subjected to acid hydrolysis and the hydrolysates were analyzed for biotin in a microbiological assay with *Lactobacillus plantarum*. As anticipated from the avidin studies the E_a preparation contained biotin (2.3 nmoles/mg protein), whereas E_b contained no biotin. Furthermore, when E_a was incubated with ATP, $H^{14}CO_3^-$, and $MnCl_2$ and then filtered through Sephadex G-50 to separate the protein from the low molecular weight radioactive compounds, radioactivity appeared in the protein fraction indicating that 2.4 nmoles of ^{14}C-carboxy-E_a per milligram of protein was formed, an amount stoichiometric with the amount of biotin in E_a. Omission of ATP or Mn^{2+} or the prior addition of avidin prevented the formation of carboxy-E_a. Similar experiments with E_b, which contained no biotin, indicated that it was not carboxylated in a similar reaction mixture. In addition, it was established that E_b was not required for the formation of E_a-$^{14}CO_2^-$. These experiments showed that E_a contained biotin and functioned in the first partial reaction of acetyl-CoA carboxylation, reaction (15).

Since E_b was not required in the formation of carboxybiotin-E_a, it was apparent that it must function in the second partial reaction, reaction (16). To test this ^{14}C-carboxybiotin-E_a, isolated by Sephadex column chromatography, was incubated with acetyl-CoA and E_b, and the results of this experiment are indicated in Table III. ^{14}C-Malonyl-CoA was formed in good yield in the complete system and both acetyl-CoA and E_b were required for this reaction. Thus E_b appeared to catalyze a transcarboxylation, the transfer of the ^{14}C-carboxyl group from ^{14}C-carboxybiotin-E_a to acetyl-CoA, forming ^{14}C-malonyl-CoA. These preliminary experiments indicated that *E. coli* acetyl-CoA carboxylase was readily separated into two protein fractions, and the functions of these

TABLE III
REQUIREMENTS FOR CARBOXYL TRANSFER FROM E_a-$^{14}CO_2^-$ TO ACETYL-CoA

Reaction system	E_a-$^{14}CO_2^-$ (nmole)	^{14}C-Malonyl-CoA formed (nmole)	Yield (%)
Complete[a]	0.0055	0.0052	94.7
Complete	0.013	0.011	84.5
Complete	0.037	0.021	56.6
Complete minus E_b	0.037	0	0
Complete minus acetyl-CoA	0.037	0	0

[a] The complete system contained 55 mM imidazole-HCl, pH 7.5; 0.15 mM acetyl-CoA; 0.003 unit E_b; and E_a-$^{14}CO_2^-$ in the amounts indicated in a volume of 0.09 ml. Reactions were incubated for 1 min, and the malonyl-CoA was determined as acid-stable radioactivity.

two fractions could be explained on the basis of the partial reactions known to be catalyzed by this enzyme.

The further characterization of E_a was facilitated by the use of three independent assays of this fraction. The first assay, described in Table II, was based upon the overall assay of acetyl-CoA carboxylation which required the addition of excess E_b (*18*). The second assay measured the carboxylation of free (+)-biotin catalyzed by E_a (*140*). Catalysis of (+)-biotin carboxylation by undissociated avian liver acetyl-CoA carboxylase had been reported earlier (*173*). The discovery that E_a catalyzed the Mn^{2+}-ATP-dependent carboxylation of (+)-biotin [reaction

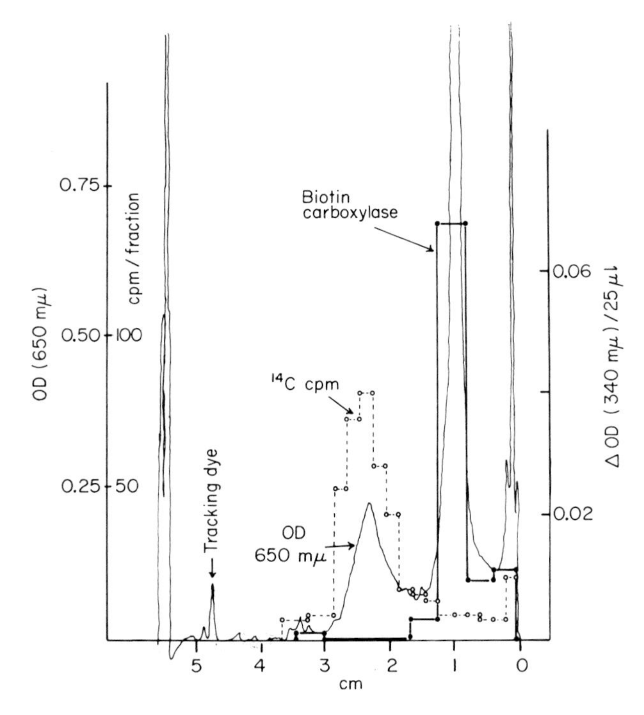

FIG. 2. Disc gel electrophoresis of E_a. Two aliquots of purified ^{14}C-biotin labeled E_a were electrophoresed on 7% acrylamide gels at pH 9.0. One gel was stained for protein, scanned at 650 nm, and then sectioned. The sections were dissolved in 30% H_2O_2 and counted. The other gel was sectioned without staining and each section was eluted with 0.05 *M* potassium phosphate buffer, pH 7.3. The eluates were assayed for biotin carboxylase (●—●), acetyl-CoA carboxylase in the presence of added E_b, and radioactivity (○--○) (*140*).

173. E. Stoll, E. Ryder, J. B. Edwards, and M. D. Lane, *Proc. Natl. Acad. Sci. U. S.* **60,** 986 (1968).

(4)] made available a spectrophotometric assay for E_a activity since the ADP formed in this reaction could be measured by a coupled pyruvate kinase-lactic dehydrogenase assay in which DPNH oxidation is monitored. The third assay of E_a made use of an *E. coli* biotin auxotroph which was grown in medium supplemented with labeled biotin (*140*). The E_a prepared from this organism contained covalently bound labeled biotin which would be detectable even if the E_a catalytic activities were lost.

Highly purified E_a, which actively catalyzed acetyl-CoA carboxylation (in the presence of E_b) and (+)-biotin carboxylation and which contained covalently bound ^{14}C-biotin, was subjected to disc gel electrophoresis. One gel was stained to identify the protein bands; a second gel, run simultaneously, was sliced and eluted, and the fractions were assayed in the two above assays and counted for ^{14}C-biotin radioactivity. Figure 2 shows that upon gel electrophoresis E_a gave rise to two major protein bands (*18*). Neither band catalyzed acetyl-CoA carboxylation either alone or in combination. Biotin carboxylase activity was associated with the slower migrating band, but this band contained very little ^{14}C-biotin radioactivity. A faster migrating band contained the ^{14}C-biotin, and this protein did not catalyze biotin carboxylation. Thus, it was apparent that disc gel electrophoresis leads to the dissociation of E_a into two protein components: one catalyzes biotin carboxylation and has been named *biotin carboxylase*, and the other contains covalently bound biotin and has been named *biotin carboxyl carrier protein* (BCCP) (*140–143*).

Sucrose density gradient centrifugation studies with E_a confirmed the fact that E_a dissociates into two components (*140*). As noted in Fig. 3 centrifugation of ^{14}C-biotin-E_a at pH 7.8 demonstrated the E_a complex; acetyl-CoA carboxylase activity (with added E_b), biotin carboxylase activity, and ^{14}C-biotin radioactivity coincided in fractions 10–18. This complex had a sedimentation coefficient of approximately 9 S. Of major interest was the slower sedimenting peak of ^{14}C radioactivity which suggested that E_a was dissociating. Centrifugation studies at pH 6.5 indicated that E_a tends to aggregate, since at this pH a sedimentation coefficient of approximately 35 S was noted (*174*). On the other hand, centrifugation at pH 9.0 led to complete dissociation of E_a into its two components (Fig. 4): one component contained essentially all the covalently bound ^{14}C-biotin and had a sedimentation coefficient of 1.3 S and no catalytic activity in the two catalytic assays; the second component catalyzed free (+)-biotin carboxylation and had a sedimentation

174. A. M. Nervi and A. W. Alberts, unpublished data (1970).

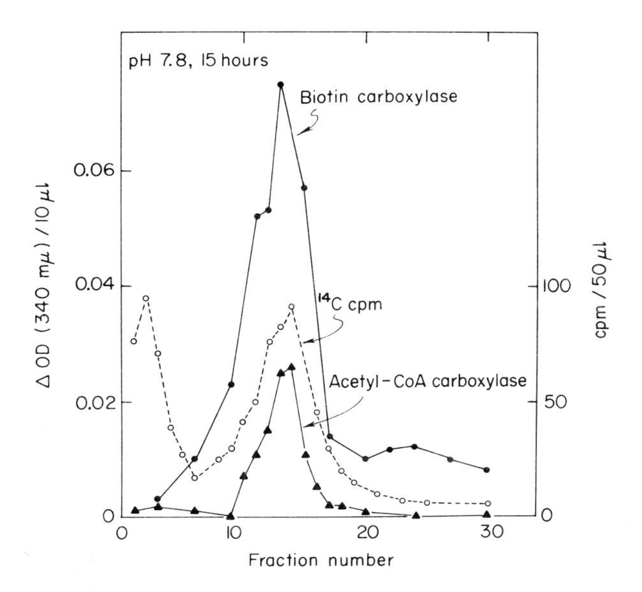

FIG. 3. Sucrose density gradient of ^{14}C-biotin-E_a at pH 7.8. ^{14}C-biotin-E_a (0.1 mg, 1000 cpm, 0.08 unit acetyl-CoA carboxylase, 0.15 unit biotin carboxylase) in 100 μl 0.02 *M* potassium phosphate buffer and 20% glycerol was layered over a 5–20% sucrose gradient containing 0.02 *M* potassium phosphate buffer pH 7.8, and 20% glycerol and centrifuged for 15 hr at 5°. Aliquots were counted (○), assayed for acetyl-CoA carboxylase in the presence of E_b (▲), and biotin carboxylase (●).

coefficient of 5.4 S, but it contained little ^{14}C-biotin (*140*). Thus these experiments demonstrated the dissociation of E_a into the same kinds of subunits that were detected on disc gel electrophoresis.

The proposed functions of the three subunits in the partial reactions of acetyl-CoA carboxylase are indicated in reactions (18) and (19):

$$\text{ATP} + \text{HCO}_3^- + \text{BCCP} \xrightleftharpoons{\text{Mn}^{2+}\text{, biotin carboxylase}} \text{CO}_2^-\text{-BCCP} + \text{ADP} + \text{P}_i \qquad (18)$$

$$\text{CO}_2^-\text{-BCCP} + \text{CH}_3\text{CO-SCoA} \xrightleftharpoons{\text{transcarboxylase}} {}^-\text{O}_2\text{CCH}_2\text{CO-SCoA} + \text{BCCP} \qquad (19)$$

Reaction (18) indicates the Mn^{2+}-ATP-dependent carboxylation of BCCP that is catalyzed by biotin carboxylase. The transcarboxylase subunit (E_b) catalyzes the transfer of the carboxyl group from carboxy-BCCP to acetyl-CoA forming malonyl-CoA in reaction (19).

2. *Biotin Carboxyl Carrier Protein*

The discovery that *E. coli* acetyl-CoA carboxylase consists of three functionally dissimilar subunits immediately suggested that this enzyme might be an ideal system for studying the enzymic mechanism of biotin-

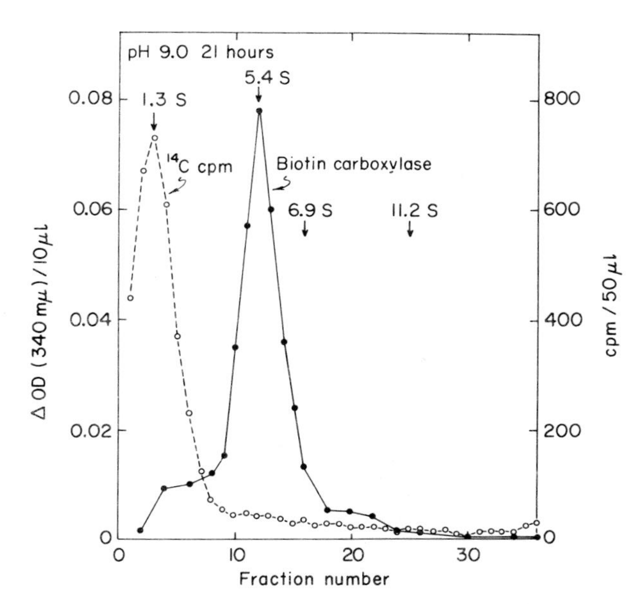

FIG. 4. Sucrose density gradient centrifugation of ^{14}C-biotin-E_a at pH 9.0. Procedure same as in experiment of Fig. 3 except buffer was 0.02 *M* tris-HCl, pH 9.0.

dependent carboxylation reactions. It was first necessary to obtain large quantities of the three subunits, BCCP, biotin carboxylase, and transcarboxylase (E_b). Purification of ^{3}H-BCCP from 50 to 100 pound batches of *E. coli* labeled with ^{3}H-biotin led to the isolation of four pure radioactive fractions *(174)* which are demonstrated in the disc gel electrophoresis pattern in Fig. 5. Prolonged dialysis at pH 9 was used during

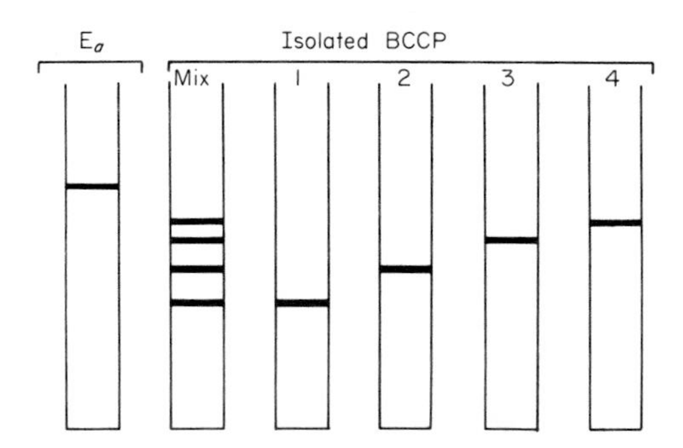

FIG. 5. Diagrammatic representation of disc gel electrophoresis of E_a and isolated BCCP fractions. Aliquots of ^{3}H-biotin-labeled E_a and the isolated BCCP fractions were electrophoresed in 15% polyacrylamide gels at pH 9. Also included is a gel containing a mixture of the BCCP fractions. Gels were stained for protein and sectioned and counted. Shown is the localization of ^{3}H radioactivity in each gel.

this procedure to completely dissociate the E_a complex. On the other hand, disc gel electrophoresis of an E_a preparation that had not been subjected to dialysis presented an entirely different pattern, a single radioactive band which migrated slower than any of the four proteins isolated from the large-scale purification (Figs. 2 and 5). Assay of the four radioactive fractions (1–4) in the reconstituted acetyl-CoA carboxylase reaction indicated that they were all equally active, exhibiting similar K_m and V_{max} values. However, the slower migrating fraction, isolated directly from E_b, had a K_m approximately 100 times lower. These experiments suggested that the isolation procedure was yielding four forms of BCCP which were less active than that present in E_a. Although it was apparent that BCCP isolated directly from E_a was far more active, this method of preparation was tedious, and only small quantities of this form could be processed. The four other forms of BCCP have been isolated in larger quantities and the fastest migrating form (fraction 1 in Fig. 5) has been crystallized [Fig. 6 (*141*)]. This form, which has been designated $BCCP_S$ since it is the smallest of the isolated forms, is available in substrate quantities, and it has been extensively characterized. The amino acid compositions of $BCCP_S$ (*141*) and of fraction 2 BCCP (*174*) are shown in Table IV; $BCCP_S$ contains 82 amino acid residues with a molecular weight of 9065, while fraction 2 BCCP (Fig. 5) contains 14 additional amino acid residues and has a molecular weight

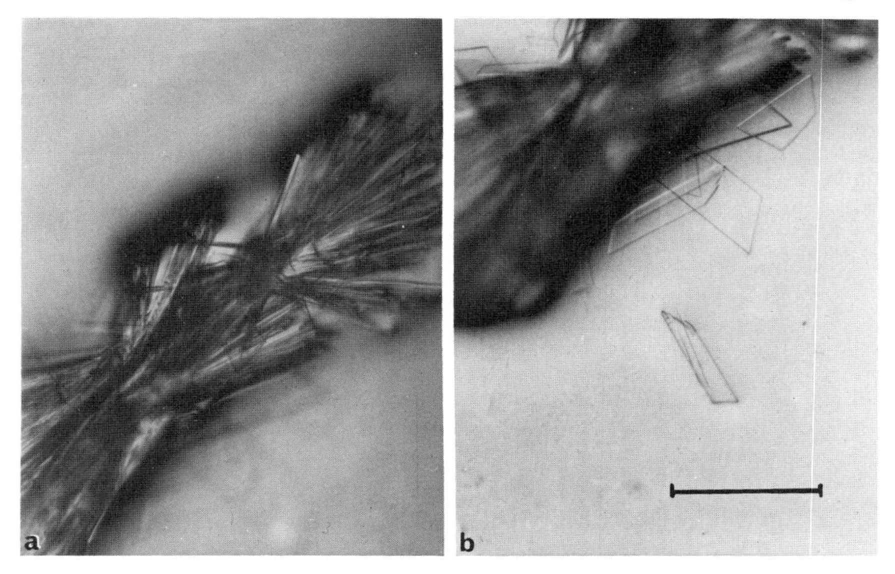

FIG. 6. Crystalline $BCCP_S$ from *E. coli*. The line represents 0.2 μ. (a and b) Different views of the crystals.

TABLE IV
AMINO ACID COMPOSITION OF BCCP

Amino acid	*E. coli* acetyl-CoA carboxylase Fraction 1[a] (residues/mole)	*E. coli* acetyl-CoA carboxylase Fraction 2 (residues/mole)	*P. shermanii* methylmalonyl-CoA oxalacetate transcarboxylase[b] (residues/mole)
Lysine	5	5	8
Histidine	1	1	2
Arginine	2	2	2
Aspartic acid	6	6	8
Threonine	4	5	5
Serine	5	6	4
Glutamic acid	11	13	8
Proline	5	8	12
Glycine	6	6	11
Alanine	7	12	9
Half-cystine	1	1	1
Valine	11	12	9
Methionine	4	5	2
Isoleucine	7	7	5
Leucine	3	3	6
Tyrosine	1	1	2
Phenylalanine	3	3	3
Tryptophan	0	0	0
Total	82	96	97

[a] Fraction 1 has been designated as $BCCP_S$ (*143*).
[b] From Gerwin *et al.* (*175*).

of 10,267. Also shown in Table IV is the amino acid composition of a BCCP obtained from a different biotin enzyme, methylmalonyl-CoA-oxalacetate transcarboxylase, from *Propionibacterium shermanii* (*175*), and this will be discussed below.

In order to obtain the most active form of BCCP (designated $BCCP_L$) an isolation procedure was devised that initially produced pure E_a which was then subjected to Sephadex G-200 chromatography and preparative disc gel electrophoresis (*143*). Analytical disc gel electrophoresis of the ^3H-$BCCP_L$ obtained by this procedure is demonstrated in Fig. 7 which shows that the preparation migrated as a single protein in this system. All the ^3H radioactivity of the preparation migrated with the protein. For comparison, gel electrophoresis of $BCCP_S$ is also shown in Fig. 7. The preparation of $BCCP_L$ was free of all other forms of BCCP, and

175. B. I. Gerwin, B. E. Jacobson, and H. G. Wood, *Proc. Natl. Acad. Sci. U. S.* **64,** 1315 (1969).

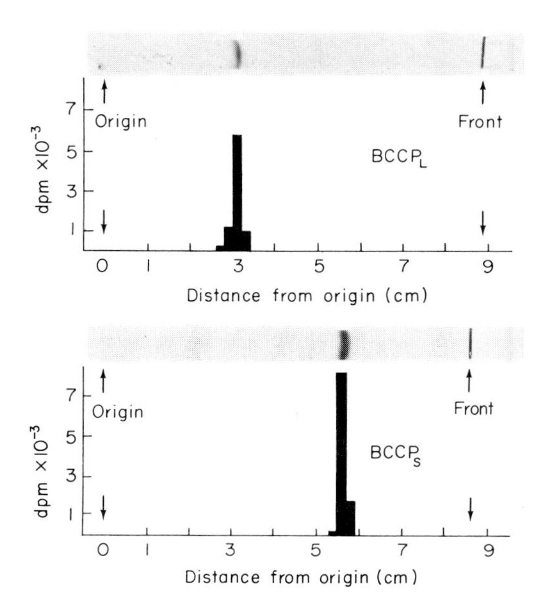

FIG. 7. Disc gel electrophoresis of purified ^{3}H-BCCP$_L$ and ^{3}H-BCCP$_S$.

it was free of biotin carboxylase and transcarboxylase activities. The limited availability of BCCP$_L$ has hampered the complete characterization of this protein. However, a minimum molecular weight of approximately 20,000 has been derived by SDS disc gel electrophoresis of BCCP$_L$ (*143*). The patterns of BCCP$_L$ and BCCP$_S$ (molecular weight 9065) on SDS gel electrophoresis are shown in Fig. 8. The molecular weight of BCCP$_L$ was determined by comparison in the SDS gel with proteins of known molecular weights. Attempts to determine the molecular weight of native BCCP$_L$ on a calibrated Sephadex G-75 column gave variable results with molecular weight values ranging from 20,000 to 45,500. Unless glycerol was included in the elution buffer, the peak of recovered BCCP exhibited multiple bands on disc gel electrophoresis somewhat similar to those shown in Fig. 5. These preliminary experiments suggest that native BCCP$_L$ probably exists as a dimer. It was apparent that BCCP$_L$ tended to break down during many normally used protein characterization procedures and that glycerol prevented the breakdown to some extent.

The biotin content of BCCP$_L$ and BCCP$_S$ as determined by two different methods are shown in Table V (*143*). As can be seen, BCCP$_L$ has an average of one mole of biotin per 22,000 g of protein while BCCP$_S$ has one mole of biotin per 9,300 g of protein, thus confirm-

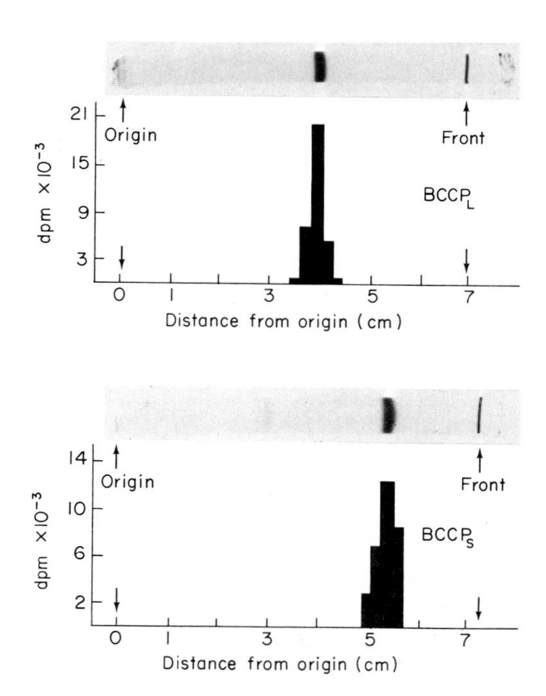

FIG. 8. Sodium dodecyl sulfate gel electrophoresis of ^{3}H-BCCP$_L$ and ^{3}H-BCCP$_S$.

ing the minimum molecular weight data determined by SDS gel electrophoresis.

The isolation of several different forms of BCCP from *E. coli* extracts raised the question as to the form of BCCP which is present in these cells *in vivo*. This was resolved by carrying out disc gel electrophoresis and SDS gel electrophoresis on freshly prepared crude extracts of *E. coli* grown on ^{3}H-biotin and comparing the patterns of radioactivity obtained with these extracts with the migration of pure ^{3}H-BCCP$_L$ and ^{3}H-BCCP$_S$ (*143*). It is apparent in Figs. 9 and 10 that the great majority of protein-

TABLE V
BIOTIN CONTENT OF BCCP$_L$ AND BCCP$_S$

Preparation	Method	Biotin content (g protein/mole biotin)
BCCP$_L$	Carboxylation	21,500
BCCP$_L$	Pronase-avidin	22,200
BCCP$_S$	Carboxylation	9,600
BCCP$_S$	Pronase-avidin	9,000

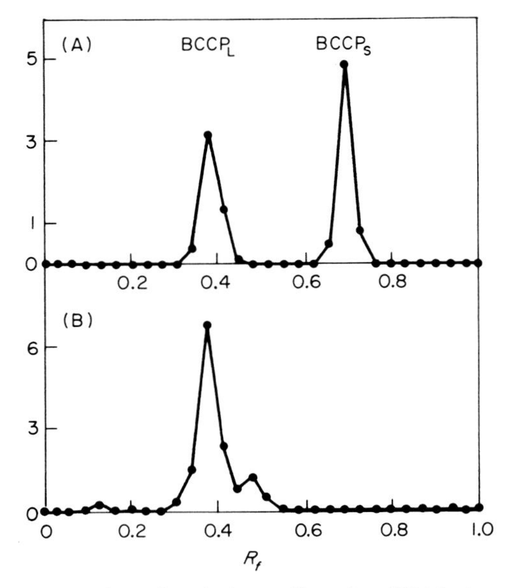

FIG. 9. Disc gel electrophoresis of the radioactive ^{3}H-biotin-containing proteins in crude *E. coli* extract: (A) mixture of ^{3}H-BCCP$_L$ and ^{3}H-BCCP$_S$ and (B) freshly prepared *E. coli* extract.

bound ^{3}H-biotin in crude extracts migrates as BCCP$_L$ in both systems. A trace of faster migrating material was always noted, but no BCCP$_S$ was detected in freshly prepared crude extracts. Dialysis of crude extracts, however, led to the appearance of several faster moving bands of ^{3}H-labeled protein (*143*). The finding that BCCP$_L$ is the major biotin protein in fresh extracts of *E. coli* and that it is the most active of the isolated biotin proteins indicates that it is the native form of *E. coli* BCCP. The multiple smaller, less active forms of BCCP probably result from proteolytic cleavage of the native molecule (*141, 143*). Although the conditions for formation of the various smaller forms have not been adequately delineated, their formation appears to be related to the purification procedures utilized (*176*). In spite of the fact that BCCP$_L$ is the native protein, most of the experiments discussed below were done with BCCP$_S$ because larger quantities of that form were available.

3. *Biotin Carboxylase*

Biotin carboxylase, originally separated from fraction E$_a$, has been extensively purified (*144, 176*), and it was recently crystallized (*146*).

176. R. R. Fall, unpublished data (1971).

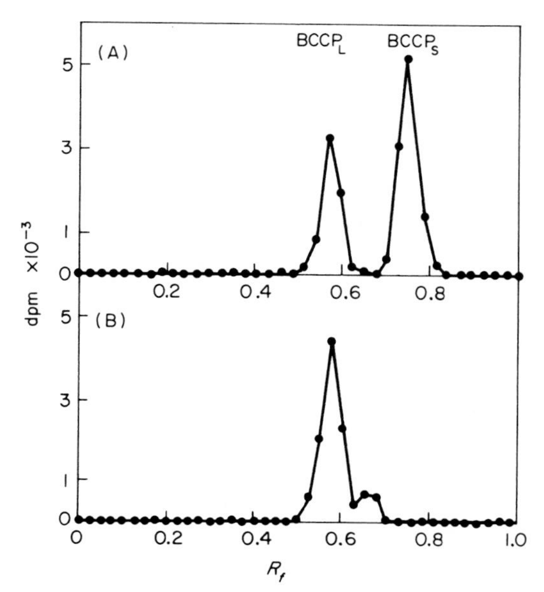

FIG. 10. Sodium dodecyl sulfate gel electrophoresis of the radioactive ^{3}H-biotin-containing proteins in crude *E. coli* extract: (A) mixture of ^{3}H-BCCP$_L$ and ^{3}H-BCCP$_S$ and (B) freshly prepared *E. coli* extract.

The molecular weight (*144*) of this acetyl-CoA carboxylase component was calculated to be 95,000 from the Stokes radius (35 Å), the sedimentation coefficient ($s_{20,w} = 5.7$ S), and an estimated partial specific volume of 0.75. The sedimentation coefficient of 5.7 S determined with the pure protein confirms the earlier value of 5.4 S determined in sucrose density gradient centrifugation (Fig. 4) with biotin carboxylase dissociated from E$_a$ (*140*). Biotin carboxylase dissociates in SDS gel electrophoresis into subunits with a molecular weight of 51,000 daltons, indicating that the protein molecule is composed of two identical subunits (*144*).

4. *Transcarboxylase*

The transcarboxylase component of acetyl-CoA carboxylase (originally designated E$_b$) has been purified about 1700-fold (*142*). The molecular characteristics have not been extensively studied because the preparation is not yet pure. However, a minimal molecular weight of 90,000 was determined by filtration of this protein through a calibrated Sephadex G-200 column (*85*). A tendency of the protein to aggregate was noted in that some preparations had molecular weights of 180,000 and higher. SDS gel electrophoresis indicated a minimal molecular weight of

45,000–50,000, suggesting that this protein also exists as a dimer of two subunits per mole.

5. *Reconstitution of Acetyl-CoA Carboxylase*

Isolation of the individual components of the *E. coli* acetyl-CoA carboxylase, BCCP, biotin carboxylase, and transcarboxylase has permitted studies of the roles of these individual proteins in the overall reaction and in the two partial reactions [reactions (18) and (19)]. Thus the absolute requirements for BCCP, biotin carboxylase, and transcarboxylase for the carboxylation of acetyl-CoA to form malonyl-CoA was readily demonstrated (*141–143*). The availability of substrate quantities of $BCCP_S$ and $BCCP_L$ (*141, 143*) has allowed the determination of the K_m values of these biotin proteins in the reaction catalyzed by biotin carboxylase [reaction (18)] or that catalyzed by transcarboxylase [reaction (19)]. Experiments have shown that $BCCP_L$ has a K_m of 0.2–0.4 μM in both reactions, whereas $BCCP_S$ has a K_m of approximately 25 μM in these reactions. At high concentrations (+)-biotin can replace BCCP as a substrate for biotin carboxylase, and it has a K_m of 84–170 mM in this reaction (*141, 144*). Thus far it has not been possible to replace BCCP with (+)-biotin in the overall reaction (*140*). However, (+)-biotin can be carboxylated by ATP and HCO_3^- with biotin carboxylase (*140, 141, 144*) and by malonyl-CoA with transcarboxylase (*145*). Several analogs of (+)-biotin have been tested in the biotin carboxylase reaction (*144*); the most active of these were biocytin and *o*-heterobiotin which were 20% as active as (+)-biotin. Ethanol and other solvents enhance the rate of carboxylation of (+)-biotin by biotin carboxylase (*141*). However, the biological significance of this observation is diminished by the finding that ethanol has no effect on the carboxylation of BCCP (*141*). The mechanism of action of biotin carboxylase is not understood. It obviously contains sites for Mn^{2+}-ATP, HCO_3^-, and BCCP since it catalyzes the carboxylation of BCCP or free (+)-biotin. Preliminary studies with the pure enzyme indicate that it catalyzes in the absence of biotin both ATP-$^{32}P_i$ and ATP-^{14}C-ADP exchanges, but the rates of these reactions are not compatible with β-γ P–O cleavage prior to carboxybiotin formation (cf. *10*). These experiments did not rule out a possible tightly bound intermediate that requires the presence of an acceptor for its release.

The mechanism of the transcarboxylase-catalyzed partial reaction [reaction (19)] is also not well understood. Preliminary studies indicated that this protein functioned in the carboxyl transfer from CO_2^--BCCP to acetyl-CoA (*18*). The unequivocal demonstration that transcarboxylase catalyzes reaction (19) was made possible by the availability of substrate quantities of ^{3}H-BCCP$_S$ and the highly purified transcarboxylase which contained no biotin carboxylase (*143*). Incubation of ^{14}C-malonyl-CoA with ^{3}H-BCCP$_S$ and transcarboxylase led to the formation of ^{14}C-carboxy-^{3}H-BCCP$_S$ which, as demonstrated in Fig. 11, was readily separated from the radioactive substrate by filtration through Sephadex G-50. In this experiment approximately 80% of the added BCCP$_S$ was carboxylated. The ^{14}C co-chromatographing with ^{3}H-BCCP was identified as $^{14}CO_2^-$-BCCP by its acid lability which is characteristic of this compound; ^{14}C-malonyl-CoA is unaffected by acid. There was no ^{14}C-radioactivity associated with ^{3}H-BCCP when the transcarboxylase was omitted nor was there any in the void volume when BCCP$_S$ was omitted. In addition the transcarboxylase catalyzes the transcarboxylation from malonyl-CoA to free (+)-biotin forming 1′-*N*-carboxybiotin, a model for the second partial reaction (*145*). The K_m

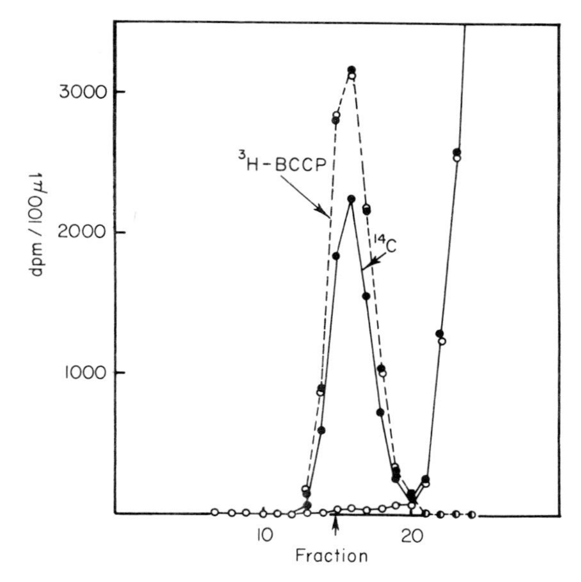

FIG. 11. Carboxylation of BCCP$_S$ by transcarboxylase. ^{3}H-BCCP$_S$ was incubated for 15 min at 30° with 1,3-^{14}C-malonyl-CoA at pH 7.0 in presence or absence of transcarboxylase. After cooling rapidly in an ice bath, $^{14}CO_2^-$-^{3}H-BCCP was separated from 1,3-^{14}C-malonyl-CoA on a Sephadex G-50 column (0.9 × 25 cm) at pH 9.0. Fractions of 0.5 ml were collected and aliquots were counted for both ^{3}H and ^{14}C. Acid-labile radioactivity was determined by adjusting aliquots to 0.02 *M* HCl, drying the samples, and counting. ^{14}C-radioactivity without HCl (●—●), ^{14}C-radioactivity with HCl (○—○), ^{3}H-BCCP radioactivity without HCl (● - - - ●), ^{3}H-BCCP radioactivity with HCl (○ - - - ○).

for (+)-biotin in this reaction is 3 mM, considerably lower than the 84–170 mM reported as the K_m for (+)-biotin in the reaction catalyzed by biotin carboxylase (*141, 144*). However, transfer of the carboxyl-group from carboxybiotin to acetyl-CoA could not be detected (*145*).

From the experiments described it is apparent that the transcarboxylase component of acetyl-CoA carboxylase catalyzes carboxyl transfer between malonyl-CoA and BCCP. Further experimentation demonstrated that the transcarboxylase contains binding sites for acetyl-CoA and/or malonyl-CoA (*142*). These studies were complicated by the properties of the transcarboxylase; i.e., purified transcarboxylase is very unstable in low ionic strength and tends to precipitate out of solution. However, it was possible to incubate transcarboxylase with either ^{14}C-acetyl-CoA or ^{14}C-malonyl-CoA and then separate enzyme-bound substrate from the starting substrate by filtration through Sephadex G-50 (*142*). This is illustrated in Fig. 12 which demonstrates the binding of 2-^{14}C-malonyl-CoA by transcarboxylase. It is apparent that a radioactive peak coincided with transcarboxylase activity in the void volume. The amount of radioactivity associated with the transcarboxylase was

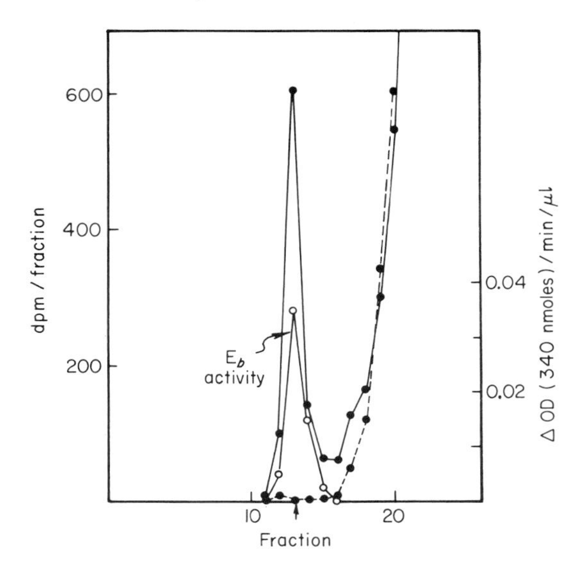

FIG. 12. Binding of 2-^{14}C-malonyl-CoA to transcarboxylase. Transcarboxylase (E_b) was incubated at 30° for 15 min at pH 7.0 with 2-^{14}C-malonyl-CoA. 2-^{14}C-Malonyl-CoA-transcarboxylase was separated from 2-^{14}C-malonyl-CoA on a Sephadex G-50 column (0.9 × 25 cm) at pH 6.7 and fractions of 0.4 ml were collected. Aliquots were counted and assayed for transcarboxylase activity. ^{14}C-Malonyl-CoA radioactivity with transcarboxylase present (●—●); ^{14}C-malonyl-CoA radioactivity with transcarboxylase absent (●- - -●); transcarboxylase (E_b) activity (○—○).

TABLE VI
BINDING OF MALONYL-CoA TO TRANSCARBOXYLASE[a]

2-^{14}C-Malonyl-CoA (M)	Acetyl-CoA (M)	E_b (mg)	2-^{14}C-Malonyl-CoA-E_b (mμmoles)
4.0×10^{-5}		0.6	0.0292
1.2×10^{-4}		0.6	0.0775
2.0×10^{-4}		0.6	0.0918
4.0×10^{-4}		0.6	0.1305
2.0×10^{-4}	2.0×10^{-4}	0.6	0.0545
2.0×10^{-4}		0.2	0.0394

[a] 2-^{14}C-Malonyl CoA-E_b was prepared as described in Fig. 12 except that the quantities of 2-^{14}C-malonyl-CoA, acetyl-CoA, and transcarboxylase (E_b) indicated above were used.

dependent upon the concentration of ^{14}C-malonyl-CoA and transcarboxylase as shown in Table VI. Addition of unlabeled acetyl-CoA decreased the amount of ^{14}C-malonyl-CoA bound to the transcarboxylase, suggesting that these two acyl-CoA derivatives are bound at a common site. Association of acetyl-CoA with transcarboxylase was shown in a similar manner. Unfortunately, transcarboxylase activity was rapidly lost in buffers of low ionic strength which were required in these binding experiments. Thus, although these experiments demonstrated the binding site(s) for both malonyl-CoA and acetyl-CoA, the extreme lability of the transcarboxylase did not permit further characterization of the binding site(s).

The results presented indicate that transcarboxylase interacts with one of its substrates, acetyl-CoA, and with the product of the reaction, malonyl-CoA. The other substrate of the transcarboxylase reaction, CO_2^--BCCP, also interacts with transcarboxylase. Avidin has been shown to inhibit the ATP-dependent carboxylation of BCCP (*18*), but the presence of transcarboxylase had no detectable effect on the inhibition. On the other hand, the reactivity of CO_2^--BCCP with avidin was greatly influenced by the presence of transcarboxylase. Thus, as illustrated in Fig. 13, the extent of inhibition by avidin of CO_2^--BCCP was greatly enhanced by the presence of transcarboxylase at the levels of avidin tested. When CO_2^--BCCP was incubated with 0.04 units of avidin, 90% of the CO_2^--BCCP remained active as a carboxyl donor to acetyl-CoA in reaction (19), whereas when transcarboxylase was present during the exposure of CO_2^--BCCP to avidin, only 13% of CO_2^--BCCP remained active. These experiments suggest that the biotin of CO_2^--BCCP in the presence of transcarboxylase is more available for binding by

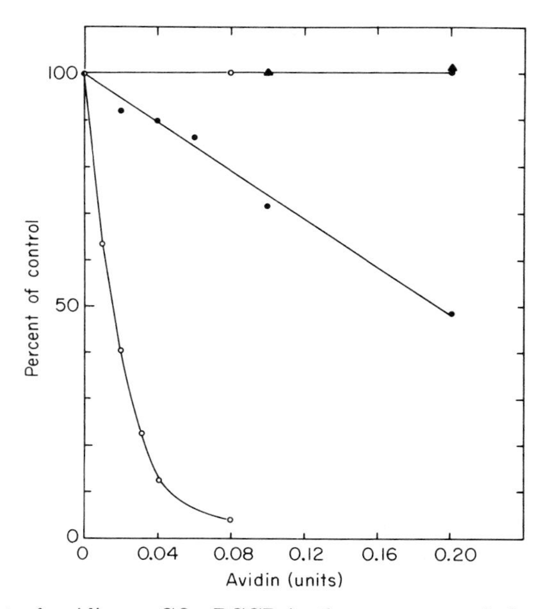

FIG. 13. Effect of avidin on CO_2^--BCCP in the presence and absence of transcarboxylase. In the middle curve (●), $^{14}CO_2^-$-BCCP was incubated at pH 7.5 with the indicated units of avidin in a volume of 50 μl. After 1 min at 25°, 0.1 mg of biotin was added. Then 0.002 unit of transcarboxylase and 15 mμmoles of acetyl-CoA were added in a final volume of 90 μl and the reaction was incubated 1 min. Malonyl-CoA formation was determined as described (*18*). Lower curve (○) is the same as the middle curve except that both $^{14}CO_2^-$-BCCP and transcarboxylase were incubated with avidin before the addition of biotin. Upper curve (▲) is the same as middle curve except only transcarboxylase was incubated with avidin. $^{14}CO_2^-$-BCCP was added after biotin. Also included on the upper curve are the controls for the middle (-●-) and lower (-○-) curves. In these experiments avidin was incubated with biotin for 1 min before the addition of $^{14}CO_2^-$-BCCP and transcarboxylase.

avidin. Transcarboxylase also influences the rate of CO_2^--BCCP decarboxylation (*142*). As has been shown with a number of biotin enzymes as well as with free (+)-biotin, CO_2^--biotin-enzyme and CO_2^--(+)-biotin spontaneously decarboxylate in the absence of carboxyl acceptors (*1, 4, 69*). This decarboxylation is both temperature and pH dependent, high temperature and low pH favoring decarboxylation. Free carboxybiotin is considerably more stable than carboxybiotin-enzyme (*9*). The stability of CO_2^--BCCP is intermediate between the carboxylated free and enzyme-bound biotin (*85*). Figure 14 illustrates the effect of increasing concentrations of transcarboxylase on the $T_{1/2}$ of CO_2^--BCCP. At pH 7.5 and 33°, CO_2^--BCCP exhibits a $T_{1/2}$ of 27.5. The $T_{1/2}$

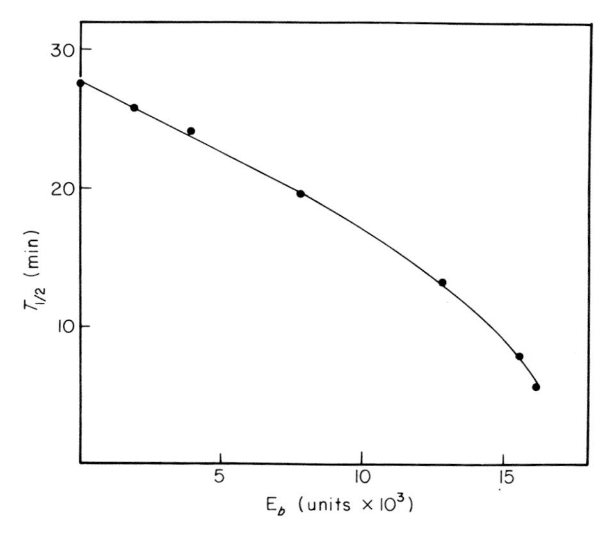

FIG. 14. Effect of transcarboxylase on the stability of CO_2^--BCCP. $^{14}CO_2^-$-BCCP was incubated at 33° in 0.32 ml with the quantity of transcarboxylase indicated at pH 7.5; 50 µl aliquots were removed at various times and the amount of $^{14}CO_2^-$-BCCP remaining was determined by the addition of 0.03 µmole acetyl-CoA and 0.002 unit of transcarboxylase and assaying for the formation of malonyl-CoA (*18*). $T_{1/2}$ was calculated from a semilog plot of the percent $^{14}CO_2^-$-BCCP remaining at the different times.

however, was drastically decreased in the presence of transcarboxylase; with 0.016 unit of transcarboxylase present, the $T_{½}$ dropped to 6 min. Thus, in the absence of the acceptor, acetyl-CoA, the transcarboxylase component of acetyl-CoA carboxylase catalyzes the decarboxylation of CO_2^--BCCP, reaction (20).

$$CO_2^-\text{-BCCP} + H_2O \xrightleftharpoons{\text{transcarboxylase}} HCO_3^- + \text{BCCP} \quad (20)$$

Methylmalonyl-CoA-oxalacetate transcarboxylase, another biotin-enzyme, has also been dissociated into nonidentical subunits (*175, 177, 178*). One of the subunits has a sedimentation coefficient of 1.3 S and contains the amino acids listed in Table IV and approximately one mole of biotin. Although the roles of the subunits of this enzyme have not been defined, it is obvious that the biotin containing subunit is comparable to BCCP of acetyl-CoA carboxylase. In light of these findings it is tempting to postulate that all biotin-enzymes might be composed of

177. F. Ahmad, B. Jacobson, and H. G. Wood, *JBC* **245,** 6486 (1970).
178. B. Jacobson, B. I. Gerwin, F. Ahmad, P. Waegell, and H. G. Wood, *JBC* **245,** 6471 (1970).

at least three different subunits: a biotin protein such as BCCP, a biotin carboxylase, and a subunit specifying the acceptor molecule. It is also possible that various biotin-enzymes in the same organism will contain similar BCCP and biotin carboxylase subunits and will differ from each other only in containing different transcarboxylase subunits which confer acceptor specificity on the enzyme complexes.

6. *Subunit Structure of Avian Liver Enzyme and Wheat Germ Enzyme*

The discovery of functional subunits in *E. coli* acetyl-CoA carboxylase stimulated examinations of the subunit structure of the avian liver acetyl-CoA carboxylase (*128, 168*), and the wheat germ acetyl-CoA carboxylase (*179*). Binding studies with the avian liver carboxylase (*128*) have revealed one tight binding site for citrate (K_D, 2–3 $\times 10^{-6}$ M) and one site for acetyl-CoA (K_D, 4 $\times 10^{-6}$ M) per protomer (molecular weight 410,000). The biotin content of the enzyme, as determined microbiologically with *Lactobacillus plantarum* after acid hydrolysis, is 0.93 mole/mole of protomer (*128*).

A similar conclusion, that there is one binding site each for acetyl-CoA and citrate per mole of protomer, was reached by Maragoudakis and Hankin (*180*) using kinetic data. By the same method they also showed that there is one binding site per protomer for ATP and one site for HCO_3^-, presumably the covalently bound biotin.

Dissociation of the protomer with sodium dodecyl sulfate gives rise to subunits with a molecular weight of 114,000 as determined by sedimentation equilibrium or 110,000 as determined by electrophoresis in SDS-polyacrylamide gels (*128, 168*). Since there is only one biotin prosthetic group per 410,000 molecular weight, it is apparent that the 110,000 molecular weight subunits are nonidentical. However, it has not yet been possible to determine the functions of the individual subunits since they are denatured by the dissociating conditions. An attractive hypothesis, based on the fact that there is one binding site each per 410,000 molecular weight for acetyl-CoA, ATP, citrate, and HCO_3^-, is that there are four functionally nonidentical subunits per protomer; one comparable to biotin carboxylase of *E. coli* and containing the ATP site; one comparable to the transcarboxylase component of the *E. coli* system; a BCCP component containing covalently bound biotin and therefore the HCO_3^- site; and a regulatory subunit which contains the

179. P. F. Heinstein and P. K. Stumpf, *JBC* **244**, 5374 (1969).
180. M. E. Maragoudakis and H. Hankin, *JBC* **246**, 348 (1971).

citrate allosteric effector site, comparable to the regulatory subunit of aspartyl transcarbamylase (*181*).

The wheat germ acetyl-CoA carboxylase, studied by Heinstein and Stumpf (*179*), is apparently intermediate in its properties between the animal enzyme, which can only be dissociated under denaturing conditions, and the *E. coli* enzyme, which is readily dissociated. Thus the plant system has been dissociated into two active protein components: one has a sedimentation coefficient of 7.3 S and appears to correspond to the transcarboxylase, and the other has a sedimentation coefficient of 9.4 S and appears to correspond to fraction E_a of the *E. coli* system. Thus far, it has not been possible to resolve this fraction into its BCCP and biotin carboxylase components.

G. Regulation

Acetyl-CoA carboxylase catalyzes the first committed step in fatty acid synthesis. It has therefore been the subject of extensive studies in several laboratories (*5, 10, 46, 126, 136, 137, 167, 168*) in order to delineate the mechanism of control of this enzyme. Several mechanisms have been proposed for the regulation of acetyl-CoA carboxylase and these have been reviewed recently by Vagelos (*182*). These include allosteric regulation, especially by tricarboxylic acids, notably citrate and isocitrate (*5, 10, 119, 128, 131, 136, 168*); feedback inhibition by the end product of fatty acid biosynthesis, palmityl-CoA (*183*); control of rates of synthesis and degradation of acetyl-CoA carboxylase (*135, 184*); and modulation between holoenzyme and apoenzyme affected by availability of biotin (*185–188*).

The most striking effect on animal acetyl-CoA carboxylase is allosteric activation by citrate (or isocitrate), and therefore this effect has been studied in the greatest detail. As noted above activation by citrate is associated with polymerization of the inactive protomeric form to the active polymeric form. Kinetic analyses of the enzyme from rat liver

181. E. R. Stadtman, "The Enzymes," 3rd ed., Vol. I, p. 397, 1970.
182. P. R. Vagelos, *Current Topics Cellular Regulation* **4**, 119 (1971).
183. P. W. Majerus and P. R. Vagelos, *Advan. Lipid Res.* **5**, 1 (1967).
184. P. W. Majerus and E. Kilburn, *JBC* **244**, 6254 (1969).
185. K. Dakshinamurti and P. R. Desjardins, *Can. J. Biochem.* **46**, 1261 (1968).
186. K. Dakshinamurti and P. R. Desjardins, *BBA* **176**, 221 (1969).
187. R. Jacobs, E. Kilburn, and P. W. Majerus, *JBC* **245**, 6462 (1970).
188. P. R. Mistry and K. Dakshinamurti, *ABB* **142**, 292 (1971).

(*155*), avian liver (*128*), and bovine adipose tissue (*189*) have established that the primary effect of the tricarboxylic acid activators is on the maximal velocity of the enzyme. Experiments have attempted to determine how the conformational changes induced by the binding of the allosteric effector markedly increase the efficiency of the enzyme.

As has been noted, acetyl-CoA carboxylation involves two partial reactions: the ATP-dependent carboxylation of biotin covalently linked to protein [reaction (15)] and the transfer of the carboxyl group from biotin to acetyl-CoA forming malonyl-CoA [reaction (16)]. The exchange reaction between ATP and either ^{32}P-orthophosphate or ^{14}C-ADP was used to assay the first partial reaction, and the exchange reaction between ^{14}C-acetyl-CoA and malonyl-CoA was used to assay the second partial reaction. Using these two exchange reactions, Matsuhashi *et al.* (*134*) demonstrated that citrate markedly activated the rates of both partial reactions catalyzed by rat liver acetyl-CoA carboxylase. These findings were subsequently confirmed with the avian liver carboxylase (*127, 128*).

Direct measurements of the kinetics of each partial reaction, i.e., the carboxylation of biotin-enzyme and the subsequent transcarboxylation to acetyl-CoA, were not possible because of the rapidity of the reactions when substrate amounts of enzyme are utilized. Lane's laboratory has circumvented these problems by the use of model compounds (*173*). Thus the carboxylation of free (+)-biotin, a reaction catalyzed by avian liver acetyl-CoA carboxylase, was used as a model for the first partial reaction. This carboxylation proceeds at a much slower rate than the carboxylation of biotin-enzyme, and therefore the effects of citrate could be tested directly. As expected, the rate of carboxylation of (+)-biotin was greatly enhanced by citrate. However, citrate also caused a 5- to 10-fold increase in K_m for (+)-biotin. This increase in K_m for (+)-biotin has been interpreted to result from a tighter binding of the biotin prosthetic group on the enzyme in the presence of citrate which thereby inhibits access of free (+)-biotin.

The carboxylation of acetyl pantetheine, an analog of acetyl-CoA that is carboxylated two orders of magnitude more slowly than acetyl-CoA, was used as a model for the second partial reaction. Transfer of the carboxyl group from carboxybiotin-enzyme to acetyl pantetheine was also greatly stimulated by citrate. Citrate had no effect on the K_m for acetyl pantetheine, suggesting that activation does not affect the binding of the acceptor. Thus the studies with model compounds con-

189. J. Moss, M. Yamagishi, A. K. Kleinschmidt, and M. D. Lane, *Federation Proc.* **28**, 1548 (1969).

firm the earlier observations (*127, 128, 134*) that citrate enhances the rate of both partial reactions.

Evidence has been presented that citrate induces conformational changes in the region of the biotin prosthetic group (*168, 190*). Avidin rapidly inactivates acetyl-CoA carboxylase irreversibly by binding to biotin. However, in the presence of citrate the enzyme is completely resistant to avidin inhibition. Acetyl-CoA partially protects against avidin inhibition, and in the presence of avidin citrate and acetyl-CoA act synergistically. These results have been interpreted as indicating that there are conformational changes induced by citrate in the region of the biotin prosthetic group. Further experiments implicating the region of the biotin prosthetic group as the site of citrate action relate to the rate of decarboxylation of 3-^{14}C-malonyl-CoA and ^{14}C-carboxy-biotin-enzyme. Both decarboxylations were stimulated by citrate and this stimulation was greatly enhanced by the presence of acetyl-CoA. Acetyl-CoA, in the absence of citrate, was without effect, again demonstrating the synergistic effect of citrate plus acetyl-CoA on the enzyme. Lynen *et al.* (*191*) have also noted that acetyl-CoA serves as an activator of rat liver acetyl-CoA carboxylase. These experiments have been used to explain the citrate effect as one involving conformational changes in the vicinity of the biotin prosthetic group.

An interesting observation relating to the mode of action of citrate has been made by Swanson *et al.* (*192*). These authors have found that treatment of rat liver acetyl-CoA carboxylase with trypsin produces an enzyme that is partially active in the absence of citrate. These findings have been confirmed and extended by Iritani *et al.* (*193*). Treatment of partially purified rat liver acetyl-CoA carboxylase with trypsin (4 μg/ml) for 10 min resulted in an enzyme that was 30–50% as active as enzyme treated with citrate. Enzyme treated with trypsin had an increased sedimentation constant similar to that of enzyme treated with citrate. However, palmityl-CoA, which is known to inactivate and dissociate acetyl-CoA carboxylase, was much less effective as an inhibitor of the trypsin-treated enzyme, and it did not affect its state of aggregation. These results suggest that trypsin might modify the citrate site or a site closely related to it so that the enzyme remains in an active form (*155, 193*). An alternative explanation is that animal acetyl-CoA carboxylase

190. E. Ryder, C. Gregolin, H. C. Chang, and M. D. Lane, *Proc. Natl. Acad. Sci. U. S.* **55,** 1148 (1966).

191. F. Lynen, M. Matsuhashi, S. Numa, and E. Schweizer, *in* "The Control of Lipid Metabolism" (J. K. Grant, ed.), p. 43. Academic Press, New York, 1963.

192. R. F. Swanson, W. M. Curry, and H. S. Anker, *BBA* **159,** 390 (1968).

193. N. Iritani, S. Nakanishi, and S. Numa, *Life Sci.* **8,** 1157 (1969).

might contain a regulatory subunit which contains the citrate binding site. In the absence of citrate the "regulatory" subunit would prevent activity; whereas, when citrate reacted with this subunit, it would presumably undergo a conformational change causing both polymerization and activation of the enzyme. Treatment of the enzyme with trypsin would either remove this subunit or cleave off a fragment, thereby rendering this inhibitory subunit inactive.

Acknowledgments

The unpublished experimental work from this laboratory presented here and the preparation of this article have been assisted by grants from the National Institutes of Health (RO1-HE-10406) and the National Science Foundation (GB-5142X).

3

Transcarboxylase

HARLAND G. WOOD

I. Introduction

Transfer reactions such as transmethylation, transamination, transphosphorylation, transhydrogenation, and those catalyzed by CoA transferase, transaldolase, and transketolase have been found in many different organisms and tissues but thus far transcarboxylation has not been reported in any organism other than the propionic acid bacteria (*1*).

1. The author is not aware of extensive efforts made to find this enzyme or other transcarboxylases in other organisms. A cursory examination was made with

Transcarboxylase [methylmalonyl-CoA:pyruvate carboxytransferase (EC 2.1.3.1)] catalyzes the following reaction:

$$CH_3 \cdot CH(COO^-) \cdot COSCoA + CH_3 \cdot CO \cdot COO^- \rightleftharpoons CH_3 \cdot CH_2 \cdot COSCoA + {}^-OOC \cdot CH_2 \cdot CO \cdot COO^- \quad (1)$$

The carboxyl transfer is reversible and oxalacetate serves as carboxyl donor not only to propionyl-CoA but also to acetyl-CoA, butyryl-CoA, and acetoacetyl-CoA yielding malonyl-CoA, ethylmalonyl-CoA, and presumably acetomalonyl-CoA, respectively (*2*). On the other hand, the keto acid requirement is specific, only pyruvate and oxalacetate are active.

A. Historical Background

The discovery of transcarboxylation (*3*) was closely linked to elucidation of the mechanism of formation of succinate and propionate and required more than 35 years to resolve. Virtanen (*4*) in 1923 proposed that succinate is formed by cleavage of glucose to C_4 and C_2 compounds. He had observed with cell suspensions of propionibacteria fermenting glucose that succinate and acetate were the major products and there was little or no formation of CO_2 or other C_1 compounds. Prior to this time it had been considered by Thunberg and others that succinate is formed by condensation of two molecules of acetate. Virtanen, Kluyver, and others adopted the C_4 and C_2 cleavage to explain metabolism by microorganisms because they believed that heterotrophic forms do not utilize CO_2. If the fermentation of glucose yielded acetate and a C_1 compound via two C_3 compounds and the formation of succinate occurred by acetate condensation, then equivalent C_1 compounds should accumulate, but this was not observed.

Van Niel (*5*) likewise observed the formation of succinate by the propionic acid bacteria, but he considered this acid to be derived from aspartate and other compounds of the yeast extract. The latter was

animal tissue to detect formation of ^{14}C-methylmalonyl-CoA from propionyl-CoA and ^{14}C-oxalacetate. Some indication of the reaction was found in dog skeletal muscle (*2*), but subsequent unpublished experiments failed to demonstrate the enzyme in this tissue.

2. H. G. Wood and R. Stjernholm, *Proc. Natl. Acad. Sci. U. S.* **47,** 289 (1961).
3. R. W. Swick and H. G. Wood, *Proc. Natl. Acad. Sci. U. S.* **46,** 28 (1960).
4. A. I. Virtanen, *Soc. Sci. Fennica, Commentatimes Phys.-Math.* **1,** No. 36, 1 (1923); **2,** No. 20, 1 (1925).
5. C. B. van Niel, Ph.D. Dissertation, Delft, Holland (1928).

included in the complex growth medium along with the substrate, glucose, glycerol, or lactate.

The discovery that propionic acid bacteria utilize CO_2 (*6*) removed previously imposed limitations relative to the metabolism of C_1 compounds and opened the way for consideration of new mechanisms. It was found in glycerol fermentations by the propionic acid bacteria that the succinate produced was almost equal to the CO_2 utilized. Based on these observations and the isolation of pyruvate from glucose and glycerol fermentations, Wood and Werkman (*7*) in 1938 proposed that CO_2 may combine with pyruvate and be the source of succinate. Shortly thereafter, carbon isotopes became available for use in biochemical studies, and in general the results obtained supported this proposal. The propionate fermentation then was considered to occur as shown in Fig. 1 (*8*). The scheme was based on the fact that the propionic acid bacteria were found to decarboxylate succinate to propionate and CO_2 (*9–11*) and $^{13}CO_2$ was found to be incorporated into the carboxyl groups of succinate and propionate (*12*).

A feature of the mechanism of Fig. 1 is that the formation of propionate involves a turnover of CO_2. Wood and Leaver (*8*) in 1958 determined the turnover of CO_2 and to their surprise found that the scheme presented in Fig. 1 was inadequate. The fermentations were done in the presence of $^{14}CO_2$, and since succinate is a symmetrical molecule there is an equal chance for decarboxylation of the fixed $^{14}CO_2$ or of the

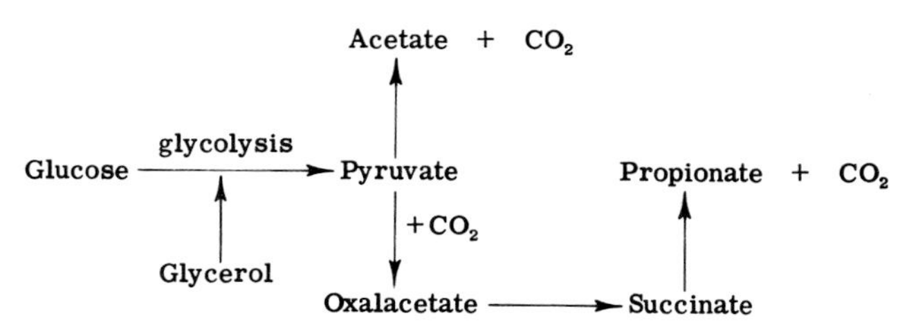

FIG. 1. Mechanism generally postulated (*8*) for the propionate fermentation prior to the discovery of transcarboxylation. Succinyl-CoA and propionyl-CoA were considered to be intermediates but are not shown in this simplified scheme.

6. H. G. Wood and C. H. Werkman, *BJ* **30,** 48 (1936).
7. H. G. Wood and C. H. Werkman, *BJ* **32,** 1262 (1938).
8. H. G. Wood and F. W. Leaver, *BBA* **12,** 207 (1953).
9. C. H. Werkman and H. G. Wood, *Advan. Enzymol.* **2,** 135 (1942).
10. E. A. Delwiche, *J. Bacteriol.* **56,** 811 (1948).
11. A. T. Johns, *J. Gen. Microbiol.* **5,** 1337 (1951)
12. H. G. Wood, C. H. Werkman, A. Hemingway, and A. O. Nier, *JBC* **139,** 365 and 377 (1941).

carboxyl which arises from the ^{12}C substrate. Accordingly, for each 100 mmoles of propionate produced there should be 50 mmoles of $^{12}CO_2$ produced in addition to that produced along with the acetate. The observed dilution of the $^{14}CO_2$ by $^{12}CO_2$ produced from the fermented substrate was found to be far less than this quantity (*8*). However, when lactate-3-^{14}C, pyruvate-2-^{14}C, or glycerol-2-^{14}C was fermented, an approximately equal distribution of ^{14}C was found in the 2 and 3 positions of the propionate and succinate (*13*) which was consistent with the role of a symmetrical C_4 dicarboxylic acid as a precursor of propionate. Furthermore, it was observed that washed cells of propionic acid bacteria catalyzed a rapid randomization of ^{14}C of propionate-1,3-^{14}C to both the 2 and 3 carbons of propionate in the absence of added fermentable substrates but the ^{14}C specific activity of the C-1 position of the propionate remained unchanged (*14*). There was no fixation of CO_2 during this randomization. The question then arose whether the randomization of ^{14}C in the products of the propionate fermentation was the result of an intramolecular rearrangement of propionate such as illustrated in Fig. 2A or whether it occurred via an intermolecular mechanism which in effect involved the reversible transfer of the carboxyl of succinate in a bound from (X–C_1) as shown in Fig. 2B. In the former case there would be no dilution of isotope of the carboxyl of propionate, and in the mechanism of Fig. 2B the dilution would be slight if the amount of

$$\underset{H_2C\text{—}{}^{14}CH_3}{{}^{14}COOH} \rightleftharpoons \underset{H_2C\text{—}{}^{14}CH_2}{\overset{O}{\overset{\|}{{}^{14}C}}} \rightleftharpoons \underset{H_3C\text{—}{}^{14}CH_2}{{}^{14}COOH}$$

(A)

$$X\text{—}C_1 + {}^{14}CH_3 \cdot CH_2 \cdot {}^{14}COOH \rightleftharpoons HOOC \cdot {}^{14}CH_2 \cdot CH_2 \cdot {}^{14}COOH + X$$

$$\updownarrow$$

$$HOOC \cdot {}^{14}CH_2 \cdot CH_3 + X\text{—}{}^{14}C_1$$

(B)

FIG. 2. Randomization of isotope from C-3 to C-2 of propionate without loss of ^{14}C from the carboxyl of propionate: (A) via an intramolecular transfer; (B) via reversible intermolecular transfer of the carboxyl of succinate in a bound form designated as X–C_1. The latter would cause little dilution of the carboxyl group if there was only a trace of succinate, but the ^{14}C of C-3 is distributed in both C-3 and C-2 and the average ^{14}C- specific activity is halved.

13. F. W. Leaver, H. G. Wood, and R. Stjernholm, *J. Bacteriol.* **70,** 521 (1955).
14. H. G. Wood, R. Stjernholm, and F. W. Leaver, *J. Bacteriol.* **72,** 142 (1956).

succinate in the cell were small and there were equilibration with a large amount of propionate-^{14}C via exchange through reaction (B).

The usual tracer techniques could not be used to differentiate between the two theoretical mechanisms of Fig. 2. These mechanisms could be differentiated, however, by use of ^{13}C and mass analysis of the products. Such an analysis was carried out by Pomerantz (*15*). Propionate was synthesized from ^{13}C so that all three carbons of a single molecule contained ^{13}C. This labeled propionate was mixed with unlabeled propionate and submitted to the randomization by the cells. The mass distribution of the resulting propionate molecules was then compared with that of a portion of the mixture which had not been exposed to the cells. It is evident that there would be no change in the mass of the propionate molecules if the randomization occurred by the intramolecular mechanism (A) whereas by the intermolecular mechanism (B) there would be interchange of ^{13}C of the carboxyl group between the labeled molecules of propionate and the unlabeled molecules and the mass spectrum of the resulting mixture would be altered from that of the original mixture. Pomerantz (*15*) obtained results which were in full accord with the mechanism of carboxyl transfer as illustrated in Fig. 2B (*16*).

The above results indicated that there was decarboxylation of succinate to propionate but CO_2 was not the product. Phares *et al.* (*17*) had likewise obtained results which were in accord with this view. The mechanism of the propionate fermentation could then be visualized as shown in Fig. 3. This mechanism would involve transcarboxylation rather than decarboxylation to free CO_2 and it would account for the random-

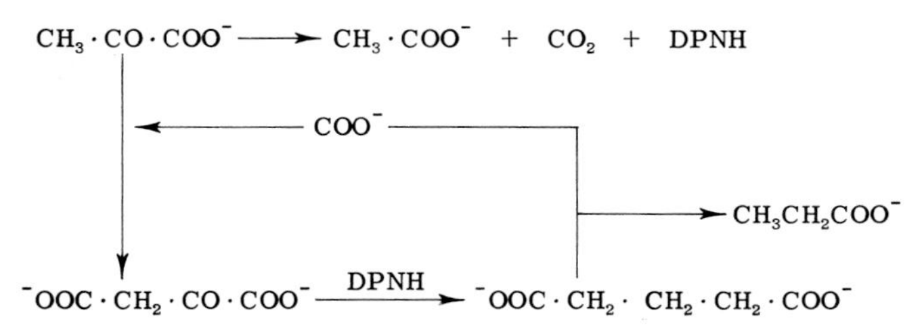

FIG. 3. A working hypothesis for formation of propionate by transcarboxylation in the propionate fermentation. For simplicity the role of CoA esters is omitted.

15. S. H. Pomerantz, *JBC* **231,** 505 (1958).

16. Present information indicates that the randomization of ^{14}C occurs by the following reversible reactions (see Figs. 4 and 5).

Propionate + succinyl-CoA ⇌ propionyl-CoA + succinate
Succinyl-CoA ⇌ (*R*)-methylmalonyl-CoA ⇌ (*S*)-methylmalonyl-CoA
E-biotin + (*S*)-methylmalonyl-CoA ⇌ E-biotin-CO_2 + propionyl-CoA

17. E. F. Phares, E. A. Delwiche, and S. F. Carson, *J. Bacteriol.* **71,** 604 (1956).

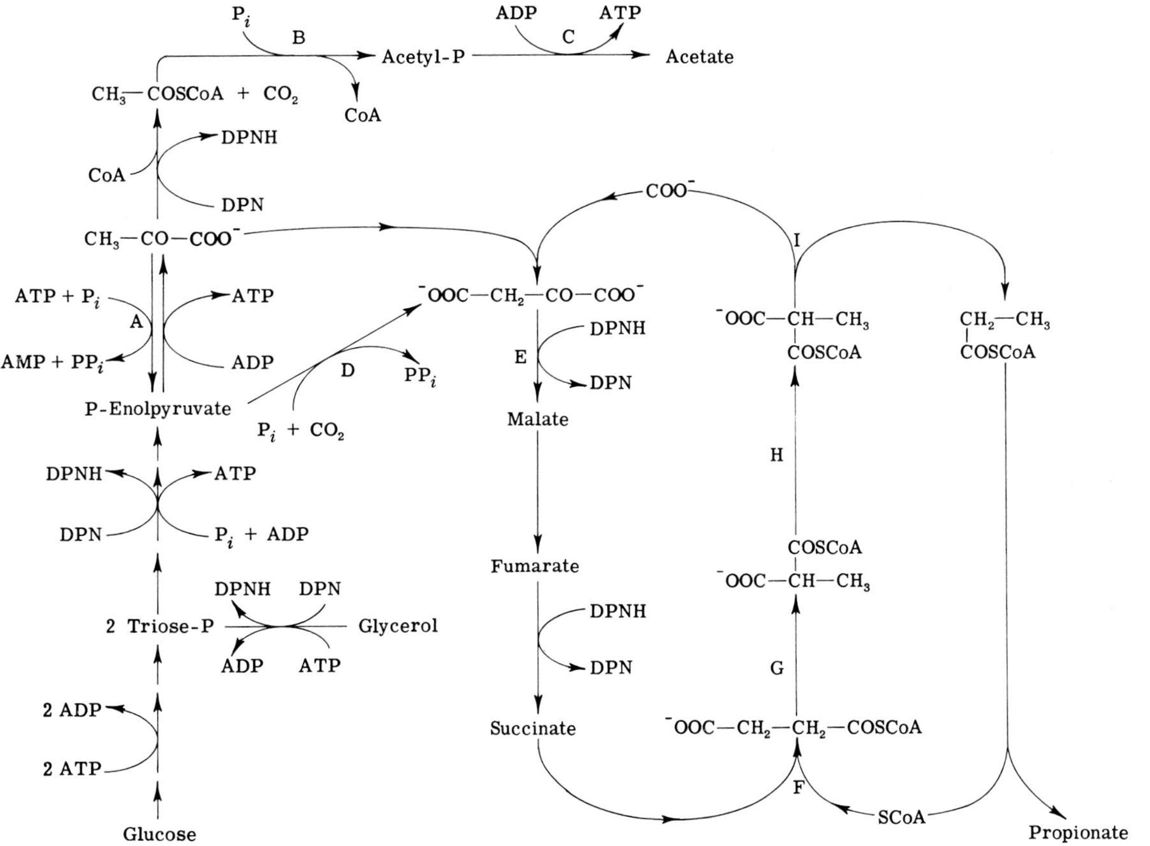

FIG. 4. Mechanism of the propionate fermentation. The following enzymes have been purified from propionibacteria: A, pyruvate, phosphate dikinase (*21*); B, phosphotransacetylase (*22*); C, acetyl kinase (*22*); D, carboxytransphosphorylase (*23*); E, malate dehydrogenase (*22*); F, CoA transferase (*22*); G, methylmalonyl-CoA mutase (*24*); H, methylmalonyl-CoA racemase (*25*); and I, transcarboxylase (*26*).

ization of the isotope in propionate from lactate-3-^{14}C or lactate-2-^{14}C or glycerol-2-^{14}C. Swick and Wood, therefore, undertook a search for an enzyme which would catalyze the formation of propionate and oxalacetate from pyruvate and succinyl-CoA and which did not involve CO_2. In a short time the enzyme was found in propionibacteria (*3*). The studies by Phares *et al.* (*17*), Whitely (*18*), Lardy and Adler (*19*), and Flavin and Ochoa (*20*) had indicated the probable role of a CoA ester. It was found (*3*) that methylmalonyl-CoA rather than succinyl-CoA is the direct reactant as shown in reaction (1) and that succinyl-CoA is converted to methylmalonyl-CoA by methymalonyl-CoA mutase.

B. The Role of Transcarboxylase in the Propionic Acid Fermentation

The major reactions of the propionate fermentation and the role of transcarboxylase are shown in Fig. 4 (*21–26*). The formation of propionate from pyruvate involves two interlinked cycles, one involving the carboxyl transfer by transcarboxylase (I) and the other the SCoA transfer by CoA transferase (F). No ATP is required in the cycles since the carboxylation of pyruvate is by transcarboxylation rather than by the fixation of CO_2 and the thioester linkage is preserved by CoA transfer. The net result of the interlinked cycles is the reduction of the carbonyl group of pyruvate yeilding propionate with the consumption of two DPNH. Seven reactions are involved in the two cycles which bring about this conversion. For simplicity, the arrows of Fig. 4 are shown in

---

18. H. R. Whitely, *Proc. Natl. Acad. Sci. U. S.* **39,** 772 (1953).
19. H. A. Lardy and J. Adler, *JBC* **219,** 933 (1956).
20. M. Flavin and S. Ochoa, *JBC* **229,** 965 (1957).
21. H. J. Evans and H. G. Wood, *Proc. Natl. Acad. Sci. U. S.* **61,** 1448 (1968); *Biochemistry* **10,** 721 (1971).
22. S. H. G. Allen, R. W. Kellermeyer, R. L. Stjernholm, and H. G. Wood, *J. Bacteriol.* **87,** 171 (1964).
23. H. G. Wood, J. J. Davis, and J. M. Willard, "Methods in Enzymology," Vol. 13, p. 297, 1969; H. G. Wood, J. J. Davis, and J. M. Willard, *Biochemistry* **8,** 3145 (1969).
24. R. W. Kellermeyer and H. G. Wood, "Methods in Enzymology," Vol. 13, p. 207, 1969; P. Overath, E. R. Stadtman, G. M. Kellerman, and F. Lynen, *Biochem. Z.* **336,** 77 (1962).
25. S. H. G. Allen, R. W. Kellermeyer, and H. G. Wood, "Methods in Enzymology," Vol. 13, p. 194, 1969; P. Overath, G. M. Kellerman, F. Lynen, H. P. Fritz, and H. J. Keller, *Biochem. Z.* **335,** 500 (1962).
26. H. G. Wood, B. Jacobson, B. I. Gerwin, and D. B. Northrop, "Methods in Enzymology," Vol. 13, p. 215, 1969.

the direction of propionate formation, although most of the reactions are readily reversible (*27*).

It is beyond the scope of this review to discuss the other enzymes of the propionic acid bacteria. Nevertheless, it is of interest to note that CO_2 is fixed into oxalacetate by carboxytransphosphorylase (D) with formation of inorganic pyrophosphate and that pyruvate is converted to P-enolypyruvate by pyruvate, phosphate dikinase (A) also with formation of inorganic pyrophosphate. Reeves (*30*) has recently found both of these enzymes in *Entamoeba histolytica*.

The following question arises: If propionate is formed by transcarboxylation, why does the propionate fermentation yield a net fixation of CO_2 in contrast to most fermentations? It is apparent from Fig. 4 that net fixation of CO_2 will occur if either the CoA transferase (F), the mutase (G), the racemase (H), or the transcarboxylase (I) becomes rate limiting so that succinate accumulates. Net fixation of CO_2 is observed with glycerol for this reason and also because the oxidation of glycerol generates two DPNH which are sufficient to provide for the reduction in the propionate cycle. Since there is no requirement to generate DPNH by oxidation of the pyruvate to acetate and CO_2, there is little formation of CO_2 from this source to mask the uptake of CO_2 via the carboxytransphosphorylase reaction.

The following question also arises: If propionate is formed by transcarboxylation why is so much labeled CO_2 incorporated in the carboxyl of propionate? In the presence of $^{13}CO_2$ (*12*) or $^{14}CO_2$ (*8*) the isotope concentration of the carboxyl of the propionate was almost half that of the CO_2. This result would indicate that practically all the propionate had arisen from succinate that had been formed via CO_2 fixation. If

27. Propionate is metabolized by animal tissue in large part by the reverse of the reactions of Fig. 4 as demonstrated by Ochoa and his co-workers [see review by Y. Kaziro and S. Ochoa (*28*)]. The major difference is that propionyl-CoA is introduced into the system by a CO_2 fixation reaction to yield methylmalonyl-CoA.

$$CO_2 + ATP + \text{propionyl-CoA} \rightleftharpoons ADP + P_i + \text{methylmalonyl-CoA} \quad (2)$$

The enzyme catalyzing this reaction like transcarboxylase is a biotin enzyme. Both enzymes are specific: propionyl-CoA carboxylase will not carboxylate propionyl-CoA with oxalacetate and transcarboxylase will not carboxylate propionyl-CoA with ATP and CO_2. Following formation of methylmalonyl-CoA from propionyl-CoA, the sequence of metabolism is via succinate to oxalacetate and then to P-enolpyruvate via P-enolpyruvate carboxykinase. Thus, propionate is glycogenic in animals and the isotope of propionate-2-^{13}C is randomized equally (*29*) in carbons 1, 2, 5, and 6 of the glucose unit because the pathway involves succinate as an intermediate.

28. Y. Kaziro and S. Ochoa, *Advan. Enzymol.* **26,** 283 (1964).

29. V. Lorber, N. Lifson, W. Sakami, and H. G. Wood, *JBC* **183,** 531 (1950).

30. R. E. Reeves, *JBC* **243,** 3202 (1968); *BBA* **220,** 346 (1970).

propionate is formed entirely by transcarboxylation as shown in Fig. 4 this concentration of isotope could only occur in the propionate if there is reversible exchange of $^{14}CO_2$ with the β-COO^- of the oxalacetate via the carboxytransphosphorylase reaction (D) (*31*).

Thus far, acetyl-CoA carboxylase has not been found in propionibacteria. Since transcarboxylase catalyzes the carboxylation of acetyl-CoA (*2*) by oxalacetate very effectively (about one-half the rate of carboxylation of propionyl-CoA), it appears likely that the malonyl-CoA required for fatty acid synthesis may be synthesized by transcarboxylation.

II. Molecular Properties of Transcarboxylase

A. Purification

The methods of purifying transcarboxylase from crude extracts of propionibacteria have been described in detail (*26*). Briefly, the procedure involves breaking the cells in an Eppenbach Colloid Mill in phosphate buffer, centrifugation to remove the debris, and adsorption of the transcarboxylase on DEAE-cellulose. The DEAE-cellulose with adsorbed enzyme is washed with 0.1 *M* phosphate. The enzyme is then eluted with 0.3 *M* phosphate at pH 6.8. The enzyme is precipitated from the phosphate buffer with $(NH_4)_2SO_4$, dialyzed, and then subjected to chromatography on a cellulose phosphate column equilibrated with 0.05 *M* phosphate,

31. Galivan and Allen (*32*) have isolated a biotin enzyme from *Micrococcus lactilyticus* which catalyzes the decarboxylation of methylmalonyl-CoA to propionyl-CoA and CO_2. If methylmalonyl-CoA were decarboxylated to CO_2 and propionyl-CoA rather than metabolized by transcarboxylation, all the oxalacetate would be formed by CO_2 fixation and the high content of ^{14}C from $^{14}CO_2$ would be accounted for. However, this would require a CO_2 turnover which does not occur (*8*). In addition, this pathway would pose energy problems in the fermentation of glycerol since only via the pyruvate kinase pathway of Fig. 4 is there a net generation of ATP. There is a possibility that inorganic pyrophosphate might be utilized as an energy source (*33, 34*). It also is possible that ATP is generated in some manner not shown in Fig. 4, such as by anaerobic oxidative phosphorylation during the reduction of the fumarate to succinate (*35*).

32. J. H. Galivan and S. H. G. Allen, *ABB* **126,** 838 (1968).

33. H. G. Wood, J. J. Davis, and H. Lochmüller, *JBC* **241,** 5692 (1966).

34. H. Baltscheffsky, *Acta Chem. Scand.* **21,** 1973 (1967); M. Baltscheffsky, *BBRC* **28,** 270 (1967).

35. D. R. Sanadi and A. L. Fluharty, *Biochemistry* **2,** 523 (1963); D. W. Haas, *BBA* **92,** 433 (1964).

pH 6.8. The column is developed with 0.05–0.3 *M* phosphate, pH 6.8. The enzyme is eluted with 0.3 *M* phosphate.

Recently it has been found that it is desirable to collect 20 ml fractions of the eluate in tubes containing 1 ml of saturated $(NH_4)_2SO_4$ (prepared from the crystallized salt and neutralized). The enzyme is more stable, and the recovery of the enzymic activity is better under these conditions (*36*). The fractions containing transcarboxylase are precipitated at 80% $(NH_4)_2SO_4$ saturation. The dialyzed enzyme is placed on a TEAE-cellulose column equilibrated with 0.05 *M* phosphate, pH 6.8, and is eluted with 0.15 *M* phosphate and 0.225 *M* phosphate. The eluates are collected in $(NH_4)_2SO_4$ solution as above. Gel filtration on Sepharose 2B may be utilized as an alternative to purification on TEAE-cellulose. Transcarboxylase with 40 IU/mg is obtained in about 30% overall yield. About 250 mg of transcarboxylase is obtained from 300 g (weight weight) of cells which is the yield from about 100 liters of medium (*26*).

Transcarboxylase preparations frequently consist of two sedimenting forms with values of $s_{20,w}$ ~18 S and ~16 S (*37*), although preparations exhibiting only one or the other have been obtained on occasion. The ~18 S form is observed if the precaution is taken to collect the fractions from the columns in saturated $(NH_4)_2SO_4$ as described above.

B. Molecular Weight, Metal Content, and Biotin Content and Linkage to Enzyme

The molecular weight of transcarboxylase with $s_{20,w}$ = 16 S has been estimated to be 670,000 ± 40,000 by the Archibald method assuming a partial specific volume of 0.75 (*26*). Recent reevaluation of the molecular weight by the Archibald method has confirmed the above value, and determination by sedimentation equilibrium has given a value of 700,000 ± 30,000 (*38*).

It has been found (*38*) that the ~16 S species of transcarboxylase arises from the ~18 S species by loss of a ~6 S biotin, Co, Zn subunit. This subunit is described below. Using transcarboxylase labeled with ^{3}H-biotin the biotin content was found to be 1.6 μg/mg of protein in the ~18 S species, and 1.29 μg/mg in the ~16 S species. The radioactivity

36. B. Jacobson, F. Ahmad, and H. G. Wood, unpublished results (1971).

37. B. Jacobson, B. I. Gerwin, F. Ahmad, P. Waegell, and H. G. Wood, *JBC* **245**, 6471 (1970).

38. N. M. Green, R. C. Valentine, N. G. Wrigley, F. Ahmad, B. Jacobson, and H. G. Wood, unpublished results (1971); F. Ahmad, B. Jacobson, H. G. Wood, R. C. Valentine, M. Green, and N. Wrigley, *Fed. Proc.* **30**, 33 (1971).

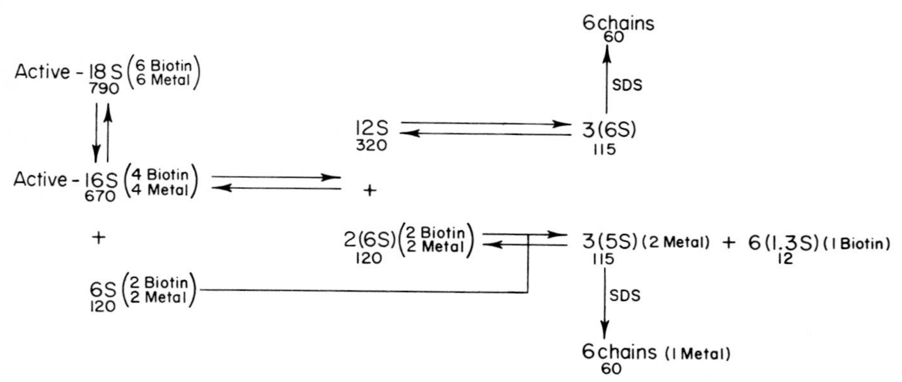

FIG. 5. Present interpretation of the dissociation of transcarboxylase to subunits, their molecular weights, and reassociation of the subunits to active forms of the enzyme. There are two active forms of the enzyme, ~18 S and ~16 S. The ~16 S form of the enzyme has lost one ~6 S biotin, Co, Zn subunit. Further dissociation of the ~16 S form of the enzyme yields an inactive ~12 S subunit and two additional inactive, ~6 S biotin, Co, Zn subunits. The latter reaction is reversible with formation of ~16 S and ~18 S active forms of the enzyme. The three ~6 S biotin, Co, Zn subunits dissociate to three ~5 S metal dimers and six 1.3 S biotin carboxyl carrier proteins. The ~12 S subunit dissociates to three ~6 S dimers which do not contain metal or biotin. On treatment with dodecyl sulfate, peptide chains of 60,000 molecular weight form from the ~6 S dimer and ~5 S metal dimer. The determined molecular weights (10^{-3}) are shown for each subunit. Transcarboxylase is made up of 18 peptide chains with a total molecular weight of 790,000 ($12 \times 60{,}000 + 6 \times 12{,}000$).

was in accord with these values. The moles of biotin per mole of ~16 S enzyme, calculated on the basis of a molecular weight of 6.7×10^5, was 3.5 (*38*). The 6 S biotin, Co, Zn subunit has a molecular weight of 1.2×10^5 as shown in Fig. 5. Accordingly, the ~18 S species would have a molecular weight of 7.9×10^5 and the moles of biotin per mole of ~18 S species were calculated to be 5.2. It has previously been reported (*26*) that transcarboxylase contains six biotins. Apparently this is the case for the 18 S species, and the 16 S species contains four biotins.

Transcarboxylase contains tightly bound cobalt and zinc. Estimations of the amount of cobalt and zinc were made from the radioactivity of transcarboxylase isolated from bacteria grown in medium-containing $^{60}Co^{2+}$ or $^{65}Zn^{2+}$ and also by determination of the metals by atomic absorption (*39*). A sum of six metals was found per mole of transcarboxylase, and this apparently is for the 18 S species; the 16 S species

39. D. B. Northrop and H. G. Wood, *JBC* **244,** 5801 (1969).

contains four atoms. The two metals are mutually interchangeable. The cobalt varied from 1.8 to 4.4 g-atoms and the zinc reciprocally from 2.2 to 4.7. The ratio of cobalt to zinc may depend on the composition of metals of the growth medium. Transcarboxylase does not require metals in addition to the bound metals, and EDTA at 0.1 *M* concentration does not inhibit the reaction.

It is of interest that pyruvate carboxylase from yeast has recently been found to contain zinc (*40*). Thus far only those biotin enzymes which have ketoacids as substrates have been found to contain metals. The first example was observed by Scrutton *et al.* (*41*) in pyruvate carboxylase of chicken liver which contains four atoms of manganese per mole

Like all other biotin enzymes so far studied [see Knappe's review (*42*)] and articles in the present volume (*43*), the biotin in transcarboxylase is amide bound to the ε-amino of a lysyl group of the protein. Carboxylation is by formation of an amide with the 1′ nitrogen of the biotin (*44*). This fact was established by the procedure illustrated in Fig. 6. The transcarboxylase reaction in common with those of the other biotin enzymes (*42*, *43*) involves two successive half-reactions [(a) and (b) of Fig. 6]. By use of 3-^{14}C-methylmalonyl-CoA the enzyme was labeled by means of half-reaction (a) and then was separated from the 3-^{14}C-methylmalonyl-CoA and propionyl-CoA by Sephadex filtration. Reaction (b) then could be shown to occur through the use of this ^{14}C-labeled protein and demonstration of the formation of ^{14}C-oxalacetate from pyruvate (*44*). The carboxy amide bond of the biotin of transcarboxylase is unstable, having a half-life of about 2 min at 30° at pH 6.8 (*44*). For comparison with the lability of other carboxylated biotin enzymes, see Knappe (*42*). The carboxylated biotin of transcarboxylase was stabilized by conversion to the methyl ester. Treatment of the resulting protein with pronase yielded ^{14}C-carbomethoxybiocytin (Fig. 6). This established that the biotin is in amide linkage with the ε-amino group of lysine. Treatment with biotinidase and then diazomethane and identification of the 1′-*N*-^{14}C-carbomethoxybiotin methyl ester established that the carboxyl is in a 1′-*N* amide linkage with the biotin. Similiar procedures have been used to establish the structure of the carboxylated enzyme of other biotin enzymes (*42*, *43*).

40. M. C. Scrutton, M. Young, and M. F. Utter, *JBC* **245**, 6220 (1970).
41. M. C. Scrutton, M. F. Utter, and A. S. Mildvan, *JBC* **241**, 3480 (1966).
42. J. Knappe, *Ann. Rev. Biochem.* **39**, 757 (1970).
43. See Chapter 2 by Alberts and Vagelos in this volume of "The Enzymes."
44. H. G. Wood, H. Lochmüller, C. Riepertinger, and F. Lynen, *Biochem. Z.* **337**, 247 (1963).

$$CH_3-CH(^{14}COO^-)-COSCoA + E\text{-Biotin} \rightleftharpoons CH_3-CH_2-COSCoA + E\text{-Biotin-}^{14}COO^-$$

(a)

$$E\text{-Biotin-}^{14}COO^- + CH_3-CO-COO^- \rightleftharpoons E\text{-Biotin} + {}^-OO^{14}C-CH_2-CO-COO^-$$

(b)

$$E\text{-Biotin-}^{14}COO^- \xrightarrow{CH_2N_2} {}^{14}C\text{-Carbomethoxyprotein}$$

↓ pronase

^{14}C-Carbomethoxybiocytin + α-Amino acids

↓ biotinidase

^{14}C-Carbomethoxybiotin + Lysine

↓ CH_2N_2

1′-*N*-^{14}C-Carbomethoxybiotin methyl ester

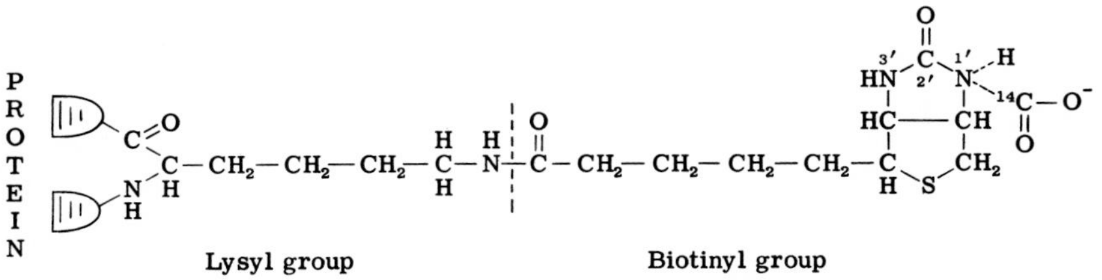

FIG. 6. Method of establishing the linkage of biotin in transcarboxylase and of the transferred carboxyl in E-biotin-COO⁻ by use of 3-^{14}C-methylmalonyl-CoA and half-reaction (a). The E-carboxyl biotin is unstable and is stabilized by esterification using diazomethane. The linkages were determined by digestion with pronose and biotinidase and identification of the products (*44*). The two dotted lines from the 1′-*N* of the biotinyl group indicate either an H or a carboxyl group may be at this position.

C. Subunits of Transcarboxylase and Reconstitution of Active Enzyme

1. *Dissociation, Distribution of Biotin, Cobalt, and Zinc in Subunits, Molecular Weight of Subunits*

Transcarboxylase dissociates in a complex fashion to inactive subunits (Fig. 5) and is conveniently monitored by the loss in enzymic activity. Alkaline pH, low ionic strength, monovalent ions, low protein concentration, and elevated temperature favor dissociation (*37*). When the enzymically active 18 S species is dissociated in 0.05 *M* tris-HCl at pH 8 at 0°, the ~18 S species disappears rapidly with formation of the active ~16 S species and the ~6 S biotin, Co, Zn subunit (*36, 38*). Increasing amounts of inactive subunits of $s_{20,w}$ values of ~12 S, ~6 S, ~5 S, and 1.3 S are observed with time. As shown in Fig. 5 there are two types of inactive ~6 S subunits. One is formed from the ~12 S subunit and does not contain metals or biotin. The other ~6 S subunit contains the biotin, cobalt, and zinc. This ~6 S subunit, referred to below as the ~6 S biotin subunit, dissociates further to a 1.3 S subunit which contains all of the biotin (*45*) and a ~5 S subunit which contains the cobalt and zinc. The dissociation of the ~12 S subunit and ~6 S biotin subunit is retarded by 20% glycerol (*37*).

Figure 7 shows the results obtained by glycerol density gradient centrifugation of partially dissociated transcarboxylase containing ^{3}H-biotin. The ^{3}H-labeled enzyme was isolated from propionibacteria grown on medium containing ^{3}H-biotin. It is seen that the enzymic activity was confined to the small amount of residual ~16 S component. The radioactivity (biotin) was present in the ~16 S component, the ~5 S and ~6 S mixture, and in the 1.3 S biotin carboxyl carrier protein (*46*). The 1.3 S subunit constitutes only about 10% of the protein of the complete transcarboxylase, and therefore the amount in the individual fractions containing the 1.3 S subunit was below the limits of accurate determination. When similar experiments were done with transcarboxylase containing ^{60}Co or ^{65}Zn which was isolated from propionibacteria grown in medium containing these radioactive metals, results similar to those of Fig. 7 were obtained (*47*) except that the radioactivity was confined to the ~16 S component and the ~6 S, ~5 S mixture. The ~12 S component thus does not contain biotin, cobalt, or zinc.

Under more drastic conditions of dissociation at pH 9 in the absence of glycerol both the ~16 S and ~12 S components disappear and only the ~5 S, ~6 S mixture and the 1.3 S biotin carboxyl carrier proteins are present (Fig. 8) (*46*). When dissociated materials similar to those of Figs. 7 and 8 are investigated by sedimentation velocity using Schlieren optics, a 1.3 S peak is not observed because of its low concentration and rapid diffusion (*37*).

As shown in Fig. 8B the 1.3 S biotin carboxyl protein is reincorporated into the ~6 S components when the mixture is brought to pH 5.2 in

45. B. I. Gerwin, B. Jacobson, and H. G. Wood, *Proc. Natl. Acad. Sci. U. S.* **64,** 1315 (1969).

46. F. Ahmad, B. Jacobson, and H. G. Wood, *JBC* **245,** 6486 (1970).

47. F. Ahmad, D. Lygre, B. Jacobson, and H. G. Wood, unpublished results (1971).

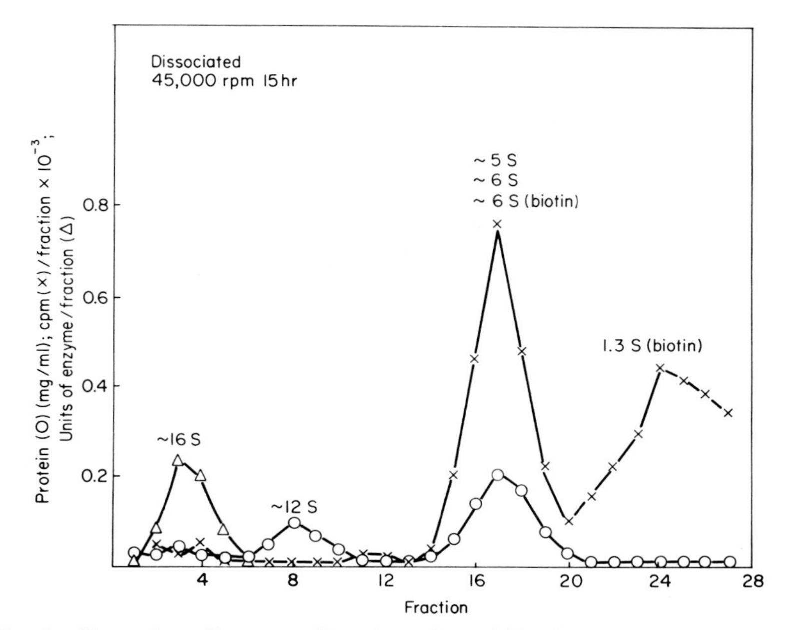

FIG. 7. Glycerol gradient centrifugation of partially dissociated transcarboxylase for isolation of ~12 S and ~6 S (biotin, Co, Zn containing) subunits. Transcarboxylase (specific activity 27.6) containing ^{3}H-biotin was passed through a Sephadex G-50 column equilibrated with 20% glycerol in tris-HCl, pH 8.4, with an ionic strength of 0.05 at 4°. The enzyme was allowed to dissociate for 24 hr at 4° and for 42 hr at 25° until the specific activity decreased to 0.47. The gradient contained glycerol (10–30%) in tris-HCl, pH 8.0, with an ionic strength of 0.05.

acetate buffer (*46*). Previously, reconstitution of active enzyme had not been accomplished from dissociated enzyme in which the ~16 S and ~12 S components were completely dissociated but recently such reconstitution has been observed (*36*).

The present information concerning the molecular weight of the components of transcarboxylase is summarized in Fig. 5. The 1.3 S biotin carboxyl carrier protein has a molecular weight of 11,700 ± 1,000 and is composed of a single chain of 97 amino acids. Alanine is the N-terminal residue and the protein contains one biotin and eight lysines (*45*). Its function obviously resembles that of the acyl carrier protein of the multiple enzyme systems for fatty acid synthesis. Alberts *et al.* (*48*) have isolated a 1.3 S biotin-containing subunit from acetyl-CoA car-

48. A. W. Alberts, A. M. Nervi, and P. R. Vagelos, *Proc. Natl. Acad. Sci. U. S.* **63**, 1319 (1969).

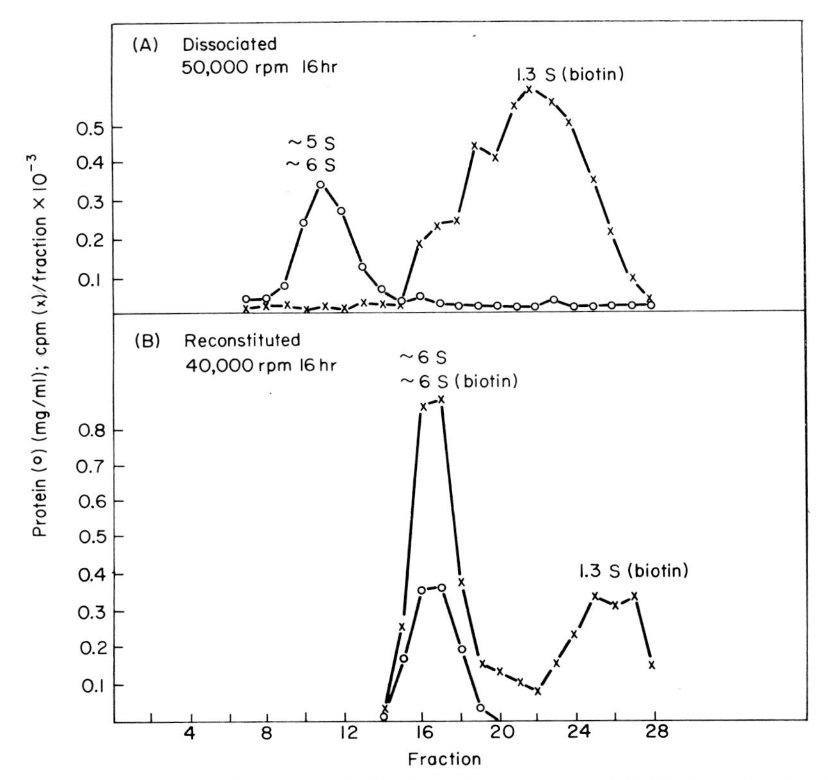

FIG. 8. Reincorporation of 1.3 S biotin subunit into ~6 S biotin subunit. (A) Transcarboxylase (specific activity 24.7) containing ^{3}H-biotin was passed through Sephadex G-50 equilibrated with tris-HCl, pH 9.0, with an ionic strength of 0.05 at 2–4°. Then the enzyme was allowed to dissociate for 16 hr at 0° and for 24 hr at 25° until the specific activity had decreased to 0.03. The gradient was in 5–20% sucrose. (B) The dissociated transcarboxylase was adjusted to pH 5.2 by the addition of 0.5 *M* acetate pH 4.4 at 0° and held at 0° for 20 hr. Gradient centrifugation was in 5–20% sucrose.

boxylase of *Escherichia coli*. It has a similar function. Thus far the 1.3 S component has not been reported from other biotin enzymes.

The molecular weight of other subunits of transcarboxylase has been determined recently by sedimentation equilibrium (*38*) using fractions isolated by glycerol gradient centrifugation. The molecular weight of the ~12 S component was 320,000 ± 20,000 and of the ~6 S component from the ~12 S component 115,000 ± 10,000. These results indicate that the ~12 S component is made up of three ~6 S components. The remainder of the transcarboxylase apparently is made up of three ~6 S biotin subunits of molecular weight 120,000 ± 12,000. Each of these contain two biotins and 2 g-atoms of cobalt plus zinc. The ~6 S biotin subunit dissociates to the ~5 S metal subunit (115,000) and the biotin carboxyl carrier protein.

Molecular weight determination by acrylamide gel electrophoresis of the components arising from ^{3}H-biotin-labeled transcarboxylase dissociated in sodium dodecyl sulfate (*38*) gave one major nonradioactive component with a molecular weight of ~60,000 and a minor radioactive component of molecular weight of 12,000 (*38*). There were other small peaks present, but these may have arisen from minor contaminants of the transcarboxylase. Gel patterns from separated ~6 S and ~12 S subunits likewise gave single major peaks with molecular weights of ~60,000. It thus appears that the ~12 S component (~320,000) is made up of six single chains (360,000), two of each in the ~6 S subunit (~115,000). The cobalt or zinc is in each chain of ~60,000 with two in each of the ~5 S subunits. Each of the six biotins are in 1.3 S subunits. Thus, the ~6 S biotin, cobalt, zinc subunit contains four chains and there are 18 single chains in transcarboxylase with a total sum of 790,000 ($12 \times 60{,}000 + 6 \times 12{,}000$), the latter agrees with the 790,000 calculated from the molecular weight of the ~16 S active species (670,000) plus that of one 6 S biotin subunit (120,000).

2. *Reconstitution of Enzymically Active Transcarboxylase from Enzymically Inactive Subunits*

Reassociation of inactive subunits to enzymically active transcarboxylase is favored by high ionic strength, polyvalent ions, acid pH, elevated protein concentration, and low temperature (*37*); thus far it has been accomplished only with the ~12 S subunit and the ~6 S biotin, metal subunit (*46*). For this purpose fractions containing the ~12 S component from experiments comparable to that of Fig. 7 were used and also those containing the ~5 S, ~6 S mixture which includes the native ~6 S biotin subunit. Reconstituted ~6 S biotin subunit was obtained from the fractions of experiments comparable to Fig. 8B. The formation of active enzyme from combinations of the ~12 S subunits and ~6 S biotin subunit is shown in Fig. 9. The transcarboxylase which was dissociated to obtain these subunits had a specific activity of about 26. Using the maximum value of Fig. 9, the specific activity of the reconstituted enzyme is about 26 μmoles of oxalacetate formed per minute per milligram of ~12 S protein. Calculations on the basis of the protein of both the ~12 S and ~6 S material are not feasible since the latter contains inactive ~6 S material from the ~12 S component as well as the ~6 S biotin fraction. Since the ~12 S component is about

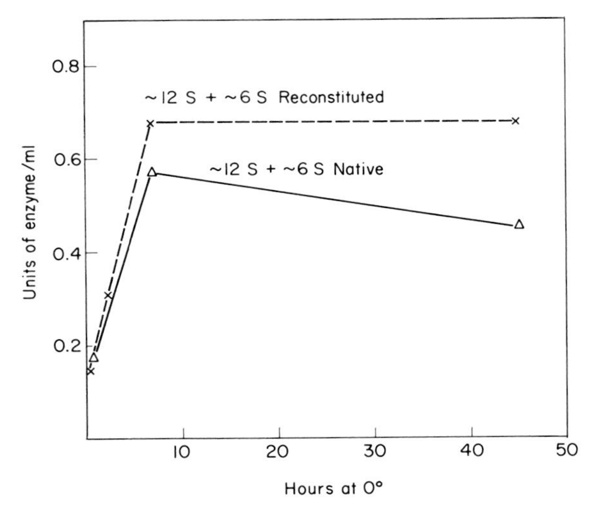

FIG. 9. Formation of enzymically active transcarboxylase from a combination of inactive ~12 S and native ~6 S biotin subunits or reconstituted ~6 S biotin subunits. Reconstitutions were done in the presence of 0.75 *M* potassium phosphate buffer, pH 6.8 (*37*). Concentrations of the subunits during reassociation were ~12 S, 26.6 μg/ml; native ~6 S biotin subunit, 146 μg/ml; reconstituted ~6 S biotin subunit, 56.0 μg/ml. Reassociations were done at 0°. There was no activity with single components.

one-half the molecular weight of the enzyme prior to dissociation, the specific activity on the basis of the complete enzyme would equal about 13, or one-half that of the original enzyme. The $s_{20,w}$ value of the reconstituted enzyme is ~16 S to ~18 S.

When transcarboxylase is dissociated in the presence of glycerol and the subunits are not separated, reconstitution to active enzyme is readily accomplished by simply replacing the 0.05 μ tris-HCl, pH 8, with 0.15 to 0.2 *M* phosphate buffer, pH 6.8, using a Sephadex column (*37*). The resulting transcarboxylase has an $s_{20,w}$ value of ~16 S to ~18 S. On the other hand, transcarboxylase which is dissociated in the absence of glycerol in 0.05 μ tris-HCl, pH 8, does not reconstitute as readily. It requires high ionic strength polyvalent ions such as 0.75 to 1.0 *M* phosphate or an acid pH of about 5 (with acetate buffer). Surprisingly, the resulting active transcarboxylase has an $s_{20,w}$ value of ~25 S (*37*).

Alberts and Vagelos (*49*) and Alberts *et al.* (*48*) have resolved acetyl-CoA carboxylase from *E. coli* into subunits somewhat similar to those of transcarboxylase. They (*49*) separated the acetyl-CoA carboxylase into two fractions, E_a and E_b which were inactive but when combined

49. A. W. Alberts and P. R. Vagelos, *Proc. Natl. Acad. Sci. U. S.* **59,** 561 (1968).

catalyzed malonyl-CoA synthesis; E_a, which contained the bound biotin, became carboxylated when ATP and bicarbonate were added as substrates. When carboxylated E_a was combined with E_b and acetyl-CoA, the fixed CO_2 was transferred to acetyl-CoA yielding malonyl-CoA. Fraction E_a with ATP and CO_2 also catalyzed the carboxylation of free biotin. Fraction E_a was further resolved into a 1.3 S biotin carboxyl carrier protein and a 5.4 S biotin-free protein (*48*). The latter subunit with ATP and CO_2 likewise catalyzed the carboxylation of free biotin. Dimroth *et al.* (*50*) have crystallized the biotin-free subunit derived from E_a and shown that the crystalline form catalyzes the carboxylation of free biotin. They designated this subunit biotin carboxylase. It has a molecular weight of 100,000 and is composed of two 50,000 subunits. Thus, acetyl-CoA carboxylase from *E. coli* and transcarboxylase have very similar subunit structures, i.e., a subunit of about 100,000 which combines with the biotin carboxyl carrier protein. Each also requires a third subunit, E_b, corresponding to the 12 S subunit.

We have been unable to carboxylate free biotin with transcarboxylase. The failure may result because free biotin cannot compete with bound biotin for the binding site of the biotin ring. The evidence for such sites in transcarboxylase will be considered in Section III. Perhaps with the subunit free of the carboxyl carrier protein such carboxylation may be possible.

Thus far, reconstitution of the intact molecule of acetyl-CoA carboxylase from the resolved subunits has not been reported.

D. Electron Microscopy of Transcarboxylase, Its Subunits, and Reconstituted Transcarboxylase

Transcarboxylase has been examined in collaboration with Green *et al.* (*38*) by electron microscopy using negative staining with sodium silicotungstate. The enzyme frequently appears as a complex structure resembling a head with two ears with a silhouette like Mickey Mouse (Fig. 10). The "head" varies from oval shape (80 $\times$ 100 Å) to almost circular shape (100 Å diam). Usually no subunit structure is evident within the "head" of the intact enzyme. The "ears" appear as circular profiles ($\sim$50 Å in diameter and 20–40 Å between them). There are numerous variations from the Mickey Mouse appearance, e.g., "heads" with one "ear" and occasionally with three "ears" as well as "heads" without "ears." The three "ears" are usually seen as a halo at one end of the

50. P. Dimroth, R. B. Guchhait, E. Stoll, and M. D. Lane, *Proc. Natl. Acad. Sci. U. S.* **67**, 1353 (1970).

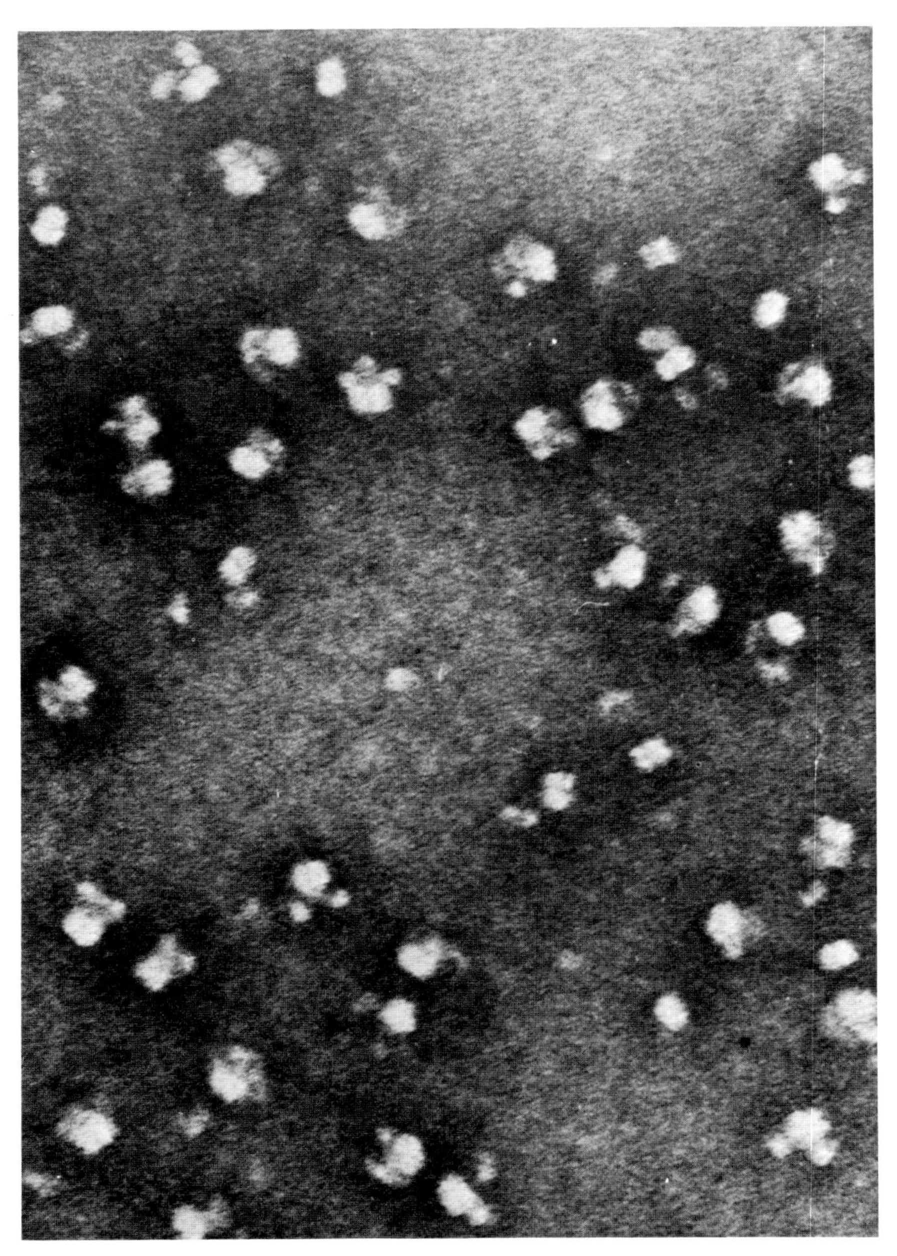

FIG. 10. Electron micrograph of native transcarboxylase. Examples are seen with a "head" and two "ears" which has been designated Mickey Mouse. Also seen are "heads" with no "ears," with single "ears." Occasionally a head is seen with what appears to be three indistinct ears grouped at one end of the head. The "head" is about 100 Å in diameter, the "ears" about 50 Å. ×500,000.

"head" of which examples are present in Fig. 10. Smaller fragments also are present which apparently arise by dissociation during preparation of the grid. It is not possible to determine whether the profiles with fewer than three "ears" have lost these structures or whether the missing "ears" are hidden by a "head."

Fractions containing the ~12 S subunit and those containing the native ~6 S biotin, Co, Zn subunit from experiments such as shown in Fig. 7 have been examined in the electron microscope, and, in addition, the ~6 S subunit obtained from the ~12 S subunit has been studied (*38*). Examples of these structures are shown in Fig. 11. The ~12 S subunit (top frame) gives two profiles which may be interpreted as different views of a single structure. Most frequently the ~12 S subunit is seen as a rectangular shape (70 × 100 Å), and it often appears divided into two halves and occasionally into four parts. The second view of the ~12 S is as a circular or hexagonal form (100 Å in diameter). The circular form appears to have a center hole. It is proposed (Fig. 13) that the subunits of the "head" are arranged in such a way as to form a cylinder with a hole in the center.

The ~6 S biotin, Co, Zn subunit obtained as outlined in Fig. 7 is shown in the middle frame of Fig. 11. The overall dimensions of these silhouettes are 55 × 90 Å, and the shape appears to be nearly cylindrical. The occasional sharp circular profile of the "ears" as seen in Fig. 10 is not observed in the isolated or dissociated material, but this view would only be evident if supported perpendicular to the grid as by attachment to the "head" unit. The free unit most likely would orient on the grid horizontally and present the 55 × 90 Å profile.

The ~6 S subunit which arises from the ~12 S subunit is rather ill defined (lower frame of Fig. 11). Some appear to be approximately square.

It has been possible by treatment with avidin to observe the central hole of the "head" with the intact enzyme (*38*). The excess avidin was removed with carboxymethyl cellulose. It is seen in some silhouettes of Fig. 12 that there are six units surrounding the "head," three of which are believed to be "ears" alternating with three avidins. In other structures one of the "ears" is missing. The avidins apparently have induced the enzyme to adsorb to the film with its threefold axis normal to the surface so that the central hole of the "head" is visible.

A photograph of models of the enzyme as proposed by Green *et al.* (*38*) is shown in Fig. 13. Figure 13A presents three views (1a, 1b, and 1c) of the ~6 S biotin, Co, Zn subunit. Number 2 is a view of the ~12 S subunit and No. 3 is a view of avidin. The small white spheres on the ~6 S biotin, Co, Zn subunit indicate the biotins and on the avidin

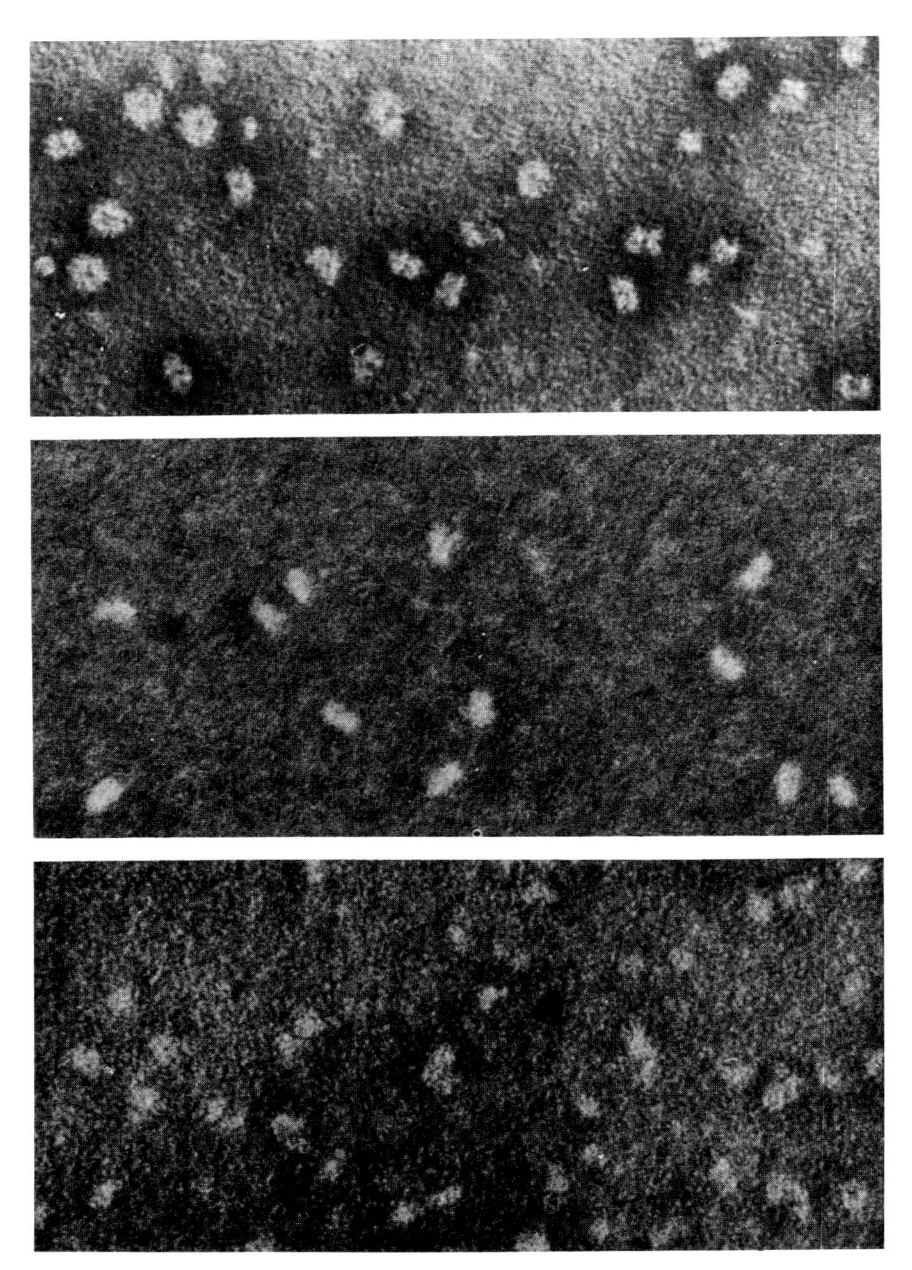

FIG. 11. Electron micrographs of transcarboxylase subunit, ×500,000. Upper frame is ~12 S subunits ("head") showing circular (100 Å diam) and rectangular (70 × 100 Å) profiles, middle frame is ~6 S biotin, Co, Zn subunits ("ears") showing their cylindrical (55 × 90 Å) profiles, lower frame ~6 S fragments from dissociation of ~12 S subunits in 20% glycerol containing 0.05 ionic strength tris-HCl, pH 8.0.

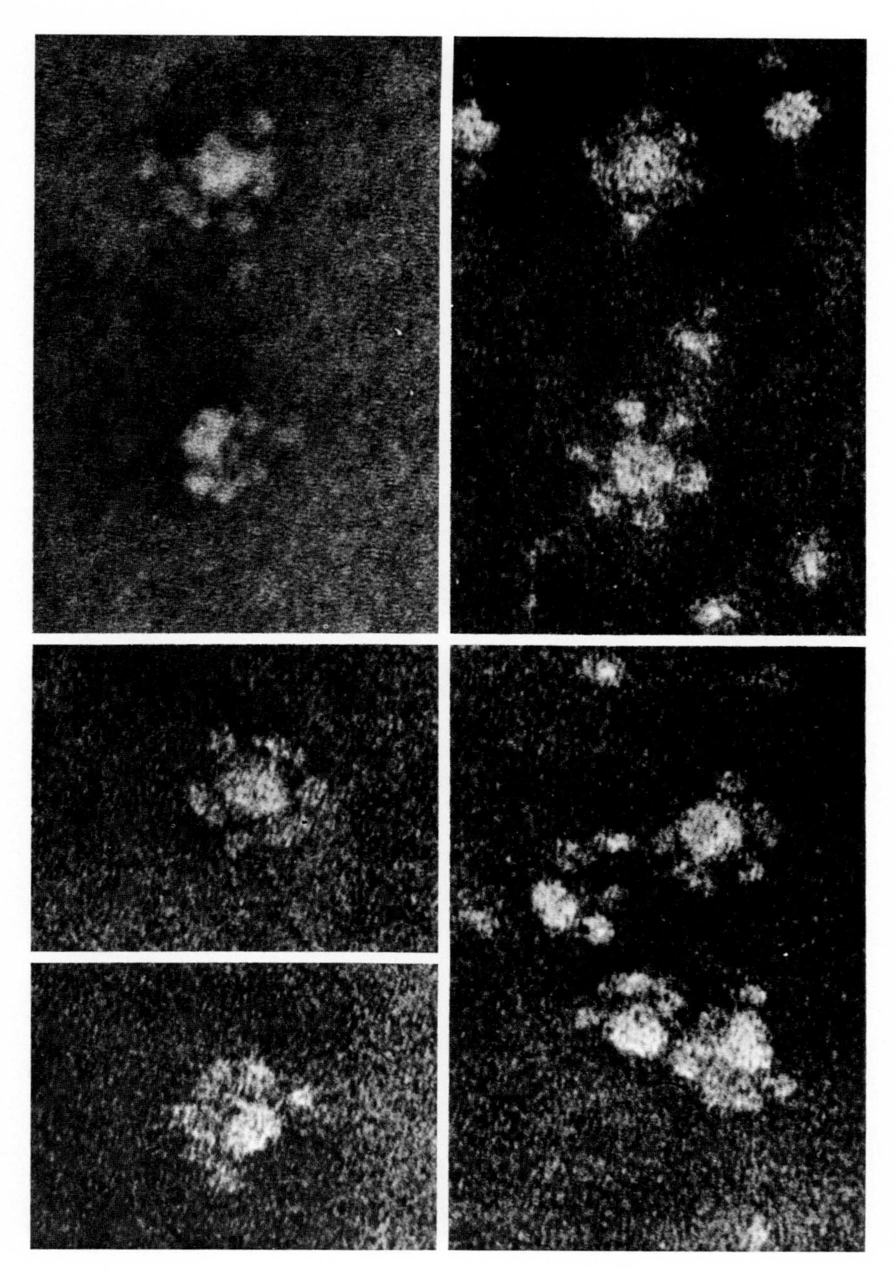

FIG. 12. Electron micrograph of transcarboxylase treated with avidin and the excess removed with carboxymethyl cellulose. In some profiles there are six units surrounding the "head," three of which are "ears" alternating with three avidins. In others, there are two "ears" with avidin between them and one avidin at the opposite end of each of the two ears. The central hole in the "head" is visible. ×750,000.

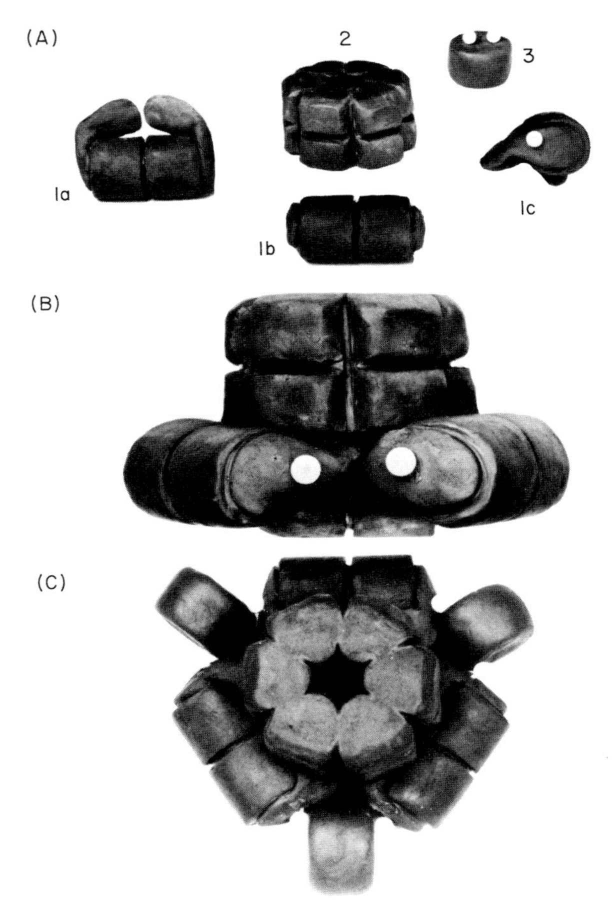

Fig. 13. Models of transcarboxylase. (A) Models of the subunits of transcarboxylase and of avidin; 1a, 1b, and 1c are different views of the ~6 S biotin, Co, Zn subunit. The cylindrical central structure is the ~5 S metal dimer (see Fig. 5) combined at each end with the 1.3 S biotin carboxyl carrier proteins. The biotin site is indicated by the white spheres. Number 2 is the ~12 S subunit or "head." It is shown in cyclic (or C_3) symmetry. The horizontal groove represents a fold of the peptide chain of the 60,000 molecular weight monomer. Number 3 is avidin with two of its four combining sites for biotin indicated by the white spheres. (B) Side view of the ~18 S form of the enzyme with an unfolded portion of the peptide chain of the carboxyl carrier protein in asymmetric combination with one end of the ~12 S subunit. One of the three "ears" is partially obscurred in the lower rear of the structure. (C) Top view of the ~18 S form of the enzyme in combination with three molecules of avidin. The avidin molecules are combined with the biotin of the carboxyl carrier proteins located at the ends of the long axis of the "ears."

binding sites for the biotin. The 1.3 S biotin subunits are shown on each end of the cylindrical ~5 S metal subunit. The protruding flaps represent an unfolded peptide chain of protein by which the ~6 S biotin, Co, Zn subunit is attached to the ~12 S subunit.

The six subunits of the "head" are shown arranged in cyclic (C_6 or C_3) symmetry. They could be in dihedral (D_3) symmetry. Either structure would be consistent with the observed geometery. However, the asymmetric attachment of the "ears" to one face only of the "head" as seen in Fig. 10 makes the C_6 symmetry the most likely interpretation. It accounts for all the appearances and properties of the enzyme except the dissociation of the ~12 S subunit to three ~6 S subunits. A C_6 oligomer of identical subunits would not be expected to dissociate to an intermediate stable dimer since all the intersubunit bonds would be expected to be identical. However, there may be mutual distortion as observed in insulin (*51*) which leads to stability as dimers and C_3 symmetry.

Figure 13B presents a side view of the "head" with the three "ears" attached. Examples corresponding to this model are seen in Fig. 10 where the "ears" are seen as a halo at one end of the "head." If the third "ear" in back is lost giving the 16 S form of the enzyme, it is evident that a Mickey Mouse could result on negative staining as observed in Fig. 10. In addition "heads" with single "ears" or none are seen in Fig. 10.

Figure 13C presents the model in which avidin has reacted with the ~18 S, three-"eared" form of the enzyme. When avidin is attached, the top view of the "head" is observed and the less densely stained center becomes apparent as seen in the top frame of Fig. 11. The "head" of the intact enzyme appears larger than that of the isolated ~12 S subunit, perhaps because of contributions to the "head" by the "ears." Examples of ~16 S form of the enzyme with two "ears" are also seen in Fig. 12. In this case there likewise are three avidins attached, one at each end of the two "ears" and one between the "ears" suggesting that the biotin carrier protein is located at opposite ends of the long axis of each "ear." With a single "ear" two avidins would be expected to attach.

The horizontal groove shown in Fig. 13 represents a fold in the 60,000 molecular weight peptide chain and accounts for the fact that the rectangular view of the ~12 S subunit (Fig. 11, top frame) sometimes appears divided both on a horizontal and perpendicular axis.

51. M. J. Adams, T. L. Lundell, E. J. Dodson, G. G. Dodson, M. Vi Jayan, E. N. Baker, M. M. Harding, D. C. Hodgkin, B. Rimmer, and S. Sheat, *Nature* **224**, 491 (1969).

In Section III below, evidence is presented that the biotin interacts with two different catalytic sites. Although not adequately indicated by the model, the biotin is considered to be positioned on a flexible peptide chain of about 10 amino acid residues (40 Å) between the "ear" and the "head." This permits oscillation of the biotin ring via the lysine-valeric acid side chain between the keto-acid-metal site on the "ears" and the CoA-ester site on the "head" (see Fig. 15).

It is possible using a mixture of isolated ~12 S and the ~6 S subunits to reconstitute active enzyme (Fig. 9). Examination of this type of reconstituted enzyme by electron microscopy has shown the presence of the Mickey Mouse form of the enzyme approximately comparable to that of the original enzyme (Fig. 10). As noted previously when transcarboxylase is dissociated in the absence of glycerol and then reconstituted, a ~25 S species is observed which is enzymically active. The ~25 S enzyme does not present a clear picture in the electron microscope, but there are indications of a set of "ears" at both ends of the "head." Possible explanations are considered by Green *et al.* (*38*).

III. Catalytic Properties of Transcarboxylase

A. Assay and General Properties of the Enzyme

Transcarboxylase is assayed in the forward direction by linking it with malate dehydrogenase and determining the decrease in DPNH by absorption at 340 nm. The reactions are as follows.

$$\begin{array}{c}\text{Methylmalonyl-CoA} + \text{pyruvate} \rightleftharpoons \text{propionyl-CoA} + \text{oxalacetate} \\ \text{Oxalacetate} + \text{DPNH} \rightleftharpoons \text{malate} + \text{DPN}\end{array} \tag{3}$$

The procedure is described in detail by Wood *et al.* (*26*). The enzyme may also be assayed in the reverse direction by linking it with lactate dehydrogenase.

$$\begin{array}{c}\text{Oxalacetate} + \text{propionyl-CoA} \rightleftharpoons \text{pyruvate} + \text{methylmalonyl-CoA} \\ \text{Pyruvate} + \text{DPNH} \rightleftharpoons \text{lactate} + \text{DPN}\end{array} \tag{4}$$

The assay is done using the same conditions as described for the forward reaction. However, since there is nonenzymic decarboxylation of oxalacetate to pyruvate, a correction is made by subtracting the results of blank assays containing the complete system minus propionyl-CoA (*52*).

Transcarboxylase has a broad pH optimum; there is little change in

52. D. B. Northrop, *JBC* **244**, 5808 (1969).

activity between pH 5.5 and 7.8 (*2*). The enzyme is stabilized to a considerable extent by polyvalent anions such as phosphate, 0.25 *M*, pH 6.8. On storage at 2° or —20° there is some loss of activity, but the enzyme can be reactivated by placing it in 1.5 μ $(NH_4)_2SO_4$ solution at 25° (*37*). The enzyme does not dissociate in the cold.

Transcarboxylase is inhibited strongly by avidin (*2, 3*). Inhibition by SH reagents is weak. Chloromercuribenzoate (10^{-5} *M*) at 0° caused 24% inactivation after incubation for 20 min and iodoacetate and *N*-ethylmaleimide (10^{-4} *M*) had little and no effect (*53*). The reaction is strongly inhibited by oxalate, α-ketobutyrate, CoA, propionyl pantetheine, and β-methyl oxalacetate (*52, 54*). It is not inhibited by metal chelators (*53*).

Transcarboxylase appears to be highly specific for the keto acid component: α-Ketobutyrate, α-ketovalerate, α-ketoglutarate, and β-ketoglutarate do not replace pyruvate as the acceptor for the carboxyl group of methylmalonyl-CoA (*2*) and β-methyloxalacetate does not serve as a carboxyl donor to propionyl-CoA (*52*). The specificity for the CoA esters is broad. Acetyl-CoA, butyryl-CoA, and acetoacetyl-CoA serve as carboxyl acceptors from oxalacetate at one-half, one-tenth, and one-fortieth the rate observed with propionyl-CoA as the acceptor (*2*). Malonyl-CoA and ethylmalonyl-CoA serve as carboxyl donors to pyruvate at one-half and one-seventh the rate observed with methylmalonyl-CoA (*2*). The enzyme is specific for the (S) isomer of methylmalonyl-CoA, the type which is formed by propionyl-CoA carboxylase. The (R) isomer is produced by methylmalonyl-CoA mutase (*25*).

B. Equilibria and the Free Energy of the Complete and Partial Reactions

The equilibrium of the transcarboxylase reaction expressed as total analytical concentrations at pH 6.5 and 30° was found to be 1.9 ± 0.1 (*2*). Since the transcarboxylase contained racemase the (S) isomer of methylmalonyl-CoA was in equilibrium with the (R) isomer and the above equilibrium constant is for the combined transcarboxylase and racemase reactions. The equilibrium of the racemase reaction is 1 (*25*). Thus, for the transcarboxylation reaction alone the equilibrium is as follows:

$$K'_{\text{anal}} = \frac{(\text{pyruvate}_T)(S\text{-methylmalonyl-CoA}_T)}{(\text{oxalacetate}_T)(\text{propionyl CoA}_T)} = 1.0 \pm 0.1 \qquad (5)$$

53. H. G. Wood, S. H. G. Allen, R. Stjernholm, and B. Jacobson, *JBC* **238,** 547 (1963).

54. D. B. Northrop and H. G. Wood, *JBC* **244,** 5820 (1969).

The $\Delta F'_{anal}$ therefore is 0.0 kcal at pH 6.5 [see Wood *et al.* (*33*) for definitions]. Since all species are ionized at pH 6.5 and there were no metals present which strongly bind the ions, the ionic equilibrium constant is the same.

$$K'_{ionic} = \frac{(\text{pyruvate}^-)(S\text{-methylmalonyl-CoA}^-)}{(\text{oxalacetate}^{2-})(\text{propionyl-CoA})} = 1.0 \pm 0.1 \quad (6)$$

The equilibrium also has been determined (*44*) for the partial reaction:

$$\text{Methylmalonyl CoA-3-}^{14}\text{C}^- + \text{E-biotin} \rightleftharpoons \text{propionyl-CoA} + \text{E-biotin-}^{14}\text{COO}^- \quad (7)$$

For this purpose methylmalonyl-CoA-3-^{14}C of known specific activity was incubated with transcarboxylase (E-biotin). When the methylmalonyl CoA-3-^{14}C reacts with E-biotin an equivalent amount of E-biotin-$^{14}COO^-$ is formed which is "acid labile." It was assumed that there was an equivalent formation of propionyl-CoA. This assumption was confirmed by the determination of the decrease in methylmalonyl-CoA. The initial E-biotin concentration was considered equal to the biotin of the enzyme since all of its biotins were shown to be subject to carboxylation in the presence of an excess of methylmalonyl-CoA.

The equilibrium of the partial reaction at pH 7.0 and 2° was found to be

$$\frac{(\text{E-biotin-COO}^-)(\text{propionyl-CoA})}{(\text{E-biotin})(S\text{-methylmalonyl-CoA}^-)} = 35 \pm 4 \quad (8)$$

the $\Delta F'_{anal}$ therefore is —1.9 kcal at pH 7.0 and 2°.

The $\Delta F'_{anal}$ has been calculated for the hydrolysis of E-biotin-COO^- by the following equations (*26*):

		$\Delta F'_{anal}$ (kcal)
E-biotin-COO^- + propionyl-CoA	$\rightleftharpoons$ E-biotin + methylmalonyl-CoA^-	1.9
(*S*)-Methylmalonyl-CoA^- + $pyruvate^-$	$\rightleftharpoons$ $oxalacetate^{2-}$ + propionyl-CoA	0.0
$Oxalacetate^{2-}$ + H_2O	$\rightleftharpoons$ $pyruvate^-$ + HCO_3^-	−6.3
Sum: E-biotin-COO^- + H_2O	$\rightleftharpoons$ E-biotin + HCO_3^-	−4.4

This value of —4.4 kcal places the E-biotin-COO^- at the lower end of the scale of energy-rich bonds. The value may be in some error because the equilibria were determined at different temperatures.

The effect of temperature has been determined on the rate of decarboxylation of E-biotin-COO^-. The decarboxylation was measured by determining the decrease in transferable ^{14}C in the following reaction which was linked with malate dehydrogenase.

$$\text{E-biotin-}^{14}\text{COO}^- + \text{pyruvate}^- \rightleftharpoons \text{E-biotin} + \text{oxalacetate-}^{14}\text{C}^{2-} \quad (9)$$

$$\text{Oxalacetate-}^{14}\text{C}^{2-} + \text{DPNH} \rightleftharpoons \text{malate-}^{14}\text{C}^{2-} + \text{DPN} \quad (10)$$

From the slope of an Arrhenius plot of the rate constants of the decarboxylation at different temperatures, the activation energy was calculated and found to be remarkably high, 26.6 kcal. The rate of decarboxylation increases about five times for each 10° rise in temperature and the half-lives at 0°, 10°, 20°, and 30° were 260, 48, 9.9, and 2 min, respectively, in phosphate buffer at pH 6.8. Other carboxylated biotin enzymes are likewise very labile.

C. Role of Cobalt and Zinc in the Catalysis

Rather conclusive evidence has been presented by Scrutton, Mildvan, and Utter (*41, 55, 56*) that Mn has a specific role at the ketoacid site of the biotin enzyme, pyruvate carboxylase from avian liver. They investigated the longitudinal proton relaxation rate of water bound to the manganese of the enzyme and showed that it was markedly reduced in the presence of pyruvate or oxalacetate and was nil with oxalate present. In addition, Mildvan and Scrutton (*56*) showed that pyruvate carboxylase caused the broadening of the methyl proton signal of pyruvate and that oxalate caused reversal of this effect. Cobalt is much less paramagnetic than manganese, thus the proton relaxation rate of water could not be studied with transcarboxylase. The effect of transcarboxylase on the nuclear magnetic signal of the protons of pyruvate has been investigated and it was observed (*39*) that the presence of the enzyme caused a marked broadening of the signal and that this effect was reversed when oxalate was added in increasing concentration. A proposed role of cobalt in the catalysis of transcarboxylase is shown in Fig. 14 which is modeled after that proposed by Mildvan *et al.* (*55*) for manganese in pyruvate carboxylase.

D. Kinetics and the Reaction Mechanism

Transcarboxylase has been found to have unusual kinetic properties. Mechanisms in which the enzyme acts alternately as an acceptor and donor of a transferred group have been designated *Ping-Pong* by Cleland (*57*) and may be outlined as follows:

$$\begin{array}{ccccccccc} & A & & P & & B & & Q & \\ & \downarrow & & \uparrow & & \downarrow & & \uparrow & \\ \hline E & & (EA:E'P) & & E' & & (E'B:EQ) & & E \end{array} \quad (11)$$

55. A. S. Mildvan, M. C. Scrutton, and M. F. Utter, *JBC* **241,** 3488 (1966).
56. A. S. Mildvan and M. C. Scrutton, *Biochemistry* **6,** 2978 (1967).
57. W. W. Cleland, *BBA* **67,** 104 (1967).

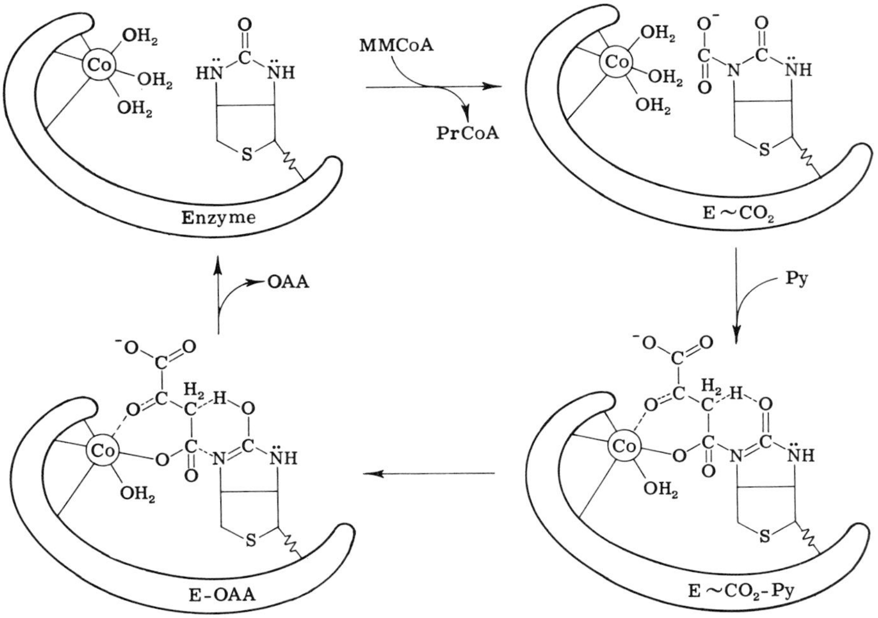

FIG. 14. Suggested role of cobalt in the transcarboxylase reaction. The scheme details the keto acid portion of the reaction. The mechanism of participation of the CoA esters is not shown. The biotin is carboxylated by methylmalonyl-CoA, the pyruvate combines with the cobalt which is considered to aid in the departure of a proton from the pyruvate and transfer of the carboxyl from the carboxylated biotin to form oxalacetate. The reactions are shown in only one direction although all steps are reversible. If zinc and cobalt can occupy common metal sites, then the mechanism also applies to the participation of zinc in the reaction. Abbreviations: PrCoA is propionyl-CoA, Py is pyruvate, OAA is oxalacetate, and MMCoA is methylmalonyl-CoA.

Here A, B, P, and Q represent substrates and products which for the transcarboxylase reaction are methylmalonyl-CoA, pyruvate, propionyl-CoA, and oxalacetate; E and E′ represent E-biotin and E-biotin-COO^-. The central feature of the standard Ping-Pong mechanism from the view of kinetic theory is the release of the first product (P) before the addition of the second substrate (B). Accordingly, the mechanism is restricted exclusively to binary enzyme–substrate complexes. Northrop (*52*) found initial velocity patterns typical of a standard Ping-Pong mechanism for both the forward and reverse reaction of transcarboxylase. However, for product inhibition he found that transcarboxylase does not follow standard Ping-Pong kinetics. It was observed that propionyl-CoA and pyruvate were noncompetitive and propionyl-CoA and methylmalonyl-CoA were competitive. In the standard Ping-Pong mechanism [Eq. (11)] it is assumed that A and Q combine at a common site on one enzyme form E, and that P and B also combine at this site but with a different form of the enzyme, E′. Consequently, product inhibition patterns between either A and Q or P and B would be competitive, and product inhibition between A and P or B and Q which combine with different forms of the enzyme would be noncompetitive. The results obtained with transcarboxylase gave exactly the opposite pattern, P and B (propionyl-CoA and methylmalonyl-CoA) competitive.

Dead-end inhibition by substrate analogies was also inconsistent with standard Ping-Pong kinetics. It is predicted for standard Ping-Pong kinetics that an inhibitor combining with both forms of the enzyme would yield noncompetitive inhibition. Coenzyme A was found to combine

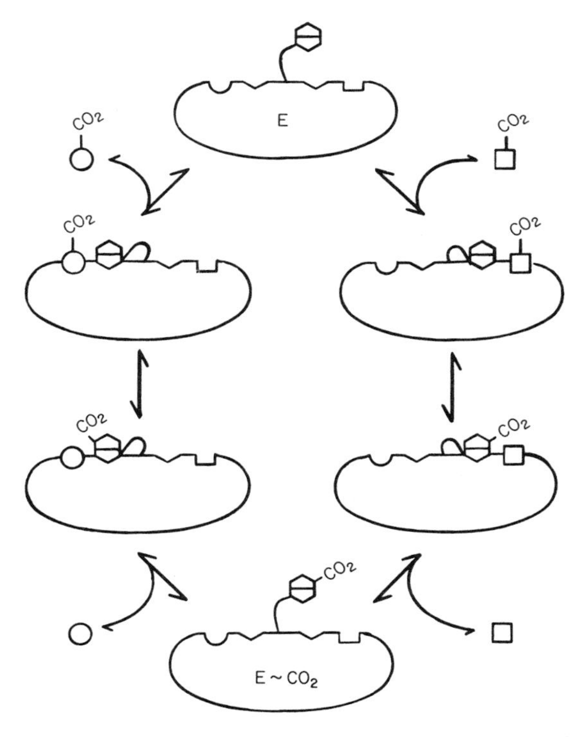

FIG. 15. Pictorial model of the transcarboxylase reaction. Free circle, pyruvate; carboxylated circle, oxalacetate; free square, propionyl-CoA; carboxylated square, methylmalonyl-CoA; hexagonal structure, biotin; carboxylated hexagonal structure, carboxylbiotin. E is one of possibly six reactive enzyme centers of transcarboxylase and the two substrate sites may be on different subunits, one associated with the ~12 S component and the other with the ~6 S biotin, Co, Zn subunit.

with both the free and carboxylated form of the enzyme, yet it was competitive with methylmalonyl-CoA or propionyl-CoA and uncompetitive with pyruvate or oxalacetate. Inhibition by α-ketobutyrate produced a reverse set of patterns.

It appears from these results that there are two groups of mutually competitive reactants: methylmalonyl-CoA, propionyl-CoA, and free coenzyme A; and pyruvate, oxalacetate, and α-ketobutyrate. The data indicate that the enzyme has two independent binding sites. Northrop has proposed a mechanism for transcarboxylase shown in Fig. 15. He assumed that the substrate and products combine equally well with either the carboxylated or the uncarboxylated form. Thus, the reaction is not restricted to either an exclusive formation of binary complexes or to a compulsory formation of a ternary complex. The term *hybrid Ping-Pong* was introduced by Northrop to designate this type of mechanism. Rate equations were derived for the hybrid Ping-Pong mechanism using rapid equilibrium and steady state kinetic theory. All the observed data were shown to fit the rate equations. Furthermore, the equilibrium constant for the transcarboxylase reaction was calculated from the kinetic data and gave a value of 1.0 which agrees with that found by direct chemical determination as described above.

In addition, Northrop and Wood (*54*) investigated the kinetics of the exchange reaction between ^{14}C-propionyl-CoA and methylmalonyl-CoA and also between ^{14}C-pyruvate and oxalacetate. It was found that α-ketobutyrate inhibited the pyruvate–oxalacetate exchange but stimulated the propionyl-CoA–methylmalonyl-CoA exchange. On the other hand,

TABLE I

K_m AND K_i VALUES OF SUBSTRATES AND INHIBITORS OF TRANSCARBOXYLASE

Substrate	K_m ($\times 10^{-4}$ M)	Inhibitors	K_i ($\times 10^{-4}$ M)
Pyruvate	7.7[a]	Pyruvate[c]	8.9[a]
Acetyl-CoA	5.0[b]	α-Ketobutyrate	25[a]
Propionyl-CoA	0.34[a]	CoA	6.3[a]
Butyryl-CoA	2.5[b]	Propionyl-CoA[c]	0.49[a]
Malonyl-CoA	0.35[b]	Propionyl pantetheine	3–7[a]
S-Methylmalonyl-CoA	0.044[a]	Oxalate	0.018[a]
Oxalacetate	0.63[a]	β-Methyloxalacetate	3[a]
		S-Methylmalonyl-CoA[c]	0.04[a]
		Oxalacetate	0.70[a]

[a] From Northrop (*52*).
[b] From Wood *et al.* (*53*).
[c] Product inhibition.

oxalate inhibited both exchange reactions and in addition gave an unusual set of kinetic patterns in the overall reaction. These results are inconsistent with the usual standard Ping-Pong mechanism. The model in Fig. 15 shows in addition to the covalent linkage of biotin by the side chain that there are two sites for binding the biotin ring, one adjacent to each substrate site. The results can be shown to fit this model if it is assumed that the presence of either α-ketobutyrate or oxalate alters the rate of migration of biotin between the two sites. It was proposed that oxalate increases the binding of biotin at the keto-acid (metal) site and α-ketobutyrate has the opposite effect.

This migration of biotin is in agreement with the results found with acetyl-CoA carboxylase (*48–50*) in which two different proteins catalyze the two half-reactions and biotin is bound to one and migrates to the other.

The K_m and K_i values of substrates and inhibitors of transcarboxylase are given in Table I.

ACKNOWLEDGMENT

The author was supported by Grants 2TO1-GM 35, GM-11839, and GM-13971 from the National Institutes of Health.

4

Formation of Oxalacetate by CO_2 Fixation on Phosphoenolpyruvate

MERTON F. UTTER • HAROLD M. KOLENBRANDER

I. Introduction

This chapter deals with the three enzymes known to catalyze the carbon dioxide fixation reaction in which phosphoenolpyruvate (PEP) is converted to oxalacetate (OAA). These enzymes include: (1) orthophosphate:oxalacetate carboxylase (phosphorylating) (EC 4.1.1.31), commonly known as phosphoenolpyruvate carboxylase and which we will refer to as PEPC; (2) GTP:oxalacetate carboxylase (transphosphorylating) (EC 4.1.1.32), commonly referred to as phosphoenolpyruvate carboxykinase and which we note as PEPCK; and (3) pyrophosphate: oxalacetate carboxylase (transphosphorylating) (EC 4.1.1.38), generally known as phosphoenolypyruvate carboxytransphosphorylase and referred to here as PEPCTrP.

These enzymes catalyze the following reactions in which water, a nucleoside diphosphate, and orthophosphate are the respective phosphate

$$PEP + CO_2 + H_2O \xrightarrow{\text{PEPC}} OAA + P_i \tag{1}$$

$$PEP + CO_2 + GDP \overset{\text{PEPCK}}{\rightleftarrows} OAA + GTP \tag{2}$$

$$PEP + CO_2 + P_i \overset{\text{PEPCTrP}}{\rightleftarrows} OAA + PP_i \tag{3}$$

acceptors. The reactions are not intended to imply that CO_2 is the actual active species for all three enzymes or to suggest that GDP is the only viable phosphate acceptor in reaction (2). Specific details relative to each of these reactions will be considered in the sections which follow, but, in summary, the reaction catalyzed by PEPC probably uses HCO_3^- as the active substrate while CO_2 is probably the species involved in reactions catalyzed by PEPCK and PEPCTrP. In the case of PEPCK, the preferred phosphate acceptor for the enzyme obtained from most animal species is GDP or IDP, while the enzymes from most bacterial species and yeast favor ADP.

Since these three enzymes all catalyze the formation of OAA from PEP, they might reasonably be expected to have similar physiological functions. However, the primary function of PEPCK *in vivo* is probably catalysis of the formation of PEP from OAA (a gluconeogenic function), while PEPC and PEPCTrP appear to catalyze the carboxylation of PEP to OAA (an anaplerotic function).

Both gluconeogenic and anaplerotic reactions are so vital to living organisms that it is not surprising that these enzymes are widely distributed in nature. PEPC is found in most if not all plants, in many genera of bacteria, and is notably absent from animal tissues and probably from yeast and fungi. PEPCK appears to be widely distributed in both the plant and animal kingdoms. Thus far, PEPCTrP is an exception which has been found only in the propionic acid bacteria and the protozoan *Entamoeba histolytica*.

In bacteria and also in many plants, PEPC probably serves an anaplerotic function, but in the tropical grasses such as sugar cane, maize, and sorghum this enzyme may play a role in a photosynthetic carbon dioxide fixation process. The latter possibility has some very interesting implications which are discussed in the following section of this chapter.

In the propionic acid bacteria, the anaplerotic role of PEPC is assumed by PEPCTrP but synthesis of dicarboxylic acids in species from which PEPC is absent, is usually catalyzed by pyruvate carboxylase. This enzyme is described in detail in another chapter of this volume.

In contrast to PEPC, PEPCK is very widely distributed and its main physiological role is probably gluconeogenic. The gluconeogenic conversion of OAA to PEP, of course, presupposes an adequate supply of OAA. The OAA could be, and in animal tissues and several bacterial species probably is, provided via the pyruvate carboxylase-catalyzed conversion of pyruvate to OAA. In a few other cases, PEPCK is found in species where dicarboxylic acids can be generated via the glyoxylate shunt.

The general plan of presentation will be to describe in more detail each of the three enzymes in this group. Each enzyme will be discussed in terms of (a) its physical and structural properties, (b) the mechanism of the reaction catalyzed, (c) factors which may regulate the reaction, and (d) its distribution and probable physiological function.

II. Phosphoenolpyruvate Carboxylase (PEPC)

A. Discovery, Distribution, and Physiological Role

Carboxylation of PEP to yield OAA and inorganic phosphate (P_i) was first observed in extracts of spinach leaves and reported by Bandurski and Greiner in 1953 (*1*). In further studies, Bandurski (*2*) purified the enzyme some tenfold and established the basic properties of the reaction and the enzyme. Shortly thereafter Vennesland and her associates (*3, 4*) showed that PEPC is widely distributed in plants. Jackson and Coleman (*5*) listed 12 common plants ranging from alfalfa to watermelon which contain significant amounts of this enzyme. There is little doubt that this list includes only a very small fraction of a potential

1. R. S. Bandurski and C. M. Greiner, *JBC* **204,** 781 (1953).
2. R. S. Bandurski, *JBC* **217,** 137 (1955).
3. T. T. Tchen and B. Vennesland, *JBC* **213,** 533 (1955).
4. M. Mazelis and B. Vennesland, *Plant Physiol.* **32,** 591 (1958).
5. W. A. Jackson and N. T. Coleman, *Plant Soil* **11,** 1 (1959).

list of plants containing the enzyme. These authors also showed that the enzyme is widely distributed among the tissues of any single plant; for example, in the snap bean the enzyme occurs in the leaves, stem, seed, and roots.

Almost from the time of its discovery, PEPC has been assigned a physiological role as the catalyst for a "dark" CO_2 fixation reaction in certain types of plants. It has been recognized for some time (*6*) that the leaves of Crassulacean and some other succulent plants accumulate large amounts of carboxylic acids, notably malic acid, as a result of CO_2 fixation reactions which occur in the dark. This process is accompanied by the disappearance of carbohydrate. During subsequent light periods, at least some of the accumulated acids are reconverted to carbohydrate. Saltman *et al.* (*7*) and Walker (*8*) showed that PEPC was present in the plants concerned and that the general properties of this reaction were consistent with dark CO_2 fixation. As noted above, the enzyme also occurs in parts of the plants other than the leaves and in a wide variety of plants which do not exhibit marked dark fixation. In these cases the PEPC presumably plays an anaplerotic role in replenishing dicarboxylic acids.

More recently, it has been suggested that in some types of plants PEPC may be closely associated with photosynthetic CO_2 fixation. The major route of photosynthetic CO_2 fixation in temperate plants and algae is generally accepted to be the cycle proposed by Bassham and Calvin (*9*) in which ribulosediphosphate carboxylase plays the key role in CO_2 fixation. However, short-term labelling experiments with $^{14}CO_2$ with sugar cane and related tropical grasses by Kortschak *et al.* (*10*) and by Hatch and Slack (*11*) are consistent with an initial CO_2 fixation into a dicarboxylic acid. When Slack and Hatch (*12*) examined the enzymic makeup of tropical grasses as compared with other types of plants such as wheat, oats, and silver beet, they found that the PEPC activity was an order of magnitude higher in the tropical grasses. Conversely, the level of ribulosediphosphate carboxylase activity was an order of magnitude lower. When the rates of photosynthetic CO_2 fixation were considered, the potential PEPC activity was severalfold higher than the

6. H. B. Vickery, *JBC* **205**, 369 (1953).
7. P. Saltman, G. Kunitaki, H. Spolter, and C. Stits, *Plant Physiol.* **31**, 464 (1956).
8. D. A. Walker, *BJ* **67**, 73 (1957).
9. J. A. Bassham and M. Calvin, "The Path of Carbon in Photosynthesis." Prentice-Hall, Englewood Cliffs, New Jersey, 1957.
10. H. P. Kortschak, C. E. Hartt, and G. O. Burr, *Plant Physiol.* **40**, 209 (1965).
11. M. D. Hatch and C. R. Slack, *BJ* **101**, 103 (1966).
12. C. R. Slack and M. D. Hatch, *BJ* **103**, 660 (1967).

observed reaction rate. As further support, Baldry *et al.* (*13*) have shown that isolated chloroplasts from sugar cane can carry out a PEP-stimulated CO_2 fixation reaction and that disrupted chloroplasts appear to contain PEPC. These authors suggested formation of OAA by a chloroplast membrane-bound PEPC, followed by a photoreduction of the OAA to malate. Slack and Hatch (*12*) also reported that PEPC was associated with the chloroplasts in maize. It must be noted, however, that the rate of CO_2 fixation by the isolated chloroplasts was rather low. The additional problem of how net synthesis of carbohydrates can be achieved by the formation of a dicarboxylic acid from PEP and CO_2 must also be considered. In animal tissues and bacteria carbohydrate synthesis is effected by decarboxylation of dicarboxylic acids to form PEP, but this pathway could not account for a *net* production of carbohydrate in the proposed scheme. Hatch and Slack (*14*) have proposed that the answer to this dilemma is a transcarboxylation reaction whereby the β-COOH of OAA would be transferred to ribulosediphosphate. Unfortunately, no such transcarboxylation reaction has yet been demonstrated although a somewhat related transcarboxylation reaction involving OAA and propionyl-CoA has been thoroughly documented in the propionic acid bacteria (*15*).

Black and his associates (*15a*) have recently proposed another role for PEPC in bermuda grass and possibly tropical grasses. On the basis of the distribution of enzymes between mesophyll and bundle sheath cells in bermuda grass leaves and short-term $^{14}CO_2$ fixation with these leaves, these authors offer the following hypothesis. The mesophyll cells contain PEPC and malate dehydrogenase, while the bundle sheath cells contain RuDP carboxylase and malate enzyme. Initial fixations of CO_2 in malate would occur in the mesophyll cells via PEPC and malate dehydrogenase. The malate would then be transported to the bundle sheath cells where it would be converted to CO_2, pyruvate, and NADPH by malate enzyme and the CO_2 would then be fixed in phosphoglycerate by RuDP carboxylase. Thus the primary role of PEPC would be concerned with the transport of CO_2 and perhaps of NADPH, and the pathway of CO_2 would be as follows: CO_2, OAA, malate, CO_2, phosphoglycerate. This interesting suggestion provides a role for PEPC without posing the awkward problem of explaining net synthesis of carbohydrate from PEP as the starting substrate.

13. C. W. Baldry, C. Bucke, and J. Coombs, *BBRC* **37,** 828 (1969).
14. M. D. Hatch and C. R. Slack, *Ann. Rev. Plant Physiol.* **21,** 141 (1970).
15. H. G. Wood, Chapter 3, this volume.
15a. B. C. Mayne, G. E. Edwards, and C. C. Black, personal communication (1971); R. H. Brown, and C. C. Black, *Plant Physiol.* **47,** 199 (1971).

As indicated by the above paragraphs, the presence of PEPC in many plants has been known since the middle 1950's, but reports of its presence in bacteria did not appear until 1958. Suzuki and Werkman (*16*) noted its presence in the autotroph, *Thiobacillus thioparus*, and the following year Amarsingham (*17*) reported that the enzyme was also present in an heterotroph, *Escherichia coli*. Since that time there have been reports of its presence in many bacterial genera, including *Psuedomonas* (*18*), *Salmonella* (*19*), *Nitrosomonas* (*20*), *Streptococcus* (*21*), *Acetobacter* (*22, 23*), *Ferrobacillus* (*24*), and *Brevibacterium* (*25*). PEPC has also been observed in algae (*26*) and protozoa (*27, 28*). Perhaps the most noteworthy fact about the distribution of PEPC, is its apparent absence from animal tissues, yeasts, and fungi. Many of these tissues and cells have been shown to contain pyruvate carboxylase, an enzyme which probably fulfills the same metabolic function as PEPC. There are no convincing reports that any single cell contains both PEPC and pyruvate carboxylase.

B. Basic Nature of the PEPC-Catalyzed Reaction

Bandurski (*2*) showed that the reaction stoichiometry depicted in reaction (1) is correct and that the enzyme requires a divalent metal ion for activity. This metal requirement obtains for all species of the enzyme observed. In almost all cases the highest rate of activity is observed when Mg^{2+} is used. Divalent manganese is usually less effective but still active and Co^{2+} has occasionally been found to be active. Bandurski (*2*) also noted that under many conditions thiol reagents were required for, or stimulated, PEPC activity. Again, this has been a very consistent finding.

The pH optimum for most of the different species of PEPC lies on the alkaline side of neutrality and can be as high as 8.5 or 9 (*2, 29*)

16. I. Suzuki and C. H. Werkman, *ABB* **76**, 103 (1958).
17. C. R. Amarsingham, *Federation Proc.* **18**, 181 (1959).
18. P. J. Large, D. Peel, and J. R. Quayle, *BJ* **85**, 243 (1962).
19. T. S. Theodore and E. Englesberg, *J. Bacteriol.* **88**, 946 (1964).
20. P. S. Rao and D. J. D. Nicholas, *BBA* **124**, 221 (1966).
21. V. F. Lachica and P. A. Hartman, *Can. J. Microbiol.* **15**, 61 (1969).
22. G. W. Claus, M. L. Orcutt, and R. T. Belly, *J. Bacteriol.* **97**, 691 (1969).
23. M. Benziman, *J. Bacteriol.* **98**, 1005 (1969).
24. G. A. Diu, I. Suzuki, and H. Lees, *Can. J. Microbiol.* **13**, 1413 (1967).
25. H. Ozaki and J. Shiio, *J. Biochem.* (*Tokyo*) **66**, 297 (1969).
26. J. R. Kates and R. F. Jones, *Physiol. Plantarum* **18**, 1022 (1965).
27. P. M. L. Siu, *Comp. Biochem. Physiol.* **23**, 785 (1967).
28. E. Ohmann and F. Plhák, *European J. Biochem.* **10**, 43 (1969).
29. T. E. Smith, *ABB* **125**, 178 (1968).

in the case of the spinach and potato enzymes, but the optimum pH for the enzyme isolated from *Brevibacterium flavum* (*25*) is reported to be 6.5.

The enzyme is usually assayed in crude extracts by the PEP-dependent fixation of $^{14}CO_2$. The OAA formed is either trapped during the reaction by coupling with NADH and malate dehydrogenase or immediately following the reaction by formation of its 2,4-dinitrophenylhydrazone. Preparations which are relatively free from interfering reactions are ordinarily assayed by the former method.

The identification of the enzyme responsible for the PEP-dependent fixation of CO_2 by crude extracts is not always a simple matter because of the basic similarity of the PEPC-, PEPCK-, and PEPCTrP-catalyzed reactions. The situation is further complicated by the possibility that OAA formation from PEP and CO_2 may result from a combination of pyruvate kinase and pyruvate carboxylase activity. One series of operations which should permit the identification of one of the four above enzymes follows:

(A) Demonstration of a PEP-dependent CO_2-fixing reaction. If positive results are not observed, none of the above enzymes is involved.

(B) Determination of whether or not the CO_2 fixation is dependent on the presence of ADP or IDP. (The extract will necessarily have to be freed from small molecules by dialysis or gel filtration.)

- (1) If the CO_2-fixing reaction is not nucleotide dependent, it is probably catalyzed by PEPC or PEPCTrP.
  - (a) PEPC will not carry out a $^{14}CO_2$–OAA exchange even in the presence of PP_i; therefore, if this reaction is not observed, PEPC is probably responsible.
  - (b) If $^{14}CO_2$–OAA exchange occurs it is necessary to show that it results from PEPCTrP and not OAA decarboxylase (*30*). A PEPCTrP-catalyzed exchange reaction will be dependent on the presence of PP_i (*31*); this will not be so for OAA decarboxylase.
- (2) If the CO_2 fixation is dependent on ADP or IDP, either PEPCK or pyruvate carboxylase might be responsible.
  - (a) Pyruvate carboxylase always contains biotin and the reaction should therefore be inhibited by preincubation with avidin.
  - (b) Pyruvate carboxylase should be able to fix CO_2 with ATP and pyruvate at least as readily as with PEP and ADP.

30. L. O. Krampitz and C. H. Werkman, *BJ* **35,** 595 (1941).
31. H. G. Wood, J. J. Davis, and J. M. Willard, *Biochemistry* **8,** 3145 (1969).

The problem of identification could become somewhat more complicated in the instances where both PEPCK and PEPC may be present or where both pyruvate carboxylase and PEPCK are found together in the same cells. Positive identification will usually depend upon isolation of the relatively pure enzyme.

C. Structural Studies of PEPC

Of the many varieties of PEPC detected in various sources, only five types have been subjected to extensive purification and examination of physical and structural properties. These five include peanut (*32*), potato (*29*), spinach (*33*), *E. coli* (*34*), and *Salmonella typhimurium* (*35*) and represent species of the enzyme from three plants and two heterotrophic bacteria. The two bacterial species are very sensitive to metabolite regulation and may not be representative of a considerable group of other bacterial enzymes with quite different regulatory properties. This aspect will be discussed in the following section.

The structural data available from the studies of these five types of the enzyme are summarized in Table I. Maruyama *et al.* (*32*) purified PEPC from germinating peanut cotyledon. Sedimentation velocity studies with their preparation, approximately 2700-fold purified and having a specific activity of about 50, showed a major peak with $s_{20,w}$ of 13.9 and a small amount of a faster moving component. The molecular weight of the major species was estimated at 350,000 using gel filtration on Sephadex G-200. Proton relaxation rate studies on the binding of Mn^{2+} to the enzyme (*36*) suggest six metal ion binding sites; this finding is consistent with six subunits.

Smith purified PEPC from potato (*29*). The preparation had a low specific activity (1–2 units/mg) compared with other types of the enzyme presented in Table I and consequently the structural data are subject to greater error. The molecular weight of the potato PEPC was estimated to be 265,000 on the basis of its sedimentation behavior in sucrose density gradients; an $s_{20,w}$ value of about 10 was observed.

PEPC from spinach has very recently been purified to at least 90%

32. H. Maruyama, R. L. Easterday, H. C. Chang, and M. D. Lane, *JBC* **241,** 2405 (1966).
33. T. Nowak, H. Miziorko, M. Bayer, and A. S. Mildvan, private communication (1970).
34. T. E. Smith, *ABB* **128,** 611 (1968).
35. P. Maeba and B. D. Sanwal, *JBC* **244,** 2549 (1969).
36. R. S. Miller, A. S. Mildvan, H. C. Chang, R. L. Easterday, H. Maruyama, and M. D. Lane, *JBC* **243,** 6030 (1968).

TABLE I
STRUCTURAL PROPERTIES OF PEP CARBOXYLASE FROM VARIOUS SOURCES

Source	Most prevalent species			Other species	Reference
	$s_{20,w}$	Molecular weight	Estimated No. of subunits		
Peanut	14	350,000	6[a]		(*32*)
Potato	10	265,000	—		(*29*)
Spinach	22	750,000	12[b]		(*33*)
E. coli	12–13	380,000	4[c]	Dimer, monomer	(*34*)
S. typhimurium	12	200,000	4[d]	Aggregates	(*35*)

[a] Based on Mn^{2+} binding sites as estimated by water proton relaxation rates (*36*).
[b] Based on Mn^{2+} binding sites and electron micrographic examination.
[c] Based on apparent sedimentation constant of monomer in urea and larger species detected in presence of substrates and inhibitors.
[d] Based on monomer as estimated by gel electrophoresis of protein treated with sodium dodecylsulfate.

purity by Nowak *et al.* (*33*). This enzyme has an approximate molecular weight of 750,000. Proton relaxation rate studies indicate 12 tight binding sites for Mn^{2+}. Figures which may be hexagonal bilayers are observed in electron micrographs of negatively stained preparations. Both observations suggest 12 subunits. The general morphology of the enzyme is very reminiscent of that of glutamine synthetase from *E. coli* (*37*).

Smith has also purified PEPC from *E. coli* (*34*). The most active fraction had a specific activity of approximately six, but no estimate of its purity was made. Canovas and Kornberg (*38*) have obtained preparations of slightly higher specific activity from the same source. Using sucrose density gradients, Smith (*34*) estimated a variety of sedimentation coefficients and probable molecular weights which depended on substances added during the centrifugation. In the absence of any added substances, the *E. coli* enzyme had an $s_{20,w}$ of 8.4. The $s_{20,w}$ value was changed to 12–13 by additions of substrate and effectors such as acetyl-CoA and aspartate, and, in the presence of urea, the apparent $s_{20,w}$ observed was about 5.8. Smith suggested that these data are consistent with a protomeric unit of molecular weight 94,000 which can form dimers or tetramers in the presence of reactants. The approximate molecular weight of the tetramer would be 380,000.

Maeba and Sanwal (*35*) have purified PEPC from *S. typhimurium* to

37. R. C. Valentine, B. H. Shapiro, and E. R. Stadtman, *Biochemistry* **7**, 2143 (1968).
38. J. L. Canovas and H. L. Kornberg, *Proc. Roy. Soc.* **B165**, 189 (1966).

homogeneity as determined by polyacrylamide gel electrophoresis. The enzyme shows a marked tendency toward aggregation; hence, $s_{20,w}$ values vary widely depending on the protein concentrations used in the determinations. At dilute concentrations the $s_{20,w}$ value is about 12 and the molecular weight, determined by sucrose-gradient methods, is 198,000. Treatment of the enzyme with sodium dodecyl sulfate produces a species with a molecular weight of just under 50,000 as determined by disc electrophoresis, and the minimum molecular weight as calculated from amino acid composition is 49,980. These observations suggest that PEPC from *S. typhimurium* is a tetramer.

The data are too limited and too imprecise to permit confident comments about the structure of the various PEP carboxylases, but as indicated in Table I, there is a possibility that the plant varieties of the enzyme may be hexagonal or bihexagonal and that bacterial types of the enzyme, at least the highly regulated examples, may be tetramers. It should be strongly emphasized, however, that this picture could change radically either with additional data on new species or more precise data on the species listed in Table I.

D. Kinetic and Regulatory Properties of PEPC from Different Sources

The regulation of PEPC from various sources has received considerable attention since the initial report by Canovas and Kornberg (*39*) that PEPC from *E. coli* was strongly activated by acetyl-CoA. Most of the currently available information about the regulatory properties of PEPC from various sources is summarized in Table II (*22–25, 28, 35, 38–44*).

Apparently PEP carboxylases may be classified into at least three groups on the basis of regulatory properties. In the first group, as exemplified by the enzymes from *E. coli* and *S. thyphimurium*, PEPC responds to activators such as acetyl-CoA and fructose 1,6-diphosphate and to inhibitors such as aspartate in a highly complex manner; cooperative interactions with the substrate, PEP, appear to be involved. The evidence strongly suggests control sites which are not identical

39. J. L. Canovas and H. L. Kornberg, *BBA* **96,** 169 (1965).
40. K. Izui, *J. Biochem.* (*Tokyo*) **68,** 227 (1970).
41. B. D. Sanwal and P. Maeba, *BBRC* **22,** 194 (1966).
42. B. D. Sanwal and P. Maeba, *JBC* **241,** 4557 (1966).
43. K. Izui, T. Nishikido, K. Ishihara, and H. Katsuki, *J. Biochem.* (*Tokyo*) **68,** 215 (1970).
44. I. P. Ting, *Plant Physiol* **43,** 1919 (1968).

TABLE II
METABOLITE EFFECTORS OF PEP CARBOXYLASES

Class	Source	Activators	Inhibitors	Reference
1	*E. coli*	Acetyl-CoA, FDP	L-Aspartate L-Malate	(*38, 39, 40*)
	S. typhimurium	Acetyl-CoA, FDP, CDP	L-Aspartate	(*35, 41, 42*)
	Brevibacterium flavum	Acetyl-CoA, FDP	L-Aspartate	(*25*)
	Ferrobacillus ferroxin	Acetyl-CoA	L-Aspartate	(*24*)
	Thiobacillus thiooxidans	Acetyl-CoA	Not inhibited by aspartate	(*43*)
2	Corn roots	—	L-Malate	(*44*)
	Euglena gracilis	—	Citrate, OAA	(*28*)
	Acetobacter xylinum	—	Succinate, ADP	(*23*)
	Acetobacter suboxydans	—	Aspartate	(*22*)
3	Spinach	—	—	(*43*)
	Wheat, potato, etc.[a]	—	—	

[a] There have been no reports of any activators or inhibitors for the wheat, potato, and several other species of this enzyme.

to the substrate-binding sites. In addition to the well-documented cases of the enzymes from *E. coli* and *S. typhimurium* (*38, 42*), there have been reports that the enzymes from *Brevibacterium flavum* (*25*) and *Ferrobacillus ferroxin* (*24*) should be classified in this first group.

One apparent anomaly exists. Izui *et al.* (*43*) reported that PEPC from *Thiobacillus thiooxidans* is activated by acetyl-CoA but not by fructose 1,6-diphosphate and, surprisingly, that the enzyme is not inhibited by aspartate. These results suggest that this species of the enzyme may be different in control properties from any other yet studied.

Another group of PEP carboxylases, derived from plant roots, protozoa, and bacteria, have been reported to be inhibited by various di- or tricarboxylic acids but not to respond to aceytl-CoA or other activators. They are summarized in the second section of Table II. It is interesting to note that these four types of PEPC differ in their response to the inhibitors. Succinate is the most effective with the enzyme from *Acetobacter xylinum* (*23*), aspartate with the enzyme from *Acetobacter suboxydans* (*22*), and citrate with the enzyme from *Euglena gracilis* (*28*). Perhaps these differences can eventually be explained in terms of specific metabolic regulations, but at present this type of inhibition is not well understood. There is little evidence at the moment on the important question of whether such inhibitors act at a "second," or noncatalytic, site or exert product inhibition because of their relationship to oxalacetate. Relatively high concentrations of the acids have been used in some cases, and it is possible that they exert their effects by interactions with Mg^{2+} and that the observations are of limited physiological significance.

A third group of enzymes do not appear to be influenced by any metabolite effectors. The intensity of the search has varied widely but it seems fairly safe to say that members of this group do not respond to positive effectors and, at least not dramatically, to the negative effectors that have been tested. The division between groups two and three, however, is very difficult to establish.

Canovas and Kornberg (*38*) showed that the acetyl-CoA–induced stimulation of the *E. coli* enzymes was due to not only a large increase in the V_{max} of the reaction but also to a lowering of the apparent K_m for PEP from 5.5 to 0.6 mM. The approximate concentration of acetyl-CoA required for half-maximal activation (K_a) was 0.14 mM. Canovas and Kornberg (*38*) found little evidence for cooperative kinetics with either PEP or acetyl-CoA, but Izui *et al.* (*45*) noted that the PEP kinetics became somewhat cooperative when aspartate, an inhibitor of the reaction, was added. The aspartate effect was confirmed by Smith (*46*) who showed that the Hill coefficient for PEP changed from approximately 1 in the absence of aspartate to 1.3–1.5 in its presence. Izui (*40*) has attempted to elucidate the interrelationships of PEP, acetyl-CoA, and aspartate. His results are summarized in Table III. In general, acetyl-CoA and aspartate have opposing effects; e.g., acetyl-CoA increases and aspartate decreases the apparent affinity of the enzyme for PEP. In the absence of effectors, little or no cooperativity of PEP is noted, but in the presence of certain concentrations of acetyl-CoA some evidence supporting a cooperative interaction is observed; aspartate has a larger effect. In the presence of both acetyl-CoA and aspartate (not shown), the cooperative effects on PEP are especially marked. Smith (*46*) has pointed out that in a formal sense the relationship of aspartate and acetyl-CoA is essentially competitive.

It is not possible to quote single kinetic constants for PEP, acetyl-CoA, and aspartate because of the high degree of interaction among them, but ranges can be given. These are: K_m for PEP, from 0.6 mM in the presence of high concentrations of acetyl-CoA to 20 mM or higher in its absence; K_a for acetyl-CoA, from 0.05 mM in the presence of high concentrations of PEP to 0.35 mM at low PEP concentrations; K_i for aspartate, from 0.17 mM in the absence of acetyl-CoA to 1.3 mM in the

45. K. Izui, A. Iwatani, T. Nishikido, H. Katsuki, and S. Tanaka, *BBA* **139,** 188 (1967).

46. T. E. Smith, *ABB* **137,** 512 (1970).

TABLE III
SUMMARY OF KINETIC PROPERTIES OF PEP CARBOXYLASE FROM *E. coli*[a]

Kinetics of	Parameter	Effect of increasing		
		Acetyl-CoA	Aspartate	PEP
PEP	Affinity	↑	↓	—
	V_{max}	↑	↔	—
	Cooperativity	↑↓[b]	↑	—
Acetyl-CoA	Affinity	—	↓	↑
	Cooperativity	—	↑	↓
Aspartate	Affinity	↓	—	↓
	Cooperativity	↑	—	↑

[a] Adapted from Izui (*40*).
[b] Increased at low concentrations of acetyl-CoA, then decreased at higher concentrations.

presence of high concentrations of acetyl-CoA [data taken from Izui (*40*) and Sanwal and Maeba (*41*)]. The effects of acetyl-CoA and aspartate on the apparent K_m for bicarbonate are relatively small; the values range from 2 to 5 m*M* (*40*).

Izui (*40*) has pointed out that the regulatory effects observed with the *E. coli* enzyme are consistent with an interpretation involving three conformational forms of the enzyme, T_i, T, and R. Using this notation T_i is the form which binds aspartate and is catalytically inactive, T is active and binds no effector, and R is the form which is active in the presence of acetyl-CoA.

Sanwal and Maeba (*41*) reported that PEPC from *S. typhimurium* is activated by fructose 1,6-diphosphate (FDP) as well as by acetyl-CoA. Izui *et al.* (*43*) found that this is also true for the *E. coli* enzyme. FDP appears to activate the enzyme by decreasing its apparent K_m for PEP; this effect, parallel to that of acetyl-CoA, is observed at FDP concentrations of 1–5 m*M*. The two effectors differ, however, in their apparent interactions with aspartate. FDP cannot relieve aspartate inhibition while acetyl-CoA can.

Izui *et al.* (*43*) also examined the specificity of L-aspartate as an inhibitor of the *E. coli* enzyme. Using the ability of various compounds to stabilize the enzyme against heat, they found that, at a concentration of 5 m*M*, L-malate and fumarate were almost as effective as aspartate. Succinate and citrate were also quite effective, but OAA was much less so. In a direct test of catalytic activity, OAA was also much less inhibitory than aspartate.

Corwin and Fanning (*47*) have reported a somewhat different picture of the PEP kinetics of the *E. coli* enzyme. They observed a very complex relationship between PEP concentration and initial velocity. In a plot of initial velocity vs. substrate concentration, they observed a sigmoidal segment (0–6 m*M* PEP), a plateau (6–9 m*M* PEP), and a hyperbolic segment (above 9 m*M* PEP). They noted that their results resemble those observed with systems, described by Teipel and Koshland (*48*), which exhibit negative cooperativity. Such kinetic behavior is consistent with a multisite enzyme, but other explanations are possible. In their investigation Corwin and Fanning (*47*) used the cyclohexylammonium salt of PEP, which Izui *et al.* (*43*) have subsequently shown to give an apparently sigmoidal relationship because this cation is itself an activator. Substitution of the potassium salt of PEP removed the apparent sigmoidicity. Smith (*46*) has also pointed out that the enzyme assay used by Corwin and Fanning, which is based on the fixation of $^{14}CO_2$, may also be subject to error under some conditions.

Sanwal and Maeba carried out extensive studies on the regulatory properties of PEPC from *S. typhimurium*. They found that the enzyme is stimulated by acetyl-CoA and FDP (*41*, *49*), and that various nucleoside phosphates can act as activators (*42*). The most active were CDP, GTP, and CMP. Acetyl-CoA increased the apparent affinity of the enzyme for these activators, an observation which is consistent with a concerted activation mechanism for the two types of effectors. Acetyl-CoA exerts its effect in a manner generally similar to that discussed for the *E. coli* enzyme, that is, by increasing both the V_{max} and the affinity of the enzyme for PEP. In the absence of effectors, the PEP kinetics of the *Salmonella* enzyme appear to be cooperative (*35*) although the Hill coefficient varies from 1.3 to 2. The addition of either acetyl-CoA or CDP changes the kinetic pattern toward the more common hyperbolic one. FDP activates the enzyme in a different manner; the V_{max} is changed without decreasing the cooperative kinetics of PEP. As was true with the *E. coli* enzyme, the addition of aspartate appears to increase the apparent K_m for PEP and probably the Hill coefficient for that substrate. Inhibition at reasonably low concentrations appears to occur only with L-aspartate and L-malate (*49*).

The regulatory patterns for PEP carboxylases from *E. coli* and *S. typhimurium* are generally similar even though some differences in detail can be cited. Probably the most important question which can be asked about both cases is whether the various metabolite effectors are acting at

47. L. M. Corwin and G. R. Fanning, *JBC* **243**, 3517 (1968).
48. J. Teipel and D. E. Koshland, Jr., *Biochemistry* **8**, 4656 (1969).
49. P. Maeba and B. D. Sanwal, *BBRC* **21**, 503 (1965).

"second," i.e., noncatalytic sites. The sheer number and diversity of the known effectors plus their complex interactions would lead one to believe that these phenomena can be explained only by invoking special regulatory sites, but direct evidence supporting this view is not plentiful. Maeba and Sanwal (*35*) have shown that the inhibitory patterns observed with oxalacetate are very different from those observed with aspartate. It is therefore unlikely that aspartate is acting solely as a product inhibitor. Izui *et al.* (*43*) have presented similar observations and arguments for the *E. coli* enzyme. Different, and perhaps more convincing, evidence is provided by Sanwal *et al.* (*50*) who made the interesting finding that polyanions (polylysine, protamine, etc.) and nonpolar solvents (dioxane, etc.) activate the *Salmonella* enzyme. The polyanions appear to bind to the enzyme forming activated complexes which still respond to regulation by aspartate and FDP. However, when the enzyme is activated by dioxane, it no longer responds to aspartate and FDP and may be said to be desensitized. The latter observation is certainly consistent with the concept that the effectors act at sites other than those occupied by the substrates and products. Smith (*46*) has studied the effects of NaCl and urea on the degree of activation achieved by acetyl-CoA. Under appropriate conditions, these substances effect a marked increase in the ratio of activity in the presence of acetyl-CoA to the activity in its absence. This behavior supports the suggestion that the kinetic and regulatory properties of the enzyme are being separated by these substances.

Regulation of PEPC by acetyl-CoA and aspartate or malate fits well with the apparent anaplerotic role of this enzyme. Acetyl-CoA acts as a feed-forward control promoting its own oxidation by increasing the amount of oxalacetate and consequently promoting the reactions of the tricarboxylic acid cycle. Aspartate and malate are obvious candidates for feedback inhibition. It is intriguing that many of the same control features are observed in other species where the anaplerotic function of PEPC is carried out by a different enzyme, pyruvate carboxylase. The pyruvate carboxylase isolated from animal tissues requires an acyl-CoA for activity; acetyl-CoA is the most effective in terms of concentration required, but propionyl-CoA and a few other acyl-CoA compounds are also able to activate the enzyme (*51*). Canovas and Kornberg (*38*) have shown that *E. coli* PEPC has a remarkably similar relative specificity for acyl-CoA activators. Pyruvate carboxylase (*52*) from animal tissues is not inhibited by malate or aspartate but the varie-

50. B. D. Sanwal, P. Maeba, and R. A. Cook, *JBC* **241,** 5177 (1966).
51. M. C. Scrutton and M. F. Utter, *JBC* **242,** 1723 (1967).
52. M. C. Scrutton and M. R. Young, Chapter 1, this volume.

ties of the enzyme obtained from yeast, *Arthrobacter globiformis*, and *Bacillus stearothermophilus* are all activated by acetyl-CoA and inhibited by aspartate. The control features of these latter varieties of pyruvate carboxylase are remarkably parallel to those presented here for the PEPC from *E. coli* or *S. typhimurium*. To carry the analogy further, species of pyruvate carboxylase have been reported which do not respond to activators but are inhibited by aspartate. These enzymes correspond to the Class 2 PEP carboxylases listed in Table II. One species of pyruvate carboxylase (from a pseudomonad) appears to be unaffected by either activators or inhibitors, and it corresponds to the Class 3 PEP carboxylases listed in Table II. An apparently parallel development of various control mechanisms is evident for these two very different enzymes which carry out the same physiological function.

The PEP carboxylases isolated from *Brevibacterium flavum* and *Ferribacillus ferroxidin* are reported to have regulatory properties similar to those found in the enzymes from *E. coli* and *S. typhimurium*. However, they have not been studied extensively.

The K_m values for PEP and HCO_3^- for different species of PEPC are presented in Table IV. The data are reported as ranges of values. As noted earlier, the K_m for PEP for the enzymes from *E. coli* and *S. typhimurium* is markedly decreased by the activators, acetyl-CoA and CDP. When species of the enzyme not subject to activation are examined, the observed values of the K_m for PEP tend to approach the activated values observed with the enzymes which are subject to regulation. Fewer values are available for HCO_3^- but with the exception of the

TABLE IV
K_m VALUES FOR PEP AND BICARBONATE FOR PEP CARBOXYLASE

Source	PEP K_m (mM)	Bicarbonate K_m (mM)	Reference
Peanut	0.63	0.31	(*32*)
Pseudomonas AM1	0.33		(*18*)
Euglena gracilis	1.6		(*28*)
Kalanchoe	0.15	0.22	(*8*)
Spinach	0.5		(*2*)
Potato	0.1	0.2–0.4	(*29*)
Thiobacillus thiooxidans	—	0.3	(*16*)
Acetobacter xylinum	1.0		(*23*)
E. coli	0.6–20[a]	2–5	(*38, 40*)
S. typhimurium	0.3–15[a]		(*35*)

[a] Low values obtained in presence of activators; high values in absence of activators.

E. coli enzyme, these appear to be in the $10^{-4} M$ region; Izui *et al.* (*43*) reported a value about tenfold higher for the enzyme from *E. coli*.

E. Mechanism of Action of PEPC

Perhaps the most striking difference between PEPC and the other two enzymes considered in this chapter is the essential irreversibility of the PEPC-catalyzed reaction. Maruyama *et al.* (*32*), for example, were unsuccessful in their attempts to form PEP from OAA by coupling PEPC with pyruvate kinase and lactate dehydrogenase. PEPC also differs from PEPCK and PEPCTrP in that it does not appear to catalyze any partial reactions; PEPC does not catalyze an exchange of CO_2 with oxalacetate (*2, 23, 53*), pyruvate with oxalacetate (*32*), or by inference, pyruvate with PEP (*32*). These negative findings appear to eliminate all three of the stepwise mechanisms shown in reactions (4)–(6) because they involve enzyme-bound pyruvate, CO_2, and phosphate, respectively.

$$\text{PEP} + \text{Enz} \rightleftarrows \text{Enz-Pyr} + \text{P}_i \tag{4A}$$

$$\text{Enz-Pyr} + \text{CO}_2 \rightleftarrows \text{Enz} + \text{oxalacetate} \tag{4B}$$

$$\text{PEP} + \text{CO}_2 + \text{Enz} \rightleftarrows \text{Enz-CO}_2 + \text{Pyr} + \text{P}_i \tag{5A}$$

$$\text{Enz-CO}_2 + \text{Pyr} \rightleftarrows \text{Enz} + \text{oxalacetate} \tag{5B}$$

$$\text{PEP} + \text{Enz} \rightleftarrows \text{Enz-P} + \text{Pyr} \tag{6A}$$

$$\text{Enz-P} + \text{CO}_2 \rightleftarrows \text{Enz-CO}_2 + \text{P}_i \tag{6B}$$

$$\text{Enz-CO}_2 + \text{Pyr} \rightleftarrows \text{Enz} + \text{oxalacetate} \tag{6C}$$

If the mechanism represented by either (4) or (6) were correct, a Ping-Pong kinetic relationship would be expected between PEP and CO_2. With at least one species of the enzyme, this does not seem to be the case. Izui (*40*) found that alterations in the concentration of PEP had no effect on the apparent K_m of the *E. coli* enzyme for bicarbonate.

Failure to demonstrate any partial reactions led Lane and his associates to propose that, as shown in Fig. 1, the PEPC-catalyzed reaction involves a concerted mechanism with a cyclic transition state (*32*). This mechanism accounts for the failure to observe exchange reactions and, in accord with the earlier finding of Tchen *et al.* (*54*), the production of the keto form of OAA. The proposed mechanism also requires that bicarbonate be the reacting form of carbon dioxide; the transition intermediate shown could not be obtained if CO_2 were the reacting species.

53. H. Maruyama and M. D. Lane, *BBA* **65**, 207 (1962).
54. T. T. Tchen, F. A. Loewus, and B. Vennesland, *JBC* **213**, 547 (1955).

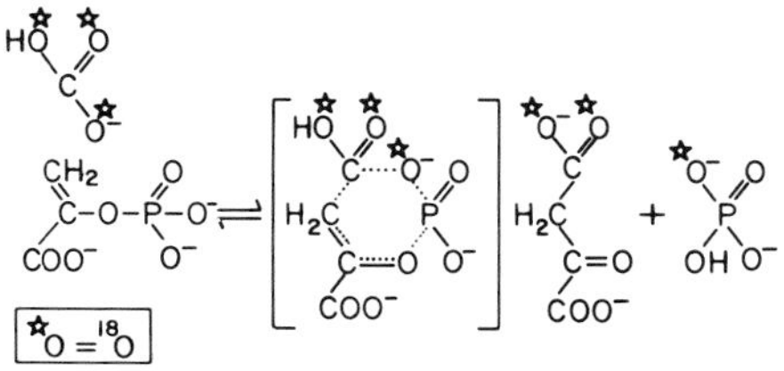

FIG. 1. Proposed concerted reaction mechanism for PEPC.

Furthermore, if ^{18}O-labeled bicarbonate is used in the reaction, ^{18}O should appear in both OAA and phosphate in a 2 to 1 ratio. Maruyama *et al.* (*32*) carried out these ^{18}O experiments using PEPC from peanut cotyledons. They found ^{18}O in both malate, which was treated as a trapped form of OAA, and in inorganic phosphate. As predicted, the ratio was approximately 2 to 1. Labeling experiments of this sort must contend with the rapid exchange of ^{18}O from the bicarbonate into the water. Maruyama *et al.* tried to minimize this process by using short incubation times, large amounts of enzyme, and an alkaline pH. Under these conditions, they observed as much as 70% of the theoretical incorporation of ^{18}O into phosphate. The ^{18}O level of the water never exceeded 5% of the concentration found in the products. Maruyama *et al.* concluded that H_2O was probably not an intermediate in the reaction; it would have to be if CO_2 were the reacting species. The proposed mechanism is similar to the one advanced by Kaziro *et al.* (*55*) for propionyl-CoA carboxylase which is also based on experiments using ^{18}O-bicarbonate.

Waygood *et al.* (*56*) using the PEPC from maize, have obtained results which disturb the satisfying picture presented by Maruyama *et al.* The former investigators examined the reacting species of carbon dioxide in the PEPC-catalyzed reaction using a technique originally described by Krebs and Roughton (*57*) in their study of the carbon dioxide species produced by decarboxylase reactions. The technique takes advantage of the fact that reaction (7) requires approximately one minute to

$$H_2O + CO_2 \rightleftarrows [H_2CO_3] \rightleftarrows H^+ + HCO_3^- \tag{7}$$

reach equilibrium at temperatures below 15° if the initial reactants are H_2O and CO_2 (*58*). Addition of carbonic anhydrase greatly accelerates the attainment of equilibrium. In experiments with CO_2-fixing enzymes

55. Y. Kaziro, L. F. Hass, P. D. Boyer, and S. Ochoa, *JBC* **237**, 1460 (1962).
56. E. R. Waygood, R. Mache, and C. K. Tan, *Can. J. Botany* **47**, 1455 (1969).
57. H. A. Krebs and F. J. W. Roughton, *BJ* **43**, 550 (1948).
58. C. Faurholt, *J. Chim. Phys.* **21**, 400 (1925).

such as PEPC the technique involves initiation of the reaction with either CO_2 or bicarbonate and measurement of the *initial* reaction rate in the presence and absence of carbonic anhydrase. If the enzyme under consideration has a high turnover number and if CO_2 is both the provided and the reacting species, the initial rate will be fast and become slower as the effective CO_2 concentration is decreased by equilibrium with the other species. The addition of carbonic anhydrase to this system will inhibit the rate of the reaction by catalyzing the equilibrium between the species of carbon dioxide. If bicarbonate is the active species and the reaction is initiated with CO_2, an initial lag will be observed. The lag can be eliminated by the addition of carbonic anhydrase, but the effects are less dramatic than those observed when CO_2 is the substrate.

Waygood *et al.* (*56*), in studies with PEPC from maize, obtained data suggesting that CO_2 is the reacting species. This result clearly contradicts that of Maruyama *et al.* (*32*). It is not possible to resolve the apparent discrepancy at present because the difference may result from the difference in either the species of the enzyme or the assay method employed. It is interesting to note that Cooper *et al.* (*59*), in considerably more extensive studies with the carbonic anhydrase method, reported that PEPCTrP (*Propionibacterium*) and PEPCK (chicken liver) both use CO_2 as the reacting species while pyruvate carboxylase (chicken liver) uses bicarbonate. Apparently the method is capable of distinguishing between the two species of substrate.

In view of the uncertainties already referred to, it would be very useful to examine the same species of PEPC using both assay methods. It may be, as suggested by Cooper *et al.* (*59*), that interpretations of the reacting species can be rationalized in terms of localized conditions at the active site. For example, the results observed by Maruyama *et al.* could also be explained on the basis of CO_2 being the reacting species if it was assumed that the CO_2 molecule was in rapid localized equilibrium with a water molecule which also participated in the reaction.

Miller *et al.* (*36*) used NMR techniques to measure the effect of PEPC and Mn^{2+} on the water proton relaxation rates. Their observations strongly support a PEPC–Mn^{2+} complex. The addition of PEP to the PEPC–Mn^{2+} system produces a transient lowering of the Mn^{2+} enhancement which suggests that the metal may form a bridge between enzyme and substrate. The situation is very similar to that observed with PEPCK. Recent studies with PEPC from spinach (*33*) yielded similar results, but the number of binding sites for Mn^{2+} appeared to be 12 instead of the six observed with the peanut enzyme.

59. T. G. Cooper, T. T. Tchen, H. G. Wood, and C. R. Benedict, *JBC* **243,** 3857 (1968).

In earlier experiments with the peanut enzyme, Maruyama *et al.* (*32*) showed that decreasing the pH from 8 to 7 resulted in a concomitant lowering of the apparent K_m values for PEP and Mg. This observation not only suggested the presence of a dissociable group with a pK between 7 and 8 (perhaps an imidazole group) but also that PEP and Mg were bound together to that group; a metal bridge between the enzyme and PEP is consistent with this finding.

Using specifically labeled 3-^{3}H-PEP, Rose *et al.* (*60*) recently investigated the stereochemistry of several reactions in which PEP is carboxylated. They found that PEPCK (chicken liver), PEPCTrP (*Propionibacterium*), and PEPC (peanut and *Acetobacter*) all catalyze stereochemically related reactions. In each case the appropriate addition occurs from the same side of the enzyme-bound PEP plane as will be discussed in more detail in the section dealing with PEPCK.

III. Phosphoenolpyruvate Carboxykinase (PEPCK)

A. Discovery, Distribution, and Physiological Role

Phosphoenolpyruvate carboxykinase (GTP:oxalacetate-carboxylase (transphosphorylating): EC 4.1.1.32) was initially found in chicken liver by Utter and Kurahashi (*61*) in 1953. The enzyme catalyzes the reversible reaction shown in reaction (8) in which CO_2 is not intended to represent any particular molecular species of this reactant and NDP and NTP represent undesignated nucleoside di- and triphosphates, respectively.

$$\text{PEP} + \text{CO}_2 + \text{NDP} \rightleftarrows \text{OAA} + \text{NTP} \quad (8)$$

Relatively high concentrations of PEPCK activity have been found in all forms of liver and kidney which have been examined. PEPCK is also found in other vertebrate tissues including mammary, certain types of muscle, and adipose tissue. The enzyme is also widely distributed outside the vertebrates. Many, perhaps most, varieties of plants, a number of genera of bacteria, yeast and fungi, flatworms, roundworms, insects, and mollusks all contain it but compose only a partial list of those organisms which do.

60. I. A. Rose, E. L. O'Connell, P. Noce, M. F. Utter, H. G. Wood, J. M. Willard, T. G. Cooper, and M. Benziman, *JBC* **244**, 6130 (1969).

61. M. F. Utter and K. Kurahashi, *J. Am. Chem. Soc.* **75**, 758 (1953); *JBC* **207**, 787 (1954).

The initial discovery of PEPCK was interpreted to be an answer to the problem of CO_2 fixation and dicarboxylic acid synthesis in liver, but subsequent findings strongly support the thesis that its primary importance involves catalyzing the synthesis of PEP as part of the carbohydrate synthesis pathway. Although all of the isotopic, enzymic, and physiological evidence supporting this thesis is circumstantial in nature, it is impressive in its bulk and consistency. More than 30 years ago, Hastings and co-workers (*62, 63*) used carbon isotopes to show that glycogen synthesis from pyruvate involved CO_2 fixation and a symmetrical dicarboxylic acid intermediate. The involvement of PEPCK as the agent for catalyzing PEP formation is consistent with their findings. On the other hand, the involvement of PEPCK in an anaplerotic role (OAA synthesis) is contraindicated by the enzyme's requirement for relatively high concentrations of bicarbonate (K_m about 10–20 mM); the estimated intracellular levels of bicarbonate [5–8 mM in man (*64*)] are much lower. Furthermore, the K_m values for bicarbonate for those enzymes believed to play a CO_2-fixing role in metabolism are much lower. As mentioned in the preceding section, the K_m for bicarbonate for most species of PEPC is 1 mM or less and the corresponding value for pyruvate carboxylase is about the same (*52*).

The strongest argument supporting the role of PEPCK in carbohydrate synthesis is based on a large amount of physiological evidence. For example, PEPCK is found in large amounts in the gluconeogenic tissues, such as liver and kidney, and it is absent or present in much lower amounts in nongluconeogenic tissues. Furthermore, the observed alterations in PEPCK activity following dietary or endocrine manipulation are consistent with its involvement in the gluconeogenic pathway. The activity of PEPCK is increased up to sixfold in rat liver following fasting, administration of glucocorticoids, or the induction of diabetes by pancreatectomy, alloxan, or mannoheptulose (*65–70*). All of these treat-

62. A. K. Solomon. B. Vennesland, F. W. Klemperer, J. M. Buchanan, and A. B. Hastings, *JBC* **140,** 171 (1941).

63. B. Vennesland, A. K. Solomon, J. M. Buchanan, R. D. Cramer, and A. B. Hastings, *JBC* **142,** 371 (1942).

64. J. L. Gamble, "Extracellular Fluid." Harvard Univ. Press, Cambridge, Massachusetts, 1947.

65. E. Shrago, H. A. Lardy, R. C. Nordlie, and D. O. Foster, *JBC* **238,** 3188 (1963).

66. J. W. Young, E. Shrago, and H. A. Lardy, *Biochemistry* **3,** 1687 (1964).

67. S. R. Wagle and J. Ashmore, *BBA* **74,** 564 (1963).

68. S. R. Wagle and J. Ashmore, *JBC* **239,** 1289 (1964).

69. L. Reshef, R. W. Hanson, and F. J. Ballard, *JBC* **245,** 5979 (1970).

70. A. J. Garber and R. W. Hanson, *JBC* **246,** 589 (1971).

ments promote gluconeogenesis, and the observed increase in the activity of PEPCK following such treatments suggests that this enzyme is involved in the overall pathway. The observations of Williamson and co-workers (*71*) obtained by using the crossover technique go somewhat further and suggest that under certain circumstances the rate of glucose synthesis is limited by the PEPCK-catalyzed reaction.

Studies involving PEPCK in bacterial systems have also provided support for the proposed role of this enzyme in carbohydrate synthesis. Mutants of *S. typhimurium* (*72*) and *E. coli* (*73*) which show little or no PEPCK activity were unable to use succinate as their sole carbon source. Other mutants of these same two species, which lacked PEPC activity but contained normal levels of PEPCK activity, were unable to synthesize dicarboxylic acids (*19, 74*). The combination of these related studies makes an excellent case, at least for these two bacterial species, in support of the concept that PEPC functions in dicarboxylic acid synthesis and cannot be replaced by PEPCK and that PEPCK is required for carbohydrate synthesis.

Studies conducted with other bacterial species also support the involvement of PEPCK in glucose synthesis. A mutant of *Neurospora crassa* which lacked PEPCK was incapable of growing on acetate but could grow on sucrose (*75*). PEPCK activity was induced in the wild strain of this organism when it was grown on acetate. Teraoka *et al.* (*76*) found a four- to sevenfold higher level of PEPCK activity in *E. coli* when the organism was grown on aspartate or malate rather than glucose. Ruiz-Amil *et al.* (*77*) observed that the level of PEPCK activity in yeasts is high in cells grown on malate, acetate, pyruvate, or aspartate and low in cells grown on glucose. In fact, glucose may repress PEPCK synthesis in yeasts (*78, 79*), *E. coli* (*80*), and *Tetrahymena pyriformis* (*81*).

71. J. R. Williamson, *Advan. Enzyme Regulation* **5**, 229 (1967).
72. G. Carrillo-Castaneda and M. V. Ortega, *J. Bacteriol.* **102**, 524 (1970).
73. A. W. Hsie and H. V. Rickenberg, *BBRC* **25**, 676 (1966).
74. J. M. Ashworth and H. L. Kornberg, *Proc. Roy. Soc.* **B165**, 179 (1966).
75. R. B. Flavell and J. R. S. Fincham, *J. Bacteriol.* **95**, 1063 (1968).
76. H. Teraoka, T. Nishikido, K. Izui, and H. Katsuki, *J. Biochem.* (*Tokyo*) **67**, 567 (1970).
77. M. Ruiz-Amil, G. deTorrontegui, E. Palacian, L. Catalina, and M. Losada, *JBC* **240**, 3485 (1965).
78. G. deTorrontegui, E. Palacian, and M. Losada, *BBRC* **22**, 227 (1966).
79. J. J. Cazzulo, L. M. Claisse, and A. O. M. Stoppani, *J. Bacteriol.* **96**, 623 (1968).
80. E. Shrago and A. L. Shug, *ABB* **130**, 393 (1969).
81. E. Shrago, W. Brech, and K. Templeton, *JBC* **242**, 4060 (1967).

B. The Reactions Catalyzed and Their Properties

1. *Formation of OAA or PEP*

As indicated in reaction (8), the PEPCK-catalyzed reaction is readily reversible. The apparent K_{eq} of the reaction as written is 2.69 at 30°. The determination was conducted using inosine nucleotides (*82*). The ratio of the rates of PEP and OAA formation vary somewhat according to the experimental conditions and the previous treatment of the enzyme but, in general, the rate of PEP formation is two to eight times as fast as that of OAA formation (*83, 84*).

2. *Pyruvate Formation from OAA*

Utter and Kurahashi (*82*) observed that PEPCK from chicken liver also catalyzes the formation of pyruvate from OAA when IDP (or GDP) is present instead of ITP. As indicated, reaction (9) has thus far been

$$\text{Oxalacetate} \xrightarrow{\text{IDP or GDP}} \text{pyruvate} + CO_2 \tag{9}$$

experimentally irreversible. This reaction, presumably characteristic of many or all species of PEPCK, has been studied only with the enzymes isolated from chicken and pig liver (*85*) and from yeast (*86*). The decarboxylation of OAA to pyruvate could result from a combination of the reactions shown in reaction (10) but it has been impossible to demonstrate this reaction using the chicken liver enzyme (*84*) and although highly purified PEPCK from yeast does catalyze reaction (10), this occurs only under conditions which inhibit pyruvate formation, i.e.,

$$\text{Oxalacetate} + \text{NTP} \underset{}{\overset{Mn^{2+}}{\rightleftarrows}} \text{PEP} + \text{NDP} + CO_2 \tag{10A}$$

$$\text{PEP} + \text{NDP} \overset{Mg^{2+}}{\rightleftarrows} \text{pyruvate} + \text{NTP} \tag{10B}$$

$$\text{Oxalacetate} \rightleftarrows \text{pyruvate} + CO_2 \tag{10C}$$

in the presence of metal ions (*86*). The formation of PEP from OAA as

82. M. F. Utter and K. Kurahashi, *JBC* **207,** 821 (1954).
83. H. C. Chang, H. Maruyama, R. S. Miller, and M. D. Lane, *JBC* **241,** 2421 (1966).
84. P. S. Noce, Ph.D. Thesis, Case Western Reserve University, Cleveland, Ohio, 1968.
85. H. C. Chang and M. D. Lane, *JBC* **241,** 2413 (1966).
86. J. J. B. Cannata and A. O. M. Stoppani, *JBC* **238,** 1919 (1963).

catalyzed by both chicken liver and yeast PEPCK requires Mn^{2+} but the pyruvate kinase-like activity of yeast PEPCK requires Mg^{2+}. Since catalysis of pyruvate formation from OAA by both the chicken liver and yeast PEPCK is inhibited by the presence of metal ions, it seems unlikely that it occurs via the reactions summarized in reaction (10). It is more likely that observation of reaction (9) in most or all species of PEPCK and of reaction (10) in certain species of PEPCK reflects the basic reaction mechanism of the enzyme. We will return to this point in Section V. The finding by Chang and Lane (*85*) that metal ions stimulate OAA decarboxylation by the pig liver enzyme is clearly at variance with the inhibitory effects of metal ions in the case of the enzymes from yeast (*86*) and chicken liver (*84*), but the basis of the discrepancy is not apparent.

It is not possible to say whether the pyruvate-forming reaction of PEPCK is of physiological importance. The mechanism of OAA decarboxylation is not well understood in many cells although OAA decarboxylases have been reported in bacteria (*30*) and in mitochondria (*87*).

3. *Nucleoside Phosphate Specificity*

In contrast with the other two enzymes discussed in this chapter, PEPCK has an absolute requirement for a nucleoside diphosphate to serve as a phosphate acceptor in the conversion of PEP to OAA. With a few exceptions, such as the enzyme from *Arthrobacter*, the specificity for nucleoside diphosphates falls neatly into the pattern: adenine nucleotides for the enzyme from bacterial species and plants and inosine or guanine nucleotides for the enzyme from higher forms. The enzyme from the protozoan, *Tetrahymena pyriformis*, may be an intermediate; adenine derivatives are more effective than those of inosine or guanine although the latter are able to function. A somewhat similar case may also exist with the enzyme isolated from the thermophilic organism, *Bacillus stearothermophilus*. This enzyme responds to ATP, GTP, TTP, and UTP in that order of effectiveness (*88*). The nucleotide specificity of a number of varieties of PEPCK is summarized in Table V (*4, 76, 89–104*). These

87. L. M. Corwin, *JBC* **234,** 1338 (1959).
88. J. J. B. Cannata, personal communication (1970).
89. K. Kurahashi, R. J. Pennington, and M. F. Utter, *JBC* **226,** 1059 (1957).
90. R. S. Bandurski and F. Lipmann, *JBC* **219,** 741 (1956).
91. R. C. Nordlie and H. A. Lardy, *JBC* **238,** 2259 (1963).
92. M. D. Lane, H. C. Chang, and R. S. Miller, "Methods in Enzymology," Vol. 13, p. 270, 1969.
93. J. W. Simpson and J. Awapara, *Comp. Biochem. Physiol.* **12,** 457 (1964).
94. H. Vanden Bossche, *Comp. Biochem. Physiol.* **31,** 789 (1969).

TABLE V
ACTIVE NUCLEOTIDES FOR VARIOUS SPECIES OF PEPCK

Source	Active nucleotides	Reference
Chicken	Inosine or guanine	(*89*)
Lamb	Inosine or guanine	(*90*)
Guinea pig	Inosine	(*91*)
Pig	Inosine or guanine	(*92*)
Mollusks	Inosine or guanine	(*93*)
Ascaris suum	Inosine or guanine	(*94*)
Ascaris lumbricoides	Inosine	(*95*)
Haemonchus contortus	Inosine	(*96*)
Fasciola hepatica	Inosine	(*97*)
Hymenolepis diminuta	Inosine or guanine	(*98*)
Echinococcus granulosis	Inosine	(*99*)
Tetrahymena pyriformis	Adenine, guanine, or inosine	(*100*)
Rhodopseudomonas spheroides	Adenine	(*101*)
Arthrobacter globiformis	Guanine or inosine	(*102*)
Pasteurella pestis	Adenine	(*103*)
Escherichia coli	Adenine	(*76*)
Yeast	Adenine	(*104*)
Higher plants	Adenine	(*4*)

requirements refer to the PEP-producing reaction. The nucleotide specificity for the pyruvate-producing reaction has not been widely investigated, but it appears to correspond to that found for the PEP-producing reaction. IDP and GDP are active with the enzymes isolated from chicken liver (*84*) and pig liver (*85*) while the yeast enzyme requires ADP (or ATP) (*86*).

4. *Metal Requirements*

In addition to the nucleotide requirement, PEPCK also appears to have an absolute requirement for a divalent metal ion in the reaction which produces PEP or OAA [reaction (8)]. The ion of choice is

95. H. J. Saz and O. L. Lescure, *Comp. Biochem. Physiol.* **22,** 15 (1967).
96. C. W. Ward, P. J. Schofield, and I. L. Johnstone, *Comp. Biochem. Physiol.* **26,** 537 (1968).
97. R. K. Prichard and P. J. Schofield, *Comp. Biochem. Physiol.* **24,** 773 (1968).
98. L. M. Prescott and J. W. Campbell, *Comp. Biochem. Physiol.* **14,** 491 (1965).
99. M. Agosin and Y. Repetto, *Comp. Biochem. Physiol.* **14,** 299 (1965).
100. E. Shrago and A. L. Shug, *BBA* **122,** 376 (1966).
101. K. Uchida and G. Kikuchi, *J. Biochem.* (*Tokyo*) **60,** 729 (1966).
102. E. S. Bridgeland and K. M. Jones, *BJ* **104,** 9P (1967).
103. C. L. Baugh, J. W. Lanham, and M. J. Surgalla, *J. Bacteriol.* **88,** 553 (1964).
104. J. J. B. Cannata and A. O. M. Stoppani, *JBC* **238,** 1208 (1963).

usually Mn^{2+} although other divalent ions (Mg^{2+}, Zn^{2+}, Co^{2+}, and Cd^{2+}) can substitute for it with lesser degrees of efficiency.

Although Mn^{2+} is the single most effective cation in activating PEPCK-catalyzed PEP formation, Noce and Utter (*105*) found that with GTP as the phosphorylating moiety, the reaction velocity was highest when both Mn^{2+} and Mg^{2+} were present. One possible explanation is that Mn^{2+} reacts directly with the enzyme and Mg^{2+} acts by complexing with GTP. Other investigators, in order to achieve maximal enzymic activity, have also used a mixture of Mn^{2+} and Mg^{2+} in assay systems for PEPCK from (a) pig liver mitochondria (*85*), (b) the cytosol of rat liver (*106*), and (c) the mitochondria and cytosol of guinea pig liver (*107*).

As noted above (see Section III,B,2), no metal ions are required in the decarboxylation of OAA to pyruvate as catalyzed by the PEPCK from yeast and chicken liver. In fact, metal ions inhibit the reaction. However, in both these systems, the reaction is inhibited by EDTA and other chelating agents (*84, 86*). These observations suggest that a relatively tightly bound metal ion participates in the pyruvate formation reaction, but attempts to show that PEPCK from chicken liver is a metalloprotein have not met with success.

5. *pH Optimum*

Optimal pH conditions for PEPCK-catalyzed reactions are listed in Table VI (*83, 91, 92, 105, 107–110*). It is apparent from the data that the decarboxylation reaction (PEP formation) has a higher optimum pH than the carboxylation and the CO_2–OAA exchange reactions which this enzyme also catalyzes. The apparent discrepancy in the pH optimum of the different reactions has not been explained but it may reflect the availability of the substrate, CO_2. As the pH is increased the actual concentration of this species, at any fixed total concentration of bicarbonate, will decrease. It is possible that this is at least partially responsible for the apparent decrease in carboxylation activity which is observed as the pH is increased. For better comparison, the V_{max} values with CO_2 as the variable substrate should be determined at different pH values.

105. P. S. Noce and M. F. Utter, unpublished observations (1968).
106. D. O. Foster, H. A. Lardy, P. D. Ray, and J. Johnston, *Biochemistry* **6,** 2120 (1967).
107. D. D. Holten and R. C. Nordlie, *Biochemistry* **4,** 723 (1965).
108. H. M. Kolenbrander and M. F. Utter, unpublished observations (1970).
109. M. F. Utter, K. Kurahashi, and I. A. Rose, *JBC* **207,** 803 (1954).
110. J. J. B. Cannata and A. O. M. Stoppani, *Nature* **200,** 573 (1963).

TABLE VI
pH Optimum for PEPCK-Catalyzed Reactions

	pH Optimum			
Enzyme source	CO_2–OAA exchange	Carboxylation	Decarboxylation	Reference
Guinea pig liver				
Soluble			8.2	(*107*)
Mitochondrial		6.6	7.8–8.5	(*91, 107*)
Pig liver	6.6	6.6	6.6–8.0	(*83, 92*)
Chicken liver		6.2	7.0–8.0	(*105, 108, 109*)
Yeast	6.1	5.4		(*110*)

6. *Effect of Thiol Reagents on the Enzyme*

It has been established in studies with PEPCK from several sources that this enzyme requires an –SH group(s) for maximum activity in catalyzing the OAA or PEP formation reactions. It has been demonstrated that a sixfold excess of *p*-hydroxymercuribenzoate produces a 98–100% inhibition of the chicken liver enzyme and that *N*-ethylmaleimide and sodium tetrathionate produce similar inhibitions. Several reagents such as 2-mercaptoethanol, reduced glutathione, and dithioerythritol are able to reverse the effect of these inhibitors with 2-mercaptoethanol being the most effective (*105*). PEPCK obtained from pig liver or from yeast is also almost completely inhibited by *p*-hydroxymercuribenzoate and in both cases the inhibition can be reversed by reduced glutathione (*85, 104*).

Studies with PEPCK isolated from sheep kidney mitochondria indicate the presence of two –SH residues in the enzyme (*111*). One of the residues, reactive toward *N*-ethylmaleimide, is protected by IDP or ITP. The other –SH group was protected by Mn^{2+}.

The pyruvate-forming reaction is affected differently by the presence of reagents such as 2-mercaptoethanol which severely inhibit it even at relatively low (1 m*M*) concentrations (*105*). As will be noted in the next section, PEPCTrP catalyzes a pyruvate-forming reaction (from PEP) which also is inhibited by 2-mercaptoethanol. As with PEPCK, however, the major reactions catalyzed by PEPCTrP are promoted by the presence of the thiol compound.

C. Physical Properties

PEPCK isolated from pig, rat, and chicken liver and from sheep kidney appears to have a molecular weight of approximately 70,000 (*85,*

108, 111, 112), while the enzyme isolated from *Bacillus stearothermophilus* appears to be about twice as large (*88*) and the yeast enzyme about four times as large (*113*). The molecular weight of the pig liver enzyme, determined by sedimentation equilibrium, is reported to be 73,300 ± 2,600 (*85*). This value is in the same range as those determined for the rat and chicken liver enzymes by less elegant means; the rat liver enzyme has a molecular weight of 74,500 as determined by gel filtration on Sephadex G-100 (*112*) and the value for the chicken liver enzyme was found to be 68,000 (*108*) using the gel electrophoresis method described by Weber and Osborn (*114*). Using sedimentation-diffusion analysis and electrophoresis in the presence of sodium dodecyl sulfate, Barns estimated the molecular weight of the sheep kidney enzyme to be 72,000 (*111*). In marked contrast, Cannata, using gel filtration, observed that the molecular weight of the *B. stearothermophilus* enzyme is about 130,000 (*88*), and, using sedimentation velocity, found that the molecular weight of yeast PEPCK is 252,000 (*113*). Perhaps the *B. stearothermophilus* enzyme is a dimer and the yeast enzyme a tetramer, but the evidence currently available is insufficient to substantiate this conclusion. Additional study is needed to clarify the significance of these differences.

The amino acid composition of the pig liver (*85*) and the yeast (*113*) enzymes are decidedly different. The percentages of proline, glycine, methionine, and arginine are all ($>2\%$) lower and the percentages of lysine, tyrosine, and isoleucine are significantly ($>2\%$) larger in the yeast enzyme. The total percentage of amino acid residues accounted for by glutamic acid and aspartic acid and by lysine, histidine, and arginine is very nearly the same in the enzymes from either source (20% acidic amino acids and 14% basic amino acids); similar results were obtained with the sheep liver enzyme (*111*). Amino acid composition data for the rat and chicken liver enzymes are not available. However, the pI reported for the rat liver enzyme is 5.04 (*112*). This value, obtained using the electrofocusing technique, is consistent with an excess of acidic amino acid residues.

D. Reaction Mechanism

Early studies of the PEPCK mechanism demonstrated that the keto form of OAA is the product of the reaction (*115*). In this experiment,

111. R. J. Barns, Ph.D. Thesis, University of Adelaide, Australia, 1971.
112. F. J. Ballard and R. W. Hanson, *JBC* **244,** 5625 (1969).
113. J. J. B. Cannata, *JBC* **245,** 792 (1970).
114. K. Weber and M. Osborn, *JBC* **244,** 4406 (1969).
115. J. L. Graves, B. Vennesland, M. F. Utter, and R. J. Pennington, *JBC* **223,** 551 (1956).

carried out in D_2O, the OAA was reduced rapidly to malate by added malate dehydrogenase. If the enol form of OAA were the initial product, the malate produced should have contained at least one atom of non-exchangeable deuterium but this was not observed.

The one carbon species involved in the PEPCK-catalyzed reaction appears to be CO_2. Cooper *et al.* (*59*) used the carbonic anhydrase method to study this reaction in a manner similar to that which they used to study the PEPC-catalyzed reaction as described in the preceding section. The stereochemistry of the addition of CO_2 to PEP has recently been investigated by Rose *et al.* (*60*). Using specifically labeled 3-^{3}H-phosphoenolpyruvate, the addition was shown to occur from the *si* side (*116*) of the plane of the enzyme-bound PEP. When PEP containing tritium at position H_A was used, 98% of the tritium was released by the fumarase-catalyzed reaction; fumarase catalyzes a *trans*-elimination reaction (*117*). If the CO_2 had added from the *re* side, the tritium would have been quantitatively retained (see Fig. 2).

Chang *et al.* (*83*) studied the kinetics of various possible exchange reactions catalyzed by PEPCK from pig liver. They found that the rate of the ITP-dependent OAA-$^{14}CO_2$ exchange was much more rapid than the rates of PEP and OAA formation. The relative rates at pH 6.8 were 30 for the exchange reaction to 8.3 to 1 for PEP and OAA formation, respectively. Under the same conditions these authors were unable to find evidence for significant exchange of GDP-8-^{14}C with GTP, ^{32}PEP with ITP, or PEP-1-^{14}C with OAA. They concluded that failure to observe these possible exchange reactions made it unlikely that a phosphoryl-enzyme was an intermediate in the exchange reaction between CO_2 and OAA. From the faster rate of the CO_2–OAA exchange as compared to the PEP and OAA formation reactions, they also concluded that

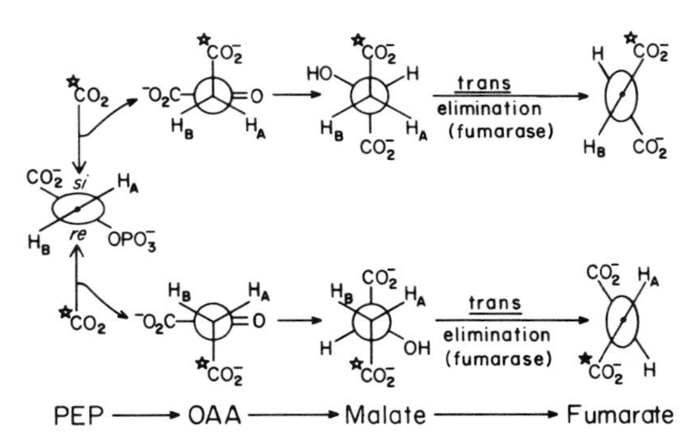

FIG. 2. Stereospecificity of carboxylation of PEP.

116. K. R. Hanson, *JACS* **88**, 2731 (1966).
117. F. A. L. Anet, *JACS* **82**, 994 (1960).

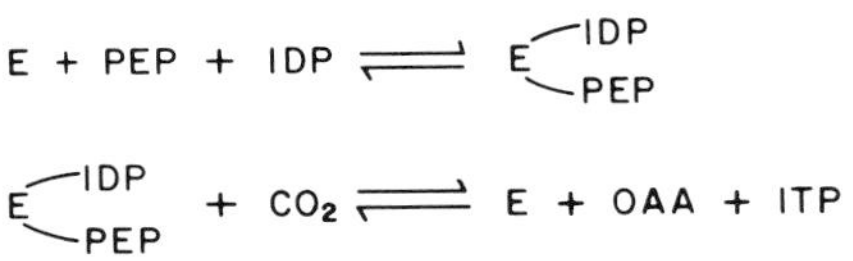

$$E + PEP + IDP \rightleftharpoons E\genfrac{}{}{0pt}{}{IDP}{PEP}$$

$$E\genfrac{}{}{0pt}{}{IDP}{PEP} + CO_2 \rightleftharpoons E + OAA + ITP$$

FIG. 3. Possible partial reactions catalyzed by PEPCK.

the overall process might be formulated as shown in Fig. 3. The nature of the events undergone by the ternary (and possibly quaternary) complexes is not specified here, but the first reaction shown in Fig. 3 must be rate limiting. PEP has been found to be an effective inhibitor of the CO_2-OAA exchange reaction but not of the overall PEP formation reaction.

In a later and more comprehensive kinetic study of the mechanism of action of pig liver PEPCK, Miller and Lane (*118*) proposed that initial steady state velocity studies and binding studies involving the same enzyme (*36*) are consistent with the mechanism shown in Fig. 4. The most significant features of this mechanism are (a) the formation of a central complex consisting of enzyme, Mn^{2+}, PEP, IDP, and CO_2 by what the authors call a mixed ordered-random addition of components; (b) PEP must bind to the enzyme before either IDP or CO_2 to yield a kinetically active form of the enzyme; (c) the binding of PEP proceeds in a random manner with respect to Mn^{2+} to form the ternary enzyme–Mn^{2+}–PEP complex; and (d) the ternary enzyme–Mn^{2+}–PEP complex may bind either IDP or CO_2 to form a quaternary complex, which then binds the third substrate to form the postulated central complex. The situation is too complex to consider this as anything more than a suggestion, particularly as far as the chemical events are concerned. Miller *et al.* (*36*) also carried out an extensive series of binding studies with the same enzyme using a variety of techniques including equilibrium dialysis, electron paramagnetic resonance, and the proton

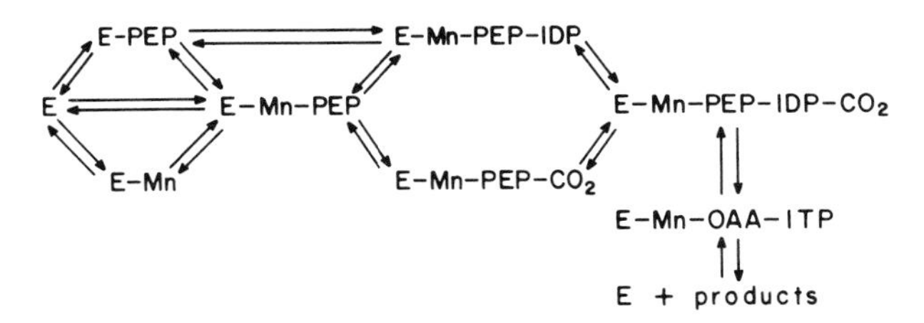

FIG. 4. Possible mechanism of PEPCK-catalyzed reaction.

118. R. S. Miller and M. D. Lane, *JBC* **243**, 6041 (1968).

relaxation rate of water. These data were used in part to formulate the mechanism shown in Fig. 4. Perhaps the most important findings emerging from these studies are that Mn^{2+} binds directly to the enzyme in an apparent 1 to 1 ratio and that this metal ion facilitates the binding of PEP. It may be that the Mn^{2+} serves as a bridge between PEP and the enzyme.

Proton relaxation rate studies (*36*) of the binary complex between PEPCK and Mn^{2+} suggest that two ligands of the protein displace two water molecules from the outer coordination sphere of the metal, leaving four molecules of water in the octahedral hydration shell. The effects on the enhancement of the proton relaxation rate of water which result from the addition of PEP and IDP to the PEPCK–Mn^{2+} complex suggest that PEP replaces two of the remaining four water molecules and IDP replaces one. Only one water molecule remains in the hydration sphere of the central carboxylation complex. The central decarboxylation complex is similar with two of the water molecules replaced by ITP and one by OAA. It should be emphasized, as it was by the authors, that these interpretations are based on the assumption that the enhancement values are influenced primarily by the number of water molecules in the hydration shell although it is known that other factors can also influence enhancement values. Nevertheless, these interpretations permit the authors to propose the central complexes as shown in Fig. 5. In experiments in which the PEPCK-catalyzed formation of PEP and IDP was conducted in $H_2{}^{18}O$, no ^{18}O excess could be detected in the products (*118*). These findings are consistent with the postulated representations of the central complexes shown in Fig. 5, but other explanations are also possible. In the next section, a mechanism involving an enzyme–enolpyruvate complex is proposed for PEPCTrP which could also explain the observations with PEPCK.

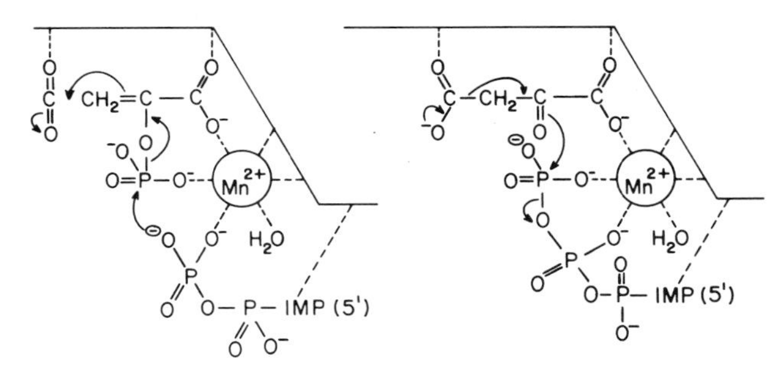

FIG. 5. Suggested central complexes for carboxylation and decarboxylation reactions catalyzed by PEPCK.

Felicioli *et al.* (*119*) have recently reported kinetic studies with PEPCK isolated from chicken liver mitochondria. On the basis of initial velocity and product inhibition studies of the OAA exchange reaction, these authors suggested an ordered kinetic mechanism in which ITP is added before OAA and the products are released in the following order: CO_2, PEP, and IDP. Careful examination of the data suggests that other interpretations are possible. Preliminary kinetic evidence with PEPCK isolated from *E. coli* also suggests that the free enzyme binds the nucleotide first followed by the remaining substrates (*120*).

The mechanism of the PEPCK-catalyzed reaction cannot be said to be well understood at present, and the same applies to the other two enzymes considered in this chapter. In particular, the chemistry of the reactions is still not clear. For example, it is not obvious how the mechanisms proposed above for the decarboxylation reaction also lead to the formation of pyruvate in the presence of IDP.

E. Regulation

Because PEPCK catalyzes the first step in the gluconeogenic series of reactions leading from dicarboxylic acids to glucose, it would appear to be a likely candidate for regulation. Indeed, as mentioned earlier, Williamson (*71*) suggested that under certain circumstances this reaction is the rate-limiting step in glucose synthesis in perfused rat liver. There is a large body of evidence to support the suggestion that, at least in some species, the activity of the PEPCK-catalyzed reaction may be regulated by variations in the amount of active enzyme, which, in turn, is under the influence of a wide variety of physiological factors. Controls operative at the enzymic level, i.e., fine controls, are less well established although there is some evidence to suggest that the enzyme may be regulated by the amount of substrates available. These two aspects of regulation are discussed below but are preceded by a consideration of a special feature of PEPCK, i.e., its intracellular location and the implications on control properties which arise therefrom.

1. *Intracellular Location of PEPCK*

The intracellular location of PEPCK varies from one species to another and may, within a given species, vary from one tissue to another.

119. R. A. Felicioli, R. Barsacchi, and P. L. Ipata, *European J. Biochem.* **13,** 403 (1970).
120. J. A. Wright and B. D. Sanwal, *JBC* **244,** 1838 (1969).

PEPCK is a predominantly soluble enzyme in adult rat liver (*91, 121*), and also in mouse and hamster liver (*91*). In contrast the enzyme is primarily or wholly located in the mitochondria of chicken and pigeon liver (*122, 123*) and rabbit liver (*91*). In pigeon liver, PEPCK appears to be located within the mitchondrial matrix (*124*). In pig (*125*), guinea pig (*107*), and human liver (*126, 127*), PEPCK is found in both the cytosol and mitochondria with perhaps a slightly greater amount in the latter. Recent experiments with yeast (*128*) indicate that the enzyme is extramitochondrial in that organism.

It is interesting that the intracellular location of the enzyme in rat liver changes from primarily mitochondrial in the fetal rat to primarily soluble in the adult (*121*). Within the intestinal mucosa of the adult rat, the enzyme appears to be mitochondrial (*129*), while in the adipose tissue 90% of the PEPCK activity is found in the cytosol (*130*).

Since the physiological significance of the PEPCK-catalyzed reaction may be related to the decarboxylation of the OAA formed from pyruvate via pyruvate carboxylase, it is of interest to know the relative location of these two enzymes. According to most workers, pyruvate carboxylase is a predominantly or wholly mitochondrial enzyme in most animal tissues (*52*). Consequently, it would appear that in tissues such as rat liver, where PEPCK is extramitochondrial, these paired enzymes are located in different cellular compartments. Walter *et al.* (*131*) have proposed that in such cases the OAA formed within the mitochondria is transported to the PEPCK outside in the form of malate or aspartate. As a result the pathway from pyruvate to PEP is considerably longer than when the two enzymes are located in the same intracellular compartment and consequently it is conceivable that the control systems might be different for the mitchondrial and cytosol forms of PEPCK. Nordlie *et al.* (*132*) suggested earlier that the two forms of the rat liver

121. F. J. Ballard and R. W. Hanson, *BJ* **104,** 866 (1967).
122. M. F. Utter, *Ann. N. Y. Acad. Sci.* **72,** 451 (1959).
123. W. Gevers, *BJ* **103,** 141 (1967).
124. C. Landriscina, S. Papa, P. Coratelli, L. Mazzarella, and E. Quagliariello, *BBA* **205,** 136 (1970).
125. K. R. Swiatek, K. L. Chao, H. L. Chao, M. Cornblath, and T. Tildon, *BBA* **206,** 316 (1970).
126. W. Brech, E. Shrago, and D. Wilken, *BBA* **201,** 145 (1970).
127. O. Wieland, E. Evertz-Prüsse, and B. Stukowski, *FEBS Letters* **2,** 26 (1968).
128. R. K. Yamazaki and M. F. Utter, unpublished observations (1970).
129. J. W. Anderson, *BBA* **208,** 165 (1970).
130. F. J. Ballard, R. W. Hanson, and G. A. Leveille, *JBC* **242,** 2746 (1967).
131. P. Walter, V. Paetkau, and H. A. Lardy, *JBC* **241,** 2523 (1966).
132. R. C. Nordlie, F. E. Verricchio, and D. D. Holten, *BBA* **97,** 214 (1965).

enzyme have different properties. Recent immunological experiments conducted by Ballard and Hanson (*112*) support the contention that these two forms of the enzyme are not identical proteins.

2. *Factors Affecting the Concentration of PEPCK*

We have already mentioned that PEPCK activity appears to respond to certain dietary and hormonal alterations. In fetal rat liver, maintained in organ culture, the activity of PEPCK is induced by several agents. Glucagon, isoproterenol, and cyclic AMP all produced an approximately 3-fold stimulation of activity (*133*). Insulin had no effect on the enzyme but did block the action of these three agents. Yeung and Oliver (*134*) found that intraperitoneal injections of cyclic AMP produced increased PEPCK activity in fetal rat liver, but the glucocorticoid, triamcinolone, when administered *in utero* did not lead to development of the enzyme (*135*). The activity of PEPCK in adult rat liver is also increased by the administration of an analog of cyclic AMP, N^6,$O^{2'}$-dibutyryl cyclic AMP, but hydrocortisone does not induce the enzyme (*136*). Reshef and Hanson (*137*) obtained similar results with rat liver, but found that neither dibutyryl cyclic AMP nor glucagon induced PEPCK in rat adipose tissue even though the liver cytosol and adipose tissue enzymes are immunochemically identical proteins (*112*). On the other hand, catecholamines appear to induce PEPCK activity in fetal rat liver (*138*) and in both liver and adipose tissue of the adult (*137*). In contrast, Shrago *et al.* (*65*) found that hydrocortisone induced PEPCK activity in adult rat liver, but Lardy *et al.* (*139*) noted that hydrocortisone treatment did not enhance PEPCK activity in either the mitochondrial or cytosol fraction of guinea pig liver. From the foregoing it seems fair to say that (a) a number of treatments and agents can alter the apparent PEPCK activity at least in rat liver and (b) the nature and basis of these changes is still obscure, particularly when other tissues and species are considered.

In some species at least, only the soluble form of PEPCK responds to dietary and hormonal manipulations. In rat liver, where the enzyme is located primarily in the cytosol, fasting induces a 2.5-fold increase

133. W. D. Wicks, *JBC* **244,** 3941 (1969).
134. D. Yeung and I. T. Oliver, *Biochemistry* **7,** 3231 (1968).
135. D. Yeung, R. S. Stanley, and I. T. Oliver, *BJ* **105,** 1219 (1967).
136. W. D. Wicks, F. T. Kenney, and K-L. Lee, *JBC* **244,** 6008 (1969).
137. L. Reshef and R. W. Hanson, personal communication (1970).
138. D. Yeung and I. T. Oliver, *BJ* **108,** 325 (1968).
139. H. A. Lardy, D. O. Foster, E. Shrago, and P. D. Ray, *Advan. Enzyme Regulation* **2,** 39 (1964).

in PEPCK activity (*65*). In contrast, the level of PEPCK activity in pigeon liver, where the enzyme is exclusively mitochondrial, is not altered by fasting (*123*). The cytosol form of guinea pig liver PEPCK appears to be increased during fasting while the mitochondrial form is not (*70*). In contrast, Swiatek *et al.* (*125*) reported that both the mitochondrial and soluble activities were increased in the liver of fasted pigs. Unfortunately, alternative explanations cannot be excluded because the authors did not normalize their data on the basis of criteria such as enzyme units per total liver or per amount of DNA.

3. *Factors Affecting the Activity of PEPCK*

In spite of the wealth of reports relating to possible regulation of PEPCK by physiological factors, there is relatively little definitive information concerning factors which might directly regulate the catalytic activity of this enzyme. One approach to the problem is to examine the reported K_m values for the various substrates of PEPCK in the hope that clues concerning possible regulatory mechanisms might be obtained. In Table VII K_m values for substrates for a variety of species of PEPCK are listed. The most striking fact which emerges from such an examination is the relatively high K_m values which have been reported for OAA. These values are in the range of 0.1–1 mM while the intracellular concentration of OAA is probably about two orders of magnitude lower (*70, 140–142*). This discrepancy is not lightly passed over. Attempts to find an activator of PEPCK which could lower the apparent K_m for OAA and thereby obviate the discrepancy have not been successful (*108*), but recent findings suggest that the previously reported K_m values for OAA may be erroneously high. Ballard (*143*) recently reported that the K_m for OAA of PEPCK isolated from rat, chicken, and sheep liver is in the range of 9–30 μM, depending on the source of the enzyme, the nature of the metal ion employed, and the presence of malate. Studies conducted in our laboratory have produced somewhat similar results but also seem to indicate that under some conditions, in which both Mn^{2+} and Mg^{2+} are present, Michaelis-Menten kinetics are not observed and hence that the apparent K_m is a function of the OAA concentration range examined (*144*). This situation obviously requires further study,

140. G. Kalnitsky and D. F. Tapley, *BJ* **70,** 28 (1958).

141. O. Wieland and G. Loeffler, *Biochem. Z.* **339,** 204 (1963).

142. H. V. Henning, B. Stempf, B. Ohly, and W. Seubert, *Biochem. Z.* **344,** 274 (1966).

143. F. J. Ballard, *BJ* **120,** 809 (1970).

144. H. M. Kolenbrander, Y. B. Chiao, and M. F. Utter, unpublished observations (1970).

TABLE VII
K_m OF PEPCK FOR ITS SUBSTRATES[a]

Enzyme source	K_m values						Reference
	CO_2	HCO_3^-	GDP[b]	PEP	GTP[c]	OAA	
Guinea pig liver			4×10^{-4}	1.4×10^{-4}	1.4×10^{-3}	2×10^{-3}	*(107)*
Guinea pig liver					2×10^{-3}	1×10^{-3}	*(91)*
Pig liver		$15–25 \times 10^{-3}$	6×10^{-5}	$1–3 \times 10^{-4}$	1.6×10^{-4}	1.5×10^{-4}	*(83)*
Pig liver		15×10^{-3}	5.4×10^{-5}	6.6×10^{-4}			*(125)*
R. spheroides						2×10^{-3}	*(101)*
Yeast		5×10^{-3}					*(104)*
Yeast						7.7×10^{-3}	*(86)*

[a] All values are reported as molar quantities.
[b] Or IDP.
[c] Or ITP.

but it seems likely that the activity of PEPCK at physiological levels of OAA may be far higher than previously supposed and further that alterations in the concentration of OAA may serve as a regulatory device for this enzyme.

The intracellular concentration of GTP is in the 0.1 m*M* range (*145*). The apparent K_m for PEPCK for this compound is in the same general range but perhaps a little higher. It is possible that PEPCK activity is also influenced by the relative phosphorylation states of adenine and guanine nucleotides. Such evidence is fragmentary, but Noce and Utter (*105*) noted that the velocity of PEP formation was stimulated by ATP. Gevers (*123*) found that PEP synthesis is inhibited in pigeon liver mitochondrial suspensions containing agents, such as dinitrophenol, which decrease the steady state ATP–ADP ratio. Garber and Ballard (*146*) noted that the ATP–ADP ratio has a profound influence on the rate of PEP synthesis in guinea pig liver mitochondria. When the ATP–ADP ratio was increased from 0.5 to 0.7, the rate of PEP synthesis was approximately tripled. (The intramitochondrial concentration of GTP was also tripled by this increase.) More recently, Philippidis and Ballard (*147*) reported that ether anesthesia resulted in an approximately 30% reduction in the ATP–ADP ratio in fetal rat liver with a concomitant 10-fold reduction in the rate of ^{14}C-incorporation from pyruvate-3-^{14}C into glycogen; PEPCK may have been involved in the decrease.

In controlling PEPCK activity, NADH may also play a role. Wright and Sanwal (*120*) noted that PEPCK from *E. coli* is allosterically inhibited by NADH. NAD^+ was not an inhibitor. Garber and Ballard (*146*) observed that an increase in the NADH–NAD^+ ratio resulted in a decreased rate of PEP synthesis in guinea pig liver mitochondria. The rate decreased approximately 2.5-fold when the NADH–NAD^+ ratio was increased from 0.05 to 0.2. This effect of NADH can be interpreted to be a result of a decrease in the concentration of OAA which thereby decreases apparent PEPCK activity, but other interpretations are possible. Kolenbrander and Utter (*108*) could not produce a significant change in the rate of PEP synthesis by experimentally manipulating the NADH–NAD^+ ratio with PEPCK purified from chicken liver mitochondria.

Although it is possible that other effectors of PEPCK have not yet been detected, extensive studies with the enzyme from chicken liver (*108*) have involved testing coenzyme A derivatives (acetyl, propionyl, butyryl, malonyl, methylmalonyl, acetoacetyl, β-hydroxybutyryl, suc-

145. B. Chance, B. Schoener, K. Krejci, W. Ruessmann, W. Wesemann, H. Schnitger, and T. Buecher, *Biochem. Z.* **341,** 325 (1965).

146. A. J. Garber and F. J. Ballard, *JBC* **245,** 2229 (1970).

147. H. Philippidis and F. J. Ballard, *BJ* **120,** 385 (1970).

cinyl, and palmityl), two carnitine derivatives (L-palmityl and DL-acetyl), four cyclic nucleotide monophosphates (adenylic, cytidylic, guanylic, and thymidylic), intermediates in the citric acid cycle, amino acids, fructose 1,6-diphosphate, lactate, quinolinate, and phthalate as potential effector compounds. All of these compounds were without effect when tested at limiting concentration of OAA.

One of the most interesting suggestions concerning a possible control mechanism for PEPCK has come from the observed effect of tryptophan on gluconeogenesis in rat liver. Gluconeogenesis was inhibited following administration of this amino acid and the crossover apparently occurred between OAA and PEP which suggested that the PEPCK activity had decreased (*148*). However, when the PEPCK activity was tested *in vitro* following this tryptophan treatment, increased activity was observed (*149*). The paradox appears to be explained by the formation of quinolinate during the metabolism of tryptophan. A ferrous complex of quinolinate is a strong inhibitor of PEPCK, but it may be sufficiently diluted during the manipulation of the tissue in preparing for the *in vitro* PEPCK assay so that the enzyme is left in a metal-activated state (*106*) which is approximately twice as active as the nonactivated enzyme [cf. Veneziale *et al.* (*150*) for details]. Quinolinate appears to be less effective in inhibiting gluconeogenesis from lactate in guinea pig and has no effect in perfused pigeon liver (*151*). Thus, despite its interesting features, the tryptophan-quinolinate mechanism may not be of general application.

IV. Phosphoenolpyruvate Carboxytransphosphorylase (PEPCTrP)

A. Discovery, Distribution, and Physiological Role

In 1961, Siu *et al.* (*152*) reported finding a new CO_2-fixing enzyme in *Propionibacterium shermanii* which they named *phosphoenolpyruvic carboxytransphosphorylase*. This enzyme catalyzes reaction (11). Although this reaction appears to be very similar to the one catalyzed by

$$\text{PEP} + CO_2 + P_i \rightleftarrows \text{oxalacetate} + PP_i \tag{11}$$

148. P. D. Ray, D. O. Foster, and H. A. Lardy, *JBC* **241**, 3904 (1966).
149. D. O. Foster, P. D. Ray, and H. A. Lardy, *Biochemistry* **5**, 563 (1966).
150. C. M. Veneziale, P. Walter, N. Kneer, and H. A. Lardy, *Biochemistry* **6**, 2129 (1967).
151. H. D. Soling, B. Willms, and J. Kleineke, *Z. Physiol. Chem.* **351**, 291 (1970).
152. P. M. L. Siu, H. G. Wood, and R. L. Stjernholm, *JBC* **236**, PC21 (1961).

PEPC, the characteristics of the two systems are different in several important respects in addition to the obvious difference in phosphate acceptor molecules. The PEPCTrP-catalyzed reaction is readily reversible (the rate of OAA formation is about seven times as rapid as the rate of PEP formation) while the PEPC-catalyzed reaction is not. Furthermore, PEPCTrP catalyzes the formation of pyruvate from PEP in the absence of CO_2 [see reaction (12)]. It has, thus far, been im-

$$PEP + P_i \rightarrow \text{pyruvate} + PP_i \tag{12}$$

possible to demonstrate reversibility of the pyruvate formation reaction. The relative rates of OAA and pyruvate formation from PEP depend not only on the availability of CO_2 but also on the presence of thiols, metal ions, and other factors. As discussed in the previous section, PEPCK also catalyzes a pyruvate formation reaction but in that case OAA is the substrate. This example is one of a number of partial similarities between PEPCK and PEPCTrP.

The discovery of PEPCTrP in the propionic acid bacteria solved a long-standing mystery concerning the CO_2-fixing reaction in these bacteria; Wood and Werkman (*153*) had discovered CO_2 fixation in these bacteria 25 years earlier. Very probably PEPCTrP is the enzyme responsible for the net fixation of CO_2 observed when succinate accumulates in propionic acid bacteria grown on glycerol. Presumably, the accumulation of succinate reflects an enhancement of the normal anaplerotic role of PEPCTrP in this genus of bacteria, but an entirely satisfactory explanation for the accumulation of succinate instead of propionate is still not available.

Thus far, this enzyme has been reported in only one other source. Reeves (*154*) has presented convincing evidence of its presence in the amoeba, *Entamoeba histolytica*. As mentioned earlier, it is possible that in certain species PEPCTrP may have been identified erroneously as PEPC. It is at least arguable that this *might* be true for the "PEPC" of tropical grasses. In addition to their reports of finding PEPC activity in these grasses, Hatch and Slack (*155*) have also noted the existence of pyruvate, phosphate dikinase in these species. The latter enzyme catalyzes reaction (13). This reaction appears to be closely related to one

$$\text{Pyruvate} + ATP + P_i \rightleftarrows PEP + AMP + PP_i \tag{13}$$

catalyzed by PEP synthetase, an enzyme discovered earlier by Cooper

153. H. G. Wood and C. H. Werkman, *BJ* **30,** 48 (1936).
154. R. E. Reeves, *BBA* **220,** 346 (1970).
155. M. D. Hatch and C. R. Slack, *BJ* **112,** 549 (1969).

TABLE VIII
ENZYMES INVOLVED IN PEP AND OAA SYNTHESIS IN VARIOUS ORGANISMS

Species	PEP formed by	Reference	OAA formed by	Reference
E. coli	PEP Synthase	(*156*)	PEPC	(*39*)
S. typhimurium	PEP Synthase	(*157*)	PEPC	(*49*)
P. shermanii	Pyruvate, phosphate dikinase	(*158*)	PEPCTrP	(*152*)
E. histolytica	Pyruvate, phosphate dikinase	(*159*)	PEPCTrP	(*154*)
Tropical grasses	Pyruvate, phosphate dikinase	(*155*)	"PEPC"	(*11*)
A. xylinum	Pyruvate, phosphate dikinase	(*160*)	PEPC	(*23*)

and Kornberg (*156*) in *E. coli* [reaction (14)]. Reactions (13) and (14)

$$\text{Pyruvate} + \text{ATP} \rightleftharpoons \text{PEP} + \text{AMP} + \text{P}_i \quad (14)$$

appear to have the same relationship to each other as the pair of reactions catalyzed by PEPCTrP and PEPC. The reactions catalyzed by pyruvate, phosphate dikinase and PEPCTrP both use P_i as a phosphate-acceptor molecule while those catalyzed by PEP synthetase and PEPC employ a water molecule as the acceptor. Data indicating the distribution of these four enzymes are presented in Table VIII (*11, 23, 39, 49, 152, 154–160*). In the first four species listed in Table VIII, there is a perfect correlation between the PEP- and OAA-forming enzymes; either both enzymes use P_i as a substrate and form PP_i as a product or neither does. With tropical grasses the presence of the phosphate dikinase and the fact that "PEPC" has not been rigorously characterized in these species suggests that the enzyme might really be PEPCTrP. However, this argument is weakened by the finding of Benziman and Eizen (*160*) of pyruvate dikinase in *Acetobacter xylinum* along with PEPC (*23*). In this case, PEPC has been purified some 200-fold, does not exhibit a $^{14}CO_2$–OAA exchange reaction in the presence of PP_i, and thus seems to be a true PEPC (*161*).

The function, if any, of reaction (12) as catalyzed by PEPCTrP is not clear, but, as shown below, coupling it with reaction (13) could serve as a mechanism for forming PP_i. The physiological role of PP_i is not well understood, but Baltscheffsky (*162*) has suggested that PP_i may serve as an energy donor at a pre-adenosine phosphate level in both

156. R. A. Cooper and H. L. Kornberg, *Proc. Roy. Soc.* **B168,** 263 (1967).
157. R. A. Cooper and H. L. Kornberg, *BBA* **141,** 211 (1967).
158. H. J. Evans and H. G. Wood, *Proc. Natl. Acad. Sci. U. S.* **61,** 1448 (1968).
159. R. E. Reeves, *JBC* **243,** 3202 (1968).
160. M. Benziman and N. Eizen, *JBC* **246,** 57 (1971).
161. M. Benziman, private communication (1970).
162. H. Baltscheffsky, *Acta Chem. Scand.* **21,** 1973 (1967).

bacterial chromatophores and animal mitochondria. Nordlie *et al.* (*163*) have also suggested that glucose-6-phosphatase can use PP_i as a phosphorylating agent under certain conditions.

$$\text{Pyruvate} + \text{ATP} + P_i \rightarrow \text{PEP} + \text{AMP} + PP_i \quad (13)$$

$$\text{PEP} + P_i \rightarrow \text{pyruvate} + PP_i \quad (12)$$

$$\overline{\text{ATP} + 2P_i \rightarrow \text{AMP} + 2PP_i}$$

B. General Characteristics and Mechanism of the PEPCTrP-Catalyzed Reaction

In 1961, PEPCTrP was first described, but with the exception of a very few recent studies, all work with this enzyme has originated in one laboratory. Wood and his associates have published an extensive series of investigations describing this enzyme and the reaction it catalyzes (*31, 164–167*). The enzyme has been obtained in crystalline form (*165*) and has a specific activity for OAA of about 20–25 μmoles/min/mg protein at 25° which compares closely with values found for PEPC and PEPCK.

PEPCTrP catalyzes the reversible carboxylation of PEP to form OAA [reaction (11)] and the irreversible formation of pyruvate from PEP [reaction (12)]. The relative rates of the two reactions depend upon a number of factors including the presence of or prior treatment with thiols. For example, Davis *et al.* (*166*) found that the rates (μmoles/min/mg protein) of OAA and pyruvate formation were 12.1 and 1.5, respectively, when the enzyme was diluted in 1 m*M* mercaptoethanol and assayed in 60 μ*M* thiol. These rates were nearly the same (11.2 and 1.8, respectively) when the enzyme was diluted in the absence of thiol but assayed in the presence of 1 m*M* mercaptoethanol. In sharp contrast, dilution and assay of the enzyme in the absence of thiols resulted in relative rates of OAA and pyruvate formation of 2.6 and 2.9, respectively. Apparently, the presence of thiols strongly promotes the formation of OAA but inhibits pyruvate formation. The thiol effect is reversible and time dependent. For example, if the reaction is started in the absence of a thiol and the latter is then added, the rate of reaction changes gradually (over a period of 2–3 min) to one characteristic of that observed in the presence of a thiol. A conformational change is plausible.

Davis *et al.* (*166*) noted that Mg^{2+}, Co^{2+}, or Mn^{2+} can satisfy the div-

163. R. C. Nordlie, "The Enzymes," 3rd ed., Vol. IV, p. 543, 1971.
164. P. M. L. Siu and H. G. Wood, *JBC* **237,** 3044 (1962).
165. H. Lochmüller, H. G. Wood, and J. J. Davis, *JBC* **241,** 5678 (1966).
166. J. J. Davis, J. M. Willard, and H. G. Wood, *Biochemistry* **8,** 3127 (1969).
167. J. M. Willard, J. J. Davis, and H. G. Wood, *Biochemistry* **8,** 3137 (1969).

alent metal ion requirement for either OAA or pyruvate formation, but both reactions are inhibited by 1 m*M* EDTA even if high concentrations of Mg^{2+} are provided. From this and subsequent studies by Willard *et al.* (*167*) and Wood *et al.* (*31*), the thesis that PEPCTrP requires two types of metal ions has developed. Both a Type I metal ion, exemplified by Mg^{2+}, Co^{2+}, or Mn^{2+} which is readily dissociated, and a Type II metal ion, which is tightly bound but subject to inhibition by EDTA and a wide variety of other chelating agents (*167*), appear to be involved. The EDTA treatment is far more effective in the presence of mercaptans, which again suggests that thiols induce a conformational change. After treatment with EDTA and dialysis, the enzyme is inactive in either the OAA or pyruvate formation reactions. The addition of Mg^{2+} alone cannot reactivate either reaction, but the OAA-forming reaction can be reactivated by low concentrations (1–10 μM) of Co^{2+}, Zn^{2+}, or Mn^{2+}. The pyruvate-forming reaction can also be reactivated by the same metal ions, but much higher concentrations (0.5 m*M*) are required; only Cu^{2+} is effective at low concentrations. Cu^{2+} is relatively ineffective in the OAA-forming reaction. Obviously this is a very complex situation which is not yet fully elucidated. As a possible explanation, Wood and associates (*31, 166, 167*) suggested that EDTA inhibits PEPCTrP, a metalloprotein, by binding to the metal on the protein and that activity is restored when EDTA is removed by the addition of another metal ion (e.g., Co^{2+}). This picture is based upon failure of attempts to prepare an apoenzyme by treatment with chelators and to demonstrate specific binding of radioactive Co^{2+} after EDTA treatment. The role of Cu^{2+} in restoring the pyruvate formation reaction is somewhat more obscure, but Davis *et al.* (*166*) suggested that a heavy metal ion may be required for the formation of pyruvate. A simplified pictorial representation of this interpretation of these results is shown in Fig. 6. In this diagram, the Type II metal is shown as Me^{2+} and Ⓜe is pictured as the additional heavy metal. The figure shows a conformational change, promoted by thiols, which leads to a greater susceptibility to EDTA inhibition. An EDTA-inhibited form of the enzyme is shown to be reactivated to the OAA and pyruvate forming conformations by Co^{2+} and Cu^{2+}, respectively.

Attempts to demonstrate that PEPCTrP is a metalloprotein have been equivocal thus far. As mentioned in an earlier section, the situation is much the same with PEPCK; chelators inhibit the reaction, but no firm evidence for a metalloprotein is available.

Wood *et al.* (*31*) have investigated the various partial reactions which PEPCTrP might catalyze. Their results are summarized in Table IX. As noted earlier, in the presence of PP_i and Mg^{2+} $^{14}CO_2$ exchanges read-

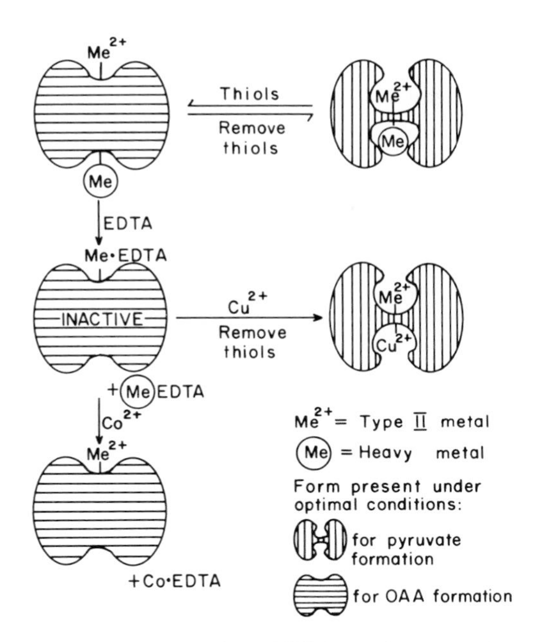

FIG. 6. Inhibition of PEPCTrP by EDTA and restoration of activity by metal ions.

ily with the β-COOH of OAA. When this exchange is examined using the EDTA-treated enzyme, no requirement for a Type II metal can be demonstrated. Further investigations by Wood *et al.* (*31*) indicate that the CO_2–OAA exchange reaction is the only partial reaction which can be demonstrated with PEPCTrP. The reaction between P_i and PP_i or between P_i and PEP can be detected only if all other substrates and both types of metals are present, that is, under conditions which permit

TABLE IX
TEST FOR PARTIAL REACTIONS WITH PEPCTrP[a]

Partial reaction	Substrate requirements	Metal requirements	
		Metal I	Metal II
$CO_2 \rightleftarrows$ OAA	PP_i	Mg^{2+}	Not required
$P_i \rightleftarrows PP_i$	PEP, CO_2	Mg^{2+}	Co^{2+}
$P_i \rightleftarrows$ PEP	PP_i, CO_2	Mg^{2+}	Co^{2+}
Pyruvate $\rightarrow$ OAA	No incorporation		
Pyruvate $\rightarrow$ PEP	No incorporation		

[a] Data taken from Wood *et al.* (*31*).

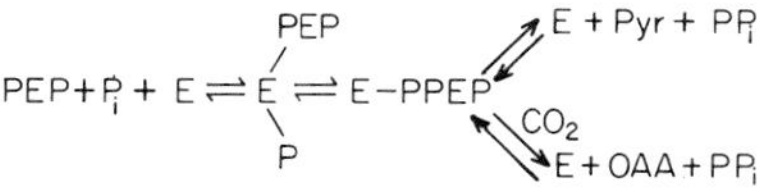

FIG. 7. Reaction sequence for PEPCTrP-catalyzed reaction involving pyrophosphopyruvate as an intermediate.

the overall PEP-OAA interconversion [reaction (11)] to occur. Thus far it has been impossible to show that ^{14}C-pyruvate can be incorporated into OAA or PEP even if all reaction components are present.

Failure to demonstrate any partial reactions other than the CO_2–OAA exchange eliminates several possible mechanisms from serious further consideration. Wood *et al.* (*31*) suggested that the sequence summarized in Fig. 7 is consistent with the foregoing experimental observations. The carbon dioxide species, shown as CO_2, is consistent with evidence presented by Cooper *et al.* (*59*) who used the carbonic anhydrase method (discussed in Section II) of their study. The main feature of this mechanism is the formation of an enzyme-bound pyrophosphopyruvate (E-PPEP) as a common intermediate in both the pyruvate- and OAA-forming pathways. The proposed mechanism appears to require the incorporation of ^{18}O from H_2O into PP_i when the pyrophosphopyruvate intermediate is converted to pyruvate and PP_i as shown in Fig. 8. When Singleton and Wood (*168*) tested this possibility they found no ^{18}O in the PP_i. This failure is not in accord with the mechanism proposed in Fig. 7, but it can be argued (rather weakly) that the water molecule involved in the reaction is sequestered and consequently not in equilibrium with the water of the medium. However, recent observations tend to focus attention on other possible mechanisms. Somewhat earlier, Rose *et al.* (*60*) examined the stereospecificity of the PEPCTrP-catalyzed OAA formation by using specifically labeled ^3H-PEP. The results indicated that carboxylation occurred predominantly from the *si* side of the plane in the PEPCTrP-catalyzed reaction and almost entirely from the *si* side in the PEPC- and PEPCK-catalyzed reactions. In more recent

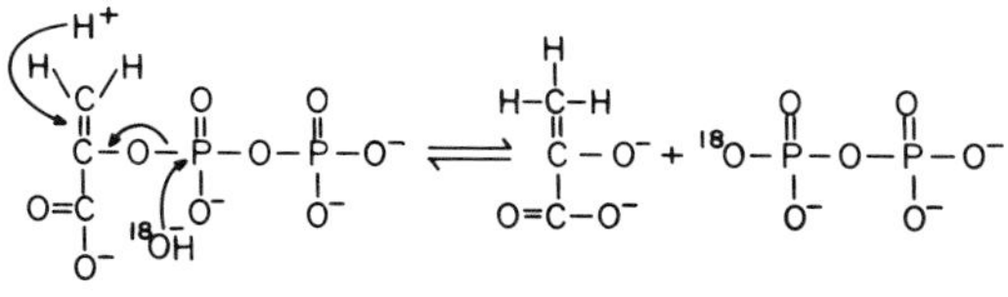

FIG. 8. Pyrophosphopyruvate intermediate in the PEPCTrP-catalyzed reaction.

168. R. Singleton and H. G. Wood, private communication (1971).

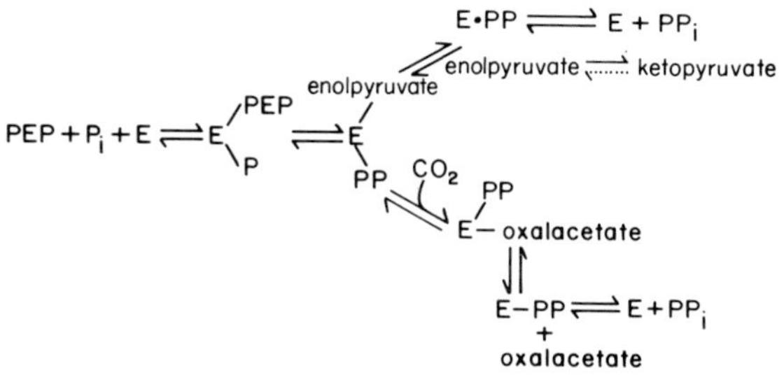

$$\mathrm{PEP} + \mathrm{P_i} + \mathrm{E} \rightleftharpoons \mathrm{E}{<}^{\mathrm{PEP}}_{\mathrm{P}} \rightleftharpoons \mathrm{E}{<}^{\mathrm{enolpyruvate}}_{\mathrm{PP}}$$

$$\mathrm{E}{<}^{\mathrm{enolpyruvate}}_{\mathrm{PP}} \rightleftharpoons \mathrm{E{\cdot}PP} + \mathrm{enolpyruvate};\quad \mathrm{E{\cdot}PP} \rightleftharpoons \mathrm{E} + \mathrm{PP_i};\quad \mathrm{enolpyruvate} \rightleftharpoons \mathrm{ketopyruvate}$$

$$\mathrm{E}{<}^{\mathrm{enolpyruvate}}_{\mathrm{PP}} + CO_2 \rightleftharpoons \mathrm{E}{<}^{\mathrm{PP}}_{\mathrm{oxalacetate}} \rightleftharpoons \mathrm{E{-}PP} + \mathrm{oxalacetate};\quad \mathrm{E{-}PP} \rightleftharpoons \mathrm{E} + \mathrm{PP_i}$$

FIG. 9. Proposed reaction sequence for PEPCTrP-catalyzed reaction with enolpyruvate intermediate.

studies with PEPCTrP, Rose and Willard (*169*) found conditions under which the formation of OAA is almost completely stereospecific, and, very interestingly, have also shown that the pyruvate formed from PEP in the PEPCTrP-catalyzed reaction arises in a relatively nonstereospecific manner. The latter results support the formation of an enolpyruvate as the reaction product and suggest that enzyme-bound enolpyruvate might play a central role as pictured in Fig. 9. This mechanism requires that OAA and PP_i be released in an ordered fashion; PP_i must be the last product released because CO_2 does not exchange with OAA in the absence of PP_i. In order to explain the lack of reversibility of the pyruvate formation reaction at least one step in the conversion of the enzyme-enolpyruvate intermediate to ketopyruvate must be essentially irreversible, but the lack of reversibility could be explained by the presence of only low concentrations of enolpyruvate under existing experimental conditions. In order that the mechanism be consistent with the observed failure to incorporate ^{18}O into PP_i, it is also necessary to postulate the formation of PP_i and enolpyruvate from PEP and P_i (all enzyme bound) by some mechanism which does not involve water. Figure 10 represents one possibility. Further experimental work is necessary to establish the validity of this mechanism, but it does have a number of attractive features. An analogous mechanism, based upon an enzyme-bound enolpyruvate, can be proposed for the PEPCK-catalyzed reaction.

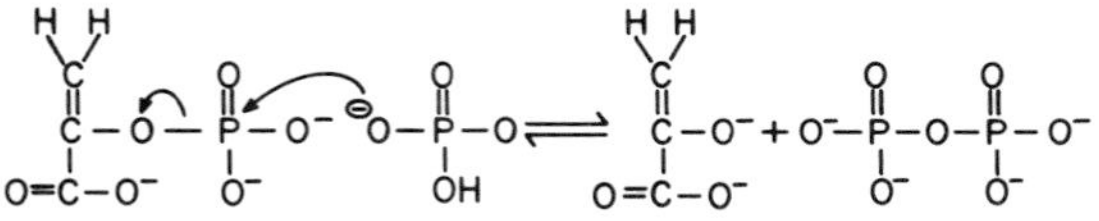

$$\mathrm{H_2C{=}C(COO^-){-}O{-}PO(O^-){-}O^-} + \mathrm{^-O{-}PO(OH){-}O^-} \rightleftharpoons \mathrm{H_2C{=}C(COO^-){-}O^-} + \mathrm{^-O{-}PO(O^-){-}O{-}PO(O^-){-}O^-}$$

FIG. 10. Enolpyruvate intermediate in the PEPCTrP-catalyzed reaction.

169. I. A. Rose and J. M. Willard, private communication (1971).

C. Kinetic Parameters, Effectors, and Inhibitors of PEPCTrP

The kinetic parameters for the substrates of the three reactions catalyzed by PEPCTrP are summarized in Table X (*165, 166, 170*). In most cases, the values shown were obtained from secondary plots and represent limiting values. The K_m for PEP is considerably lower in the pyruvate than in the OAA formation reaction, and the K_m for Mg^{2+} is probably also lower. Bicarbonate strongly influences the apparent K_m for PEP in the pyruvate formation reaction (*166*); it rises severalfold as the HCO_3^- level is increased to 10 m*M*. The kinetic interaction of substrates with each other and the interaction of products with substrates have not been studied extensively enough to permit mechanistic interpretations. It is worth noting, however, that moderate concentrations of the substrate, P_i, strongly inhibit pyruvate formation (*166*) and that moderate concentrations of the product, PP_i, strongly inhibit OAA formation even in the presence of excess Mg^{2+} (*165*).

Positive effectors for PEPCTrP have not been detected, but many substances inhibit the reaction including common buffers such as imidazole, tris-Cl, glycylglycine, and maleate (*165*). That 10 m*M* L-malate inhibits OAA formation by 50% is of more interest (*165*). Although this observation is reminiscent of the L-malate inhibition of certain varieties of PEPC, the specificity and mechanism of action of this inhibitor with PEPCTrP are not clear. The physiological significance of this possible feedback inhibition cannot be estimated at present.

TABLE X
Kinetic Parameters of PEPCTrP[a]

Reaction	Component	K_m (m*M*)
OAA formation	PEP	0.14
	P_i	0.8
	CO_2	1.9
	Mg^{2+}	1.2
PEP formation	OAA	0.47
	PP_i	0.22
Pyruvate formation	PEP	0.04
	P_i	0.66
	Mg^{2+}	0.63

[a] Based on data from Wood *et al.* (*165, 166, 170*).

170. M. Haberland, J. M. Willard, and H. G. Wood, private communication (1971).

D. Structure and Catalytic Activities of Different Forms of PEPCTrP

Lochmüller *et al.* (*165*) reported that crystalline PEPCTrP from *P. shermanii* had an $s_{20,w}$ of 15.2 (8 mg protein per milliliter). Using the Archibald method and assuming a partial specific volume for the protein of 0.75, they calculated a molecular weight of 430,000 ± 30,000. The $s_{20,w}$ value of the protein was decreased to 6.5–7 by treatment with 6 *M* urea, suggesting that it was dissociated into a smaller form.

Considerable additional evidence has accumulated since then which suggests that the 15 S species is a tetramer, that the 7 S species is a monomer, and that a dimeric species of about 10 S also exists. By using negative staining techniques Valentine *et al.* (*171*) found tetrameric forms in electron micrographs of the glutaraldehyde-stabilized 15 S species. Recent determinations of the molecular weight of the 15 S species (using the short column method of Yphantis) yield a value of 440,000 (*170*). The supernatant liquid above the 15 S crystals contains 7 S as well as 15 S material. The approximate molecular weight of the 7 S component is 110,000–115,000 suggesting that it bears a monomeric–tetrameric relationship to the 15 S species. Material obtained by treating the 15 S crystals with guanidine-HCl in the presence of thiols also appears to be the monomeric form, and, as discussed below, other material with an $s_{20,w}$ of about 10 S and a molecular weight approximating 230,000 can be detected under certain conditions. An active 7 S particle can also be formed from 15 S material by prolonged dialysis at low ionic strengths, but the reverse process has not been observed.

The catalytic activities of the species described present some surprising possibilities. Haberland *et al.* (*170*) have investigated the reacting species of PEPCTrP using the method of Cohen *et al.* (*172*). This method involves sedimenting the enzyme through its reaction mixture and determining the rate of sedimentation of the catalytically active species by optical means which employ the scanner attachment of the analytical ultracentrifuge. In this instance, PEPCTrP-catalyzed OAA formation is coupled with reduction by malate dehydrogenase, and the ultracentrifuge cell is scanned frequently at 350 n*M* to determine the extent of NADH oxidation at each position in the cell. The $s_{20,w}$ values for reacting species can be determined from such results. A catalytically active 10 S species was the predominant form detected in addition to an active 15 S species

171. R. M. Valentine, J. M. Willard, M. Haberland, and H. G. Wood, unpublished results (1970).

172. R. Cohen, B. Giraud, and A. Messiah, *Biopolymers* **5**, 203 (1967).

when an OAA formation assay was conducted with 15 S material. When the same experiment was run with 7 S material, an active 10 S species was again the predominant form but some 7 S material was also detected. Separate assays of the 15 S–10 S mixture and of the 7 S–10 S mixture showed an approximate ratio of 4 to 1 in the relative specific activities of these mixtures. These findings suggest that 10 S species with different catalytic capabilities are obtained from the 15 S and 7 S starting materials. The results also indicate that the monomeric, dimeric, and tetrameric forms of PEPCTrP all have catalytic activity in the OAA formation reactions. Experiments utilizing sucrose density gradients have also shown that PEPCTrP can exist as three different species.

When the 15 S and 7 S species were examined separately under conditions optimal for pyruvate formation (i.e., in the absence of CO_2 and thiol), the only catalytically active forms observed were the respective starting materials. This result was not altered by addition of thiol. Addition of CO_2 has been found to be essential, although not sufficient in itself, to effect conversion of the 15 S or 7 S species to 10 S material.

The results of these studies and those discussed above (*170*) are summarized in Table XI and the accompanying diagram.

TABLE XI

Molecular Weights, Sedimentation Characteristics, Catalytic Activities, and Interconversions of Different Forms of PEPCTrP

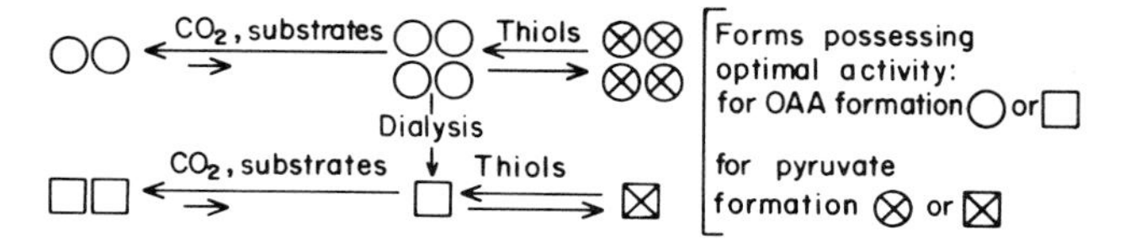

			Relative catalytic activities	
$s_{20,w}$	Molecular weight	Origin	OAA (optimal) (μmoles/min/mg)	Pyruvate (optimal) (μmoles/min/mg)
15	440,000[a]	Crystals	24	6
	430,000[b]	Crystals	24	6
10	230,000[a]	From 15 S	24	(Not done)
		From 7 S	6	(Not done)
7	116,000[a]	(From dialyzed crystals)	6	3
	109,000[a]	Guanidine-HCl + thiol	—	—

[a] By short-column method of Yphantis.

[b] By Archibald approach to equilibrium method.

V. Concluding Remarks

In the preceding sections, the three enzymes which fix CO_2 on PEP to form OAA have been discussed individually and only occasional attempts have been made to point out similarities and differences among them. At this juncture it seems appropriate to ask how much justification there is for considering these enzymes as a single group. The question can be put in terms of physiological function, structure, or mechanism.

According to present views the physiological roles of PEPC and PEPCTrP are very similar, i.e., to provide dicarboxylic acids. Both enzymes appear to be related in function to pyruvate carboxylase which is found in another group of species. In contrast, the primary physiological role of PEPCK appears to be catalysis of PEP formation. The PEP produced is subsequently utilized in the synthesis of various carbohydrates. The closest metabolic analogies to PEPCK are PEP synthase and pyruvate, phosphate dikinase both of which also catalyze the synthesis of PEP. However, the latter two enzymes require pyruvate as the substrate rather than the dicarboxylic acid, OAA, required by PEPCK.

Comparison of the controls of the three enzymes by metabolite effectors also shows few similarities among the three enzymes. The varieties of PEPC obtained from Enterobacteriaciae are subject to both positive and negative effectors but, while several other species of the enzyme are inhibited by certain dicarboxylic acids, no activators have been reported (cf. Table II). PEPCTrP from *P. shermanii*, is also inhibited by malate, and thus may resemble one type of PEPC, but this seems to exhaust the similarities. On the other hand, a striking set of analogies may be drawn between the various types of controls which affect PEPC and pyruvate carboxylase.

Structural studies of the various enzymes under discussion have not advanced to the stage where meaningful comparisons can be made among the three enzymes in the group or, for that matter, among the various types of the same enzyme obtained from different sources. Present evidence does suggest that PEPCTrP and PEPC are multisubunit enzymes which contain four or more subunits. This is probably true also for PEPCK from yeast, but the species of this enzyme obtained from animal tissues appear to be much smaller and are probably monomeric. It is likely that the definitive decision concerning the relationships of these enzymes will need to be based on structural evidence including primary sequence and other detailed information, but this lies in the future. At present our efforts at comparison must be based mainly on mechanistic considerations.

In terms of mechanism of action, PEPCTrP and PEPCK appear to be much more closely related to each other than either is to PEPC. This holds for the $\Delta G°$ of the three reactions; the value for PEPC (wheat germ) has been calculated to be —7.2 kcal (*54*) while the values for PEPCK and PEPCTrP, under roughly similar conditions, have been calculated to be —0.1 and —0.3 kcal, respectively (*173*). The exact values depend upon the pH, metal ion concentrations, and other assumptions, but it is clear that in this respect PEPC does not resemble the other two enzymes.

As mentioned earlier (*32*), there is evidence that PEPC uses HCO_3^- as the one carbon substrate while PEPCK and PEPCTrP use CO_2 (*59*). These findings again link the latter two enzymes together. In addition, PEPCK and PEPCTrP resemble each other in (a) appearing to be metalloproteins, (b) their response to thiols, (c) catalyzing a CO_2–OAA exchange reaction, and (d) catalyzing pyruvate formation under certain conditions. The source of the pyruvate has usually been OAA in the PEPCK-catalyzed reaction and PEP in the PEPCTrP-catalyzed reaction, but this may mainly result from the experimental conditions employed. As shown in Fig. 11, it is possible to write a skeletal reaction mechanism which would apply to both PEPCK and PEPCTrP. Enzyme-bound enolpyruvate is the key intermediate in this mechanism. By using the appropriate phosphate derivative it is possible to explain all of the various reactions catalyzed by the two enzymes. In both cases, the $^{14}CO_2$–OAA exchange would involve reaction D (see Fig. 11) and would require the presence of the appropriate phosphate derivative. Pyruvate formation from PEP would involve reactions A, B, and C and pyruvate formation from OAA would require D and C (see Fig. 11). Thus far, PEPCTrP has not been shown to catalyze pyruvate formation from OAA, a reaction which should occur if the above mechanism is correct for this enzyme, but the experimental design, with OAA as

```
                              E–P–P–X ⇌ E+PP–X
                                 +
                          C // enolpyruvate ⇌ ketopyruvate
                  P–X    P–P–X
                 A /   B  /
E+PEP+X–P ⇌ E   ⇌   E
                   \      \
                  PEP  enolpyruvate
                          D \\ CO2
                             /P–P–X   E–PPX ⇌ E+PP–X
                            E        ⇌        +
                             \OAA             OAA
```

FIG. 11. Possible common mechanism for PEPCK- and PEPCTrP-catalyzed reactions: X = —OH for PEPCTrP, X = GMP for PEPCK from animals, and X = AMP for PEPCK from yeast and bacteria.

173. H. G. Wood, J. J. Davis, and H. Lochmüller, *JBC* **241**, 5692 (1966).

the substrate, has not favored the demonstration of pyruvate formation. PEPCK from yeast does catalyze the formation of pyruvate from PEP, but this reaction has not been reported for PEPCK from animal tissues. Again, this may reflect the particular experimental conditions employed.

The enolpyruvate mechanism shown in Fig. 11 can be adapted to the PEPC-catalyzed reaction as well, but the various observations mentioned above, which distinguish this enzyme from the other two, suggest that the basic mechanism may be different. Maruyama *et al.* (*32*) have suggested a concerted mechanism for PEPC, and Miller *et al.* (*36*) postulated that analogous concerted mechanisms might apply to PEPCK and pyruvate kinase as well. In each case it is suggested that PEP undergoes nucleophilic attack on its phosphoryl group, tautomerizes, and accepts a positively charged group on its C-3 atom. The proposed mechanism for PEPCK and PEPC is shown in Fig. 12 and includes the possible role of the divalent metal ion in the reaction. The PEPCTrP-catalyzed reaction could easily be written in a similar manner although analogous metal-binding data are not yet available, but it may be a mistake to try to force all three enzymes into the same mechanistic mold. Both the enolpyruvate and concerted mechanisms have attractive features but neither is supported by unequivocal or even very strong evidence.

At present, the sum of the available data would lead one to say that the three enzymes discussed in this chapter are probably only related in a formal sense through the apparent similarities of the reactions which they catalyze. This statement may have to be modified if the mechanisms

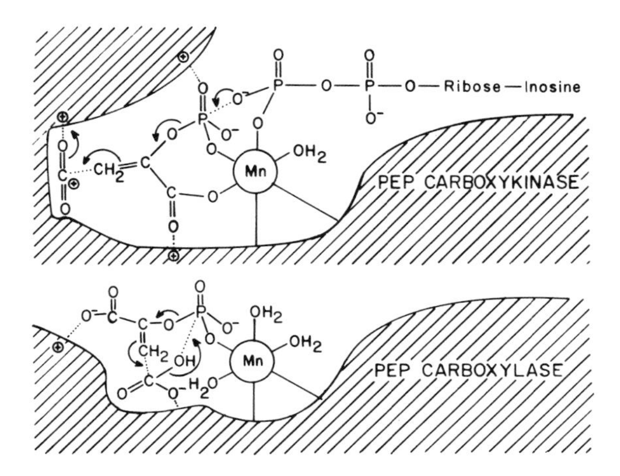

FIG. 12. Postulated concerted mechanisms for PEPCK- and PEPC-catalyzed reactions.

of the reactions catalyzed by PEPCK and PEPCTrP turn out to be very closely related.

ACKNOWLEDGMENTS

The authors express their sincere appreciation to M. Bayer, M. Benziman, J. J. B. Cannata, Y. B. Chiao, M. Haberland, R. W. Hanson, A. S. Mildvan, H. Miziorko, P. S. Noce, T. Nowak, R. E. Reeves, I. A. Rose, R. Singleton, J. M. Willard, and H. G. Wood for permission to cite some of their unpublished results. We also want to acknowledge the helpful suggestions offered by M. Haberland, P. S. Noce, R. Singleton, and H. G. Wood who read the manuscript and gave constructive criticism. Finally, we recognize the support of the U. S. Atomic Energy Commission through AEC Contract No. AT-1242 (MFU) and the Central College Research Council (HMK).

5

Ribulose-1,5-Diphosphate Carboxylase

MARVIN I. SIEGEL • MARCIA WISHNICK • M. DANIEL LANE

I. Introduction

The identification by Calvin and his colleagues of phosphoglyceric acid as the first stable radioactive product formed during brief exposure of algae to $^{14}CO_2$ (*1–3*) led ultimately to the discovery that the carboxylation of ribulose diphosphate is the initial step in the photosynthetic carbon cycle (*1–3*). Radioactivity appeared first in the carboxyl carbon of 3-phosphoglycerate (*1–3a*) and subsequently distributed equally be-

1. M. Calvin and A. A. Benson, *Science* **107,** 476 (1948).
2. A. A. Benson, J. A. Bassham, M. Calvin, T. C. Goodale, V. A. Haas, and W. Stepka, *JACS* **72,** 1710 (1950).
3. J. A. Bassham, A. A. Benson, and M. Calvin, *JBC* **185,** 781 (1950).
3a. E. W. Fager, J. L. Rosenberg, and H. Gaffron, *Federation Proc.* **9,** 535 (1950).

tween the 2 and 3 positions (*4, 5*), suggesting that a cyclic process was involved (*6, 7*). It was also observed that ribulose diphosphate itself became labeled early (*5, 6, 8–11*) and that illumination caused a decrease in its steady state concentration and a concomitant rise in the concentration of 3-phosphoglycerate (*6, 8–12*). Moreover, when the supply of CO_2 was cut off, the ribulose diphosphate level increased and the level of 3-phosphoglycerate decreased correspondingly (*6, 10, 11*). These results were consistent (*7, 12, 13*) with the cyclic precursor-product relationship shown in reaction (1):

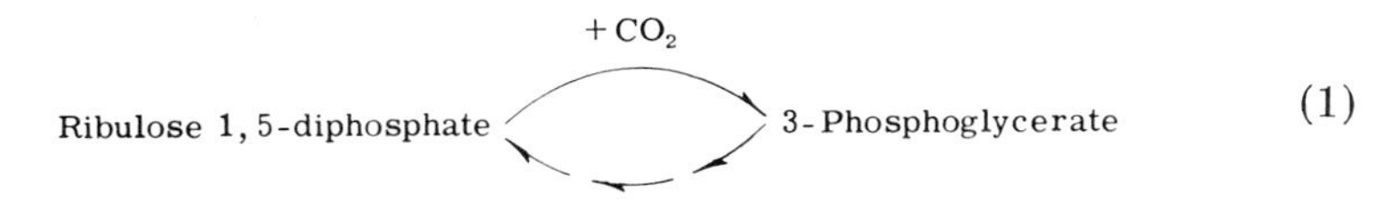

Careful analysis of the distribution of ^{14}C label in the carbon skeletons of the sugars and of changes in pool sizes of intermediates during light–dark transitions led to the formulation of the now well-known photosynthetic carbon reduction cycle (*1–8, 10–12, 14–22*).

Simultaneous reports from Calvin's (*7*) and Horecker's (*23*) laboratories revealed that cellfree preparations of both algae and spinach, respectively, carried out apparently similar pentose phosphate-dependent

4. A. A. Benson and M. Calvin, *Ann. Rev. Plant Physiol.* **1,** 25 (1950).
5. A. A. Benson, *Z. Elecktrochem.* **56,** 848 (1952).
6. J. A. Bassham, A. A. Benson, L. D. Kay, A. Z. Harris, A. T. Wilson, and M. Calvin, *JACS* **76,** 1760 (1954).
7. J. R. Quayle, R. C. Fuller, A. A. Benson, and M. Calvin, *JACS* **76,** 3610 (1954).
8. A. A. Benson, *JACS* **73,** 2971 (1951).
9. M. Calvin, *Harvey Lectures* **46,** 218 (1952).
10. M. Calvin, *JCS* p. 1895 (1956).
11. M. Calvin, *Science* **135,** 879 (1962).
12. M. Calvin and P. Massini, *Experientia* **8,** 445 (1952).
13. B. L. Horecker and A. H. Mehler, *Ann. Rev. Biochem.* **24,** 207 (1955).
14. M. Calvin, J. A. Bassham, A. A. Benson, V. H. Lynch, C. Ouellet, L. Schov, W. Stepka, and N. E. Tolbert, *Symp. Soc. Exptl. Biol.* **5,** 284 (1951).
15. A. A. Benson, J. A. Bassham, M. Calvin, A. G. Hall, H. E. Hirsch, S. Kawaguchi, V. Lynch, and N. E. Tolbert, *JBC* **196,** 703 (1952).
16. M. Calvin, *Federation Proc.* **13,** 697 (1954).
17. M. Calvin, J. R. Quayle, R. C. Fuller, J. Mayaudon, A. A. Benson, and J. A. Bassham, *Federation Proc.* **14,** 188 (1955).
18. A. T. Wilson and M. Calvin, *JACS* **77,** 5948 (1955).
19. J. Mayaudon, A. A. Benson, and M. Calvin, *BBA* **23,** 342 (1957).
20. B. R. Rabin, D. F. Shaw, N. G. Pon, J. M. Anderson, and M. Calvin, *JACS* **8,** 2528 (1958).
21. J. A. Bassham and M. R. Kirk, *BBA* **43,** 447 (1960).
22. J. A. Bassham and M. Kirk, *Atomlight* **45,** 1 (1965).
23. A. Weissbach, P. Z. Smyrniotis, and B. L. Horecker, *JACS* **76,** 3611 (1954).

carboxylations. Quayle *et al.* (*7*) found that a compound tentatively identified as ribulose diphosphate (*15*) was carboxylated by cellfree algal extracts in the presence of $H^{14}CO_3^-$ giving rise to carboxyl-labeled 3-phosphoglycerate. Furthermore, labeled intermediates between ribulose diphosphate and phosphoglycerate could not be detected. Weissbach *et al.* (*23*) found that spinach extracts known to contain ribose-5-phosphate isomerase and 5-phosphoribulokinase supported a ribose 5-phosphate-, ATP-, and Mg^{2+}-dependent $^{14}CO_2$ fixation which also led to the formation of carboxyl-labeled phosphoglycerate. It was demonstrated (*24, 25*) that under these conditions, but in the absence of added HCO_3^-, a pentose diphosphate accumulated almost quantitatively from ribose 5-phosphate. This enzymic product, which appeared to be the sole substrate for the carboxylase, was characterized and unequivocally identified as D-ribulose 1,5-diphosphate (*26*). It is now well established that ribulose diphosphate carboxylase is present in most, if not all, plants and photosynthetic microorganisms.

In the mid-1950's the isolation and partial characterization of ribulose diphosphate carboxylase from spinach leaves was reported from the laboratories of Horecker (*27*), Ochoa (*25*), and Racker (*28*). The carboxylase preparation obtained by Horecker's group (*27*) was shown to be nearly homogeneous and to be relatively large, the principal component having a sedimentation coefficient of about 17 S. Subsequently, it became apparent (*29*) that the carboxylase was identical to the principal 17–18 S component of leaf "fraction I protein" isolated earlier by Wildman and Bonner (*30*). It is interesting that this protein, i.e., ribulose diphosphate carboxylase, constitutes approximately 16% of the protein of the spinach leaf (*30, 31*) and is localized within the chloroplast (*32, 33*). Ribulose diphosphate carboxylase is widely distributed in plants and bacteria having been isolated from spinach (*27, 31, 33–35a*), spinach beet (*36,*

24. J. Hurwitz, A. Weissbach, B. L. Horecker, and P. Z. Smyrniotis, *JBC* **218,** 769 (1956).
25. W. B. Jakoby, D. Brummond, and S. Ochoa, *JBC* **218,** 811 (1956).
26. B. L. Horecker, J. Hurwitz, and A. Weissbach, *JBC* **218,** 785 (1956).
27. A. Weissbach, B. L. Horecker, and J. Hurwitz, *JBC* **218,** 795 (1956).
28. E. Racker, *ABB* **69,** 300 (1957).
29. R. W. Dorner, A. Kahn, and S. G. Wildman, *JBC* **229,** 945 (1957).
30. S. G. Wildman and J. Bonner, *ABB* **14,** 381 (1947).
31. J. M. Paulsen and M. D. Lane, *Biochemistry* **5,** 2350 (1966).
32. J. W. Lyttleton and P. O. Ts'o, *ABB* **73,** 120 (1958).
33. R. B. Park and N. G. Pon, *JMB* **3,** 1 (1961).
34. P. W. Trown, *Biochemistry* **4,** 908 (1965).
35. N. G. Pon, *ABB* **119,** 179 (1967).
35a. M. Wishnick and M. D. Lane, "Methods in Enzymology," Vol. 23, p. 570, 1971.

37), rice (*38*), tobacco (*39–41*), Chinese cabbage (*42, 43*), tomato (*44, 45*), oat (*46*), clover (*47*), navy bean (*48*), soybean (*49*), and wheat (*50*) among the higher plants; *Chlamydomonas reinhardi* (*43*), *Chlorella ellipsoidea* (*51, 52*), and *Euglena gracilis* (*53*) among the green algae; *Plectonema boryauum* (*43*) and *Anacystis nidulans* (*53*) among the blue-green algae; from the purple sulfur bacterium *Chromatium* strain D (*43, 53, 54*); from the green sulfur bacteria *Rhodopseudomonas palustris* (*53*) and *R. spheroides* (*52, 53*); from the purple nonsulfur bacterium *Rhodospirillum rubrum* (*52, 53*); and from the chemotrophic bacteria *Thiobacillus neapolitanas* (*55*), *T. thioparus* (*55*), and *T. novellus* (*56*), and *Hydrogenomonas eutropha* and *H. facilis* (*57, 58*).

Homogeneous preparations of ribulose diphosphate carboxylase have been obtained from several plant (*31, 35, 35a, 37, 38, 50, 59–62*) and

---

36. J. P. Thornber, S. M. Ridley, and J. L. Bailey, *BJ* **96,** 29c (1965).
37. S. M. Ridley, J. P. Thornber, and J. L. Bailey, *BBA* **140,** 62 (1967).
38. T. Akazawa, K. Saio, and N. Sugiyama, *BBRC* **20,** 114 (1965).
39. J. A. Bassham, *Advan. Enzymol.* **25,** 39 (1963).
40. G. van Noort and S. G. Wildman, *BBA* **90,** 309 (1964).
41. N. Kawashima, *BBRC* **38,** 119 (1970).
42. R. Haselkorn, H. Fernandez-Moran, F. J. Kieras, and E. F. J. van Bruggen, *Science* **150,** 1598 (1965).
43. F. J. Kieras and R. Haselkorn, *Plant Physiol.* **43,** 1264 (1968).
44. W. R. Andersen, G. F. Wildner, and R. S. Criddle, *ABB* **137,** 84 (1970).
45. R. S. Criddle, B. Dau, G. E. Kleinkopf, and R. C. Huffaker, *BBRC* **41,** 621 (1970).
46. M. W. Steer, B. E. S. Gunning, T. A. Graham, and D. J. Carr, *Planta* **79,** 254 (1968).
47. J. W. Lyttleton, *BJ* **64, 70** (1956).
48. W. H. Jyung, *Plant Physi*ol. **45, 7** (1970).
49. G. Bowes and W. L. Ogren, *Plant Physiol.* **45, 7** (1970).
50. T. Sugiyama and T. Akazawa, *J. Biochem.* (*Tokyo*) **62,** 474 (1967).
51. T. Sugiyama, C. Matsumoto, T. Akazawa, and S. Miyachi, *ABB* **129,** 597 (1969).
52. T. Akazawa, K. Sato, and T. Sugiyama, *ABB* **132,** 255 (1969).
53. L. E. Anderson, G. B. Price, and R. C. Fuller, *Science* **161,** 482 (1968).
54. J. Gibson and B. A. Hart, *Biochemistry* **8,** 2737 (1969).
55. R. D. MacElroy, E. J. Johnson, and M. K. Johnson, *ABB* **127,** 310 (1968).
56. M. I. Aleem and E. Huang, *BBRC* **20,** 515 (1965).
57. G. D. Kuehn and B. A. McFadden, *Biochemistry* **8,** 2403 (1969).
58. G. D. Kuehn and B. A. McFadden, *Biochemistry* **8,** 2394 (1969).
59. T. Sugiyama, N. Nakayama, M. Ogawa, T. Akazawa, and T. Oda, *ABB* **125,** 98 (1968).
60. T. Sugiyama, T. Akazawa, N. Nakayama, and Y. Tanaka, *ABB* **125,** 107 (1968).
61. N. Kawashima, *Plant Cell Physiol.* **10,** 31 (1969).
62. K. E. Moon and E. O. P. Thompson, *Australian J. Biol. Sci.* **22,** 463 (1969).

microbial sources (*58, 63*). The spinach enzyme, which is monodisperse by hydrodynamic criteria (sedimentation velocity and equilibrium centrifugation) and by polyacrylamide gel electrophoresis, is free of carbohydrate (*31*), phosphoriboisomerase (*31*), 5-phosphoribulokinase (*31*), 3-phosphoglycerokinase (*64*), NAD^+ or $NADP^+$-linked 3-phosphoglyceraldehyde dehydrogenase (*64*), and oxalacetate-ribulose diphosphate transcarboxylase (*64*). Thus, it appears that ribulose diphosphate carboxylase (or fraction I protein) is not a multienzyme complex containing several enzymes of the photosynthetic carbon cycle as had been suggested earlier (*39, 65–67*).

II. Molecular Properties

A. Native Enzyme

Ribulose diphosphate carboxylases from higher plants have complex quaternary structures as evidenced by their appearance in the electron microscope (*33, 37, 42, 46, 50, 68, 69*), high molecular weights (*31, 34–37, 43, 46*), numbers of substrate, cofactor, inhibitor, and antibody binding sites (*35a, 70–77*), and the fact that they are composed of nonidentical subunits (*35a, 44, 45, 50, 62, 73, 76, 78, 79*). An electron micro-

63. L. E. Anderson and R. C. Fuller, *JBC* **244,** 3105 (1969).
64. A. Rutner and M. D. Lane, unpublished observations (1968).
65. S. G. Wildman, G. van Noort, and W. Hudson, *Plant Physiol.* **36,** Suppl., xix (1961).
66. D. W. Kupke, *JBC* **237,** 3287 (1962).
67. L. Mendiola and T. Akazawa, *Biochemistry* **3,** 174 (1964).
68. A. A. Yasnikov, N. V. Volkova, B. N. Bernshtein, A. S. Okanenko, and T. A. Reingard, *Biochemistry* **32,** 191 (1967).
69. D. Branton and R. B. Park, *J. Ultrastruct. Res.* **19,** 283 (1967).
70. N. Kawashima and T. Mitake, *Agr. Biol. Chem.* (*Tokyo*) **33,** 539 (1969).
71. N. Kawashima and S. G. Wildman, *Proc. 11th Intern. Botan. Congr., Seattle, 1969* p. 5 (1969).
72. M. Wishnick and M. D. Lane, *JBC* **244,** 55 (1969).
73. A. C. Rutner, *BBRC* **39,** 923 (1970).
74. M. Wishnick, M. D. Lane, and M. C. Scrutton, *JBC* **245,** 4939 (1970).
75. M. Wishnick, *Federation Proc.* **29,** 531 (1970).
76. T. Sugiyama and T. Akazawa, *Biochemistry* **9,** 4499 (1970).
77. M. Wishnick, Ph.D. Thesis, New York University (1970).
78. T. Akazawa and T. Sugiyama, *Abstr. 11th Intern. Botan. Congr., Seattle, 1969* Vol. 2, p. 1 (1969).
79. A. C. Rutner and M. D. Lane, *BBRC* **28,** 531 (1967).

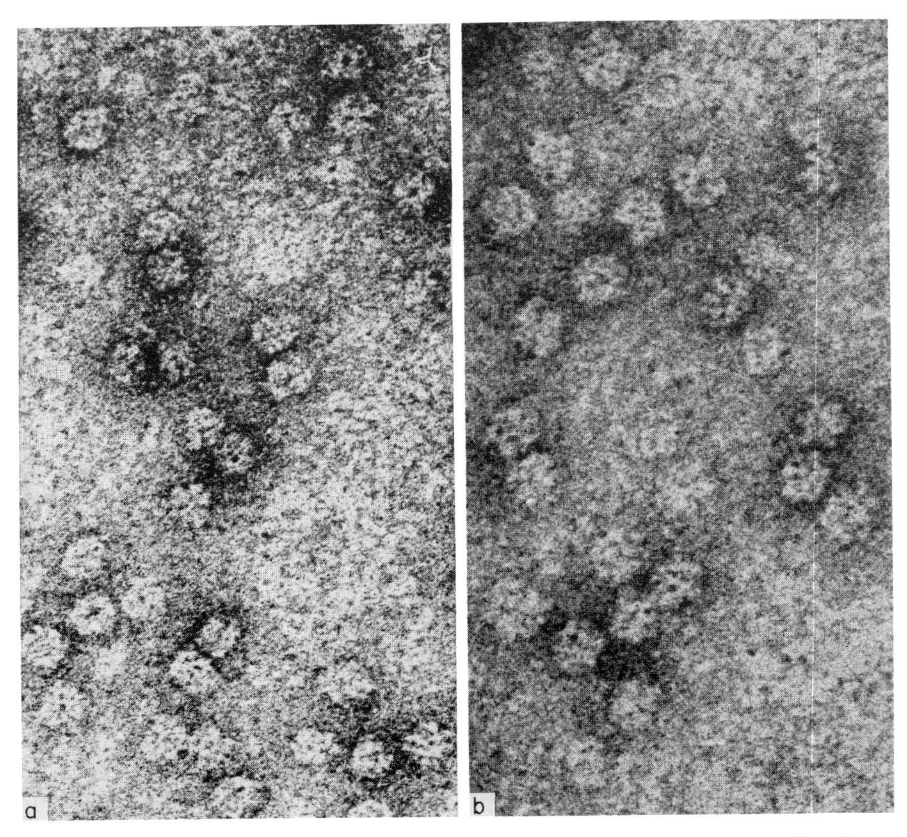

FIG. 1. Electron micrographs of spinach leaf ribulose-1,5-diphosphate carboxylase. Staining was carried out with uranyl acetate. Magnification: (a) 520,000×, (b) 658,000×. (Electron micrographs by Dr. A. K. Kleinschmidt.)

graph of the carboxylase from spinach leaves revealing its complex "raspberrylike" structure is shown in Fig. 1.

The hydrodynamic properties of the ribulose diphosphate carboxylases from various plants and microorganisms are summarized in Table I. The higher plant enzymes have similar sedimentation coefficients and molecular weights, the most reliable estimates of molecular weight determined by sedimentation equilibrium being approximately 560,000 (*31*, *34–37*, *43*). The frictional coefficient ratio (f/f_0) for the spinach carboxylase calculated from appropriate molecular parameters is 1.11 (*31*). This value is within the usual range for spherical protein molecules having average hydration characteristics, i.e., about 0.2 g of water per gram of protein, and is consistent with its shape as revealed by electron microscopy (Fig. 1).

While the algal carboxylases (*51–53*) appear to have complex high molecular weight structures similar to those of higher plants, the bacterial enzymes exhibit a high degree of variation and some appear to be considerably smaller. Molecular weights as low as 68,000–120,000 have been reported for ribulose diphosphate carboxylase from *Rhodospirillum rubrum* (*52, 63*), 260,000–360,000 for the *Rhodopseudomonas spheroides* carboxylase (*52, 53*), and as high as 550,000 for the enzyme from *Hydrogenomonas facilis* (*57*).

The amino acid compositions of several ribulose diphosphate carboxylases have been reported; these include the enzymes from spinach (*61, 79*), spinach beet (*37*), oat (*46*), tobacco (*61*), and two *Hydrogenomonas* species (*57*). Since these composition data have been cataloged in a recent review by Kawashima and Wildman (*80*) only the amino acid composition of the most widely studied of these enzymes, namely, the spinach leaf carboxylase, will be given here (Table II). These data reveal no unique structural features. The carboxylases from several higher plants (*50, 59, 64*) have approximately 95 half-cystine residues per molecule (560,000 daltons, Table I), all of which appear to be cysteinyl residues since titrations with *p*-CMB (*50, 59*) and Ellman's reagent (*64*) indicate a similar number of free sulfhydryl groups. In contrast, the carboxylases from *Hydrogenomonas eutropha* and *H. facilis* have 97 and 93 half-cystine residues per molecule, respectively, only 35–37 of which appear to be cysteinyl residues (*57*).

Homogeneous spinach ribulose diphosphate carboxylase contains tightly bound copper, i.e., copper which is not removed by gel filtration in the presence of 50 m*M* ethylenediaminetetracetate (EDTA) (*35a, 81*). The enzyme has a typical Cu(II) EPR spectrum ($g\perp \sim g_{\max} = 2.09$). Analyses by atomic absorption spectrophotometry, EPR, and neutron activation indicate the presence of 1 g-atom of copper per mole (560,000 g) of carboxylase. The bound copper appears to be present as Cu(II) and can be removed by incubation with cyanide in the presence of EDTA as has been observed with other cupro-proteins (*74*). Copper-free ribulose diphosphate carboxylase readily rebinds Cu^{2+} (1 g-atom/mole of enzyme), but other metal ions, with the possible exception of Ni^{2+}, do not exhibit high affinity for the copper binding site (*74*). Formation of a copper–dithiothreitol complex appears to facilitate the rebinding of copper as might be expected if the Cu^{2+} binding site involves vicinal sulfhydryl ligands. The role of the metal ion remains obscure.

Homogeneous spinach carboxylase has an absorbancy ratio, $A_{280\ \mathrm{nm}}/$

80. N. Kawashima and S. G. Wildman, *Ann. Rev. Plant Physiol.* **21**, 325 (1970).
81. M. Wishnick, M. D. Lane, M. C. Scrutton, and A. S. Mildvan, *JBC* **244**, 5761 (1969).

TABLE I
HYDRODYNAMIC PROPERTIES OF RIBULOSE DIPHOSPHATE CARBOXYLASES

Source	Sedimentation coefficient (s)	pH	Protein conc. (mg/ml)	Partial specific volume ($\bar{v}$)	Molecular weight $\times 10^{-5}$	Ref.
Spinach	21.0[a]	8.3	0.0954–0.954	0.74	5.57 ± 0.16[d]	(*31*)
	18.57[b]	7.4	1–10	0.73	5.15 ± 0.10[d]	(*34*)
	18.7[a]	7.73	3.61	0.73	5.59[d]	(*35*)
Spinach beet	18.29[a]	7.6, 7.9	0.3–2.0	0.737	5.47–5.83[d]	(*36*)
	18.30[a]	7.6, 7.9	0.3–2.0	0.744	5.61–5.85[d]	(*37*)
Chinese cabbage	17.0[b]	—	—	—	5.11[d]	(*43*)
Oat	18.2[a]	—	—	0.73	5.7[e]	(*46*)
Chlorella ellipsoidea	18.4[b]	—	—	—	—	(*51*)
	—	7.5	—	—	~4.7[f]	(*52*)
Anacystis nidulans	19.5[c]	8.5	—	—	~6.6[f]	(*53*)
Euglena gracilis	19[c]	8.5	8.5	—	~6.6[f]	(*53*)
Chromatium strain D	18[c]	8.5	33	—	~6.6[f]	(*53*)
Rhodopseudomonas spheroides	14.5[c]	8.5	—	—	~3.6[f]	(*53*)
	—	7.5	—	—	~2.6[f]	(*52*)
Rhodopseudomonas palustris	12[c]	8.5	—	—	~3.6[f]	(*53*)
Rhodospirillum rubrum	6.2[a]	7.5	—	—	~0.68[f]	(*52*)
	—	7.5	—	0.725	~1.20[f]	(*63*)
Thiobacillus thioparus	17[c]	—	—	—	—	(*55*)
Thiobacillus neapolitanus	17[c]	—	—	—	—	(*55*)
Hydrogenomonas eutropha	20	7.0	30	0.729	5.15[d]	(*57*)
Hydrogenomonus facilis	20	7.0	30	0.726	5.51[d]	(*57*)

[a] $\overset{\circ}{s}_{20,w}$.
[b] $s_{20,w}$.
[c] s_{rel}.
[d] M_{SE}.
[e] M_{SD}.
[f] Determined by gel filtration on Sephadex columns.

TABLE II

AMINO ACID COMPOSITION OF "NATIVE" SPINACH RIBULOSE DIPHOSPHATE CARBOXYLASE AND RIBULOSE DIPHOSPHATE CARBOXYLASE SUBUNITS FROM SEVERAL PLANT SOURCES

Amino acid	"Native" spinach carboxylase residues/mole (560,000 daltons[a])	"Heavy" chain (mole ratio relative to phenylalanine)					"Light" chain (mole ratio relative to phenylalanine)				
		Spinach	Spinach	Spinach	Spinach beet	Tobacco	Spinach	Spinach	Spinach	Spinach beet	Tobacco
Phenylalanine	207	1.00	1.00	1.00	1.00	1.00	1.00	1.00	1.00	1.00	1.00
Lysine	238	1.18	1.04	1.13	1.15	1.07	1.21	1.19	1.09	1.44	1.77
Histidine	139	0.67	0.66	0.77	0.62	0.62	0.45	0.53	0.54	0.21	0.42
Arginine	277	1.44	1.42	1.48	1.26	1.44	0.99	0.96	1.10	0.58	0.96
Aspartate	416	2.18	2.35	2.20	2.00	2.27	2.14	2.31	2.20	1.44	2.88
Threonine	282	1.75	1.63	1.73	1.41	1.45	1.19	1.29	1.23	0.77	1.60
Serine	140	0.81	0.87	0.70	0.93	0.86	0.77	0.78	0.64	0.91	1.51
Glutamate	452	2.20	2.42	2.28	2.30	2.52	2.26	2.77	2.71	2.17	4.14
Proline	257	1.13	1.17	1.12	1.06	1.13	1.56	1.56	1.56	1.43	1.64
Glycine	430	2.10	2.43	2.33	2.22	2.44	1.15	1.28	1.07	1.58	2.33
Alanine	391	2.16	2.21	2.18	2.13	2.23	0.86	0.98	0.79	1.09	1.63
Valine	332	1.70	1.52	1.67	1.57	1.49	1.17	1.15	1.12	1.40	1.36
Methionine	92	0.42	0.35	0.39	0.33	0.36	0.46	0.43	0.45	0.30	0.41
Isoleucine	173	0.88	0.69	0.92	0.88	0.76	0.56	0.47	0.66	0.70	0.84
Leucine	413	2.09	2.19	2.16	1.91	2.16	1.63	1.81	1.78	1.54	2.10
Tyrosine	220	0.92	0.91	0.94	0.76	0.85	1.57	1.43	1.74	0.99	1.92
Tryptophan	133	—	—	—	—	—	—	—	—	—	—
Half-cystine	95	—	—	—	—	—	—	—	—	—	—
Ref.	(*64*)	(*79*)	(*61*)	(*76*)	(*62*)	(*61*)	(*79*)	(*61*)	(*76*)	(*62*)	(*61*)

[a] See Table I and Paulsen and Lane (*31*).

$A_{260\text{ nm}}$, of 1.85 (*31*, *35a*); the relation between $A_{280\text{ nm}}$, and refractometrically determined protein concentration (c, mg/ml) being $c = 0.61$ ($A^{1\text{ cm}}_{280\text{ nm}}$) at pH 8.3 (*31*, *35a*). The interaction of ribulose 1,5-diphosphate with the enzyme leads to a characteristic ultraviolet difference spectrum (*82–84*). Two positive absorption peaks, one broad and one sharp with maxima at 268 and 288 nm, respectively, as well as a negative band with a minimum at 298 nm are observed. These peaks disappear upon addition of HCO_3^- and Mg^{2+} as a result of enzymic conversion of ribulose diphosphate to phosphoglycerate. The appearance of these bands in the difference spectrum probably reflects a change in the environment(s) of the aromatic amino acid residues resulting from a ribulose diphosphate-induced conformational change in the enzyme. It is interesting to note that 2-carboxy-D-ribitol 1,5-diphosphate, an analog of the proposed six-carbon intermediate in the reaction and potent inhibitor of the carboxylase (see Sections III,B,4 and III,C), elicits a similar difference spectrum to that of ribulose diphosphate upon interaction with the enzyme (*85*).

B. Subunits

Ribulose diphosphate carboxylase from spinach leaves was first shown by Rutner and Lane (*79*) to be composed of two types of subunits which differ markedly in size and amino acid composition. These observations have more recently been confirmed with the leaf carboxylases from tobacco (*61*, *85a*), spinach beet (*62*), barley (*45*), wheat (*50*), and tomato plants (*44*). Interestingly, the enzymes from two bacterial species (*Hydrogenomonas eutropha* and *H. facilis*) are composed of subunits of only one size (*57*). Ribulose diphosphate carboxylases from higher plants can be dissociated into subunits with sodium dodecyl sulfate (*35a*, *41*, *50*, *79*), 4 *M* urea (*78*), 8 *M* urea (*57*, *62*, *76*), or 6 *M* guanidine·HCl (*57*, *64*). After blocking sulfhydryl groups by aminoethylation (*41*, *57*, *79*), alkylation (*50*, *62*, *76*), or sulfitolysis (*45*), the subunits are resolved on Sephadex G-100 or G-200 columns (*41*, *45*, *50*, *62*, *79*) or by polyacrylamide gel electrophoresis (*57*, *73*, *76*). The molecular weights of the "heavy" and 'light" chains determined by the sodium dodecyl sulfate–polyacrylamide gel electrophoresis method were found to be 54,000–60,000 and 12,000–16,000, respectively (*73*, *76*).

82. P. W. Trown and B. R. Rabin, *Proc. Natl. Acad. Sci. U. S.* **52,** 88 (1964).
83. B. R. Rabin and P. W. Trown, *Nature* **202,** 1290 (1964).
84. M. Wishnick and M. D. Lane, unpublished observations (1969).
85. M. I. Siegel and M. D. Lane, unpublished observations (1970).
85a. N. Kawashima and S. G. Wildman, *BBRC* **41,** 1463 (1970).

From the large difference in amino acid composition of the resolved heavy and light polypeptide chains (see Table II) it is evident that the subunits are nonidentical. In the case of the spinach enzyme, for example, the heavy chain has a phenylalanine:tyrosine:glycine ratio of approximately 1:1:2, whereas the ratio for the light chain is approximately 1:1.6:1. Using the selective tritium labeling technique of Matsuo *et al.* (*86*, *87*) for C-terminal amino acids, native spinach carboxylase was found to contain both C-terminal valine and tyrosine (*76*). The carboxyl terminal amino acids of the resolved heavy and light polypeptide chains are valine and tyrosine, respectively (*76*). A tentative C-terminal sequence (shown below) of the light chain has recently been reported (*76*).

–Phe–Leu–(Tyr, Thr)–Tyr (CO_2H)

Comparison of the amino acid compositions of the heavy chains from the spinach, tobacco, and spinach beet enzymes reveals a remarkable similarity. On the other hand, the compositions of the light chains from the same sources differ markedly. The two polypeptide chains are synthesized in different subcellular compartments and at different rates which indicates that they are products of different structural genes, (*41*, *45*, *88*, *89*). The heavy chain appears to be synthesized in the chloroplast, whereas the light chain appears to be assembled on "cytoplasmic-type" ribosomes (*41*, *45*).

C. Composite Quaternary Structure of the Plant Enzyme

As indicated earlier, spinach leaf ribulose diphosphate carboxylase has a molecular weight of 560,000 (Section II,A) and is composed of nonidentical subunits of 56,000 and 14,000 daltons (Section II,B). The most reliable data available (*73*) on the mass ratio of the two types of subunits in the native enzyme, taken together with their molecular weights, indicate their presence in a 1:1 molar ratio in the enzyme complex. The simplest structure consistent with these data is a molecular complex composed of eight large subunits (56,000 daltons) and eight small subunits (14,000 daltons). The observation that each carboxylase molecule possesses eight antibody-binding sites is compatible with such

86. H. Matsuo, Y. Fujimoto, and T. Tatsuno, *BBRC* **22**, 69 (1966).
87. H. Matsuo, Y. Fujimoto, and T. Tatsuno, *Chem. & Pharm. Bull.* (*Tokyo*) **15**, 716 (1967).
88. R. M. Smillie, D. Graham, M. R. Dwyer, A. Grieve, and N. F. Tobin, *BBRC* **28**, 604 (1967).
89. J. J. Armstrong, S. J. Surzycki, B. Moll, and R. P. Levine, *Biochemistry* **10**, 692 (1971).

an octameric structure (*70, 71*). Furthermore, the fact that the enzyme has eight binding sites for ribulose 1,5-diphosphate (at low ionic strength, Section III,B,2) and eight sites for the proposed transition state analog, 2-carboxy-D-ribitol 1,5-diphosphate (see Sections III,B,4 and III,C), also supports a structure composed of eight elements. Preliminary investigations in Criddle's laboratory (*90*) indicate that the large subunit possesses the ribulose diphosphate binding site. These results suggest that the heavy and light polypeptide chains may be paired giving rise to a basic protomer, eight of which constitute the ribulose diphosphate carboxylase molecule. A structure of the type proposed is compatible with the complex degree of organization evident in electron micrographs of the native carboxylase (Fig. 1; *33, 37, 42, 46, 50, 68, 69*).

III. Catalytic Process

A. Reaction

Ribulose diphosphate carboxylase catalyzes the divalent cation-dependent carboxylation of D-ribulose 1,5-diphosphate to form 3-phosphoglycerate according to reaction (2) (*17, 25, 27, 35a, 91, 92*).

$$O{=}C^{*}{=}O \quad \begin{array}{ll} CH_2{-}O{-}PO_3^{2-} & (1) \\ C{=}O & (2) \\ H{-}C{-}OH & (3) \\ H{-}C{-}OH & (4) \\ CH_2{-}O{-}PO_3^{2-} & (5) \end{array} + HOH \xrightarrow{Mg^{2+}} \begin{array}{c} CH_2{-}O{-}PO_3^{2-} \\ {}^{-}O_2C^{*}{-}CH{-}OH \\ CO_2^{-} \\ H{-}C{-}OH \\ CH_2{-}O{-}PO_3^{2-} \end{array} + 2H^{+} \qquad (2)$$

D-Ribulose 1,5-diphosphate — D-(−)-3-Phosphoglycerate

Investigators in several laboratories using soluble carboxylase preparations identified D-(—)-3-phosphoglycerate as the sole carboxylation product and showed that the stoichiometry was 2 moles of phosphoglycerate formed per mole of bicarbonate fixed (*17, 25, 27, 91, 92*). It was subsequently established that carbon dioxide per se, rather than bicarbonate or carbonate, is the "active species" involved in the carboxylation mechanism (*93*). On the basis of the distribution of label in phosphoglycerate

90. R. S. Criddle, personal communication (1970).
91. A. Weissbach and B. L. Horecker, *Federation Proc.* **14**, 302 (1955).
92. E. Racker, *Nature* **175**, 249 (1955).
93. T. G. Cooper, D. Filmer, M. Wishnick, and M. D. Lane, *JBC* **244**, 1081 (1969).

formed enzymically from $H^{14}CO_3^-$ or $[1-^{14}C]$-ribulose diphosphate, it appeared that addition of CO_2 occurred at carbon-2 of ribulose diphosphate and that cleavage took place between carbons-2 and -3 (*27*). $H^{14}CO_3^-$ was incorporated exclusively into the carboxyl carbon of phosphoglycerate (*1–3, 27, 94*) while label from $[1-^{14}C]$-ribulose diphosphate appeared in the C-3 position (*27*). The site of cleavage remained in question, however, since these results did not indicate into which of the two identical product molecules ^{14}C label had been incorporated. Cleavage at the carbon-2–carbon-3 bond was unequivocally established by Müllhofer and Rose (*95*). ^2H from ^2H_2O, which ultimately becomes affixed to carbon-2 of phosphoglycerate (*94, 96*), appeared only in phosphoglycerate molecules derived from $[2-^{14}C]$-ribulose diphosphate and not in those derived from $[4-^{14}C]$-ribulose diphosphate. $[2-^{14}C]$-Phosphoglycerate with ^2H label in the 2 position was distinguished from $[2-^{14}C]$-phosphoglycerate with ^1H in the 2 position by conversion to $[2-^{14}C]$-glycolate and exploitation of the kinetic hydrogen isotope effect of the glycolate oxidase-catalyzed reaction.

The carboxylation reaction [reaction (2)] is for all practical purposes irreversible; attempts to demonstrate the reverse reaction under conditions favorable for its detection have been unsuccessful (*25, 27*). This is consistent with the high negative free energy change, $\Delta G'^{\circ} = -12.4$ kcal/mole, calculated for the forward reaction from empirical free energy of formation data (*39*). The fact that protons are generated in the reaction at pH 8, where carboxylase activity is optimal, must contribute substantially to the high apparent equilibrium constant for the reaction.

B. Kinetics and Specificity

The pH optimum (10 m*M* Mg^{2+}) for the reaction catalyzed by ribulose diphosphate carboxylase from spinach leaves was found by Weissbach *et al.* (*27*) to be between 7.8 and 8.0. pH optima in this region have been described for most of the carboxylases isolated from other plant and bacterial sources (*51, 52, 58, 63*). A shift in the pH optimum from 8.5 to 7.7 by increasing the Mg^{2+} concentration from 1.8 to 45.4 m*M* has been reported by Bassham and co-workers (*97*) and by Su-

94. V. H. Simon, H. D. Dorrer, and A. Trebst, *Z. Naturforsch.* **19b,** 734 (1964).
95. G. Müllhofer and I. A. Rose, *JBC* **240,** 1341 (1965).
96. F. Fiedler, G. Müllhofer, A. Trebst, and I. A. Rose, *European J. Biochem.* **1,** 395 (1967).
97. J. A. Bassham, P. Sharp, and I. Morris, *BBA* **153,** 898 (1968).

giyama *et al.* (*52, 98, 99*). Although the basis for this shift is not understood, it is possible that the $CO_2 \rightleftharpoons H_2CO_3 \rightleftharpoons HCO_3^- \rightleftharpoons CO_3^{2-}$ equilibrium is shifted toward CO_3^{2-} at elevated Mg^{2+} concentrations, thereby decreasing the concentration of CO_2, the "active species" in the reaction (see Section III,B,1). This may be compensated for by decreasing the pH of the reaction mixture.

An activation energy of 16.9 kcal/mole has been reported (*27*) for the reaction catalyzed by the spinach enzyme over a temperature range of 4–47°; this corresponds to a Q_{10} of approximately two. A recent report from Wildman's laboratory (*100*) suggests that the tobacco leaf carboxylase undergoes a slow reversible cold inactivation without a change in polymeric state.

The molecular activity of the homogeneous spinach ribulose diphosphate carboxylase, which is typical of most of the plant carboxylases, is 1340 moles of ribulose diphosphate carboxylated (2680 moles of 3-phosphoglycerate formed) per minute per mole of enzyme at pH 7.9 (*31, 35a*). If the eight ribulose diphosphate binding sites per enzyme molecule (*35a, 74*) represent active sites, the catalytic center activity (160 molecules per active center per minute) is low in comparison to other types of carboxylases, e.g., the biotin-dependent enzymes [see Table VII in Moss and Lane (*101*)]. The kinetic constants (K_m values) for ribu-

TABLE III
K_m VALUES FOR SUBSTRATES OF RIBULOSE-1,5-DIPHOSPHATE CARBOXYLASES FROM VARIOUS SOURCES

Enzyme source	Temp. (°C)	pH	K_m: Total "CO_2" (mM)	K_m: RuDP (mM)	K_m: Mg^{2+} (mM)	Ref.
Spinach leaves	30	7.8	22	0.12	1.1	(*31*)
	25	7.7	11	0.25	—	(*27*)
	25	7.8	20	—	—	(*28*)
Chinese cabbage	25	8.0	30	0.1	—	(*43*)
Rhodospirillum rubrum	25	8.1	12.5–65?	0.08	—	(*53, 63*)
Hydrogenomonas facilis	30	8.0	4.2	0.24	1.4	(*58*)
Hydrogenomonas eutropha	30	8.0	—	0.12	—	(*58*)

98. T. Sugiyama, N. Nakayama, and T. Akazawa, *ABB* **126,** 737 (1968).
99. T. Sugiyama, N. Nakayama, and T. Akazawa, *BBRC* **30,** 118 (1968).
100. N. Kawashima, S. Singh, and S. G. Wildman, *BBRC* **42,** 664 (1971).
101. J. Moss and M. D. Lane, *Advan. Enzymol.* **35,** 321 (1971).

lose diphosphate, Mg^{2+}, and total "CO_2" for ribulose diphosphate carboxylases from various sources are summarized in Table III.

1. "CO_2"

Numerous studies have revealed (Table III; *27, 28, 31, 35a, 43*) that ribulose diphosphate carboxylases from plants have remarkably high K_m values for total "CO_2," i.e., ranging from 11–30 mM at the optimal pH of about 8 and at saturating Mg^{2+} concentration. Under these conditions the principal species of "CO_2" in solution is bicarbonate.

A discrepancy exists between the high CO_2 fixation rates obtained with intact plant cells, or carefully prepared chloroplasts at low total "CO_2" concentrations, and the relatively low carboxylation rates catalyzed by isolated ribulose diphosphate carboxylase at similar total "CO_2" concentrations (*21, 28, 102–106*). A solution buffered at pH 7.9 and 30° in equilibrium with air ($pCO_2 = 0.23$ mm Hg) has a total "CO_2" concentration of 0.5 mM (7 μM $CO_2 + H_2CO_3$ and 0.5 mM HCO_3^-), conditions under which CO_2 fixation of intact plant cells is rapid. Even at saturating "CO_2" concentrations, but particularly at physiological "CO_2" concentrations, the CO_2 fixation capacity of the carboxylase does not account for the rates observed with intact leaves (*21, 28, 102–106*).

At physiological pH values "CO_2" exists both as dissolved CO_2 and its hydrated forms (HCO_3^- and CO_3^{2-}), the predominant species being HCO_3^- and CO_2. Enzymic carboxylation reactions are known in which either CO_2 or HCO_3^- is the active species [see Section IV,A in Moss and Lane (*101*) and Edsall (*107, 108*)]. An understanding of the carboxylation mechanism requires knowledge of the species which participates in the reaction. The active species of "CO_2" utilized in the reaction catalyzed by spinach leaf ribulose diphosphate carboxylase was determined (*93*) by taking advantage of the fact that the hydration–dehydration step in the equilibria shown in reaction (3) is comparatively slow and can be made

$$CO_2 + H_2O \underset{\text{slow}}{\rightleftharpoons} H_2CO_3 \underset{\text{fast}}{\rightleftharpoons} H^+ + HCO_3^- \underset{\text{fast}}{\rightleftharpoons} H^+ + CO_3^{2-} \quad (3)$$

rate-limiting relative to the rate of carboxylation of ribulose diphosphate

102. J. Coombs and C. W. Baldry, *Nature* **228,** 1349 (1970).
103. M. D. Hatch and C. R. Slack, *Ann. Rev. Plant Physiol.* **21,** 141 (1970).
104. E. Lutzko and M. Gibbs, *Plant Physiol.* **44,** 295 (1969).
105. A. Peterkofsky and E. Racker, *Plant Physiol.* **36,** 409 (1961).
106. C. R. Slack and M. D. Hatch, *BJ* **103,** 660 (1967).
107. J. T. Edsall, *Harvey Lectures* **62,** 191 (1966–1967).
108. J. T. Edsall, *NASA Spec. Publ.* **NASA SP-188,** 21 (1969).

(*101, 107–109*). The rate of incorporation of ^{14}C label into 3-phosphoglycerate was found to be faster when CO_2 was the initially labeled species than when HCO_3^- was labeled initially (*93*). ^{14}C incorporation rates were equalized by addition of carbonic anhydrase which causes essentially instantaneous equilibration of ^{14}C label between all species of "CO_2." Thus, molecular CO_2, rather than H_2CO_3, HCO_3^-, or CO_3^{2-} appears to be the active species in the carboxylation of ribulose diphosphate.

This finding offers a partial explanation for the exceptionally high K_m value for CO_2 in the ribulose diphosphate carboxylase-catalyzed reaction (*27, 28, 31, 35a, 43, 53, 63*). Correction of the Michaelis constant to the concentration of the active species of "CO_2" results in a K_m for CO_2 of approximately 0.45 m*M*. In the two other instances in which CO_2 has been shown (*110*) to be the active species in carboxylation reactions, high K_m values for "CO_2" were also observed (*111, 112*). In general, those carboxylations in which HCO_3^- has been shown to be the active species have been found to have considerably lower K_m values for "CO_2" (*110, 113, 114*).

In view of the finding that CO_2 is the active carboxylating species in the reaction, the report (*98*) suggesting a homotropic response to bicarbonate in the ribulose diphosphate carboxylase reaction must be viewed with caution. The actual substrate (CO_2 rather than HCO_3^-) concentrations used in these calculations may be in error by as much as 50-fold since it was assumed that all species of "CO_2," i.e., CO_2, H_2CO_3, and HCO_3^-, were active in the carboxylation reaction. In view of this, the earlier data (*115*) concerning the metal activation of ribulose diphosphate carboxylase and the role of CO_2 or a CO_2–Mg^{2+} complex in this activation should be reevaluated.

2. *Ribulose Diphosphate*

Ribulose diphosphate carboxylase from spinach appears to be absolutely specific for D-ribulose 1,5-diphosphate. None of the compounds listed in the accompanying tabulation serves as substrates for the enzyme as measured by standard $^{14}CO_2$ fixation assay procedures.

109. D. Filmer and T. G. Cooper, *J. Theoret. Biol.* **29**, 131 (1970).
110. T. G. Cooper, T. T. Tchen, H. G. Wood, and C. R. Benedict, *JBC* **243**, 3857 (1968).
111. H. Lochmuller, H. G. Wood, and J. J. Davis, *JBC* **241**, 5678 (1966).
112. T. G. Cooper and C. R. Benedict, *Plant Physiol.* **43**, 788 (1968).
113. Y. Kaziro, L. F. Hass, P. D. Boyer, and S. Ochoa, *JBC* **237**, 1460 (1962).
114. H. Maruyama, R. L. Easterday, H. C. Chang, and M. D. Lane, *JBC* **241**, 2405 (1966).
115. N. G. Pon, B. R. Rabin, and M. Calvin, *Biochem. Z.* **338**, 7 (1963).

Compound	Ref.	Compound	Ref.
Ribulose 1-phosphate	(*84*)	3-Phosphoglycerate	(*23*)
Ribulose 5-phosphate	(*27*)	Fructose 1,6-diphosphate	(*23*)
Ribose 5-phosphate	(*27*)	Pyruvate	(*23*)
Glucose 6-phosphate	(*27*)	Vinyl phosphate	(*116*)
Glucose 1,6-diphosphate	(*27*)	2-Phosphoglycolate	(*116*)
Fructose 1-phosphate	(*27*)	Hydroxypyruvate	(*116*)
Sedoheptulose 1,7-diphosphate	(*27*)	Acetol phosphate	(*117*)
Glycolaldehyde phosphate	(*23, 27*)	Glyceraldehyde 3-phosphate	(*84*)
Dihydroxyacetone phosphate	(*64*)		

The K_m values for ribulose diphosphate, summarized in Table III, are similar for the carboxylases from a variety of plant and bacterial sources. In the case of the spinach enzyme, ribulose diphosphate becomes inhibitory at concentrations greater than 0.7 mM (*31, 35a*).

Wishnick *et al.* (*35a, 74*) have found that the spinach carboxylase has eight tight binding sites for ribulose diphosphate at low ionic strength; the dissociation constant, K_D, for these sites is 1 μM. Divalent metal ion is not necessary for binding of ribulose diphosphate. At high ionic strength (0.25 M tris-Cl, pH 8) only four of these sites are available to ribulose diphosphate. It is significant that the carboxylase has eight tight binding sites for 2-carboxy-D-ribitol diphosphate, a potent inhibitor of the reaction and an analog of the proposed 6-carbon intermediate in the reaction [see Section III,C and Wishnick *et al.* (*74*)]. Tight binding of this compound is observed only in the presence of divalent metal ion (Mg^{2+} and Mn^{2+}) (*35a, 74*).

3. *Divalent Metal Ion*

The ribulose diphosphate carboxylase-catalyzed reaction exhibits an absolute requirement for divalent metal ion activation (*27, 31, 35a, 115*), which in the case of the spinach enzyme can be satisfied by Mg^{2+}, Mn^{2+}, Co^{2+}, or Ni^{2+} (*27, 31, 115*). Although Mg^{2+} is the most effective activator (K_m = 1.1 mM, Table III), the V_{max} for Mn^{2+} is 56% that with Mg^{2+} and the K_m is 0.04 mM (*31*). The formation of a dissociable enzyme–Mn^{2+} binary complex has been shown (*118*) by electron paramagnetic resonance and by measurements of the effect of this species on the longitudinal nuclear magnetic relaxation rate ($1/T_1$) of water protons. Furthermore, ter-

116. E. W. Fager, *BJ* **57,** 264 (1954).
117. W. R. Weimar, M. I. Siegel, and M. D. Lane, unpublished observations (1970).
118. M. Wishnick, M. D. Lane, and A. S. Mildvan, unpublished observations (1969).

nary complexes have been demonstrated by gel filtration (*72, 74*) between enzyme, Mg^{2+} or Mn^{2+}, and 2-carboxy-D-ribitol diphosphate or the ribulose 1,5-diphosphate–cyanide adduct; these compounds are thought to be analogs either of the transition state or of the proposed 6-carbon carboxylated intermediate in the reaction (see Sections III,B,4 and III,C).

4. *Inhibitors*

Spinach leaf ribulose diphosphate carboxylase is reversibly inhibited by cyanide at low concentration; 10^{-5} *M* and 10^{-4} *M* cyanide inhibit the carboxylation reaction 51 and 91%, respectively (*72*). Inhibition by cyanide ($K_i = 1.6 \times 10^{-5}$ *M*) is uncompetitive with respect to ribulose 1,5-diphosphate and is of mixed character with respect to Mg^{2+} and HCO_3^-. This and other kinetic evidence suggest that cyanide combines readily with enzyme–ribulose 1,5-diphosphate complex but not with enzyme.

The inference that enzyme–ribulose 1,5-diphosphate complex reacts with cyanide to form an inactive ternary complex is supported by binding studies with the use of the gel filtration technique. In short-term binding experiments, ribulose diphosphate uniformly labeled with ^{14}C is tightly bound to the carboxylase (1 mole of ribulose diphosphate per mole of enzyme) only in the presence of cyanide, and ^{14}C-cyanide is bound to the carboxylase (1 mole of cyanide per mole of enzyme) only in the presence of ribulose diphosphate. A ternary complex of enzyme, ribulose 1,5-diphosphate, and cyanide in a mole ratio of 1:1:1 is indicated. This ternary complex may result from cyanohydrin formation resulting from the reaction of cyanide with the 2-keto group of ribulose 1,5-diphosphate before or after interaction of this substrate with the enzyme. Although the presence of divalent metal ion is not essential for ^{14}C-ribulose diphosphate binding in the presence of cyanide or for ^{14}C-cyanide binding in the presence of ribulose diphosphate (*74, 75*), binding experiments by gel filtration carried out in the presence of Mg^{2+} or Mn^{2+} give rise to complexes containing the "ribulose diphosphate–cyanide adduct" and metal ion in equimolar concentration (*74, 75*). It has been suggested (*74*) that the "ribulose diphosphate adduct" is an analog of the proposed (*16, 20, 92*) 6-carbon carboxylated intermediate (2-carboxy-3-keto-D-ribitol 1,5-diphosphate, see Section III,C) in the carboxylase-catalyzed reaction.

2-Carboxy-D-ribitol 1,5-diphosphate, which may be a transition state analog (*11, 20, 74, 92*) and resembles the carboxylated intermediate proposed by Calvin (*16*), is a potent and essentially irreversible inhibitor of the carboxylase-catalyzed reaction (*72, 85*). Maximal inhibition requires preliminary incubation of the enzyme with the inhibitor and divalent metal ion, e.g., Mg^{2+} (*85*); nearly complete inhibition is obtained with a 10-fold stoichiometric excess of inhibitor over carboxylase in the preliminary incubation (*85*). While ribulose diphosphate added to the preliminary incubation mixture completely prevents inhibition by carboxyribitol diphosphate, once the inhibitor is bound, inhibition cannot be reversed by even a 10^4-fold excess of ribulose diphosphate over inhibitor in the assay reaction mixture (*85*). Tight binding of carboxyribitol diphosphate ($K_D < 1\ \mu M$) to the carboxylase occurs only in the presence of divalent metal ions, e.g., Mg^{2+} or Mn^{2+}. The complex formed contains equimolar amounts of inhibitor and divalent metal ion when isolated by gel filtration. The spinach carboxylase has eight tight binding sites for carboxyribitol diphosphate, the same number of binding sites it possesses for the substrate, ribulose diphosphate (*74*).

Ribulose diphosphate itself becomes inhibitory at concentrations exceeding 0.7 mM (*31, 35a*). Orthophosphate and $(NH_4)_2SO_4$ are competitive inhibitors with respect to ribulose diphosphate and have K_i values of 4.2 and 8.1 mM, respectively (*31, 35a*). 3-Phosphoglycerate is a noncompetitive inhibitor ($K_i = 8.3$ mM) with respect to ribulose diphosphate and a competitive inhibitor ($K_i = 9.5$ mM) with respect to HCO_3^- (*31, 35a*). Glyceraldehyde 3-phosphate shows a similar pattern of inhibition; it is competitive ($K_i = 22$ mM) with respect to HCO_3^- and noncompetitive ($K_i = 19$ mM) with respect to ribulose diphosphate (*31, 35a, 84*). The carboxylase is inhibited 63% by 0.5 mM arsenite in the presence of 0.4 mM British anti-Lewisite (BAL; 2,3-dimercaptopropanol). In the absence of BAL, the observed inhibition is 19%, while BAL alone causes no inhibition (*35a, 84*). At levels as high as 10 mM azide, hydroxylamine, and oxalate are without effect on the carboxylase reaction (*35a, 84*). Ribulose diphosphate carboxylase is inhibited by $HgCl_2$ and *p*-chloromercuribenzoate (*27, 59, 119*), as well as by iodoacetamide (*60, 82, 83, 120, 121*).

C. Mechanistic Considerations

The basic elements of the ribulose diphosphate carboxylase-catalyzed reaction which must be accounted for in the formulation of a mechanism are: (a) development of a nucleophilic center at carbon-2 of ri-

119. T. Sugiyama, T. Akazawa, and N. Nakayana, *ABB* **121**, 522 (1967).
120. B. R. Rabin and P. W. Trown, *Proc. Natl. Acad. Sci. U. S.* **51**, 497 (1964).
121. J. H. Argyroudi-Akoyunoglou and G. Akoyunoglou, *Nature* **213**, 287 (1967).

bulose diphosphate, (b) intramolecular dismutation in which oxidation at carbon-3 occurs at the expense of reduction at carbon-2, (c) addition of CO_2 at carbon-2, and (d) cleavage of the C-2–C-3 bond. Attempts to determine the sequence of these events have been hampered by the irreversible nature of the overall reaction and inability to detect partial reactions by isotopic exchange or other means. For example, the carboxylase does not catalyze isotope exchange between ribulose diphosphate and $[^{14}C]$-3-phosphoglycerate (*122*), ribulose diphosphate and $H^{14}CO_3^-$ (*122*), 3-phosphoglycerate and $H^{14}CO_3^-$ (*25*), ribulose diphosphate and ^3H_2O, 3-phosphoglycerate and ^3H_2O, or H_2O and $[3\text{-}^3H]$-ribulose diphosphate (*70, 123*).

Although water protons do not exchange with substrate, a water proton is incorporated at carbon-2 of the molecule of 3-phosphoglycerate arising from carbon atoms 1 and 2 of ribulose diphosphate and from CO_2 [*70, 95, 123*; see reaction (2)]. The reciprocal experiment with $[3\text{-}^3H]$-ribulose 1,5-diphosphate showed that ^3H was not incorporated into 3-phosphoglycerate but rather appeared in the solvent water; the release of ^3H was dependent upon the occurrence of the overall carboxylation reaction and occurred at approximately the same rate.

It has been established that molecular CO_2, rather than HCO_3^- or CO_3^{2-}, is the active carboxylating species in the carboxylase-catalyzed reaction (*24*). Carbon dioxide is a linear symmetrical molecule in which electrons are delocalized toward the two oxygen atoms rendering the carbon atom particularly susceptible to nucleophilic attack. Carbon-2 of ribulose diphosphate, to which CO_2 ultimately becomes affixed, is an electrophilic center [see (II)]; therefore, it would be an unfavorable site for carboxylation

$$\begin{array}{l} CH_2{-}O\text{Ⓟ} \\ \vert \\ C{=}O \\ \vert \\ R \end{array} \begin{array}{c} (1) \\ (2) \\ (3-5) \end{array} \longleftrightarrow \begin{array}{l} CH_2{-}O\text{Ⓟ} \\ \vert \\ {}^{\oplus}C{-}O^{\ominus} \\ \vert \\ R \end{array} \qquad (4)$$

(I) (II)

without prior rearrangement to a center with nucleophilic character. This chemical requirement would be satisfied by ene-diol formation as suggested by Calvin (*16*) or by isomerization to the corresponding 3-keto pentose as suggested by Racker (*92*) prior to carboxylation as illustrated in reactions (5a) and (5b). In either case the appropriate nucleophilic center at carbon-2 would be developed and carboxylation would

122. A. C. Rutner, M. Wishnick, and M. D. Lane, unpublished observations (1968).
123. J. Hurwitz, W. B. Jakoby, and B. L. Horecker, *BBA* **22,** 194 (1956).

be expected to lead to the six-carbon intermediate (VII) proposed by Calvin (*16*) or a closely related transition state.

$$\underset{\text{(III)}}{\begin{array}{c}\mathrm{CH_2{-}O\,ⓟ}\ (1)\\ \mathrm{C{=}O}\ (2)\\ \mathrm{H{-}C{-}OH}\ (3)\\ \mathrm{R}\ (4\text{-}5)\end{array}} \underset{(a)}{\rightleftharpoons} \underset{\text{(IV)}}{\begin{array}{c}\mathrm{CH_2{-}O\,ⓟ}\\ \mathrm{C{-}OH}\\ \|\\ \mathrm{C{-}O{-}H}\\ \mathrm{R}\end{array}} \underset{(b)}{\rightleftharpoons} \underset{\text{(V)}}{\begin{array}{c}\mathrm{CH_2{-}O\,ⓟ}\\ \mathrm{H{-}C{-}OH}\\ \mathrm{C{=}O}\\ \mathrm{R}\end{array}} \quad (5)$$

$$\text{(IV)},\ \text{(V)} \ \underset{(c)}{\overset{-\mathrm{H}^{\oplus}}{\rightleftharpoons}} \ \underset{\text{(VI)}}{\mathrm{O{=}C{=}O}\ \ \begin{array}{c}\mathrm{CH_2{-}O\,ⓟ}\\ {}^{\ominus}\mathrm{C{-}OH}\\ \mathrm{C{=}O}\\ \mathrm{R}\end{array}} \ \underset{(d)}{\rightleftharpoons} \ \underset{\text{(VII)}}{\begin{array}{c}\mathrm{CH_2{-}O\,ⓟ}\\ {}^{\ominus}\mathrm{O_2C{-}C{-}OH}\\ \mathrm{C{=}O}\\ \mathrm{R}\end{array}}$$

$$\text{(VII)} \xrightarrow[(e)]{+\mathrm{HOH}} \underset{\text{(VIII)}}{\begin{array}{c}\mathrm{CH_2{-}O\,ⓟ}\\ {}^{\ominus}\mathrm{O_2C{-}C{-}OH}\\ \mathrm{H}\end{array} + \begin{array}{c}\mathrm{CO_2^{\ominus}}\\ \mathrm{R}\end{array}} + \mathrm{H}^{\oplus}$$

While lack of proton exchange between the carbon-3 hydrogen of ribulose diphosphate and water does not rule out ene-diol or 3-keto intermediates (*96*), it does preclude their formation in the absence of the other components (CO_2 and Mg^{2+}) needed to complete the reaction. The occurrence of a large (4- to 6-fold) kinetic isotope effect in the carboxylation reaction with [3-^{3}H]-ribulose diphosphate as substrate indicates that the enolization step [reaction (5a)] is followed by a fast and irreversible step (*96*). Since there is discrimination (*70, 123*) against ^{3}H incorporation from ^3H_2O into 3-phosphoglycerate [reaction (5e)], this step must also be slow. The effect with ^2H_2O on the net reaction rate is smaller than predicted by ^3H_2O incorporation (*96*), thus it appears that reaction (5e) is not rate determining in the overall process. The large discrimination against [3-^{3}H]-ribulose diphosphate suggests (*96*) that enolization [reaction (5a)] is rate limiting in the overall reaction.

An appropriate chemical model for the mechanism proposed for the enzymic carboxylation has been described by Cramer and Proske (*124*).

124. F. Cramer and B. Proske, *Angew. Chem.* **68**, 120 (1956).

As illustrated in reaction (6), CO_2 adds to sodio benzoin methyl ether (IX) alpha to the carbonyl group forming a new C–C bond at the position which corresponds to carbon-2 of 3-ketoribitol 1,5-diphosphate (V).

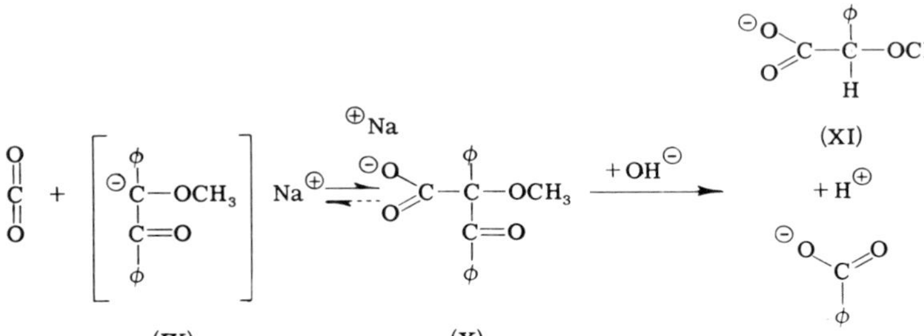

$$O{=}C{=}O + \left[{}^{\ominus}C(\phi)(OCH_3){-}C({=}O){-}\phi\right] Na^{\oplus} \rightleftharpoons {}^{\ominus}O_2C{-}C(\phi)(OCH_3){-}C({=}O){-}\phi \;\; ({}^{\oplus}Na) \xrightarrow{+OH^{\ominus}} {}^{\ominus}O_2C{-}CH(\phi){-}OCH_3 + H^{\oplus} + {}^{\ominus}O_2C{-}\phi \quad (6)$$

(IX) (X) (XI) (XII)

Upon treatment with base the carboxylation adduct (X) was cleaved between the "hydroxymethyl" and carbonyl carbon atoms giving rise to mandelic acid methyl ether (XI) and benzoate (XII).

Further support for the participation of a six-carbon carboxylated intermediate (VII) or transition state in the ribulose diphosphate carboxylase–catalyzed reaction is derived from studies with 2-carboxy-D-ribitol 1,5-diphosphate (XIII) and the ribulose 1,5-diphosphate–cyanide adduct (XIV), analogs of the proposed six-carbon intermediate. The chemical structures of these compounds are compared.

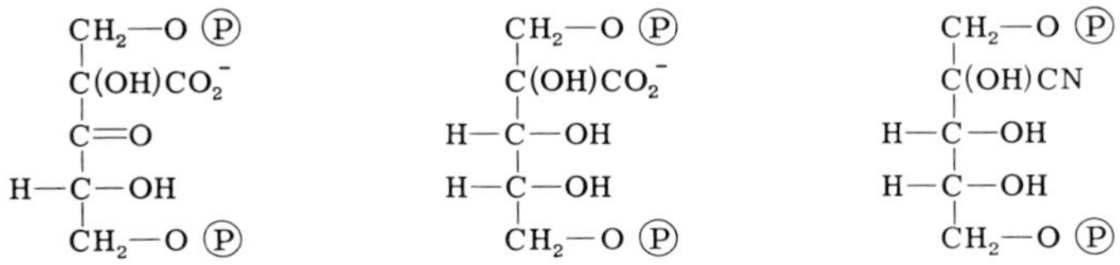

Proposed six-carbon carboxylated intermediate (VII)	Carboxy-D-ribitol 1,5-diphosphate (XIII)	D-Ribulose 1,5-diphosphate-cyanide adduct (XIV)
$CH_2{-}O\,Ⓟ$	$CH_2{-}O\,Ⓟ$	$CH_2{-}O\,Ⓟ$
$C(OH)CO_2^-$	$C(OH)CO_2^-$	$C(OH)CN$
$C{=}O$	$H{-}C{-}OH$	$H{-}C{-}OH$
$H{-}C{-}OH$	$H{-}C{-}OH$	$H{-}C{-}OH$
$CH_2{-}O\,Ⓟ$	$CH_2{-}O\,Ⓟ$	$CH_2{-}O\,Ⓟ$

(7)

Preliminary incubation of the carboxylase with a 10-fold stoichiometric excess (over enzyme) of carboxyribitol diphosphate (final concentration in the assay, $2 \times 10^{-7}\ M$) in the presence of Mg^{2+} causes 80–90% inhibition of the carboxylation reaction (*85*). Although $10^{-3}\ M$ ribulose diphosphate present during the preliminary incubation prevents this inhibitory effect, the inhibition is not reversed by $10^{-3}\ M$ ribulose diphosphate in the assay. Tight binding ($K_D < 1$ μM) of carboxyribitol diphosphate to ribulose diphosphate carboxylase is observed only in the presence of divalent metal ions, e.g., Mg^{2+} or Mn^{2+}. The complexes formed contain equimolar amounts of the inhibitor and divalent metal ion when isolated by gel filtration (*74*). Similarly, incubation of carboxylase with ribulose diphosphate and cyanide, particularly in the presence of Mg^{2+}, leads to the formation of catalytically inactive complexes which contain equimolar amounts of enzyme, ribulose diphosphate, cyanide, and Mg^{2+} (*72, 74*).

The requirement for Mg^{2+} in the tight binding of carboxyribitol 1,5-diphosphate to "native" ribulose 1,5-diphosphate carboxylase suggests the involvement of the divalent metal ion in the formation or stabilization of the six-carbon intermediate (VII), which may result from condensation of ribulose 1,5-diphosphate with CO_2. A stabilizing role for Me^{2+} is further suggested by the increase in the affinity of the native enzyme for carboxyribitol 1,5-diphosphate in the presence of Mg^{2+} which is observed in the gel filtration studies (*74*). Formation of a dissociable enzyme–Mn^{2+} complex has been shown by measurements of the effect of this species on the longitudinal nuclear magnetic relaxation rate ($1/T_1$) of water protons (*118*).

Wolfenden (*125*) has proposed that unusually tight binding of an inhibitor might be characteristic of transition state analogs on the basis of absolute reaction rate theory. Many apparent examples of this phenomenon have been described, e.g., 2-phosphoglycolate for triosephosphate isomerase (*125*), oxalate for lactate dehydrogenase (*126*) and pyruvate carboxylase (*127*), and pyrrole 2-carboxylate for proline racemase (*128*). It should be noted that the affinity of the ribulose 1,5-diphosphate–cyanide adduct (XIV) and of carboxyribitol 1,5-diphosphate (XIII) for ribulose 1,5-diphosphate carboxylase is at least an order of magnitude greater than the affinity of this enzyme for ribulose 1,5-disphosphate (*74*).

D. Light Activating Factor

A small chromophore-containing protein, light activating factor (LAF), found in the extracts of green plant cells and chloroplasts, has been

125. R. Wolfenden, *Nature* **223,** 704 (1969).

126. W. B. Novoa, A. D. Winer, A. J. Glaudy, and G. W. Schwert, *JBC* **234,** 1143 (1959).

127. A. S. Mildvan, M. C. Scrutton, and M. F. Utter, *JBC* **241,** 3488 (1966).

128. G. J. Cardenale and R. H. Abeles, *Biochemistry* **7,** 3970 (1968).

isolated in pure form from tomato leaves by Wildner and Criddle (*44, 129*). Illumination of purified ribulose diphosphate carboxylase in the presence of this factor results in a 3- to 4-fold activation of the carboxylase catalyzed reaction. The action spectrum for light activation of the ribulose diphosphate carboxylation reaction exhibits a maximum at 325 nm with shoulders at 550 and 750 nm. As might be anticipated LAF has an absorption spectrum with a maximum at 325 nm which coincides with that of the action spectrum (*129*). It is conceivable that the tightly bound copper (*35a, 81*) in the ribulose diphosphate carboxylase molecule (1 g-atom of Cu^{2+} per mole) may be involved in the transmission of the light signal from LAF to the carboxylase, since Cu^{2+} is a good electron acceptor being readily reduced to Cu^{+} (*129*).

The "factor" is heat stable and can be extensively purified by selective extraction of leaf tissue with ethanol; further fractionation with ammonium sulfate leads to preparations which appear homogeneous by acrylamide gel electrophoresis. The molecular weight of LAF, estimated from the dependence of its electrophoretic mobility on acrylamide gel concentration, is in the range of 4000–8000 (*129*).

The suggestion has been made (*129*) that LAF serves a regulatory function by modulating the rate of the photosynthetic carbon cycle through light-induced effects on the carboxylase. Although there have been numerous investigations (*6, 8–12, 130, 131*) with intact plant cells or chloroplasts which demonstrate the dependence of CO_2 fixation on dark–light transitions, evidence implicating LAF-mediated control of this fundamental process is lacking.

Acknowledgments

Investigations conducted in the authors' laboratory were supported by a research grant from the National Institutes of Health, USPHS (AM 14575).

129. G. F. Wildner and R. S. Criddle, *BBRC* **37**, 952 (1969).
130. O. Warburg, G. Krippahl, and W. Schroeder, *Z. Naturforsch.* **96**, 667 (1954).
131. J. F. Turner, C. C. Black, and M. Gibbs, *JBC* **237**, 577 (1962).

6

Ferredoxin-Linked Carboxylation Reactions

BOB B. BUCHANAN

I. Introduction

The carboxylation reactions described in this chapter were not known when the previous edition of this treatise was published 10 years ago. The reactions under discussion had been proposed prior to that date, but evidence for their existence was not obtained for many years.

An important historical development leading to final documentation of the reactions was the review of Lipmann (*1*) in 1946. Lipmann discussed, on thermodynamic grounds, the feasibility of a reductive carboxylation of what is now known as acetyl-CoA. He envisaged that the formation of pyruvate in this manner would offer to photosynthetic

1. F. Lipmann, *Advan. Enzymol.* **6**, 231 (1946).

or chemosynthetic cells an ideal mechanism of incorporating CO_2 into key metabolic intermediates.

Early attempts to achieve a net synthesis of pyruvate from CO_2 and a two-carbon unit, however, met with no success. Several investigators (*2, 3*) were able to show a rapid exchange between ^{14}C-bicarbonate and the carboxyl group of pyruvate in cellfree extracts of certain nonphotosynthetic bacteria but could not demonstrate an incorporation into pyruvate of acetyl-CoA [or a precursor, acetyl phosphate, which, in the presence of phosphotransacetylase (*4*), is converted to acetyl-CoA]. Not until 1959 was a synthesis of pyruvate from a two-carbon unit and CO_2 achieved in an enzymic reaction. In that year, Mortlock and Wolfe (*5*) reported that the nonphysiological reductant, sodium dithionite, could drive at a low rate the synthesis of pyruvate from acetyl phosphate and CO_2 in cellfree extracts of the fermentative bacterium *Clostridium butylicum*. But there was no indication that dithionite could be replaced by a physiological reductant or that the reaction was of physiological significance.

There was for years general agreement that pyruvate synthesis at best was of questionable importance in CO_2 assimilation. An appraisal in 1963 of the evidence for reversibility of the phosphoroclastic and other α-decarboxylation reactions led Wood and Stjernholm (*6*) to conclude ". . . the utilization of CO_2 by the reversal of α-decarboxylation is of little practical significance in the heterotrophic assimilation of CO_2, at least in the organisms so far studied."

Further progress on pyruvate synthesis was not made until ferredoxin had been isolated and certain key properties determined (*7, 8*). Accordingly, when Tagawa and Arnon (*8, 9*) reported that the redox potential of ferredoxin was 100 mV more reducing than that of nicotinamide-adenine dinucleotides, the possibility arose that the strong reducing power of ferredoxin might be used in pyruvate synthesis. However, there was at the time no experimental evidence that ferredoxin could participate directly as a reductant in this or any other enzymic reaction concerned with carbon assimilation. [The indirect participation of ferredoxin in CO_2 assimilation by way of nicotinamide adenine dinucleotides—with

2. J. Wilson, L. O. Krampitz, and C. N. Workman, *BJ* **42,** 598 (1948).
3. R. S. Wolfe and D. J. O'Kane, *JBC* **215,** 637 (1955).
4. E. R. Stadtman and H. A. Barker, *JBC* **184,** 769 (1950).
5. R. P. Mortlock and R. S. Wolfe, *JBC* **234,** 1657 (1959).
6. H. G. Wood and R. Stjernholm, *Bacteria* **3,** 41 (1963).
7. L. E. Mortenson, R. C. Valentine, and J. E. Carnahan, *BBRC* **7,** 448 (1962).
8. K. Tagawa and D. I. Arnon, *Nature* **195,** 537 (1962).
9. K. Tagawa and D. I. Arnon, *BBA* **153,** 602 (1968).

an attendant drop of 100 mV in reducing potential—was not in doubt because ferredoxins were by then known to act as electron carriers in the reduction of nicotinamide adenine dinucleotides by illuminated chloroplasts (*10, 11*) and by cellfree bacterial extracts (*12–14*).]

In 1964, Bachofen *et al.* (*15*) obtained the first evidence that reduced ferredoxin can drive the reductive synthesis of pyruvate from acetyl-CoA and CO_2 in cellfree extracts of the fermentative (heterotrophic) bacterium *Clostridium pasteurianum* [Eq. (1)]:

$$\text{Acetyl-CoA} + CO_2 + \text{ferredoxin}_{red} \rightarrow \text{pyruvate} + \text{CoA} + \text{ferredoxin}_{ox} \quad (1)$$

Named *pyruvate synthase*, Eq. (1) was the first demonstration of ferredoxin as a direct reductant in CO_2 assimilation. Pyruvate synthase achieves with reduced ferredoxin a reversal of the "phosphoroclastic" splitting of pyruvate [known in fermentative organisms (*16, 17*) since 1944 and found in 1962 to depend on ferredoxin (*7*)] and thus represents the reductive carboxylation that Lipmann had proposed in 1946.

Later work has demonstrated that reduced ferredoxin can promote the synthesis of α-keto acids other than pyruvate (Fig. 1). Each ferredoxin-linked carboxylation involves a reductive carboxylation of an acyl-CoA derivative to an α-keto acid [Eq. (2)] and appears to be catalyzed by a specific enzyme.

$$\text{Acyl-CoA} + \text{ferredoxin}_{red} + CO_2 \rightarrow \alpha\text{-keto acid} + \text{CoA} + \text{ferredoxin}_{ox} \quad (2)$$

Although discovered in fermentative bacteria (*15*) and later found to be widely distributed therein, ferredoxin-linked carboxylation is not peculiar to this group (Table I) (*17a*). Ferredoxin-linked carboxylation has been demonstrated also in numerous photosynthetic anaerobes; but, despite the presence of ferredoxin, it has not been found in photosynthetic organisms that evolve oxygen (i.e., algae and higher plants) or in aerobic nonphotosynthetic cells. A common feature of all organisms showing

10. M. Shin, K. Tagawa, and D. I. Arnon, *Biochem. Z.* **338,** 84 (1963).
11. M. Shin and D. I. Arnon, *JBC* **240,** 1405 (1965).
12. R. C. Valentine, W. J. Brill, and R. S. Wolfe, *Proc. Natl. Acad. Sci. U. S.* **48,** 1856 (1962).
13. B. B. Buchanan and R. Bachofen, *BBA* **162,** 607 (1968).
14. B. B. Buchanan and M. C. W. Evans, *BBA* **180,** 123 (1969).
15. R. Bachofen, B. B. Buchanan, and D. I. Arnon, *Proc. Natl. Acad. Sci. U. S.* **51,** 690 (1964).
16. H. J. Koepsell, M. J. Johnson, and J. S. Meek, *JBC* **154,** 535 (1944).
17. R. S. Wolfe and D. J. O'Kane, *JBC* **205,** 755 (1953).
17a. B B. Buchanan and D. I. Arnon, *Advan. Enzymol.* **33,** 120 (1970).

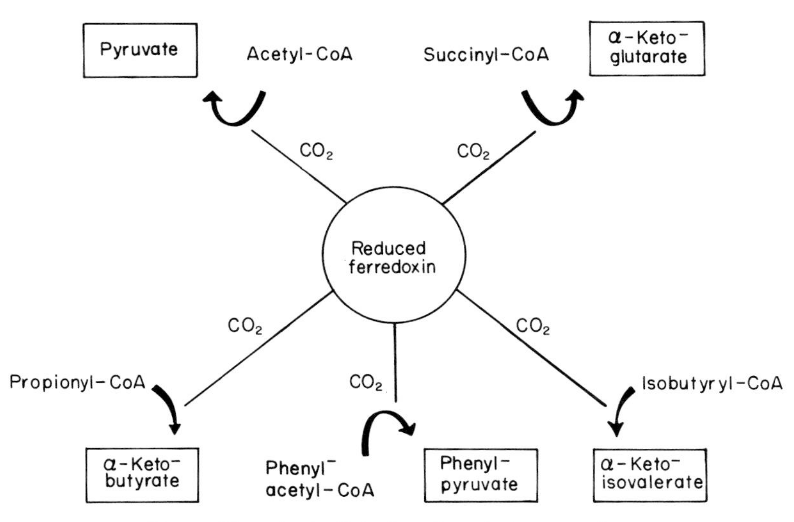

Fig. 1. Ferredoxin-linked carboxylation reactions.

ferredoxin-linked carboxylation is a requirement for an anaerobic environment for growth.

All α-keto acids shown to be synthesized via a ferredoxin-linked carboxylation are important intermediates in the biosynthesis of amino acids. Moreover, the formation of α-keto acids by ferredoxin-dependent carboxylations also constitutes a mechanism for CO_2 fixation. There is now considerable evidence that carboxylation reactions driven by reduced ferredoxin are essential for the operation of two new cyclic mechanisms for the assimilation of CO_2: the reductive carboxylic acid cycle of photo-

TABLE I
Occurrence of Ferredoxin and Ferredoxin-Linked Carboxylation Enzymes in Different Types of Organisms

Organism	Ferredoxin[a]	Ferredoxin-linked carboxylation enzymes
Bacteria		
Fermentative (anaerobic)	+	+
Photosynthetic (anaerobic)	+	+
Aerobic	+	−
Plants	+	−
Animals	−	−

[a] Ferredoxin is distributed ubiquitously in photosynthetic cells (bacterial and plant) but occurs only in certain groups of fermentative and aerobic bacteria (*17a*).

synthetic bacteria (*18, 19*) and the reductive monocarboxylic acid cycle of fermentative bacteria (*20*).

This chapter summarizes current evidence for the ferredoxin-linked carboxylation reactions and presents evidence pertaining to their physiological significance, with particular reference to the new carbon cycles.

II. Enzymes Catalyzing Ferredoxin-Linked Carboxylation Reactions

Each of the ferredoxin-linked carboxylation enzymes tested can oxidize (by a reversal of the carboxylation reaction) a specific α-keto acid to reduced ferredoxin, CO_2, and an acyl-CoA derivative—one carbon shorter than the original α-keto acid [see Eq. (2)]. Each preparation catalyzes, in addition, an exchange between CO_2 (added as bicarbonate) and the α-keto acid formed in the synthetic reaction.

The synthetic, oxidative, and exchange reactions appear to be separate activities catalyzed by a single protein which is specific for each α-keto acid. Thiamine pyrophosphate (TPP) is required as cofactor for each of the three activities.

In this section the properties and functions of the individual carboxylation enzymes will be described. Because of its importance historically, pyruvate synthase will be discussed first.

A. Pyruvate Synthase

Pyruvate synthase [Eq. (1)] is widely distributed in anaerobic organisms (Table II); it has been found in all photosynthetic bacteria (*18, 19, 21, 22*) examined and in several types of fermentative bacteria (*15, 23–28*).

18. M. C. W. Evans, B. B. Buchanan, and D. I. Arnon, *Proc. Natl. Acad. Sci. U. S.* **55,** 928 (1966).
19. B. B. Buchanan, M. C. W. Evans, and D. I. Arnon, *Arch. Mikrobiol.* **59,** 32 (1967).
20. R. K. Thauer, E. Rupprecht, and K. Jungermann, *FEBS Letters* **8,** 304 (1970).
21. B. B. Buchanan, R. Bachofen, and D. I. Arnon, *Proc. Natl. Acad. Sci. U. S.* **52,** 839 (1964).
22. M. C. W. Evans, *BBRC* **33,** 146 (1968).
23. I. G. Andrew and J. G. Morris, *BBA* **97,** 176 (1965).
24. S. Raeburn and J. C. Rabinowitz, *BBRC* **18,** 303 (1965).
25. J. R. Stern, *in* "Non-Heme Iron Proteins" (A. San Pietro, ed.), p. 199. Antioch Press, Yellow Springs, Ohio, 1965.
26. E. Heer and R. Bachofen, *Arch. Mikrobiol.* **54,** 1 (1966).
27. H. Bothe, *Progr. Photosyn. Res.* **3,** 1483 (1969).
28. H. H. Keller, manuscript in preparation.

TABLE II
DISTRIBUTION OF FERREDOXIN-LINKED CARBOXYLATION ENZYMES IN PHOTOSYNTHETIC AND FERMENTATIVE BACTERIA

Carboxylation reaction	Photosynthetic bacteria			Fermentative bacteria		
	Green sulfur	Purple sulfur	Purple nonsulfur	Clostridia	Rumen	Sulfur
Pyruvate synthase	*Chlorobium thiosulfatophilum* *Chloropseudomonas ethylicum*	*Chromatium*	*Rhodospirillum rubrum*	*Clostridium pasteurianum* *Clostridium kluyveri* *Clostridium acidi-urici* *Clostridium butyricum*	—	*Desulfovibrio desulfuricans*
α-Ketoglutarate synthase	*Chlorobium thiosulfatophilum* *Chloropseudomonas ethylicum*	—	*Rhodospirillum rubrum*	—	*Bacteroides ruminicola*	—
α-Ketobutyrate synthase	—	*Chromatium*	—	*Clostridium pasteurianum*	—	*Desulfovibrio desulfuricans*
α-Ketoisovalerate synthase	—	—		—	*Peptostreptococcus elsdenii*	—
Phenylpyruvate synthase	*Chlorobium thiosulfatophilum* *Chloropseudomonas ethylicum*	*Chromatium*	—	—	—	—

A procedure has been published for the assay and partial purification of pyruvate synthase from the photosynthetic bacteria *Chlorobium thiosulfatophilum* and *Chromatium* (*29*). The *C. thiosulfatophilum* enzyme, purified 8- to 10-fold, is free of ferredoxin, α-ketoglutarate synthase, and phosphotransacetylase; the enzyme is not stable at any stage of purification.

A similar (if not identical) enzyme, named *pyruvate ferredoxin oxidoreductase,* has been purified to homogeneity from the purine-fermenting bacterium *Clostridium acidi-urici* (*24, 30, 31*). This enzyme contains an iron-sulfur chromophore (*30, 31*) which may couple directly to ferredoxin in the synthesis (or breakdown) of pyruvate.

Molecular weight: Pyruvate synthase from *C. thiosulfatophilum* shows a molecular weight of 290,000 based on density gradient centrifugation and gel filtration (*32*). The enzyme can be separated from the closely associated α-ketoglutarate synthase in the centrifugation procedure (or, alternatively, by DEAE-Sephadex chromatography), thus providing direct evidence that the two enzymic activities result from different proteins.

Effect of TPP: Pyruvate synthase from *C. thiosulfatophilum* shows no response to added TPP unless treated specifically to remove bound TPP. Dialysis of the enzyme against an alkaline ammonium sulfate solution releases sufficient TPP so that a requirement for the cofactor can be demonstrated. Pyruvate synthase from another photosynthetic bacterium, *Rhodospirillum rubrum,* by contrast, shows a stimulation by TPP in initial cellfree extracts (*19*). The highly purified enzyme from the fermentative bacterium *C. acidi-urici* contains tightly bound TPP which could not be dissociated (*30, 31*).

pH optima: Pyruvate synthase from *C. thiosulfatophilum* shows a pH optimum at 6.2; at pH 5.9 and 6.5 the activity is about three-fourths that observed at pH 6.2 (*33*). *Rhodospirillum rubrum* pyruvate synthase has a pH optimum range of 7.0–7.5.

Reversibility: The pyruvate synthase preparations from both *C. acidi-urici* (*24, 30*) and *C. thiosulfatophilum* have been shown to catalyze (in the presence of CoA and ferredoxin) an oxidative decarboxylation of

29. B. B. Buchanan and D. I. Arnon, "Methods in Enzymology," Vol. 13, p. 170, 1969.

30. S. Raeburn and J. C. Rabinowitz, *in* "Non-Heme Iron Proteins" (A. San Pietro, ed.), p. 189. Antioch Press, Yellow Springs, Ohio, 1965.

31. K. Uyeda and J. C. Rabinowitz, *Federation Proc.* **26,** 561 (1967).

32. U. Gehring and D. I. Arnon, manuscript in preparation.

33. B. B. Buchanan, M. C. W. Evans, and D. I. Arnon, *in* "Non-Heme Iron Proteins" (A. San Pietro, ed.), p. 175. Antioch Press, Yellow Springs, Ohio, 1965.

pyruvate to acetyl-CoA, CO_2, and reduced ferredoxin. Ferredoxin can be replaced by FAD or by the nonphysiological dye triphenyltetrazolium in pyruvate oxidation by the *C. acidi-urici* enzyme (*30*). With this preparation DPN was inactive as acceptor.

Effect of ferredoxins from different sources: Native ferredoxins are no more effective with pyruvate synthase preparations from *C. thiosulfatophilum* and *Chromatium* than are the ferredoxins obtained from other photosynthetic bacteria or from certain fermentative bacteria. Spinach leaf ferredoxin is appreciably less effective (*33*).

CO_2-pyruvate exchange: All preparations of pyruvate synthase tested catalyze an exchange between CO_2 and the carboxyl group of pyruvate (*3, 24, 29, 30, 33*). Like pyruvate synthesis, the CO_2-pyruvate exchange reaction requires CoA, but at a lower level. The exchange reaction also requires TPP. With the partly purified pyruvate synthase from *C. thiosulfatophilum,* pyruvate synthesis and CO_2-pyruvate exchange have different pH optima: The optimal pH for pyruvate synthesis is 6.2, whereas CO_2-pyruvate exchange is most rapid at pH 7.2 (*33*).

Physiological role: There is substantial evidence with whole cells that pyruvate synthase is important in the assimilation of exogenous acetate and carbon dioxide in photosynthetic (*34, 35*) and fermentative (*36*) bacteria. [Acetate is activated to the coenzyme A thioester in an ATP-dependent reaction prior to assimilation (*29*).] The main products formed from acetate and carbon dioxide assimilated via pyruvate synthase are amino acids, particularly alanine which is formed directly by transamination. Other amino acids such as aspartate may be derived from pyruvate following additional carboxylation steps (*18*).

As shown below, the ferredoxin-dependent synthesis of alanine from acetate and carbon dioxide by way of pyruvate is independent of phosphoenolpyruvate—the compound previously considered the main precursor of pyruvate in both photosynthetic and nonphotosynthetic cells (*37*).

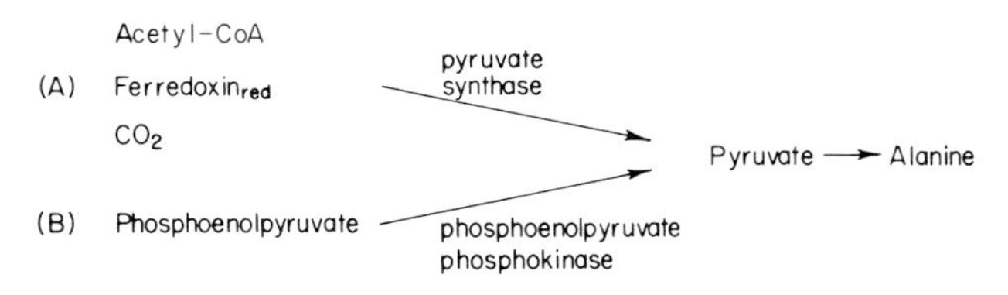

34. C. Cutinelli, G. Ehrensvärd, L. Reio, E. Saluste, and R. Stjernholm, *Arkiv Kemi* 3, 315 (1951).

35. D. S. Hoare and J. Gibson, *BJ* **91,** 546 (1964).

36. N. Tomlinson, *JBC* **209,** 597 (1954).

37. A. Meister, "Biochemistry of the Amino Acids," 2nd ed., Vol. 2. Academic Press, New York, 1965.

In addition to its role in the synthesis of alanine and other amino acids, pyruvate synthase is a key reaction in the reductive carboxylic acid cycle of photosynthetic bacteria and the reductive monocarboxylic acid cycle of fermentative bacteria (see below).

Apart from its role in carbon assimilation, pyruvate synthase has a different but important function in certain fermentative (nonphotosynthetic) bacteria such as *C. acidi-urici* and *C. pasteurianum*. In these organisms, the enzyme operates mainly in reverse for the oxidation (rather than the synthesis) of pyruvate derived from fermentation.

B. α-Ketoglutarate Synthase

α-Ketoglutarate synthase [Eq. (3)] was discovered in green photosyn-

$$\text{Succinyl-CoA} + CO_2 + \text{ferredoxin}_{red} \rightarrow \alpha\text{-ketoglutarate} + \text{CoA} + \text{ferredoxin}_{ox} \quad (3)$$

thetic bacteria (*38*) and for several years its occurrence in nonphotosynthetic bacteria was not known. The important finding by Allison and Robinson (*39*) of α-ketoglutarate synthase in fermentative bacteria of the rumen (Table II) shows that this enzyme, like pyruvate synthase, occurs also in nonphotosynthetic anaerobes.

Procedures for the assay (*29*) and purification (*32*) of α-ketoglutarate synthase from the photosynthetic bacterium *C. thiosulfatophilum* have been described. α-Ketoglutarate synthase is exceedingly oxygen sensitive, but it can be stabilized under anaerobic conditions by a combination of dimercaptopropanol and $FeSO_4$. The enzyme, purified 120-fold, was found to be closely associated with pyruvate synthase; the two could be separated by either density gradient centrifugation or DEAE-Sephadex chromatography, thus providing conclusive evidence that the two activities result from separate enzymes.

Molecular weight: α-Ketoglutarate synthase from *C. thiosulfatophilum* shows a molecular weight of 240,000 based on density gradient centrifugation and gel filtration (*32*).

Effect of TPP: α-Ketoglutarate synthase from *C. thiosulfatophilum* shows no response to added TPP. By contrast, the enzyme from another photosynthetic bacterium, *R. rubrum*, shows a requirement for TPP without appreciable purification (*19*, *29*).

pH optima: α-Ketoglutarate synthase from *C. thiosulfatophilum* shows a pH optimum between 7.5 and 7.8 with a relatively sharp drop on the acid side (*32*); for *R. rubrum*, the pH optimum is 7.5 (*29*).

38. B. B. Buchanan and M. C. W. Evans, *Proc. Natl. Acad. Sci. U. S.* **54,** 1212 (1965).
39. M. J. Allison and I. M. Robinson, *J. Bacteriol.* **104,** 63 (1970).

Reversibility: α-Ketoglutarate synthase purified 120-fold from *C. thiosulfatophilum* catalyzes, in addition to a reductive synthesis of α-ketoglutarate, an oxidative decarboxylation of α-ketoglutarate with ferredoxin as acceptor. The enzyme can use triphenyltetrazolium dye, FMN, or FAD in lieu of ferredoxin, but it cannot use NAD or NADP.

Effect of ferredoxins from different sources: Native ferredoxin is no more effective in the synthesis (or the oxidation) of α-ketoglutarate by α-ketoglutarate synthase from *C. thiosulfatophilum* than are ferredoxins from other photosynthetic bacteria. The ferredoxins from spinach leaves, blue–green algae, and fermentative bacteria are much less effective (*32*).

CO_2-α-ketoglutarate exchange: Purified *C. thiosulfatophilum* α-ketoglutarate synthase catalyzes an exchange between CO_2 and α-ketoglutarate analogous to the exchange between CO_2 and pyruvate catalyzed by pyruvate synthase (*32*). The exchange reaction, like that of pyruvate, requires a low level of CoA (1×10^{-5} *M*). The CoA requirement for α-ketoglutarate-CO_2 exchange disappears when the reaction is carried out in the presence of reduced ferredoxin or the nonphysiological substitutes, reduced methyl viologen and sodium dithionite. The mechanism of action of either CoA or reduced ferredoxin in the exchange reaction is not known.

With the α-ketoglutarate synthase purified from *C. thiosulfatophilum*, α-ketoglutarate synthesis and CO_2-α-ketoglutarate exchange have different pH optima: The optimal pH for α-ketoglutarate synthesis is 6.2, whereas CO_2-α-ketoglutarate exchange shows an optimum between pH 7.2 and 7.8 (*32*).

Physiological role: α-Ketoglutarate synthase provides a new mechanism for assimilation of externally supplied succinate and CO_2 by both photosynthetic (*40*) and nonphotosynthetic (*39*) anaerobes. Precursors of succinate such as propionate may also be assimilated by this mechanism. [Like acetate, succinate is activated to the coenzyme A thioester in an ATP-dependent reaction prior to assimilation (*29*).] The principal products formed from succinate and carbon dioxide are amino acids—especially glutamate which is derived directly from α-ketoglutarate by transamination.

The biosynthesis of glutamate by the α-ketoglutarate synthase reaction (A) is independent of isocitrate (B) derived from the citric acid cycle (*41*, *42*).

40. K. Shigesada, K. Hidaka, H. Katsuki, and S. Tanaka, *BBA* **112,** 182 (1966).
41. H. A. Krebs, *Prix Nobel* (1953).
42. H. A. Krebs and J. M. Lowenstein, *Metab. Pathways* **1,** 129 (1960).

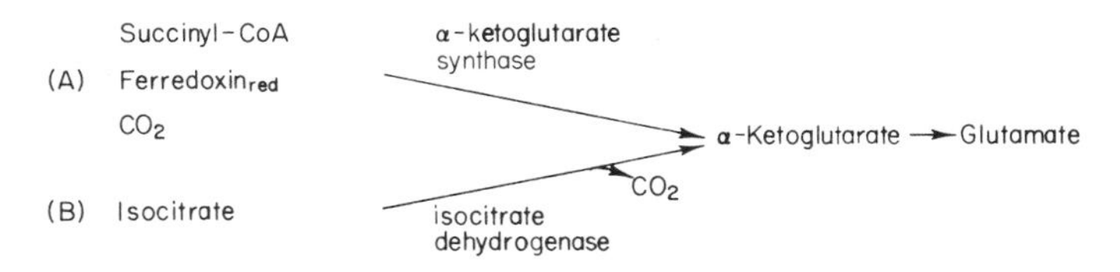

In addition to its role in biosynthesis of glutamate, α-ketoglutarate synthase is a key enzyme in the reductive carboxylic acid cycle of bacterial photosynthesis.

C. α-Ketobutyrate Synthase

α-Ketobutyrate synthase catalyzes a ferredoxin-dependent reductive carboxylation of propionyl-CoA to α-ketobutyrate [Eq. (4)].

$$\text{Propionyl-CoA} + \text{ferredoxin}_{red} + CO_2 \rightarrow \alpha\text{-ketobutyrate} + \text{CoA} + \text{ferredoxin}_{ox} \quad (4)$$

The enzyme is present in preparations from both photosynthetic and fermentative bacteria (*43*) (Table II) but has not been purified.

α-Ketobutyrate synthase appears to be protein-separate from pyruvate and α-ketoglutarate synthases. This conclusion is based primarily on the observation (*43*) that a preparation from *C. thiosulfatophilum*, showing high pyruvate and α-ketoglutarate synthase activities, had no α-ketobutyrate synthase activity.

Effect of ferredoxins from different sources: The α-ketobutyrate synthase preparation from the photosynthetic bacterium *Chromatium* was active with all ferredoxins tested (native ferredoxin as well as ferredoxins from spinach leaves, fermentative, and other photosynthetic bacteria) with no large differences in activity.

Physiological role: In *Chromatium* cells, α-ketobutyrate synthase is constitutive; cellfree preparations show the same enzyme content regardless of the carbon source used for growth.

α-Ketobutyrate synthase appears to be important in a new pathway for the biosynthesis of α-aminobutyrate and isoleucine (*43*). The new pathway is independent of threonine and threonine deaminase—the key enzyme in the mechanism previously considered to account for the biosynthesis of α-aminobutyrate and isoleucine (*37*). In the new pathway (A), propionate and carbon dioxide replace threonine (B) as carbon source for formation of the α-ketobutyrate required in the synthesis of α-aminobutyrate or isoleucine.

43. B. B. Buchanan, *JBC* **244**, 4218 (1969).

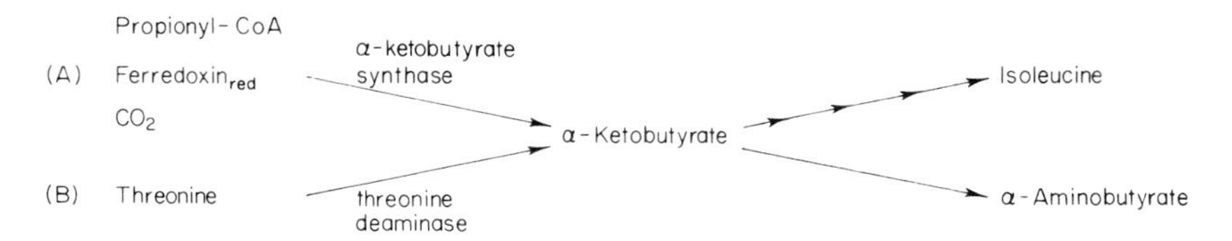

Aside from occurrence in several organisms of the key enzyme, α-ketobutyrate synthase (Table II), evidence for the new path of isoleucine synthesis has been obtained mainly for the photosynthetic bacterium *Chromatium*. Growth experiments (*43*) showed that *Chromatium* cells assimilated ^{14}C-propionate into a variety of nonvolatile compounds, particularly amino acids. Of the compounds labeled, isoleucine was the most prominent and accounted for 16% of the ^{14}C-propionate assimilated (Fig. 2). Threonine, a precursor of isoleucine in the known pathway, showed less than one-fifth the activity of isoleucine; and asparate, a known precursor of threonine in other organisms (*37*), was only slightly labeled. The high labeling in isoleucine, relative to threonine, supports the conclusion that in *Chromatium* a significant part of the α-ketobutyrate needed for the synthesis of isoleucine (and α-aminobutyrate) is formed via α-ketobutyrate synthase.

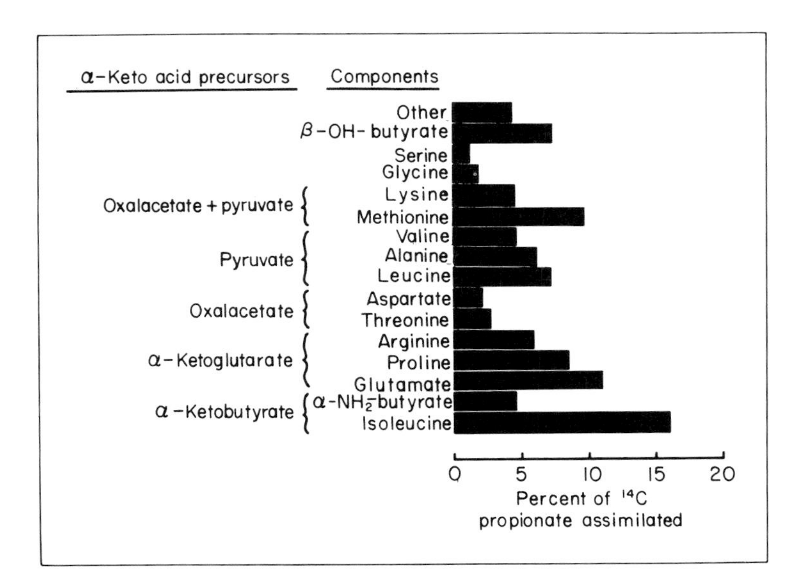

FIG. 2. Relative activities of ^{14}C components of *Chromatium* cells grown with ^{14}C-propionate.

D. α-Ketoisovalerate Synthase

A ferredoxin-dependent reductive carboxylation of isobutyryl coenzyme A to α-ketoisovalerate [herein referred to as the α-ketoisovalerate synthase reaction, Eq. (5)] was found by Allison and Peel (*44, 44a*) in cellfree extracts from the fermentative bacterium *Peptostreptococcus elsdenii*. The enzyme has not been purified.

$$\text{Isobutyryl-CoA} + \text{ferredoxin}_{\text{red}} + CO_2 \rightarrow \alpha\text{-ketoisovalerate} + \text{CoA} + \text{ferredoxin}_{\text{ox}} \quad (5)$$

Physiological role: α-Ketoisovalerate is known to be converted by transamination to valine (*37*). Accordingly, α-ketoisovalerate synthase appears to be important in a new path of valine biosynthesis (*44a*).

As with other ferredoxin-linked carboxylations which lead to amino acids, the α-ketoisovalerate synthase mechanism (A) for valine biosynthesis is independent of the mechanism previously established (B). The carbon for α-ketoisovalerate is derived from isobutyrate and carbon dioxide in the new mechanism rather than from α,β-dihydroxyisovalerate (which is produced from α-acetolactate) in the previously established path (*37*).

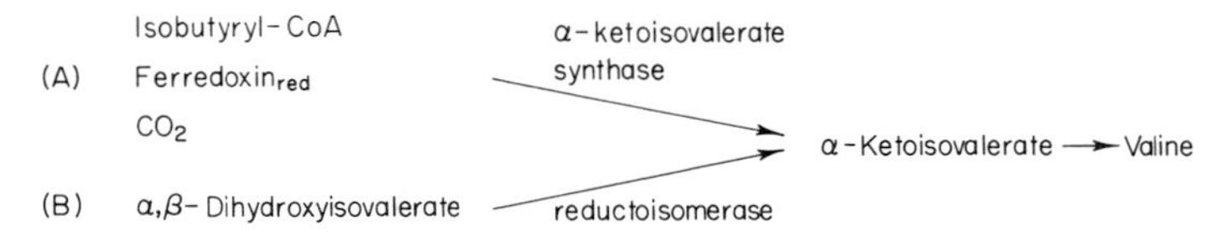

α-Ketoisovalerate synthase has so far been found only in preparations from the rumen bacterium, *P. elsdenii*. Whether the ferredoxin-dependent route of valine biosynthesis is used by photosynthetic or other fermentative bacteria is an open question.

E. Phenylpyruvate Synthase

The phenylpyruvate synthase reaction was first proposed by Allison and Robinson (*45*) on the basis of experiments with whole cells of *Chromatium* and *R. rubrum*. These authors described ^{14}C labeling data which were consistent with a synthesis of phenylalanine via phenylpyruvate from a condensation of phenylacetate and carbon dioxide.

44. M. J. Allison and J. L. Peel, *Bacteriol. Proc.* **68,** 142 (1968).
44a. M. J. Allison and J. L. Peel, *BJ* **121,** 431 (1971).
45. M. J. Allison and I. M. Robinson, *J. Bacteriol.* **93,** 1269 (1967).

Direct evidence for phenylpyruvate synthase in a cellfree system and its dependence on ferredoxin, however, has been obtained only recently. Gehring and Arnon (*46*) have obtained evidence for a ferredoxin-dependent reductive carboxylation of phenylacetyl coenzyme A to phenylpyruvate [Eq. (6)].

$$\text{Phenylacetyl-CoA} + \text{ferredoxin}_{red} + CO_2 \rightarrow \text{phenylpyruvate} + \text{CoA} + \text{ferredoxin}_{ox} \quad (6)$$

Phenylpyruvate synthase has not been purified. The enzyme is closely associated with, and has not been separated from, pyruvate synthase. However, phenylpyruvate synthase appears to be distinct from pyruvate synthase. This conclusion is based on (1) the presence of the pyruvate (but not the phenylpyruvate enzyme) in *C. pasteurianum*, (2) differences in the stability of phenylpyruvate and pyruvate synthase in *Chromatium* cells, and (3) differential requirements of TPP. In cellfree extracts of *Chromatium*, phenylpyruvate synthesis (and phenylpyruvate-CO_2 exchange) requires TPP; pyruvate synthase in these extracts requires TPP only after endogenous TPP is removed by special treatment (*33*, *46*).

pH optimum: The optimum pH for phenylpyruvate synthesis is 8.3–8.7 with a fairly sharp drop on both sides.

Effect of ferredoxins from different sources: Ferredoxins from photosynthetic bacteria (*Chromatium*, *C. thiosulfatophilum*, and *C. ethylicum*), aerobic bacteria (*Azotobacter vinelandii*), and fermentative bacteria (*C. pasteurianum*) showed a relative effectiveness of 4:2:1. Ferredoxin from spinach leaves showed nearly no effect (*46*).

CO_2-phenylpyruvate exchange: The exchange reaction, like other α-keto acid–CO_2 exchange reactions, requires a low level of CoA. The CoA requirements appear to be specific and cannot be satisfied by other SH reagents (*46*).

The phenylpyruvate synthesis and CO_2-phenylpyruvate exchange activities of *Chromatium* phenylpyruvate synthase show different pH optima. The optimal pH for synthesis is 8.3–8.7, whereas CO_2-phenylpyruvate exchange shows an optimum of 7.2. At their respective pH optima, the rate of the exchange reaction is three times greater than the rate of phenylpyruvate synthesis (*46*).

Physiological role: The biosynthesis of phenylalanine by the phenylpyruvate synthase mechanism (A) extends to aromatic amino acids a ferredoxin route of biosynthesis; as for the ferredoxin-mediated pathways of amino acid synthesis described above, the new phenylalanine pathway (A) is independent of the mechanism—the shikimate pathway (*37*)—previously established (B).

46. U. Gehring and D. I. Arnon, *JBC* **246**, 4518 (1971).

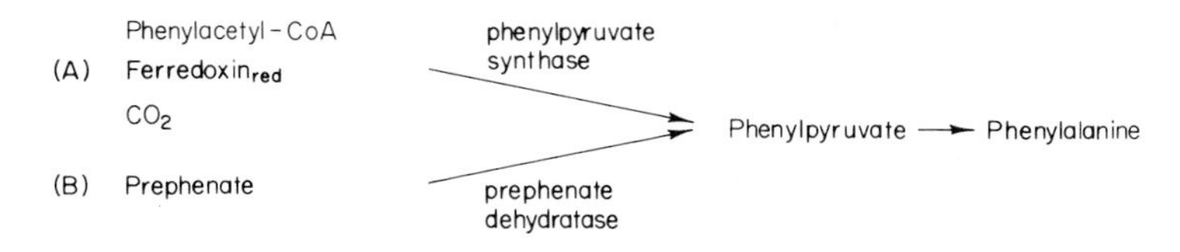

Although phenylpyruvate synthase has been found only in preparations from photosynthetic sulfur bacteria (*Chromatium*, *C. thiosulfatophilum*, and *C. ethylicum*, Table II), the ^{14}C experiments of Allison and associates with whole cells would indicate that the enzyme is present in rumen bacteria (*47*) as well as in the nonsulfur photosynthetic bacterium *R. rubrum* (*45*).

III. Reductive Carboxylic Acid Cycle of Bacterial Photosynthesis

Apart from a role in the biosynthesis of amino acids, the ferredoxin-dependent syntheses of pyruvate [Eq. (1)] and α-ketoglutarate [Eq. (3)] form the basis for the reductive carboxylic acid cycle—a new pathway of CO_2 assimilation in bacterial photosynthesis (*18, 19*). The new cycle is independent of the reductive pentose phosphate cycle (*48*) and provides another cyclic mechanism for CO_2 assimilation that continuously regenerates an acceptor for CO_2. One complete turn of the reductive carboxylic acid cycle (Fig. 3) incorporates four molecules of CO_2 and results in the net synthesis of oxalacetate, which is itself an intermediate in the cycle. Thus, beginning with one molecule of oxalacetate, one complete turn of the reductive carboxylic acid cycle will regenerate it and yield, in addition, a second molecule of oxalacetate formed by the reductive fixation of four molecules of CO_2.

The carboxylations of the reductive carboxylic acid cycle include, apart from the pyruvate and α-ketoglutarate synthase reactions, isocitrate dehydrogenase (*49, 50*), which catalyzes reversibly the carboxylation of α-ketoglutarate to isocitrate [Eq. (7)], and phosphoenolpyruvate carboxylase (*51*), which catalyzes the carboxylation of phosphoenolpyruvate to oxalacetate [Eq. (8)].

47. M. J. Allison, *BBRC* **18**, 30 (1965).
48. J. A. Bassham and M. Calvin, "The Photosynthesis of Carbon Compounds." Benjamin, New York, 1962.
49. S. Ochoa and E. Weisz-Tabori, *JBC* **159**, 245 (1954).
50. J. Moyle, *Biochem. J.* **63**, 552 (1956).
51. R. S. Bandurski and C. M. Greiner, *JBC* **204**, 781 (1953).

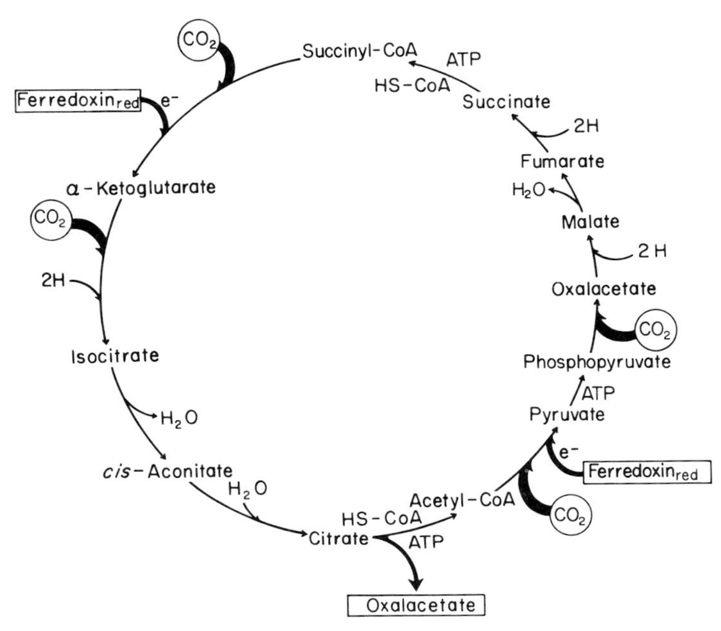

FIG. 3. Complete reductive carboxylic acid cycle of bacterial photosynthesis. Reversibility of the reactions is not indicated.

$$\alpha\text{-Ketoglutarate} + CO_2 + NADPH_2 \rightleftharpoons \text{isocitrate} + NADP \qquad (7)$$

$$\text{Phosphoenolpyruvate} + CO_2 \rightarrow \text{oxalacetate} + P_i \qquad (8)$$

A variant of the complete reductive carboxylic acid cycle is the "short" reductive carboxylic acid cycle (Fig. 4) which, in one turn, incorporates two molecules of CO_2 and yields one molecule of acetate. The complete cycle (Fig. 3) and the short cycle (Fig. 4) have the same sequence of reactions from oxalacetate to citrate. Thus, beginning again with oxalacetate, a complete turn of the short reductive carboxylic acid cycle would result in the regeneration of the oxalacetate and the synthesis of acetyl coenzyme A from two molecules of CO_2.

In its overall effect, the short reductive carboxylic acid cycle (Fig. 4) which generates acetyl-coenzyme A from two molecules of CO_2 is a reversal of the Krebs citric acid cycle, which degrades acetyl coenzyme A to two molecules of CO_2 (*41*, *42*). A basic distinction between the two cycles is that the reductive carboxylic acid cycle is endergonic in nature and hence must be linked with energy-yielding reactions which in this instance are the photoreduction of ferredoxin and photophosphorylation (*52*). Moreover, although several reversible enzyme reac-

52. D. I. Arnon, *Physiol. Rev.* **47**, 317 (1967).

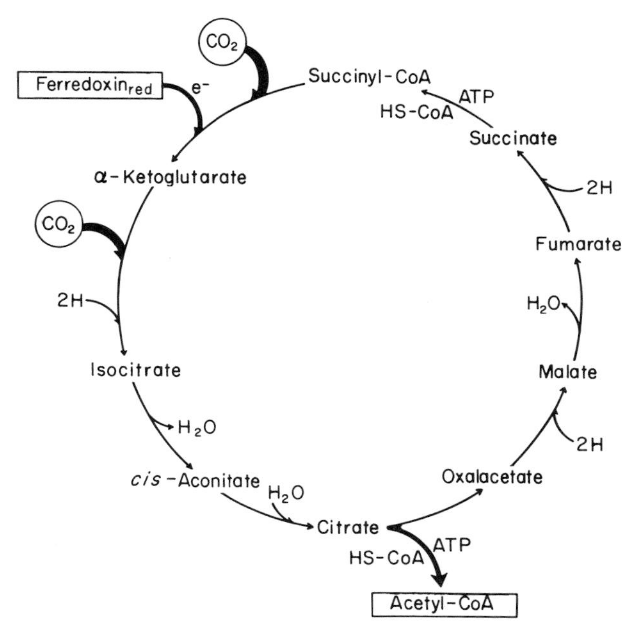

FIG. 4. Short reductive carboxylic acid cycle of bacterial photosynthesis. Reversibility of the reactions is not indicated.

tions of the citric acid cycle function also in the reductive carboxylic acid cycle, only the reductive cycle has the pyruvate and α-ketoglutarate synthases which, by reversing two steps that are irreversible in the citric acid cycle, permit the reductive carboxylic acid cycle to function as a pathway for CO_2 assimilation.

The operation of the reductive carboxylic acid cycle at the expense of radiant energy involves bacterial photophosphorylation and photoreduction of ferredoxin. The evidence for the latter in subcellular preparations from photosynthetic bacteria is not nearly so extensive as in chloroplasts. Evans and Buchanan (*53*), using chlorophyll-containing particles from *C. thiosulfatophilum*, were able to show a formation of reduced ferredoxin that was strictly dependent on light and on an added electron donor, such as sodium sulfide. Furthermore, Buchanan and Evans (*14*) have shown that in cellfree preparations of *C. thiosulfatophilium* photoreduced ferredoxin can serve as an electron donor for the reduction of NAD and thus provide a supply of $NADH_2$ (or $NADPH_2$ which is formed less effectively) for the operation of the reductive carboxylic acid cycle. The reduced ferredoxin (and nicotinamide-adenine dinucleo-

53. M. C. W. Evans and B. B. Buchanan, *Proc. Natl. Acad. Sci. U. S.* **53,** 1420 (1965).

TABLE III
ACTIVITIES OF ENZYMES OF THE REDUCTIVE CARBOXYLIC ACID CYCLE IN EXTRACTS OF *C. thiosulfatophilum* AND *R. rubrum*[a]

Enzyme	Enzymic activity (μmole/mg protein/hr)	
	C. thiosulfatophilum	*R. rubrum*
Acetyl-CoA synthetase	0.8	24
Pyruvate synthase	0.2	0.06
Phosphoenolpyruvate synthase	2.3	0.7
Phosphoenolpyruvate carboxylase	4.8	6.0
Malate dehydrogenase	37	159
Fumarate hydratase	118	128
Succinate dehydrogenase	0.85	1.2
Succinyl-CoA synthetase	1.6	3.3
α-Ketoglutarate synthase	0.4	0.012
Isocitrate dehydrogenase	102	70
Aconitate hydratase	3.1	7.3
Citrate lyase	0.15	0.17

[a] Data from Evans *et al.* (*18*) and Buchanan *et al.* (*19*).

tides) required for operation of the cycle may also be supplied, independently of light, by hydrogen gas (*13, 21, 54*) and hydrogenase—an enzyme native to photosynthetic bacteria.

The operation of the reductive carboxylic acid cycle (*18, 19*) in bacterial photosynthesis rests on the identification in cellfree extracts of *C. thiosulfatophilum* and *R. rubrum* of the enzymes listed in Table III that are required to catalyze the sequence of reactions shown in Fig. 2. The activities of the individual enzymes (in μmole per milligram protein per hour) ranged from 0.012 for α-ketoglutarate synthase to 159 for malate dehydrogenase. Since these measurements were made to establish the presence of these enzymes in cellfree extracts without a systematic search for optimal experimental conditions, they give no definitive information about the relative activities of these enzymes *in vivo*.

Other evidence includes phosphoenolpyruvate synthase, an enzyme that catalyzes the synthesis of phosphoenolpyruvate from ATP and pyruvate [Eq. (9)] in *Escherichia coli*, as reported by Cooper and Kornberg (*55*). The presence of phosphoenolpyruvate synthase in *C. thiosul-*

$$\text{Pyruvate} + \text{ATP} \rightarrow \text{phosphoenolpyruvate} + P_i + \text{AMP} \quad (9)$$

fatophilum and *R. rubrum* (and *Chromatium*) was demonstrated by

54. P. Weaver, K. Tinker, and R. C. Valentine, *BBRC* **21,** 195 (1965).
55. R. A. Cooper and H. L. Kornberg, *BBA* **104,** 618 (1965).

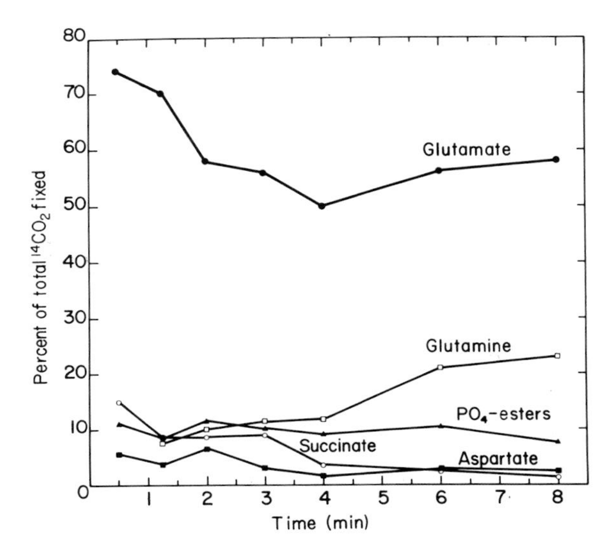

FIG. 5. Products of short-term photosynthesis by the green sulfur bacterium *Chlorobium thiosulfatophilum*.

Buchanan and Evans (*56*). The equilibrium of the phosphoenolpyruvate synthase reaction lies far on the side of phosphoenolpyruvate formation and would thus favor the operation of the reductive carboxylic acid cycle. A similar effect would also result from the irreversibility of phosphoenolpyruvate carboxylase (*51*) [Eq. (1)].

The reductive carboxylic acid cycle appears to function as a biosynthetic pathway that is particularly well suited to provide carbon skeletons for the amino acids that are the main products of photosynthesis in *C. thiosulfatophilum* (Fig. 5) (*18, 35*) and other bacteria (*57–59*). Thus, the reactions of the new cycle supply α-ketoglutarate for the synthesis of glutamate, oxalacetate for aspartate, and pyruvate for alanine. A principal product of the short cycle is acetyl coenzyme A which would be used for biosynthetic reactions, particularly the synthesis of fatty acids.

The new reductive carboxylic acid cycle invites comparison with the reductive pentose phosphate cycle (*48*) which hitherto has been regarded as the sole cyclic mechanism for CO_2 assimilation and which has been

56. B. B. Buchanan and M. C. W. Evans, *BBRC* **22,** 484 (1966).
57. M. Losada, A. V. Trebst, S. Ogata, and D. I. Arnon, *Nature* **186,** 753 (1960).
58. R. C. Fuller, R. M. Smillie, E. C. Sisler, and H. L. Kornberg, *JBC* **236,** 2140 (1961).
59. D. S. Hoare, *BJ* **87,** 284 (1963).

TABLE IV
COMPARISON OF REQUIREMENTS FOR ATP AND REDUCTANT IN THE TOTAL SYNTHESIS OF α-KETOGLUTARATE (α-KG) AND 3-PHOSPHOGLYCERATE (PGA) BY ENZYMES OF OR RELATED TO THE REDUCTIVE CARBON CYCLES OF PHOTOSYNTHESIS

	Equivalents of ATP		Equivalents of reductant			
			Ferredoxin[a]		NAD(P)[b]	
Carbon cycle	α-KG	PGA	α-KG	PGA	α-KG	PGA
Pentose phosphate	15	8	0	0	8	5
Carboxylic acid	5	4	3	2	5	3

[a] Based on a stoichiometry of two electrons per molecule of bacterial ferredoxin.
[b] Includes also the reduced flavin needed for reduction of fumarate by succinic dehydrogenase in the reductive carboxylic acid cycle.

reported in *C. thiosulfatophilum* (*60*), *R. rubrum* (*61*), and *Chromatium* (*58, 62*).

A noteworthy distinction between the two cycles is the requirement for ATP for the assimilation of CO_2 (Table IV). The synthesis of α-ketoglutarate and 3-phosphoglycerate (two important metabolic intermediates in photosynthetic cells) by the carboxylic acid cycle requires, respectively, one-third and one-half the ATP needed for their synthesis by the pentose cycle. This lower requirement for ATP in carbon assimilation results in all probability from the direct participation of ferredoxin in CO_2 fixation by the carboxylic acid cycle. Accordingly, the pentose phosphate cycle, which does not invoke ferredoxin as a direct reductant, shows a greater requirement for ATP.

Until recently, it was not possible to assess the importance of the new carboxylic acid cycle in relation to the pentose cycle. The recent demonstration of the inhibition of cellular photosynthesis by low levels of the inhibitor fluoroacetate led Sirevåg and Ormerod (*63, 64*) to conclude that in *C. thiosulfatophilum* CO_2 assimilation occurs largely via the reductive carboxylic acid cycle while the reductive pentose phosphate cycle, if functional, is of minor significance. Such a quantitative assessment of the relative importance of the new cycle in *R. rubrum* cannot yet be made, but pertinent to this point is the demonstration by Shigesada *et al.* (*40*) that whole cells of *R. rubrum* convert anaerobically in the light

60. R. M. Smillie, N. Rigopoulos, and H. Kelly, *BBA* **56,** 612 (1962).
61. L. Anderson and R. C. Fuller, *Plant Physiol.* **42,** 497 (1967).
62. E. Latzko and M. Gibbs, *Plant Physiol.* **44,** 295 (1969).
63. R. Sirevåg and J. G. Ormerod, *Science* **169,** 186 (1970).
64. R. Sirevåg and J. G. Ormerod, *BJ* **120,** 399 (1970).

$^{14}CO_2$ and ^{14}C-succinate into glutamate by a condensation that can now be explained by the operation of α-ketoglutarate synthase. [Similar experiments with ^{14}C-acetate and $^{14}CO_2$ by Cutinelli *et al.* (*34*) with *R. rubrum* and Tomlinson (*36*) with *Clostridium kluyveri* had previously revealed a condensation that can now be explained by the operation of pyruvate synthase—another key enzyme of the reductive carboxylic acid cycle.] Moreover, the $^{14}CO_2$ experiments of Yoch and Lindstrom (*65*) are consistent with the primary operation of α-ketoglutarate synthase in CO_2 assimilation by a related purple photosynthetic bacterium, *Rhodopseudomonas palustris*.

IV. Reductive Monocarboxylic Acid Cycle of Fermentative Metabolism

The fermentative bacterium *C. kluyveri* can convert CO_2 directly to formate in C_1 metabolism (*66*). Until the recent work of Thauer *et al.* (*20*) no mechanism was known to account for this conversion in *C. kluyveri*.

The new evidence shows that CO_2 can be converted to formate via a cyclic mechanism of CO_2 fixation (Fig. 6). Designated the reductive monocarboxylic acid cycle, the new cycle uses the power of reduced ferredoxin for the reduction in two steps of CO_2 to formate and leads to a regeneration of the CO_2 acceptor, acetyl coenzyme A. The formate produced by the cycle may be used [following conversion to the formyltetrahydrofolate derivative (*20*)] in a variety of processes, such as thymine biosynthesis, that depend on a C_1 unit.

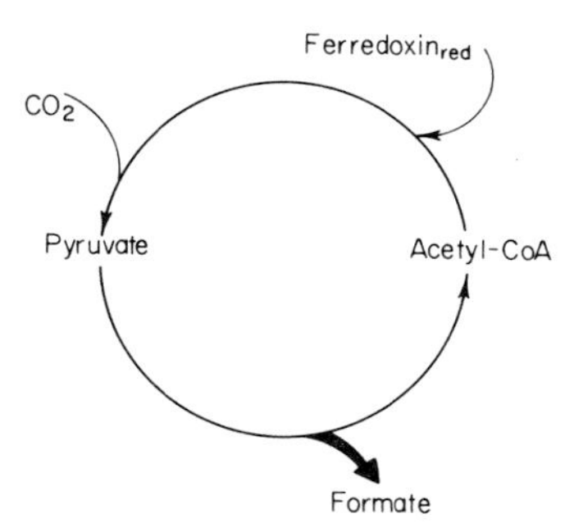

FIG. 6. Reductive monocarboxylic acid cycle of fermentative metabolism. Reversibility of the reactions is not indicated.

65. D. C. Yoch and E. S. Lindstrom, *BRRC* **28**, 65 (1967).
66. K. Jungermann, R. K. Thauer, and K. Decker, *European J. Biochem.* **3**, 351 (1968).

Evidence for the reductive monocarboxylic acid cycle in *C. kluyveri* consists of: (a) demonstration in cellfree extracts of the two component enzymes, pyruvate synthase [Eq. (1)] and pyruvate-formate lyase (*67, 68*) [Eq. (10)] and (b) demonstration (in a cellfree preparation) of a

$$\text{Pyruvate} + \text{CoA} \rightarrow \text{acetyl-CoA} + \text{formate} \tag{10}$$

production of formate from CO_2 and reduced ferredoxin in the presence of acetyl coenzyme A. Such a conversion would be predicted from the cycle shown in Fig. 6.

A feature of the reductive monocarboxylic acid cycle of fermentative bacteria, which distinguishes it from the reductive carboxylic acid cycle of bacterial photosynthesis, is the lack of a requirement for ATP. The lack of an ATP requirement shows that the reducing power of ferredoxin suffices to drive the monocarboxylic acid cycle in the direction of formate synthesis. Reduced ferredoxin could be supplied by either hydrogen gas or $NADH_2$ (*69, 70*). [The effectiveness of $NADH_2$ as a reductant for ferredoxin in the syntheses of pyruvate or the evolution of hydrogen has been observed for preparations from *C. kluyveri* but not for preparations from other organisms.]

The monocarboxylic acid cycle has been described only for the fermentative bacterium, *C. kluyveri*. It seems likely that other anaerobes, including photosynthetic bacteria, could use this mechanism to convert CO_2 to a formate derivative functional in C_1 metabolism.

V. Concluding Remarks

The discovery of a role for ferredoxin in the reductive synthesis of pyruvate in photosynthetic and nonphotosynthetic anaerobic bacteria has led to demonstration of other ferredoxin-linked carboxylations in these organisms. Each of the reactions involves a reductive carboxylation of an acyl coenzyme A derivative to an α-keto acid. Present evidence indicates that each reaction is catalyzed by a specific enzyme.

The new ferredoxin-dependent reactions function in the assimilation by bacterial cells of CO_2 and organic acids such as acetate or succinate. The α-keto acids formed are key metabolic intermediates particularly

67. T. C. Chase and J. C. Rabinowitz, *J. Bacteriol.* **96,** 1065 (1968).

68. J. Knappe, J. Schacht, W. Müchel, T. Höpner, H. Vetter, and R. Edenharder, *European J. Biochem.* **11,** 316 (1969).

69. G. Gottschalk and A. A. Chowdhury, *FEBS Letters* **2,** 342 (1969).

70. R. K. Thauer, K. Jungermann, E. Rupprecht, and K. Decker, *FEBS Letters* **4,** 108 (1969).

important in the synthesis of amino acids. The synthesis of amino acids by the ferredoxin-dependent reactions involves, in each case, a new pathway.

The significance of the new ferredoxin-linked carboxylations pertains not only to amino acid synthesis per se but also to mechanisms of CO_2 fixation. Two of the carboxylations—pyruvate and α-ketoglutarate synthase—form the basis for two new cycles of carbon dioxide assimilation: the reductive carboxylic acid cycle of bacterial photosynthesis and the reductive monocarboxylic acid cycle of fermentative metabolism. The operation of both of these cycles is dependent on the strong reducing potential of ferredoxin.

The ability of both fermentative and photosynthetic bacteria (but not plants) to use ferredoxin as a reductant in CO_2 assimilation may be important from the standpoint of evolution. A hypothesis favored by this laboratory holds that photosynthesis developed first in photosynthetic anaerobes such as *Chlorobium* which had evolved from fermentative organisms of the *Clostridium* type (Fig. 7). Bacterial photosynthesis would then have been followed by algal and higher plant photosynthesis which added oxygen to the earth's atmosphere. This view—advanced by Arnon *et al.* (*71, 72*) on the basis of a comparative analysis of certain metabolic features of the pertinent organisms—is supported by the findings on ferredoxin-linked carbon assimilation summarized here. This hypothesis has also received important support from recent comparative

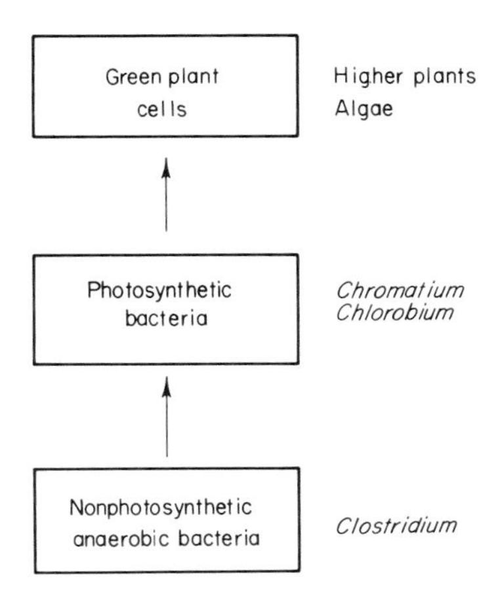

FIG. 7. A current view of the evolutionary development of photosynthesis.

71. D. I. Arnon, F. R. Whatley, and M. B. Allen, *Science* **127,** 1026 (1958).
72. D. I. Arnon, M. Losada, M. Nozaki, and K. Tagawa, *Nature* **190,** 601 (1961).

studies on amino acid composition and sequence of ferredoxins (*73–77, 17a*).

Prior to the accumulation of the earth's atmosphere of oxygen derived from plant photosynthesis, the metabolic energy available to anaerobic cells was limited to the relatively low amount released by fermentation. The addition of oxygen to the earth's atmosphere made possible the important development of respiration and the complete oxidation of organic substrates to carbon dioxide and water via the citric acid cycle (*41, 42*)—a development that increased the yield of ATP 19-fold over that obtained anaerobically in glucose breakdown by glycolysis alone.

In such a course of biochemical evolution, the reductive carboxylic acid cycle would represent a primitive biosynthetic pathway that has survived to this day in photosynthetic bacteria. The striking resemblance of certain features of the anaerobic "short" cycle (for formation of acetyl coenzyme A from CO_2) to the aerobic citric acid cycle (for oxidation of acetyl coenzyme A to CO_2) may therefore be of evolutionary significance. It seems possible that with the advent of atmospheric oxygen the earlier anaerobic reductive carboxylic acid cycle (used as a mechanism for reductive CO_2 assimilation) was converted to the citric acid cycle that serves as a mechanism for the oxidation of acetyl coenzyme A to CO_2 and water. In this view, the present use of reactions of the citric acid cycle for biosynthesis (*41, 42*) would be analogous to an earlier stage in evolution when these reactions were part of the reductive carboxylic acid cycle that functioned solely for biosynthesis.

Note added in proof: Recent reports (*78, 79*) on the purification and properties of pyruvate synthase ("pyruvate-ferredoxin oxidoreductase") from *Clostridium acidi-urici* provide strong evidence that the enzyme contains an iron–sulfur chromophore (absorption maximum, 400 nm) that is reduced (bleached) by reduced ferredoxin in the synthesis of pyruvate or by pyruvate plus CoA in the breakdown of pyruvate. This finding raises the possibility that an iron–sulfur group may be a general feature of the ferredoxin-linked carboxylation enzymes.

73. M. Tanaka, T. Nakashima, A. Benson, H. Mower, and K. T. Yasunobu, *Biochemistry* **5,** 1666 (1966).
74. H. Matsubara, R. M. Sasaki, and R. K. Chain, *Proc. Natl. Acad. Sci. U. S.* **57,** 439 (1967).
75. S. Keresztes-Nagy, F. Perini, and E. Margoliash, *JBC* **244,** 981 (1969).
76. A. M. Benson and K. T. Yasunobu, *JBC* **244,** 955 (1969).
77. B. B. Buchanan, H. Matsubara, and M. C. W. Evans, *BBA* **189,** 46 (1969).
78. K. Uyeda and J. C. Rabinowitz, *JBC* **246,** 3111 (1971).
79. K. Uyeda and J. C. Rabinowitz, *JBC* **246,** 3120 (1971).

7

Amino Acid Decarboxylases

ELIZABETH A. BOEKER • ESMOND E. SNELL

I. Introduction

In a series of fundamental papers dating back to 1940, Gale and his co-workers (*1*) established the existence of six inducible bacterial enzymes catalyzing the α-decarboxylation of amino acids and the fact that several of these enzymes required an unidentified coenzyme. The discovery of mammalian and plant decarboxylases was proceding during this same period (*2*). α-Decarboxylases catalyzing reaction (1) and specific for more than half of the 20 commonly occurring amino acids, as well as several less common ones, are now known (*3, 4*). In addition,

1. E. F. Gale, *Advan. Enzymol.* **6,** 1 (1946).
2. H. Blaschko, *Advan. Enzymol.* **5,** 67 (1945).
3. O. Schales, "The Enzymes," 1st ed., Vol. 2, p. 216, 1951.
4. B. M. Guirard and E. E. Snell, *Compr. Biochem.* **15,** 138 (1964).

one β-decarboxylase which catalyzes reaction (2) (*5*), one decarboxylation-dependent transaminase catalyzing reaction (3) (*6*), and three α-de-

$$RCH\overset{+}{N}H_3COO^- + H \longrightarrow RCH_2\overset{+}{N}H_3 + CO_2 \tag{1}$$

$$^-OOCCH_2CH\overset{+}{N}H_3COO^- + H^+ \longrightarrow CO_2 + CH_3CH\overset{+}{N}H_3COO^- \tag{2}$$

carboxylation-dependent condensing enzymes that catalyze reaction (4) (*7*) have now been identified.

$$R\overset{\overset{R'}{|}}{C}\cdot\overset{+}{N}H_3COO^- + R''COCOO^- + H^+ \longrightarrow CO_2 + R\overset{\overset{R'}{|}}{C}=O + R''CH\overset{+}{N}H_3COO^- \tag{3}$$

$$\underset{\underset{^+NH_3}{|}}{RCHCOO^-} + R'\underset{\underset{O}{\|}}{C}-SCoA + H^+ \longrightarrow CO_2 + \underset{\underset{^+NH_3}{|}}{RCH}-\underset{\underset{O}{\|}}{C}-R' + CoASH \tag{4}$$

Following the discovery of the six bacterial decarboxylases, Bellamy and Gunsalus (*8*) and shortly thereafter Baddiley and Gale (*9*) showed that the coenzyme required by four of these enzymes (tyrosine, lysine, arginine, and ornithine decarboxylase) was a phosphorylated derivative of pyridoxal, later shown to be pyridoxal 5′-phosphate (*10, 11*). However, Gale was unable to achieve resolution and reconstitution of the remaining two enzymes, glutamate and histidine decarboxylase, and also observed that boiled preparations of them did not serve as cofactors for lysine or tyrosine apodecarboxylase (*1*); he concluded that they must not be PLP (*12*) dependent. Shukuya and Schwert (*13, 14*) used spectrophotometric techniques to demonstrate that purified glutamate decarboxylase does depend on tightly bound PLP; Strausbauch and Fischer (*15*) have now determined the structure of its PLP binding site. Histidine decarboxylase from a *Lactobacillus*, however, is not dependent upon PLP (*16*); instead the pure enzyme contains covalently bound, functional

5. A. Meister, H. A. Sober, and S. V. Tice, *JBC* **189,** 577 (1951).
6. G. B. Bailey and W. B. Dempsey, *Biochemistry* **6,** 1526 (1967).
7. E. E. Snell and S. J. DiMari, "The Enzymes," 3rd ed., Vol. II, p. 355, 1970.
8. W. D. Bellamy and I. C. Gunsalus, *J. Bacteriol.* **50,** 95 (1945).
9. J. Baddiley and E. F. Gale, *Nature* **155,** 727 (1945).
10. D. Heyl, E. Luz, S. A. Harris, and K. Folkers, *JACS* **73,** 3436 (1951).
11. J. Baddiley and A. P. Mathias, *JCS* p. 2583 (1962).
12. Abbreviations used in this article include PLP, pyridoxal 5′-phosphate; PMP, pyridoxamine 5′-phosphate; and PxyP, phosphopyridoxyl residue.
13. R. Shukuya and G. W. Schwert, *JBC* **235,** 1649 (1960).
14. R. Shukuya and G. W. Schwert, *JBC* **235,** 1653 (1960).
15. P. H. Strausbauch and E. H. Fischer, *Biochemistry* **9,** 233 (1970).
16. J. Rosenthaler, B. M. Guirard, G. W. Chang, and E. E. Snell, *Proc. Natl. Acad. Sci. U. S.* **54,** 152 (1965).

pyruvate residues (*17–19*). It is not known whether other bacterial histidine decarboxylases (including that studied by Gale) are PLP or pyruvate dependent, especially since the specific mammalian histidine decarboxylase appears to require PLP as coenzyme (*20, 21*). *S*-Adenosylmethionine decarboxylase from *Escherichia coli* also appears to be a pyruvate enzyme (*22*).

The discovery that pyruvate replaces PLP as an essential prosthetic group in certain amino acid decarboxylases demonstrates an unexpected diversity in this group of enzymes. It is no longer safe to assume that because a crude decarboxylase preparation is inhibited by carbonyl reagents it must be PLP dependent.

II. Metabolic Importance of the Decarboxylases

A. Bacterial Decarboxylases

Most of the six inducible enzymes originally described by Gale (*1*) have proved to be readily available and amenable to purification; four have been purified to homogeneity and extensively characterized. Despite this, their metabolic function remains unknown, and hypotheses concerning it remain untested. Since all six have low pH optima and are produced only in an enriched, acidic growth medium, Gale (*1*) suggested that amine formation might represent a mechanism for counteracting acidic conditions. Guirard and Snell (*4*), also considering the low pH necessary for maximum induction, proposed that decarboxylation might also serve to control the intracellular pCO_2, which would be very low if unregulated.

A nonspecific decarboxylase which acts on hydrophobic L-amino acids occurs in *Proteus vulgaris* (*23*). It is also inducible, but its formation is unaffected by the pH of the culture. It has a neutral pH optimum and a relatively low specific activity (see Table I), suggesting that its function may differ from that of the other inducible decarboxylases.

Several of the bacterial decarboxylases have biosynthetic functions. Aspartic acid β-decarboxylase provides one of the biosynthetic routes to alanine, and diaminopimelic acid decarboxylase catalyzes the last reaction of lysine biosynthesis in bacteria, certain fungi, blue–green and

17. W. D. Riley and E. E. Snell, *Biochemistry* **7**, 3520 (1968).
18. W. D. Riley and E. E. Snell, *Biochemistry* **9**, 1485 (1970).
19. P. A. Recsei and E. E. Snell, *Biochemistry* **9**, 1492 (1970).
20. R. Håkanson, *Biochem. Pharmacol.* **12**, 1289 (1963).
21. R. Håkanson, *European J. Pharmacol.* **1**, 34 (1967).
22. R. B. Wickner, C. W. Tabor, and H. Tabor, *JBC* **245**, 2132 (1970).
23. L. Ekladius, H. K. King, and C. R. Sutton, *J. Gen. Microbiol.* **17**, 602 (1957).

green algae, and vascular plants (*24*). The products of both adenosylmethionine decarboxylase and ornithine decarboxylase are required for synthesis of polyamines (*25*) which are essential growth factors for organisms which do not synthesize them, such as *Hemophilus parainfluenzae* (*26*). In addition to the inducible ornithine decarboxylase, which is not produced by cultures of *E. coli* grown on minimal media, Morris and Pardee (*27, 28*) have found a distinct, biosynthetic enzyme which catalyzes the same reaction. Normally present at relatively low levels, its production is derepressed fourfold during putrescine limitation (*29, 30*).

Morris and Pardee (*28*) have also identified an unusual biosynthetic pathway which operates in *E. coli* when arginine supplementation limits the synthesis of ornithine from glutamate. Under these conditions, most organisms use either arginase or arginine desimidase and ornithine transcarbamylase (*31–33*) to convert arginine to ornithine, the precursor of putrescine and the polyamines. Since none of these enzymes is present in *E. coli,* this organism, grown in an arginine-containing medium, would become deficient in polyamines except for the presence of arginine decarboxylase and agmatine ureohydrolase, which together convert arginine directly to putrescine. As with ornithine decarboxylase, the biosynthetic arginine decarboxylase is distinct from the inducible enzyme (*28*) and is derepressed during putrescine limitation (*29, 30*).

An unusual type of PLP-dependent α-decarboxylation reaction [Eq. (4)], in which the carboxyl group of the amino acid is replaced by the acyl group of an acyl-CoA, operates in the biosynthesis of δ-aminolevulinic acid from glycine and succinyl-CoA (*34, 35*), of 3-ketosphinganine (an intermediate in sphingolipid biosynthesis) from serine and palmityl-CoA (*36*) and of 7-keto-8-aminopelargonic acid (an intermediate in biotin biosynthesis), from alanine and pimelyl-CoA (*37*).

Amino acid decarboxylases also participate in energy-producing se-

24. H. J. Vogel, *in* "Evolving Genes and Proteins" (V. Bryson and H. J. Vogel, eds.), p. 25. Academic Press, New York, 1965.

25. H. Tabor and C. W. Tabor, *Pharmacol. Rev.* **16,** 245 (1964).

26. E. J. Herbst and E. E. Snell, *JBC* **181,** 47 (1949).

27. D. R. Morris and A. B. Pardee, *BBRC* **20,** 697 (1965).

28. D. R. Morris and A. B. Pardee, *JBC* **241,** 3129 (1966).

29. H. Tabor and C. W. Tabor, *JBC* **244,** 2286 (1969).

30. D. R. Morris and C. M. Jorstad, *J. Bacteriol.* **101,** 731 (1970).

31. M. E. Jones, *Science* **140,** 1373 (1963).

32. V. Stalon, F. Ramos, A. Piérard, and J. M. Wiame, *BBA* **139,** 91 (1967).

33. R. H. Davis, M. B. Lawless, and L. Port, *J. Bacteriol.* **102,** 299 (1970).

34. D. Shemin, C. S. Russell, and T. Abramsky, *JBC* **215,** 613 (1955).

35. B. F. Burnham and J. Lascelles, *BJ* **87,** 462 (1963).

36. E. E. Snell, S. J. DiMari, and R. N. Brady, *Chem. Phys. Lipids* **5,** 116 (1970).

37. M. A. Eisenberg and C. Star, *J. Bacteriol.* **96,** 1291 (1968).

quences of reactions. For example, crude extracts of *Peptococcus glycinophilus,* an anaerobic bacterium which utilizes glycine as its sole carbon source (*38*), convert glycine to acetate with the overall stoichiometry shown in Eq. (5) (*39, 40*):

$$2\ H_3N^+CH_2COO^- + 2\ DPN^+ + ADP + P_i + H_2O \rightarrow 2\ CO_2 + 2\ NH_4^+ + CH_3COOH + 2\ DPNH + ATP \quad (5)$$

The first reaction in this sequence, reaction (6),

$$H_3N^+CH_2COO^- + DPN^+ + \text{tetrahydrofolic acid} \rightarrow CO_2 + NH_4^+ + DPNH + \text{methylene-tetrahydrofolic acid} \quad (6)$$

is catalyzed by a complex of four enzymes, one of which is a PLP-dependent glycine decarboxylase (*41–44*). A similar pathway for catabolism of glycine appears to exist in rat liver mitochondria (*45*).

Another decarboxylase which functions in an energy-yielding pathway occurs in a *Pseudomonad* which can utilize α-substituted amino acids as a carbon (*46*) or nitrogen source (*6*). Since α-substituted amino acids have no α-hydrogen, these organisms produce an enzyme which carries out first, decarboxylation, and second, transamination. In this decarboxylation-dependent transamination, the α-carboxyl assumes the role normally played by the α-hydrogen, as shown in Eq. (3).

B. Mammalian and Plant Decarboxylases

Like their bacterial counterparts, the mammalian ornithine and adenosylmethionine decarboxylases are required for the synthesis of polyamines. Although the precise biological role of polyamines is still unknown (*25*), it is now clear that the levels of spermidine (*47–50*) and at least some of the enzymes responsible for its synthesis increase signifi-

38. B. P. Cardon and H. A. Barker, *J. Bacteriol.* **52,** 629 (1946); *ABB* **12,** 165 (1947).
39. R. D. Sagers and I. C. Gunsalus, *J. Bacteriol.* **81,** 541 (1961).
40. S. M. Klein and R. D. Sagers, *J. Bacteriol.* **83,** 121 (1962).
41. R. D. Sagers and S. M. Klein, *Federation Proc.* **24,** 505 (1965).
42. S. M. Klein and R. D. Sagers, *Bacteriol. Proc.* P192 (1966).
43. S. M. Klein and R. D. Sagers, *JBC* **241,** 197 and 206 (1966).
44. S. M. Klein and R. D. Sagers, *JBC* **242,** 297 and 301 (1967).
45. T. Yoshida and G. Kikuchi, *ABB* **139,** 380 (1970).
46. H. G. Aaslestaad and A. D. Larson, *J. Bacteriol.* **88,** 1296 (1964).
47. W. G. Dykstra and E. J. Herbst, *Science* **149,** 428 (1965).
48. C. M. Caldarera, B. Barbirolli, and G. Moruzzi, *BJ* **97,** 84 (1965).
49. J. Kostyo, *BBRC* **23,** 150 (1966).
50. A. Raina, J. Jänne, and M. Siimes, *BBA* **123,** 197 (1966).

cantly during rapid growth. A dramatic increase of ornithine decarboxylase is the first enzymic response so far known in regenerating rat liver (*51–54*), in developing chick embryos (*51*), in intact rat livers treated with growth hormone (*51, 55–57*), in chick embryo epidermis in response to epidermal growth factor (*58*), in chick oviducts in response to estrogen (*59*), and in the prostate glands of castrated rats in response to testosterone (*60*). Adenosylmethionine decarboxylase activity was examined only in the last case; it too increased (*60*).

Mammalian glutamate decarboxylase appears to be of primary importance in the central nervous system. Its substrate, glutamic acid, is thought to be an excitatory transmitter in cerebral neurons, while its product, γ-aminobutyric acid, is apparently an inhibitory transmitter (*61–63*). Several studies point to a synaptosomal localization of glutamate decarboxylase in brain (*64–68*), and Kravitz *et al.* (*69*) observed that the level of glutamate decarboxylase activity was 10 times higher in the inhibitory axons of lobsters than in the excitatory axons.

Dihydroxyphenylalanine decarboxylase (*70*) also plays an important role in the central nervous system; this relatively nonspecific enzyme catalyzes the formation of 5-hydroxtryptamine (serotonin) as well as dihydroxyphenylethylamine (dopamine), a precursor of epinephrine and

51. D. Russell and S. Snyder, *Proc. Natl. Acad. Sci. U. S.* **60,** 1420 (1968); *Endocrinology* **84,** 223 (1969).

52. J. Jänne and A. Raina, *Acta Chem. Scand.* **22,** 1349 (1968).

53. N. Fausto, *BBA* **190,** 193 (1969).

54. T. R. Schrock, N. J. Oakman, and N. L. R. Bucher, *BBA* **204,** 564 (1970).

55. J. Jänne, A. Raina, and M. Siimes, *BBA* **166,** 419 (1968).

56. J. Jänne and A. Raina, *BBA* **174,** 769 (1969).

57. D. H. Russell, S. H. Snyder, and V. J. Medina, *Endocrinology* **86,** 1414 (1970).

58. M. Stastny and S. Cohen, *BBA* **204,** 578 (1970).

59. S. Cohen, B. W. O'Malley, and M. Stastny, *Science* **170,** 336 (1970).

60. H. G. Williams-Ashman, A. E. Pegg, and D. H. Lockwood, *Advan. Enzyme Regul.* **7,** 291 (1969).

61. D. R. Curtis and J. C. Watkins, *Pharmacol. Rev.* **17,** 347 (1965).

62. E. Roberts and K. Kuriyama, *Brain Res.* **8,** 1 (1968).

63. K. Krnjevíc, *Nature* **228,** 119 (1970).

64. H. Weinstein, E. Roberts, and T. Kakefuda, *Biochem. Pharmacol.* **12,** 503 (1963).

65. L. Salganicoff and E. De Robertis, *J. Neurochem.* **12,** 287 (1965).

66. F. Fonnum, *BJ* **106,** 401 (1968).

67. K. Kuriyama, B. Sisken, J. Ito, D. G. Simonsen, B. Haber, and E. Roberts, *Brain Res.* **11,** 412 (1968).

68. M. J. Neal and L. L. Iverson, *J. Neurochem.* **16,** 1245 (1969).

69. E. A. Kravitz, P. B. Molinoff, and Z. W. Hall, *Proc. Natl. Acad. Sci. U. S.* **54,** 778 (1965).

70. T. L. Sourkes, *Pharmacol. Rev.* **18,** 53 (1966).

norepinephrine. The actions of these neurohormones have been reviewed by Kety (*71*), by Hornykiewicz (*72*), and by Molinoff and Axelrod (*73*). α-Methyldihydroxyphenylalanine, a potent inhibitor of this decarboxylase, is used therapeutically as an antihypertensive agent (*74, 75*) and the substrate, dihydroxyphenylalanine, is used in the treatment of Parkinson's disease, in which reduced levels of brain dopamine are observed clinically (*76*).

Although dihydroxyphenylalanine decarboxylase will act on relatively high concentrations of histidine at a very slow rate (*77–79*), a distinct histidine decarboxylase, present in several mammalian tissues (*80*), is more important in the formation of histamine. Histamine is involved in several physiological processes including gastric secretion, peripheral circulation, allergic and similar hypersensitivity reactions, and certain types of rapid growth (*81*). In the rat gastric mucosa, histidine decarboxylase decreases during starvation and increases in response to feeding or the injection of gastrin (*82–85*). Although other hormones cause a similar increase, these effects are apparently indirect and are mediated by gastrin (*83, 86*). Histidine decarboxylase also increases in epithelial tissues during reparative growth (*87*) and as a result of grafting (*88*). Fetal rats and hamsters show a peak of histidine decarboxylase activity just before birth (*21, 89*). In four tumors, the levels of histidine and ornithine decarboxylase appear to show an inverse correlation (*51*).

71. S. S. Kety, *Pharmacol. Rev.* **18,** 787 (1966).
72. O. Hornykiewicz, *Pharmacol. Rev.* **18,** 925 (1966).
73. P. B. Molinoff and J. Axelrod, *Ann. Rev. Biochem.* **40,** 465 (1971).
74. T. L. Sourkes and H. R. Rodriguez, *Med. Chem.* **7,** 151 (1967).
75. C. A. Stone and C. C. Porter, *Advan. Drug Res.* **4,** 71 (1967).
76. G. C. Cotzias, M. H. Van Woert, and L. M. Schiffer, *New Engl. J. Med.* **276,** 374 (1967).
77. H. Weissbach, W. Lovenberg, and S. Udenfriend, *BBA* **50,** 177 (1961).
78. W. Lovenberg, H. Weissbach, and S. Udenfriend, *JBC* **237,** 89 (1962).
79. J. G. Christenson, W. Dairman, and S. Udenfriend, *ABB* **141,** 356 (1970).
80. D. Aures, R. Håkanson, and W. G. Clark, *in* "Handbook of Neurochemistry" (A. Lajtha, ed.), Vol. IV, p. 165. Plenum, New York, 1970.
81. G. Kahlson and E. Rosengren, *Physiol Rev.* **48,** 155 (1968).
82. G. Kahlson, E. Rosengren, D. Svahn, and R. Thunberg, *J. Physiol.* (*London*) **174,** 400 (1964).
83. D. Aures, R. Håkanson, and A. Schauer, *European J. Pharmacol.* **3,** 217 (1968).
84. Y. Kobayashi and D. V. Maudsley, *Brit. J. Pharmacol.* **32,** 428P (1968).
85. S. H. Snyder and L. Epps, *Mol. Pharmacol.* **4,** 187 (1968).
86. L. R. Johnson, R. S. Jones, D. Aures, and R. Håkanson, *Am. J. Physiol.* **216,** 1051 (1969).
87. R. W. Schayer and O. H. Ganley, *Am. J. Physiol.* **197,** 721 (1959).
88. T. C. Moore and R. W. Schayer, *Transplantation* **7,** 99 (1969).
89. R. Håkanson, *European J. Pharmacol.* **1,** 42 (1967).

Cysteine sulfinic acid decarboxylase catalyzes one reaction in the degradation of cysteine. In liver, cysteine is first oxidized to cysteine sulfinic acid, which is either transaminated and desulfinated to pyruvate or decarboxylated to hypotaurine and oxidized to taurine. Since labeled cysteine gives rise to hypotaurine and not to cysteic acid, the final oxidation step does not precede decarboxylation (*90, 91*). The finding that the decarboxylase acts on cysteine sulfinic acid approximately six times faster than on cysteic acid is consistent with this (*92, 93*).

The function of α-aminomalonic acid decarboxylase, also found in liver, is not known; it may catalyze the last reaction in an alternate pathway of glycine biosynthesis [for discussion, see Thanassi and Fruton (*94*)]. Degradation of glycine to CO_2, NH_3, and methylene tetrahydrofolate in liver mitochondria (*45*) provides "one-carbon units" for a variety of biosynthetic reactions, and presumably it requires a PLP-dependent decarboxylase similar to that demonstrated in *Peptococcus glycinophilus* (*42–44*). The essential role of two atypical decarboxylases, δ-aminolevulinic acid synthetase and 3-ketosphinganine synthetase, in providing intermediates for porphyrin and sphingolipid biosynthesis, respectively, has been referred to earlier (Section II,A).

The decarboxylation of valine and leucine in the newly opened buds of lilies (*Arum maculatum*), hawthorne (*Crataegus monogyna*), and mountain ash (*Sorbus aucuparia*) serves a unique function. These species are insect-pollinated; the insects are attracted by the odor of the volatile amines produced by decarboxylation (*95, 96*).

In barley, the levels of arginine decarboxylase are elevated during K^+ deficiency; Smith (*97*) has suggested that the organic base produced may have a sparing effect on K^+ requirements.

III. Distribution and General Properties of Decarboxylases

Table I summarizes some of the information available for decarboxylases from all sources (*6, 13, 16–19, 22, 23, 35, 42–44, 79, 89, 92, 94,*

90. J. Awapara, *JBC* **203**, 183 (1953).
91. J. Awapara and W. J. Wingo, *JBC* **203**, 189 (1953).
92. D. B. Hope, *BJ* **59**, 497 (1955).
93. A. N. Davison, *BBA* **19**, 66 (1956).
94. J. W. Thanassi and J. S. Fruton, *Biochemistry* **1**, 975 (1962).
95. E. Simon, *J. Exptl. Botany* **13**, 1 (1962).
96. M. Richardson, *Phytochemistry* **5**, 23 (1966).
97. T. A. Smith, *Phytochemistry* **2**, 241 (1963).

98–122). Where possible, data are presented for the most highly purified or most extensively characterized preparations, but in many cases only preliminary studies have been performed, and the data must be considered provisional. Studies of mammalian decarboxylases have been particularly hindered by their instability and their very low specific activities in crude extracts; even in the most favorable cases, this value is orders of magnitude less than that of most bacterial enzymes, as inspection of Table I shows. Other than a single report that legumes possess weak decarboxylases for several amino acids (*123*), no systematic search for plant decarboxylases has been undertaken, and relatively few are known.

S-Adenosylmethionine decarboxylase from *E. coli* has been obtained in essentially homogeneous form (*22*), although the specific activity seems low in comparison to other purified decarboxylases. The purified enzyme, which catalyzes reaction (7), does not contain PLP but does contain 1–2 pyruvate residues per mole. It is inhibited by carbonyl re-

98. C. W. Tabor, "Methods in Enzymology," Vol. 5, p. 756, 1962.
99. S. L. Blethen, E. A. Boeker, and E. E. Snell, *JBC* **243,** 1671 (1968).
100. D. R. Morris, W. H. Wu, D. Applebaum, and K. L. Koffron, *Ann. N. Y. Acad. Sci.* **171,** 968 (1970).
101. W. H. Wu and D. R. Morris, manuscript in preparation.
102. A. Novogrodsky and A. Meister, *JBC* **239,** 879 (1964).
103. S. S. Tate and A. Meister, *Biochemistry* **7,** 3240 (1968).
104. S. S. Tate and A. Meister, *Biochemistry* **8,** 1660 (1969).
105. P. J. White and B. Kelly, *BJ* **96,** 75 (1965).
106. P. H. Strausbauch, E. H. Fischer, C. Cunningham, and L. P. Hager, *BBRC* **28,** 525 (1967).
107. A. D. Homola and E. E. Dekker, *Biochemistry* **6,** 2626 (1967).
108. G. W. Chang and E. E. Snell, *Biochemistry* **7,** 2005 (1968).
109. C. R. Sutton and H. K. King, *ABB* **96,** 360 (1962).
110. K. Soda and M. Moriguchi, *BBRC* **34,** 34 (1969).
111. E. S. Taylor and E. F. Gale, *BJ* **39,** 52 (1945).
112. D. Applebaum and D. R. Morris, manuscript in preparation.
113. H. M. R. Epps, *BJ* **38,** 242 (1944).
114. P. R. Sundaresan and D. B. Coursin, "Methods in Enzymology," Vol. 18A, p. 509, 1970.
115. A. E. Pegg and H. G. Williams-Ashman, *BJ* **108,** 533 (1968).
115a. J. Jänne and H. G. Williams-Ashman, *BBRC* **42,** 222 (1971).
116. J. P. Susz, B. Haber, and E. Roberts, *Biochemistry* **5,** 2870 (1966).
117. R. Håkanson, *European J. Pharmacol.* **1,** 383 (1967).
118. A. Raina and J. Jänne, *Acta Chem. Scand.* **22,** 2375 (1968).
119. O. Schales, V. Mims, and S. Schales, *ABB* **10,** 455 (1946).
120. H. Beevers, *BJ* **48,** 132 (1951).
121. L. Fowden, *J. Exptl. Botany* **5,** 28 (1954).
122. S. J. Norton and Y. T. Chen, *BBA* **198,** 610 (1970).
123. L. Ambe and K. Sohonie, *J. Sci. Ind. Res. (India)* **18C,** 135 (1959); *CA* **54,** 4792a (1960).

TABLE I
PROPERTIES OF THE BETTER CHARACTERIZED AMINO ACID DECARBOXYLASES (CARBOXY-LYASES)

Source	Best substrate[a]	Other substrates	Fold purification and purity (%)	Specific activity (μmoles/min/mg)	Optimum pH	K_m (mM)	Cofactor	K_{PLP}[b] (μM)	References
A. Bacterial Decarboxylases									
Escherichia coli	Adenosyl-methionine	—	800X, >95%	0.02	7.4	0.09	Pyruvate Mg^{2+}	—	(*22, 98*)
Pseudomonas sp.	α-Aminoiso-butyric acid	Isovaline and others	130X, >95%	10	8.0–8.5	8	PLP	5	(*6*)
E. coli	Arginine (inducible)	Canavanine	40X, >95%	400	5.2	0.65	PLP	0.027	(*99*)
E. coli	Arginine (bio-synthetic)	—	1200X, >95%	16	8.4	0.03	PLP	0.5	(*100, 101*)
Alcaligenes faecalis	Aspartic acid (β-COOH)	Cysteine sulfinic acid and others	90X, >95%	125	5.0	0.64	PLP	200	(*102–104*)
E. coli	Diaminopimelic acid	None known	200X, ca. 70%	10	6.8	0.17	PLP	25[c]	(*105*)
E. coli	Glutamic acid	α-Methylene glutamic acid and others	>10X, >95%	110	3.8	0.54	PLP	—	(*13, 106, 107*)
Peptococcus glycinophilus	Glycine (P_1)	None known	60X, ca. 80%	47	7.0	32	PLP	4.6	(*42–44*)
Rhodopseudomonas spheroides	Glycine + succinyl-CoA (→ δ-aminolevulinate)	None known	9X	0.03	7.8–8.0	0.28 (Gly)	PLP	ca. 4.0	(*35*)
Lactobacillus 30a	Histidine	None[d]	200X, >95%	70	4.7–5.2	0.9	Pyruvate	—	(*16–19, 108*)

Proteus vulgaris	Leucine, valine	Norvaline, isoleucine, and others	28X	2.8	6.5	10 (Leu) 25 (Val)	PLP	4.5 (Leu) 0.35 (Val)	(*23*, *109*)
Bacterium cadaveris	Lysine	*S*-Aminoethylcysteine, 5-hydroxylysine	60X, >95%	86	5.8	0.37	PLP	—	(*110*)
Clostridium septicum	Ornithine (inducible)	—	—	—	5.2	4	PLP	—	(*111*)
E. coli	Ornithine (biosynthetic)	—	1300X, >70%	90	8.0	3.9	PLP	0.5	(*100*, *112*)
Streptococcus faecalis	Tyrosine	Dihydroxyphenylalanine, possibly others	100X	35[e]	5.5	2.3[f]	PLP	—	(*113*, *114*)
B. Mammalian Decarboxylases									
Rat prostate	Adenosylmethionine	—	510X	0.028	7.0	0.05	? Inhibited by carbonyl reagents	—	(*115*, *115a*)
Rat liver	Aminomalonic acid	None known	2X	0.062	5.5–6.5	14	PLP	—	(*94*)
Dog liver	Cysteine sulfinic acid	Cysteic acid	—	0.0013[g]	—	—	PLP	—	(*92*)
Hog kidney	Dihydroxyphenylalanine	5-Hydroxytryptophan, other aromatic amino acids	330X, >99%	8.67	8.7[h]	0.19	PLP	0.09[i]	(*79*)

TABLE I (*Continued*)

Source	Best substrate[a]	Other substrates	Fold purification and purity (%)	Specific activity (μmoles/min/mg)	Optimum pH	K_m (mM)	Cofactor	K_{PLP}[b] (μM)	References
Mouse brain	Glutamic acid	—	160X	0.68	6.4–7.2	3–8	PLP	—	(*116*)
Fetal rat	Histidine	None known	200X	0.0015	6.0	1.80[j]	PLP	ca. 0.25	(*89, 117*)
Regenerating rat liver	Ornithine	—	3X	0.044	7.4–8.1	0.2	PLP	—	(*118*)
C. Plant Decarboxylases[k]									
Barley seedlings	Arginine	—	10X	—	6–8	0.75	?	—	(*97*)
Barley roots	Glutamic acid	γ-Methylene glutamic acid	—	—	5–6	3–10	PLP	—	(*119–121*)

[a] All of the decarboxylases listed are specific for L-amino acids except diaminopimelic acid decarboxylase (which acts on the D center of *meso*-diaminopimelic acid) and glycine decarboxylase.

[b] Apparent K_m for PLP, usually determined from plots of 1/v vs. 1/PLP.

[c] K_{PLP} has been determined only for the enzyme from *L. arabinosus* (*122*).

[d] Several compounds [e.g., 1-methylhistidine, (1,2,4-triazole)-3-alanine, and β-(2-pyridyl)alanine] are decarboxylated at a rate about 1% that of histidine (*108*) if present at sufficiently high concentrations.

[e] Expressed on a dry weight basis.

[f] K_m for dihydroxyphenylalanine; tyrosine is not sufficiently soluble.

[g] Per milligram wet weight of tissue, assayed at pH 6.6.

[h] On all substrates except dihydroxyphenylalanine which is unstable at alkaline pH. Assays with dihydroxyphenylalanine as substrate were conducted at pH 7.0.

[i] Determined for the guinea pig liver enzyme, which, unlike the hog kidney enzyme, is resolved by dialysis against EDTA [see Christenson *et al.* (*79*)].

[j] At pH 6; the apparent K_m varies with pH.

[k] Other plant decarboxylases are known but have not been included since only their pH optima are known (see text).

agents (*22*) but apparently only at concentrations higher than those that inhibit the mammalian enzyme (*115*). It is not clear whether the mammalian enzyme is a PLP enzyme, a pyruvate enzyme, or dependent upon some unidentified carbonyl compound. Crude preparations of the *E. coli* enzyme are activated 30-fold or more by Mg^{2+} but not by putrescine. The rat prostate enzyme, by contrast, is activated at least 10-fold by putrescine but is not affected by Mg^{2+} (*115*). Stimulation by putrescine does not result solely from removal of the decarboxylation product of adenosylmethionine by a transferase catalyzing reaction (8) because, while partially purified [80×, (*115*)] preparations of the mammalian enzyme catalyze the overall reaction (9) [i.e., reaction (7) + (8)], more highly purified preparations [500×, (*115a*)] are still acti-

$$S\text{-Adenosylmethionine} \rightarrow CO_2 + S\text{-methyladenosylhomocysteamine} \quad (7)$$

$$S\text{-Methyladenosylhomocysteamine} + \text{putrescine} \rightarrow \text{spermidine} + \text{thiomethyladenosine} \quad (8)$$

vated by putrescine but catalyze only reaction (7). It is not known whether additional purification removes a separate transferase catalyzing reaction (8), or simply inactivates one site or subunit of a decarboxylase complex that normally catalyzes the overall reaction (9).

$$S\text{-Adenosylmethionine} + \text{putrescine} \rightarrow CO_2 + \text{spermidine} + \text{thiomethyladenosine} \quad (9)$$

Aminomalonic acid decarboxylase was first observed in silkworms (*124*) and has been purified twofold from rat liver (*94*); further attempts to purify it produced unstable preparations. It is stimulated by PLP and inhibited by both cyanide and hydroxylamine; the activity is apparently not a secondary consequence of another enzyme since aspartic acid and several other amino acids are not decarboxylated by the liver enzyme (*94*). Aminomalonic acid is also decarboxylated by the bacterial β-aspartate decarboxylase (*125*).

Arginine decarboxylase has been found in several species of bacteria (*1*), in barley, and probably in other plants (*97*), but not in mammals (*60*). Both the inducible (*99, 126–129*) and the biosynthetic (*28, 100, 101*) enzymes from *E. coli* have been extensively characterized. These two enzymes catalyze the same reaction but otherwise differ in nearly

124. K. Shimura, H. Nagayama, and A. Kikuchi, *Nature* **177**, 935 (1956).
125. A. G. Palekar, S. S. Tate, and A. Meister, *Biochemistry* **9**, 2310 (1970).
126. G. Melnykovich and E. E. Snell, *J. Bacteriol.* **76**, 518 (1958).
127. E. A. Boeker and E. E. Snell, *JBC* **243**, 1678 (1968).
128. E. A. Boeker, E. H. Fischer, and E. E. Snell, *JBC* **244**, 5239 (1969).
129. E. A. Boeker, E. H. Fischer, and E. E. Snell, *JBC* **246**, 6776 (1971).

every respect. They are produced under entirely different growth conditions and have pH optima separated by three pH units; their molecular weights and subunit structures also differ (see Table II). Double diffusion in agar of the biosynthetic enzyme and antibodies directed against the inducible enzyme does not result in a zone of precipitation, and the biosynthetic enzyme is not inhibited by antibody concentrations which cause 80% inhibition of the inducible enzyme. While the inducible enzyme does not require metal ions for activity, the biosynthetic enzyme has an absolute requirement for Mg^{2+} ($K_m = 1.2$ mM); the dependence on Mg^{2+} concentration is cooperative and has a Hill coefficient of 1.8. Mn^{2+} is 40% as effective as Mg^{2+} and also has a Hill coefficient of 2. Neither enzyme shows cooperative substrate interactions.

Arginine decarboxylase from barley seedlings (*97*), like the biosynthetic enzyme from *E. coli*, has an alkaline pH optimum but does not require metal ions for activity. It is inhibited by carbonyl trapping reagents, but attempts to resolve it or to demonstrate stimulation by PLP were unsuccessful.

Aspartic acid β-decarboxylase has been found in several species of bacteria [see Kakimoto *et al.* (*130*) and Tate and Meister (*130a*) for references] and purified to homogeneity from *Achromobacter* sp. (*131*), *Alcaligenes faecalis* (*102, 103*), and *Pseudomonas dacunhae* (*130, 131a*). To the extent that directly comparable information is available, all three enzymes are quite similar. The *P. dacunhae* and *A. faecalis* enzymes have very similar amino acid compositions (*130, 132*) and probably have identical subunit structures, although there is conflicting evidence on this point (see Section VI). All three have similar spectra, are stimulated by keto acids, and catalyze a slow transamination reaction in addition to β-decarboxylation (*102, 131a–133*; see Section IV,C). Chibata *et al.* (*131a*) found that the *P. dacunhae* enzyme, unlike the *A. faecalis* enzyme, exhibits cooperative kinetics unless α-ketoglutarate is present during the reaction. Tate and Meister (*132*), however, were unable to find evidence of cooperative interactions and suggested that the results of Chibata *et al.* may have been affected by the serum albumin present in their reaction mixtures. Since Tate and Meister were also able to

130. T. Kakimoto, J. Kato, T. Shibatani, N. Nishimura, and I. Chibata, *JBC* **244,** 353 (1969).

130a. S. S. Tate and A. Meister, *Advan. Enzymol.* **35,** 503 (1971).

131. E. M. Wilson, *BBA* **67,** 345 (1963).

131a. I. Chibata, T. Kakimoto, J. Kato, T. Shibatani, and N. Nishimura, *BBRC* **32,** 375 (1968).

132. S. S. Tate and A. Meister, *Biochemistry* **9,** 2626 (1970).

133. E. M. Wilson and H. L. Kornberg, *BJ* **88,** 578 (1963).

demonstrate hybrid formation between the *P. dacunhae* and *A. faecalis* enzymes, these two decarboxylases must be closely homologous. Aspartate β-decarboxylase is the subject of a recent extensive review (*130a*).

α-Decarboxylation of aspartic acid (to yield β-alanine) has been reported to occur in legume bacteria (*134–136*), but the activity is extremely weak and the enzyme has not been characterized sufficiently to eliminate the possibility that β-alanine formation is not direct.

Cysteine sulfinic acid decarboxylase is found most commonly in mammalian livers (*92, 93, 137, 138*) and also in brain, where much of it is localized in the crude mitochondrial and nerve ending particles (*138a*); it has not been purified extensively. It decarboxylates cysteic acid approximately one-sixth as fast as cysteine sulfinic acid, is stimulated by PLP and protected by reducing agents, and disappears from livers during vitamin B_6 deficiency at a rate significantly faster than several other PLP enzymes.

α-Dialkyl amino acid transaminase has been found in several soil bacteria (*139*), isolated from a pseudomonad of unknown species (*46, 140*), and purified to homogeneity from *P. fluorescens* (*6, 141*). It catalyzes transamination of α-substituted amino acids by labilizing the α-carboxyl group instead of the more usual α-hydrogen which is not present in these substrates; the overall stoichiometry is shown in reaction (3). Pyruvate ($K_m = 2$ mM) and α-ketobutyrate are equally effective as amino group acceptors; α-ketovalerate, glyoxalate, and α-ketoisocaproate are less effective. In addition to decarboxylation-dependent transamination of several substituted amino acids, the enzyme catalyzes a slow α-hydrogen-dependent transamination of L-alanine (see Section IV,A).

Diaminopimelic acid decarboxylase is a biosynthetic enzyme found in bacteria and other organisms which synthesize lysine from aspartic acid β-semialdehyde (*24, 142*). In all species examined, its formation is repressed both in enriched media and in late stages of growth, and it

134. A. I. Virtanen and T. Laine, *Enzymologia* **3**, 266 (1937).
135. D. Billen and H. C. Lichstein, *J. Bacteriol.* **58**, 215 (1949).
136. W. E. David and H. C. Lichstein, *Proc. Soc. Exptl. Biol. Med.* **73**, 216 (1950).
137. H. Blaschko, *BJ* **36**, 571 (1942).
138. F. Chatagner, H. Tabechian, and B. Bergeret, *BBA* **13**, 313 (1954).
138a. L. K. Kaczmarek, H. C. Agrawal, and A. N. Davison, *BJ* **119**, 45P (1970).
139. W. B. Dempsey, *J. Bacteriol.* **97**, 182 (1969).
140. H. G. Aaslestad, P. J. Bouis, Jr., A. T. Phillips, and A. D. Larson, *in* "Pyridoxal Catalysis" (E. E. Snell *et al.*, eds.), p. 479. Wiley (Interscience), New York, 1968.
141. G. B. Bailey, O. Chotamangsa, and K. Vuttivej, *Biochemistry* **9**, 3243 (1970).
142. W. E. Gutteridge, *BBA* **184**, 366 (1969).

is unaffected by diaminopimelic acid supplementation (*122, 143–147*). Since the substrate is *meso*-diaminopimelic acid and the product is L-lysine (*105, 148*), the enzyme is a D-amino acid decarboxylase; it is the only one known. It has been purified from both *E. coli* (*105*) and *Lactobacillus arabinosus* (*122*); acrylamide gel electrophoresis of the most highly purified preparation showed one intense band, comprising about 70% of the protein, and three lighter bands. The enzyme requires PLP and a reducing agent for full activity and stability. Mutant strains of *E. coli* are known in which this enzyme shows a reduced affinity for PLP; this change is reflected in a nutritional requirement for either lysine or vitamin B_6, but not both (*148a*).

Dihydroxyphenylalanine decarboxylase is widely distributed in mammals; it has been partially purified from guinea pig kidney (*78, 149–151*), from beef (*152*) and ox (*153*) adrenals, from rat liver (*154*), and more recently obtained as an essentially homogeneous protein from hog kidney (*79*). In addition to dihydroxyphenylalanine, preparations of the enzyme decarboxylate *o*-, *m*-, and *p*-tyrosine as well as tryptophan and 5-hydroxytryptophan quite rapidly; histidine is decarboxylated only very slowly. The concept that a single enzyme is responsible (*78, 149, 153–155*) is consistent with the substrate specificity of the pure hog kidney enzyme (*79*), but the possibility that other enzymes may also decarboxylate certain of these substrates *in vivo* remains open. The most active of the mammalian decarboxylases, dihydroxyphenylalanine decarboxylase, is not readily resolved but is inhibited by carbonyl reagents and is stimulated by PLP under the appropriate conditions. Even though the hog kidney enzyme is not resolved during purification and contains approximately 1 mole of PLP per 112,000 g of protein, its activity is significantly increased (up to 5-fold) by added PLP (*79*); the nature of this effect is not understood.

143. D. L. Dewey and E. Work, *Nature* **169,** 533 (1952).
144. D. L. Dewey, D. S. Hoare, and E. Work, *BJ* **58,** 523 (1954).
145. J-C. Patte, T. Loviny, and G. N. Cohen, *BBA* **58,** 359 (1962).
146. P. J. White, B. Kelly, A. Suffling, and E. Work, *BJ* **91,** 600 (1964).
147. D. P. Grandgenett and D. P. Stahly, *J. Bacteriol.* **96,** 2099 (1968).
148. D. S. Hoare and E. Work, *BJ* **61,** 562 (1955).
148a. A. J. Buchari and A. L. Taylor, *J. Bacteriol.* **105,** 908 (1971).
149. C. T. Clark, H. Weissbach, and S. Udenfriend, *JBC* **210,** 139 (1954).
150. E. Werle and D. Aures, *Z. Physiol. Chem.* **316,** 45 (1959).
151. W. Lovenberg, J. Barchas, H. Weissbach, and S. Udenfriend, *ABB* **103,** 9 (1963).
152. J. H. Fellman, *Enzymologia* **20,** 366 (1959).
153. P. Hagen, *Brit. J. Pharmacol.* **18,** 175 (1962).
154. J. Awapara, R. P. Sandman, and C. Hanley, *ABB* **98,** 520 (1962).
155. E. Rosengren, *Acta Physiol. Scand.* **49,** 364 (1960).

Tyrosine decarboxylase, the bacterial counterpart of dihydroxyphenylalanine decarboxylase, is found primarily in *Streptococcus faecalis* (*156*) but has been observed in one pseudomonad (*157*) and a few lactobacilli (*158, 159*). In the most recently reported purification, Epps (*113*) showed that the *S. faecalis* enzyme also decarboxylates dihydroxyphenylalanine at one-sixth the rate of tyrosine. The enzyme is readily resolved and crude preparations of the apodecarboxylase are frequently used to assay for PLP (*114*). Slow bacterial decarboxylation of phenylalanine (*157*) and tryptophan (*160*) has been observed; the possibility that tyrosine decarboxylase is responsible for these activities remains open.

Glutamic acid decarboxylase is widely distributed in higher plants (*119–121, 161–165*). The enzyme has not been purified but is known to be PLP dependent and to decarboxylate γ-methyleneglutamic acid in addition to glutamic acid.

In mammals, the enzyme was thought to be confined to the central nervous system until recently when low levels of what may be a distinct enzyme were detected in mouse kidneys and several other organs (*166*). The enzyme from mouse brain is PLP dependent, stabilized by reducing agents, and has been partially purified (*116, 167*). It is inhibited by a variety of carbonyl trapping reagents, such as aminooxyacetic acid, as well as by Cl^- and several other anions. In contrast, the kidney enzyme is not stimulated by PLP, is apparently unaffected by semicarbazide, is activated by aminooxyacetic acid, and is apparently activated rather than inhibited by Cl^-. Glutamate decarboxylase has also been partially purified from the central nervous system of lobsters (*168*); this enzyme is clearly PLP dependent and is unaffected by anions, but it has an almost absolute requirement for K^+.

The bacterial glutamate decarboxylase has been purified to homo-

156. E. F. Gale, *BJ* **34,** 846 (1940).
157. G. R. Seaman, *J. Bacteriol.* **80,** 830 (1960).
158. V. A. Lagerborg and W. E. Clapper, *J. Bacteriol.* **63,** 393 (1952).
159. A. W. Rodwell, *J. Gen. Microbiol.* **8,** 224, 233, and 238 (1953).
160. C. Mitoma and S. Udenfriend, *BBA* **37,** 356 (1960).
161. K. Okunuki, *Botan. Mag.* (*Tokyo*) **51,** 270 (1937).
162. O. Schales and S. Schales, *ABB* **11,** 155 (1946).
163. M. Ohno and K. Okunuki, *J. Biochem.* (*Tokyo*) **51,** 313 (1962).
164. C. Nations, *Can. J. Botany* **45,** 1917 (1967).
165. C. Nations and R. M. Anthony, *Can. J. Biochem.* **47,** 821 (1969).
166. B. Haber, K. Kuriyama, and E. Roberts, *Biochem. Pharmacol.* **19,** 1119 (1970).
167. E. Roberts and D. G. Simonsen, *Biochem. Pharmacol.* **12,** 113 (1963).
168. P. B. Molinoff and E. A. Kravitz, *J. Neurochem.* **15,** 391 (1968).

geneity from *Clostridium perfringens* (*169*) and from several strains of *E. coli* (*13–15, 106, 107, 170–172*). The effects of cultural conditions and genetic constitution on the production of the *E. coli* enzyme have been analyzed, and both a structural and a regulatory gene have been identified and mapped (*173–177*). The properties of the purified enzyme are discussed in detail in the succeeding sections.

Glycine decarboxylase has been detected in avian (*178*) and mammalian (*179*) livers and in *E. coli* (*180*) as well as in *Peptococcus glycinophilus* (*42–44, 181*) where it has been examined in detail. Since the mechanism of the reaction catalyzed by this complex of four enzymes differs from that of the other decarboxylases in many important respects, the entire system is discussed in Section IV,B.

Many bacteria produce small quantities of histamine from histidine, presumably by the action of *histidine decarboxylase*, but this activity varies greatly from species to species and from strain to strain within a single species for reasons that are completely unknown (*1, 181a,b*). Among the bacteria tested, the decarboxylase is found most commonly among intestinal organisms (*181c*). Among the richest sources, in which its level is greatly increased by culturing in the presence of L-histidine, are certain lactobacilli (*159*), micrococci (*182*), and some strains of *E. coli* (*1, 181a*). The enzyme has been purified to homogeneity and crystallized from a *Micrococcus* (*183*) and from *Lactobacillus* 30a (*16–18, 108, 184*). The apparently pure enzyme from the latter source contains

169. I. Cozzani, A. Misuri, and C. Santoni, *BJ* **118,** 135 (1970).
170. R. Shukuya and G. W. Schwert, *JBC* **235,** 1658 (1960).
171. J. A. Anderson and H. W. Chang, *ABB* **110,** 346 (1965).
172. P. H. Strausbauch and E. H. Fischer, *Biochemistry* **9,** 226 (1970).
173. Y. S. Halpern and H. E. Umbarger, *J. Gen. Microbiol.* **26,** 175 (1961).
174. Y. S. Halpern, *BBA* **61,** 953 (1962).
175. M. Marcus and Y. S. Halpern, *J. Bacteriol.* **93,** 1409 (1967).
176. M. Lupo, Y. S. Halpern, and D. Sulitzeanu, *ABB* **131,** 621 (1969).
177. M. Lupo and Y. S. Halpern, *BBA* **206,** 295 (1970); *J. Bacteriol.* **103,** 382 (1970).
178. D. A. Richert, R. Amberg, and M. Wilson, *JBC* **237,** 99 (1962).
179. I. C. Gunsalus, R. B. Morton, M. J. Namtvedt, and R. D. Sagers, *Abstr. Papers, 148th ACS Meeting, Chicago* 51C (1964).
180. J. D. Pitts and G. W. Crosbie, *BJ* **83,** 35P (1962).
181. M. L. Baginsky and F. M. Huennekens, *BBRC* **23,** 600 (1966).
181a. M. T. Hanke and K. K. Koessler, *JBC* **50,** 131 (1922).
181b. O. Ehrismann and E. Werle, *Biochem. Z.* **318,** 560 (1947).
181c. A. H. Eggerth, *J. Bacteriol.* **37,** 205 (1939).
182. S. R. Mardashev and L. A. Siomina, *Dokl. Akad. Nauk SSSR* **156,** 465 (1964).
183. S. R. Mardashev, L. A. Semina, V. N. Prozorovskii, and A. M. Sokhina, *Biokhimiya* **32,** 761 (1967).
184. G. W. Chang and E. E. Snell, *Biochemistry* **7,** 2012 (1968).

no PLP (*16, 108*) and is not dependent upon added PLP; this is probably true for the *Micrococcus* enzyme as well (*185*). The enzyme is inhibited by high concentrations of carbonyl reagents (*16, 17, 185*); this sensitivity results from the presence of pyruvate as a catalytically essential prosthetic group (*17–19*). The enzyme has an unusually broad pH optimum centered around pH 4.8; its action is discussed more fully in Section IV,B.

Mammalian histidine decarboxylase is widely distributed although activities are generally low. Because histidine is decarboxylated inefficiently by dihydroxyphenylalanine decarboxylase (*q.v.*) some confusion exists in the older literature concerning the distribution of the specific histidine decarboxylase. However, the latter enzyme has been observed in mast cell tumors (*77*), fetal rat tissues (*21, 89, 186*), hamster placenta (*186*), rat bone marrow (*187*), rat stomach (*82, 83, 188*), and possibly in other vertebrate stomachs (*189, 190*) although conflicting evidence exists on this point (*191*). Unlike the purified bacterial enzyme, the mammalian enzyme is almost certainly PLP dependent since it loses activity when dialyzed and activity is restored by PLP (*20, 21, 117*). However, because of the low activity of the mammalian enzyme, this point will require reexamination when more highly purified preparations become available. The K_m value of the mammalian decarboxylase decreases regularly between pH 4 and 9 indicating that only the unprotonated form of the substrate is active (*21*). This behavior contrasts with that of the purified enzyme from *Lactobacillus* 30a, where the K_m exhibits minimal values between pH 4.5 and 6.0, then increases rapidly with further increase in pH (*19*).

Histidine decarboxylase activity is found also in spinach and some other plants (*192*); however, its distribution in plants has not been widely investigated.

Lysine decarboxylase has been detected in lactobacilli (*159*) and purified to homogeneity from *Bacterium cadaveris* (*110*) and *E. coli*

185. S. R. Mardashev, L. A. Siomina, N. S. Dabagov, and N. A. Gonchar, *in* "Pyridoxal Catalysis" (E. E. Snell *et al.*, eds.), p. 451. Wiley (Interscience), New York, 1968.

186. R. Håkanson, *European J. Pharmacol.* **1,** 383 (1967).

187. R. Håkanson, *Experientia* **20,** 205 (1964).

188. R. Håkanson and C. Owman, *Biochem. Pharmacol.* **15,** 489 (1966).

189. W. Lorenz, S. Halbach, M. Gerant, and E. Werle, *Biochem. Pharmacol.* **18,** 2625 (1969).

190. W. W. Noll and R. J. Levine, *Biochem. Pharmacol.* **19,** 1043 (1970).

191. D. Aures, W. Davidson, and R. Håkanson, *European J. Pharmacol.* **8,** 100 (1969).

192. E. Werle and A. Raub, *Biochem. Z.* **318,** 538 (1947).

(*193–195*). Although the enzyme has not yet been fully characterized, it clearly resembles the inducible arginine decarboxylase of *E. coli* in several respects (see Sections V,B and VI). The enzymes are readily separable, however, and antibodies directed against arginine decarboxylase do not cross react with lysine decarboxylase.

Regarding *neutral amino acid decarboxylases*, crude extracts of *Proteus vulgaris* decarboxylate valine, leucine, norvaline, isoleucine, and α-aminobutyric acid at roughly the same rate (*23, 109*). A single inducible enzyme is apparently responsible for all five activities since the pH profile of each is identical and the relative activity levels for each substrate are independent of the inducer used; the enzyme has been partially purified but is rather unstable. It is slowly inactivated when leucine is the substrate; this is probably caused by resolution of the bound coenzyme since excess PLP restores activity.

Decarboxylation of valine and leucine has also been observed in three plant species (*95, 96*). The pH optimum (7.5) and pattern of hydroxylamine inhibition are similar for both reactions, which are probably catalyzed by a single enzyme. Extracts of *Ecballium elaterium* decarboxylate both alanine and serine (*196, 197*); the reactions have pH optima of 5.5 and 6.5, respectively, and may result from different enzymes.

Ornithine decarboxylase is widely distributed in both bacteria (*1, 159*) and mammals (see Section II). Like arginine decarboxylase, both a biosynthetic and an inducible ornithine decarboxylase are produced in *E. coli* (*27, 28*). The biosynthetic enzyme has been extensively purified and is currently being characterized (*100, 112*); the best preparations show two to four bands after acrylamide gel electrophoresis. The apparent molecular weight changes with pH indicating that the enzyme forms an associating–dissociating system. Like the biosynthetic arginine decarboxylase, it is inhibited by physiological concentrations of putrescine and spermidine, but it does not require metal ions for activity. The inducible ornithine decarboxylase has not been purified from any source; it is reported to be extremely unstable (*111*).

The mammalian ornithine decarboxylase is also unstable (*115, 118*); recent evidence (*198*) shows that thiols such as dithiothreitol preserve the activity of the rat prostate enzyme and suggests that, in the ab-

193. I. H. Sher and M. F. Mallette, *ABB* **53,** 354 (1954).
194. A. Maretzki and M. F. Mallette, *J. Bacteriol.* **83,** 720 (1962).
195. D. Sabo, H. Waron, E. A. Boeker, and E. H. Fischer, manuscript in preparation.
196. P. Dunnill and L. Fowden, *J. Exptl. Botany* **14,** 237 (1963).
197. O. J. Crocomo and L. Fowden, *Phytochemistry* **9,** 537 (1970).
198. J. Jänne and H. G. Williams-Ashman, *JBC* **246,** 1725 (1971).

sence of thiols, the enzyme polymerizes to an inactive species. The enzymic activity is also dependent on PLP and is inhibited by the polyamines, putrescine, spermidine, and spermine.

IV. Mechanism of Action

The mechanism of PLP catalyzed reactions, of which these decarboxylation reactions comprise one class, has been extensively reviewed in previous volumes of this treatise (*7*, *198a*) and elsewhere (*4*, *199*, *200*).

A. PLP-Dependent α-Decarboxylation

The experiments of Mandeles *et al.* (*201*) with bacterial tyrosine, lysine, and glutamate decarboxylase established that decarboxylation in ^2H_2O introduces only one atom of deuterium on the α-carbon of the product; i.e., the α-hydrogen atom of the substrate amino acid does not exchange with the medium. The monodeuteroamine produced is one of a pair of enantiomers; the opposite enantiomer can be obtained by decarboxylation of the α-deutero-amino acid in H_2O (*202*). Belleau and Burba (*203*) demonstrated that decarboxylation proceeds with retention of configuration by synthesizing both enantiomers of α-monodeutero-tyramine with known configurations and using monoamine oxidase to compare them to the enzymically produced enantiomers. These results suggest that the α-hydrogen atom of the amino acid substrate is not required for α-decarboxylase action. In accord with this suggestion dihydroxyphenylalanine decarboxylase does in fact decarboxylate the α-methyl analogs of three of its substrates approximately 1% as fast as the substrates themselves (*204*). Furthermore, nonenzymic pyridoxal-catalyzed decarboxylation of α-methyl-α-amino acids has been observed (*205*), and Aaslestad and Larson (*46*) and Bailey and Dempsey (*6*)

198a. A. E. Braunstein, "The Enzymes," 2nd ed., Vol. 2, Part A, p. 113, 1960.
199. E. E. Snell, *Vitamins Hormones* **16,** 77 (1958).
200. W. P. Jencks, "Catalysis in Chemistry and Enzymology." McGraw-Hill, New York, 1969.
201. S. Mandeles, R. Koppelman, and M. E. Hanke, *JBC* **209,** 327 (1954).
202. S. Mandeles and M. E. Hanke, *Abstr. Papers, 124th ACS Meeting, Chicago,* 55C (1953).
203. B. Belleau and J. Burba, *JACS* **82,** 5751 and 5752 (1960).
204. H. Weissbach, W. Lovenberg, and S. Udenfriend, *BBRC* **3,** 225 (1960).
205. G. D. Kalyankar and E. E. Snell, *Biochemistry* **1,** 594 (1962).

have isolated an enzyme from soil bacteria which decarboxylates similar substrates.

Rothberg and Steinberg (*206*) observed that decarboxylation in $H_2{}^{18}O$ does not lead to stoichiometric enrichment of the CO_2 produced with ^{18}O and thus eliminated the possibility of hydrolytic cleavage of the carbon–carbon bond with subsequent release of the CO_2 as bicarbonate. These experiments also eliminated the possibility that an amino-acyl–enzyme intermediate of the type shown below was formed prior to the release of CO_2.

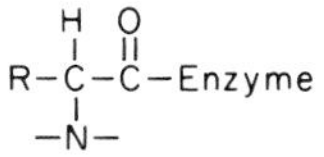

All of these results are consistent with the mechanism of PLP-catalyzed α-decarboxylation originally proposed by Metzler *et al.* (*207*) and independently by Westheimer [see (*201*)]. This mechanism, modified to include transaldimination between the enzyme–PLP complex and the substrate, is shown in Fig. 1. Two sets of observations indicate that transaldimination must be included. Fischer *et al.* (*208*) showed that

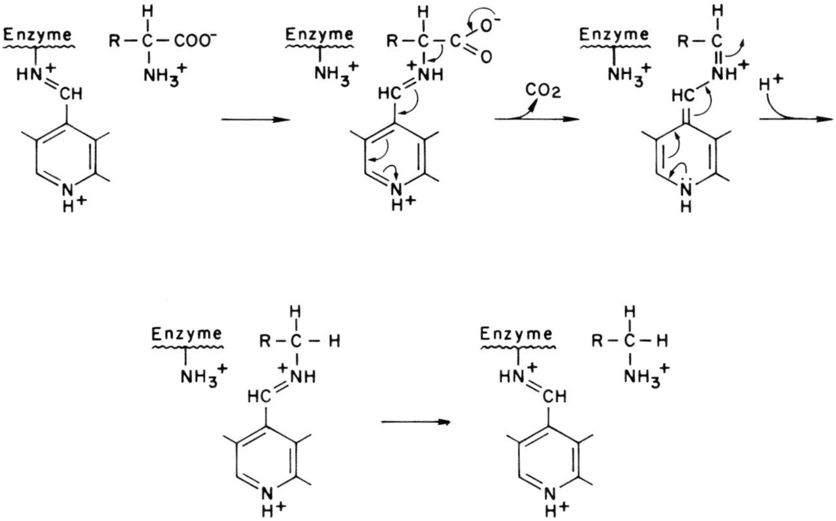

FIG. 1. Postulated mechanism for participation of PLP in enzymic α-decarboxylation of amino acids.

206. S. Rothberg and D. Steinberg, *JACS* **79,** 3274 (1957).
207. D. Metzler, M. Ikawa, and E. E. Snell, *JACS* **76,** 648 (1954).
208. E. H. Fischer, A. B. Kent, E. R. Snyder, and E. G. Krebs, *JACS* **80,** 2906 (1958).

in phosphorylase PLP is linked to the ε-amino group of a specific lysyl residue by a labile covalent bond which behaves toward $NaBH_4$ under certain conditions as an azomethine bond. Similar results have since been obtained for every PLP enzyme examined, and specifically for glutamate (*15, 171*), β-aspartate (*133*), arginine (*99, 129*), and lysine (*195*) decarboxylase. In the presence of substrate, however, borohydride reduction of certain PLP enzymes yields the reduced adduct of PLP with its amino acid substrate (*209*). Thus enzyme–substrate (ES) formation requires transaldimination, and Cordes and Jencks (*210*) demonstrated that this provides an advantage in model systems where the Schiff base of PLP and methylamine is about 30 times more reactive toward semicarbazide than is free PLP; i.e., transaldimination proceeds more rapidly than does *de novo* Schiff base formation.

Snell (*199*) suggested that reaction specificity in enzymic systems might be achieved by specific binding of two of the three groups surrounding the α-carbon atom of the substrate in such a way that productive labilization of only one of the three bonds (which are all weakened in model systems) could occur. Dunathan (*211*) has proposed that this effect is achieved by limiting the potentially free rotation around the substrate α-carbon to a single orientation in which the one α-carbon bond in a plane perpendicular to the plane of the conjugated system of the Schiff base is labile. The observation that decarboxylation does not cause exchange of the α-hydrogen atom is consistent with such proposals. A more rigorous test of part of Dunathan's hypothesis was constructed by Bailey and co-workers (*6, 141*) using the decarboxylation-dependent α-dialkylamino acid transaminase [see Eq. (3)]. The natural substrates of this enzyme, α-aminoisobutyric acid and α-aminoisovaleric acid, have two alkyl substituents on the α-carbon. The enzyme is weakly active on amino acids having a single alkyl substituent such as alanine or α-aminomethylmalonic acid, but it is not active on amino acids having no alkyl substituents such as glycine or α-aminomalonic acid. Bailey and co-workers therefore concluded that a single alkyl substituent is both necessary and sufficient to orient the substrate in the active site and reasoned that the two isomers of alanine should undergo different reactions since one would have its α-hydrogen in the reactive position and should be transaminated without decarboxylation while the other would undergo the normal decarboxylation-dependent transamination. Using labeled substrates, they demonstrated that the predicted reactions, (10) and (11), do in fact occur:

209. E. A. Malakhova and Yu. M. Torchinsky, *Dokl. Akad. Nauk SSSR* **161,** 1224 (1965).
210. E. H. Cordes and W. P. Jencks, *Biochemistry* **1,** 773 (1962).
211. H. C. Dunathan, *Proc. Natl. Acad. Sci. U. S.* **55,** 712 (1966).

$$CH_3-\underset{NH_3^+}{\overset{^{14}COO^-}{C}}-H + CH_3COCOO^- \xrightarrow{H^+} {}^{14}CO_2 + CH_3CHO + CH_3\underset{NH_3^+}{C}HCOO^- \tag{10}$$

D-Alanine

$$CH_3-\underset{NH_3^+}{\overset{H}{C}}-{}^{14}COO^- + CH_3COCOO^- \longrightarrow CH_3CO^{14}COO^- + CH_3\underset{NH_3^+}{C}HCOO^- \tag{11}$$

L-Alanine

This experiment clearly suggests that the orientation of the substrate determines which of the α-carbon bonds becomes reactive, but, of course, it does not indicate what that orientation actually is.

B. Glycine Decarboxylase

The mechanism of the reaction catalyzed by glycine decarboxylase differs from that of the other α-decarboxylases in several important respects. Klein and Sagers (*41–44*) have shown that four proteins are required for the overall reaction (see Fig. 2): P_1, which has a molecular weight of approximately 125,000 and contains two moles of PLP; P_2, a heat-stable protein with a molecular weight of 10,000, which is apparently the initial electron acceptor; P_3, which has a molecular weight of approximately 120,000, contains 1 mole of FAD, and carries out the

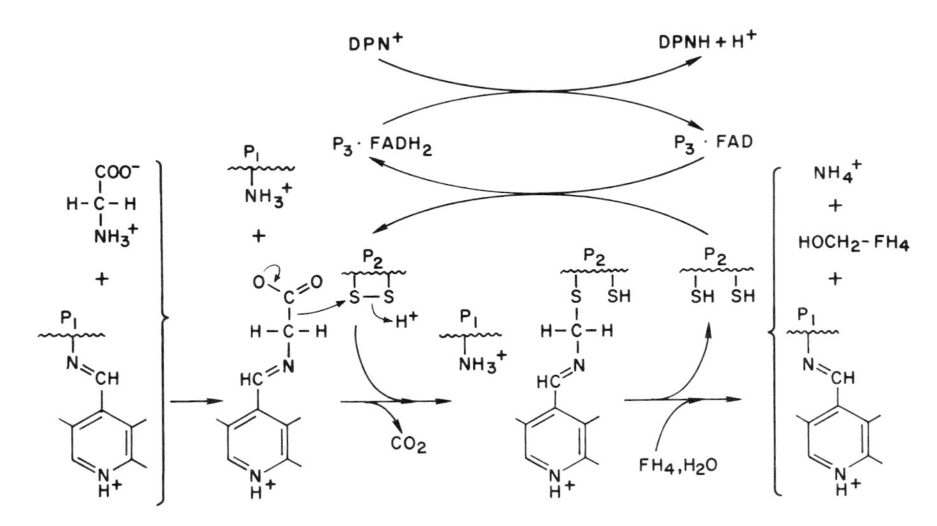

Fig. 2. Postulated role of PLP in the enzymic decarboxylation of glycine.

final transfer of electrons to DPN; and P_4, which is less well characterized but evidently catalyzes the final transfer of the glycine α-carbon to tetrahydrofolic acid.

P_1 alone catalyzes neither CO_2 release nor the exchange of labeled CO_2 with glycine; P_1 and P_2 together catalyze the exchange reaction and the entire system is required for CO_2 release. Since P_2 is required for CO_2 exchange, decarboxylation must be preceded or accompanied by reduction of P_2. Reduction could precede decarboxylation only if an α-hydrogen were first released from the glycine-PLP Schiff base; simultaneous reduction and decarboxylation requires only that the α-carboxyl group be labile. Since the complete system, including tetrahydrofolate, is required for reduction of P_3, the oxidation of P_2 must not precede the transfer of the α-carbon of glycine to tetrahydrofolate. The mechanism shown in Fig. 2 has been adapted from that suggested by Baginsky and Huennekens (*181*) to account for the observed sequence of reactions.

C. PLP-Dependent β-Decarboxylation

The reaction catalyzed by aspartic acid β-decarboxylase is fundamentally different from α-decarboxylation. The substituent on the α-carbon atom first labilized by the enzyme is the α-hydrogen atom: β-Decarboxylation resembles transamination and β-elimination reactions in this sense (*7*, *199*). In a series of papers beginning in 1951, Meister and his group (*5*, *102*, *104*, *212*, *213*) have unequivocally demonstrated that subsequent to β-decarboxylation this enzyme catalyzes either the formation of alanine and enzyme-bound PLP or (relatively infrequently) pyruvate and enzyme-bound PMP, and they have formulated the mechanism shown in Fig. 3 to account for this. The catalytic duality apparently arises from a lack of specificity in the protonation of the ketimine Schiff base (D in Fig. 3).

The transamination reaction (D $\rightarrow$ G $\rightarrow$ H in Fig. 3) is not specific in the sense that any of several keto acids will contribute to the steady state level of the ketimine Schiff base (D in Fig. 3); excess keto acid prevents the inactivation of the enzyme (which otherwise results from PMP formation during transamination) and allows β-decarboxylation to proceed with maximum efficiency. In the absence of added keto acid, steady state levels of pyruvate and the PMP form of the enzyme are first established at the expense of the PLP form, and the rate of β-de-

212. J. S. Nishimura, J. M. Manning, and A. Meister, *Biochemistry* **1**, 442 (1962).
213. A. Novogrodsky, J. S. Nishimura, and A. Meister, *JBC* **238**, PC1903 (1963).

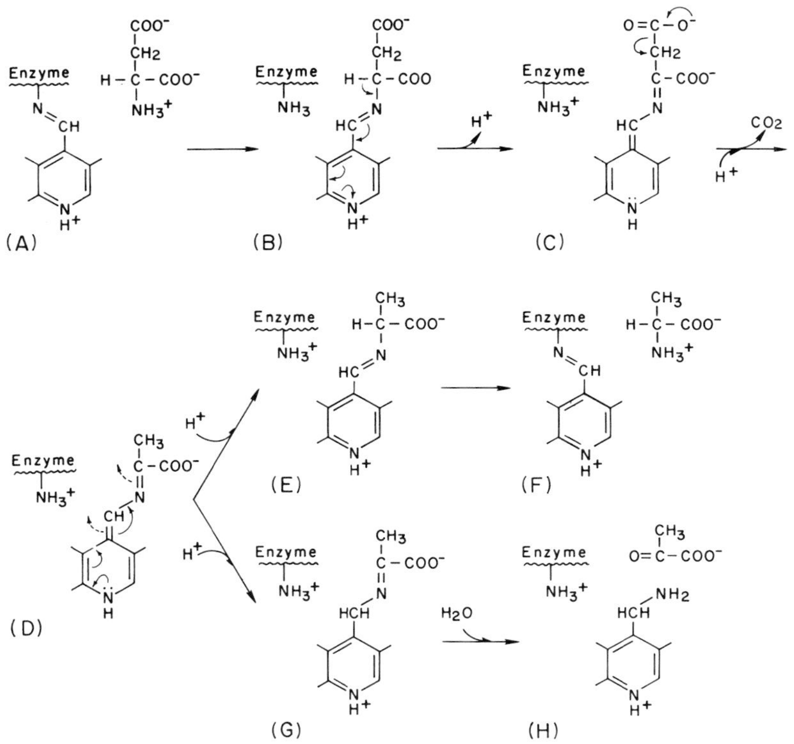

FIG. 3. Postulated role of PLP in the β-decarboxylation of aspartate.

carboxylation declines rapidly. When the steady state is achieved, the rate of β-decarboxylation is approximately 30% of the rate in the presence of excess keto acid, and Tate and Meister (*104*) have shown that the concentration of pyruvate formed under these conditions is approximately two-thirds of the molar concentration of the enzyme. Furthermore, if lactate dehydrogenase is added to remove the pyruvate as rapidly as it is formed, all of the enzyme is converted to the PMP form and β-decarboxylation ceases; addition of α-ketoglutarate restores full activity.

The activity of aspartate β-decarboxylase is clearly subject to very sensitive regulation by the level of α-keto acid; Novogrodsky and Meister (*102*) have suggested that other PLP enzymes may be controlled by similar mechanisms, but no evidence for this has yet been found. Tate and Meister (*104*) have also suggested that aspartate β-decarboxylase is, in a sense, an allosteric enzyme, i.e., that separate sites exist for substrate and keto acid which can be occupied simultaneously. Such separate sites are inferred because (1) the apparent K_m for aspartic acid is, within experimental error, the same in the presence of 1 mM α-ketoglutarate as in its absence, suggesting that the substrate and the keto acid do not compete for the same site; (2) the presence of a competitive inhibitor, β-cyanoalanine, does not interfere with the binding of pyruvate to the enzyme, as measured under equilibrium conditions by gel filtration; and (3) pyruvate is also bound by the apoenzyme, indicating that neither PLP nor PMP is required for formation of the binding site. Although separate binding sites for the substrate and the keto acid are not a necessary consequence of the mechanism shown in Fig. 3, these results provide highly suggestive evidence for such sites.

In addition to transamination and β-decarboxylation, aspartate β-decarboxylase also catalyzes the analogous desulfination of cysteine sulfinic acid (*214*), β elimination of Cl^- from β-chloroalanine (*215*), and α-decarboxylation of α-aminomalonic acid (*125*). All but the last reaction require the initial loss of an α-hydrogen and are consistent with the hypothesis (*211*) that the labile α-carbon substituent is determined by the orientation of the substrate. Studies in ^3H_2O have shown that aminomalonic acid is decarboxylated stereospecifically to S-^{3}H-glycine; assuming that inversion does not occur during addition of the proton, the carboxyl group lost corresponds to the β-carboxyl of aspartic acid (*125*). Additional recent studies using specifically labeled ^{14}C-carboxyl aminomalonic acids (prepared by oxidation of L-serine-1-^{14}C and L-serine-3-^{14}C) have eliminated the possibility that inversion occurs and thus confirm this conclusion (*215a*). This can be reconciled with Dunathan's hypothesis only if, of the possible binding modes for aminomalonic acid shown in Fig. 4, the mode in which the α-hydrogen occupies the usual β-carboxyl binding site (C in Fig. 4) does not occur.

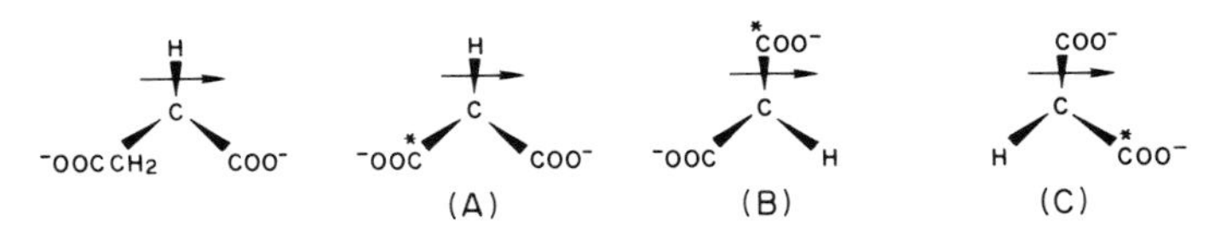

FIG. 4. Possible modes of binding of L-aspartate and of aminomalonate by aspartate β-decarboxylase. The α-amino group lies below the α-carbon atom in each instance. The asterisk indicates the carboxyl group which corresponds to the β-carboxyl group of aspartate, and the arrow indicates the bond which, according to Dunathan's hypothesis (*211*), would be labilized in each case.

214. K. Soda, A. Novogrodsky, and A. Meister, *Biochemistry* **3**, 1450 (1964).
215. S. S. Tate, N. M. Relyea, and A. Meister, *Biochemistry* **8**, 5016 (1969).
215a. A. G. Palekar, S. S. Tate, and A. Meister, *Biochemistry* **10**, 2180 (1971).

D. Pyruvate-Containing Decarboxylases

Histidine decarboxylase of *Lactobacillus* 30a is completely inhibited by reduction with $NaBH_4$ or by reaction with phenylhydrazine. When tritiated $NaBH_4$ was used, tritiated lactate was isolated from the acid-hydrolyzed enzyme; when [^{14}C]-phenylhydrazine was employed, the labeled phenylhydrazone of pyruvylphenylalanine was isolated from an enzymic digest (pronase + chymotrypsin) of the derivatized enzyme (*17*). These findings identified the essential carbonyl group of the enzyme as occurring in a pyruvate residue covalently bound to phenylalanine; five such residues occur per molecule (MW 190,000) of enzyme and arise in a yet-unknown fashion from precursor serine residues (*17, 18*). On reduction with $NaBH_4$ in the presence of ^{14}C-histidine, label is fixed to protein, and following acid hydrolysis of the labeled protein N^2-(1-carboxyethyl)[^{14}C]-histidine and larger amounts of N^1-(1-carboxyethyl)[^{14}C]-histamine were isolated. These data show that both histidine and its decarboxylation product, histamine, form Schiff bases during the enzymic reaction with these essential pyruvyl residues (*19*), presumably as part of a catalytic cycle. These data provide the experimental support for the mechanism shown in Fig. 5, which was originally suggested (*17*) largely on the basis of analogy with PLP-dependent decarboxylation reactions. An appropriate Lewis acid near the carbonyl

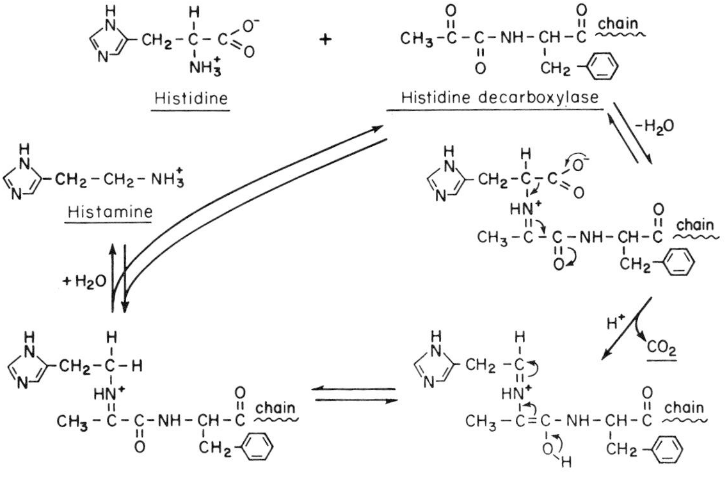

Fig. 5. Postulated role of the pyruvyl moiety of histidine decarboxylase in the decarboxylation of histidine.

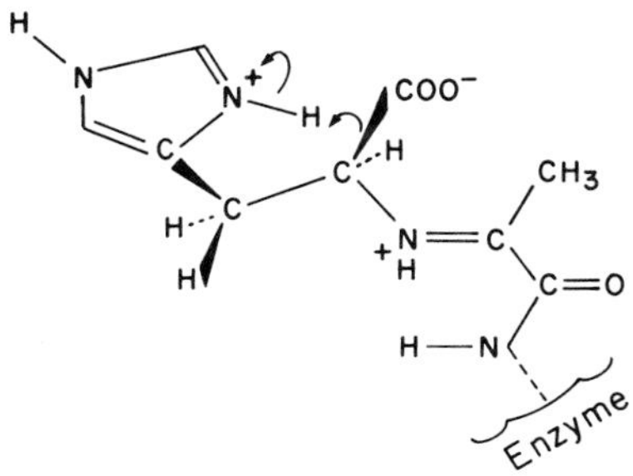

FIG. 6. A speculative catalytic role for the imidazole group of histidine during its decarboxylation by the bacterial histidine decarboxylase.

oxygen of the amide bond (Fig. 5) might, if present, contribute to the action of the protonated imine bond in weakening the bond to the α-carboxyl group. An attractive possibility, in view of the specificity of histidine decarboxylase, is that the protonated imidazole group of the substrate itself provides a proton that assists in elimination of the carboxyl group, as shown in Fig. 6. Kinetic data of Recsei and Snell (*19*), however, indicated very little change in V_m from pH 2.9 to 7.6 and thus argue against such a mechanism.

V. The Active Site of Amino Acid Decarboxylases

A. ABSORPTION SPECTRA OF PLP-DEPENDENT DECARBOXYLASES

A prominent absorption maximum at 415–430 nm is characteristically present in all four of the purified α-decarboxylases so far examined at their optimum pH. Arginine and lysine decarboxylase each have a maximum at 420 nm which is independent of pH between 5.0 and 8.0 (*99, 110*). The PLP enzyme of the glycine decarboxylase complex (P_1) has been examined only at its optimum pH, 7.0, where it has a maximum at 430 nm (*43, 44*). Glutamate decarboxylase is maximally active between pH 3 and 5 (*13, 106*) and has an absorption maximum at 415 nm in this pH range. This maximum decreases rapidly if the pH is increased from 5 to 6, and a new maximum at 340 nm appears (*14, 106, 171*). In all four enzymes, $NaBH_4$ reduction of the species absorbing at 415–430 nm causes inactivation of the enzyme (*15, 44, 99, 171, 195*); the glutamate decarboxylase species which absorbs at 340 nm is resistant to $NaBH_4$ reduction (*171*).

In contrast to the α-decarboxylases, aspartic acid β-decarboxylase has a pH independent absorption maximum at 360 nm (*102, 130, 131, 133, 216*); this species is reduced by $NaBH_4$ (*133*).

Although the structures of the absorbing species in these enzymes have not been established, similar species have been identified in model systems (*217, 218*), phosphorylase (*208, 219*), aspartate aminotransferase (*220*), and bovine plasma albumin plus PLP (*221*). By analogy with these results, it is probable that the species absorbing at 340, 360, and 420 nm are, respectively, an uncharged Schiff's base or a substituted aldamine (Fig. 7A), an unsubstituted aldimine with no internal hydrogen bonding (Fig. 7B), and a hydrogen-bonded aldimine (Fig. 7C). The spectra of PLP and related compounds have been discussed in detail by Johnson and Metzler (*218*). Since spectra can be modified extensively depending upon the microenvironment of the PLP (*219*), assertions concerning the exact forms that correspond to a given spectral band must be evaluated with caution.

Although PLP is optically inactive, it is bound asymmetrically to both aspartic acid β-decarboxylase and glutamate decarboxylase, since the optical rotatory dispersion curve of the former exhibits a positive Cotton effect at 360 nm (*222*) and the circular dichroism curve of the latter has a maximum at 420 nm at pH 4.6 (*223, 224*). Similar asym-

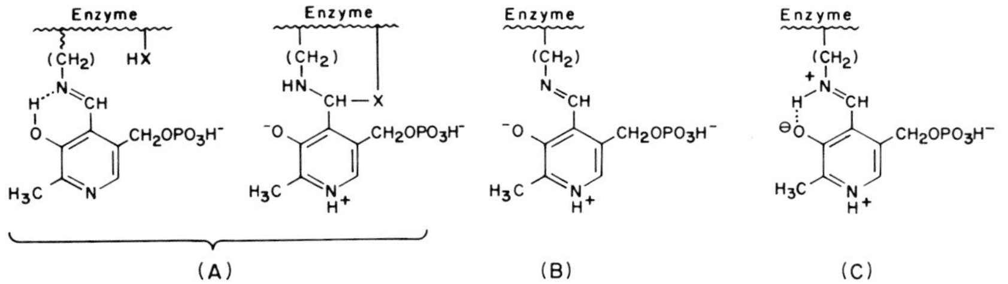

FIG. 7. Possible structures of PLP adducts having absorption maxima at (A) 330–340, (B) 360–370, and (C) 410–430 nm.

216. I. Chibata, T. Kakimoto, J. Kato, T. Shibatani, and N. Nishimura, *BBRC* **26,** 662 (1967).
217. D. E. Metzler, *JACS* **79,** 485 (1957).
218. R. J. Johnson and D. E. Metzler, "Methods in Enzymology," Vol. 18A, p. 433, 1970.
219. S. Shaltiel and M. Cortijo, *BBRC* **41,** 594 (1970).
220. W. T. Jenkins and I. W. Sizer, *JBC* **234,** 1179 (1959).
221. W. B. Dempsey and H. N. Christensen, *JBC* **237,** 1113 (1962).
222. E. M. Wilson and A. Meister, *Biochemistry* **5,** 1166 (1966).
223. B. S. Sukhareva and Yu. M. Torchinsky, *BBRC* **25,** 585 (1966).
224. T. E. Huntley and D. E. Metzler, *BBRC* **26,** 109 (1967).

metric binding to other PLP-dependent decarboxylases can be tentatively assumed.

B. The PLP Binding Site

Since $NaBH_4$ reduction was introduced (*208*) as a method of fixing the PLP residue to the peptide chain, the primary structure of that portion of the PLP binding site near the azomethine bond to PLP has become readily accessible. These structures have been determined for several PLP enzymes, including several decarboxylases. The extent to which such peptides include those amino acid residues which comprise the "active site" of the enzyme is unknown. It is clear that PLP is also bound to the apoenzyme by ionic and hydrophobic forces, and that the azomethine bond is broken by transimination during formation of the enzyme–substrate complex. It is not clear, after ES formation has occurred, whether this same portion of the peptide chain also provides other binding groups for the coenzyme or for the substrate, or contains any of the catalytic residues of the enzyme. Such peptide sequences from a variety of PLP enzymes have been compared elsewhere (*7*); the corresponding sequences from the glutamate (*15*), arginine (*128, 129*), and lysine (*195*) decarboxylases of *E. coli* are shown in Fig. 8. These are the only sequences known for PLP enzymes which are from the same source and which catalyze a single type reaction. It is perhaps not surprising, therefore, that they are also the only sequences which show any obvious

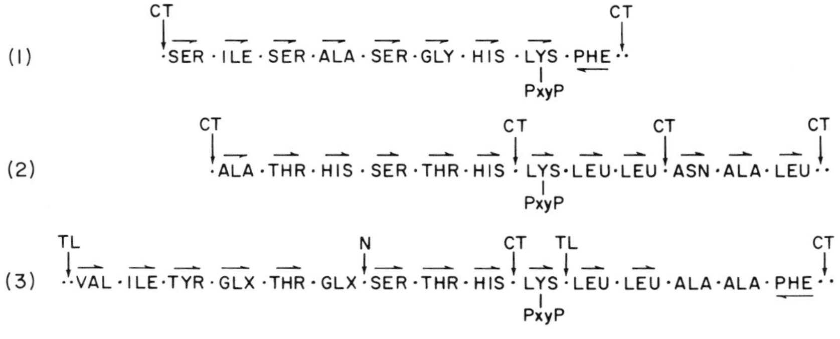

(1) ·SER·ILE·SER·ALA·SER·GLY·HIS·LYS·PHE··
(LYS–PxyP; CT cleavage before SER and after PHE)

(2) ·ALA·THR·HIS·SER·THR·HIS·LYS·LEU·LEU·ASN·ALA·LEU··
(LYS–PxyP; CT cleavage before ALA, after HIS, after LEU·LEU, after LEU)

(3) ··VAL·ILE·TYR·GLX·THR·GLX·SER·THR·HIS·LYS·LEU·LEU·ALA·ALA·PHE··
(LYS–PxyP; TL before VAL, N before SER, CT after HIS, TL after LYS, CT after PHE)

Fig. 8. Amino acid sequences of the peptide chains of (1) glutamate, (2) arginine, and (3) lysine decarboxylases from *E. coli* near the lysine residue which undergoes azomethine formation with PLP. CT, chymotrypsin; TL, thermolysin; N, Nagarse; and PxyP, the phosphopyridoxyl residue resulting from $NaBH_4$ reduction of PLP on the enzyme. Sequence data were obtained with (←) carboxypeptidase, (⇀) subtractive Edman degradations, and (→) the dansyl-Edman procedure.

homology; by the same token, however, it is difficult to assess the significance of this homology.

Perhaps the most noteworthy feature of the three peptides is the histidine residue which precedes the (PxyP)-lysyl residue. Since the optimum pH of all three enzymes is less than 6.0, the imidazole ring would presumably be protonated; if it is, indeed, part of the active site it could serve one of at least three functions: (1) it could donate a proton to the substrate α-carbon in the ketimine Schiff base formed during decarboxylation, or, (2) by ion pair formation, it could orient the α-carboxyl group, or (3) provide an accessory binding site for the PLP residue.

Two observations pertinent to the first two possibilities were made by Kalyankar and Snell (*205*) in a study of nonenzymic pyridoxal-catalyzed decarboxylations. Using α-aminoisobutyric acid, an α-substituted amino acid, to minimize competing reactions, they observed that metal ions inhibit decarboxylation and suggested that neutralization of the charge on the α-carboxyl by the metal ion might be partly responsible. Thus a histidine residue which formed an ion pair with the carboxyl group might in fact be expected to hinder decarboxylation. Dunathan's hypothesis and the evidence which supports it (Section IV,A) clearly indicate that the α-carboxyl group is specifically oriented, but the specific binding of any one of the α-carbon substituents would achieve this; the amino acid side chain is perhaps a better candidate than the α-carboxyl itself.

Kalyankar and Snell also observed that in the nonenzymic reaction the protonation of the ketimine Schiff base formed after decarboxylation was not specific; i.e., that isopropylamine and pyridoxal or acetone and pyridoxamine were formed about equally. This is essentially similar to the nonspecificity of the final steps of aspartic acid β-decarboxylation ($D \rightarrow F$ and $D \rightarrow H$ in Fig. 3). The α-decarboxylases, however, are specific, and they do not catalyze transamination, which implies that a specifically oriented proton donor such as histidine probably plays a part in the enzymic catalysis.

VI. Subunit Structure

The subunit structures of β-aspartate, glutamate, arginine, and histidine decarboxylases have been investigated quite thoroughly, and molecular weight data are available for some of the others. Much of this information is summarized in Table II.

TABLE II
MOLECULAR WEIGHTS AND SUBUNIT STRUCTURES OF PURIFIED AMINO ACID DECARBOXYLASES

		Native enzyme			Subunits			
Substrate	Source	Molecular weight	$s^{\circ}_{20,w}$	Residues of cofactor	Molecular weight	Number	Type	References
		A. PLP-Dependent Decarboxylases						
α-Aminoisobutyric acid	*P. fluorescens*	150,000	8.4 S[a]	—	—	—	—	(*6*)
Arginine	*E. coli* (inducible)	820,000	23.3 S	10	82,000	10	Probably identical	(*127, 128*)
Arginine	*E. coli* (biosynthetic)	280,000–300,000	—	—	70,000–75,000	—	—	(*101*)
Aspartic acid	*A. faecalis*	675,000	18.9 S[b]	12	58,000	12	Probably identical	(*225*)
Diaminopimelic acid	*E. coli*	200,000	5.4 S[c]	—	—	—	—	(*105*)
Dihydroxyphenylalanine	Hog kidney	112,000	5.8 S	1[d]	—	—	—	(*79*)
Glutamic acid	*E. coli*	300,000	12.7 S	6	50,000	6	Probably identical	(*172*)
Glycine	*P. glycinophilus* (P_1)	125,000	—	2	—	—	—	(*43*)
Lysine	*B. cadaveris*	ca. 10^6	21.1 S	10	82,000[e]	—	—	(*110*)
		B. Pyruvate-containing Decarboxylases						
Adenosylmethionine	*E. coli*	113,000	—	1–2	15,000	8	Noniden-tical (?)	(*22*)
Histidine	*Lactobacillus* 30a	190,000	9.2 S	5	29,700 9,000	5 5	Non-identical	(*18, 184*)

[a] At 5.0 mg/ml.
[b] At 0.4 mg/ml.
[c] At 1.7 mg/ml.
[d] Despite presence of this firmly bound PLP, the enzyme is further stimulated (up to 5-fold) by added PLP (*79*).
[e] Determined for the enzyme from *E. coli* (*195*).

Equilibrium ultracentrifugation of the *A. faecalis* aspartic acid β-decarboxylase at pH 5.5 yields a molecular weight of 675,000; electron micrographs show that this species has sixfold symmetry (*225*). If the enzyme is resolved of its PLP and the pH is raised to 8, the apoenzyme dissociates to dimers with an observed molecular weight of 115,000 (*103, 225*). Reassociation can be brought about either by the addition of PLP [or certain PLP analogs (*226*)] at pH 8 or by a decrease in the pH in the absence of PLP. The enzyme binds 1 mole of PLP per 50,000–60,000 g by a variety of criteria (*102, 104, 130, 131, 222*); the species observed in 5 *M* guanidine has a molecular weight of 58,000 and shows no indications of heterogeneity (*225*). The number of peptides produced by cyanogen bromide cleavage or tryptic digestion is consistent with 12 identical polypeptide chains with a molecular weight of 58,000 (*225*).

For the aspartic acid β-decarboxylase from *P. dacunhae*, Kakimoto *et al.* (*130, 227*) have calculated a molecular weight of 800,000 from its sedimentation and diffusion coefficients and obtained a similar value from the Archibald method of equilibrium ultracentrifugation. On the basis of molecular weight and sulfhydryl group determinations in 5 *M* guanidine·HCl with and without prior reduction of the enzyme, these workers concluded that the enzyme has eight subunits, each consisting of two polypeptide chains linked by a disulfide bond. However, Tate and Meister (*132*), who have also examined the *P. dacunhae* enzyme, found no evidence of disulfide bonds; they observed that its dissociation behavior, appearance in the electron microscope, and sedimentation coefficients were essentially identical to those of the enzyme from *A. faecalis*. Tate and Meister also took advantage of the fact that both enzymes dissociate reversibly to dimers and showed that they will form a seven-membered hybrid set if reassociated together.

It seems almost certain therefore that aspartic acid β-decarboxylase consists of 12 subunits, which are probably identical, and dissociates readily and reversibly to six dimers. Bowers *et al.* (*225*) pointed out that, in general, six isologous dimers can be symmetrically assembled only as a hexagonal array or as the six edges of a tetrahedron, and they

225. W. F. Bowers, V. B. Czubaroff, and R. H. Haschemeyer, *Biochemistry* **9**, 2620 (1970).

226. Pyridoxal 5′-phosphate has a major effect on conformation near the active center of many PLP enzymes. The activity of various PLP analogs, in playing both the structural and catalytic role of the coenzyme, is not treated in this review but has been summarized recently elsewhere [E. E. Snell, *Vitamins Hormones* **28**, 265 (1970)].

227. T. Kakimoto, J. Kato, T. Shibatani, N. Nishimura, and I. Chibata, *JBC* **245**, 3369 (1970).

suggested that the latter possibility is indicated by their electron micrographs.

Glutamate decarboxylase also forms an associating–dissociating system. Strausbauch and Fischer (*172*) have shown that both the sedimentation coefficient and the diffusion coefficient are severely concentration dependent, and equilibrium ultracentrifugation at both pH 4.5 and 7.0 shows clear evidence of dissociation at low protein concentrations. No search for mild conditions leading to a specific dissociation has been undertaken, but a slowly sedimenting species has been observed during sedimentation velocity experiments at pH 7. The best estimate for the molecular weight of the native species is 300,000; it binds 1 mole of PLP per 50,000–55,000 g. The subunit molecular weight obtained by equilibrium ultracentrifugation in 6 *M* guanidine·HCl is 50,000 and is essentially identical to the value of 47,000 obtained by acrylamide gel electrophoresis in sodium dodecyl sulfate. No evidence for nonidentical subunits was obtained in either experiment. The observed N- and C-terminal residues are methionine (0.4 moles/50,000 g) and threonine (0.8 moles/50,000 g), respectively. The PLP binding site is unique, and the number of tryptic peptides is also consistent with identical subunits (*15, 172*). Electron micrographs show in some orientations a hexagonal array of subunits (*228*); more extensive investigations (*228a*) show that most images are triangular in outline and appear to be superpositions of two triangles, one rotated 60° with respect to the other. These results strongly indicate that the undissociated glutamate decarboxylase is a hexamer with dihedral symmetry, with the subunits packed in an approximate octahedral arrangement.

The inducible arginine decarboxylase of *E. coli* (*99, 127–129*) also dissociates under mild conditions. At pH 5.2 and 0.1 *M* Na^+, the molecular weight observed by equilibrium ultracentrifugation is 820,000; this species dissociates reversibly to five dimers with an observed molecular weight of 160,000. Dissociation occurs only when the concentration of monovalent cations is less than 0.05 *M* and the pH is greater than 6.5. It is inhibited by divalent cations but does not require resolution of the enzyme-bound PLP. A five-membered ring is clearly visible in electron micrographs, and five association states can be observed simultaneously in sedimentation velocity experiments performed under the appropriate conditions. The enzyme binds 1 mole of PLP per 82,000 g. Equilibrium untracentrifugation in denaturing solvents and acrylamide gel electrophoresis in sodium dodecyl sulfate are both consistent with

228. A. S. Tikhonenko, B. S. Sukhareva, and A. E. Braunstein, *BBA* **167,** 476 (1968).

228a. C. M. To, *J. Mol. Biol.* **59,** 215 (1971).

identical subunits with an observed molecular weight of 74,000–86,000. The N- and C-terminal residues (0.7 mole of methionine and 1.0 mole of alanine per 82,000 g, respectively) are both apparently unique, as is the amino acid sequence at the PLP binding site. Isoelectric focusing of the reassociated dimers gives a single species, and the number of tryptic peptides is also consistent with identical subunits. The five subunits (dimers) visible in the electron micrographs (*127*) show no evidence of a fine structure (*229*). Two types of quaternary structure are consistent with these results: a double pentamer of dimers, or a decamer in which each dimer forms a single, closely packed subunit. The first structure requires that the interactions between dimers be heterologous; the second is consistent with isologous interactions of two different types. The importance of quaternary structure to activity of this enzyme is emphasized by the fact that its dimeric form has the same PLP content and ultraviolet spectrum as the decameric form but is very much less active (*127*) or inactive in catalyzing decarboxylation of arginine (*230*).

The subunit structure of lysine decarboxylase may resemble that of arginine decarboxylase quite closely. The native enzyme is an associating–dissociating system (*193*) and has a molecular weight on the order of 10^6 (*110*). The subunits of the *E. coli* arginine and lysine decarboxylases are not separated by acrylamide gel electrophoresis in sodium dodecyl sulfate (*195*).

Although the data shown in Table II for the biosynthetic arginine decarboxylase of *E. coli* are not complete, it is very clear that the subunit structures of the two arginine decarboxylases are not similar. The subunits of these two enzymes can be separated by acrylamide gel electrophoresis in sodium dodecyl sulfate (*101*).

The pyruvate-dependent histidine decarboxylase presents a unique subunit structure. The native enzyme has a molecular weight of 190,000 and a sedimentation coefficient of 9.2 S; these values decrease after carboxymethylation to about 19,000 and 1.1 S, respectively, consistent with dissociation to 10 subunits (*184*). Since only 5 pyruvate residues are present per 190,000 daltons, subunits of two different types were indicated (*17*), and, following carboxymethylation of the enzyme phenylhydrazone, nonidentical subunits were readily separated by Sephadex

229. G. B. Haydon, *J. Ultrastruct. Res.* **25**, 349 (1968).

230. Arginine and its analogs induce association of the dimeric to the decameric form of arginine decarboxylase (*127*); hence, the reduced activity shown by the dimer may result only from the decamer formed during assay. This interpretation is supported by the fact that L-ornithine does not induce such reassociation and is decarboxylated slowly by decameric arginine decarboxylase (at about 1% the rate of arginine) but not by the dimeric form of the enzyme.

or DEAE-chromatography (*18*). The pyruvate-containing subunit has a molecular weight of about 29,700; pyruvylphenylalanine is N-terminal, tyrosine is C-terminal, and two cysteine residues are present per chain. The second subunit has a molecular weight of 9,000, contains neither pyruvate nor cysteine, and has a serine at both the N- and C-terminus. There are apparently five subunits of each type in the native enzyme; reconstitution of activity from separated subunits has not been achieved, and nothing is known of the arrangement of subunits within the native enzyme.

8

Acetoacetate Decarboxylase

IRWIN FRIDOVICH

I. Introduction

World War I caused a great increase in the demand for acetone, which was used at the time in the manufacture of cordite and of airplane wing dope. Since the available sources of acetone were inadequate, fermentation processes for the conversion of carbohydrate to acetone were explored (*1–5*). Microorganisms were isolated which were capable of

1. J. H. Northrop, L. H. Ashe, and R. R. Morgan, *J. Ind. Eng. Chem.* **11,** 723 (1919).
2. H. B. Speakman, *J. Soc. Chem. Ind.* (*London*) **38,** 155 (1919).
3. J. H. Northrop, L. H. Ashe, and J. K. Senior, *JBC* **39,** 1 (1919).
4. H. B. Speakman, *JBC* **41,** 319 (1920).
5. J. Reilly, W. J. Hickinbottom, F. R. Henley, and A. C. Thaysen, *BJ* **14,** 229 (1920).

reasonably efficient conversion of cheap sources of carbohydrate, such as maize, to acetone and butanol, by anaerobic fermentations. It was soon observed that although growth of the organisms was most vigorous at slightly alkaline pH the production of acetone occurred only at slightly acid pH. When the acidification, which occurred naturally during these fermentations, was prevented by the addition of lime, no acetone was produced. This led to the discovery that acetic acid was produced in the culture medium and that the yield of acetone could be increased by adding acetic acid to the cultures (*5*). Two moles of exogenous acetic acid gave rise to approximately one mole of acetone. Acetate-1-^{13}C was shown to give rise to labeled acetone in cultures of these bacteria, and a mechanism was proposed which involved the condensation of acetate to acetoacetate followed by its decarboxylation to acetone (*6*). Ethyl acetoacetate had been observed to augment acetone production by active cultures (*5*). Suspension of these microorganisms were found to catalyze the decarboxylation of acetoacetate (*7*). This activity increased with the age of the culture from which the cells had been harvested; thus, it may be inferred that the decarboxylase was induced by some change in the culture medium rather than being constitutive to the microorganism. All modern work on the acetoacetic decarboxylase may be traced to Davies (*8–10*), who first studied the enzyme in cellfree extracts of *Clostridium acetobutylicum* and who achieved a partial purification of the enzyme from that source.

II. Properties of the Enzyme

A. Assay of the Enzyme

The earliest assays of acetoacetate decarboxylase were based upon manometric measurements of the CO_2 released (*9–12*). A more convenient spectrophotometric assay was based upon the observation that at pH 6.0 the molar extinction coefficients for acetoacetate at 270 and 210 nm are 55.0 and 420, respectively, whereas the corresponding values for acetone are 28.3 and 3.9. The decarboxylation of acetoacetate at this

6. H. G. Wood, R. W. Brown, and C. H. Werkman, *AB* **6,** 243 (1945).
7. M. J. Johnson, W. H. Peterson, and E. B. Fred, *JBC* **101,** 145 (1933).
8. R. Davies and M. Stephenson, *BJ* **35,** 1320 (1941).
9. R. Davies, *BJ* **36,** 582 (1942).
10. R. Davies, *BJ* **37,** 230 (1943).
11. H. W. Seeley and P. J. Van DeMark, *J. Bacteriol.* **59,** 381 (1950).
12. H. W. Seeley, "Methods in Enzymology," Vol. 1, p. 624, 1955.

pH is therefore accompanied by a change in molar extinction of —26.7 at 270 nm and of —416.1 at 210 nm (*13*). Because the K_m for acetoacetate is very high ($K_m = 0.08\,M$ at pH 6.0), the assay based upon changes in extinction observed at 270 nm is sufficiently sensitive for most purposes. The assay based upon observations at 210 nm is much more sensitive but can ordinarily be used only in pure systems because of the intense absorption at this wavelength by proteins and by many other classes of compounds. The absorbtivity of acetoacetate is largely due to the enol form, some of which persists even at low pH, because of stabilization by intramolecular hydrogen bonding (*14*). Boric acid was observed to react reversibly with acetoacetate to yield a 1:1 adduct which exhibits intense absorbance at 250 nm (*15*). Because boric acid does not inhibit acetoacetic decarboxylase, this phenomenon provides the basis of a spectrophotometric assay of β-decarboxylation of enhanced sensitivity (*15*). An assay which allows detection of acetoacetate decarboxylase activity on cellulose acetate electrophoretograms has also been devised (*16*).

Because acetoacetate decarboxylase is inhibited by a wide variety of monovalent anions (*13*) the assay medium should be free of such ions, and the buffering species should consequently be chosen with care. Potassium phosphate does not inhibit the enzyme, and it has been commonly used. A unit of enzyme has arbitrarily been designated as that amount which produces an absorbancy change at 270 nm of 1.0 per 100 sec when acting $0.03\,M$ lithium acetoacetate in $0.10\,M$ potassium phosphate at pH 5.9 and at 30° (*17*). The lithium salt of acetoacetate, which is much less hygroscopic than the sodium or potassium salts, is prepared from ethyl acetoacetate by a modification (*18*) of the method of Hall (*19*).

B. Purification of the Enzyme

The earliest studies on the purification of the acetoacetate decarboxylase from *Clostridium acetobutylicum* were performed by Davies (*10*).

13. I. Fridovich, *JBC* **238,** 592 (1963).
14. L. F. Fieser and M. Fieser, "Organic Chemistry," p. 218. Heath, Boston, Massachusetts, 1956.
15. S. Neece and I. Fridovich, *ABB* **108,** 240 (1964).
16. I. Fridovich, *Anal. Biochem.* **7,** 371 (1964).
17. B. Zerner, S. M. Coutts, F. Lederer, H. H. Waters, and F. H. Westheimer, *Biochemistry* **5,** 813 (1966).
18. I. Fridovich, *JBC* **243,** 1043 (1968).
19. L. M. Hall, *Anal. Biochem.* **3,** 75 (1962).

This enzyme has subsequently been extensively purified from this organism (*20*) and crystallized in the form of thin hexagonal plates (*17*). The decarboxylase has also been purified from *Clostridium madisonii* (*11*, *21*). Cells were harvested in late log phase (after three days at 37°) because this enzyme does not appear in these cultures until growth begins to plateau. The cell paste was acetone powdered, and the resultant light tan powder which is stable to storage at —20° was extracted with dilute phosphate buffer. The enzyme was precipitated from the resultant yellowish, cellfree extract by adjusting the pH to 3.9 by cautious addition of acetic acid. This is the most important single step of the entire procedure because it accomplishes an approximately 50-fold purification in good yield and greatly reduces the volume. It is also the most critical step of the procedure because lowering the pH only a few tenths of a unit below pH 3.9 results in irreversible inactivation.

The active precipitate was redissolved in dilute phosphate buffer at pH 5.9 and was then fractionated by salting out with ammonium sulfate. Chromatography on DEAE-cellulose was the last step of the purification procedure. Crystallization was accomplished at room temperature out of ammonium sulfate solutions of the enzyme. Several lines of evidence indicated that the crystalline enzyme was homogeneous (*17*). Thus, the specific activity of the enzyme was unaffected by repeated recrystallization; it migrated as a single protein component upon starch gel electrophoresis, it was monodisperse in terms of sedimentation velocity, and end group analysis indicated only methionine as the N-terminus (*22*).

C. Latency

Although the crystalline enzyme was apparently homogeneous its specific activity varied from batch to batch (*17*). This variability was explained by the discovery that half of the acetoacetate decarboxylase activity in extracts of *Clostridium acetobutylicum* was latent (*23*). The latent decarboxylase was converted to an expressed activity by an irreversible first-order process which had a half-life of 53 min at 50° and an Arrhenius energy of activation of 25 kcal/mole (*23*). Latent and expressed decarboxylase co-purify and cosediment in a sucrose density gradient (*23*) and are generated in parallel in cultures of the microorga-

20. G. A. Hamilton and F. H. Westheimer, *JACS* **81**, 2277 (1959).
21. R. F. Colman, Ph.D. Thesis, Harvard University, 1962.
22. F. Lederer, S. M. Coutts, R. A. Laursen, and F. H. Westheimer, *Biochemistry* **5**, 823 (1966).
23. M. S. Neece and I. Fridovich, *JBC* **242**, 2939 (1967).

nism (*24*). Once it has undergone the conversion to the expressed form, the activity of the originally latent decarboxylase is indistinguishable from that of the expressed decarboxylase (*23*). Since the percentage of residual latent decarboxylase is a function of how long and at what temperatures a given preparation has been aged, the specific activity must be expected to vary within a twofold range. The rate of conversion of latent to expressed decarboxylase was increased approximately two-fold as the pH was decreased from 6.9 to 4.8. Only heat treatment was found to accomplish the activation of latent decarboxylase. Exposure of the latent decarboxylase to urea, ethanol, or proteolytic enzymes was ineffectual. Reversible inhibitors of the enzyme such as monovalent anions, 2-oxopropane sulfonate, or acetylacetone were without effect on the rate of activation of the latent decarboxylase (*23*). Active enzyme was irreversibly inactivated when treated with borohydride in the presence of substrate (*25*) or of substrate analogs (*18*, *26*). This treatment had no effect on the latent sites. Furthermore, irreversible inactivation of the expressed decarboxylase by this treatment had no effect on the activation of the residual latent activity (*23*). The only differences which have yet been detected between enzyme which is half expressed and half latent and that which has been fully activated by heating, aside from the doubling of specific activity, relate to the effects of *p*-chloromercuriphenyl sulfonate (*26*), and this will be discussed in Section III,C.

The molecular basis of the latency of this enzyme and of the process of conversion of latent to expressed activity remains unknown, as does the biological rationale for the existence of latent activity. Latent and expressed decarboxylases were not distinguishable by any of the steps of the procedure used in the purification. Thus, the first extract of the acetone-powdered cells contained as much latent activity as did the fully purified enzyme, provided that the purification was performed rapidly and at low temperature. Latency was not an artifact of acetone powdering since sonicates of cells which had not been exposed to acetone exhibited the same phenomenon and to the same degree. Attempts to separate latent from expressed decarboxylase by centrifugation in a sucrose density gradient and by disc gel electrophoresis were also unsuccessful. It appears most likely, therefore, that expressed and latent sites coexist in equal amounts in every molecule of the native enzyme and that the expression of latent sites depends upon a conformational transition.

24. A. P. Autor, Ph.D. Thesis, Duke University, 1970.
25. I. Fridovich and F. H. Westheimer, *JACS* **84**, 3208 (1962).
26. A. P. Autor and I. Fridovich, *JBC* **245**, 5214 (1970).

D. Molecular Weight, Subunits, and Amino Acid Composition

The acetoacetate decarboxylase was found by the equilibrium sedimentation method and a calculated partial specific volume of 0.73 to have a molecular weight of 340,000 ± 10,000 (*27*). The enzyme was dissociated into its constituent subunits by exposure to pH 2.15, to 4.0 *M* urea or to 6.0 *M* guanidinium chloride. The subunit weight was found to be 29,000 ± 1,000. It follows that the enzyme is composed of 12 subunits. Since methionine has been found to be the only amino terminal amino acid and lysine was the only residue detected at the carboxyl terminus and since the number of tryptic peptides detectable were slightly less than the number of arginine plus lysine residues per subunit (*27*), it may be inferred that these subunits are identical. Enzymic activity was reversibly lost at 4° in 4.0 *M* urea at pH 8.0 and in the presence of β-mercaptoethanol. Sedimentation velocity measurements have indicated (*27, 28*) that under these conditions the decarboxylase was dissociated into dimers which were capable of reforming the active dodecameric enzyme when the denaturing stress was removed. In contrast, the dissociation into monomers has not yet been successfully reversed. In view of the coexistence of equal numbers of latent sites and expressed sites, it is tempting to speculate that the dimers formed under the conditions that cause reversible dissociation were composed of one subunit in the expressed state and one in the latent condition.

The amino acid composition of this enzyme has been reported (*22*). The isoelectric point which would be deduced from the amino acid composition lies close to neutrality, yet the experimentally determined isoelectric point was 4.9. It appears likely that the discrepancy resulted from the binding of buffer anions to the enzyme. The absorption spectrum of the enzyme exhibits the usual protein absorption band at 280 nm and an additional broad band of low intensity at 320 nm. The molar extinction coefficient at 280 nm is 34.0×10^4 (*22*), which may be compared to the value of 30×10^4, which may be calculated on the basis of its content of tyrosine plus tryptophan and on the contribution made by these residues to the absorption spectra of proteins (*29*). There have been no indications of the existence of nonamino acid prosthetic groups, and no explanation is currently available for the source of the 320-nm absorption band. Since this band was eliminated by denaturation of the

27. W. Tagaki and F. H. Westheimer, *Biochemistry* **7**, 895 (1968).
28. W. Tagaki and F. H. Westheimer, *Biochemistry* **7**, 891 (1968).
29. D. Wetlaufer, *Advan. Protein Chem.* **17**, 303 (1962).

enzyme (*22*), it may be related to a charge transfer interaction between amino acid functionalities.

E. Stability of the Enzyme

The enzyme is stable at 25° in the pH range 4–9 (*22*). It is rather easily inactivated by denaturants such as urea, guanidinium chloride, and sodium dodecyl sulfate (*22, 27*). The enzyme is quite resistant to thermal inactivation and may be incubated at 70° in 0.10 *M* phosphate buffer at pH 5.9 for an hour without perceptible loss of activity (*23*). When subjected to temperatures above 70°, the enzyme exhibited a biphasic irreversible inactivation, the rapid phase of which was more responsive to changes in temperature than was the slow phase. ΔH_a for the rapid phase of the thermal inactivation was found to be 49 kcal/mole. An explanation for the time course of this thermal inactivation has been proposed (*30*). Acetylacetone, which is a potent reversible inhibitor of the enzyme, also serves to protect it against thermal inactivation (*23, 30*). Solutions of the enzyme are stable to storage in the frozen state at —20° for several months.

III. Catalytic Properties

A. Mechanism

Numerous and painstaking studies of the decarboxylation of β-keto acids in general and of acetoacetate in particular and of the catalysis of these decarboxylations by amines have led to the proposal that the catalytic amine forms a Schiff-base salt with the β-carbonyl group of the acid. It is further proposed that this center of positive change then facilitates withdrawal of electrons from the carboxylate group and its subsequent release as CO_2. The decarboxylated product would then be liberated by hydrolysis of the Schiff base. This mechanism for the non-enzymic decarboxylation of acetoacetate and the evidence which supports it has been well reviewed (*31*). The decarboxylation catalyzed by aceto-acetic decarboxylase similarly involves the formation of a Schiff base between the carbonyl of the substrate and a specific ε-amino group on the enzyme. This assertion is supported by several lines of evidence.

30. A. P. Autor and I. Fridovich, *JBC* **245**, 5223 (1970).
31. F. H. Westheimer, *Proc. Chem. Soc.* p. 253 (1963).

Schiff-base formation between the keto acid and the enzyme would entail obligatory exchange of the carbonyl oxygen with the oxygen of water concomitant with the decarboxylation. This exchange has been demonstrated (*32*). Schiff-base intermediates are prone to reduction by borohydride. This reaction has been used to trap and thus to demonstrate Schiff-base intermediates in a number of enzyme-catalyzed reactions (*33–38*). Acetoacetate decarboxylase was not affected by borohydride alone but was rapidly and irreversibly inactivated when treated with borohydride in the presence of acetoacetate (*25*). Furthermore, when the enzyme was inactivated by treatment with borohydride in the presence of 3-^{14}C-acetoacetate, radioactivity was introduced onto the protein. One equivalent of ^{14}C was introduced per 50,000 g of enzyme, and this label could not be removed by exhaustive dialysis. Acid hydrolysis of the labeled and dialyzed enzyme, followed by two-dimensional paper chromatography of the digest, yielded a single radioactive spot which was located by radioautography (*25*) and which was identified as ϵ-*N*-isopropyllysine (*39*). Acetopyruvate, a potent inhibitor of this enzyme, prevented both its inactivation by borohydride plus acetoacetate and its labeling by radioactive acetoacetate. Furthermore, acetone was much less effective than acetoacetate, both in supporting the inactivation and as a vehicle for labeling of the enzyme (*39*). Proteolytic digestion of the labeled enzyme gave a single radioactive peptide whose sequence was found to be –Glu–Leu–Ser–Ala–Tyr–Pro–*Lys–Lys–Leu– (*40*). The asterisked lysine is the site of Schiff-base formation and of borohydride reduction. Acetoacetate decarboxylase contains approximately 20 lysine residues per subunit (*22*). The conclusion that the active site of acetoacetate decarboxylase contains a unique lysine residue whose ϵ-amino group is intimately involved in the catalytic process appears inescapable.

The mechanism which has been proposed for the action of acetoacetate decarboxylase is as follows (*41*):

32. G. A. Hamilton and F. H. Westheimer, *JACS* **81,** 6332 (1959).
33. W. T. Jenkins and I. W. Sizer, *JACS* **79,** 2655 (1957).
34. Y. Matsuo and D. M. Greenberg, *JBC* **234,** 507 (1959).
35. B. L. Horecker, S. Pontremoli, C. Ricci, and T. Cheng, *Proc. Natl. Acad. Sci. U. S.* **47,** 1949 (1961).
36. J. M. Ingraham and W. A. Wood, *JBC* **240,** 4146 (1965).
37. R. G. Rosso and E. Adams, *JBC* **242,** 5524 (1967).
38. J. G. Shedlarski and C. Gilvarg, *JBC* **245,** 1362 (1970).
39. S. Warren, B. Zerner, and F. H. Westheimer, *Biochemistry* **5,** 817 (1966).
40. R. A. Laursen and F. H. Westheimer, *JACS* **88,** 3426 (1966).
41. W. Tagaki and F. H. Westheimer, *Biochemistry* **7,** 901 (1968).

$$E{-}NH_2 + CH_3{-}\overset{O}{\overset{\|}{C}}{-}CH_2{-}\overset{O}{\overset{\|}{C}}{-}O^- + H^+ \underset{k_{-1}}{\overset{k_1}{\rightleftharpoons}} CH_3{-}\underset{\underset{\underset{E}{|}}{\overset{+}{NH}}}{\underset{\|}{C}}{-}CH_2{-}\overset{O}{\overset{\|}{C}}{-}O^- + H_2O \quad (1)$$

$$CH_3{-}\underset{\underset{\underset{E}{|}}{\overset{+}{NH}}}{\underset{\|}{C}}{-}CH_2{-}\overset{O}{\overset{\|}{C}}{-}O^- \underset{k_{-2}}{\overset{k_2}{\rightleftharpoons}} CH_3{-}\underset{\underset{\underset{E}{|}}{NH}}{\underset{|}{C}}{=}CH_2 + CO_2 \quad (2)$$

$$CH_3{-}\underset{\underset{\underset{E}{|}}{NH}}{\underset{|}{C}}{=}CH_2 + H^+ \underset{k_{-3}}{\overset{k_3}{\rightleftharpoons}} CH_3{-}\underset{\underset{\underset{E}{|}}{\overset{+}{NH}}}{\underset{\|}{C}}{-}CH_3 \quad (3)$$

$$CH_3{-}\underset{\underset{\underset{E}{|}}{\overset{+}{NH}}}{\underset{\|}{C}}{-}CH_3 + H_2O \underset{k_{-4}}{\overset{k_4}{\rightleftharpoons}} CH_3{-}\underset{\underset{O}{\|}}{C}{-}CH_3 + H^+ + E{-}NH_2 \quad (4)$$

Reactions (1) and (4) of this mechanism were demonstrated by the finding that the enzyme catalyzes the exchange of the carboxyl oxygen of acetoacetate and of acetone with water (*32*). The Schiff-base salt produced by reaction (3) was trapped by borohydride reduction (*25, 39*). In addition, the enzyme has been observed to catalyze the exchange of deuterium between acetone and water as required by reactions (4) and (3) (*41*). Under identical conditions, but in the absence of enzyme, this exchange was too slow to be observed and amino acetonitrile, which is a catalyst for the decarboxylation of acetoacetate (*42*), was less effective than the enzyme in catalyzing this deuterium exchange by a factor of 50,000 (*41*).

B. Kinetic Properties

The K_m for acetoacetate is approximately $0.008\,M$ (*10, 13, 18, 23*). Optimal activity was observed at pH 5.9, and the effects of pH on activity could be rationalized in terms of two activity-limiting ionizable groups on the enzyme whose pK_a were approximately 5.5 and 6.5 (*21, 31*). Since the addition of ethanol to 30% shifted these to slightly lower values, it was concluded that the activity-limiting ionizable groups were cationic

42. J. P. Guthrie and F. H. Westheimer, *Federation Proc.* **26**, 562 (1967).

acids. Through the use of a reporter group covalently attached to the lysine residue at the active site, it was shown that the environment at the active site was such as to cause a lowering of pK_a by at least three units (*43*). It is therefore likely that one of the activity-limiting ionizations detected on the basis of kinetic measurements resulted from the ϵ-amino group of the active site lysine.

Arrhenius plots of V_m as a function of temperature were linear, and the energy of activation calculated from the slope was 11.4 kcal/mole (*18*). The anomalous increase in Q_{10} with rising temperature, which was reported by Davies (*10*), was probably because of unrecognized latency which resulted in the appearance of more enzymic activity at higher temperatures, owing to the conversion of latent to expressed decarboxylase. The effect of temperature on K_m for acetoacetate was also considered, and ΔH was found to be 7.3 kcal/mole (*18*).

The enzyme was found to exhibit a high degree of specificity; having no action upon a variety of α- and β-keto acids other than acetoacetate. Among the compounds tested were acetopyruvate, acetone dicarboxylate, oxalacetate, α-ketoglutarate, and pyruvate (*10*). Phenyl acetoacetate was found to serve as a substrate for acetoacetate decarboxylase (*44*). Its K_m at pH 5.9 and at 26° was 0.010 M. The enzymic decarboxylation of this compound was conveniently followed at 290 nm where the molar extinction change was 30 (*45*).

C. Inhibition Studies

1. *2-Oxopropane Sulfonate*

Acetoacetate decarboxylase is reversibly inhibited by a number of compounds some of which function as substrate analogs while others inhibit by reacting with the Schiff bases which are intermediates in the catalytic process. The sulfonic acid analog of acetoacetate, i.e., 2-oxopropane sulfonate appears to be an ideal virtual substrate. Thus, it was competitive with respect to acetoacetate and its K_i was numerically equal to the K_m for acetoacetate over a wide range of temperature (*18*). In addition, 2-oxopropane sulfonate was found to mimic the action of acetoacetate in supporting the inactivation of the enzyme by borohydride (*18, 23*).

43. P. A. Frey and F. H. Westheimer, *Federation Proc.* **29**, 461 (1970).
44. M. S. Neece, Ph.D. Thesis, Duke University (1967).
45. M. S. Neece and I. Fridovich, unpublished observations (1966).

2. *β-Diketones*

β-Diketones such as acetopyruvate (*10, 21, 46*) and acetylacetone (*18, 30*) are powerful inhibitors of the decarboxylase. These compounds do not support the inactivation of the enzyme by borohydride, but rather they protect the enzyme against the inactivation caused by borohydride in the presence of acetoacetate (*18, 25, 31, 39, 46*). Acetopyruvate, which inhibits competitively with respect to acetoacetate and whose K_i at pH 5.9 and at 30° was 1×10^{-7} *M*, has been shown to react with the decarboxylase and with model amino compounds such as β-amino acetonitrile to yield enamines whose electron-rich carbon–carbon double bonds absorb intensely in the ultraviolet and are not reducible by borohydride (*46*). The formation of an enamine not reducible by borohydride had also been proposed in explanation of the ability of acetylacetone to inhibit the decarboxylase and to protect it against inactivation by borohydride (*18*). The formation of such enamines appear to be catalyzed by the enzyme (*46*) and do not appear to involve the prior formation of Schiff-base intermediates (*26*). A variety of diketones were investigated for their abilities to inhibit the decarboxylase (*26*). β-Diketones were strongly inhibitory, but α- or γ-diketones were not. Considering acetylacetone as the simplest β-diketone, several generalizations are possible concerning structural requirements for inhibition of the enzyme. Thus, without deleterious effect, one of the methyl groups may be replaced by a carboxyl as in acetopyruvate or may be substituted with a bulky group as in benzoyl acetone. The methylene group can also be substituted as in 3-benzylidene-2,4-pentanedione, again without effect. It was also possible to replace one of the carbonyl groups by another electron-withdrawing functionality, as in benzene sulfonylacetone, and still retain the ability to inhibit the decarboxylase. Substitution on both methyl groups, as in dibenzoyl methane, did eliminate inhibitory activity. It thus appears that the essential feature of the inhibitory β-diketones is $CH_3\text{–}\overset{\overset{\displaystyle O}{\|}}{C}\text{–}CH_2\text{–}R$ where R is some electron-withdrawing group. It follows that in the case of inhibitory β-diketones it is the carbonyl group adjacent to the unsubstituted methyl group which is involved in reacting with the essential amino group on the enzyme to yield on enamine.

Whereas acetopyruvate acted as a classic competitive inhibitor (*46*), acetyl acetone (*18*) and benzoyl acetone (*26*) exhibited some intercept

46. W. Tagaki, J. P. Guthrie, and F. H. Westheimer, *Biochemistry* **7**, 905 (1968).

effects when their inhibitory action was analyzed according to Lineweaver and Burk (*47*). This implies that these compounds exhibited some affinity for enzyme–substrate or enzyme–product complexes in addition to a predominant affinity for the free enzyme. As the temperature was raised the intercept effect seen with acetyl acetone decreased, and at temperatures above 50° it was entirely competitive with respect to acetoacetate (*18*).

3. *Monovalent Anions*

Acetoacetate decarboxylase was reversibly inhibited by a variety of monovalent anions (*13*). Association of a single anion with the sensitive site on the enzyme resulted in complete inactivation. This anion binding site behaved like a cationic acid with an effective pK_a of about 5.8. The thermodynamic parameters for the association of inhibitory anions with the sensitive site on the enzyme were determined and were related to the sizes of the monovalent anions. The data so obtained were consistent with the proposal that binding to the sensitive site involved a migration of the inhibitory anion from the aqueous phase into a relatively hydrophobic environment (*13*). Model studies of the effects of monovalent anions in facilitating the migration of a cationic dye, such as pararosaniline, from an aqueous into a chloroform phase have been performed as have studies of the effects of changes in temperature on the behavior of this model system (*48*). This simple two-phase system faithfully duplicated the behavior of the enzyme with respect to the effects of changes in temperature. It is therefore tempting to propose that the ϵ-amino group at the active site of the enzyme is very close to a hydrophobic crevice and that the binding of monovalent anions to that group in its protonated form or to the enzyme–substrate or enzyme–product Schiff-base salts involves migration of the resultant ion pair into that hydrophobic crevice.

At pH 7.0 and above, monovalent anions were entirely noncompetitive inhibitors, but as the pH was lowered the inhibition became increasingly uncompetitive (*44*). At pH 5.0 the inhibition by monovalent anions was virtually completely uncompetitive. Mixed inhibition studies were performed to determine whether the inhibitors 2-oxopropane sulfonate, acetylacetone, and nitrate acted at identical or at independent and noninteracting sites (*18*). This is of some interest because at pH 5.9 these inhibitors are competitive, mixed competitive–noncompetitive, and mixed

47. H. Lineweaver and D. Burk, *JACS* **56,** 658 (1934).
48. I. Fridovich, unpublished observations (1965).

noncompetitive–uncompetitive, respectively. The results clearly indicated that all three inhibitors acted at the same site.

4. *Hydrogen Cyanide*

HCN inhibits acetoacetic decarboxylase and apparently does so by reacting with the Schiff-base compounds formed at the active site in the course of the catalytic process (*10, 25, 30, 31*). Thus, the inhibition caused by the presence of HCN in reaction mixtures developed with perceptible slowness, and incubation of enzyme or of substrate with HCN prior to addition of the missing component had no effect. HCN also protected the enzyme against inactivation by borohydride in the presence of substrate. Carbonyl compounds, other than substrate, which are capable of forming Schiff bases with the active site amino group, should generate a site for the addition of HCN. An inhibitory synergism between such carbonyl compounds and HCN was therefore to be expected. This inhibitory synergism has been observed and has been used to gain information about the relative abilities of carbonyl compounds to form Schiff bases at the active site of the decarboxylase (*26*). Because it can react with enzyme–substrate or enzyme–product complexes but not with free enzyme, HCN should act as an uncompetitive inhibitor of the decarboxylase. HCN should thus increase the ordinate intercepts but not the slopes of the lines generated by plotting the kinetic data on reciprocal coordinates. This was almost the case (*26*). But, while increasing intercepts, HCN caused some anomalous decrease in slopes. Such a result could be explained on the basis of a reaction of HCN with the substrate to form an adduct which was more inhibitory than HCN itself.

5. *Borohydride*

When an alkaline solution of borohydride was slowly infused into solutions buffered at pH 5.9, a balance between the infusion and the rapid acid-catalyzed hydrolysis of borohydride was achieved within a few seconds. Under such conditions of a contrived constant concentration of borohydride, the inactivation of the decarboxylase was entirely dependent upon the presence of substrate or of other compounds capable of generating Schiff bases at the active site and, in the presence of such compounds, was a first-order process (*18, 26*). The first-order rate constant for inactivation under these conditions depended upon the proportion of the enzyme which was in the form of the reducible Schiff base and upon the intrinsic rate of reaction of the particular Schiff base with borohydride. The rate of inactivation by a fixed level of borohydride increased with the concentration of 2-oxopropane sulfonate in a hyperbolic fashion. Graphing of such data on reciprocal coordinates generated straight lines from which K'_m and V'_m could be obtained. In this context V'_m refers to the rate of inactivation which was observed in the presence of saturating concentrations of 2-oxopropane sulfonate and K'_m to the concentration of 2-oxopropane sulfonate which supported half of the maximal rate of inactivation by borohydride. The K'_m thus obtained for 2-oxopropane sulfonate was equal to the K_i derived from measurements of its ability to inhibit the decarboxylation of acetoacetate in a competitive fashion (*18*).

Several ketones were compared in this way for their abilities to support the inactivation of the enzyme by borohydride (*26*). The role of a negative charge in the binding of small molecules to the active site of this enzyme was underscored by the observation that K'_m for acetone at 0° and at pH 5.9 was 0.20 M, whereas that for the sulfonate of acetone was 0.002 M. On the other hand, the V'_m observed with acetone was five times larger than the V'_m seen with 2-oxopropane sulfonate. This probably relates to a small electrostatic repulsion of borohydride from the vicinity of the Schiff base by the sulfonate group.

The unusual efficacy of borohydride as a reductant of the Schiff-base salts generated at the active site of acetoacetic decarboxylase must be noted and explained. Thus, a steady state level of borohydride of 2.4×10^{-7} M, maintained by the continuous infusion technique, resulted in 90% inactivation of 5.2×10^{-7} M enzyme in 8 min at 0° when 0.0178 M 2-oxopropane sulfonate was present (*18*). The sensitivity of this enzyme to inhibition by monovalent anions offers the basis for an explanation of the effectiveness of borohydride as a reductant at its active site. Thus, if borohydride were bound to the active site as a monovalent anion, its effective concentration at that site would be enormously increased and its subsequent action as a reductant at that site would be proportionately enhanced. There are several reasons for seriously considering this proposition. Monovalent anions are predominantly uncompetitive inhibitors of the decarboxylase at pH 5.9 and therefore exhibit greater affinity for enzyme–substrate or enzyme–product complexes than for the free enzyme. Considered as a monovalent anion, borohydride should likewise have greater affinity for these complexes than for the free enzyme.

Furthermore, the negative charge on the carboxylate group of the enzyme–substrate Schiff-base salt should repel borohydride, whereas there would be no such electrostatic repulsion in the case of the enzyme–product Schiff-base salt. Hence, borohydride should primarily bind to and should therefore preferentially reduce the enzyme–product Schiff-base salt. That the group introduced onto the ε-amino at the active site, through the action of borohydride plus acetoacetate, was isopropyl rather than butyryl is in keeping with this line of reasoning. Monovalent anions effectively protect the enzyme against inactivation by borohydride in the presence of substrate (*23*), 2-oxopropane sulfonate (*18*), or acetone (*26*). This ability of monovalent anions to deny borohydride access to the reducible site on the enzyme is most readily understood in terms of competition between borohydride and other monovalent anions for binding to that site. Finally, decreasing the temperature enhanced the ability of monovalent anions to inhibit acetoacetic decarboxylase and the magnitude of this effect, quantitatively expressed in terms of thermodynamic parameters, was related to the size of the monovalent anion (*13*). Lowering the temperature similarly increased the effectiveness of borohydride as a reductant of Schiff-base salts formed at the active site of the enzyme (*18*). Borohydride has an ionic radius which is very similar to that for chloride, and the ΔH which describes the effects of temperature on its ability to inactivate the enzyme by reduction was —11.5 kcal/mole (*18*). This may be compared to the ΔH of —11.8 kcal/mole which was obtained from the effects of temperature on the ability of chloride to inhibit the decarboxylase.

6. *p-Chloromercuriphenyl Sulfonate*

When corrected to a subunit weight of 29,000 (*27*), the results of amino acid analysis of performic acid oxidized enzyme indicated 2.6 half-cystine residues per subunit. Of these, 1.9 reacted rapidly with Ellman's reagent at pH 8.0 in the presence of 4.0 *M* guanidinium chloride, whereas only 0.88 per subunit reacted rapidly in the absence of guanidinium chloride (*22*). The effect of *p*-chloromercuriphenyl sulfonate on the activity of acetoacetate decarboxylase has been investigated and correlated with the extent of mercaptide formation (*26*). When the enzyme was mixed with a slight excess of this reagent at pH 5.9 and at 25°, two sulfhydryls per subunit reacted within 1 min. This reaction was followed in terms of mercaptide formation at 235 nm. There followed a much slower increase in absorbance which, if resulting from mercaptide formation, signified the reaction of 0.3 SH per subunit per hour. Under identical conditions native decarboxylase rapidly lost 30% of its activity, whereas fully activated decarboxylase lost 50% of its activity. These rapid losses of activity were reversed by cysteine. There were also much slower losses of activity which correlated with the slow phase of mercaptide formation and which were not reversed by the addition of cysteine. Titration of the decarboxylase with *p*-chloromercuriphenyl sulfonate demonstrated that one SH per subunit could be converted to the mercaptide without significant loss of activity, whereas derivitization of the second caused the partial inactivation already noted above. It follows that, of the two sulfhydryls per subunit of enzyme, the one which is more reactive with *p*-chloromercuriphenyl sulfonate is not at all essential for activity, while the second may also be converted to the mercaptide with only partial inactivation of the enzyme.

9

Aldose–Ketose Isomerases

ERNST A. NOLTMANN

I. Introduction

Aldose–ketose isomerases catalyze the interconversion of isomeric aldo and keto sugars by causing the migration of a carbon-bound hydrogen between carbons 1 and 2. These enzymes may be classified in two groups according to their action on free or on phosphorylated monosaccharide substrates. Those acting on free sugars appear to be confined mainly to microorganisms, whereas some of those acting on phosphorylated substrates are common to all living organisms. Prominent among the latter are glucose-6-phosphate isomerase, triosephosphate isomerase, and ribose-5-phosphate isomerase. Although all three catalyze essential steps in the pathways of glycolytic or oxidative metabolism of carbohydrates, they received little attention from enzymologists until approximately ten years ago. This lack of interest is exemplified by the sparsity of quantitative information in Topper's chapter in the second edition of this treatise (*1*): At that time the only published molecular property of any of the enzymes in this class was a crude estimate for the molecular weight of yeast glucose-6-phosphate isomerase (*2*).

This chapter covers the period since Topper's review (*1*) and refers to older work only when relevant to the current discussion. Other comprehensive reviews on aldose–ketose isomerases have not appeared, but Rose has discussed mechanistic aspects of this class of enzymes in 1962 (*3*), 1966 (*4*), and 1970 (*5*). In addition, anyone interested in aldose–ketose isomerization should not miss Speck's excellent and historically colorful account of the work dealing with its nonenzymic counterpart, the Lobry de Bruyn–Alberda van Ekenstein transformation (*6*).

II. Glucose-6-phosphate Isomerase

D-Glucose 6-phosphate ⇄ D-Fructose 6-phosphate

A. Occurrence and Function

Glucose-6-phosphate isomerase (EC 5.3.1.9), also called Lohmann's isomerase after its discoverer (*7*) (or more conveniently PGI, i.e., phos-

1. Y. J. Topper, "The Enzymes," 2nd ed., Vol. 5, p. 429, 1961.
2. E. Noltmann and F. H. Bruns, *Biochem. Z.* **331**, 436 (1959).
3. I. A. Rose, *Brookhaven Symp. Biol.* **15**, 293 (1962).
4. I. A. Rose, *Ann. Rev. Biochem.* **35**, 23 (1966).
5. I. A. Rose, "The Enzymes," 3rd ed., Vol. II, p. 281, 1970.
6. J. C. Speck, Jr., *Advan. Carbohydrate Chem.* **13**, 63 (1958).
7. K. Lohmann, *Biochem. Z.* **262**, 137 (1933).

phoglucose isomerase), is ubiquitously and abundantly present in nature. It has been isolated in crystalline form from skeletal muscle of rabbit (*8*), beef (*9*), and pig (*10*), from baker's (*11*) and brewer's yeast (*12*) and from peas (*13*). In addition to the references quoted by Topper (*1*), numerous reports have appeared describing various extents of purification for glucose-6-phosphate isomerases of different origins such as rat kidney (*14*), human and dog liver (*15*), *Aspergillus niger* (*16*), bovine mammary gland (*17*) and intestinal mucosa (*18*), *Aerobacter aerogenes* (*19*), various *Trypanosoma* species (*20*), and human erythrocytes (*20a, 20b*) and skeletal muscle (*21*). The enzyme has also been studied or at least shown to be present in *Schistosoma mansoni* (*22*), *Mycobacterium tuberculosis* (*23*), lingcod (*24*) and carp muscle (*25*), *Escherichia coli* (*26*), bovine milk (*27*), various *Entamoeba* species (*28*), as well as in the Spanish (*29*), Indian (*30*), and American honeybee (*31*). These examples illus-

8. E. A. Noltmann, *JBC* **239,** 1545 (1964).
9. E. A. Noltmann, unpublished experiments (1964).
10. R. K. Scopes, unpublished experiments [H. Muirhead, personal communication (1970)].
11. H. Klotzsch and H. U. Bergmeyer, *Angew. Chem.* **72,** 920 (1960).
12. Y. Nakagawa and E. A. Noltmann, *JBC* **240,** 1877 (1965).
13. Y. Takeda, S. Hizukuri, and Z. Nikuni, *BBA* **146,** 568 (1967).
14. A. N. Wick, D. R. Drury, H. I. Nakada, and J. B. Wolfe, *JBC* **224,** 963 (1957).
15. M. N. Lipsett, R. B. Reisberg, and O. Bodansky, *ABB* **84,** 171 (1959).
16. K. Singh, *Can. J. Biochem. Physiol.* **37,** 927 (1959).
17. A. Baich, R. G. Wolfe, and F. J. Reithel, *JBC* **235,** 3130 (1960).
18. F. Alvarado, *Enzymologia* **26,** 12 (1963).
19. K. Matsushima and F. J. Simpson, *Can. J. Microbiol.* **11,** 967 (1965).
20. E. L. Risby, T. M. Seed, and J. R. Seed, *Exptl. Parasitol.* **25,** 101 (1969).
20a. J. E. Smith and M. McCants, *BBA* **171,** 332 (1969). Although these authors described their purification procedure as superior to that of Tsuboi *et al.* (*20b*), direct comparison of the two methods is not possible since Smith and McCants failed to indicate the temperature at which they made their activity measurements.
20b. K. K. Tsuboi, J. Estrada, and P. B. Hudson, *JBC* **231,** 19 (1958).
21. N. D. Carter and A. Yoshida, *BBA* **181,** 12 (1969).
22. E. Bueding and J. A. MacKinnon, *JBC* **215,** 507 (1955).
23. F. Bastarrachea, D. G. Anderson, and D. S. Goldman, *J. Bacteriol.* **82,** 94 (1961).
24. G. B. Martin and H. L. A. Tarr, *Can. J. Biochem. Physiol.* **39,** 297 (1961).
25. R. K. Scopes and C. Gosselin-Rey, *J. Fisheries Res. Board Can.* **25,** 2715 (1968).
26. L. Klungsöyr and A. Endresen, *BBA* **92,** 378 (1964).
27. G. V. Heyndrickx, *Enzymologia* **27,** 209 (1964).
28. F. Montalvo and R. E. Reeves, *Exptl. Parasitol.* **22,** 129 (1968).
29. A. Sols, E. Cadenas, and F. Alvarado, *Science* **131,** 297 (1960).
30. S. Ray, N. C. Pant, and K. N. Mehrotra, *Indian J. Entomol.* **30,** 37 (1968).
31. S. A. Saunders, R. W. Gracy, K. D. Schnackerz, and E. A. Noltmann, *Science* **164,** 858 (1969).

trate the widespread occurrence of this isomerase, although the compilation should not be considered as complete.

Comparison of enzyme levels in various types of rabbit skeletal muscle (*32*) resulted in the finding that in so-called white muscle the concentration of glucose-6-phosphate isomerase was up to nine times as high as in red muscle, in contrast to hexokinase of which six times as much was found in red muscle as in the white species. These differences were attributed to the different physiological functions of the two muscle types (*32*).

The basic function of glucose-6-phosphate isomerase, *viz.*, catalyzing an obligatory step in glycolysis, is obvious. Its potential for exerting a regulatory influence on carbohydrate metabolism, however, must be considered uncertain. Despite its strategic location near the branching point of several pathways utilizing glucose 6-phosphate, three properties would appear to make this enzyme difficult to control: (a) its equilibrium constant being close to unity, (b) its ubiquitous presence in relatively high concentration, and (c) its high catalytic efficiency. Nevertheless, the finding that several intermediates of the oxidative pentose phosphate cycle [6-phosphogluconate (*33*), erythrose 4-phosphate (*34*), and sedoheptulose 7-phosphate (*35*)] are potent competitive inhibitors has repeatedly led to suggestions involving glucose-6-phosphate isomerase in control mechanisms. In fact, in his first report on the inhibition of this enzyme by 6-phosphogluconate, Parr (*35a*) hypothesized that it "could conceivably play a part in the regulation of carbohydrate metabolism" and suggested (*36*) that it may be involved in the "mechanism responsible for the partitioning of carbohydrate between the direct oxidative pathway and the glycolytic route." Stadtman listed the enzyme among those that are "subject to allosteric control" (*36a*), and Racker considered the inhibitory effect of pentose phosphate cycle intermediates on glucose-6-phosphate isomerase an example of "negative feedback inhibition" (*37*). On the other hand, Rose and Rose were skeptical about

32. I. G. Burleigh and R. T. Schimke, *BBRC* **31**, 831 (1968).
33. C. W. Parr, *BJ* **65**, 34P (1957).
34. E. Grazi, A. DeFlora, and S. Pontremoli, *BBRC* **2**, 121 (1960).
35. R. Venkataraman and E. Racker, *JBC* **236**, 1876 (1961).
35a. C. W. Parr, *Nature* **178**, 1401 (1956).
36. In order to appreciate the significance of these quotes one must realize that they were made in 1956, i.e., about parallel in time to Umbarger's first note on feedback inhibition [*Science* **123**, 848 (1956)] and more than a decade before the phrase *metabolic regulation* became a household word among biochemists.
36a. E. R. Stadtman, *Advan. Enzymol.* **28**, 41 (1966).
37. E. Racker, "Mechanisms in Bioenergetics." Academic Press, New York, 1965.

the possibility of the enzyme becoming rate limiting since there are no concrete data available to support this concept (*38*).

In a study of a mutant strain of *Salmonella typhimurium* lacking glucose-6-phosphate isomerase (*39*), the rate of growth and of glucose utilization was found to be 20% of that of the wild-type organism. This was also the value found for the fraction of glucose utilized by the wild-type strain via the oxidative pentose phosphate cycle. On gluconate growth of the mutant strain was as fast as on glucose, but substrate utilization occurred via the Entner–Doudoroff pathway. Thus the oxidative pentose phosphate cycle could not be adapted to the increased need of the mutant strain, and, at least for this *Salmonella* species, absence of the glucose-6-phosphate isomerase step does not lead to a compensatory increase of the alternate oxidative pathway. Similarly, the physiological significance of inhibition only of the reverse reaction (fructose 6-phosphate → glucose 6-phosphate) by fructose 1-phosphate and fructose 1,6-diphosphate (*40*) is difficult to assess in view of the considerable turnover capacity of glucose-6-phosphate isomerase.

Another interesting aspect of this enzyme is its use as a diagnostic tool in medicine. Elevated glucose-6-phosphate isomerase levels in patients with cancer were reported by Bodansky (*41*) and simultaneously by Bruns and collaborators (*42, 43*). The latter also reported dramatically increased levels of this enzyme (*44*) in the serum of patients with hepatitis. These studies, together with those on aldolase in the serum of

38. I. A. Rose and Z. B. Rose, *Comprehensive Biochem.* **17,** 93 (1969).
39. D. G. Fraenkel and B. L. Horecker, *JBC* **239,** 2765 (1964).
40. J. Zalitis and I. T. Oliver, *BJ* **102,** 753 (1967).
41. O. Bodansky, *Cancer* **7,** 1191 and 1200 (1954).
42. F. Bruns and K. Hinsberg, *Biochem. Z.* **325,** 532 (1954).
43. F. H. Bruns and W. Jacob, *Klin. Wochschr.* **32,** 1041 (1954).
44. In the early clinical work the enzyme was usually referred to as phosphohexose isomerase since Bodansky felt that a more specific name had to await unambiguous identification of the product of the mannose-6-phosphate isomerase reaction [O. Bodansky, *JBC* **202,** 829 (1953)]. Although it is understandable that Bodansky retained the original name for his continuing studies on this enzyme even after the evidence he requested had been provided (see Section III,A), the choice of the designation phosphohexose isomerase by authors who entered the field in the late 1960's [e.g., (*20, 30, 32, 44a*)] would appear to be hardly justifiable. After Slein's explicit description (*45*) of two enzymes under the heading of phosphohexose isomerases (plural), continued use of this name for one of the two will only serve to confuse those not intimately familiar with the difference.

44a. J. C. Detter, P. O. Ways, E. R. Giblett, M. A. Baughan, D. A. Hopkinson, S. Povey, and H. Harris, *Ann. Human Genet.* (*London*) **31,** 329 (1968).

45. M. W. Slein, "Methods in Enzymology," Vol. 1, p. 299, 1955.

patients with progressive muscular dystrophy (*46, 47*) and on glutamate-oxalacetate transaminase in patients with myocardial infarction (*48*), opened the field of modern clinical enzymology which experienced an explosive growth in the late 1950's [e.g., (*49*)]. The medical aspects of glucose-6-phosphate isomerase were reviewed by Bodansky and Schwartz in 1966 (*50*).

B. Molecular Architecture

Studies on glucose-6-phosphate isomerase at the molecular level are the current subject of vigorous research in a number of laboratories. Since much of the work is just beginning to be published, a substantial amount of new information can be expected in the near future. Most of the data available at present are for the enzyme crystallized from rabbit skeletal muscle, but individual properties have also been reported for glucose-6-phosphate isomerases from cow mammary gland (*17*), young peas (*13*), human skeletal muscle (*21*), and brewer's yeast (*51, 52*).

1. *Molecular Weight and Related Properties*

A first estimate of 145,000 daltons for the molecular weight was obtained for the enzyme from brewer's yeast (*2*). This value was based on three measurements of the sedimentation coefficient ($s^0_{20,w} = 7.6$ S) and the diffusion coefficient ($D^0_{20,w} = 5.1$ F) and by assuming a value of 0.749 for $\bar{V}$, an unsatisfactory premise by today's standards. Nevertheless, this figure appears to be within 20% of the present best estimate for yeast glucose-6-phosphate isomerase (*52*). Shortly after the value of 145,000 was published (*2*), Baich *et al.* reported a molecular weight of 48,000 daltons for the mammary gland isomerase which was determined by the Archibald method (*17*). This value was later revised to 125,000

46. J. A. Sibley and A. L. Lehninger, *J. Natl. Cancer Inst.* **9,** 303 (1949).
47. G. Schapira, J. C. Dreyfus, and F. Schapira, *Semaine Hop. Paris* **29,** 1917 (1953); J. C. Dreyfus and G. Schapira, *Compt. Rend. Soc. Biol.* **147,** 1145 (1953).
48. J. S. LaDue, F. Wróblewski, and A. Karmen, *Science* **120,** 497 (1954).
49. E. Schmidt, F. W. Schmidt, H. D. Horn, and U. Gerlach, *in* "Methods of Enzymatic Analysis" (H. U. Bergmeyer, ed.), 1st ed., p. 651. Academic Press, New York, 1963.
50. O. Bodansky and M. K. Schwartz, "Methods in Enzymology," Vol. 9, p. 568, 1966.
51. Y. Nakagawa, Ph.D. Dissertation, University of California, Riverside (1966).
52. T. D. Kempe and E. A. Noltmann, *Federation Proc.* **30,** 1122 (1971).

TABLE I

SOME PHYSICAL PARAMETERS OF GLUCOSE-6-PHOSPHATE ISOMERASES FROM VARIOUS SOURCES

Parameter	Enzyme source			
	Rabbit muscle[a] [Ref. (56)]	Human muscle [Ref. (21)]	Bovine mammary gland [Ref. (54)]	Peas [Ref. (13)]
$s^0_{20,w}$	7.19 S	[6.1 S at c = 0.4%]	7.09 S	[6.8 S at c = 0.65%]
$D^0_{20,w}$	5.15×10^{-7} cm^2 sec^{-1}			
M_{equil}	132,000 ($c \rightarrow 0$)	134,000 (c = 0.03%)	125,000 (c = ?)	[110,000 by Archibald method]
$\bar{V}$ used in MW calculations (method in parentheses)	0.740 ml g^{-1} (pycnometry)	0.734 ml g^{-1} (amino acid composition)	0.75 ml g^{-1} (pycnometry)	0.720 ml g^{-1} (arbitrary assumption)
$(dn/dc)_{10°}$	1.80×10^{-3} dl g^{-1}			
f/f_0	1.23			
$(A_{280\ nm})^{0.1\%}_{1\ cm}$	1.32	1.2	0.83[b]	0.875
$pH^{1°,\ I\rightarrow 0}_{isoelectric}$	8.5 [Ref. (62)]			
$pH^{30°,\ I\rightarrow 0}_{isoionic}$	7.47 [Ref. (62)]			

[a] From a small number of isolated analyses Carter and Yoshida have estimated $s_{20,w}$ values ranging from 6.2 to 6.5 S (at c = approximately 0.4%) and molecular weights ranging from 122,000 to 139,000 daltons by high-speed equilibrium sedimentation (at c = 0.02–0.03%) for three chromatographic forms of the rabbit muscle enzyme (63).

[b] This value is calculated from the molecular extinction coefficient given in the original paper by Baich *et al.* (17) which has apparently been erroneously transcribed to the handbook article by Reithel (54) without recalculation for the revised molecular weight.

daltons (*53, 54*). In the meantime, however, Wieland and Pfleiderer in their review on multiple forms of enzymes (*55*) cited the difference in molecular weight between 145,000 (*2*) and 48,000 daltons (*17*) for yeast and mammary gland glucose-6-phosphate isomerase, respectively, as a classic example in support of their concept of "heteroenzymes."

At present, a molecular weight of 132,000 daltons appears to be well established for rabbit muscle glucose-6-phosphate isomerase (*56*). An earlier estimate gave a range from 130,000 to 140,000 (*57*), and a value of 130,000 had been used in previous work from the reviewer's laboratory (*58–60*). Molecular weights of 110,000 and approximately 120,000 daltons have been obtained for the pea enzyme (*13*) and the three isoenzymes from brewer's yeast, respectively (*52, 61*). The available physical parameters are summarized in Table I (*13, 17, 21, 54, 62, 63*).

2. *Amino Acid Composition*

The amino acid composition of rabbit muscle glucose-6-phosphate isomerase does not possess any unusual features. For a detailed appraisal the reader may refer to the paper by Pon *et al.* (*56*). A remarkable aspect, however, is the close similarity in the amino acid compositions of the rabbit and the human muscle enzyme (Table II). With the exception of arginine, none of the differences in the content of the individual amino acid goes beyond a reasonable experimental deviation, and for some (glycine, alanine, valine, and phenylalanine) the agreement is better than can normally be expected from duplicate analyses of one and the same protein. Since the amino acid data of Pon *et al.* are backed by more than 500 individual analyses (*56*), while those of Carter and Yoshida have been derived from a comparatively small number of 20-hr hydrolyzates only (*21*), a reinvestigation of the arginine content of the human muscle enzyme might be desirable in order to decide whether there is in fact a difference. If there should be none, this reviewer hesitates to speculate on the genetic implications.

53. M. C. Hines and R. G. Wolfe, *Biochemistry* **2,** 770 (1963).
54. F. J. Reithel, "Methods in Enzymology," Vol. 9, p. 565, 1966.
55. T. Wieland and G. Pfleiderer, *Advan. Enzymol.* **25,** 329 (1963).
56. N. G. Pon, K. D. Schnackerz, M. N. Blackburn, G. C. Chatterjee, and E. A. Noltmann, *Biochemistry* **9,** 1506 (1970).
57. E. A. Noltmann, "Methods in Enzymology," Vol. 9, p. 557, 1966.
58. G. C. Chatterjee and E. A. Noltmann, *European J. Biochem.* **2,** 9 (1967).
59. G. C. Chatterjee and E. A. Noltmann, *JBC* **242,** 3440 (1967).
60. J. E. D. Dyson and E. A. Noltmann, *JBC* **243,** 1401 (1968).
61. T. D. Kempe and E. A. Noltmann, unpublished experiments (1970).
62. J. E. D. Dyson and E. A. Noltmann, *Biochemistry* **8,** 3533 (1969).
63. A. Yoshida and N. D. Carter, *BBA* **194,** 151 (1969).

TABLE II
COMPARISON OF THE AMINO ACID ANALYSES[a] OF 20-HR HYDROLYZATES (6 N HCl) OF HUMAN AND RABBIT SKELETAL MUSCLE GLUCOSE-6-PHOSPHATE ISOMERASES

Amino acid residue	Glucose-6-phosphate isomerase (number of residues per subunit[c]) — Five-times crystallized from rabbit muscle [Ref. (*56*)]	Glucose-6-phosphate isomerase (number of residues per subunit[c]) — Chromatographically purified from human muscle [Ref. (*21*)][b]
Aspartic acid	57.9	58.9
Threonine	38.7	39.6
Serine	33.2	32.9
Glutamic acid	63.4	67.3
Proline	24.9	22.8
Glycine	44.0	44.5
Alanine	42.4	42.5
Valine	30.5	30.8
Methionine	14.3	13.3
Isoleucine	31.2	27.7
Leucine	57.5	55.8
Tyrosine	12.2	13.3
Phenylalanine	30.3	29.8
Lysine	41.2	39.1
Histidine	23.6	21.8
Arginine	21.2	32.3

[a] Amino acid recoveries after 20 hr of hydrolysis in 6 N HCl have been chosen for this table in order to allow direct comparison of the analyses performed by Pon *et al.* (*56*) and those by Carter and Yoshida (*21*) who derived their data only from 20-hr hydrolyzates. It should be emphasized that the individual values given here are therefore not necessarily an accurate reflection of the true amino acid composition. For the complete amino acid composition of the rabbit muscle enzyme, refer to Pon *et al.* (*56*).

[b] The data in this column are recalculated from Table II of the paper by Carter and Yoshida (*21*). The original table contains computational errors for the values of tryptophan, ammonia, and total number of residues (not listed here) (personal communications from Dr. Carter and Dr. Yoshida).

[c] For the purpose of this comparison, identical subunit molecular weights of 66,000 daltons ($= 0.5 \times 132{,}000$) have been assumed for both the rabbit and the human muscle enzyme. This assumption is considered justifiable since the corresponding molecular weights from equilibrium sedimentation analyses have been found to be 132,400 ± 1,200 (*56*) and 134,000 ± 10,000 (*21*), respectively.

3. *Subunit Structure*

Various lines of physical evidence appear to leave no doubt that muscle and yeast glucose-6-phosphate isomerases are dimeric proteins with identical or nearly identical subunits. Molecular weight data in

TABLE III
SEDIMENTATION COEFFICIENTS AND MOLECULAR WEIGHTS OF GLUCOSE-6-PHOSPHATE ISOMERASE SUBUNITS[a]

Dissociating agent	Method	Sedimentation coefficient[b] of the experimentally observed molecular species (Svedbergs)			Molecular weight (daltons)
		Dimer	Monomer	Random coil	
Sodium dodecyl sulfate	Velocity centrifugation	6.73[c]	4.13[c]		64,400[d]
	Polyacrylamide gel electrophoresis				64,000 [60,000][e]
Maleic anhydride[f]	Velocity and equilibrium sedimentation	6.8	3.9		63,900
Urea (8 M)[g]	Velocity sedimentation		4.0		
Propionic acid (1 M)	Velocity sedimentation			1.6–1.7	
Guanidine hydrochloride (4.5–6 M)[h]	Velocity and equilibrium sedimentation	4.55		1.63	65,500 [70,000][i] [61,000][j]
	X-ray crystallography				~60,000[k]

[a] If not indicated otherwise, data in this table refer to subunits of the rabbit muscle enzyme (*64, 65*).

[b] Sedimentation coefficients are given as $s_{20,w}$ except when designated otherwise.

[c] Apparent sedimentation coefficient, $s_{20°,b}^{0.45\%}$ Tris-NaCl (pH 7.85), determined in the presence of 3.3 mM sodium dodecyl sulfate.

[d] Molecular weight is calculated from the relationship $s/(M^{2/3})$ = constant, based on the Scheraga-Mandelkern equation and on the assumption that dissociation occurs without change in shape. A molecular weight of 132,000 daltons has been assumed for the native dimeric enzyme. At higher ratios of sodium dodecyl sulfate to protein (see footnote *c*) the indicated relationship apparently does not hold. This may be taken to suggest additional conformational alterations.

[e] Isoenzymes of brewer's yeast glucose-6-phosphate isomerase (*52*).

[f] Degree of dissociation is dependent on the extent of modification.

[g] Dissociation in 8 M urea is not complete; prolonged exposure of the enzyme to the denaturant leads to aggregation.

[h] In 4.5–6 M guanidine hydrochloride the enzyme undergoes a reversible transition between two species with different sedimentation coefficients, both of which are present side by side over the range of 4.5–5.5 M guanidine hydrochloride (*65*). Whether the 4.5 S species represents a monomer, a dimer, or a mixture of the two, has not yet been determined unequivocally.

[i] Determined by Yoshida and Carter (*63*).

[j] Human muscle enzyme, determined by Carter and Yoshida (*21*).

[k] Pig muscle enzyme, determined by H. Muirhead (*66*).

support of this contention are summarized in Table III (*21, 52, 63–66*). Prior to the discovery that the yeast enzyme could be resolved into several isoenzymes, some preliminary experiments suggested the possibility of a tetrameric structure (*66a*). A full understanding of these data will have to await further detailed study of the dissociation behavior of the individual yeast isoenzymes.

Chemical studies on the subunit structure of the rabbit muscle enzyme are not yet completed. Peptide maps prepared from tryptic digests of native and carboxymethylated enzyme gave fewer spots than expected. However, peptide maps from tryptic digests of fully carbamylated enzyme yielded 20–22 ninhydrin-positive spots; this finding is in good agreement with two identical subunits, each containing 21 arginine residues (*67*). Carboxyl terminal analyses by carboxypeptidase digestion, by tritium exchange, and by hydrazinolysis coupled with dinitrophenylation indicate glutamine as the terminal and isoleucine as the penultimate residues. Quantitation of the carboxypeptidase experiments yielded 1.7–2.1 glutamine residues per enzyme molecule (*67*). Considerable efforts to identify amino terminal residues by standard methods were unsuccessful suggesting that terminal α-amino groups may be blocked. This was recently shown to be the case by isolation of acetylalanine from a pronase digest as the only chromatographic species containing both acetate and an amino acid. Identification and quantitation of alanine and acetic acid were accomplished by amino acid analysis and gas chromatography, respectively, yielding 1.6–1.9 acetylalanine residues per enzyme molecule of 132,000 daltons (*67a*). Thus, both the carboxyl terminal and the amino terminal analyses are in agreement with a dimeric enzyme composed of two identical subunits. Chemical data on other glucose-6-phosphate isomerases are not yet available.

4. *Multiple Forms of the Enzyme*

a. Isoenzymes of Glucose-6-phosphate Isomerase from Brewer's Yeast. Shortly after the brewer's yeast enzyme had been isolated in crystalline form (*12*), homogeneity studies indicated that the crystalline prepa-

64. M. N. Blackburn, T. D. Kempe, and E. A. Noltmann, *Abstr. Pacific Slope Biochem. Conf., La Jolla, California, 1970* p. 52 (1970).

65. M. N. Blackburn, Ph.D. Dissertation, University of California, Riverside (1971); M. N. Blackburn and E. A. Noltmann, manuscript in preparation (1971).

66. H. Muirhead, personal communication (1970).

66a. J. E. Robbins, S. L. Lowe, and F. J. Reithel, *Federation Proc.* **26**, 275 (1967).

67. G. T. James and E. A. Noltmann, unpublished experiments (1970).

67a. G. T. James, Ph.D. Dissertation, University of California, Riverside, California (1971).

ration might not represent a single molecular species. Systematic chromatographic studies led to resolution of three isoenzymes, and evidence was presented that these chromatographic species exist *in vivo* and are not artifacts of the isolation procedures (*68*). Analogous but not identical isoenzymes could also be resolved from a commercial preparation (Boehringer) of the crystalline baker's yeast enzyme (*68*). This finding was recently confirmed in another laboratory (*69*). The current status of studies on the characterization of the brewer's yeast isoenzymes may be summarized as follows:

(a) They can be reproducibly isolated, chromatographed, and rechromatographed in the presence or absence of thiols without changes in their properties (*61*). They therefore fulfill the essential criterion for the definition of isoenzymes, i.e., they are stable, isolatable protein species which do not undergo interconversion among each other under "physiological" conditions.

(b) Their molecular weights are identical (approximately 120,000 daltons) (*61*) and, by the Hedrick and Smith method (*70*) of polyacrylamide gel electrophoresis, they behave strictly as charge isomers (*52*). This is in agreement with their chromatographic elution pattern on ion exchange columns (*51, 68*) and with their mobilities in cellulose acetate electrophoresis. The latter yield apparent isoelectric points of 5.4, 5.2, and 5.0 for the brewer's yeast isoenzymes A, B, and C, respectively (*51*).

(c) Their overall amino acid compositions are too similar to pinpoint any differences in their primary structure (*51*).

(d) Polyacrylamide gel electrophoresis in the presence of sodium dodecyl sulfate yields indistinguishable subunit molecular weights of approximately 60,000 daltons (*52*).

b. Pseudoisoenzymes of Glucose-6-phosphate Isomerase from Rabbit Muscle. In contrast to the yeast enzyme, all evidence gathered in the reviewer's laboratory on five-times crystallized glucose-6-phosphate isomerase from rabbit skeletal muscle indicated homogeneity within the precision of the applied methods (*56*). This premise was challenged, however, by Yoshida and Carter who reported (*63*) the resolution of a commercial preparation (Calbiochem) of the crystalline rabbit muscle isomerase into three "isoenzymes." They claimed that two of the three chromatographic species (the amount of the third fraction, considered to be an artifact produced from "Isoenzyme II," was insufficient for

68. Y. Nakagawa and E. A. Noltmann, *JBC* **242,** 4782 (1967).
69. L. A. Gonzalez de Galdeano and H. Simon, *Z. Physiol. Chem.* **351,** 1113 (1970).
70. J. L. Hedrick and A. J. Smith, *ABB* **126,** 155 (1968).

more detailed study) have different amino acid compositions (*71*) and yield different peptide maps. They therefore suggested that different structural genes are involved in producing the "isoenzymes" (*63*). Numerous attempts in the reviewer's laboratory to reproduce the chromatographic separation pattern on carboxymethyl Sephadex in the system described by Yoshida and Carter (*63*) have failed to yield fractions corresponding to their "isoenzymes" from either crude muscle extracts, or partially purified preparations, or two- and five-times crystallized muscle enzyme from different rabbit strains (*72*, *73*). However, several chromatographic forms of the rabbit muscle enzyme could be separated with a sodium (phosphate) gradient about one order of magnitude in concentration below that used by Yoshida and Carter. Yet the enzyme species obtained in this manner did not remain single species on rechromatography; in fact, they were interconvertible by treatment with either dithiothreitol or exposure to oxygen. Performing the entire purification procedure in the presence of dithiothreitol yielded only one enzyme species. Amino acid analysis of any chromatographic species gave values indistinguishable within the precision of the analysis and essentially equal to that reported (*56*) for the five-times crystallized enzyme (*72*, *73*). The only property by which the individual chromatographic species could be distinguished was a different accessibility to *p*-mercuribenzoate and variation in the extent of oxidation of their sulfhydryl groups. Exhaustive analysis of the various chromatographic species with respect to immediately available–SH groups [*p*-mercuribenzoate assay of the native protein (*59*, *74*)], total –SH content (*p*-mercuribenzoate assay of sodium dodecyl sulfate–denatured enzyme), and total half-cystine content [measured as cysteic acid after performic acid oxidation (*75*)] yielded differences between the chromatographic forms in the available –SH groups but not in the total half-cystine content (*73*). It must therefore be concluded that the different chromatographic species represent different states of the *same* enzyme protein and *not*, as implied by Yoshida and Carter (*63*), genuinely different "isoenzymes" with genet-

71. It is interesting to note that the amino acid compositions reported for the two "isoenzymes" (*63*) show larger differences than those that exist between the values given by the same authors for the human muscle enzyme (*21*) and those obtained by Pon *et al.* for the five-times crystallized rabbit muscle enzyme (*56*) (see Section II,B,2 above).

72. G. T. James, M. N. Blackburn, J. M. Chirgwin, A. M. Register, K. D. Schnackerz, and E. A. Noltmann, *Abstr. Pacific Slope Biochem. Conf., La Jolla, California, 1970* p. 51 (1970).

73. M. N. Blackburn, J. M. Chirgwin, G. T. James, T. D. Kempe, T. F. Parsons, A. M. Register, K. D. Schnackerz, and E. A. Noltmann, *JBC* **247,** 1170 (1972).

74. P. D. Boyer, *JACS* **76,** 4331 (1954).

75. S. Moore, *JBC* **238,** 235 (1963).

ically different primary structures. To this extent, they should be considered artifacts, and their designation as "pseudoisoenzymes" (*76, 77*) seems appropriate.

c. Other Multiple Electrophoretic Enzyme Species. Several other cases of multiple forms of glucose-6-phosphate isomerase have been described, e.g., in human blood serum (*78*), tissues of mice (*79, 80*) and fish (*25*), erythrocytes, leukocytes and platelets of normal and genetically variant humans (*44a, 81*) and rabbits (*82*). In most instances, evidence for defining the various forms in terms of different, genetically determined isoenzymes has been exclusively derived from activity staining patterns of multiple electrophoretic bands with little emphasis on possible artifacts created by the technique itself or by the preceding tissue work-up and storage procedures (*83*). Although this reviewer does not wish to dispute the existence of genetically distinct isoenzymes of glucose-6-phosphate isomerase in all instances, he feels that the alternative, *viz.*, the existence of pseudoisoenzymes, especially in the form of various oxidation states as elaborated upon in the preceding section, has been largely ignored not only for the case discussed here but also for many other enzymes (e.g., see Sections V,B and VII,B,3).

5. *Conformational States of Native and Denatured Glucose-6-phosphate Isomerase from Rabbit Muscle*

Electrometric and spectrophotometric titration studies (*62*) as well as ultraviolet difference spectroscopy (*84*), of the native and the denatured enzyme, indicated considerable differences—depending upon the denaturant used—in the type and number of accessible amino acids side

76. R. J. Wieme, *in* "Homologous Enzymes and Biochemical Evolution" (N. van Thoai and J. Roche, eds.), p. 19. Gordon & Breach, New York, 1968. In his assessment of the isoenzyme concept, Wieme attributed the term *pseudoisoenzymes* to Kirkman and Hanna (*77*). Although these authors had admirably taken the then unfashionable stand to warn against the practice of uncritically equating multiple activity bands in electrophoretic systems with genetically different isoenzymes, they apparently had not specifically used the term *pseudoisoenzymes* in their paper to describe multiple enzyme forms that may originate from "uncontrolled conditions of extraction and storage" (*77*).

77. H. N. Kirkman and J. E. Hanna, *Ann. N. Y. Acad. Sci.* **151**, 133 (1968).

78. M. K. Schwartz and O. Bodansky, *Am. J. Med.* **40**, 231 (1966).

79. N. D. Carter and C. W. Parr, *Nature* **216**, 511 (1967).

80. R. J. DeLorenzo and F. H. Ruddle, *Biochem. Genet.* **3**, 151 (1969).

81. L. I. Fitch, C. W. Parr, and S. G. Welch, *BJ* **110**, 56P (1968).

82. S. G. Welch, L. I. Fitch, and C. W. Parr, *BJ* **117**, 525 (1970).

83. In one study, for example, the preparation of enzyme extracts from tissue culture cells involves "trypsinizing the cells from the glass surface" (*80*).

84. J. E. D. Dyson and E. A. Noltmann, *Biochemistry* **8**, 3544 (1969).

TABLE IV
NUMBER OF MAXIMALLY ACCESSIBLE AMINO ACID SIDE CHAINS OF RABBIT MUSCLE GLUCOSE-6-PHOSPHATE ISOMERASE UNDER VARIOUS DENATURING CONDITIONS[a]

	Average number of amino acid side chains per subunit of 66,000 daltons[b]			
Condition of denaturation	Sulfhydryl	Imidazolyl	Indolyl	Phenolic hydroxyl
Total amino acid analysis	6	24	12	12
Native enzyme	2	14	5	0
Sodium dodecyl sulfate (0.5%)	6		8	0
pH 2		24	9	6
pH 13			9	12
Urea (8 *M*, pH 7)			10	9
Urea (8 *M*, pH 2.9)			12	11

[a] Data from Dyson and Noltmann (*62, 84*).
[b] Values are calculated in terms of average numbers per subunit. It should be kept in mind, however, that more than one conformational species may exist at some denaturing conditions.

chains (Table IV); thus, several different levels of structural integrity were proposed (*84*). Although the quantitative aspect of these interpretations will be affected by whether only one or more than one molecular species is present under a particular set of denaturation conditions, the experimental results show clearly that glucose-6-phosphate isomerase must possess a very high stability within its core structure which enables it to withstand severe denaturing conditions without complete exposure of all tryptophyl or tyrosyl residues (*84*). Furthermore, ultracentrifuge data (cf. Table II) indicated that the denatured enzyme still retains a considerable degree of structure in 8 *M* urea or 5 *M* guanidine hydrochloride (*65*). Another interesting observation was that acid and alkaline denaturation resulted in the titration of normally inaccessible histidyl or tyrosyl residues. On the basis of the parallel nature of exposure of these residues and loss of enzymic activity, it was speculated that juxtaposed histidine–tyrosine pairs may play a role in maintaining the catalytically active conformation of the enzyme (*62*). The validity of such a hypothesis will obviously have to await analysis of the three-dimensional structure.

6. *Amino Acid Residues at the Active Site*

The first clue to the nature of a catalytically functional group came from deuterium exchange experiments of Rose and O'Connell (*85*) which

85. I. A. Rose and E. L. O'Connell, *JBC* **236**, 3086 (1961).

led them to suggest that a base with no exchangeable hydrogens, i.e., a carboxylate group or the nonprotonated nitrogen of an imidazole group, is involved in the glucose-6-phosphate isomerase reaction. Further support for an imidazole was provided by kinetic studies of the enzyme as a function of pH by Hines and Wolfe (*53*) and an extension of this approach by Dyson and Noltmann (*60*) who included temperature dependence of the kinetic parameters as an additional variable (*vide infra*). Beyond its agreement with the postulated role of histidine, this latter work also provided evidence for the additional participation of a lysine side chain amino group in the catalytic process (*60*).

Chemical modification of the rabbit muscle enzyme with parallel loss of enzymic activity by dye-sensitized photooxidation (*58*), reaction with organic mercurials (*59*), and carboxamidomethylation (*86*) was in agreement with the proposed role of histidine and also rendered cysteine and methionine unlikely as catalytically critical amino acid residues. Full protection of histidine from alkylation by iodoacetamide, with complete preservation of enzymic activity, in the presence of the competitive inhibitor 6-phosphogluconate (*86*) lends further support to the argument for a histidyl residue at the active site. Recent work by Rose (*87*) involving affinity labeling of glucose-6-phosphate isomerase with isotopic 1,2-anhydromannitol 6-phosphate (*87a*) and liberation of the label by dilute alkali or hydroxylamine led him to suggest that a carboxyl group had reacted with the substrate analog as in the case of triosephosphate isomerase (see Section VII,B,4). Thus, the question as to which residue is responsible for the kinetically operable pK of 6.75 (*60*) cannot at present be answered unambiguously.

Although the assignment of the pK of 9.3, arrived at from the kinetic studies (*60*), to the ϵ-amino group of a lysyl residue was reasonable because both cysteine and tyrosine were not compatible with the

86. K. D. Schnackerz and E. A. Noltmann, *JBC* **245**, 6417 (1970).

87. I. A. Rose, *8th Intern. Congr. Biochem., Symp. III, Postersession 1, Interlaken, Switzerland, 1970;* I. A. Rose and E. L. O'Connell, *Federation Proc.* **30**, 1158 (1971).

87a. The epoxide affinity label was initially presumed to be 1,2-anhydroglucitol 6-phosphate (*87*). Subsequent detailed studies with the enzymically prepared substrate analog, however, revealed that the compound used in the alkylation experiments was a mixture of both 1,2-anhydromannitol and 1,2-anhydroglucitol phosphates and that, paradoxically, it was the manno compound that produced alkylation and inactivation, with half-lives of less than 0.5 hour and 99.9% inactivation occurring within 60 hours (reported by I. A. Rose at the 62nd Annual Meeting of the American Society of Biological Chemists, San Francisco, June 1971). Since then, Rose and collaborators have been able to prepare the gluco-epoxide compound and have found it also to inactivate glucose-6-phosphate isomerase, but at a rate 100 times slower than that observed with the manno compound (I. A. Rose, personal communication, August 1971).

thermodynamic (*60*) and the spectrophotometric titration data (*62*), it was made entirely on the basis of circumstantial evidence. Carboxamidomethylation of the enzyme at pH 8.55 resulted in modification of lysine. However, the binding constants for substrate and competitive inhibitor, derived from quantitation of their protective effect against alkylation, were one to two orders of magnitude greater than the corresponding kinetic constants (*86*). It was therefore desirable to test a more specific modifying reagent for lysine. Pyridoxal 5′-phosphate was found to serve this purpose ideally (*88*). It reacted stoichiometrically with the enzyme to form a Schiff base displaying the spectral properties typical for pyridoxal 5′-phosphate–protein interaction. After reduction with sodium borohydride, the absorption spectrum showed a maximum at 325 nm, characteristic for ϵ-amino pyridoxyllysine, and N^6-pyridoxyllysine was identified in acid hydrolysates of the reduced glucose-6-phosphate isomerase–pyridoxal 5′-phosphate complex. A point of interest was the stoichiometry which, by assuming the molar absorptivity coefficient of N^6-phosphopyridoxyllysine (*89*) held also for protein-bound pyridoxal 5′-phosphate, was found to be two mole equivalents per mole of dimeric enzyme for 100% loss of enzymic activity (*90*). Data are currently being sought to substantiate this stoichiometry by measuring the binding of isotopically labeled pyridoxal 5′-phosphate.

C. Catalytic Properties

1. *Assay Methods*

Glucose-6-phosphate isomerase has been mainly assayed by two methods: (a) in the forward direction by colorimetric estimation of fructose 6-phosphate in a chemical-stop assay (for detailed descriptions, see, e.g., references *8, 45, 50, 54*) and (b) in the reverse direction by a continuously recording spectrophotometric assay with glucose-6-phosphate dehydrogenase as the coupling enzyme (for precise conditions, see, e.g., references *8, 57, 60, 91*). A convenient continuously recording pH-stat assay for the forward reaction, utilizing phosphofructokinase as

88. K. D. Schnackerz and E. A. Noltmann, *Abstr. Pacific Slope Biochem. Conf., Seattle, Washington, 1969* p. 19 (1969).
89. E. H. Fischer, A. W. Forrey, J. L. Hedrick, R. C. Hughes, A. B. Kent, and E. G. Krebs, *Proc. Intern. Symp. Chem. Biol. Aspects Pyridoxal Catal., 1962* p. 543. Macmillan, New York, 1963.
90. K. D. Schnackerz and E. A. Noltmann, *Biochemistry* **10**, 4837 (1971).
91. S. E. Kahana, O. H. Lowry, D. W. Schulz, J. V. Passonneau, and E. J. Crawford, *JBC* **235**, 2178 (1960).

indicator enzyme, has been described by Dyson and Noltmann (*92*). Other multiple-couple assays have been reported for the forward reaction [e.g., (*91, 93*)], but their usefulness is restricted by the requirement for several highly purified coupling enzymes. Also, complications may arise in kinetic studies from interference by the large number of intermediates formed in the sequence of reactions. A potentially useful method for assaying both the forward and the reverse reaction should be an adaptation of Takasaki's spectropolarimetric assay for glucose and mannose isomerization (*94, 95*) to the isomerases acting on the phosphorylated hexoses.

A recent paper by Wurster and Hess (*96*) begins with the categorical statement that, "The two enzyme system composed of glucosephosphate isomerase and glucose-6-phosphate dehydrogenase from yeast does *not* [italics added] follow the reaction sequence fructose-6-phosphate → glucose-6-phosphate → 6-phosphogluconate. . . ."

This gives the erroneous impression that the standard coupled assay of the isomerase with the dehydrogenase is invalid. Although their basic premise (*viz.*, the isomerase reacts with α-glucose 6-phosphate, whereas the yeast dehydrogenase reacts exclusively with β-glucose 6-phosphate; therefore, the overall rate will be affected by the rate of α to β conversion) is correct (*vide infra*), the implied conclusion that the routine assay is not a proper measure of the isomerase activity is unwarranted. The emphasis of their experimental approach is on conditions of absolute and relative concentrations of both glucose-6-phosphate isomerase and dehydrogenase, which may be employed in stopped-flow kinetic studies, but which are normally never approached in standard spectrophotometric assays. Their data show specifically that at a ratio of 28 for dehydrogenase/isomerase, no difference exists between the measured and the theoretical rate for glucose-6-phosphate isomerase up to rates that are equivalent to producing an absorbance change at 340 nm of 0.6/min/ml reaction mixture (in a routine assay the rate would normally be 0.1 to 0.2 $\Delta A_{340\ \text{nm}}$ per min). Even at a rate equivalent to 2.5 $\Delta A_{340\ \text{nm}}$ per min, the measured rate is still less than 10% slower than the calculated theoretical rate. Thus, ironically, the data of Wurster and Hess (*96*) do in fact *prove* that the standard coupled assay *is* a valid measurement of glucose-6-phosphate isomerase activity.

It should be stressed, however, that the results of Wurster and Hess

92. J. E. Dyson and E. A. Noltmann, *Anal. Biochem.* **11**, 362 (1965).
93. E. Racker, *JBC* **167**, 843 (1947).
94. Y. Takasaki, *Agr. Biol. Chem.* (*Tokyo*) **31**, 309 (1967).
95. Y. Takasaki, *Agr. Biol. Chem.* (*Tokyo*) **31**, 435 (1967).
96. B. Wurster and B. Hess, *Z. Physiol. Chem.* **351**, 1537 (1970).

reiterate the importance of experimentally verifying the conditions for a valid enzymic assay. For the coupled assay of glucose-6-phosphate isomerase with glucose-6-phosphate dehydrogenase this aspect has been elaborated upon previously. It was emphasized that sufficient amounts of the auxiliary enzyme must be added "to produce a low steady state level of glucose-6-phosphate in the coupled system, i.e. *conditions where the isomerase velocity is equal to the glucose-6-phosphate dehydrogenase velocity*" [italics added] (*8, 91*), and this was established to be the case for the entire pH range studied (*60*).

2. *Kinetic Parameters*

a. Michaelis Constants. Values of K_m for the substrates of both the forward and the reverse reaction have been reported for glucose-6-phosphate isomerase from numerous laboratories. The reviewer has chosen to give only a representative sample of K_m data which are compiled in Table V (*2, 13, 17, 19, 22, 28, 30, 40, 51, 53, 60, 82, 91, 97, 98*). Rationale for the selection has been (a) to represent those values that have been derived from detailed kinetic studies, (b) to point out discrepancies where they exist for the isomerase from a single source, and (c) to give examples of K_m values for various phylogenetic species. Some of the differences may only be the result of different assay conditions employed by different investigators. This is especially true for values based on assays involving the resorcinol method (*99*) for which it was initially assumed that fructose 6-phosphate would give only 60% the color value of fructose routinely used as standard, whereas in fact the color produced by both is equal; this was found after fructose 6-phosphate of sufficient purity became available for comparison [e.g., (*8, 91*); for a more detailed discussion of this aspect, see reference *100*].

Most of the K_m values are of the order of 10^{-4} *M*, those for fructose 6-phosphate usually one-half to two-thirds of those for glucose 6-phosphate. A notable exception are the constants reported by Kahana *et al.* (*91*) which are approximately one order of magnitude below all others. Attempts have been made (*40, 60*) to rationalize these low values on the basis of the assay conditions employed by Kahana *et al.* who used fluorometric and multiple-couple spectrophotometric assays and also applied extensive corrections to their raw data in order to arrive at the figures quoted in Table V.

97. M. Salas, E. Viñuela, and A. Sols, *JBC* **240,** 561 (1965).
98. B. Wurster and F. Schneider, *Z. Physiol. Chem.* **351,** 961 (1970).
99. J. H. Roe, *JBC* **107,** 15 (1934).
100. M. K. Schwartz and O. Bodansky, *Anal. Biochem.* **11,** 48 (1965).

TABLE V

SELECTED VALUES[a] OF MICHAELIS CONSTANTS FOR GLUCOSE 6-PHOSPHATE AND FRUCTOSE 6-PHOSPHATE AND OF INHIBITOR CONSTANTS FOR 6-PHOSPHOGLUCONATE REPORTED FOR GLUCOSE-6-PHOSPHATE ISOMERASES OF VARIOUS ORIGIN

	K_m		K_i	
Enzyme source	Glucose-6-P (mM)	Fructose-6-P (mM)	6-phosphogluconate (mM)	Ref.
Rabbit muscle	0.31	0.17	0.16	(*60*)
	0.8	0.12		(*40*)
	0.03	0.01	0.005	(*91*)
Bovine mammary gland	0.12	0.07		(*53*)
	0.57			(*17*)
Rabbit liver	0.6	0.12		(*40*)
Rabbit erythrocytes				
wild-type		0.21	0.075	(*82*)
B9 mutant		0.74	0.20	(*82*)
Yeast	0.3	0.15	0.015	(*97*)
	0.7			(*2*)
	1.5[b]	0.23[b]		(*51*)
	0.8	0.11		(*98*)
Peas	0.27		0.013	(*13*)
Honeybee	2.0			(*30*)
Entamoeba species		0.06–0.19	0.013–0.028	(*28*)
Aerobacter aerogenes	0.36			(*19*)
Schistosoma mansoni		0.1		(*22*)

[a] For selection criteria see text. Most of the constants in this table have been measured at 30° (or at room temperature) and at pH values between 7 and 8. For precise experimental conditions, the reader should refer to the original publications.

[b] Average of K_m values determined for three isoenzymes.

b. Substrate Specificity. Glucose-6-phosphate isomerase appears to be one of the very few enzymes that possess an absolute or nearly absolute specificity for a single substrate-product pair. It neither acts on phosphorylated sugars other than glucose 6-phosphate and fructose 6-phosphate nor on any free sugars except in those cases where arsenate appears to take the place of phosphate, as discussed in Section VIII,A. For the five-times crystallized enzyme from rabbit muscle, maximal molecular activities of 7.7×10^4 and 13.7×10^4 per min (molecular weight, 132,000 daltons), corresponding to 590 and 1040 μmoles/min/mg for the forward and reverse reaction, respectively, have been measured at pH 8.5, 30° and 0.12 ionic strength (*60*). Similar proportions have been found for the mammary gland enzyme (*53*).

Whether the enzyme also isomerizes glucose 6-sulfate is an interesting problem which has not yet been solved. Both the isomerases from mammary gland and from yeast have been reported to act on the sulfate ester (*53, 54, 100a*), but it was not certain whether observed changes in optical rotation were solely the result of mutarotation or of mutarotation and isomerization (*100a*).

c. Effect of pH. Detailed studies on the effect of pH on the catalyzed reaction have been conducted with the enzyme from bovine mammary gland (*53*) and from rabbit muscle (*60*). Hines and Wolfe (*53*) measured K_m and V_{max} as a function of pH at 30° and calculated the following apparent ionization constants. Free enzyme (average values from combined data of the forward and the reverse reaction): $pK_1 = 6.0$, $pK_2 = 9.1$; enzyme–substrate complex, forward reaction: $pK_1 = 6.0$, $pK_2 = 9.5$; enzyme–substrate complex, reverse reaction: $pK_1 = 6.8$, $pK_2 = 9.8$. For the same temperature (30°), Dyson and Noltmann (*60*) obtained these values: $pK_1^{E} = 6.75$, $pK_2^{E} = 9.3$; $pK_1^{ES(f)} = 6.4$, $pK_2^{ES(f)} = 9.9$; $pK_1^{ES(r)} = 6.9$, $pK_2^{ES(r)} = 9.4$. In addition, the latter also measured the effect of temperature on the apparent ionization constants which permitted estimation of the heats of ionization for the kinetically operable groups in the free enzyme. For pK_1 the heat of ionization was approximately 7,700 cal/mole which supports the proposal that this pK represents an imidazole group (*53, 60*). For pK_2 a value of approximately 16,000 cal/mole was found which was considered strong evidence for the involvement of an ϵ-amino lysyl side chain since such a large heat of ionization is incompatible with either a sulfhydryl or a phenolic hydroxyl group (*60*).

An interesting observation was recently reported by Welch *et al.* (*82*). These authors studied the pH dependence of the K_m for fructose 6-phosphate and of the K_i for 6-phosphogluconate for a glucose-6-phosphate isomerase preparation from erythrocytes of an autosomally inherited variant rabbit strain as compared with the enzyme obtained from the usual wild-type rabbit. They found different pK values for the two preparations and attributed the difference to a mutation in the variant enzyme which affects a kinetically operable histidine residue.

The implication of catalytically critical amino acid residues on the basis of kinetic studies can only be suggestive evidence since the apparent ionization constants may be a reflection of conformational changes. However, the argument has been made (*60*) that in cases where only two pK values are found within the pH range of the catalytically active enzyme, these must likely originate from groups involved in the catalytic process. If these pK values were in fact to reflect ionizations related to

100a. C. W. Carlson, S. L. Lowe, and F. J. Reithel, *Enzymologia* **33**, 192 (1967).

conformational changes instead, then ionizations of those residues involved in catalysis must not affect the catalytic process in the working pH range, which appears to be unlikely for the majority of enzymes (*60*). For rabbit muscle glucose-6-phosphate isomerase, gross conformational changes apparently do not occur within the pH range of the kinetically determined pK values (*62*). The relationship of the kinetically identified amino acid side chains to the catalytic mechanism is discussed below in Section II,C,5,c.

An interesting proposal has been made regarding the rapid decrease of activity at acid pH of glucose-6-phosphate isomerase from *Escherichia coli:* It was suggested that at low pH the enzyme may be the rate-limiting regulator of glycolysis (*26*).

d. Equilibrium Constant and Thermodynamic Parameters. The equilibrium constant, K_{eq} ([fructose-6-P]/[glucose-6-P]), for glucose-6-phosphate isomerase is dependent upon temperature (*6, 91, 98*) and independent of pH within the range of pH 6–10 (*60, 91*). For 30°, K_{eq} has been found by at least four laboratories to be 0.30–0.32 for the enzyme from mammalian tissues and from yeast (*98*) both by direct measurement of the equilibrium concentrations of substrate and product and by calculation from the Haldane relationship.

From a study of K_{eq} and V_{max} as a function of temperature, the following thermodynamic parameters have been calculated for 25°, pH 8.5 and 0.12 ionic strength (*60*): K_{eq} ([fructose-6-P]/[glucose-6-P]) = 0.27; $\Delta G° = 767$ cal/mole; $\Delta H° = 2{,}570$ cal/mole; $\Delta S° = 6.0$ cal/mole/deg; ΔH^*(glucose-6-P → fructose-6-P) = 11,250 cal/mole; ΔH^*(fructose-6-P → glucose-6-P) = 9,460 cal/mole.

e. Competitive Inhibitors. Parr's original reports (*33, 35a*) that 6-phosphogluconate is a competitive inhibitor of mammalian glucose-6-phosphate isomerase were confirmed for the purified yeast enzyme (*2*) and inhibition was regularly observed whenever tested with other preparations. Values for the inhibitor dissociation constant are listed in Table V. It is notable that in most cases the K_i values seem to indicate a tighter binding of the competitive inhibitor than of either substrate, although the value of 5 μM (*91*) may be too low for the reasons indicated above (Section II,C,2,a). In contrast to the typically competitive nature of the 6-phosphogluconate inhibition for the enzyme from rabbit muscle (*60*), the inhibition of the yeast enzyme appears to be not of the conventional type. Abnormal inhibition patterns have been observed in the author's laboratory for the crystalline yeast enzyme prior to resolution into isoenzymes (*92*) as well as for the individual isoenzymes (*51*). Also,

Davis *et al.* (*101*) reported different K_i values for 6-phosphogluconate for the forward and the reverse reaction and, in addition, differences in the inhibitor constant at high and low substrate concentrations. These abnormalities have not yet found a satisfactory explanation.

The inhibition by 6-phosphogluconate is strongly dependent on pH (*60, 97, 102, 103*), and the decrease in K_i with decreasing pH has been suggested as a possible means by which temporary decreases in the intracellular pH may exert a regulatory effect through increased inhibition of the glucose-6-phosphate isomerase reaction (*102*).

A considerable number of other phosphorylated compounds have also been found to act as competitive inhibitors: erythrose-4-P, $K_i = 1\text{–}2\ \mu M$ (*34, 97*); sorbitol-6-P, $K_i = 0.025$ mM (*33, 97*); 1,5-anhydroglucitol-6-P, $K_i = 2.5$ mM (*97, 104*); mannose-6-P, $K_i = 0.46$ mM (*97*); mannitol-6-P, $K_i = 0.13$ mM (*97*); 2-deoxyglucose-6-P, $K_i = 0.7$ mM (*14, 97*); galactose-6-P, $K_i = 1.7$ mM (*97*); L-sorbose-6-P, $K_i = 4$ mM (*97*); glucosamine-6-P (*105*); fructose-1-P (reverse reaction only), $K_i = 1.5$ mM (*40*); fructose 1,6-diphosphate (reverse reaction only), $K_i = 7.5$ mM (*40*); ribose-5-P, $K_i = 0.5$ mM (*97*); ribulose-5-P, $K_i = 0.05$ mM (*97*); L-xylulose-5-P, $K_i = 0.7$ mM (*97*); sedoheptulose-7-P, $K_i = 2$ mM (*35, 97*); erythritol-4-P, $K_i = 0.04$ mM (*97*); phosphoenolpyruvate, $K_i = 1.1$ mM (*91*); ATP, $K_i = 0.4$ mM (*91*); inorganic phosphate, $K_i = 1.7$ mM (*91*).

3. *The Active Form of the Substrate in Solution*

A critical aspect in deriving a mechanism of action for glucose-6-phosphate isomerase is the assumption of which form of the substrate initially binds to the active site. Glucose can be present in aqueous solution as pyranose and furanose, each as the α and β anomer, and also in the acyclic form as the free aldehyde or its hydrate (*106*). Although little experimental data are available for phosphorylated sugars, glucose-6-phosphate and fructose 6-phosphate can also be expected to be present in solution as multiple species. While the problem of different conforma-

101. J. S. Davis, F. J. Kull, and J. E. Gander, *Abstr. 6th Intern. Congr. Biochem., New York, 1964* p. 506 (1964).

102. D. M. J. Geewater, E. D. Hanshaw, R. E. Martin, and C. W. Parr, *BJ* **97,** 12P (1965).

103. J. E. Dyson and E. A. Noltmann, *Abstr. Pacific Slope Biochem. Conf., Eugene, Oregon, 1966* p. 22 (1966); J. E. D. Dyson and E. A. Noltmann, *Federation Proc.* **26,** 390 (1967).

104. R. A. Ferrari, P. Mandelstam, and R. K. Crane, *ABB* **80,** 372 (1959).

105. J. B. Wolfe and H. I. Nakada, *ABB* **64,** 489 (1956).

106. S. J. Angyal, *Angew. Chem. Intern. Ed. Engl.* **8,** 157 (1969).

tional forms has not yet been considered in detail (however, cf. reference *5*), the question of whether the ring form or the straight-chain form of the hexose 6-phosphates are the primary substrates of the enzyme has been subject to considerable discussion.

Arguments favoring the ring form as the "true" substrate have been based on polarographic measurements according to which an equilibrium solution of D-glucose contains only 0.0026% (*107*) to 0.024% (*107a*) of the free aldehyde form (see also Section II,C,5 below) and on the assumption that for the phosphorylated hexose this amount would be larger (*53, 60, 97, 108–110, 110a*) but not by more than one or two orders of magnitude [e.g., (*53, 60, 97, 110a*)]. Since the measured K_m values for *total* substrates were found to be of the order of 10^{-4} M, it was reasoned that it would require the K_m to be between 10^{-7} and 10^{-9} M to account for these values in terms of free aldehyde (*53, 60*; footnote 8 of reference *111*), but this value for K_m was thought to be too small to receive serious consideration. More recent analyses by ORD, CD, UV, IR, and NMR spectroscopy (*112–114a*), however, suggest that the earlier values for the acyclic forms of hexose phosphates may have been far too low. The best and internally most consistent estimates, obtained by parallel measurement of UV, IR, and NMR, spectra, appear to be those of Gray and Barker (*113*) and Swenson and Barker (*113a*) which place the acyclic form of the hexose phosphates in solution at approximately 2% of the total. Based on this percentage, K_m values of 10^{-4} M for total substrate would correspond to 10^{-6} M for the straight-chain form, which is definitely within the range of accepted Michaelis constants.

4. *Anomerase Activity of Glucose-6-phosphate Isomerase*

Since even the revised estimates for the amount of straight-chain hexose phosphates in solution (see preceding section) leave 98% of the

107. J. M. Los, L. B. Simpson, and K. Wiesner, *JACS* **78**, 1564 (1956).
107a. S. M. Cantor and Q. P. Peniston, *JACS* **62**, 2113 (1940).
108. F. Lippich, *Biochem. Z.* **248**, 280 (1932).
109. A. V. Stepanov and B. N. Stepanenko, *Biokhimiya* **2**, 917 (1937).
110. J. M. Bailey, P. H. Fishman, and P. G. Pentchev, *JBC* **243**, 4827 (1968).
110a. J. M. Bailey, P. H. Fishman, and P. G. Pentchev, *Biochemistry* **9**, 1189 (1970).
111. R. W. Gracy and E. A. Noltmann, *JBC* **243**, 5410 (1968).
112. R. W. McGilvery, *Biochemistry* **4**, 1924 (1965).
113. G. R. Gray and R. Barker, *Biochemistry* **9**, 2454 (1970).
113a. C. A. Swenson and R. Barker, *Biochemistry* **10**, 3151 (1971).
114. G. Avigad, S. Englard, and I. Listowsky, *Carbohydrate Res.* **14**, 365 (1970).
114a. G. R. Gray, *Biochemistry* **10**, 4705 (1971).

total substrate in the ring form, a means must exist by which the ring form is converted into the aldehyde form required for the isomerization per se to occur (*vide infra*). The opening of the glucose ring is an obligatory step in the anomerization between the α and β forms which proceeds spontaneously at a high turnover rate [0.04 to 0.06 sec^{-1} for glucose 6-phosphate (*96*, *110*, *110a*)]. The question therefore is whether the spontaneous anomerization is fast enough to provide sufficient open chain substrate in order not to be rate limiting for the isomerization reaction or whether the isomerase itself has anomerase activity which can produce the open chain form at a rate above and beyond that supplied spontaneously. Sols and collaborators (*97*) have shown in a series of elegant experiments that glucose-6-phosphate isomerase from yeast possesses anomerase activity toward α-glucose 6-phosphate (but not toward the β anomer) and that the enzyme has therefore the inherent ability to catalyze a ring opening step. Most significantly, the anomerase activity of the isomerase was found to catalyze the conversion only to the point of ring opening, whereas the ring closure to form the β configuration occurred spontaneously (cf. also discussion in Section II,C,5,c). The relative anomeric specificity of the various enzymes acting on glucose

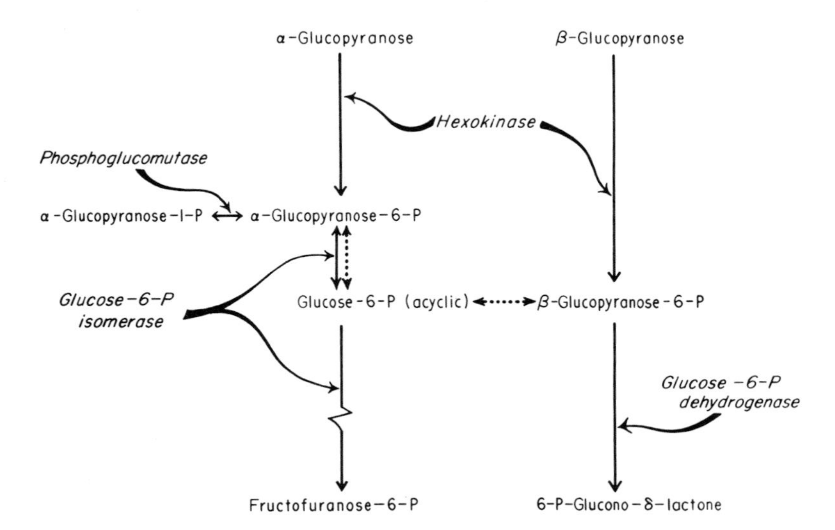

FIG. 1. Anomeric specificity of enzymes acting on glucose 6-phosphate, modified from Salas *et al.* (*97*) to include glucose 6-phosphate as an obligatory intermediate in both the enzymically catalyzed isomerase reaction and the spontaneous anomerization process (cf. Section II,C,5,c). Linear arrows denote enzymic reactions, dotted arrows spontaneous anomerization. The interrupted arrow between acyclic glucose-6-P and fructose-6-P indicates the probable existence of intermediates for which experimental evidence is not yet available.

6-phosphate is depicted in Fig. 1, modified from the paper by Salas *et al.* (*97*). Further evidence for this anomeric specificity has recently been provided by Wurster and Hess (*96*) who interpreted the occurrence of a lag period in the coupled glucose-6-phosphate isomerase–glucose-6-phosphate dehydrogenase system (at high isomerase and low dehydrogenase concentrations; see Section II,C,1) as originating from accumulation of α-glucose 6-phosphate that is not fast enough anomerized spontaneously to β-glucose 6-phosphate, the latter being the only form on which the dehydrogenase acts. [This same reasoning had previously been applied by Salas *et al.* (*97*) to the phosphoglucomutase (specific for α-glucose 1-phosphate)–glucose-6-phosphate dehydrogenase sequence.] A report to the contrary, claiming β-anomeric specificity for the glucose-6-phosphate isomerase reaction (*115*), was later shown (*116*) to have been the result of erroneous interpretation of NMR spectroscopic data.

5. *Mechanism of Action*

a. Early Proposals. Data available to date seem to leave no doubt that the basic mechanism for glucose-6-phosphate isomerase (and other aldose–ketose isomerases) involves the formation of an enediol anion intermediate (*1, 5, 6, 60*) and that it is formally analogous to the acid- or base-catalyzed Lobry de Bruyn–Alberda van Ekenstein transformation (*1, 6, 117*). At one time, a hydride shift mechanism was proposed (*53*) which was considered incompatible (*3*), however, with the temperature dependence of the isotope effect observed for the hydrogen transfer–exchange ratio (*85*).

It is interesting to note that the involvement of an enediol intermediate in aldose–ketose isomerization was originally proposed far earlier than is commonly assumed. The existence of a structure of the

115. M. S. Feather and M. J. Lybyer, *BBRC* **35,** 538 (1969).

116. M. S. Feather, V. Deshpande, and M. J. Lybyer, *BBRC* **38,** 859 (1970).

117. Since references to this transformation seem to suffer from more misspelling and incorrect abbreviation (e.g., E. E. Conn and P. K. Stumpf, "Outlines of Biochemistry," 2nd ed., p. 45. Wiley, New York, 1966; H. R. Mahler and E. H. Cordes, "Biological Chemistry," p. 421. Harper, New York, 1966) than any other name reaction, it might be worthwhile to emphasize that the designation given here is the correct one. Both [Cornelius Adrian] Lobry de Bruyn and [Willem] Alberda van Ekenstein were Dutch chemists who performed the work on carbohydrate transformations named after them at the turn of the century. Their family names "Lobry de Bruyn" and "Alberda van Ekenstein," which also appear as surnames in their publications, should be used in properly referring to the transformation they first described. The historically interested reader may wish to consult the enlightening article by Speck (*6*) for more details.

form R–C(OH)=CH(OH) in alkaline medium was first suggested by E. Fischer in 1895 (*118*), i.e., in the same year when Lobry de Bruyn and Alberda van Ekenstein published their first joint paper (*119*). At the turn of the century, Wohl and Neuberg (*120*) postulated such an enolic form as the common intermediate in the alkaline interconversion of aldoses and ketoses (see especially footnote 2 on p. 3099 of reference *120*). E. F. Armstrong (*121*) then appears to have been the first to suggest in 1904 that such an intermediate is also part of the enzymic process when he wrote ". . . the formulae ordinarily assigned to glucose, mannose and fructose are reducible to one common enolic form. *It is conceivable that this enolic form is the substance actually fermented* . . ." (*121*) [italics added]. Armstrong also had what in retrospect must be seen as a remarkable vision of a transient intermediate when he concluded that "It is to be borne in mind that ordinary glucose in solution consists almost wholly of two stereoisomeric compounds in equilibrium, together with a very small *proportion . . . of the enolic form*, the presence of which must be assumed . . . in order to explain the results arrived at by Lobry de Bruyn . . ." [italics added] (asterisk footnote on p. 522 of reference *121*).

Topper's proposal of an enzyme bound enediol intermediate (*122, 123*) is well referenced in the literature. A similar suggestion by Bruns and Okada (*124*) appears to have been quoted only once (*33*). Also, even though Rose brought out the analogy of the enzymic mechanism with chemical models in which the isomerization reaction proceeds via attack of a base on the hydrogen alpha to the carbonyl with simultaneous formation of the conjugate acid–enzyme–enolate substrate complex and subsequent return of the proton to the adjacent carbon (*3, 85*), Michaelis and Rona (*125*) proposed an essentially identical mechanism in 1912. The following is a quote from their paper (in translation): "The sugar molecule alternatingly undergoes substitution of a H-atom by a negative electron and resubstitution of the latter by a H-atom. . . . This

118. E. Fischer, *Chem. Ber.* **28**, 1145 (1895).

119. C. A. Lobry de Bruyn and W. Alberda van Ekenstein, *Rec. Trav. Chim.* **14**, 203 (1895).

120. A. Wohl and C. Neuberg, *Chem. Ber.* **33**, 3095 (1900).

121. E. F. Armstrong, *Proc. Roy. Soc.* **B73**, 516 (1904).

122. Y. J. Topper, *Federation Proc.* **15**, 371 (1956).

123. Y. J. Topper, *JBC* **225**, 419 (1957).

124. F. Bruns and S. Okada, *Nature* **177**, 87 (1956) (submitted Aug. 5, 1955). Although the quantitative aspects of this communication have some shortcomings, the proposal of an enzyme-bound enediol intermediate was timely and qualitatively correct.

125. L. Michaelis and P. Rona, *Biochem. Z.* **47**, 447 (1912).

transformation . . . may occur in such a way that the ionic species of the sugar is present in its enolic form . . ." (*125*).

Finally, the postulate of Topper (*1, 123*) that the enediol anion will have either *cis* or *trans* configuration, depending upon which aldohexose is converted to the keto isomer, was contradicted by Speck (*6*) and experimentally refuted by Rose and collaborators who have shown that *all* aldose–ketose isomerases form a *cis*-enediol anion intermediate (*5*, and references quoted therein).

b. The Path of Hydrogen in Isomerase Reactions. Considerable debate and some controversy currently surrounds the question of whether the proton migration in isomerase reactions occurs directly from the leaving carbon to the base on the enzyme and back to the accepting carbon or whether the proton is lost to the medium and then returns to the adjacent carbon. The first process is transfer, the second exchange; in different terminology, the first is *intra*molecular, the second *inter*molecular. In some enzymic isomerizations, both pathways are observed side by side. The implications of this rather anomalous behavior of an atom in an enzymic reaction were recently discussed by Rose (*5*); therefore, only what is directly related to the glucose-6-phosphate isomerase reaction will be mentioned here (however, refer also to Sections III,D,3; V,C,3; and VII,C,5).

Several laboratories have measured the relative amounts of proton transfer vs. exchange with various isolated glucose-6-phosphate isomerases and in crude multienzyme systems (*69, 85, 126–130*). The relevant data are summarized in Table VI. The transfer–exchange ratio for the spinach enzyme was found to be independent of pH (*128*) and that of both the rabbit muscle and the spinach enzyme was found to increase with decreased temperature (*85, 128*). This has been explained by Rose (*85*) to be the result of a higher activation energy for the exchange reaction compared with the transfer reaction, thus favoring direct transfer at lower temperature. In spite of their variation, the data of Table VI show clearly that the glucose-6-phosphate isomerase reaction proceeds both by intramolecular and intermolecular proton migration and that the proton is transferred directly in more than half of the catalytic turnovers. It is notable that in the systems studied, the isotope effect appears to have been approximately twice as large for the forward reaction com-

126. H. Simon, R. Medina, and G. Müllhofer, *Z. Naturforsch.* **23b,** 59 (1968).
127. C. Fedtke, Ph.D. Dissertation, University of Göttingen (1968).
128. C. Fedtke, *Progr. Photosyn. Res.* **3,** 1597 (1969).
129. H. Simon, H. D. Dorrer, and A. Trebst, *Z. Naturforsch.* **19b,** 734 (1964).
130. H. D. Dorrer, C. Fedtke, and A. Trebst, *Z. Naturforsch.* **21b,** 557 (1966).

TABLE VI
INTRAMOLECULARITY AND KINETIC ISOTOPE EFFECT OF $^3H^+$ TRANSFER IN THE GLUCOSE-6-PHOSPHATE ISOMERASE REACTION

Enzyme source	Direction of measurement	Temp. of measurement	Isotope effect k_H/k_T	Intramolecularity (%)	Reference
Rabbit muscle	Forward[a]	27°	4.4	80	(*69*)
	Reverse	27°	2.8	56	(*85*)
Yeast	Forward	27°	5.8–6.0	68–93	(*69*)
	Forward	25°	6.2	50	(*126*)
	Reverse	27°	2.9	57	(*69*)
	Reverse	22°	3.0 (assumed)	55	(*127*)
Spinach extract	Reverse	22°	3.0	70	(*127, 128*)
Chlorella extract	Reverse	25°	3–4 (assumed)	100	(*127–130*)

[a] Glucose-6-P → fructose-6-P.

pared with the reverse reaction. If it is assumed that this difference is not a methodological phenomenon due to the necessary employment of different trapping systems, the magnitude of the isotope effect can be said to be proportional to the Michaelis constant and inversely proportional to the maximal velocity of the respective direction of catalysis.

c. Current Status. Based upon the following premises, which have been elaborated upon in the preceding sections, the mechanism depicted in Fig. 2 has been proposed by Dyson and Noltmann (*60*):

(1) Two reactive amino acid side chains of the enzyme, assumed to be the protonated ε-amino group of a lysyl residue and the nonprotonated imidazole nitrogen of a histidyl residue, participate in catalysis.
(2) The isomerization reaction per se occurs while the substrate is in acyclic form.
(3) The reaction involves nucleophilic attack by a base on the α-hydrogen with formation of an enzyme-bound enediol anion as an intermediate.
(4) Both direct proton transfer to the neighboring carbon and exchange with the medium occur.
(5) The enzyme can also reversibly catalyze ring opening and closing of the α form of the substrates to and from the acyclic form but cannot produce or act upon the β form.
(6) Depending upon the supply of substrate, the enzyme will nor-

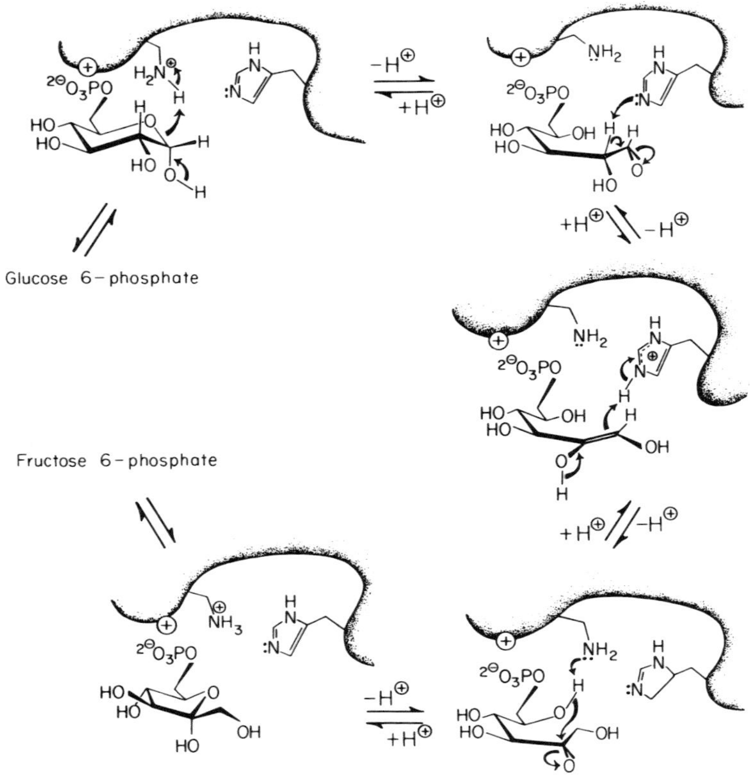

Fig. 2. Proposed dual function mechanism of action of glucose-6-phosphate isomerase, redrawn from Dyson and Noltmann (*60*) to show the conformation of the ring structures.

mally bind the ring form, but it may also act on freely available straight-chain hexose 6-phosphate.

The catalytic process is visualized to involve a true dual function mechanism (*60*) in which the ϵ-amino nitrogen of the lysyl group catalyses the ring opening step and an imidazole nitrogen is the base for nucleophilic attack on the α-hydrogen. The mechanism in Fig. 2 is shown for the forward reaction but is understood to be fully reversible.

A point deserving additional comment is the anomerase activity ascribed (*97*) to glucose-6-phosphate isomerase (see also Section II,C,4). Although perhaps a semantic point, describing the ability of ring opening as anomerase activity is actually incorrect. An anomerase, more strictly defined, catalyzes the *reversible* interconversion between the α and β forms of a sugar. However, it has been shown explicitly that glucose-6-phosphate isomerase can only perform the half-reaction, *viz.*, opening of the α-ring structure to form the straight-chain species and closing it back to the same or another corresponding α anomer, while it is unable to attack β-glucose 6-phosphate. This reviewer therefore feels that it is inaccurate to describe the isomerase as having a secondary anomerase activity. Instead, what has been designated as such is in fact the first equilibrium of the proposed dual function mechanism shown in Fig. 2.

Rose has argued on mechanistic grounds (*5, 130a*) that the anomeric specificity of various isomerases (cf. also Sections VIII,E and VIII,F) can be rationalized by the necessity to maintain *cis* configuration of the hydroxyls on C-1 and C-2 in order to facilitate formation of a *cis*-1,2-enediol intermediate. The observed specificity of glucose-6-phosphate isomerase for the α form of glucose 6-phosphate supports this postulate.

When the mechanism was described in detail (*60*), it was also pointed out that binding of such hydrophilic substrates as the hexose 6-phosphates cannot likely be very effective in a hydrophobic environment. Furthermore, it is improbable that an enediolate intermediate, which is more polar than the starting substrate and which requires charge stabilization during its transient existence, can be formed in predominantly nonpolar surroundings. Finally, exchange of protons with the medium (see preceding section) can obviously occur only when water molecules are available to participate in this exchange. Fedtke (*128*) has recently proposed that the active site of glucose-6-phosphate isomerase is buried in a hydrophobic cleft because of the relatively high intramolecularity observed in some instances (Table VI). This reviewer is somewhat reluctant to accept such reasoning, partly because of the general arguments to the opposite made above and partly because of the high catalytic turnover of the enzyme. The catalytic center activity of glucose-6-phosphate isomerases is of the order of 10^3 sec^{-1} which at neutrality is in the same range as the diffusion-controlled limiting rate of proton dissociation from a group with a pK of about 7 (*131*) (see also Section VIII,C,5). On the other hand, collision frequency alone could allow rates much in excess of 10^3 sec^{-1}, irrespective of the nature of the site, so that the question remains unresolved at the present time.

130a. K. J. Schray and I. A. Rose, *Biochemistry* **10,** 1058 (1971).
131. M. Eigen, *Angew. Chem. Intern. Ed. Engl.* **3,** 1 (1964).

III. Mannose-6-phosphate Isomerase

D-Mannose 6-phosphate ⇄ D-Fructose 6-phosphate

A. History, Occurrence, and Function

Although earlier studies on the fermentation rate of mannose as compared with glucose or fructose (*132–135*) may be interpreted in retrospect as having indicated the existence of two different hexose phosphate isomerases, the first clear enunciation of this fact must be attributed to Gottschalk (*136*). On the basis of experiments similar to those of Jephcott and Robison (*135*) in which Gottschalk found that at temperatures above 28° mannose was fermented considerably more slowly than glucose, he suggested that two separate and specific enzymes, which he designated as Isomerases I and II, were responsible for the conversion of the two aldohexose 6-phosphates to the common product fructose 6-phosphate.

$$\text{Glucopyranose-6-P} \underset{}{\overset{\text{Isomerase I}}{\rightleftarrows}} \text{Fructofuranose-6-P} \overset{\text{Isomerase II}}{\rightleftarrows} \text{Mannopyranose-6-P}$$

He furthermore concluded that at temperatures between 28° and 38° mannose 6-phosphate accumulated because Isomerase II was more susceptible to temperature inactivation than Isomerase I (*136*). In his 1947 paper (*136*), Gottschalk introduced the names *glucose-6-phosphate isomerase* and *mannose-6-phosphate isomerase* (EC 5.3.1.8) for Isomerases I and II, respectively, and predicted what was later found to be the correct reaction sequence linking the two enzymes together. The separate identity of the two hexose-6-phosphate isomerases was established in 1950 by Slein (*137*) who purified both from rabbit muscle. However, difficulties in obtaining mannose-6-phosphate isomerase preparations completely devoid of glucose-6-phosphate isomerase (*45, 137*) made it impossible at first to ascertain whether the primary reaction product of mannose 6-phosphate isomerization was glucose 6-phosphate or fructose 6-phosphate (*137, 138*).

Even when Slein concluded later from activity assays at low pH,

132. A. Slator, *JCS* **93,** 217 (1908).
133. W. J. Young, *Proc. Roy. Soc.* **B81,** 528 (1909).
134. I. S. Neuberg and C. Ostendorf, *Biochem. Z.* **221,** 154 (1930).
135. C. M. Jephcott and R. Robison, *BJ* **28,** 1844 (1934).
136. A. Gottschalk, *BJ* **41,** 276 (1947).
137. M. W. Slein, *JBC* **186,** 753 (1950).
138. L. F. Leloir, *Advan. Enzymol.* **14,** 193 (1953).

where interference by contaminating glucose-6-phosphate isomerase is minimal, that fructose 6-phosphate was the reaction product of the mannose-6-phosphate isomerase catalyzed reaction (*45*), uncertainty in the literature (*138*) remained. In fact, a well-known textbook of enzyme chemistry (*139*) described the reaction catalyzed by mannose-6-phosphate isomerase as that between mannose 6-phosphate and glucose 6-phosphate, and claims to this effect were made as late as 1961 [e.g., (*140*)]. Unequivocal proof for the reaction to be between mannose 6-phosphate and fructose 6-phosphate was provided by Noltmann and Bruns (*141*) with a mannose-6-phosphate isomerase preparation from brewer's yeast, which was free of any glucose-6-phosphate isomerase activity.

In addition to rabbit muscle and brewer's yeast, numerous other tissues and organisms have been shown to contain mannose-6-phosphate isomerase, usually at concentrations substantially lower than those of glucose-6-phosphate isomerase. Examples of mammalian tissues are: skeletal muscle of pig, rat, guinea pig, beef, and sheep (*142*); various other tissues of the rat (*140, 143, 144*); pig erythrocytes (*143, 145*); bovine intestinal mucosa (*146*); bovine testis (*146*); and human erythrocytes (*147*) and arterial tissue (*148*). Mannose-6-phosphate isomerase was also shown to occur in the honeybee (*29, 31*), in the parasite roundworm *Ascaris suis* (*146*), in the phosphomannan producing yeast *Hansanula holstii* (*149*), constitutively in *Aerobactor aerogenes* (*150*), and as an adaptive enzyme in *Pseudomonas aeruginosa* (*151*).

The function of mannose-6-phosphate isomerase in mammalian organisms appears to be primarily that of converting ingested mannose (after phosphorylation) to the more readily metabolized fructose 6-phosphate, thus channeling it into glycolysis. In animals, a functional role for the reverse catalysis (fructose 6-phosphate → mannose 6-phosphate) has

139. J. B. Neilands, and P. K. Stumpf, "Outlines of Enzyme Chemistry," 2nd ed., p. 244. Wiley, New York, 1958.
140. S. Abraham, W. M. Fitch, and I. L. Chaikoff, *ABB* **93**, 278 (1961).
141. E. Noltmann and F. H. Bruns, *Biochem. Z.* **330**, 514 (1958).
142. A. Willemsen, M.D. Dissertation, Medizinische Akademie Düsseldorf, Germany (1960).
143. F. H. Bruns, E. Noltmann, and A. Willemsen, *Biochem. Z.* **330**, 411 (1958).
144. D. E. Kizer and T. A. McCoy, *Proc. Soc. Exptl. Biol. Med.* **103**, 772 (1960).
145. F. H. Bruns and E. Noltmann, *Nature* **181**, 1467 (1958).
146. F. Alvarado and A. Sols, *BBA* **25**, 75 (1957).
147. E. Beutler and L. Teeple, *J. Clin. Invest.* **48**, 461 (1969).
148. J. E. Kirk, *J. Gerontol.* **21**, 420 (1966).
149. R. K. Bretthauer, D. R. Wilken, and R. G. Hansen, *BBA* **78**, 420 (1963).
150. M. Y. Kamel and R. L. Anderson, *J. Bacteriol.* **92**, 1689 (1966).
151. R. G. Eagon and A. K. Williams, *J. Bacteriol.* **79**, 90 (1960).

not been established. In microorganisms, however, the synthetic direction, i.e., the formation of mannose 6-phosphate from fructose 6-phosphate, is undoubtedly of importance. Mannose 6-phosphate produced in this manner can be incorporated—via mannose 1-phosphate as intermediate—into guanosine diphosphomannose which participates in numerous processes related to cell wall biosynthesis. These aspects have been discussed in more detail by Gracy (*152*).

B. Molecular Properties of the Yeast Enzyme

Studies on the molecular properties of mannose-6-phosphate isomerase have so far been reported only for the enzyme from yeast which has been isolated in homogeneous form (*153*). Thus, essentially all of the available quantitative information is based on data obtained for the enzyme isolated from Fleischmann's dry yeast, type 20-40, which is defined by the supplier (*154*) as a "bottom high-attenuating brewer's yeast of the genus and species *Saccharomyces cerevisiae*." The concentration of mannose-6-phosphate isomerase in autolysates from this yeast was found (*153*) to be nine times higher than in the so-called "dried brewer's yeast for enzyme work" of Anheuser-Busch (which has served in the past as a source for numerous enzyme preparations) and three times higher than in the most highly purified preparation previously reported (*141*). The isolated enzyme, after a 1500-fold purification, has a specific activity of approximately 800 units/mg at 30°, measured as micromoles of mannose 6-phosphate converted to fructose 6-phosphate in the coupled spectrophotometric assay with glucose-6-phosphate isomerase and glucose-6-phosphate dehydrogenase (*153*). The preparation was judged homogeneous (*155*) by the criteria of sedimentation velocity and sedimentation equilibrium ultracentrifugation, ion exchange chromatography, gel filtration,

152. R. W. Gracy, Ph.D. Dissertation, University of California, Riverside (1968).

153. R. W. Gracy and E. A. Noltmann, *JBC* **243**, 3161 (1968).

154. Standard Brands, Inc., New York.

155. Boehringer Mannheim Corp. has recently begun to market under catalog No. 15104 EPAM "crystalline" mannose-6-phosphate isomerase from yeast with a specific activity of 40 units/mg. Data on the homogeneity of this preparation are not available. It may be relevant to point out that the purification procedure of Gracy and Noltmann (*153*) at specific activities of about 40 units/mg yields two different crystalline proteins whose nature has not been identified but which have been shown to be neither mannose-6-phosphate isomerase nor glucose-6-phosphate isomerase (*152*). At present, it is not clear whether the "crystalline" enzyme from Boehringer as such has only 5% of the activity of the enzyme isolated according to Gracy and Noltmann or whether the crystals are a different protein which is contaminated by 5% mannose-6-phosphate isomerase.

TABLE VII
PHYSICAL PROPERTIES OF YEAST MANNOSE-6-PHOSPHATE ISOMERASE[a]

Physical parameter	Method	Value
Sedimentation coefficient ($s_{20,w}$)	Velocity sedimentation ($c = 0.5\%$)	3.86×10^{-13} sec
Diffusion coefficient ($D_{20,w}$)	Boundary spreading during velocity sedimentation ($c = 0.5\%$)	7.78×10^{-7} cm^2 sec^{-1}
	Gel filtration	7.62×10^{-7} cm^2 sec^{-1}
Molecular weight ($M_{s,D}$)	$s_{20,w}$; $D_{20,w}$ from boundary spreading	44,500 g mole^{-1}
	$s_{20,w}$; $D_{20,w}$ from gel filtration	45,500 g mole^{-1}
Molecular weight (M_{eq})	Equilibrium sedimentation	45,900 g mole^{-1}
Molecular weight (M_{gel})	Gel filtration	42,000 g mole^{-1}
Stokes' radius	Gel filtration	27.0 Å
Partial specific volume ($\bar{V}$)	Equilibrium sedimentation in D_2O	0.73 ml g^{-1}

[a] From Gracy and Noltmann (*153*).

and zone electrophoresis on cellulose acetate. The molecular weight was determined by a variety of methods and was found to be 45,000 daltons. Values for the measured physical parameters are listed in Table VII.

C. CHARACTERIZATION AS A ZINC METALLOENZYME

Inhibition by chelating agents and subsequent reconstitution of enzymic activity with metals led Bruns and Noltmann in 1958 to suggest that mannose-6-phosphate isomerase from pig erythrocytes is a metal–enzyme complex (*145*). This postulate was justly criticized by Vallee (*156*) on the grounds that inhibition studies alone are not sufficient criteria for the identification of metals in enzymes. The question of a possible metal involvement in mannose-6-phosphate isomerase action rearoused interest when the isolated yeast enzyme proved to be readily inhibited by metal-binding agents.

Quantitative studies by Gracy and Noltmann (*157*) with 8-hydroxyquinoline, ethylenediaminetetraacetate, *o*-phenanthroline, α,α'-dipyridyl, cysteine, and dithiothreitol showed the inhibition to be first order with respect to both enzyme and inhibitor. Moreover, the effectiveness of the various chelating agents in bringing about inhibition paralleled their ability to bind zinc, and pretreatment with metals to block their coordina-

156. B. L. Vallee, "The Enzymes," 2nd ed., Vol. 3, p. 225, 1960.
157. R. W. Gracy and E. A. Noltmann, *JBC* **243**, 4109 (1968).

tion sites prior to addition to the enzyme abolished their inhibitory potential. EDTA-induced inhibition could be reversed instantaneously by the addition of metals, and the extent of reversal or prevention of this inhibition was proportional to the ability of the added metals to form complexes with EDTA, as expressed by their stability constants. An interesting aspect of the inhibition of mannose-6-phosphate isomerase by thiol compounds is that they were found to exert their effect not as reducing agents but by virtue of their metal-binding properties. Sulfur ligands are known to form particularly stable complexes with zinc (*158*), and the inhibition of other zinc metalloenzymes by mercaptans is well established (*159–162*).

Analysis of mannose-6-phosphate isomerase samples, judged pure by criteria described above under Section III,B, yielded a metal content of 1.0 atom of zinc per enzyme molecule (molecular weight, 45,000 daltons). Atomic absorption analyses for Ca, Mg, Cu, Co, Mn, Fe, and Zn were performed on all major fractions obtained in the course of the purification procedure. Only the specific concentration of zinc increased consistently with the specific activity of the preparation, and zinc was the only metal present in stoichiometric quantities in the isolated protein. A direct correlation was found between the zinc content and the specific activity of the enzyme, with the best preparation analyzed (specific activity, 812 units/mg) containing 0.98 g-atoms of zinc per mole (*157*).

Basically, two mechanisms can account for the inhibition of a metalloenzyme by chelating agents: the inhibitor may bind the metal *in situ*, forming an inactive ternary complex, or it may physically remove the metal from the protein. These alternatives have been investigated for the reaction of EDTA with mannose-6-phosphate isomerase and the latter has been found to prevail (*163*). When the enzyme was treated with ^{14}C-EDTA until 99% of the catalytic activity was lost and then passed through Sephadex or dialyzed exhaustively, no radioactivity remained bound to the protein. Zinc analysis after extensive dialysis of EDTA-treated enzyme demonstrated a decrease in the zinc content directly proportional to the loss in enzymic activity.

Experimental data available for the yeast mannose-6-phosphate

158. B. L. Vallee and J. E. Coleman, *Comprehensive Biochem.* **12**, 165 (1964).
159. S. J. Adelstein and B. L. Vallee, *JBC* **234**, 824 (1959).
160. T. L. Coombs, J. P. Felber, and B. L. Vallee, *Biochemistry* **1**, 899 (1962).
161. S. G. Agus, R. P. Cox, and M. J. Griffin, *BBA* **118**, 363 (1966).
162. W. H. Fishman and N. K. Ghosh, *BJ* **105**, 1163 (1967).
163. As pointed out in the original paper (*157*), this same mechanism need not apply for all other inhibitors. A metalloenzyme may be inhibited by one chelating agent via formation of a ternary complex, whereas another may inactivate it by removal of the metal.

isomerase would thus appear to provide convincing evidence that it is a genuine zinc metalloenzyme (*157*). It is the first isomerase for which a metal has been established to be an integral part of the enzyme protein and so far seems to have also remained the only enzyme in this class for which direct analytical data on the stoichiometric metal content have been provided (*164*). Although no further studies on mannose-6-phosphate isomerases of mammalian origin have been made, the effects of chelating agents and metals on the isomerase activity of various tissue extracts observed in the past (*143, 144, 145*) are consistent with those found with the yeast enzyme; they suggest that mannose-6-phosphate isomerase of higher organisms may also be a metalloenzyme.

Finally, two sets of data concerning an interesting metabolic phenomenon warrant mentioning since they relate to the possible metalloenzyme nature of mannose-6-phosphate isomerase in honeybees. When Sols *et al.* observed only negligible amounts of the enzyme in these insects, they concluded this deficiency to be the cause of the well-known mannose toxicity, presumably caused by a mutational loss of the bee's ability to synthesize the isomerase (*29*). Their conclusion, however, was based upon experiments with homogenates of honeybees prepared in 5 m*M* EDTA. When extracts of the insects were obtained and assayed under conditions in which activity loss resulting from instability and possible metal binding was avoided (*31*), mannose-6-phosphate isomerase levels were found, relative to metabolically related enzymes, at ratios quite comparable to those prevailing in mammalian tissues and in several yeast species. It can therefore be argued that a metal may also be involved in the honeybee isomerase.

D. Catalytic Properties

1. *General Kinetic Parameters*

The early work on catalytic properties has been mentioned by Topper in the previous edition of this treatise (*1*) and need not be repeated here. Kinetic measurements of the basic steady state parameters were recently conducted for the isolated mannose-6-phosphate isomerase from yeast (*111*), and most of the present discussion will be concerned with these studies.

All of the kinetic assays have been made in the direction from mannose 6-phosphate to fructose 6-phosphate, either spectrophotometrically

164. Although Mildvan's elegant NMR data on D-xylose isomerase (*164a, 164b*) implicate manganese as participating in an enzyme–metal–substrate bridge complex, they do not show whether manganese is in fact part of the enzyme molecule *in vivo*.

by coupling with glucose-6-phosphate isomerase and glucose-6-phosphate dehydrogenase or, when desired conditions would interfere with the spectrophotometric measurement, with the discontinuous resorcinol assay for keto sugars as described by Roe (*99*). Kinetic constants have not yet been determined for the reverse direction (fructose-6-phosphate → mannose 6-phosphate) because of the lack of a convenient assay. Nonchelating buffers were used for the determination of all kinetic constants of the yeast enzyme, notably piperazine bisethanesulfonate for the standard assay at pH 7.15 and 0.1 ionic strength.

Assayed under initial velocity conditions, mannose-6-phosphate isomerase from yeast possesses optimum activity at pH 7.0–7.2 (*111*). Values between pH 5.5 and 6.0, reported from earlier work with crude or partially purified enzyme preparations (*45, 143*) contaminated by a large excess of glucose-6-phosphate isomerase, can be explained as apparent optima produced by the assay conditions of the colorimetric determination of fructose 6-phosphate. In this assay, contaminating glucose-6-phosphate isomerase will immediately convert any fructose 6-phosphate formed by the action of mannose-6-phosphate isomerase on mannose 6-phosphate into an equilibrium mixture between glucose 6-phosphate and fructose 6-phosphate. As the pH is decreased below pH 6.0, glucose-6-phosphate isomerase activity is more rapidly lost than mannose-6-phosphate isomerase activity, and fructose 6-phosphate is no longer converted to glucose 6-phosphate. As a consequence, an apparent pH optimum is produced which in reality merely represents that pH below which glucose-6-phosphate isomerase can no longer interfere with the assay. This phenomenon is not observed with the spectrophotometric assay as long as care is exercised in maintaining initial velocity conditions (*111*).

Other kinetic parameters are as follows (all measured at pH 7.15 in the direction of mannose 6-phosphate → fructose 6-phosphate): K_m = 1.35 mM; V_{max} = 830 μmoles/min/mg at 30°, corresponding to a molecular activity of 3.7×10^4 mole/min/mole of enzyme (MW 45,000 daltons); K_{eq} ([fructose 6-phosphate]/[mannose 6-phosphate]) = 1.03. At 30°, the enzyme was found to be completely stable between pH 6.0 and 8.5; 50% loss of activity occurred at pH 5.3 and 9.0, and total inactivation below pH 5.0 and above 9.5. Evaluation of K_m and V_{max} data as a function of pH according to Dixon (*165*) yielded pK values of 6.6 and 7.8 for the free enzyme and 6.4 and 8.1 for the enzyme–substrate complex (*111*).

164a. A. S. Mildvan, *Abstr. ACS Middle Atlantic Reg. Meeting, Newark, Delaware, 1970* p. 22 (1970).
164b. A. S. Mildvan and I. A. Rose, *Federation Proc.* **28**, 534 (1969).
165. M. Dixon, *BJ* **55**, 161 (1953).

TABLE VIII
SUBSTRATE ANALOGS AS COMPETITIVE INHIBITORS OF YEAST MANNOSE-6-PHOSPHATE ISOMERASE[a]

Inhibitor	K_i (mM)
Mannose 6-phosphate (substrate)	1.35 (K_m)
Mannitol 1-phosphate	0.80
Glucose 6-phosphate	13.6
Glucosamine 6-phosphate	10
Sorbitol 6-phosphate	2.6
6-Phosphogluconate	3.24
Erythrose 4-phosphate	8.0
Galactose 6-phosphate	65
Ribose 5-phosphate	19
Glucose 1-phosphate	33
Inorganic phosphate	86

[a] Inhibition constants determined at 30°, pH 7.15, and 0.1 ionic strength [from Gracy and Noltmann (*111*)].

Neither glucose 6-phosphate, nor mannose, fructose, glucose, or any other analog tested underwent isomerization to a measurable extent. All inhibitors found were strictly competitive in nature; their K_i values are listed in Table VIII. The following structurally related compounds did not show any inhibitory effect or they produced insignificant inhibition: mannose, fructose, glucose, GDP-mannose, mannan, *N*-acetylmannosamine, mannitol, inositol 1-phosphate, inositol 3-phosphate, fructose 1,6-diphosphate, and L-α-glycerophosphate.

2. *Effect of Zinc on the Nonenzymic Isomerization*

When mannose-6-phosphate isomerase proved to be a zinc metalloenzyme (*157*), it became of immediate interest whether zinc might have an effect on the nonenzymic interconversion of mannose 6-phosphate and fructose 6-phosphate, long known as a Lobry de Bruyn–Alberda van Ekenstein transformation (*117*). Indeed, the presence of zinc was found to increase dramatically the rate of nonenzymic isomerization of mannose 6-phosphate, compared with identical conditions of pH and ionic strength, but with sodium acetate in place of zinc acetate (*111*). The reaction was found to be of apparent first order with respect to zinc, and the stimulatory effect of zinc was abolished when *o*-phenanthroline was included in the reaction mixture. Quantitative analysis of aliquots taken during the course of the reaction yielded the same relative amount of products (glucose 6-phosphate and fructose 6-phosphate) independent of whether the reaction proceeded in the presence or absence of zinc. In addition, reactant and products could be stoichiometrically accounted for, indicating the virtual absence of side reactions. These data suggest that the nonenzymic isomerization may serve as a direct model for the enzymically catalyzed process although caution must be exercised in comparing the two. First, reaction conditions are considerably more rigorous for the nonenzymic isomerization (*111*) and, second, glucose 6-phosphate and fructose 6-phosphate also show an increased isomerization rate in the presence of zinc, although both rabbit muscle and yeast glucose-6-phosphate isomerases have definitely been shown not to contain zinc (*157*).

3. *Mechanism of Action*

The most pertinent questions one might ask regarding the catalytic mechanism of mannose-6-phosphate isomerase are perhaps those related to a comparison with its counterpart glucose-6-phosphate isomerase. Do both enzymes follow basically the same type of mechanism? What are the distinctive features of their active sites that enable them to distinguish the two hydrogens on C-1 of fructose 6-phosphate and that cause one enzyme to produce mannose 6-phosphate and the other glucose-6-phosphate? What are the respective forms of the substrates that initially bind to the enzyme surface? What is the significance of apparently only mannose-6-phosphate isomerase being a metalloenzyme? Some partial answers are available, but much quantitative work on the protein nature of both enzymes still needs to be done.

In terms of gross structure, little common ground can be found: Glucose-6-phosphate isomerases both from yeast and from skeletal muscle are dimers with subunit molecular weights of 60,000–66,000 daltons; mannose-6-phosphate isomerase from yeast has a molecular weight of only 45,000 daltons. Mannose-6-phosphate isomerase is a zinc metalloenzyme; both glucose-6-phosphate isomerases have been shown not to contain zinc (*157*). Kinetic studies of mannose-6-phosphate isomerase as a function of pH yielded pK values of 6.6 and 7.8 for the free enzyme. With due consideration for the limitations inherent in this method, a histidine residue has been suggested as the source of the ionization at pH 6.6. An assignment for the pK at 7.8 is as yet not warranted, although arguments for an NH_2-terminal α-amino group have been advanced (*111*). The histidine residue is assumed to function in a manner analogous to that proposed for glucose-6-phosphate isomerase, i.e., in its nonprotonated form as a nucleophile abstracting the activated C-2 hydrogen and effecting its transfer to C-1. This postulated function of a histidine residue is compatible with the maximal catalytic rate of isomerization experimentally observed and with the diffusion-controlled limit for the rate of proton dissociation from imidazolium ion which is considered to be of the order of 10^3 sec^{-1} at 25° and pH 7.0 (*131*, *166*). From the maximal velocity of mannose-6-phosphate isomerase, 830 units/mg at 30° (*111*), a molecular turnover rate of about 4×10^2 sec^{-1} may be estimated for 25°. If one follows the argument that for 10% proton transfer and 90% exchange with the medium (*126*) (*vide infra*) the actual proton shifts must occur ten times faster than the measured substrate turnover (*5*, especially p. 290 ff.), the resulting rate of 4×10^3 sec^{-1} can be accommodated by imidazolium catalysis, although other explanations are possible according to which intramolecular proton transfer may occur more rapidly than exchange with the solvent (*166*) (cf. also Section VII,C,5).

After zinc had been found to be an integral part of the enzyme molecule and essential for activity (*157*), subsequent experiments showed that the substrate or mannitol 1-phosphate, a competitive inhibitor, could protect the enzyme against activity loss occurring on treatment with chelating agents. This was taken to suggest that the metal must be located in, or very near, the active center, and that binding of substrate blocks the interaction of the chelating agent with the metal (*111*). Other evidence in favor of a functional role of zinc in the catalytic process—rather than in only maintaining the enzyme in an active conformation—is provided by the marked stimulation which zinc conveys to the nonenzymic isomerization (see Section III,D,2). It is also relevant to mention older studies on complex formation between carbohydrates and alkali metals or alkaline earth metals [recently reviewed by Rendleman (*167*)]. Many monosacchardies, including fructose (*168*), mannose (*169*), and glucose (*169a*), form chelates with monovalent and divalent metals. Such chelates (especially those forming five-membered rings) are quite stable, those with divalent metals more so than those with monovalent cations (*170*). This latter observation is of particular interest since the nonenzymic isomerization of mannose is much faster in the presence of alkaline earth metals than with alkali metals. Kusin (*171*), and later

166. W. P. Jencks, "Catalysis in Chemistry and Enzymology," p. 207 ff. McGraw-Hill, New York, 1969.

167. J. A. Rendleman, Jr., *Advan. Carbohydrate Chem.* **21,** 209 (1966).

168. R. H. Smith and B. Tollens, *Chem. Ber.* **33,** 1277 (1900).

169. J. K. Dale, *JACS* **51,** 2788 (1929).

169a. J. A. Rendleman, Jr., *J. Org. Chem.* **31,** 1839 (1966).

170. J. A. Mills, *BBRC* **6,** 418 (1961/62).

171. A. Kusin, *Chem. Ber.* **69,** 1041 (1936).

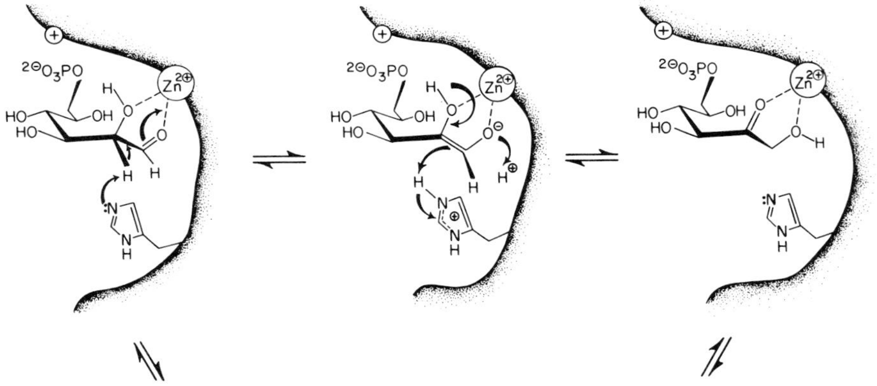

D-Mannose 6-phosphate D-Fructose 6-phosphate

FIG. 3. Proposed mechanism of action of mannose-6-phosphate isomerase, redrawn from Gracy and Noltmann (*111*).

Speck (*6*), therefore suggested that a divalent metal increases the isomerization rate by forming a chelate with the carbonyl oxygen and an adjacent hydroxyl oxygen, resulting in labilization of the hydrogen alpha to the carbonyl.

On the basis of the preceding considerations, the mechanism shown in Fig. 3 has been proposed (*111*). It shares a number of common features with the suggested mechanism for glucose-6-phosphate isomerase (*60*) (discussed in more detail in Section II,C,5,c) such as attack of a basic group on the activated hydrogen alpha to the carbonyl (*3*), the endiol anion intermediate (*123, 172*), and the requirement for absolute stereospecificity with respect to the hydrogens on C-1 of fructose 6-phosphate (*1*). It also accounts for both intramolecular hydrogen transfer and for hydrogen exchange with the medium (*126, 173*). The distinct role of zinc in the case of the mannose-6-phosphate isomerase is thought to be that of an electrophile which participates in the isomerization reaction by coordinating with the carbonyl oxygen and the adjacent C-2 hydroxyl oxygen of mannose 6-phosphate via formation of a five-membered chelate, thus exerting an electron withdrawing effect, facilitating removal of the α-hydrogen. This proposed role of zinc (*111*) places mannose-6-phosphate isomerase into the category of metalloenzymes forming ternary metal bridge complexes according to Mildvan's recent classification (*174*).

Several other aspects of the proposed mechanism deserve mentioning.

172. Y. J. Topper and D. Stetten, Jr., *JBC* **189**, 191 (1951).
173. H. Simon and R. Medina, *Z. Naturforsch.* **21b**, 496 (1966).
174. A. Mildvan, "The Enzymes," 3rd ed., Vol. II, p. 445, 1970.

If we accept the five-membered ring structure of Fig. 3 as representing the mode of interaction between the substrate and zinc, then it is obvious that the enediol intermediate must have a *cis* configuration. In fact, this steric configuration was postulated for the intermediate of the mannose-6-phosphate isomerase reaction by Bruns *et al.* (*143*) when involvement of a metal in this enzyme was first observed. However, following the line of Topper's early reasoning (*123*), it was also thought at the time that glucose-6-phosphate isomerase forms the corresponding *trans*-enediol intermediate (*143*), although this turned out to be incorrect. Rose and collaborators have developed compelling evidence that *all* aldose–ketose isomerases form enediol intermediates of the *cis* configuration (*175–177*) [discussed in more detail in Volume II of this edition (*5*)].

Another significant feature of this mechanism is that in the $ES_{(F)}$ complex, owing to coordination with a single atom (i.e., zinc), both the hydroxyl group on C-1 and the carbonyl group on C-2 are held in a fixed relationship, which places only one (and always the same) hydrogen of the two on C-1 in close proximity to the imidazole group; consequently, the same hydrogen is always labilized by the enzyme. Inversion of the hydroxyl on C-2 (*viz.*, in glucose 6-phosphate) does not permit it to coordinate with the zinc without substantial distortion of the coordination sphere. This may also account for the lower affinity of substrate analogs in which the C-2 hydroxyl is inverted and may perhaps be the ultimate reason for the absolute substrate specificity of mannose-6-phosphate isomerase (*111*). Constraint imposed by the zinc on the enediol intermediate may also be responsible for the low ratio of intramolecular hydrogen transfer to hydrogen exchange with the medium, which was found with mannose-6-phosphate isomerase to be only 0.05–0.1 under conditions for which a ratio of 0.50–0.58 was found with glucose-6-phosphate isomerase (*126, 173*). This difference may well be the result of the greater stability imposed by the metal on the $ES_{(X)}$ complex of mannose-6-phosphate isomerase. Charge neutralization of the enediol anion would make this intermediate stage less transient and thereby allow exchange to occur more easily than with glucose-6-phosphate isomerase which, lacking the metal, would form a less stable $ES_{(X)}$ intermediate (*111*).

It may finally be pointed out that the inhibition studies, which indicated tighter binding for the straight-chain analogs than for the corresponding ring compounds (Table VIII), favor the acyclic form of man-

175. I. A. Rose and E. L. O'Connell, *BBA* **42**, 159 (1960).

176. I. A. Rose, E. L. O'Connell, and R. P. Mortlock, *BBA* **178**, 376 (1969).

177. I. A. Rose, *Abstr. ACS Middle Atlantic Reg. Meeting, Newark, Delaware, 1970* p. 22 (1970).

nose 6-phosphate as the primary substrate for mannose-6-phosphate isomerase, as shown in Fig. 3. However, the possibility cannot be eliminated that also the cyclic form of mannose 6-phosphate may bind and undergo ring opening prior to the isomerization step in a manner similar to that suggested for glucose-6-phosphate isomerase (cf. Section II,C,5,c).

IV. Glucosamine-6-phosphate Isomerase

$$\text{D-Glucosamine-6-P} + H_2O \rightleftarrows \text{D-Fructose-6-P} + NH_3$$

Although its catalyzed reaction is formally that of an aldose–ketose isomerase, D-glucosamine-6-phosphate isomerase (EC 5.3.1.10) is involved in both an isomerization and a deamination step. It does not appear to have been established yet whether the hydrolytic deamination occurs spontaneously or whether it is specifically catalyzed by the enzyme (*1*). In all of the early work [discussed by Topper (*1*)] and also as late as 1969 (*178*), the enzyme has been referred to as glucosamine-6-phosphate deaminase, and emphasis has been placed on the similarity of its mechanism (*179*) to that of L-glutamine:D-fructose-6-phosphate aminotransferase (EC 2.6.1.16); the latter has also been called a transamidase (*180, 181*) or a glucosamine phosphate synthetase (*182*). In spite of their mechanistic similarity, the two enzymes are usually assumed to play opposite physiological roles: With glutamine as ammonia donor, the transferase reaction proceeds entirely in the direction of amino sugar formation, whereas the equilibrium of the isomerase reaction is on the side of ketohexose phosphate production (*179, 180, 183*) (however, *vide infra*). Leloir and Cardini (*184*) first reported that the enzyme from hog kidney is activated by *N*-acetylglucosamine phosphate. This activation has been confirmed for all subsequent preparations of the enzyme but one: Glucosamine phosphate isomerase from *Proteus vulgaris* was found by Nakada not to be affected by the *N*-acetyl analog of the substrate (*183*). Data on some catalytic properties of enzyme preparations

178. J. H. Veerkamp, *ABB* **129,** 248 (1969).
179. E. A. Davidson, "Carbohydrate Chemistry," p. 185 ff. Holt, New York, 1967.
180. L. F. Leloir and C. E. Cardini, *BBA* **12,** 15 (1953).
181. S. Ghosh, H. J. Blumenthal, E. Davidson, and S. Roseman, *JBC* **235,** 1265 (1960).
182. C. J. Bates and C. A. Pasternak, *BJ* **96,** 147 (1965).
183. H. I. Nakada, "Methods in Enzymology," Vol. 9, p. 575, 1966.
184. L. F. Leloir and C. E. Cardini, *BBA* **20,** 33 (1956).

from *Escherichia coli* (*185, 186*), hog kidney (*186*), human brain (*187*), rat connective tissue (*188*), *Proteus vulgaris* (*183*), *Bacillus subtilis* (*182*), *E. coli* K_{12} (*189*), and the adult housefly (*190*) are given in Table IX (*182–184, 187–192, 194*). Cell-free extracts from *Lactobacillus casei* have been reported to convert galactosamine 6-phosphate to tagatose 6-phosphate and ammonia (*193*). Although the enzyme responsible for this isomerization was not purified to permit specificity studies, it is conceivable that the reaction may have actually been catalyzed by a glucosamine phosphate isomerase.

The recent study by Benson and Friedman on the properties and the allosteric control of this enzyme in the housefly (*190*) deserves special mention. These authors purified the insect isomerase 580-fold, made an estimate of its molecular weight by sucrose density gradient centrifugation (154,000 daltons), and conducted a careful kinetic investigation of the effect of related metabolites on its catalyzed reaction. They found a unidirectional activation of the reverse direction (i.e., the formation of glucosamine phosphate) by glucose 6-phosphate and inorganic phosphate and, at saturating concentrations of substrates and *N*-acetylglucosamine 6-phosphate (which activates in both directions), a 3.8-fold higher rate of amination compared to deamination. Since the insects were also found not to contain a glutamine:fructose-6-phosphate aminotransferase, the authors concluded that in the housefly glucosamine-6-phosphate isomerase may play a significant role in the synthesis of the amino sugar and thus be an important link in the synthetic pathway leading to chitin (*190*).

As for other isomerases whose substrates can occur in cyclic and acyclic forms (see Sections II,C,3 and III,D,3), alternate mechanisms of action may also be considered for glucosaminephosphate isomerase. Kim and Song proposed a mechanism in which a divalent metal is chelated between the straight-chain form of the subtrate and *N*-acetylglucosamine phosphate, yielding an enzyme–substrate–metal activator complex whose enolate intermediate will be hydrated and then de-

185. J. B. Wolfe, B. B. Britton, and H. I. Nakada, *ABB* **66,** 333 (1957).
186. D. G. Comb and S. Roseman, *JBC* **232,** 807 (1958).
187. T. N. Pattabiraman and B. K. Bachhawat, *BBA* **54,** 273 (1961).
188. T. N. S. Varma and B. K. Bachhawat, *BBA* **69,** 464 (1963).
189. R. J. White and C. A. Pasternak, *BJ* **105,** 121 (1967).
190. R. L. Benson and S. Friedman, *JBC* **245,** 2219 (1970).
191. D. G. Comb and S. Roseman, "Methods in Enzymology," Vol. 5, p. 422, 1962.
192. J. S. Clarke and C. A. Pasternak, *BJ* **84,** 185 (1962).
193. T. Shiota, H. Blumenthal, M. N. Disraely, and M. P. McCann, *ABB* **96,** 143 (1962).
194. L. F. Leloir and C. E. Cardini, "Methods in Enzymology," Vol. 5, p. 418, 1962.

TABLE IX

SOME CATALYTIC PROPERTIES OF GLUCOSAMINE-6-PHOSPHATE ISOMERASE FROM VARIOUS SOURCES

Kinetic parameter	Enzyme source							
	Hog kidney [Ref. (*184*)]	*E. coli* B [Ref. (*191, 194*)]	*E. coli* K_{12} [Ref. (*189*)]	*Proteus vulgaris* [Ref. (*183*)]	*Bacillus subtilis* [Ref. (*182, 192*)]	Human brain [Ref. (*187*)]	Rat connective tissue [Ref. (*188*)]	Housefly [Ref. (*190*)]
Specific activity (units[a]/mg at 37°)	51	158	28	0.82	0.007	0.053	~0.002	0.024
pH optimum	pH 8.0–9.0	pH 7.8	pH 7.0	pH 7.2	pH 7.5–8.0	pH 8.5–8.7	pH 8.0	pH 7.1–7.3 (pH 8.3 in the presence of glucose-6-P)
K_m								
Glucosamine-6-P without Ac-Gm6P	14 m*M*	7.1 m*M*	9.0 m*M*	12.5 m*M*	3.0 m*M*	2.4 m*M*	1.8 m*M* (with Mn^{2+})	17 m*M*

with Ac-Gm6P	0.66 m*M*	1.2 m*M*			0.25 m*M* (with Mn^{2+})	0.58 m*M* (with Mn^{2+})	<0.5 m*M*
Fructose-6-P					23 m*M* (3.5 m*M* with Ac-Gm6P and Mn^{2+})		36 m*M*
$(NH_4)^+$					0.23 *M* (32 m*M* with Ac-Gm6P and Mn^{2+})		25 m*M*
K_{eq} $\left(\frac{[\text{fructose-6-P}][NH_4^+]}{[\text{glucosamine-6-P}]}\right)$	0.12–0.18 *M*						0.1–0.18 *M*

[a] Micromoles of substrate converted per minute at specified temperature.

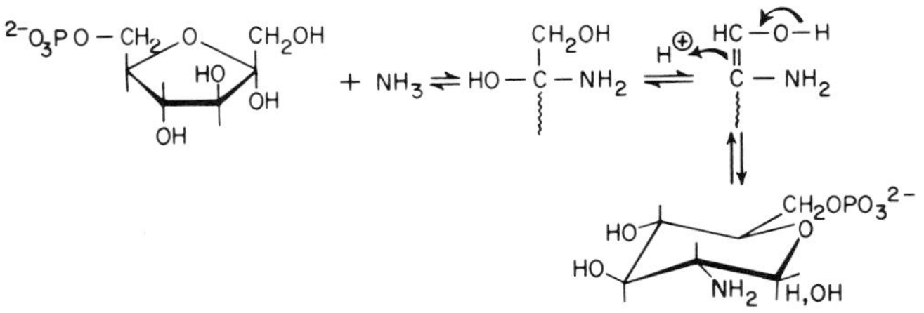

FIG. 4. Proposed mechanism of action of glucosamine-6-phosphate isomerase, modified from Davidson (*179*).

aminated (*195*). This mechanism was designed on the basis of reports that glucosamine phosphate isomerase from human brain and from rat connective tissue is activated by divalent metal ions (*187*, *188*); however, such activation has not been observed with this isomerase from other sources. The potential applicability of this mechanism may therefore be limited. The reaction sequence suggested by Davidson (*179*) is shown in Fig. 4.

As an oddity, and somewhat related to the point made below in regard to erythrose isomerase (Section VIII), it may be remarked that an enzyme designated as acetylglucosaminephosphate isomerase has both obtained and retained official status and an IUB number (EC 5.3.1.11) in spite of the fact that the initial conclusions (*184*) concerning the existence of such an enzyme had been shown to be erroneous (*186*) several years before the first IUB nomenclature report was issued (*196*).

V. Ribose-5-phosphate Isomerase

D-Ribose 5-phosphate ⇄ D-Ribulose 5-phosphate

A. GENERAL

Ribose-5-phosphate isomerase (EC 5.3.1.6) was discovered in the early 1950's by Horecker and collaborators (*197*) as a by-product of their work on the oxidative decarboxylation of 6-phosphogluconate. The en-

195. H. S. Kim and P. S. Song, *Soul Taehakkyo Nonmunjip, Chayon Kwahak, Saengnongge* **14**, 5 (1963); *Chem. Abstr.* **63**, 16691c (1965); P. S. Song, *Experientia* **20**, 211 (1964).

196. "Report of the Commission on Enzymes of the International Union of Biochemistry 1961." Pergamon, Oxford, 1961.

197. B. L. Horecker, P. Z. Smyrniotis, and J. E. Seegmiller, *JBC* **193**, 383 (1951).

zyme is usually assumed to catalyze the interconversion between ribose 5-phosphate and ribulose 5-phosphate (*1, 197*), although F. C. Knowles and Pon (*198*) have recently proposed that the first reaction product is not ribulose 5-phosphate but its dehydrated β-diketone phosphate derivative (*vide infra*). By being part of both the oxidative and the reductive pentose phosphate cycles, ribose-5-phosphate isomerase is ubiquitously present in nature and has been found wherever it has been sought. Rather than presenting selected examples of tissues in which the enzyme has been measured, the author refers the interested reader to the comprehensive review by Pon (*199*). Earlier studies on some catalytic properties of the enzyme from alfalfa (*200*) and erythrocytes (*201*) have been summarized by Topper (*1*). Investigations of the molecular properties are of more recent origin since they have had to await the development of isolation procedures capable of producing quantities of the homogeneous enzyme sufficient for physical measurement. Such procedures are now available for ribose-5-phosphate isomerase from spinach (*202, 203*) by which the enzyme has been purified to specific activities approaching that of the "classic" preparation of Axelrod and Jang from alfalfa (2400 units/mg), i.e., 2200 units/mg of protein (*202, 204–207*), all activities

198. F. C. Knowles and N. G. Pon, *JACS* **90,** 6536 (1968).

199. N. G. Pon, *Comp. Biochem.* **7,** 1 (1964).

200. B. Axelrod and R. Jang, *JBC* **209,** 847 (1954).

201. F. H. Bruns, E. Noltmann, and E. Vahlhaus, *Biochem. Z.* **330,** 483 (1958).

202. N. G. Pon and F. C. Knowles, *Federation Proc.* **27,** 790 (1968).

203. A. C. Rutner and M. D. Lane, *Abstr., 156th ACS Natl. Meeting, Atlantic City, 1968* Biol. Chem Div., Abstr. No. 146 (1968).

204. A. C. Rutner, *Biochemistry* **9,** 178 (1970).

205. F. C. Knowles, Ph.D. Dissertation, University of California, Riverside (1968).

206. In the report by N. G. Pon and F. C. Knowles (*202*), the specific activity of their preparation is actually given as 5000 units/mg at 37°. This value was based on the arbitrary assumption that 1.0 $A_{280\ nm}$ corresponds to 1 mg of protein (*205*). The reason for the adoption of this conversion factor is not clear since the ultraviolet spectrum of their isolated enzyme (*205*) yields a 280 nm/260 nm ratio of only 1.52, indicative of an unusually low content of aromatic amino acids. Rutner determined a conversion factor of 2.3 mg of protein corresponding to 1.0 $A_{280\ nm}$ based upon refractometry of the pure enzyme as the primary standard (*204*). With this absorptivity coefficient for spinach ribose-5-phosphate isomerase, the specific activity of the preparation of Pon and Knowles (*202*) is in reality only 2200 units/mg. The case is an impressive example of the danger inherent in the use of "average" absorptivity coefficients derived from "average" proteins as the sole basis for calculating the final turnover number of an isolated enzyme.

207. It is interesting to note that in spite of the very high turnover number of this enzyme and the technical disadvantage of performing activity measurements substantially above ambient temperature, investigators who purified this enzyme have chosen, apparently without exception, 37° as assay temperature. Most probably, this may have resulted from the keen competition to reach numerical values for the turnover number as high as that reported earlier by Axelrod and Jang (*200*).

having been measured at 37°. Partial purifications have also been reported (*207*) for the spinach enzyme (306 units/mg) (*208*), for the enzyme from *Echinococcus granulosus* (230 units/mg) (*209*), and for human erythrocyte ribose-5-phosphate isomerase (375 units/mg) (*210*).

B. Molecular Properties

Only few data on the physical properties have thus far been published. A rough estimate for the sedimentation coefficient of ribose-5-phosphate isomerase from *Rhodospirillum rubrum* (*211*) has been made ($s_{20,w}^{app}$ = 5.1 S) by comparing its mobility in a sucrose gradient with that of yeast enolase whose $s_{20,w}$ was taken to be 5.9 S (*212*). In the same paper, a molecular weight of 57,000 daltons was derived from the Stokes' radius (27 Å) of enolase which coeluted with the *Rhodospirillum* isomerase from Sephadex G-100, with use of an assumed partial specific volume of 0.725 ml/g (*211*). Acceptance of these values should be postponed until they have been confirmed by more rigorous methods.

Rutner has reported $s_{20,w}$ for spinach ribose-5-phosphate isomerase to be 4.10 ± 0.02 S by analytical ultracentrifuge techniques and the molecular weight to be 53,000 ± 2000 by high-speed equilibrium centrifugation (*204*). The latter value was calculated with use of a partial specific volume of 0.749 ml/g measured by the "falling-drop" technique of Linderstrøm-Lang and Lanz (*213*). Polyacrylamide gel electrophoresis in the presence of sodium dodecyl sulfate gave a value of approximately 26,000 daltons for the molecular weight, suggesting that the native enzyme is composed of two subunits (*214*). Data on the amino acid composition are not yet available, although a 280 nm/260 nm absorbance ratio of only 1.55 (*204*) has been taken to suggest an abnormally low content of aromatic amino acids. Examination of the isolated enzyme by atomic absorption spectroscopy did not yield significant amounts of bound metal, and no inhibition of enzymic activity was observed on addition of chelating agents to the assay system (*214*). It would thus appear that ribose-5-phosphate isomerase is not a metalloenzyme.

208. M. Tabachnick, P. A. Srere, J. Cooper, and E. Racker, *ABB* **74,** 315 (1958).
209. M. Agosin and L. Aravena, *Enzymologia* **22,** 281 (1961).
210. M. Urivetzky and K. K. Tsuboi, *ABB* **103,** 1 (1963).
211. L. Anderson, L. E. Worthen, and R. C. Fuller, *in* "Comparative Biochemistry and Biophysics of Photosynthesis" (K. Shibata *et al.*, eds.), p. 379. Univ. Park Press, State College, Pennsylvania, 1968.
212. B. G. Malmström, "The Enzymes," 2nd ed., Vol. 5, p. 471, 1961.
213. H. K. Schachman, "Methods in Enzymology," Vol. 4, p. 32, 1957.
214. A. C. Rutner, personal communication (1970).

Brief mention should be made of the possible existence of isoenzymes of ribose-5-phosphate isomerase. Despite the recent discovery (*215*) of "high K_m" and "low K_m" enzymic activities in *E. coli* X289 (Section V,C,2), convincing evidence for separate protein species has not yet been presented. Rutner concluded from his data that the spinach enzyme isolated by his procedure represents only a single protein species and, furthermore, that spinach leaves seem to have only one ribose-5-phosphate isomerase (*204*). F. C. Knowles, on the other hand, basing his claim upon solubility properties and behavior on ion exchange columns, believed that there are two isoenzymic forms in spinach leaves (*205*). It is possible, however, that these multiple forms are artifacts, and additional methodic studies are necessary to resolve this discrepancy.

C. Catalytic Properties

1. *Assay Methods*

Three types of assay methods have been employed for measurements of ribose-5-phosphate isomerase: (a) discontinuous chemical-stop assays of the formation of ketopentose phosphate [e.g., (*197, 200, 216*)] by the colorimetric cysteine-carbazole method of Dische and Borenfreund (*217*) or of the disappearance of ribose 5-phosphate (*201*) by the phloroglucinol reaction (*218*), (b) multiple-couple spectrophotometric assays involving measurement of the absorbance at 340 nm produced in the terminal dehydrogenase reaction [e.g., (*208, 219–221*)], and (c) direct spectrophotometric measurement of the ketopentose phosphate chromophore which has a maximum at approximately 280 nm (*198*). Both laboratories that have used this assay (*222, 223*) have calibrated it against the cysteine-carbazole assay, and in order to arrive at a molar extinction coefficient both have used an equilibrium constant of 0.32–0.33 for [ketopentose phosphate]/[ribose 5-phosphate] selected from the literature (*197, 200, 216*). They arrive, however, at different molar absorptivity

215. J. David and H. Wiesmeyer, *BBA* **208,** 56 (1970).
216. F. Dickens and D. H. Williamson, *BJ* **64,** 567 (1956).
217. Z. Dische and E. Borenfreund, *JBC* **192,** 583 (1951).
218. Z. Dische and E. Borenfreund, *BBA* **23,** 639 (1957).
219. F. Novello and P. McLean, *BJ* **107,** 775 (1968).
220. M. E. Kiely, E. L. Tan, and T. Wood, *Can. J. Biochem.* **47,** 455 (1969).
221. J. David and H. Wiesmeyer, *BBA* **208,** 45 (1970).
222. F. C. Knowles, M. K. Pon, and N. G. Pon, *Anal. Biochem.* **29,** 40 (1969).
223. T. Wood, *Anal. Biochem.* **33,** 297 (1970).

coefficients: F. C. Knowles and Pon (*222*) reported a value of 58.6 at 280 nm, and 85 may be calculated for this wavelength from Wood's data at 290 nm (*223*). Although the direct spectrophotometric assay is clearly the most desirable to use because of its ease and the absence of auxiliary coupling enzymes, its applicability at this time appears to suffer from the lack of a reliable absorptivity coefficient which would allow comparison with rate measurements of other enzymes.

The pH optima of ribose-5-phosphate isomerases from various sources have been found to be quite broad and somewhat above pH 7. A pH between 7.0 and 7.5 is usually used for activity assay.

2. *Michaelis Constants*

The K_m for ribose 5-phosphate has been measured by most laboratories concerned with ribose-5-phosphate isomerase. The following values have been reported for the enzymes from various sources: human erythrocytes, 2.2 mM (*201, 210*); *Rhodospirillum rubrum*, 0.43 mM (*224*) and 4 mM (*211*); *Aerobacter aerogenes*, 1.8 mM (*19*); *Echinococcus granulosus*, 2.7 mM (*209*); *Pediococcus pentosaceus*, 2.8 mM (*225*); yeast, 0.74 mM (*226*); and spinach, 4.6 mM (*222*) and 0.46 mM (*204*). In addition, David and Wiesmeyer (*215*) have recently suggested the presence of two distinct constitutive ribose-5-phosphate isomerase activities in wild-type *E. coli* X289. Although their data would not allow a decision as to whether or not the activities reflect two separate protein species, a distinction was possible on the basis of heat lability. One activity was found to be heat stable with an apparent K_m ranging from 1.85 to 2.59 mM, the other was heat labile with an apparent K_m between 0.13 and 0.25 mM. The heat-labile "low K_m" isomerase was very strongly inhibited by glucose 6-phosphate ($K_i^{\text{noncomp}} = 0.06$ to 0.085 mM), the "high K_m" activity less so (*215*). The isomerase pair with its different apparent affinities for the substrate ribose 5-phosphate was proposed to control the flux of ribose (via ribose 5-phosphate) either into energy metabolism or, in the presence of a plentiful supply of other carbon sources, toward the production of 5-phosphoribosyl pyrophosphate (*215, 227*). In the light of this observed "double nature" of ribose-5-phosphate isomerase in *E. coli*, it is tempting to speculate on the significance of the difference in the K_m values reported for the enzyme from spinach, *viz.*, 4.6 mM by F. C. Knowles and Pon (*222*) vs. 0.46 mM by Rutner (*204*), especially

224. L. Anderson and R. C. Fuller, *Plant. Physiol.* **43,** Suppl., S-30 (1968).
225. W. J. Dobrogosz and R. D. DeMoss, *BBA* **77,** 629 (1963).
226. H. Horitsu and M. Tomoeda, *Agr. Biol. Chem.* (*Tokyo*) **30,** 956 (1966).
227. J. David and H. Wiesmeyer, *BBA* **208,** 68 (1970).

since Knowles' isolation procedure involves a heat step whereas Rutner's does not. If the difference is not merely of methodological origin, it may be worthwhile to consider the possibility of either the presence of two isomerase species in spinach, one of which is inactivated by the heat treatment, or the transformation of a "low K_m" form to a "high K_m" form during the heat step. This problem would seem to warrant further investigation.

3. *Mechanism*

Ribose-5-phosphate isomerase has usually been thought to catalyze the interconversion of ribose 5-phosphate and ribulose 5-phosphate. The latter was proposed by Horecker *et al.* (*197*) to be the reaction product on the basis of isolation of an optically active *o*-nitrophenylhydrazone of ribulose. The interconversion is considered to proceed as a proton transfer involving the intermediate formation of an enediol anion of *cis* configuration, demonstrated by Rose and collaborators for all aldose–ketose isomerases (*3, 5, 175, 176*). Tritium exchange experiments performed with the isolated spinach enzyme (*228*) and with chloroplast extracts (*128*) showed the proton shift to occur intramolecularly without any significant proton exchange with the medium. Nothing is known at this time about how the enzyme protein participates in the catalytic process.

The nature of the catalyzed reaction has recently been questioned by F. C. Knowles and Pon who, on the basis of spectral data, postulated that the initial product of the reaction catalyzed by ribose-5-phosphate isomerase is not ribulose 5-phosphate but rather its dehydrated β-diketone derivative (*198*). They proposed the sequence of events to be as depicted in Fig. 5. Although this scheme provides a plausible explanation for the observed spectral changes which occur on addition of either acid/base or enzyme to the substrate ribose 5-phosphate, it is clearly inconsistent with an asymmetrical product required to account for Horecker's original finding of an optically active phenylhydrazine derivative of ribulose (*197*). For this to be compatible with the shown mechanism, one would have to postulate the action of an additional hydratase activity converting the β-diketone phosphate to the asymmetrical ribulose 5-phosphate. Also, the entire argument has been derived from spectral data alone (*198, 205*) and thus far only a short note has been published (*198*). Chemical evidence for the existence of the proposed intermediates is needed to determine whether a mono- or a diketone species is produced under the conditions of catalysis. In spite of its hy-

228. M. W. McDonough and W. A. Wood, *JBC* **236**, 1220 (1961).

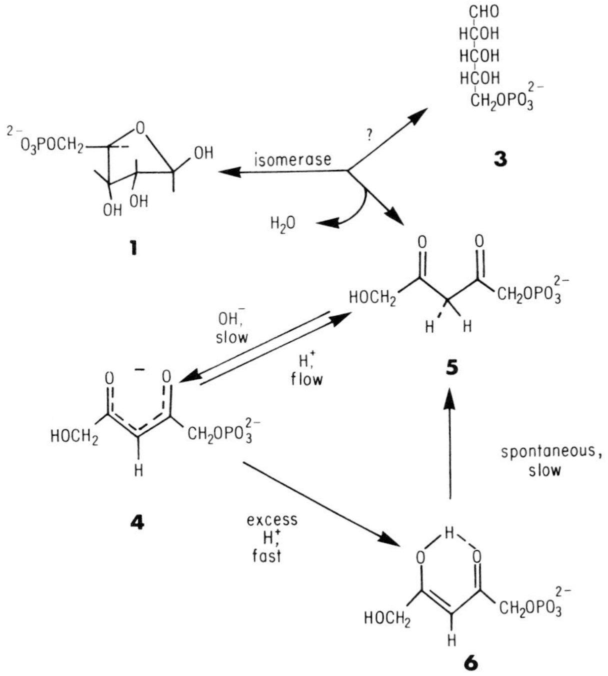

FIG. 5. Reaction scheme proposed by F. C. Knowles and Pon (*198*) for the ribose-5-phosphate isomerase catalyzed reaction: 1, ribose 5-phosphate; 3, acyclic ribose 5-phosphate; 4, enolate anion; 5, β-diketone phosphate; 6, β-hydroxy enone. The scheme is shown as presented in the original paper (*198*).

pothetical nature, the proposed mechanism is of considerable interest since, if confirmed, it would have mechanistic implications with respect to the subsequent reaction sequences of both oxidative and reductive ribose 5-phosphate metabolism.

VI. Arabinose-5-phosphate Isomerase

$$\text{D-Arabinose 5-phosphate} \rightleftarrows \text{D-Ribulose 5-phosphate}$$

D-Arabinose-5-phosphate isomerase (EC 5.3.1.13) appears to be the one enzyme of those dealt with in this chapter that was not known at the time Topper reviewed this topic for the previous edition of "The Enzymes" (*1*). It catalyzes the interconversion of D-arabinose 5-phos-

TABLE X
CATALYTIC PROPERTIES OF D-ARABINOSE-5-PHOSPHATE ISOMERASE PREPARATIONS FROM *Propionibacterium pentosaceum*[a] AND *Escherichia coli*[b]

Property	Enzyme source	
	P. pentosaceum	*E. coli*
Specific activity[c] (units[d]/mg)	5.4 (37°)	0.35 (37°)
pH optimum	pH 8.0	pH 8.0
Michaelis constant		
D-Arabinose 5-phosphate	1.98 m*M*	1.36 m*M*
D-Ribulose 5-phosphate		0.54 m*M*
$K_{eq}\left(\frac{[\text{D-ribulose 5-phosphate}]}{[\text{D-arabinose 5-phosphate}]}\right)$	0.295 (37°)	0.333 (37°?)
Compounds tested and found not to act as substrates	D-Arabinose, L-arabinose, D-ribose, D-xylose, D-ribose 5-phosphate, D-glucose 6-phosphate	D-Arabinose, D-ribulose, D-ribose, D-ribose 5-phosphate
Inhibitors	*p*-HMB	*p*-HMB, Mn^{2+}, Co^{2+}, Zn^{2+}, Cd^{2+}

[a] From Volk (*229, 231*).
[b] From Lim and Cohen (*230*).
[c] Cysteine-carbazole assay for ketopentoses (*217*).
[d] Micromoles of product formed per minute at specified temperature.

phate and D-ribulose 5-phosphate as shown by cochromatography of the reaction products, after dephosphorylation, with authentic D-arabinose and D-ribulose (*229, 230*), by quantitative assay of the ketopentose with ribitol dehydrogenase (*229*) and by demonstration of D-ribose 5-phosphate as the product of the combined action of arabinose-5-phosphate isomerase and ribose-5-phosphate isomerase on arabinose 5-phosphate as substrate (*229*). Arabinose-5-phosphate isomerase has been purified from *Propionibacterium pentosaceum* (*229*) and from *Escherichia coli* (*230*). The physiological role in both organisms is thought to be the production of arabinose 5-phosphate from ribulose 5-phosphate (*229, 230*). In *E. coli*, the enzyme may have a significant function since arabinose 5-phosphate is the precursor for 2-keto-3-deoxyoctonate, a constituent of its cell wall lipopolysaccharide (*230*). Some catalytic properties of the two preparations are listed in Table X (*229–231*).

229. W. A. Volk, *JBC* **235,** 1550 (1960).
230. R. Lim and S. S. Cohen, *JBC* **241,** 4304 (1966).
231. W. A. Volk, "Methods in Enzymology," Vol. 9, p. 585, 1966.

VII. Triosephosphate Isomerase

D-Glyceraldehyde 3-phosphate ⇄ Dihydroxyacetone phosphate

A. History and General

Triosephosphate isomerase (EC 5.3.1.1) is a fascinating example of a "forgotten" enzyme. Although it has been known for over thirty-five years, enzymologists have not subjected this enzyme to rigorous study until quite recently. Thus, in his review in 1961, Topper cites only four references to it (*1*). The neglect of triosephosphate isomerase is the more remarkable in view of the fact that it directly catalyzes a step in what is probably the best known and most studied metabolic pathway, that it can be easily isolated, and that it is perhaps the "fastest" enzyme in intermediary metabolism. In 1953, about twenty years after its discovery (*232*), the enzyme became available in crystalline form (*233*); yet a first estimate of its molecular weight (which turned out to be incorrect) was not made until 1966 (*234*). It is only while this chapter is being written that reliable measurements of the molecular weight by hydrodynamic methods are being published (*235*, *236*).

In contrast to this long dormancy, there is now fierce competition among about half a dozen laboratories in the United States and Europe investigating the active site, mechanism, and molecular architecture of this enzyme. An attempt will be made to provide an up-to-date appraisal, but by the time of publication much of this information may be superseded by newer data.

B. Molecular Properties

At the time of this writing, purification procedures have been published for the isolation of crystalline triosephosphate isomerase from five

232. O. Meyerhof and K. Lohmann, *Biochem. Z.* **273**, 413 (1934); O. Meyerhof and W. Kiessling, *Biochem. Z.* **279**, 40 (1935).

233. E. Meyer-Arendt, G. Beisenherz, and T. Bücher, *Naturwissenschaften* **40**, 59 (1953).

234. P. M. Burton and S. G. Waley, *BJ* **100**, 702 (1966).

235. W. K. G. Krietsch, P. G. Pentchev, H. Klingenbürg, T. Hofstätter, and T. Bücher, *European J. Biochem.* **14**, 289 (1970).

236. I. L. Norton, P. Pfuderer, C. D. Stringer, and F. C. Hartman, *Biochemistry* **9**, 4952 (1970).

sources: calf skeletal muscle (*237*), rabbit skeletal muscle (*236*, *238*), rabbit liver (*235*), bovine lens (*239*), and yeast (*235*). The following specific activities have been reported for these preparations with glyceraldehyde 3-phosphate as substrate (in the same order): 9,500 [26° (*237*)], 5,500 (*238*) to 7,500 [24° (*236*)], 6,400 (*235*), 7,350 (*240*), and 10,000 (*235*), all in units per milligram measured at 25°, except as indicated. The enzyme has also been isolated in crystalline form from human erythrocytes (*240a*).

1. *Molecular Weight*

Estimates for the molecular weight vary from 43,000 to 60,000. A compilation of the data from various laboratories is given in Table XI. As is often the case, some of the differences may originate from lack of measurement of the partial specific volume. Thus, Norton *et al.* used a value of 0.737 ml g^{-1} calculated from their data for the amino acid composition of the rabbit muscle enzyme (*236*); Krietsch *et al.* arbitrarily assumed 0.750 ml g^{-1} without reference to any data (*235*). A potentially very accurate determination could be made by X-ray crystallography. Unfortunately, the one value reported by this technique, *viz.*, 53,000, has been calculated on the basis of assumed "typical" values of 30% for the water content of the crystal, 30% for the salt content of the mother liquor, and 1.2 g ml^{-1} for the density of the needle-shaped crystal (*240*). Nevertheless, it is interesting to observe that many of the recent data (see Table XI) seem to converge on this value of 53,000 (*240b*). The sedimentation coefficient of the rabbit muscle enzyme was found to show a slight concentration dependence with $s^{\circ}_{20,w}$, extrapolating to 3.95 S (*236*).

2. *Subunit Structure*

Molecular weight determinations under dissociation conditions provide good evidence for the dimeric structure of the enzyme (see Table XI) (*234–236*, *240–244*). It is especially interesting to note that calcu-

237. G. Beisenherz, "Methods in Enzymology," Vol. 1, p. 387, 1955.

238. F. W. Bube, R. Czok, and I. Jäger, unpublished work (1960); quoted by R. Czok and T. Bücher, *Advan. Protein Chem.* **15**, 315 (1960).

239. P. M. Burton and S. G. Waley, *Exptl. Eye Res.* **7**, 189 (1968).

240. L. N. Johnson and S. G. Waley, *JMB* **29**, 321 (1967).

240a. E. E. Rozacky, T. H. Sawyer, R. A. Barton, and R. W. Gracy, *ABB* **146**, 312 (1971).

240b. A value of 56,000 daltons has recently been reported for the molecular weight of triosephosphate isomerase from human erythrocytes (*240a*).

241. P. Esnouf and S. G. Waley, unpublished work, quoted by Johnson and Waley (*240*) and Burton and Waley (*245*).

TABLE XI
MOLECULAR WEIGHT DATA OF NATIVE AND DISSOCIATED TRIOSEPHOSPHATE ISOMERASE

Analytical technique	Enzyme source: Rabbit skeletal muscle (daltons)	Enzyme source: Yeast (daltons)
Native enzyme		
Equilibrium sedimentation	60,000 [Ref. (*241*)] 48,000 [Ref. (*242*)] 57,000 [Ref. (*235*)] 53,000 [Ref. (*236*)]	53,000 [Ref. (*235*)]
Sucrose gradient centrifugation[a]	56,000 [Ref. (*235*)]	56,000 [Ref. (*235*)]
Gel filtration	43,000 [Ref. (*234*)] 60,000 [Ref. (*235*)]	56,000 [Ref. (*235*)]
X-ray crystallography of needle-shaped crystals[b]	53,000 [Ref. (*240*)]	
Subunit		
Equilibrium sedimentation in 1% sodium dodecyl sulfate	25,000 [Ref. (*235*)] [50,000][c]	29,500 [Ref. (*235*)] [59,000]
Equilibrium sedimentation after maleylation	29,000 [Ref. (*235*)] [58,000]	23,500 [Ref. (*235*)] [47,000]
Polyacrylamide gel electrophoresis in sodium dodecyl sulfate	26,500 [Ref. (*236*)] [53,000] 26,500 [Ref. (*244*)] [53,000]	
X-ray crystallography of hexagonal crystals[d]	26,000 [Ref. (*240*)] [52,000]	

[a] The molecular weight of rabbit liver triosephosphate isomerase by this method was also found to be 56,000 (*235*).

[b] Based on assumed values for water content of the crystal, salt content of the mother liquor, and density of the crystal (*240*), see text.

[c] Values in brackets are the respective subunit molecular weights, multiplied by 2, to indicate the molecular weight of a hypothetical dimer.

[d] Based on a measured water content of 37% in the crystal, a salt content of 27% in the mother liquor, and a density of the crystal of 1.262 (*240*).

242. P. Esnouf and J. Jesty, quoted by Coulson *et al.* (*243*).

243. A. F. W. Coulson, J. R. Knowles, and R. E. Offord, *Chem. Commun.* **1970,** p. 7 (1970).

244. J. R. Katze and J. R. Knowles, unpublished experiments, quoted by Coulson *et al.* (*243*).

lation from the X-ray data indicates that the asymmetric unit of the monoclinic needle crystals apparently represents the dimeric molecule, whereas the asymmetric unit of the hexagonal bipyramid crystals corresponds to the subunit (*240*). All of the physical data suggest that the subunits are similar or identical as far as size is concerned since only one species has been found under dissociation conditions (*236, 240, 245*). Chemical evidence for identical vs. nonidentical subunits does not seem to be unequivocal at the time of this writing. For the rabbit muscle enzyme, only one type of C-terminal residue, glutamine (*236, 245*), and only one type of N-terminal residue, alanine (*235, 245*), have been found. Also, only five different cysteine peptides have been found after carboxymethylation (*245*) and only three peptide fragments after cyanogen bromide cleavage (*246*). This compares with cysteine and methionine contents of 5 and 2 residues, respectively, calculated for a subunit molecular weight of 26,500 (*234–236, 245*). These data together with the results from active site labeling (*vide infra*) were taken to suggest that the subunits are most likely identical (*243, 245, 246*). Recent work by Krietsch *et al.* (*247*), however, indicates different subunits on the basis of hybridization experiments and variations in the peptide elution pattern from tryptic hydrolyzates of chromatographically separated isoenzyme forms.

3. *Isoenzymes*

Multiple electrophoretic or chromatographic forms of triosephosphate isomerase have been described for the crystalline enzyme from rabbit skeletal muscle (*234–236, 245, 248*) and also for the enzyme present in or isolated from pig muscle (*249, 250*), horse heart (*248*), bovine lens (*239*), human erythrocytes (*251*), and rabbit (*235*), horse (*248*), and human liver (*248*). Where individual species have been further studied (*248*), no differences in size or in kinetic parameters have been found. It has been concluded that the observed variation in charge only may be a reflection of conformational dissimilarities between the various separated species (*245, 251*). On the basis of hybridization experiments and

245. P. M. Burton and S. G. Waley, *BJ* **107,** 737 (1968).

246. J. C. MacGregor and S. G. Waley, *Abstr. Symp., 8th Intern. Congr. Biochem., Interlaken, Switzerland, 1970* p. 116 (1970).

247. W. K. G. Krietsch, P. G. Pentchev, W. Machleidt, and H. Klingenburg, *FEBS Letters* **11,** 137 (1970).

248. E. W. Lee and R. Snyder, *Federation Proc.* **29,** 898 (1970). The Michaelis and inhibitor constants in this abstract should have been given in mM instead of μM [R. Snyder, personal communication (1970)].

249. R. K. Scopes, *Nature* **201,** 924 (1964).

250. R. K. Scopes, *BJ* **107,** 139 (1968).

251. J. C. Kaplan, L. Teeple, N. Shore, and E. Beutler, *BBRC* **31,** 768 (1968).

slight differences in the elution pattern from cation exchange columns of tryptic digests which had been made of three chromatographically separable enzyme species, Krietsch *et al.* (*247*) have obtained preliminary indication that rabbit muscle triosephosphate isomerase may exist as three isoenzymes of the AA, AB, BB type, an interpretation that has also been considered by J. C. Kaplan *et al.* as an alternative to conformers (*251*). This reviewer, however, disagrees with the conclusion of Krietsch *et al.* that their presented data constitute "*strong* [italics added] evidence that the rabbit muscle [triosephosphate isomerase] isoenzymes are composed of two different polypeptide chains, coded on two cistrons" (*247, 251a*). The reader may wish to refer to the discussion on pseudoisoenzymes of rabbit muscle phosphoglucose isomerase for which experimental evidence seems to suggest that the observed chromatographic differences may merely be the result of different oxidation states of their sulfhydryl groups (Section II,B,4,b). Triosephosphate isomerase would also appear to deserve further study in this respect especially since mercaptoethanol has been found to convert the three isoenzymes into a single electrophoretic band at pH 9.5 although it had no effect at pH 6.6 (*235*). Conceivably, differently charged cysteine derivatives could also result in different elution patterns of tryptic peptides.

4. *Active Site Structure*

Studies on the active site of triosephosphate isomerase were initiated by Burton and Waley (*234*). Data from carboxymethylation and photooxidation experiments, accompanied by measurements of the loss in enzymic activity, were found (*234*) to be consistent with a histidine residue at the active center supplying the basic group which was thought to be involved in catalysis (*3*). Furthermore, reaction of the enzyme with limited amounts of diazotized sulfanilic acid also led to inactivation. This finding was taken to suggest that a lysine residue might be involved in addition to the histidine, perhaps in binding the phosphate group of the substrate (*252*).

Most of the recent progress in elucidating the active site structure, however, has come from specific labeling with synthetic substrate analogs which possess a halogen enabling covalent bonding to the protein. Hartman synthesized 1-iodo-3-hydroxyacetone phosphate, intending to link it specifically to the active site of aldolase (*253*). A more detailed

251a. More convincing data have now been provided in the full-length paper [W. K. G. Krietsch, P. G. Pentchev, and H. Klingenburg, *European J. Biochem.* **23**, 77 (1971)].

252. P. M. Burton and S. G. Waley, *BJ* **104,** 3P (1967).

253. F. C. Hartman, *Federation Proc.* **27,** 454 (1968).

study (*254*) of the initially reported inactivation (*253*) made it apparent that for aldolase the reaction is not specific for the active site but that iodohydroxyacetone phosphate inactivates this enzyme by causing the majority of its sulfhydryl groups to be oxidized to disulfides and sulfenic acids. The same reagent, however, was found to modify specifically the active site of triosephosphate isomerase by totally abolishing enzymic activity with the incorporation of 1.7 mole equivalents per mole of enzyme (*255*). After exploratory experiments with mixtures of amino acids had indicated that the SH group of cysteine was the only functional group of free amino acids which reacted with either iodo-, chloro-, or bromohydroxyacetone phosphate, a more detailed investigation of the reaction of the haloacetolphosphates with glutathione showed that only the chloro and the bromo derivative modify its sulfhydryl group by alkylation, whereas the iodo compound does so by oxidation (*256*).

It was, therefore, somewhat of a surprise when analysis of an active site peptide, obtained from a tryptic digest of triosephosphate isomerase treated with chlorohydroxyacetone phosphate yielded a glutamyl residue as the point of attack by the haloacetolphosphate (*257, 258*). As part of the active site, this glutamyl residue must possess an unusual reactivity since 1000-fold higher concentrations of the reagent failed to react with free glutamic acid (*258*). Clearly, such an extraordinary reactivity would also support its role in catalysis. Its environmental characteristics may be similar to those of the glutamic acid residue previously identified by Takahashi *et al.* (*259*) as part of the active site of ribonuclease T_1, which was found to have an unusual reactivity toward alkylation by iodoacetate. In fact, it could perhaps be concluded in retrospect that the inactivation of triosephosphate isomerase by iodoacetate observed by Burton and Waley (*234*) may have resulted from such alkylation of an essential glutamyl residue rather than from alkylation of the originally postulated imidazole group of a histidyl residue.

Closely following Hartman's initial reports (*253, 255*), communications from several laboratories appeared describing similar work on "affinity labeling" (*260*) of triosephosphate isomerase. Rose and O'Connell chose glycidol phosphate (1,2-epoxipropanol 3-phosphate) as a potential site-specific alkylating agent "because its planarity was thought

254. F. C. Hartman, *Biochemistry* **9,** 1783 (1970).
255. F. C. Hartman, *BBRC* **33,** 888 (1968).
256. F. C. Hartman, *Biochemistry* **9,** 1776 (1970).
257. F. C. Hartman, *JACS* **92,** 2170 (1970).
258. F. C. Hartman, *Biochemistry* **10,** 146 (1971).
259. K. Takahashi, W. H. Stein, and S. Moore, *JBC* **242,** 4682 (1967).
260. L. Wofsy, H. Metzger, and S. J. Singer, *Biochemistry* **1,** 1031 (1962).

to make it a reasonable analogue of the enediol form of the triosephosphates believed to be an intermediate in the catalyzed reaction" (*261*). They found that, indeed, the enzyme completely lost its activity on incorporation of one molecule of the modifier per catalytic subunit (*261*). This approach was then pursued in collaboration with Waley and Miller (*262*). Rabbit muscle triosephosphate isomerase labeled with glycidol-^{32}P was hydrolyzed with pepsin, the radioactive peptide was isolated and subjected to sequence analysis, glutamic acid was identified as the residue carrying the label, and the sequence around it was determined (*262*) (*vide infra*).

J. R. Knowles and collaborators used bromohydroxyacetone phosphate for active-site-directed inactivation first of rabbit muscle triosephosphate isomerase (*243*) and then, after difficulties arose in isolating a single radioactive peptide from various proteolytic digests (*243*), of the chicken muscle enzyme (*263*). With the ^{32}P-labeled reagent they found incorporation of 1.3 equivalents per mole of enzyme, but with ^{14}C-bromohydroxyacetone phosphate the incorporated radioactivity amounted to 2.1 equivalents (*243*); the authors therefore concluded that the inhibitor phosphate group was lost from the inactivated enzyme whereas the carbon skeleton remained attached to the protein (*243, 263*). For the modified peptide isolated from chicken muscle triosephosphate isomerase, the site of attachment was determined to be a tyrosine residue and the mode of linkage that of a phenyl ether (*263*). Coulson *et al.*, however, considered the possibility that a migration of label might have taken place because of the freely available keto group in the bromohydroxyacetone phosphate (*263*). A feasible mechanism for such an event is illustrated in Fig. 6 (*264*). This notion was subsequently borne out by isolation of a uniquely labeled dipeptide Glu*–Pro from the digest of triosephosphate isomerase, which had been inactivated by bromohydroxyacetone phosphate and reduced by borohydride immediately after alkylation in order to keep the inhibitor locked in place at the primary site of attachment (*264, 265*). Thus all laboratories concerned are now in agreement that a glutamic acid residue provides the link for attachment of the substrate analogs (*257, 258, 262, 263*) and therefore, most likely, also of the natural substrates. At the time of this writing, 15 residues have been established in the sequence of the active site peptide:

261. I. A. Rose and E. L. O'Connell, *JBC* **244,** 6548 (1969).

262. S. G. Waley, J. C. Miller, I. A. Rose, and E. L. O'Connell, *Nature* **227,** 181 (1970); J. C. Miller and S. G. Waley, *BJ* **123,** 163 (1971).

263. A. F. W. Coulson, J. R. Knowles, J. D. Priddle, and R. E. Offord, *Nature* **227,** 180 (1970).

264. J. R. Knowles, personal communication (1970).

265. Note added in proof to Coulson *et al.* (*263*).

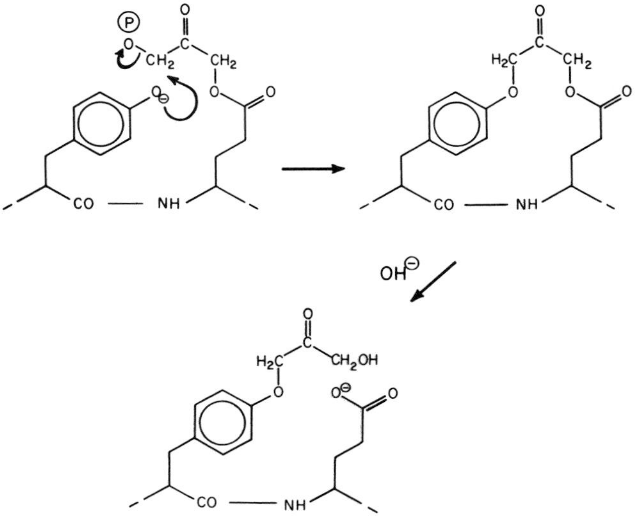

FIG. 6. Proposed mechanism for the migration of an active site label of triosephosphate isomerase from a glutamyl to a tyrosyl residue [according to J. R. Knowles (*264*)].

Trp	Val	Leu	Ala	Tyr	Glu*	Pro	Val	Trp	Ala	Ile	Gly	Thr	Gly	Lys
1	2	3	4	5	6	7	8	9	10	11	12	13	14	15

This is the sequence of the pentadecapeptide given by Hartman for rabbit muscle triosephosphate isomerase (*258*) which is a slight revision of the sequence in his preliminary communication (*266*). The same sequence was determined by Priddle and Offord (*267*) for the enzyme from chicken muscle, except that these authors place a valine rather than a tryptophan in position 1 (*267*). Identical sequences of a hexapeptide corresponding to positions 4–9 have been published by Waley *et al.* (*262*) and by Coulson *et al.* (*263*) for the enzyme from rabbit and chicken muscle, respectively; in addition, a labeled dipeptide Glu*–Pro (*vide supra*) has been identified by the latter authors (*265*).

C. CATALYTIC PROPERTIES

1. *Kinetic Parameters*

The one property that makes triosephosphate isomerase stand out among other glycolytic enzymes is its extraordinary catalytic capacity, *viz.*, 7,000–10,000 μmoles of substrate conversion in the forward direction

266. F. C. Hartman, *BBRC* **39**, 384 (1970).
267. R. E. Offord, personal communication (1970).

per minute per milligram of enzyme at 25° (*vide supra*) under routine assay conditions. Based on V_{max} data at pH 7.6, Krietsch *et al.* (*235*) arrive at molecular activities of 500,000–1,000,000 per minute at 25°, depending upon the source of the enzyme; catalytic center activities would be half these values, all turnover rates being expressed in terms of *total* substrate. In the reverse direction (dihydroxyacetone phosphate → glyceraldehyde 3-phosphate) the molecular activity for the mammalian and the yeast enzymes is approximately 50,000 per minute (*235*).

K_m values for D-glyceraldehyde 3-phosphate ranging from 0.32 to 0.5 m*M* for various mammalian isomerases (*233, 235, 239, 268, 269*) and 1.27 m*M* for the enzyme from yeast (*235*) have been reported. For dihydroxyacetone phosphate the corresponding values are 0.60–0.87 and 1.23 m*M* (*235*), after correction for inhibition by arsenate which is present in the coupled assay system of the reverse direction. The equilibrium of the triosephosphate isomerase reaction is on the side of the ketone with values for K_{eq} ([dihydroxyacetone phosphate]/[glyceraldehyde 3-phosphate]) ranging from 19 to 29 (*235, 248, 268, 270–273*), either measured directly or calculated on the basis of the Haldane relationship. The most consistent value for K_{eq} is 22 (*268, 270, 271, 273*); it appears to be independent of both temperature (25°–38°) (*270, 271, 273, 274*) and pH (6.4–7.1) (*273*). Again, all these values are expressed in terms of *total* substrate. The enzyme displays optimal activity at pH 7–8 (*235, 237*) whereby the plateau region for the yeast enzyme is somewhat narrower compared with the rabbit enzyme species (*235*).

The following compounds (values for K_i in m*M*) have been found to act as competitive inhibitors with the respective enzyme source indicated in parentheses: arsenite, 5.5 (*268*), 6.2 (*235*) (rabbit muscle), 6.3 (rabbit liver), 12.0 (yeast) (*235*); phosphate (rabbit muscle), 6 (*268*); phosphoenolpyruvate, 3.0 (rabbit muscle), 2.5 (rabbit liver), 9.0 (yeast) (*235*); and D-α-glycerophosphate, 0.12 (rabbit muscle) (*268*). This last value is taken as one-half that found with DL-α-glycerophosphate, assuming that only the D isomer is inhibitory (*268*). In addition, a comparative inhibition study of chicken and rabbit muscle triosephosphate isomerase (*275*) yielded the following data (K_i in m*M*): phosphoglycolate, 0.0068,

268. P. M. Burton and S. G. Waley, *BBA* **151,** 714 (1968).
269. A. S. Schneider, W. N. Valentine, M. Hattori, and H. L. Heins, *New Engl. J. Med.* **272,** 229 (1965).
270. O. Meyerhof and R. Junowicz-Kocholaty, *JBC* **149,** 71 (1943).
271. P. Oesper and O. Meyerhof, *AB* **27,** 223 (1950).
272. O. H. Lowry and J. V. Passonneau, *JBC* **239,** 31 (1964).
273. R. L. Veech, L. Raijman, K. Dalziel, and H. A. Krebs, *BJ* **115,** 837 (1969).
274. S. J. Reynolds, D. W. Yates, and C. I. Pogson, *BJ* **122,** 285 (1971).
275. L. N. Johnson and R. Wolfenden, *JMB* **47,** 93 (1970).

0.0060; DL-α-glycerophosphate, 0.15, 0.23; allylphosphate, 1.0, 1.3; 2-hydroxyethylphosphate, 1.4, 2.0; D-ribose 1-phosphate, 3.1, 4.8; inorganic phosphate, 5.5, 4.8; inorganic sulfate, 5.0, 5.7; succinate, 1.0, 1.4; maleate, 3.5, 1.7; malonate, 1.5, 0.18; oxalate, 6.2, 9.0; glycolate, 10, 5.4; acetate 10, 7.7 (chicken and rabbit muscle enzyme, respectively).

2. *Phosphoglycolate as a Potential Transition State Analog*

On the basis of mechanistic considerations, Wolfenden postulated that the substrate portion of an enzyme–substrate complex in the transition state (and therefore also a true transition state analog) should be bound to the enzyme considerably tighter than the substrate itself, in fact, that much more tightly than the rate of the enzymically catalyzed reaction exceeds that of its nonenzymic counterpart (*276*). In contrast to the real substrate in the transition state, the analog is not transient but remains bound to the enzyme while simulating the active Michaelis complex, thus permitting study under conditions resembling those of catalysis. 2-Phosphoglycolate was considered (*275, 276*) to fulfill these requirements as a transition state analog for the enediol anion intermediate (*3*) of triosephosphate isomerase when its inhibition constant (6 μM) was found to be about two orders of magnitude smaller than the K_m for the substrate (*vide supra*). Wolfenden's attractive hypothesis was supported by a number of experimental observations. 2-Phosphoglycolate produced an ultraviolet difference spectrum of the enzyme analogous to that with DL-α-glycerophosphate. When studied by X-ray crystallography, orthorhombic crystals of the chicken muscle isomerase soaked in phosphoglycolate were found to contract more than those treated with glycerophosphate (*275*). The postulated transition state analog protected the enzyme against inactivation by both glycidolphosphate and bromohydroxyacetone phosphate (*275*) (*vide supra*) and also against denaturation by increased temperature (*277*). Finally, the binding affinity of phosphoglycolate was found to be maximal below pH 7.35 and to change in parallel with the changing rate enhancement produced by the enzyme as catalyst, the latter expressed in terms of V_{max} for isomerization (*277*). On the other hand, recent data by Pogson and collaborators (*274, 278*), which put the K_m for the true substrate species at the same level as the K_i for 2-phosphoglycolate, would seem to call the nature of the analog in question. Further studies will have to decide whether 2-phosphoglycolate can, in fact, be viewed as a genuine transition state analog capable of "trapping the enzyme in a

276. R. Wolfenden, *Nature* **223,** 704 (1969).
277. R. Wolfenden, *Biochemistry* **9,** 3404 (1970).
278. D. R. Trentham, C. H. McMurray, and C. I. Pogson, *BJ* **114,** 19 (1969).

mockery of the normal catalytic event" (*275*) or whether it should merely be considered to be an ordinary competitive inhibitor.

3. *Enzymically Active States of the Substrates*

Detailed studies of enzyme-catalyzed reactions and of the structure of substrates present in aqueous solution make it apparent that in many cases an enzyme will not bind and act upon all structural forms in which the substrates occur. It will, instead, be specific for one of the species which may constitute only a small fraction of the total (compare Section II,C,3). In these instances the kinetic parameters of the true substrates may be significantly different from those determined in terms of total substrate concentrations; this may seriously affect the interpretation of feasible mechanisms of catalysis. Pogson and collaborators have recently shown that both glyceraldehyde 3-phosphate and dihydroxyacetone phosphate exist in aqueous solution to a significant extent (96.7 and 44%, respectively) in the form of hydrated species, i.e., as geminal diols (*274, 278*) and that the rates of their interconversion are relatively slow. They also found the enzyme to bind the aldehyde and ketone species, respectively, but to be incapable of binding the *gem*-diols (*274, 278*). The various substrate species of the triosephosphate isomerase equilibrium are shown in Fig. 7. The value of 55% keto-dihydroxyacetone is in excellent agreement with that determined by Gray and Barker (also 55%) on the basis of infrared and NMR measurements (*113*).

These results have two important consequences (*274*). First, the equilibrium constant in terms of enzymically active substrates is substantially

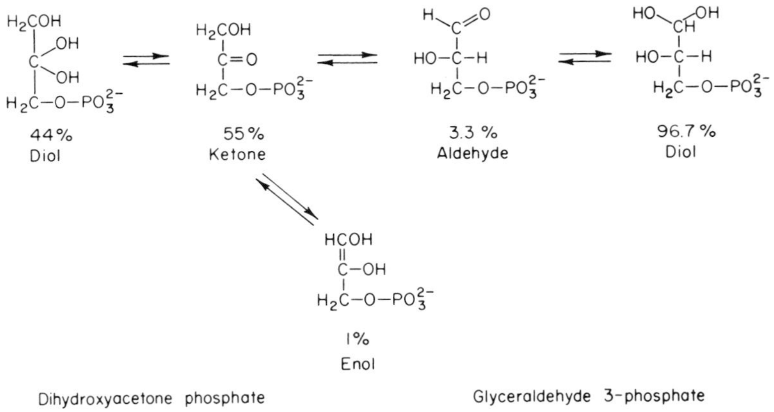

FIG. 7. Equilibrium of the triosephosphate isomerase system at 25° [from Reynolds *et al.* (*274*)].

TABLE XII
EQUILIBRIUM CONSTANT AND MICHAELIS CONSTANTS OF TRIOSEPHOSPHATE ISOMERASE RECALCULATED FOR THE ENZYMICALLY ACTIVE FORMS OF THE SUBSTRATES[a]

Parameter	Original value	Recalculated value
K_{eq}	22	367 (25°)
$\left(\frac{\text{[dihydroxyacetone phosphate]}}{\text{[glyceraldehyde 3-phosphate]}}\right)$		420 (37°)
K_m (glyceraldehyde 3-phosphate)		
Rabbit muscle enzyme	3.2×10^{-4} *M* (total)	11×10^{-6} *M* (aldehyde)
Yeast enzyme	12.7×10^{-4} *M* (total)	42×10^{-6} *M* (aldehyde)
K_m (dihydroxyacetone phosphate)		
Rabbit muscle enzyme	6.2×10^{-4} *M* (total)	3.4×10^{-4} *M* (ketone)
Yeast enzyme	12.3×10^{-4} *M* (total)	6.8×10^{-4} *M* (ketone)

[a] Condensed from a table of Reynolds *et al.* (*274*).

larger because of the disparity in the relative hydration of the two substrates. Second, the K_m values for the active forms of the substrates are considerably smaller. Table XII gives a listing of the kinetic constants recalculated by Reynolds *et al.* (*274*). It is notable that the K_m recalculated for aldehydo-glyceraldehyde 3-phosphate (11 μM) is of the same order as the K_i for 2-phosphoglycolate (6–7 μM) which Wolfenden considers to be a transition state analog (*275, 277*).

4. *Catalytic Function in Vivo*

The function of triosephosphate isomerase in mammalian organisms is believed to be twofold. In glycolysis it controls the symmetrical conversion of the two halves of glucose to lactate, in gluconeogenesis the assembly of the glucose units of liver glycogen from pyruvate (*38, 279*). Its role in gluconeogenesis has recently also been demonstrated for *Escherichia coli* when a mutant was detected which was unable to grow on either glucose or lactate as the sole carbon source and which had a full complement of enzymes except for triosephosphate isomerase (*280*). From a quantitative standpoint it is remarkable that in spite of its enormous turnover rate the enzyme does not maintain equilibrium between the two triosephosphates in rat liver (*273, 281*). This disequi-

279. H. Krebs, *Proc. Roy. Soc.* **B159**, 545 (1964).
280. A. Anderson and R. A. Cooper, *FEBS Letters* **4**, 19 (1969).
281. I. A. Rose, R. Kellermeyer, R. Stjernholm, and H. G. Wood, *JBC* **237**, 3325 (1962).

librium has been explained as resulting from low activity of the enzyme, relative to the metabolic flux, at the prevailing low physiological concentration (about 3 μM) of unbound glyceraldehyde 3-phosphate (*273*).

At one time, when crystalline liver alcohol dehydrogenase was found to isomerize triose and triosephosphate at a rate at least eight times that of ethanol oxidation it was thought to be also an isomerase (*282, 283*). It was found later (*284*), however, that the isomerase activity was merely a contamination (*285*) of the crystalline dehydrogenase which could be removed by chromatography on carboxymethyl cellulose.

5. *Mechanism of Catalysis*

Studies on the mechanism of triosephosphate isomerase began in the mid-1950's (*286, 277*), long before anything was known about the molecular properties of this enzyme. Experiments with deuterium- and tritium-labeled substrates provided evidence (a) for the stereospecificity of the reaction with respect to proton abstraction and (b) for the mechanism to involve an enzyme (conjugate acid)—substrate (enediolate) complex as intermediate (*1, 3, 286, 287*). The mechanism is presumed to be similar to that of glucose-6-phosphate isomerase (*123*) in that a basic residue of the enzyme is thought to attack the hydrogen alpha to the substrate carbonyl group and to effect its transfer to the alternate carbon via an enediol anion intermediate (*1, 3*). Rose has presented persuasive arguments that the enediol intermediate of all aldose–ketose isomerases must be in the *cis* configuration (*3, 175, 176*). Various aspects of the isomerase mechanism have been discussed by Topper in the second edition of "The Enzymes" (*1*) and by Rose in Chapter 5, Volume II of this edition (*5*) and in other places (*3, 4*) to which the reader is referred (cf. also Section II,C,5).

Most of the recent attention has been focused on two aspects of the

282. J. Van Eys and K. Proctor, *JBC* **236,** 1531 (1961).

283. R. Snyder, *Federation Proc.* **23,** 385 (1964).

284. R. Snyder and E. W. Lee, *ABB* **117,** 587 (1966).

285. This case is reminiscent of a similar example involving glycogen phosphorylase which was also claimed to possess a second unrelated activity, i.e., that of a transaminase [A. Waksman and E. Roberts, *BBRC* **12,** 263 (1963)]. In a detailed and thorough investigation J. L. Hedrick and E. H. Fischer [*Biochemistry* **4,** 1337 (1965)] provided proof that the transaminase activity was simply a contamination in the commercial phosphorylase preparation used by Waksman and Roberts, who at about the same time also retracted their original conclusion [*ABB* **109,** 522 (1965)].

286. B. Bloom and Y. J. Topper, *Science* **124,** 982 (1956).

287. S. V. Rieder and I. A. Rose, *Federation Proc.* **15,** 337 (1956); *JBC* **234,** 1007 (1959).

mechanism: the nature of the base performing the proton abstraction and attempts to prove the existence of the postulated intermediate. Efforts concerning the latter are directed toward clarification of the controversial point of whether any intramolecular proton transfer occurs in addition to exchange (*126*) or whether the isomerization proceeds entirely via proton exchange through the medium (*3*). If there is only exchange, as Rose believes (*3, 4*), differences with respect to the reaction pathway and perhaps even the nature of the base may conceivably exist as compared with glucose-6-phosphate isomerase for which both transfer and exchange have been found (*85*).

Although at the time of this writing there is no agreement on the extent of proton transfer—if any—relative to exchange, a value of 10% (*126*) taken as the upper limit (*5*) permits the following calculation: If only 10% intramolecular transfer occurs and 90% is exchange with the medium, and if the 10% transfer corresponds to the measured rate of product formation, the exchange must occur ten times as fast as the transfer (*288*). In other words, the base performing the proton abstraction must be capable of a proton dissociation rate exceeding the catalytic substrate turnover by this factor. The catalytic center activity of triosephosphate isomerase may be taken as no less than 300,000 min^{-1} at 25° (Section VII,C,1) or approximately 5000 sec^{-1}. Therefore, it follows that the base involved in proton abstraction must permit dissociation of protons at a rate of 5×10^4 sec^{-1} which at neutrality exceeds the diffusion-controlled limiting rate of proton dissociation from *any* group with a pK_a of about 7 [generally accepted to be approximately 10^3 sec^{-1} (*131, 166*)] by more than an order of magnitude. One is therefore left with two alternatives: (1) The kinetically operable pK near neutrality (*3, 277*) does not reflect a critical dissociation of the residue involved in proton abstraction. (2) The reaction may not precisely follow Rose's scheme (*3, 5*) but may instead proceed as a competition for the base proton between the solvent and the enolate ion bound to the enzyme active site; in that case any transfer to the solvent would appear as exchange, and transfer to the enolate would appear as net reaction.

The latter alternative has been proposed by Jencks (*166*), who con-

288. New data concerning the transfer versus exchange ratio have recently become available from J. R. Knowles and collaborators (*288a*) who found less than 2% tritium transfer from C-3 of dihydroxyacetone phosphate to C-2 of glyceraldehyde phosphate, thus confirming the earlier findings by Rose (*3*). This implies that in the calculation made in the text, the factor would have to be fifty instead of ten yielding a rate of proton dissociation of more than 2×10^5 sec^{-1} for the forward reaction (glyceraldehyde 3-phosphate → dihydroxyacetone phosphate) if an upper limit of 1% transfer is also assumed for this direction of isomerization.

sidered Eigen's value of 10^5 M^{-1} sec^{-1} for the rate of proton transfer from an acid with pK 7 to the enolate of acetylacetone (*131*) also to be a reasonable estimate for the rate of proton transfer to an enolate ion at the active site of an enzyme. Perhaps other possibilities such as the involvement of more than one residue at the active site may have to be considered, but at this time any proposals of this kind will be no more than speculation.

Very recent experiments from J. R. Knowles' laboratory (*288a*) have provided additional evidence regarding (a) the postulated enzyme-bound enediolate intermediate and (b) the identity of the rate-limiting step in either direction of catalysis. The following parameters were compared for the forward and the reverse reaction: primary kinetic deuterium isotope effects, isotopic discrimination in the formation of product, and solvent isotope exchange into substrate after partial reaction. The results from these experiments indicated that proton abstraction from dihydroxyacetone phosphate is only slightly faster than the rate of loss of glyceraldehyde phosphate from the enzyme, that an enzyme-enediol intermediate is obligatory, and that this intermediate partitions unequally in favor of the glyceraldehyde phosphate-enzyme complex. It was furthermore concluded that the overall reaction with glyceraldehyde phosphate as substrate is diffusion-controlled whereas the reaction from dihydroxyacetate phosphate is limited by the rate of loss of product from the enzyme.

VIII. Aldose–Ketose Isomerases Acting on Nonphosphorylated Sugars

In contrast to the isomerases acting on phosphorylated substrates, which have a narrow specificity for the aldose–ketose pair they interconvert, those acting on the free sugars are usually characterized by their ability to isomerize several substrates. This broader specificity has led to several cases in which the same enzyme has been described under more than one name. As a consequence, arguments have sometimes developed as to what should be considered as the "true" substrate of such an enzyme. For clarity in this review, all names are therefore given below in the headings when an enzyme has been designated by more than one name. Another characteristic common to the isomerases acting on nonphosphorylated substrates is that with very few exceptions they have only been found to occur in microorganisms.

288a. J. R. Knowles, P. F. Leadlay, and S. G. Maister, *Cold Spring Harbor Symp. Quant. Biol.* **36** (1972), in press.

D-Erythrose isomerase may be mentioned as somewhat of a curiosity. The preparation of a rabbit muscle extract with D-erythrose isomerase activity has been described in "Methods in Enzymology," but further purification of the enzyme was not possible because of its "extreme instability and its insolubility" (*289*). The enzyme has been reported to isomerize D-glucose as well as D-arabinose, although at less than 50% of the rate observed with D-erythrose (*289*). Except for the cited handbook article (*289*), no communication concerning this enzyme has appeared in the literature (*290*). Thus, evidence for the identity of a D-erythrose isomerase would seem to be tenuous at best, and its formal classification and the assignment of an IUB identification number (EC 5.3.1.2) must be considered as having been premature.

A highly unusual type of isomerase, which defies any classification in the conventional system, is the NAD-linked D-glucose isomerase found in *Bacillus megaterium* A-1 (*291*) and the D-glucose- and D-mannose-isomerizing enzyme of *Paracolobacterum aerogenoides* requiring both NAD and Mg^{2+} as cofactors (*292*). The latter is even more puzzling than the former since none of the other isomerases acting on glucose will also act on mannose and vice versa. Although the enzyme could not be purified because of its lability, identical response of the activity with both substrates to various inhibition and inactivation conditions was taken to suggest that only a single enzyme is involved (*292*). However, the authors have not explained how such an enzyme, when offered D-fructose—supposedly the product of the forward reaction with either D-glucose or D-mannose—as substrate, will decide whether to produce D-glucose or D-mannose in the reverse reaction. Nor has a proposal been made as to how NAD may participate as cofactor in the isomerization reaction. Since NAD-linked isomerases have so far only been reported from one laboratory, speculations concerning the role of NAD should perhaps be postponed until additional evidence is available.

A. D-Glucose Isomerase Activity

Because of its potential for industrial application, researchers in Japan have expended substantial effort to find a glucose isomerase capable of converting glucose (which is commercially produced by hy-

289. K. Uehara, "Methods in Enzymology," Vol. 5, p. 350, 1962.

290. This statement refers to the literature covered by *Chem. Abstr.* since 1958.

291. Y. Takasaki and O. Tanabe, *J. Agr. Chem. Soc. Japan* **36,** 1010 and 1013 (1962); **37,** 93 (1963).

292. Y. Takasaki and O. Tanabe, *Agr. Biol. Chem.* (*Tokyo*) **28,** 740 (1964); **30,** 220 (1966).

drolysis of sweet potato starch) to a mixture of glucose and fructose that yields a more potent sugar substitute (*293*). Numerous microorganisms were screened for D-glucose isomerizing activity (*294, 295*) but a genuine, specific D-glucose isomerase apparently does not exist. Instead, most of the enzymes in this category turned out to be either D-xylose isomerase or glucose-6-phosphate isomerase. The former will be described separately in Section VIII,F, but mention of the glucose isomerase activity of glucose-6-phosphate isomerase seems to be more germane to the discussion here than to the other aspects of this enzyme dealt with in Section II above.

Marshall and Kooi were the first to report in 1957 on an enzymically catalyzed interconversion between D-glucose and D-fructose occurring in D-xylose-grown cells of *Pseudomonas hydrophila* and requiring the presence of arsenate (*296*). Yamanaka observed the same activity with a D-xylose isomerase preparation made from D-xylose grown cells of *Lactobacillus brevis;* but, in this case, addition of arsenate was not required for the reaction (*297*). He concluded that in this microorganism D-xylose isomerase and D-glucose isomerase are one and the same enzyme. After extensive screening, Natake and Yoshimura then isolated strain HN-500 of *Escherichia intermedia* grown on D-glucose, which showed high D-glucose isomerase activity in the presence of arsenate but no activity toward D-xylose (*298*). After study of the optimal conditions for activity measurement (*299*), the enzyme was purified to apparent homogeneity (*300*) and finally established to be identical with glucose-6-phosphate isomerase (*301*). Natake concluded that *in vivo* glucose 6-phosphate—which is isomerized in the absence of arsenate—is the natural substrate, and that the activity toward glucose is an artifact presumably caused by nonenzymic formation of a glucose arsenate compound which poses as a substrate analog. This interpretation is further supported by his finding 6-phosphogluconate to act as a competitive inhibitor with respect to both phosphorylated and free glucose (the latter in the presence of arsenate) (*301*). Some of the properties of the enzyme isolated from

293. Y. Takasaki, Y. Kosugi, and A. Kanbayashi, *in* "Fermentation Advances" (D. Perlman, ed.), p. 561. Academic Press, New York, 1969.

294. A compilation of references pertaining to D-glucose isomerase activities of various microorganisms has been given by Yamanaka (*295*).

295. K. Yamanaka, *BBA* **151,** 670 (1968).

296. R. O. Marshall and E. R. Kooi, *Science* **125,** 648 (1957).

297. K. Yamanaka, *Agr. Biol. Chem.* (*Tokyo*) **27,** 271 (1963).

298. M. Natake and S. Yoshimura, *Agr. Biol. Chem.* (*Tokyo*) **28,** 505 (1964).

299. M. Natake and S. Yoshimura, *Agr. Biol. Chem.* (*Tokyo*) **28,** 510 (1964).

300. M. Natake, *Agr. Biol. Chem.* (*Tokyo*) **30,** 887 (1966).

301. M. Natake, *Agr. Biol. Chem.* (*Tokyo*) **32,** 303 (1968).

TABLE XIII
SOME CATALYTIC PROPERTIES OF THE "GLUCOSE ISOMERIZING ENZYME" FROM *Escherichia intermedia*, STRAIN NH-500[a]

Property	D-Glucose isomerase activity (in the presence of arsenate)	D-Glucose-6-phosphate isomerase activity
Specific activity (units[b]/mg)	15[c] (40°)	940 (40°)
pH optimum	7.0	7.0
K_m	1.6 M (D-glucose)	1.4 mM (D-glucose 6-phosphate)
K_{eq}	$\frac{[\text{D-fructose}]}{[\text{D-glucose}]} = 0.82$ (45°)	$\frac{[\text{D-fructose-6-P}]}{[\text{D-glucose-6-P}]} = 0.67$ (35°)
ΔH (cal mole^{-1})	15,300	8,300
K_d of glucose arsenate compound	0.9 M	—

[a] Data from Natake (*301*).
[b] Micromoles of substrate converted to product per minute at the specified temperature.
[c] At 1.6 M D-glucose and 0.8 M arsenate (*301*).

E. intermedia, strain NH-500, are summarized in Table XIII. Arsenate-inducible D-glucose isomerase activity has also been found in crystalline glucose-6-phosphate isomerase isolated from peas (*13*).

Investigating the effect of arsenate on a number of glycolytic enzymes of diverse origin (*302*), Lagunas and Sols observed in several instances that it caused these enzymes to increase their activity toward non-phosphorylated analogs of their normal phosphorylated substrates from 10- to 1000-fold over that in the absence of arsenate (*302*). The authors suggested, similarly to Natake (*301*), that the "inducing effect of arsenate may involve an interaction of the arsenate with the sugar giving an ester-like intermediate which . . . can bind specifically to the enzyme in a form suitable for attack." They also quoted precedents for the transient formation of labile arsenate esters of sugars demonstrated in phosphorolytic reactions (*303*). Lagunas explored in more detail the effect of arsenate on yeast glucose-6-phosphate isomerase, and he concluded that the enzyme plays an important role in the formation of the glucose-arsenate compound (*304*). He proposed that initial binding of the arsenate may induce a conformational change which produces the precise spatial orientation for the interacting groups to yield an esterlike

302. R. Lagunas and A. Sols, *FEBS Letters* **1,** 32 (1968).
303. M. Cohn, "The Enzymes," 2nd ed., Vol. 5, p. 179, 1961.
304. R. Lagunas, *BBA* **220,** 108 (1970).

compound between the arsenate and the C-6 hydroxyl group of the sugar (*304*).

B. D-MANNOSE ISOMERASE; D-LYXOSE ISOMERASE

D-Mannose ⇄ D-Fructose
D-Lyxose ⇄ D-Xylulose

An enzyme catalyzing the isomerization of D-mannose to D-fructose was first described by Palleroni and Doudoroff (*305*) as being present in all mutant strains of *Pseudomonas saccharophila* capable of growing with sucrose or fructose as carbon source. This enzyme has been reviewed in the previous edition of this treatise (*1*). A similar isomerase was found widely distributed in *Xanthomonas* species where it appears to be produced constitutively, independent of the carbon source in the growth medium (*306*). It has been purified 35-fold from *Xanthomonas rubrilineans* S-48. A K_m of 12 m*M* was determined for mannose; the enzyme was found to be inhibited by D-arabinose (K_i = 11 m*M*) and L-fucose (K_i = 0.71 m*M*) (*307*). Some kinetic and thermodynamic parameters have also been reported for a preparation from *Streptomyces aerocolorigenes:* pH optimum = 7.7; $K_m^{mannose}$ = 1.4 m*M*, $K_m^{fructose}$ = 2.1 m*M*; K_{eq} ([fructose]/[mannose]) = 3.0; $\Delta H = 0$; $\Delta G = -650$ cal mole^{-1}; and $\Delta S = 2.2$ cal mole^{-1} deg^{-1} (*95, 308*). Another D-mannose isomerase (EC 5.3.1.7) has been partially (60-fold) purified from *Mycobacterium smegmatis* grown on D-mannose as the sole carbon source. It was shown to act on both D-mannose (K_m = 7 m*M*) and D-lyxose, and to have a pH optimum of 7.5 and a K_{eq} ([fructose]/[mannose]) of approximately 1.9 (*309*).

Anderson and Allison obtained a 130-fold purified D-lyxose isomerase (specific activity = 7.7 units/mg at 25°) from *Aerobacter aerogenes* PRL-R3 grown on D-lyxose (*310*). The enzyme could not be induced by growth on either fructose, mannose, or glucose, pentitols, or pentoses other than D-lyxose (*310, 311*). This is in contrast to the

305. N. J. Palleroni and M. Doudoroff, *JBC* **218**, 535 (1956).
306. Y. Takasaki and O. Tanabe, *Agr. Biol. Chem.* (*Tokyo*) **28**, 601 (1964).
307. Y. Takasaki, S. Takano, and O. Tanabe, *Agr. Biol. Chem.* (*Tokyo*) **28**, 605 (1964).
308. Y. Takasaki and O. Tanabe, *Rept. Ferment. Res. Inst.* (*Chiba City, Japan*) **28**, 89 (1965).
309. A. Hey-Ferguson and A. D. Elbein, *J. Bacteriol.* **101**, 777 (1970).
310. R. L. Anderson and D. P. Allison, *JBC* **240**, 2367 (1965); R. L. Anderson, "Methods in Enzymology," Vol. 9, p. 593, 1966.
311. R. P. Mortlock and W. A. Wood, *J. Bacteriol.* **88**, 838 (1964).

isomerase from *Pseudomonas saccharophila* grown on fructose which had been shown to act on D-lyxose at 11% of the rate on D-mannose (*305*). Since the *Aerobacter* enzyme converted D-mannose at less than half the rate of D-lyxose and the K_m for D-lyxose (3.6 mM) was smaller than that for D-mannose (10 mM), Anderson and Allison concluded that the unnatural pentose D-lyxose is the natural substrate for this enzyme and named it accordingly (*310*). The enzyme was found to lose its activity on treatment with 10 mM EDTA but could be completely reactivated (after dialysis against water) by 0.1 mM Mn^{2+}; in the isomerase reaction the K_m for Mn^{2+} was determined to be 4 μM. The pH optimum was found to be 7.0 and K_{eq}([D-xylulose]/[D-lyxose]) equal to 0.23 at pH 7.0 and 25°. Gradient sedimentation in NaCl and gel filtration on Sephadex G-100 yielded rough estimates for the sedimentation coefficient (3.7–3.8 S) and for the molecular weight (40,000 daltons), respectively (*310*).

C. L-RHAMNOSE ISOMERASE; L-MANNOSE ISOMERASE

L-Rhamnose ⇄ L-Rhamnulose
L-Mannose ⇄ L-Fructose

Discovery and initial studies of this enzyme in *Escherichia coli* (*312, 313*) and *Pasteurella pestis* (*314*) have been discussed previously by Topper (*1*). Since then the enzyme has been purified as L-rhamnose isomerase (EC 5.3.1.14) from *Lactobacillus plantarum* (*315*) and *Escherichia coli* (*316*) and in *Aerobacter aerogenes* shown to be probably identical with L-mannose isomerase (*317, 318*).

Properties of the enzyme purified 250-fold from *Lactobacillus plantarum* (*315*): specific activity = 31.8 units/mg at 37°; pH optimum = 8.5; K_m (L-rhamnose) = 0.7 mM; K_m (L-mannose) = 5 mM; K_{eq} ([L-rhamnulose]/[L-rhamnose]) = 0.75. In spite of the reasonably small K_m for L-mannose, its isomerization rate was found to be only 0.6% of that with L-rhamnose (*315*). Properties of the enzyme purified 50-fold from *Escherichia coli* strain B (*316*): specific activity = 3.0 units/mg at 37°;

312. D. M. Wilson and S. Ajl, *BBA* **17,** 289 (1955); *J. Bacteriol.* **73,** 410 (1957).
313. G. Tecce and M. Di Girolamo, *Giorn. Microbiol.* **1,** 286 (1955–1956); *Chem. Abstr.* **51,** 18116h (1957).
314. E. Englesberg, *Federation Proc.* **15,** 586 (1956); *ABB* **71,** 179 (1957).
315. G. F. Domagk and R. Zech, *Biochem. Z.* **339,** 145 (1963); "Methods in Enzymology," Vol. 9, p. 579, 1966.
316. Y. Takagi and H. Sawada, *BBA* **92,** 10 (1964).
317. J. W. Mayo and R. L. Anderson, *JBC* **243,** 6330 (1968).
318. J. W. Mayo and R. L. Anderson, *J. Bacteriol.* **100,** 948 (1969).

pH optimum = 7.6; K_m (L-rhamnose) = 2.0 mM; K_m (L-rhamnulose) = 1.7 mM; K_{eq} ([L-rhamnulose]/[L-rhamnose]) = 1.5. The enzyme induced in *Aerobacter aerogenes* PRL-R3 grown on either L-mannose or L-rhamnose (*318*) was found to isomerize L-rhamnose twice as fast as L-mannose; the equilibrium has only been measured for the L-mannose isomerase reaction, K_{eq} ([L-fructose]/[L-mannose]) = 1.63 (*317*).

Reports on the enzymes from all three microorganisms mention the involvement of divalent metals, but qualitative as well as quantitative differences appear to exist. Tecce and Di Girolamo found the *E. coli* enzyme to be activated by Co^{2+} and Mn^{2+} and also by cysteine in the absence of any metal, whereas with metal present cysteine had no effect. They concluded that the isomerase is a metalloenzyme with cobalt linked to a –SH group (*319*). Takagi and Sawada, however, found the purified enzyme from the same organism to be inhibited by Co^{2+} and activated only by Mn^{2+} and suggested that it may be a metalloenzyme with Mn^{2+} linked to a –SH group (*316*). Domagk and Zech noticed no effect by Co^{2+} on the enzyme purified from *Lactobacillus plantarum* but found maximal activity at 0.1 mM Mn^{2+} and also activation by sulfhydryl compounds (*315*). In *Aerobacter aerogenes*, finally, Mayo and Anderson observed activation by Co^{2+} and strong inhibition by Mn^{2+} (*317*).

D. D-Arabinose Isomerase; L-Fucose Isomerase

D-Arabinose ⇄ D-Ribulose
L-Fucose ⇄ L-Fuculose

The early history of this enzyme has been reviewed by Topper (*1*). D-Arabinose isomerase (EC 5.3.1.3) is an interesting example of the kind of nomenclature change, with the accompanying confusion, that seems to have plagued several enzymes in this category. D-Arabinose isomerase was first demonstrated in *Escherichia coli* grown on D-arabinose and shown to act also on L-fucose (*320*). Mortlock later described the purification of a D-arabinose isomerase from *Aerobacter aerogenes*, which also isomerized L-xylose and L-fucose to their corresponding ketopentoses (*321*). The enzyme had a pH optimum at 7.0 and required Mn^{2+} for maximal activity; the K_m for D-arabinose was found to be 0.22 M. In another

319. G. Tecce and M. Di Girolamo, *Boll. Soc. Ital. Biol. Sper.* **32,** 1195 (1956); *Chem. Abstr.* **51,** 9793f (1957).
320. M. Green and S. S. Cohen, *JBC* **219,** 557 (1956).
321. R. P. Mortlock, "Methods in Enzymology," Vol. 9, p. 583, 1966.

paper (*322*), Camyre and Mortlock reported on a L-fucose isomerase from *Aerobacter aerogenes* that also act on D-arabinose and L-xylose, with little emphasis on the fact this is the *identical* enzyme thus far described as D-arabinose isomerase (*321*). The name change was introduced because L-fucose was isomerized more rapidly than D-arabinose and also because it can be considered a natural, albeit rare, sugar (e.g., reference *323*, p. 600); the latter is not true for either D-arabinose or L-xylose.

More recently, dithiothreitol was found to be a competitive inhibitor of the isomerase (*324*) with K_i values of 4.6 μM with respect to D-arabinose and 2.4 μM with respect to L-fucose. The difference may not be significant since for a genuine competitive inhibitor of a single enzyme acting on two substrates, K_i must be the same when tested with either substrate (e.g., reference *325*, p. 203). This inhibition may be explained by the stereochemical configuration of dithiothreitol on C-2 and C-3 which is the same as in all of the substrates. Topper had previously pointed out (*1*) that all of the substrates for D-arabinose isomerase have the same configuration at C-2 to C-4 when commenting on the early observation of Green and Cohen (*320*) that the *E. coli* enzyme was also acting on L-galactose and D-altrose. The enzyme was also inhibited by dithioerythritol (*324*), but this inhibition, if assumed to occur through binding as a substrate analog, is more difficult to understand since in dithioerythritol the C-2 hydroxyl is *cis* to that on C-3, in contrast to the other sugars mentioned for which it is always *trans*. It may be worthwhile to mention an aspect of inhibition by dithiothreitol that is often overlooked, *viz*., its ability to bind metals. This function has recently been shown to be its mode of inhibition in the case of the zinc metalloenzyme mannose-6-phosphate isomerase (*157*). This possibility should deserve further investigation for three reasons:

(1) Partially purified D-arabinose isomerase from *E. coli* 30_{DA} was found to be inhibited by all chelating agents tested and to be activated by divalent metal ions of which Co^{2+} was the most efficient (*326*). Di Girolamo *et al.* therefore suggested that D-arabinose isomerase may be a metalloenzyme like L-rhamnose isomerase and D-xylose isomerase (*326*).

(2) The enzyme purified from *Aerobacter aerogenes* requires Mn^{2+} for activity (*321*).

322. K. P. Camyre and R. P. Mortlock, *J. Bacteriol.* **90,** 1157 (1965).

323. R. G. Spiro, *Ann. Rev. Biochem.* **39,** 599 (1970).

324. E. J. Oliver and R. P. Mortlock, *BBRC* **36,** 24 (1969).

325. M. Dixon and E. C. Webb, "Enzymes," 2nd ed. Academic Press, New York, 1964.

326. M. Di Girolamo, L. Frontali, and G. Tecce, *Quaderni Nutr.* **18,** 131 (1959); *Chem. Abstr.* **57,** 14164h (1962).

(3) Although a *strong* chelating agent would not be expected to produce a competitive inhibition pattern, K_i values of 2–4 μM could merely be an expression of competition for Mn^{2+} between the thiol as a chelator and an enzyme-pentose compound requiring the metal to form the ternary metal–substrate bridge complex (*174*).

E. L-Arabinose Isomerase

L-Arabinose ⇄ L-Ribulose

L-Arabinose isomerase (EC 5.3.1.4) appears to be the exception among the isomerases acting on nonphosphorylated sugars in that it has so far been described under only one name. The early work on this enzyme has been reviewed in 1961 (*1*). Since then, induction of the enzyme in *Lactobacillus plantarum* (*327*), studies on purified preparations from *Aerobacter aerogenes* (*328*), *Escherichia coli* (*329*), *Clostridium acetobutyliticum* (*330*), and *Candida utilis* (*331*), and isolation in crystalline state from *Lactobacillus gayonii* (*332*) have been reported. Some catalytic properties of these L-arabinose isomerase preparations are compared in Table XIV. Schray and Rose have recently demonstrated that L-arabinose isomerase possesses anomeric specificity for the β-form of L-arabinose only (*130a*).

Detailed investigations on the molecular properties have until now only been published by Patrick and Lee for L-arabinose isomerase from *E. coli* (*329*), although data on the enzyme from *Lactobacillus gayonii* may soon be forthcoming since "large quantities of the crystalline enzyme" (*332*) are now available. The *E. coli* isomerase has been shown to be homogeneous by ultracentrifugal analysis and to be better than 98% pure by polyacrylamide gel electrophoresis (*329*). The following physical parameters have been established: $s^0_{20,w} = 12.7$ S, independent of pH from 5.0 to 7.6; no concentration dependent dissociation (between ~0.5 and 4 mg ml^{-1}) within this pH range; the molecular weight of the native enzyme was determined to be 362,000 daltons by high-speed equilibrium sedimentation (*329*). Exposure to pH 2.0 or to 8 M urea resulted in dis-

327. M. Chakravorty, *BBA* **85,** 152 (1964).

328. K. Yamanaka and W. A. Wood, "Methods in Enzymology," Vol. 9, p. 596, 1966.

329. J. W. Patrick and N. Lee, *JBC* **243,** 4312 (1968).

330. M. Tomoeda, H. Horitsu, and I. Sasaki, *Agr. Biol. Chem.* (*Tokyo*) 33, 151 (1969).

331. H. Horitsu, I. Sasaki, and M. Tomoyeda, *Agr. Biol. Chem.* (*Tokyo*) **34,** 676 (1970).

332. T. Nakamatu and K. Yamanaka, *BBA* **178,** 156 (1969).

sociation of the native enzyme into six, apparently identical, subunits of 2.85 S ($s^0_{20,w}$) with a molecular weight of 60,000 daltons by equilibrium sedimentation (*333*). Following cyanogen bromide cleavage of the enzyme, 18–19 peptides were discernible after separation by a combination of polyacrylamide gel electrophoresis, high voltage electrophoresis, and descending paper chromatography; this is in good agreement with 17 peptides to be expected on the basis of a methionine content of 16 residues per subunit of 60,000 molecular weight, determined by amino acid analysis (*329, 333*).

F. D-Xylose Isomerase; D-Glucose Isomerase

D-Xylose ⇄ D-Xylulose
D-Glucose ⇄ D-Fructose

Whereas the initial studies on D-xylose isomerase (EC 5.3.1.5) (*1*) were mainly concerned with the adaptive nature of the enzyme present in extracts prepared from microorganisms grown on D-xylose, most of the effort in the 1960's resulted only incidentally in extensive investigation of this enzyme. As mentioned at the beginning of this section, the search was actually to find a D-glucose isomerase. This is most impressively demonstrated by the fact that all three laboratories which recently reported the isolation of a crystalline D-xylose isomerase (*295, 334, 335*) initially described their work as studies on D-glucose isomerase (*297, 336–338*) until it became clear that in each case the enzyme was actually a D-xylose isomerase (e.g., see reference *339*). In fact, one laboratory still persists in calling the crystalline enzyme from *Streptomyces albus* a D-glucose isomerase (*334*). However, there have also been reports on D-xylose isomerases that do not act on D-glucose, e.g., the enzymes from *Pasteurella pestis*, strain A-1122 (*340*), and from wheat germ (*341*). Interestingly, the latter was found to utilize D-ribose.

333. J. W. Patrick and N. Lee, *JBC* **244,** 4277 (1969).
334. Y. Takasaki, Y. Kosugi, and A. Kanbayashi, *Agr. Biol. Chem. (Tokyo)* **33,** 1527 (1969).
335. G. Danno, *Agr. Biol. Chem. (Tokyo)* **34,** 1795 (1970).
336. K. Yamanaka, *Agr. Biol. Chem. (Tokyo)* **27,** 265 (1963).
337. Y. Takasaki, *Agr. Biol. Chem. (Tokyo)* **30,** 1247 (1966).
338. S. Yoshimura, G. Danno, and M. Natake, *Agr. Biol. Chem. (Tokyo)* **30,** 1015 (1966); G. Danno, S. Yoshimura, and M. Natake, *ibid.* **31,** 284 (1967).
339. K. Yamanaka, "Methods in Enzymology," Vol. 9, p. 588, 1966.
340. M. W. Slein, *JACS* **77,** 1663 (1955); "Methods in Enzymology," Vol. 5, p. 347, 1962.
341. M. H. Pubols, J. C. Zahnley, and B. Axelrod, *Plant Physiol.* **38,** 457 (1963).

TABLE XIV
CATALYTIC PROPERTIES OF L-ARABINOSE ISOMERASES PURIFIED FROM VARIOUS MICROORGANISMS

Property	Enzyme source				
	E. coli strain B/r [Ref. (*329*)]	*A. aerogenes* PRL-R3 [Ref. (*328*)]	*C. acetobutyliticum* [Ref. (*330*)]	*L. gayonii*[a] [Ref. (*332*)]	*C. utilis* [Ref. (*331*)]
Specific activity, units[b]/mg	200 (37°)	150 (30°)	0.24 (37°)[c]	5.3 (35°)[d]	8.1 (37°)[e]
pH optimum	Plateau from pH 6–8	pH 6.4–6.9	pH 7.5	pH 6–7	pH 7.0
Substrates (K_m)					
L-Arabinose	60 mM	33 mM	11 mM	55 mM	0.11 M
L-Ribulose				5 mM	
D-Fucose	+[f]	0.27 M			
D-Galactose		0.37 M		+	
D-Xylose	+[f]				+[g]
Competitive inhibitors (K_i)					
L-Arabitol	18 mM	2.3 mM	23 mM	7.5 mM	6.2 mM
L-Ribitol	6 mM	0.35 mM		6.0 mM	
L-Xylitol				38.0 mM	

Metal activators (K_m)	Mn^{2+}	Mn^{2+}	Co^{2+} (38 μM) Fe^{2+} Mn^{2+} (83 μM)	Mn^{2+} (5 μM)	Co^{2+} Mn^{2+}
K_{eq} ([L-ribulose]/[L-arabinose])			1.94 (30°, pH 7.5)		
Thermodynamic parameters					
ΔH (cal mole^{-1})			7350		8900
ΔG (cal mole^{-1})			−400 (30°)		−1100 (30°)
ΔS (cal mole^{-1} deg^{-1})			26		33

[a] Isolated in crystalline form.

[b] Units of enzymic activity in this chapter are always given in terms of the IUB definition, i.e., micromoles of substrate converted to product per minute at the specified temperature.

[c] Specific activity is given by the authors (*330*) as 1079 arbitrary units/mg of protein, one unit defined as 1 μg of ribulose produced under the conditions of assay, i.e., 30 min at 37°.

[d] The specific activity of the crystalline enzyme appears to be only about 35% of that reported previously by Yamanaka and Wood (*328*) for the partially purified preparation from the same source (14 units/mg at 35°), both values measured by the cysteine-carbazole reaction for keto sugars (*217*).

[e] Specific activity is given by the authors (*331*) as 36666.7 (!) arbitrary units per milligram of protein, with units defined as in footnote *c*.

[f] Under standard assay conditions, except for replacement of the substrate, the *E. coli* enzyme showed activity also toward D-fucose (7%), L-fucose (6%), and D-xylose (6%), compared to L-arabinose (100%), but none with D-galactose.

[g] L-Arabinose isomerase, purified from *Candida utilis* grown in sulfite pulp mill waste, acted on D-xylose at about 5% of the rate with L-arabinose.

Although mixed-substrate studies supported the concept of a single enzyme, the authors felt that higher purification of their preparation was necessary for an unequivocal decision as to whether one or two enzymes are involved (*341*). D-Xylose isomerase from wheat germ is notable in that it appears to be the first and only instance of an aldose–ketose isomerase in higher plants that acts on nonphosphorylated sugars. A D-xylose isomerase was also purified 10-fold from *Candida utilis*, but D-glucose was apparently not tested as a potential substrate (*342*). Recent work by Schray and Rose (*130a*) on the anomeric specificity of the enzyme from *Streptomyces* showed that D-xylose isomerase is specific for the α forms of D-xylose and D-glucose.

In most studies on D-xylose isomerase, enzymic activity has been measured by a discontinuous chemical-stop assay with quantitation of the produced ketopentose by the cysteine-carbazole method (*217*). Yamanaka has recently proposed a more specific, continuous spectrophotometric assay in which the isomerase is coupled to the reverse direction of the D-arabitol dehydrogenase reaction (D-xylose + NADH + $H^+ \rightarrow$ D-arabitol + NAD^+) allowing convenient measurement at 340 nm of the decrease in NADH absorbance (*343*). Obviously, this assay is dependent on the availability of a suitable D-arabitol dehydrogenase preparation. Another very intriguing method has been employed by Takasaki, *viz.*, assaying the D-glucose isomerase reaction by direct measurement at 300 nm of the change in optical rotation produced on converting D-glucose to D-fructose, employing a recording spectropolarimeter sensitive to 0.005° rotational angle (*94*). The same principle has also been used to determine kinetic parameters of D-mannose isomerase (*95*).

A number of catalytic and physical properties of three crystalline D-xylose isomerases are summarized in Table XV (*293, 295, 334, 335, 344, 345*). Yamanaka made a detailed kinetic analysis of the inhibition of the enzyme from *Lactobacillus brevis* by pentitols and D-lyxose as a function of Mn^{2+} concentration. From the inhibition pattern he concluded that the substrate as well as the inhibitors combine with the enzyme through a Mn^{2+} bridge, and he postulated a compulsory order of Mn^{2+} binding preceding the binding of the substrate (*344*). Mildvan and Rose confirmed the enzyme–Mn–xylose bridge complex by Mn^{2+} binding studies with NMR and EPR which yielded a K_d for Mn^{2+} of 8 ± 2 μM (*164b*), in good agreement with Yamanaka's K_m of 6.1 μM (*295*). They suggested that the enzyme-bound Mn^{2+} coordinates an oxygen on C-1

342. M. Tomoyeda and H. Horitsu, *Agr. Biol. Chem.* (*Tokyo*) **28,** 139 (1964).
343. K. Yamanaka, *Agr. Biol. Chem.* (*Tokyo*) **33,** 834 (1969).
344. K. Yamanaka, *ABB* **131,** 502 (1969).
345. G. Danno, *Agr. Biol. Chem.* (*Tokyo*) **34,** 1805 (1970).

TABLE XV
PROPERTIES OF CRYSTALLINE D-XYLOSE ISOMERASES

Property	Enzyme source: *L. brevis* [Refs. (*295, 344*)]	*S. albus* [Refs. (*293, 334*)]	*B. coagulans* strain HN-68 [Refs. (*335, 345*)]
Specific activity (units[a]/mg) (D-xylose as substrate)	6.4 (35°)	4.4 (70°)[b]	6.7 (40°)
pH optimum	pH 6–7 (D-xylose, D-glucose, D-ribose)	pH 8–8.5 (D-glucose)	pH 8–8.5 (D-xylose) pH 7–7.5 (D-glucose, D-ribose)
Substrates (K_m)			
D-Xylose	5 m*M*	32 m*M*	1.1 m*M*
D-Glucose	0.92 *M*	0.16 *M*	90 m*M*
D-Ribose	0.67 *M*		83 m*M*
Competitive inhibitors (K_i)			
D-Xylitol	2.7 m*M*		2.5 m*M*
D-Sorbitol		++	29 m*M*
D-Arabitol	0.13 *M*		
L-Arabitol	0.146 *M*		
D-Mannitol		++	
D-Mannose			70 m*M*
D-Lyxose	70 m*M*	+	
Metal activators (K_m)	Mn^{2+} (6.1 μ*M*)	Co^{2+} (0.18 m*M*) Mg^{2+} (1.8 m*M*)	Co^{2+} Mn^{2+} Mg^{2+}
Thermodynamic parameters (25°)			
K_{eq} ([D-fructose]/[D-glucose])		0.74	
ΔH (cal mole^{-1})		2220	
ΔG (cal mole^{-1})		180	
ΔS (cal mole^{-1} deg^{-1})		6.8	
$s^0_{20,w}$ (S)		8.0	10.2
$D_{20,w}$ (cm^2 sec^{-1})		4.0×10^{-7}	
$\bar{V}$ (ml g^{-1})		0.69	0.705
Molecular weight	230,000[c]	157,000 (from *s* and *D*)	175,000 (from *s* and viscosity) 160,000 (by gel filtration)

[a] Micromoles of substrate converted to product per minute at the specified temperature.

[b] Conditions for the standard glucose isomerase activity assay are given by Takasaki *et al.* (*334*) as 1 hr incubation at 70°C [*in words:* seventy degrees centigrade (!)].

[c] Mentioned by Mildvan and Rose without reference to experimental details (*164b*).

of D-xylose, thus facilitating proton removal at C-2 by a basic group, resulting in the formation of a *cis*-enediol (*164b*). Isotope studies had also shown the enediol intermediate in the D-xylose isomerase reaction to have the *cis* configuration (*176*). A more detailed discussion of the catalytic mechanism has recently been provided by Mildvan (*174*).

Acknowledgments

The author is grateful to his current collaborators for aiding in the preparation of this chapter through continual discussion of the issues involved. He thanks in particular J. M. Chirgwin for redrawing the mechanisms in Figs. 2 and 3 and for bringing several of the historical references to his attention. He finally acknowledges the cooperation of numerous colleagues, especially Dr. G. Danno, Dr. F. C. Hartman, Dr. J. R. Knowles, Dr. A. S. Mildvan, Dr. C. I. Pogson, Dr. I. A. Rose, and Dr. A. C. Rutner, in communicating unpublished data and making manuscripts available prior to publication.

Work in the author's laboratory dealing with studies on the phosphohexose isomerases has been generously supported by grants from the U. S. Public Health Service (AM 07203) and the National Science Foundation (GB 2236) and by Cancer Research Funds of the University of California.

10

Epimerases

LUIS GLASER

I. Introduction

Epimerases are defined for the purpose of this chapter as single proteins which catalyze the reversible inversion of the configuration at the asymmetric center of a molecule, generally a sugar (*1*), in all cases

1. All sugars, unless indicated, are of the D configuration.

without the formation of a free intermediate. Amino acids will not be discussed in this chapter.

Historically the first enzyme of this type was the discovery by Leloir (*1a*) of the UPD-D-glucose 4′-epimerase which catalyzes the interconversion of UDP-D-glucose and UDP-D-galactose. First named *galactowaldenase*, the enzyme is an obligatory step in the utilization of D-galactose by many microorganisms and by mammals, as well as being required for the synthesis of UDP-D-galactose, which is the precursor of D-galactose in a wide variety of polysaccharides, glycoprotein, and glycolipids.

The initial name *galactowaldenase*, by analogy to Walden inversion, is probably a misnomer, since the detailed mechanism of the reaction appears to involve oxidation and reduction. The most important clues to the mechanism of the reaction was provided by the observation that the liver enzyme specifically required added DPN as a cofactor (*2*), followed by the observation that the yeast enzyme catalyzing the same reaction did not require added DPN because the enzyme as isolated contained tightly bound DPN (*3, 4, 4a*).

More information is currently available about the mechanism of the UDP-D-glucose 4′-epimerase than any other enzyme of this class, and it will therefore be the main subject of this chapter. Fragmentary evidence for other nucleoside diphosphate sugar epimerases of the same type suggests that their mechanism, when investigated in detail, will be similar if not identical to that of the UDP-D-glucose 4′-epimerase. The proposed mechanism of the UDP-D-glucose 4′-epimerase is shown in Fig. 1. Although mechanisms of epimerization which involve reaction of the pyrimidine ring have been suggested several times (*6, 7*), there is at

1a. L. F. Leloir, *ABB* **33,** 186 (1951).

2. E. S. Maxwell, *JBC* **229,** 139 (1957); "The Enzymes," 2nd ed., Vol. 5, p. 443, 1961.

3. E. S. Maxwell, H. R. Szulmajster, and H. M. Kalckar, *ABB* **78,** 407 (1958).

4. E. S. Maxwell and H. R. Szulmajster, *JBC* **235,** 308 (1960).

4a. Recent data on the DPN bound to the yeast UDP-D-glucose 4′-epimerase suggest that it can react with yeast alcohol dehydrogenase but not with UDP-D-glucose dehydrogenase, indicating that it is a modified form of DPN (*5*). The DPN bound to the *E. coli* UDP-D-glucose 4′-epimerase can react with UDP-D-glucose dehydrogenase. It is therefore possible that the binding of the natural coenzyme to the yeast UDP-D-glucose 4′-epimerase differs from that of DPN.

5. H. M. Kalckar, A. U. Bertland, J. T. Johansen, and M. Ottesen, "The Role of Nucleotides for the Function and Conformation of Enzymes," Proc. Alfred Benzon Symp. No. 1, p. 247. Munskgaard, Copenhagen, 1969.

6. E. I. Budowsky, T. N. Drujhinina, O. I. Elseva, N. D. Gabrielyan, N. R. Kochetkow, M. A. Novikova, V. N. Shibaev, and O. L. Zhadanov, *BBA* **122,** 213 (1966).

7. H. R. Szulmajster, *JMB* **3,** 253 (1961).

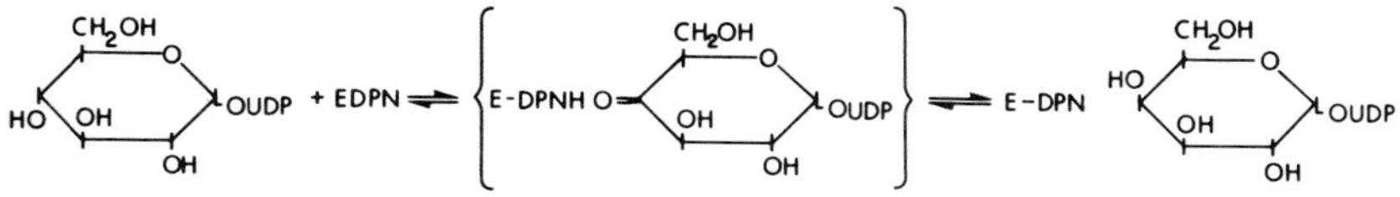

Fig. 1. Proposed mechanism of UDP-D-glucose 4′-epimerase.

present no adequate experimental evidence to support them. At the end of this chapter, epimerases, which proceed by a different mechanism, will be discussed.

II. UDP-D-Glucose 4′-Epimerase

The enzyme has been obtained in essentially homogeneous form from yeast (*8*) and from *Escherichia coli* (*9, 10*). Highly purified enzyme has been obtained from liver (*2*), wheat germ (*11*), and mammary gland (*12*), and it has been demonstrated to occur in a wide variety of cell types.

The known properties of the enzyme from yeast and *E. coli* are summarized in Table I. Complete amino acid analysis of the yeast enzyme (*8*) and the *E. coli* enzyme (*9*) have been published.

Both enzymes appear to be dimers containing 1 mole of firmly bound DPN per mole of enzyme, suggesting that in these enzymes there is a single active site per dimer. In the case of the yeast enzyme, it is not clear whether the two subunits are identical. Although quantitatively the protein contains two carboxyl terminal residues (tyrosine and leucine) (*5*) it is not yet known whether the subunits of molecular weight 60,000 are single polypeptide chains.

In the case of the *E. coli* enzyme, amino terminal analysis and peptide mapping suggest that the two subunits of molecular weight 40,000 are single chains and probably identical. This is a rather surprising observation. While a number of enzymes are known in which the number of active sites is less than the number of subunits (*13*), few examples are known where identical subunits combine to form a single active site. A related example is the binding of only 1 mole of 2,3-diphosphoglycerate

8. R. A. Darrow and R. Rodstrom, *Biochemistry* **7,** 1645 (1968).
9. D. B. Wilson and D. S. Hogness, *JBC* **244,** 2132 (1969).
10. D. B. Wilson and D. S. Hogness, *JBC* **239,** 2469 (1964).
11. D. F. Fan and D. S. Feingold, *Plant Physiol.* **44,** 599 (1969).
12. C. M. Tsai, N. Holmberg, and K. E. Ebner, *ABB* **136,** 233 (1970).
13. C. Frieden, *Ann. Rev. Biochem.* **40,** 653 (1971).

TABLE I
PROPERTIES OF UDP-D-GLUCOSE 4′-EPIMERASE FROM YEAST AND *E. coli*

E. coli UDP-D-Glucose 4′-Epimerase [Refs. (*9, 10*)]	
Molecular weight	79,000[a,b]
Subunit structure	Two subunits molecular weight 40,000[c]
DPN content	1 mole of DPN/79,000 g protein[d,e]
SH group	4 per 79,000 g protein
Yeast UDP-D-Glucose 4′-Epimerase [Ref. (*8*)]	
Molecular weight	125,000[b,f]
Subunit structure	Two subunits 60,000[g]
DPN content	1 mole of DPN/125,000 g of protein[d]
SH group	16/125,000 g of protein

[a] Sedimentation diffusion.
[b] Sedimentation equilibrium
[c] Molecular weight based on sedimentation in 6 *M* guanidine-HCl. Subunits appear identical by peptide mapping and amino terminal analysis.
[d] Determined enzymically.
[e] Determined chemically.
[f] Sucrose density centrifugation.
[g] Obtained after treatment with *p*-hydroxymercuribenzoate. Their identity is not known.

per mole of tetrameric hemoglobin located between the two β chains (*14*). It remains to be determined whether the two subunits of the *E. coli* enzyme are functionally identical; that is, whether because of the interaction with the pyridine nucleotide, the conformation of the two subunits in the functional enzyme is in fact different.

A. KINETICS AND SPECIFICITY

The enzyme, in spite of its complex mechanism, follows simple kinetics. The K_m and V_{max} for various substrates are listed in Table II. The equilibrium constant UDP-Glu/UDP-Gal = 3.0–3.5 has been determined in a number of laboratories (*2, 10, 12, 15*). The sugar and nucleotide specificity of the enzymes(s) from various sources have not been examined in very great detail nor with every one of the enzymes listed. It would appear that the enzyme from rat liver will tolerate a wide variety of substitutions in the pyrimidine moiety of UDP-Glu while retaining

14. R. Benesch and R. E. Benesch, *Nature* **221**, 618 (1968); M. F. Perutz, *Nature* **228**, 7266 (1970).
15. H. Imai, N. Morikawa, and K. Kurahashi, *J. Biochem.* (*Tokyo*) **56**, 135 (1964).

enzymic activity. It is particularly noteworthy that while the K_m for many of the analogs is much higher than for UDP-Glu, the V_{max} of the enzyme with the analogs, is in fact higher (up to a factor of 7 with 5,6-dihydro-UDP-Glu) compared to UDP-D-Glu suggesting that in the reaction with UDP-D-Glu substrate release may be rate limiting. A variety of uridine containing nucleotides, with or without sugar attached, can bind to some of these enzymes and appear to be competitive inhibitors (Table II) (*2, 6, 10–12, 15–18*).

Sugar substitutions have again not been investigated very systematically, partly because the appropriate nucleotide derivatives are not available. The *E. coli* enzyme will act on either 6-deoxy sugars (UDP-6-deoxy-D-glucose or UDP-6-deoxy-D-galactose) or pentoses (L-arabinose and D-xylose) (*16*). Results obtained with the same analogs with the yeast enzyme are somewhat ambiguous (*19*). As expected, UDP-4-deoxy-D-glucose is inactive but surprisingly so is UDP-3-deoxy-D-glucose as tested with the liver enzyme (*18*).

Additional information on both the nucleotide and sugar specificity of the yeast epimerase will be discussed below when the partial reactions of the enzyme are discussed in the section devoted to catalytic mechanism.

B. Activators

The activity of some epimerases are affected by the addition of cations. This effect was first noted with the UDP-D-Glu-NAc 4′-epimerase from *Bacillus subtilis* (*20*), but it has been most extensively studied with the UDP-D-Glu 4′-epimerase from yeast (*21–23*). The yeast enzyme is stimulated by a variety of cations of which spermidine is the most effective. The stimulation is not on the maximal velocity of the enzyme, but the cations increase the apparent affinity of the enzyme for nucleotide (*23*). Thus, at spermidine concentrations of 0.2, 0.5, and 10 m*M* the K_m for UDP-Gal is 0.6, 0.4, and 0.1 m*M*. Very high concentrations of all cations tested inhibit enzymic activity. Some forms of the yeast epimerase

16. H. Ankel and U. S. Maitra, *BBRC* **32**, 526 (1968).
17. L. Glaser and L. Davis, unpublished observations (1970).
18. N. K. Kochetkov, E. I. Budowsky, T. N. Druzhinina, N. D. Gabrielyan, I. V. Komlev, Y. Y. Kusov, and V. N. Shibaev, *Carbohydrate Res.* **10**, 152 (1968).
19. W. L. Salo, J. H. Nordin, D. R. Peterson, R. D. Bevill, and S. Kirkwood, *BBA* **151**, 484 (1968).
20. L. Glaser, *JBC* **239**, 2801 (1959).
21. R. A. Darrow and C. R. Creveling, *JBC* **239**, PC362 (1964).
22. R. A. Darrow and R. Rodstrom, *Proc. Natl. Acad. Sci. U. S.* **55**, 205 (1966).
23. R. A. Darrow and R. Rodstrom, *JBC* **245**, 2036 (1970).

TABLE II
KINETIC PROPERTIES OF UDP-D-GLUCOSE 4′-EPIMERASE

Enzyme source	Substrate	K_m or K_i	V_{max}	pH optimum	Ref.
E. coli	UDP-D-Gal	$K_m = 1.6 \times 10^{-4}\ M$	30,250[a]	7–9	(*10, 15*)
	UDP-D-Glu	$K_m = 1.0 \times 10^{-3}\ M$			(*15*)
	UDP-D-xylose	$K_m = 1.2 \times 10^{-3}\ M$			(*16*)
	UDP-L-arabinose	Active	1,600[a]		(*16*)
	UDP-D-fucose (6-deoxy-D-galactose)	$K_m = \sim 1.0 \times 10^{-4}\ M$	130[a]		(*17*)
	2′-Deoxy-UDP-D-Glu	Active			(*17*)
	dTDP-D-glucose	Active			(*17*)
Yeast	UDP-D-Gal	$K_m = 1.1 \times 10^{-4}\ M$[b]	10,900[a]		(*8*)
Liver[e]	UDP-D-Gal	$K_m = 5 \times 10^{-5}\ M$	0.66[c]	8.5–9.5	(*2*)
	UDP-D-Glu	$K_m = 7 \times 10^{-5}\ M$			(*2*)
	UDP-D-Glu	$K_m = 9 \times 10^{-5}\ M$	(1)[d]		(*6*)
	2′-Deoxy-UDP-Glu	$K_m = 50 \times 10^{-5}\ M$	(2.5)[d]		(*6*)
	5,6-Dihydro-UDP-Glu	$K_m = 80 \times 10^{-5}\ M$	(7.0)[d]		(*6*)
	6-Aza-UDP-Glu	Active			(*6*)

Source	Substrate/compound	Constant	Activity	pH	Ref.
	2-Thio-UDP-Glu	Active			
	3-*N*-Methyl-UDP-Glu	Inactive			*(6)*
	CDP-Glu	Inactive			*(6)*
	UDP-3-Deoxy-D-Glu	$K_i = 2.6 \times 10^{-3}$ *M*	Inactive		*(18)*
	UDP-4-Deoxy-D-Glu	$K_i = 4.4 \times 10^{-4}$ *M*	Inactive		*(18)*
Mammary gland[e]	UDP-D-Gal	$K_m = 3 \times 10^{-5}$ *M*	4.7[c]	8–9.0	*(12)*
	UDP-D-Glu	$K_m = 5 \times 10^{-4}$ *M*			
	UDP	$K_i = 3.5 \times 10^{-5}$ *M*			
	UTP	$K_i = 2.2 \times 10^{-4}$ *M*			
	UDPU	$K_i = 1.8 \times 10^{-5}$ *M*			
	UMP	$K_i = 5 \times 10^{-5}$ *M*			
Wheat germ[e]	UDP-D-Gal	$K_m = 1 \times 10^{-4}$ *M*	0.75[c]	9.0	*(11)*
	UDP-D-Glu	$K_m = 2 \times 10^{-4}$ *M*			
	dTDP-D-Glu	Inactive			

[a] Moles/mole of enzyme/min.
[b] The precise value depends on cation concentration (see Section II,B).
[c] μmoles/min/mg protein.
[d] Relative to value of UDP-D-Glu as 1.
[e] These enzymes require NAD for activity.

dimerize in the presence of spermidine to a form of molecular weight 250,000 (*21*, *22*). However, present in normal preparations epimerase is a second form, which is kinetically sensitive to spermidine activation but does not dimerize. Such enzyme can also be obtained by treatment of the enzyme with *p*-hydroxymercuribenzoate followed by a thiol. It would therefore appear that dimerization is not required for spermidine activation.

C. Mechanism of Catalysis

As briefly mentioned in Section I, the DPN requirement for activity of the UDP-D-glucose 4′-epimerase and other similar reactions suggest the mechanism in Fig. 1 for this reaction. This mechanism is *not* very simple, since it requires that DPN be able to accept a hydrogen from both sides of the pyranose ring of the uridine-bound hexose.

This mechanism is at present supported by the following observations:

(1) No hydrogen from the medium is incorporated into the sugar when the reaction is carried out in tritiated water (*24–26*).

(2) No ^{18}O is incorporated into the sugar from $H_2{}^{18}O$ (*25*, *26*).

(3) When DPNH-4-^{3}H is added to the liver enzyme which requires added DPN for activity, no ^{3}H was incorporated into the sugar moiety of UDP-D-Glu or UDP-D-Gal (*2*, *24*).

(4) When DPN-4-^{3}H was used with the liver enzyme no tritium was found in the UDP-D-glucose of UDP-D-galactose (*2*).

(5) No loss of hydrogen from UDP-D-glucose-4-^{3}H or UDP-D-galactose-4-^{3}H (*27*) was observed in the course of the reaction.

(6) UDP-D-glucose-4-^{3}H shows a negative isotope effect when used as a substrate for the yeast UDP-D-glucose 4′-epimerase (*28*).

(7) Addition of substrate to the *E. coli* enzyme causes the appearance of a spectral band, with absorption maximum at 345 nm suggesting the formation of enzyme-bound DPNH (*10*).

(8) A small but probably significant increase of enzyme-bound DPNH, measured enzymically, can be observed on substrate addition to the yeast enzyme amounting to about 10% of the total protein-bound pyridine nucleotide (*29*).

24. H. M. Kalckar and E. S. Maxwell, *BBA* **22**, 588 (1956).
25. A. Kowalsky and D. E. Koshland, *BBA* **22**, 575 (1956).
26. L. Anderson, A. M. Landel, and D. F. Diedrick, *BBA* **22**, 573 (1956).
27. R. D. Bevill, J. H. Nordin, F. Smith, and S. Kirkwood, *BBRC* **12**, 152 (1963).
28. R. D. Bevill, E. A. Hill, F. Smith, and S. Kirkwood, *Can. J. Chem.* **43**, 1577 (1965).
29. A. U. Bertland and H. M. Kalckar, *Proc. Natl. Acad. Sci. U. S.* **61**, 629 (1968).

TABLE III
REDUCTION OF YEAST UDP-D-GLUCOSE 4′-EPIMERASE WITH SUBSTRATE ANALOGS[a]

Active nucleotides	Inactive nucleotides
5′-UMP	5′-CMP
5′-CDP	3′-UMP
5′-TDP	
Active sugar	**Inactive sugar**
D-Glu	Sucrose
D-Gal	L-Fucose
D-Fucose	D-Ribose
D-Xylose	D-Arabinose
L-Arabinose	

[a] The compounds listed are tested in the reaction epimerase-DPN + nucleotide + sugar → epimerase-DPNH + products. Sugars were tested at 10^{-2} *M*, nucleotides at 10^{-3} *M*. Sugars are tested in the presence of 5′-UMP, nucleotides in the presence of glucose. The sugar product is not known. One mole of nucleotide remains firmly bound per mole of epimerase DPNH (*30*).

(9) The pyridine nucleotide on the enzyme can be totally reduced although very slowly by the addition of a nucleotide such as UMP and a sugar as D-galactose. This reaction, which will be discussed again in the next section, behaves as if the enzyme can slowly react with the two halves of UDP-D-glucose or UDP-D-galactose even if they are not linked to each other, but the sugar moiety is released after formation of reduced pyridine nucleotide to lead to inactive enzyme containing DPNH. The specificity of this reaction is shown in Table III. The reaction product of the sugar has not been isolated. Similar, although less extensive, observations have been made with the enzyme from *E. coli* and *Salmonella typhimurium* LT2 (*29–31*).

Several mechanisms more detailed than that in Fig. 1 can be written for this reaction on the basis of these observations. Such mechanisms attempt to explain the steric requirement for the pyridine nucleotide to react with both sides of the pyranose ring of the hexose and assume that DPNH in the enzyme is generated by oxidation of C-4 of the sugar.

Equation (1) assumes the existence of a second hydrogen acceptor on the enzyme which is in equilibrium with DPNH.

Equation (2), assumes that DPNH is formed directly from UPD-D-glucose or UDP-D-galactose. Mechanisms of this type illustrated with UDP-D-glucose-4-*d* as a substrate can either assume that different sides

30. A. U. Bertland, B. Bugge, and H. M. Kalckar, *ABB* **116,** 280 (1966).
31. H. M. Kalckar, A. U. Bertland, and B. Bugge, *Proc. Natl. Acad. Sci. U. S.* **65,** 1113 (1970).

$$\text{E}\langle^{\text{BH}}_{\text{DPN}} + \text{UDP-glucose} \rightleftharpoons \left[\text{E}\langle^{\text{BH}}_{\text{DPNH}} \quad \text{UDP-4-ketoglucose}\right] \rightleftharpoons$$

$$\text{E}\langle^{\text{B}}_{\text{DPNH}} + \text{UDP-galactose}; \quad \text{E}\langle^{\text{B}}_{\text{DPNH}} \rightleftharpoons \text{E}\langle^{\text{BH}}_{\text{DPN}} \qquad (1)$$

of the pyridine ring of DPN react with glucose and galactose [Eq. (2a)] or that the same side of the pyridine ring may react with glucose and galactose. This second situation is sterically much more difficult and requires a large conformational change presumably both in the protein and/or the hexose.

$$\text{E-DPN} + \text{UDP-D-glucose-4-}d \rightleftharpoons \text{E-DPN}(d)\ \text{UDP-4-ketoglucose} \rightarrow \text{E-DPN}(4\text{-}d) + \text{UDP-D-galactose} \qquad (2a)$$

$$\text{E-DPN} + \text{UDP-D-glucose-4-}d \rightleftharpoons \text{E-DPN}(d)\ \text{UDP-4-ketoglucose} \rightarrow \text{E-DPN} + \text{UDP-D-galactose-4-}d \qquad (2b)$$

Of these three mechanisms, Eqs. (1) and (2a) will give rise on intramolecular higher transfer in the reaction, while mechanism (2b) will give rise to an intramolecular hydrogen transfer. This has been tested with the UDP-D-glucose 4′-epimerase *E. coli* (*32*) (Fig. 2) and the hydrogen

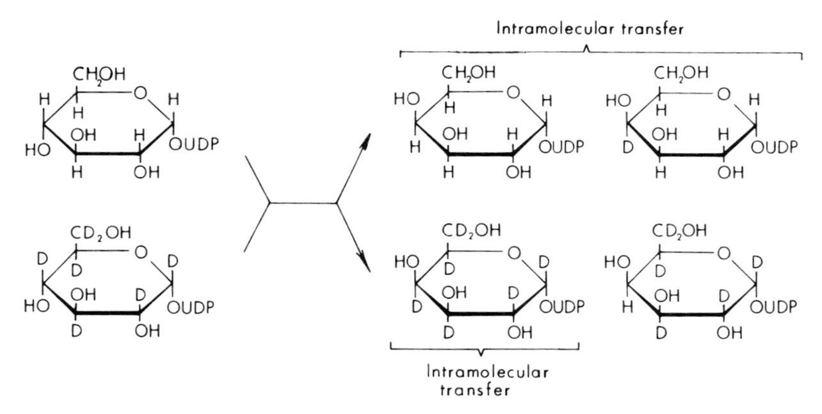

FIG. 2. Reaction products expected in the UDP-D-glucose 4′-epimerase for an intramolecular or intermolecular hydrogen transfer, using deuterated substrates. The various products can easily be distinguished by mass spectrometry of the isolated alditol acetates derived from the sugar nucleotides (*32*).

32. L. Glaser and L. Ward, *BBA* **198**, 613 (1970).

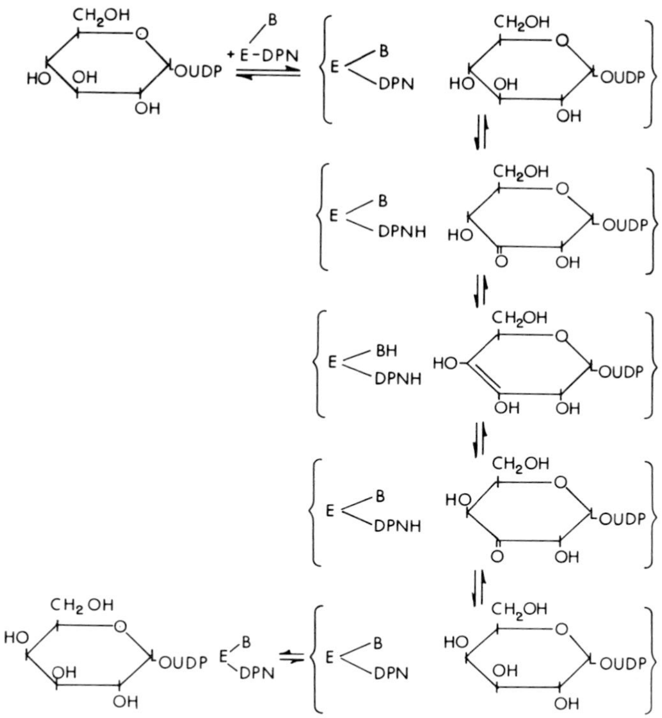

FIG. 3. Hypothetical alternative mechanism for the UDP-D-glucose 4′-epimerase.

transfer found to be strictly intramolecular, thus supporting mechanism (2b) for the reaction (*32a*). It should be noted that so far the formation of a 4-keto sugar intermediate in the reaction has not been directly demonstrated, nor has it been shown that the hydrogen at C-4 is transferred to DPN. Therefore, although unlikely, mechanisms such as that shown in Fig. 3 are still theoretically possible. This mechanism would be formally analogous to that shown for the epimerization at C-3 and C-5 glucose during L-rhamnose synthesis (see below), with the additional postulate that B_1H and B_2H in Fig. 3 do not exchange with the medium. This mechanism is shown not because the author believes it to be correct but to illustrate the uncertainty in the currently accepted mechanism.

32a. The same conclusion can be tentatively made for the liver enzyme from the fact that addition of 4-^{3}H-DPN did not yield titrated UDP-Gal and UDP-Glu with the reservation that it is not known in that case whether enzyme-bound DPN exchanges rapidly enough with free DPN to provide a critical test of this mechanism. If enzyme-bound DPN exchanges slowly with free DPN in the case of the liver enzyme, then these experiments have no bearing on mechanisms (1), (2a), or (2b).

Tentative evidence that a 4-keto sugar can be formed in this reaction as well as confirmation of the fact that only one of the hydrogens in DPNH participates in hydrogen transfer come from the observation that when *E. coli* epimerase is reduced with sodium borotritide, B-labeled DPNH can be isolated from the enzyme. When enzyme so labeled is incubated with the analog dTDP-4-keto-6-deoxy-D-glucose, the analog is reduced by the enzyme in a stoichiometric reaction to yield tritiated dTDP-6-deoxy-D-glucose and dTDP-6-deoxy-D-galactose (*33*).

Unfortunately, this does not prove that a 4-keto sugar is formed in the normal reaction since 2-ketoglucose will also reoxidize enzyme-DPNH to enzyme-DPN (*5*).

D. Effect of Bound Pyridine Nucleotide on Protein Structure

1. *Subunit Association*

Titration of the yeast UDP-D-glucose 4′-epimerase with *p*-hydroxymercuribenzoate results in loss of enzymic activity [the protein contains 16 titratable SH groups per mole (125,000 g)]. Enzymic activity is lost when half of these groups are titrated (*5, 34*). Titration of all of the available sulfhydryl groups, followed by centrifugation either in a sucrose gradient or in sedimentation velocity experiments, indicates that the enzyme has dissociated into subunits of molecular weight 60,000. Removal of the *p*-hydroxymercuribenzoate with a sulfhydryl reagent results in the formation of enzyme which requires the addition of DPN for activity. This apoenzyme is stated to behave in sucrose gradients like a protein of molecular weight 125,000 although detailed data have not been published (*22*).

The molecular weight of the *E. coli* UDP-D-glucose 4′-epimerase after removal of bound DPN has not been determined.

The data so far published on the reassociation of the yeast apoenzyme with DPN are extremely limited. The data so far available suggest that the apoenzyme requires the addition of large quantities of DPN for activity and that full activity is not regained (*22*). This interpretation may be incorrect because it fails to take into account the time dependence of the reactivation of the apoenzyme by DPN (*22*).

In Fig. 4 are shown data on the reactivation of the apoenzyme of the *E. coli* UDP-D-glucose 4′-epimerase by DPN under conditions where the reaction of apoenzyme with DPN is measured accurately before the holoenzyme formed is assayed for activity. Activation of the residual

33. G. Nelsestuen and S. Kirkwood, *Federation Proc.* **29**, 337 (1970).

34. C. R. Creveling, A. Bhadure, A. Christensen, and H. M. Kalckar, *BBRC* **21**, 624 (1965).

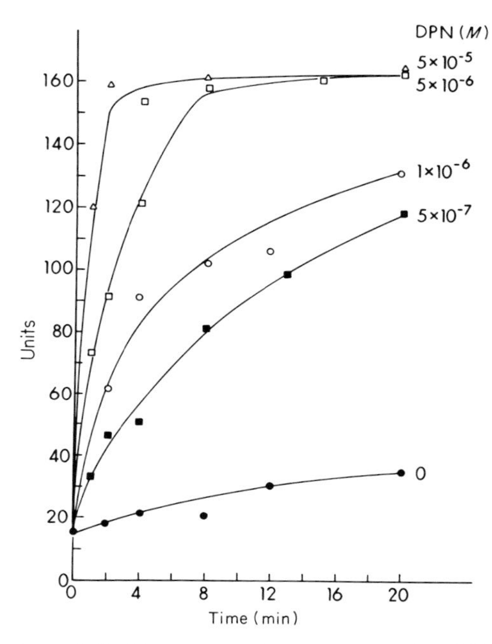

FIG. 4. Reactivation of UDP-D-glucose 4′-epimerase with DPN. *Escherichia coli* UDP-D-glucose 4′-epimerase was incubated with $3 \times 10^{-4}\,M$ *p*-hydroxymercuribenzoate for 1 hr at 25°; at the end of the reaction, 2-mercaptoethanol was added to a final concentration of $2 \times 10^{-3}\,M$, and DPN was added to the concentration indicated. Aliquots of the reactivation mixture was assayed for UDP-D-glucose 4′-epimerase at the time indicated in the presence of $5 \times 10^{-4}\,M$ ADP-ribose, which effectively prevents further reactivation by DPN.

apoenzyme during the assay is prevented by the addition of excess ADP-ribose which reacts with the apoenzyme competitively with DPN. The results suggest that activation can be described by the equation

$$\text{Apoenzyme} + \text{DPN} \rightleftharpoons [\text{apoenzyme DPN}] \xrightarrow{\text{slow}} \text{active enzyme}$$

where the binding of DPN to the apoenzyme has a dissociation constant of $\approx 10^{-6}\,M$ and the complex undergoes a slow isomerization to active enzyme. This isomerization is essentially irreversible since reactivated enzyme is not inhibited by ADP-ribose addition, which, however, prevents further activation by DPN. This interpretation is essentially the same as has been given for an analogous situation with the dTDP D-glucose oxidoreductase (*35*) (see Glaser and Zarkowsky, Chapter 16, Volume V).

It follows from this that if apoenzyme and DPN are mixed during

35. H. Zarkowsky, E. Lipkin, and L. Glaser, *JBC* **245,** (1970).

the assay, full activity may not be recovered and high concentrations of DPN may be required to observe enzymic activity.

It should be noted that the apoprotein from the yeast epimerase clearly shows a much looser structure as determined by circular dichroism and by hydrogen exchange than the native protein (*5, 29*).

2. *Effect of Pyridine Nucleotide Reduction on Protein Conformation*

As mentioned in Section II, reduction of all of the bound pyridine nucleotide on the yeast UDP-D-glucose 4′-epimerase can be accomplished by the addition to the enzyme of a free sugar in high concentration provided a nucleotide such as UMP is always present but at a relatively low concentration (see Table III). The rate of reduction is slow, and full reduction of the enzyme requires several hours of incubation. It is noteworthy that reduction of the bound pyridine nucleotide by $NaBH_4$ is also stimulated by 5′-UMP suggesting that the reactivity of the bound DPN is increased when the uridine subsite of the substrate binding site is occupied (*31*).

The enzyme containing bound DPNH shows several very interesting characteristics. It binds 5′-UMP and presumably other uridine nucleotides very tightly, and by gel filtration it can be shown that one uridine is bound per mole of enzyme (MW 125,000) (*30, 31*). This observation may explain (as for the dTDP-D-glucose oxidoreductase, see Glaser and Zarkowsky, Chapter 16, Volume V) why the presumed intermediate in the reaction UDP-4-keto-D-glucose does not occur free, since it would only occur bound to enzyme containing bound (DPNH (Fig. 1).

This tight substrate binding clearly indicates that the protein structure must change depending on whether DPN or DPNH is bound to the enzyme.

Direct evidence for a gross change in protein structure has been obtained. Thus, differences between the reduced and oxidized form of the yeast epimerase can be detected both by optical rotatory dispersion and circular dichroism as well as by measuring the rate of exchange of hydrogens in the protein backbone with the medium (*5, 29*).

The effect of DPNH reduction on the structure of the yeast UDP-D-glucose 4′-epimerase is even more complicated than the above summary would indicate. It is possible to reoxidize the inactive reduced UDP-D-glucose 4′-epimerase by leaving dilute solutions exposed to air for several days. Under these conditions active enzyme containing bound DPN is recovered with about twice the original specific activity before reduction (*29*). This surprising observation either indicates that going through the oxidation reduction cycle increases the catalytic efficiency of the enzyme or that the enzyme as prepared contains some inactive molecules which can still be reduced by free sugar in the presence of UMP and which on reoxidation regain enzymic activity. Even though these two alternatives cannot at present be distinguished, the second needs to be seriously considered in light of the observation that yeast epimerase as prepared may contain fractions of varying specific activity and slightly different properties (*8, 23*), although of equal DPN content, and presumably equal purity.

The fluorescence yield of DPNH on the yeast epimerase is close to 90% of theoretical, suggesting that the nicotinamide portion of DPN is buried in a hydrophobic region of the protein; fluorescence can be excited not only at 340 nm but also by energy transfer from aromatic residues in the protein (excitation at 280 nm) (*5, 29, 30*). The high quantum yield of DPNH fluorescence accounts for the early observation that the yeast epimerase is a fluorescent protein since, as isolated, 10–15% of the enzyme-bound DPN is in the form of DPNH.

Many epimerases require added DPN for activity, for example, the liver UDP-D-glucose 4′-epimerase. It will be of great interest to determine the effect of DPN on the structure of these proteins. In all cases where DPN is weakly bound, DPNH has been shown to be an inhibitor of the reaction and it has been suggested that the cellular DPN–DPNH ratio serves to control the velocity of these reactions *in vivo* (*36, 37*).

III. Other Epimerases That Potentially Involve Oxidation–Reduction

Table IV (*36, 39–52*) gives a list of the known nucleoside diphosphate sugar epimerases. Their mechanism has not been investigated in detail.

36. H. M. Kalckar, *Science* **150,** 305 (1965).
37. E. A. Robinson, H. M. Kalckar, and H. Troedson, *BBRC* **13,** 313 (1963).
38. V. Ginsburg, *Advan. Enzymol.* **26** (1969).
39. F. Maley and G. F. Maley, *BBA* **31,** 577 (1959).
40. D. S. Feingold, E. F. Neufeld, and W. Z. Hassid, *JBC* **235,** 910 (1960).
41. E. E. B. Smith, G. T. Mills, H. P. Bernheimer, and R. Austrian, *BBA* **29,** 640 (1958).
42. H. Ankel and R. G. Tischer, *BBA* **178,** 415 (1969).
43. B. Jacobson and E. A. Davidson, *JBC* **237,** 638 (1962).
44. E. F. Neufeld, V. Ginsburg, E. W. Putman, D. Fanshier, and W. Z. Hassid, *ABB* **69,** 602 (1957).
45. C. E. Cardini and L. F. Leloir, *JBC* **225,** 317 (1957).
46. D. G. Comb and S. Roseman, *BBA* **29,** 653 (1958).
47. C. T. Spivak and S. Roseman, "Methods in Enzymology," Vol. 9, p. 612, 1966.
48. R. Kalan and G. Avigad, *BBRC* **24,** 18 (1966).

TABLE IV
NUCLEOSIDE DIPHOSPHATE SUGAR EPIMERASES[a]

(a) Reactant	(b) Product	Ratio at equilibrium a/b	DPN requirement[b]	Source of enzyme
UDP-D-glucose	UDP-D-galactose	3.0–3.5	(+) or (B)	Various cell types[c] (see text)
UDP-D-Glu-NAc	UDP-D-Gal-NAc	2.0	(?)	*Bacillus subtilis*[c] [Ref. (*20*)]
			(+)	Liver[c] [Refs. (*20*, *39*)]
UDP-D-glucuronic acid	UDP-D-galacturonic	1	(?)	*Phaseolus aureus* [Ref. (*40*)]
				Pneumococcus type 1 [Ref. (*41*)]
				Anabaena flos aquae [Ref. (*42*)]
UDP-D-glucuronic acid	UDP-L-iduronic active	—	(+)	Connective tissue [Ref. (*43*)]
UDP-L-arabinose	UDP-D-xylose	1	(?)	*Phaseolus aureus* [Refs. (*40*, *44*)]
UDP-D-Glu-NAc	UDP + Man-NAc	See text		Liver[c] [Refs. (*45–47*)]
dTDP-D-Glu	dTDP-D-Gal	—	(?)	Sugar beets [Ref. (*48*)]
dTDP-D-Glu	dTDP-D-Man	—		*Streptomyces griseus* [Ref. (*49*)]
CDP-paratose	CDP-tyvelose[d]	1.33	(+)	*Salmonella* strains[c] [Refs. (*50*, *51*)]
GDP-D-mannose	GDP-L-galactose		(?)	*Helix pomatia*[e] [Ref. (*52*)]

[a] For an earlier review dealing with the metabolism of nucleoside diphosphate sugar, see Ginsburg (*38*).
[b] (+) DPN addition required for activity. (B) Bound DPN on the enzyme. (?) Unknown.
[c] Enzyme at least partially purified.
[d] Paratose is 3,6-dideoxy-D-ribohexose and tyvelose is 3,6-dideoxy-D-arabinohexose. Paratose and tyvelose are 2-epimers.
[e] It appears highly unlikely that this reaction is catalyzed by a single protein.

As indicated in the Table IV, a DPN requirement has been demonstrated for a number of these reactions. It would appear likely that the mechanism of these enzymes will not differ fundamentally from that of the UDP-D-glucose 4′-epimerase. Only one of these enzymes, the UDP-*N*-acetyl-D-glucosamine 2′-epimerase, requires special discussion.

A. UDP-*N*-Acetyl-D-glucosamine 2′-Epimerase

This reaction is unique in that it catalyzes a hydrolysis of the nucleotide according to the following equation:

UDP-*N*-acetyl-D-glucosamine + H_2O → UDP + *N*-acetyl-D-mannosamine

Neither free *N*-acetyl-D-glucosamine nor free UDP-*N*-acetyl-D-mannosamine is an intermediate in the reaction.

The reaction differs from other epimerases in that hydrogen from H_2O is incorporated into the product (*N*-acetyl-D-mannosamine) at C-2 (*53*), this observation would exclude DPN as a hydrogen carrier in this reaction but not other hydrogen carriers such as flavin. A hypothetical mechanism for this reaction is shown in Fig. 5.

The enzyme does not catalyze the incorporation of hydrogen from the medium into free *N*-acetyl-D-mannosamine or into UDP-*N*-acetyl-D-glucosamine (*53*).

Recently, UDP-*N*-acetyl-D-mannosamine has been synthesized chemically. This nucleotide is hydrolyzed by the enzyme yield UDP + *N*-acetyl-D-mannosamine. Surprisingly, it is found that *N*-acetyl-D-mannosamine formed in this reaction had incorporated hydrogen from the medium at C-2, and that the specific activity of the *N*-acetyl-D-mannosamine formed was the same whether UDP-*N*-acetyl-D-glucosamine or UDP-*N*-acetyl-D-mannosamine was the substrate when the reaction was carried out in ^3H water (*54*).

The authors have suggested a mechanism in which the amino sugar is transferred to the enzyme to yield a glycosyl enzyme. A 3-keto sugar is then formed with an enzyme-bound cofactor such as DPN, and epimerization at C-2 occurs by an enediol mechanism followed by reoxidation of DPNH by reduction of the 3-keto group. This scheme or

---

49. J. Baddiley, N. L. Blumson, A. Di Girolamo, and M. Di Girolamo, *BBA* **50,** 391 (1961).
50. H. Nikaido and K. Nikaido, *JBC* **241,** 1376 (1966).
51. S. Matsuhashi, *JBC* **241,** 4275 (1966).
52. E. M. Goudsmit and E. F. Neufeld, *BBRC* **24,** 730 (1967).
53. L. Glaser, *BBA* **41,** 534 (1960).
54. W. L. Salo and H. G. Fletcher, *Biochemistry* **9,** 882 (1970).

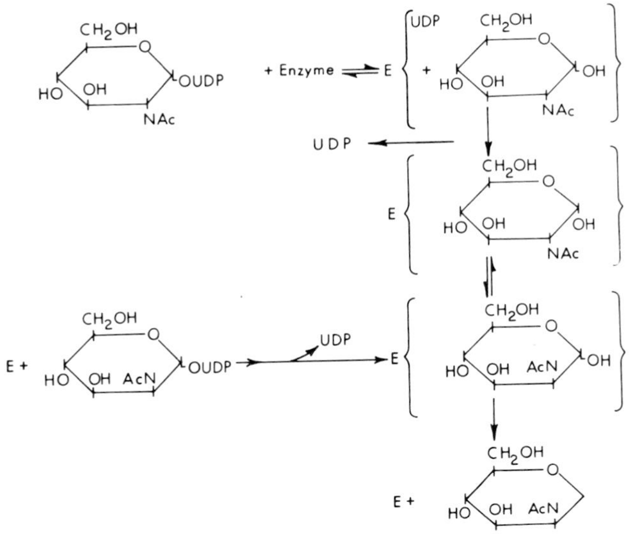

FIG. 5. Mechanism of the UDP-*N*-acetyl-D-glucosamine 2′-epimerase.

the more general one in Fig. 5 is certainly not proven. It will explain the presently available data only if the rate-limiting step in this reaction is the release of *N*-acetyl-D-mannosamine from this enzyme, in order to account for the fact that the tritium incorporation into product is the same whether UDP-*N*-acetyl-D-glucosamine or UDP-*N*-acetyl-D-mannosamine is a substrate.

If product release is the rate-limiting step in the reaction then it might preclude the one possible test of the suggested mechanism which would be the finding of an isotope effect if 3-^{2}H- or 3-^{3}H-labeled UDP-*N*-acetyl-D-glucosamine were used as a substrate. Finding of a primary isotope effect with these substrates could be used as evidence that the C–H bond at C-3 of the sugar is cleaved during the reaction, but such an isotope effect would only be seen if this step becomes rate limiting after isotope substitution.

The enzyme is extremely unstable (*47*). Until it can be stabilized and purified to homogeneity so that any cofactor bound to it can be determined, the mechanism will remain somewhat speculative (see Section IV,D for a related reaction).

B. L-RIBULOSE-5-P 4′-EPIMERASE

L-Ribulose-5-P 4′-epimerase catalyzes the epimerization of L-ribulose-5-P to D-xylulose-5-P, a homogeneous enzyme has been obtained from

TABLE V
PROPERTIES OF L-RIBULOSE-5-P 4′-EPIMERASE

Property	Enzyme source		
	E. coli[a]	*A. aerogenes*[a]	*L. plantarum*[b]
Molecular weight	103,000	114,000	?
Equilibrium (L-ribulose-5-P/D-xylulose-5-P)	—	0.54	0.83
K_m, L-ribulose-5-P	9.5×10^{-5} *M*	1×10^{-4} *M*	1.1×10^{-3} *M*
V_{max}, μmoles of L-ribulose-5-P utilized/min/mg	18.7 (37°)	12.5 (25°)	22.0 (37°)
pH optimum	7 and above	9.0	—

[a] This enzyme is homogeneous by sedimentation velocity, sedimentation equilibrium, and acrylamide gel electrophoresis and has been obtained in crystalline form.
[b] The absolute purity of this enzyme is now known, but its specific activity compares favorably with that of the other two enzymes.

Escherichia coli (*55*) and *Aerobacter aerogenes* (*56*), and highly purified preparations have been obtained from *Lactobacillus plantarum* (*57*). No added cofactors are required.

Earlier studies with the enzyme from *A. aerogenes* (*58*) indicated that no hydrogen or oxygen from the water was incorporated into the sugar, and it thus very much resembles the UDP-D-glucose 4′-epimerase.

The *E. coli* enzyme is homogeneous but has not been carefully examined for the presence of bound DPN. A very sensitive method was used to determine the presence of DPN in the enzyme from *L. plantarum* (*57*) and none was found. Nor has any DPN or other dissociable cofactor been found in a careful examination of the crystalline enzyme from *A. aerogenes* (*56*). If these enzyme(s) proceed by an oxidation–reduction mechanism, the hydrogen acceptor on the protein is not known. Some of the properties of these enzymes are shown in Table V.

IV. Epimerases Which Probably Proceed by a Keto-Enol Rearrangement

The enzymes discussed in this section catalyze the inversion of the configuration at a carbon atom adjacent to an aldehyde or keto group. Their proposed mechanism resembles that of an isomerase reaction,

55. N. Lee, J. W. Patrick, and M. Masson, *JBC* **243**, 4700 (1968).
56. J. D. Deupree and W. A. Wood, *JBC* **245**, 3988 (1970).
57. D. P. Burma and B. L. Horecker, *JBC* **231**, 1053 (1953).
58. M. W. McDonough and W. A. Wood, *JBC* **236**, 1220 (1961).

however, the proposed intermediate ene-diol is not released from the enzyme.

A. D-RIBULOSE-5-P 3′-EPIMERASE

D-Ribulose-5-P 3′-epimerase catalyzes the conversion of D-ribulose-5-P to D-xylulose-5-P. The enzyme was purified from *L. plantarum* (*59*) and obtained in homogeneous form from yeast (*60*); its properties are listed in Table VI. The enzyme has been detected in a variety of other cells (*59*), being an obligatory step in the pentose phosphate shunt.

TABLE VI
PROPERTIES OF D-RIBULOSE-5-P 3′-EPIMERASE

Property	Enzyme source	
	Yeast	*L. plantarum*
Molecular weight	45,800	—
Equilibrium constant = (D-ribulose-5-P/D-xylulose-5-P)	—	1.5
V_{max}, D-ribulose-5-P → D-xylulose-5-P (μmoles min/mg protein 25°)	250	69
pH optimum	8.0	7.5
K_m, D-xylulose-5-P	—	5×10^{-4} *M*
K_m, D-ribulose-5-P	—	1×10^{-3} *M*

This enzyme will incorporate one atom of tritium at C-3 during the reaction (*58, 59*), and an obvious mechanism for the reaction is shown in Fig. 6. It is worthwhile to point out that this mechanism assumes two groups on the enzyme which can act as hydrogen acceptors or donors for the hydrogens at C-3 either of D-xylulose-5-P or D-ribulose-5-P and

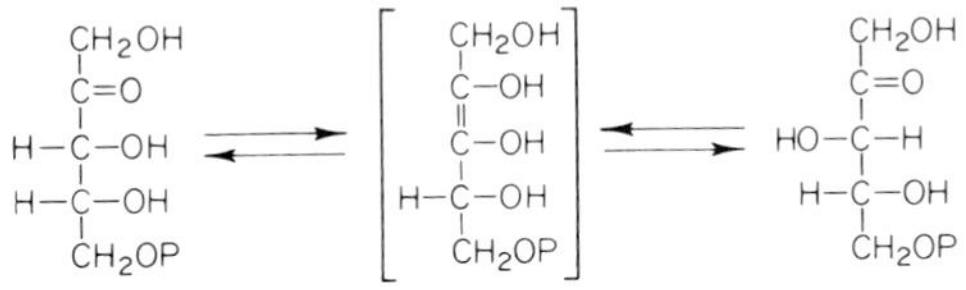

FIG. 6. Mechanism of D-ribulose-5-P 3′-epimerase. Enzyme-bound intermediates are shown in brackets.

59. J. Hurwitz and B. L. Horecker, *JBC* **223**, 993 (1956).
60. W. T. Williamson and W. A. Wood, "Methods in Enzymology," Vol. 9, p. 605, 1966.

resembles the mechanism which has been proven for the proline racemase by Cardinale and Abeles (*61*).

A detailed kinetic analysis of the rate of hydrogen incorporation from the medium, as compared to the catalytic rate, will be required to test this mechanism.

B. dTDP-L-Rhamnose Synthetase

dTDP-L-Rhamnose synthetase catalyzes the reaction

$$\text{dTDP-4-keto-6-deoxy-D-glucose} + \text{TPNH} \rightarrow \text{TPN} + \text{dTDP-L-rhamnose} + \text{H}^+$$

and requires inversion of the configuration at C-3 and C-5 of the hexose (see Fig. 7). In the overall reaction the hydrogen atoms at C-3 and C-5 of L-rhamnose are derived entirely from the solvent (*62*) and the hydrogen at C-4 is derived from TPNH (*63*).

In the overall reaction substitution of hydrogen by deuterium at C-3 gives a 3.4-fold isotope effect, and substitution of deuterium at C-5 gives a 2.0-fold isotope effect. Both of these are observed at saturating substrate levels and do not result from changes in K_m. The magnitude of both isotope effects suggest that they are primary isotope effects. Their occurrence not only substantiates the fact that the carbon hydrogen

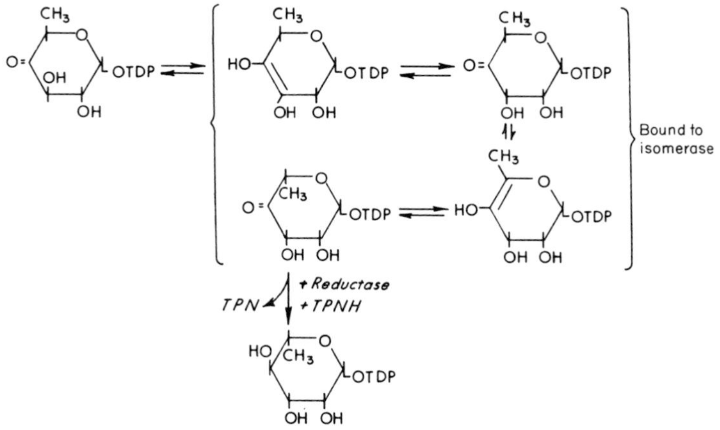

FIG. 7. Mechanism of dTDP-rhamnose synthetase. Enzyme-bound intermediates are shown in braces.

61. G. J. Cardinale and R. H. Abeles, *Biochemistry* **5**, 3971 (1968).
62. A. Melo and L. Glaser, *JBC* **243**, 1475 (1968).
63. L. Glaser, *BBA* **51**, 169 (1961).

TABLE VII
PARTIAL EXCHANGE OF THE HYDROGENS OF dTDP-4-KETO-6-DEOXY-D-GLUCOSE WITH SOLVENT[a]

Experiment	% of 4-keto-6-deoxy-D-glucose molecules having deuterium at positions indicated			
	C-3	C-3 and C-5	C-5	No deuterium
I	18.1	24	6.0	51.9
II	24.6	32.6	3.2	39.6

[a] TDP-4-keto-6-deoxy-D-glucose was incubated in D_2O with the E-II component of the TDP-L-rhamnose synthetase, and the deuterium distribution in the sugar was determined after reduction to the corresponding sugar alcohol (*17*, *62*). Longer incubation yields molecules which are fully deuterated at C-3 and C-5.

bonds at C-3 and C-5 are broken in the course of the reaction but also suggests that after deuterium substitution either of these steps can become rate limiting for the overall reaction (*17*).

The enzyme has been separated into two fractions, E-I and E-II. Neither fraction alone catalyzes any net reaction. E-I appears to contain the TPNH binding site, on the basis of heat denaturation data. E-II will catalyze the exchange of the hydrogen at C-3 and C-5 of the thymidine-linked 4-keto-6-deoxy-D-glucose with the solvent; this exchange is sequential, as shown by the fact that partial exchange in D_2O followed by mass spectrometry shows the presence of molecules containing deuterium at C-3, at C-3 and C-5, but essentially no molecules with deuterium at C-5 (Table VII). The fact that inversion at C-3 and C-5 is sequential suggests that the epimerization at C-3 induces a structural change in the enzyme that allows epimerization at C-5 to take place.

The proposed mechanism of the reaction is shown in Fig. 6. It suggests that E-II catalyzes inversion at C-3 and C-5 via corresponding enediols but that these molecules cannot be released from the enzyme. E-I then catalyzes the stereospecific reduction of the 4-keto group of E-II-bound dTDP-4-keto-6-deoxy-L-mannose. Neither of the enzymes has been obtained in pure form; thus, the proposed features have not been directly tested.

Other similar enzymes may be expected to follow the same mechanism, such as the enzyme(s) catalyzing the conversion of GDP-4-keto-6-deoxy-D-mannose to GDP-6-deoxy-L-galactose (L-fucose) which in every respect are entirely analogous to the rhamnose system (*64*).

64. V. Ginsburg, *JBC* **235**, 2196 (1960); **236**, 2389 (1961).

C. Cellobiose-2′-epimerase

Cellobiose-2′-epimerase is obtained from the culture filtrate of *Rumninococcos albus* strain 7 and catalyzes the reaction (*65*):

4-O-β-D-glucopyranosyl-D-glucose ⇌ 4-O-β-D-glucopyranosyl-D-mannose

When the reaction is carried out in D_2O one atom of deuterium is incorporated in C-2 of the reducing sugar (*66*). Free 4-O-β-D-glucopyranosyl-D-fructose (cellobiulose) is neither an intermediate in the reaction nor a substrate when added to the enzyme. The authors have proposed a stabilized carbanion as an intermediate for this reaction (*66*). At the present moment no distinction can be made between this suggestion and an enzyme-bound ene-diol as suggested for the epimerases involved in L-rhamnose biosynthesis.

D. *N*-Acetyl-D-glucosamine and *N*-Acetyl-D-glucosamine 6-P 2′-Epimerase

Two enzymes have been described which catalyze the inversion of the configuration of C-2 of an acetylated amino sugar. The first catalyzes the interconversion of *N*-acetyl-D-glucosamine-6-P and *N*-acetyl-D-mannosamine-6-P. It has been purified from *Aerobacter cloaeca* as an adaptive enzyme formed when this microorganism is grown in the presence of *N*-acetyl-D-mannosamine or *N*-acetyl-D-glucosamine. The enzyme has also been detected as an inducible enzyme in other microorganisms, although at lower levels than in *A. cloaeca* (*67*).

A similar enzyme catalyzing the interconversion of the corresponding free amino sugar has been partially purified from hog kidney (*68*); this enzyme is activated by ATP acting as an allosteric activator. ATP is not used in the reaction, nor are phosphorylated sugars intermediates in the reaction (*68*). The known properties of these enzymes are summarized in Table VIII. Their mechanism has never been studied, and one can only assume that they resemble the other enzymes in this group. It should be noted that at pH 11 and room temperature *N*-acetyl-D-glucosamine and *N*-acetyl-D-mannosamine are readily interconvertible nonenzymically (*69, 70*).

65. T. R. Tyler and J. M. Leatherwood, *ABB* **119,** 363 (1967).
66. H. Amein and J. M. Leatherwood, *BBRC* **36,** 223 (1969).
67. S. Ghosh and S. Roseman, *JBC* **240,** 1525 (1964).
68. S. Ghosh and S. Roseman, *JBC* **240,** 1531 (1965).
69. S. Roseman and D. Comb, *JACS* **80,** 3166 (1958).
70. R. Kuhn and R. Brossmer, *Ann. Chem.* **616,** 221 (1958).

TABLE VIII
PROPERTIES OF *N*-ACETYL-D-GLUCOSAMINE 2′-EPIMERASE

Property	*N*-Acetyl-D-glucosamine-6-P 2′-epimerase	*N*-Acetyl-D-glucosamine 2′-epimerase
Source	*A. cloaeca*	Hog kidney
K_m	Glu-NAc-6-P = 1.6×10^{-3} *M* Man-NAc-6-P = 2.4×10^{-3} *M*	Glu-NAc = 3.4×10^{-3} *M* Man-NAc = 3×10^{-3} *M*
V_{max} purest preparation (μmoles/mg protein/min) in direction of Glu NAc or Glu-NAc-6-P synthesis (37°)	478	5.3
Equilibrium constant	$\frac{\text{Man-NAc-6-P}}{\text{Glu-NAc-6-P}} = 0.43$	$\frac{\text{Man-NAc}}{\text{Glu-NAc}} = 0.26$
Activator and K_m	None	ATP (1×10^{-3} *M*) or dATP
pH optimum	7.6	7.6

V. Noncarbohydrate Epimerases

Three enzymes will be considered very briefly in this section, the methylmalonyl-CoA racemase, the mandelic acid racemase, and the lactic acid racemase. Amino acid racemases are discussed by Adams, Chapter 13, this volume.

A. METHYLMALONYL-COA RACEMASE

The methylmalonyl-CoA racemase has been purified from sheep liver (*71*) and has been obtained in pure form from *Propionibacterium shermanii* (*72*). The latter is a protein of molecular weight 29,000 showing unusual high stability to heat and acid. Both enzymes catalyze the exchange of a proton with the medium (*71–75*), although the published data do not clearly establish whether incorporation of a proton

71. R. Mazumder, T. Sasakawa, Y. Kaziro, and S. Ochoa, *JBC* **237,** 3065 (1962).
72. S. H. G. Allen, R. Kellermeyer, R. Stjernholm, B. Jacobson, and H. G. Wood, *JBC* **238,** 1637 (1963).
73. M. Sprecher, M. J. Clarck, and D. B. Sprinson, *JBC* **241,** 872 (1966).
74. J. Retey and F. Lynen, *Biochem. Z.* **342,** 256 (1965).
75. P. Overath, G. M. Kellerman, F. Lynen, H. P. Fritz, and H. J. Keller, *Biochem. Z.* **335,** 500 (1962).

from the medium is obligatorily involved in the reaction (*75a*). An enzyme-bound enolate has been postulated as an intermediate in this reaction (*75*).

B. Mandelic Acid Racemase

Mandelic acid racemase has been purified to homogeneity from *Pseudomonas putida* (*76*) as a protein of molecular weight 200,000 catalyzing the reaction:

$$C_6H_5-\underset{OH}{\overset{H}{C}}-COOH \rightleftharpoons C_6H_5-\underset{H}{\overset{OH}{C}}-COOH$$

The enzyme appears to be free of any bound cofactors. The K_m for mandelic acid is $4 \times 10^{-5}\, M$ and the turnover number is 1.72×10^3 sec^{-1}. Substitution of the α-hydrogen with deuterium gives a 5.4-fold isotope rate effect. Free benzoyl formic acid is not an intermediate (*76, 77*).

When tritium incorporation from the solvent is measured using D-mandelic acid as a substrate, equal isotope incorporation into D- and L-mandelic acid is observed at early time periods, indicating the formation of a symmetrical intermediate in the reaction, which the authors suggest is a carbanion (*77*). This assumption is supported by the observation that the rate of the reaction is increased by the presence of electron withdrawing groups at C-4 of the benzene ring when substituted mandelic acids are used as substrates.

When the actual rate of deuterium incorporation is measured as compared to the rate of the reaction, it is found that deuterium incorporation is considerably slower. This observation would indicate that an "intramolecular" hydrogen transfer is possible with this enzyme similar to what has been observed with phosphoglucose isomerase, where it is assumed that hydrogen is transferred from the sugar to a group on the enzyme, which can either exchange with the medium or transfer hydrogen back to either C-1 or C-2 of the sugar (*78*). It has been

75a. The proton at C-2 of methylmalonyl-CoA also exchanges nonenzymically with hydrogen of the medium especially at acid pH (*71–75*). The absolute stereochemistry of methylmalonyl-CoA has been determined (*72, 73*).

76. G. D. Hegeman, E. Y. Rosenberg, and G. L. Kenyon, *Biochemistry* **9,** 4029 (1970).

77. G. L. Kenyon and G. D. Hegeman, *Biochemistry* **9,** 4036 (1970).

78. I. A. Rose and E. L. O'Connell, *JBC* **236,** 3086 (1961).

suggested by Jencks that in the case of the phosphoglucose isomerase this observation need not represent the presence of a group on the enzyme with limited accessibility to the solvent but can be explained by the fact that for the phosphoglucose isomerase the rate of proton transfer to the solvent from a basic group on the enzyme is comparable in magnitude to the rate of catalysis (*79*). Whether this argument can be applied to the mandelic acid racemase case is not clear because of the following stereochemical requirement.

If one assumes that a single group on the enzyme can accept hydrogen from both D- and L-mandelic acid as the deuterium data would indicate, this would imply that either the steric position of the carboxyl and hydroxyl group is not fixed on the enzyme surface or that the hydrogen acceptor group on the enzyme is extremely mobile. No data are at present available to distinguish between these possibilities.

C. Lactic Acid Racemase

This enzyme has been detected in extracts of *Clostridium butylicum* ATCC 860 (*80*) and has not been purified. It catalyzes the interconversion of D- and L-lactic acid without the formation of free pyruvate or a requirement for DPN addition. No hydrogen exchange occurs with the medium, and substitution of the α-hydrogen by deuterium gives a 2-fold isotope effect (*80*). It is unclear at the moment whether an enzyme-bound DPN is involved in the reaction, and until the enzyme is purified, speculation about its mechanism seems unwarranted.

Acknowledgment

Work in the author's laboratory has been supported by NSF Grant GB-6243X and NIH Grant GMI-8405.

79. W. Jencks, "Catalysis in Chemistry and Enzymology," p. 209. McGraw-Hill, New York.
80. P. Dennis and S. S. Shapiro, *Biochemistry* **4**, 2283 (1965).

11

Cis–Trans Isomerization

STANLEY SELTZER

I. Enzymic Cis–Trans Isomerization about Carbon–Carbon Double Bonds

The subject of enzymic cis–trans isomerization has been treated previously in the second edition of "The Enzymes" by Knox (*1*) and Topper (*2*). The reader should consult these excellent reviews.

At the present time cis–trans isomerases appear to fall into two classes. The first encompass those that isomerize without double bond migration in the final product and the second include cis–trans isomerization accompanied with positional migration.

A. Geometrical Isomerization without Bond Migration

Up to now this class of cis–trans isomerases appears to require one or more sulfhydryl groups for enzymic activity. In the first subclass the

1. W. E. Knox, "The Enzymes," 2nd ed., Vol. 2, Part A, p. 253, 1960.
2. Y. J. Topper, "The Enzymes," 2nd ed., Vol. 5, p. 413, 1961.

sulfhydryl group is an integral part of the enzyme but in the second the sulfhydryl may be furnished as a coenzyme. This distinction may be specious in that only a small number of isomerases are presently known.

The majority of the isomerases in the isomerization-without-migration class are concerned with maleate or derivatives of maleic acid. These convert the substrate to fumarate or derivatives of fumaric acid and the fumarate derived from both of these products enter the tricarboxylic acid cycle. Presence of these enzymes and the equilibrium favoring the trans product (fumarate) likely account for the small amount of maleate and maleyl derivatives in nature as compared to fumarate and fumaryl derivatives (*3*).

1. *Maleate Isomerase*

Three maleate cis–trans isomerases have been extracted from different bacterial sources. In the first, Behrman and Stanier (*4*) demonstrated some time ago that a strain of *Pseudomonas fluorescens*, grown aerobically on nicotinic acid as the sole source of carbon, manufactures enzymes that oxidize and decarboxylate nicotinic acid to 2,5-dihydroxypyridine. Further enzymic oxidation opens the ring to an intermediate which is assumed to be *N*-formylmaleamic acid. The presence of maleamic acid and its hydrolysis to maleic acid could be demonstrated (*4*). The extract of *P. fluorescens* also contains an enzyme which catalyzes the cis–trans isomerization of maleate to fumarate.

In another study, Otsuka (*5*) exposed a growth medium containing maleate to air and isolated two bacteria. One of these, a *Pseudomonas* species, was found to have maleate cis–trans isomerase activity. An eightfold purification was accomplished by salt precipitation. The enzyme is reported to lose activity at this stage of purification if it is dialyzed against very dilute buffer but activity could be returned with low concentrations of glutathione or cysteine. Iodoacetate and *p*-mercuribenzoate at $10^{-5}\,M$ completely inhibit activity. Sodium fluoride inhibits activity by 19–45% between 6×10^{-4} and $3 \times 10^{-3}\,M$. Twenty percent inhibition is achieved by EDTA at about $10^{-4}\,M$ or by sodium cyanide at $6 \times 10^{-4}\,M$.

If strains of *P. fluorescens* isolated by Behrman and Stainer are grown on a medium containing maleic acid as the sole carbon source, relatively large quantities of the isomerase (about 0.5% of total protein) are

3. "Beilstein's Handbuch der organischen Chemie." Vol. II, pp. 737 and 748; Springer-Verlag, Berlin and New York, 1920 [1st Suppl., pp. 299 and 303 (1929); 2nd Suppl., pp. 631 and 641 (1942); 3rd Suppl., pp. 1890 and 1911 (1961)].

4. E. J. Behrman and R. Y. Stanier, *JBC* **277**, 923 (1957).

5. K. Otsuka, *Agr. Biol. Chem.* (*Tokyo*) **25**, 726 (1961).

furnished. The enzyme purified by Scher and Jakoby (*6*) by standard techniques is relatively unstable (*6*). It, however, appears to be homogeneous as shown by acrylamide disc gel electrophoresis and ultracentrifugation. Crystallization could be accomplished but is accompanied by a 50% loss in activity. As further indications of instability, a solution of the enzyme loses 10% of its activity overnight when frozen. At 4°, however, it is reported that 70% of the activity is lost after three months when stored in 10% glycerol containing 10 m*M* thioglycerol. Half of the lost activity is recovered if the solution is treated with fresh thioglycerol.

The molecular weight of the enzyme is 74,000 as determined by sedimentation velocity experiments. The amino acid composition has been determined, and from this the partial specific volume has been computed to be 0.74 ml/g.

The pH optimum for the enzyme catalyzed rate is 8.4 with half-maximal rates at 6.7 and 9.4. The turnover number is 18,000 moles/mole of enzyme per minute. This isomerase also requires nonspecific mercaptans for activity, but this appears to be needed to keep a functional sulfhydryl group on the enzyme reduced. The half-maximal rate near pH 9.4 lends further support to the notion of a functional sulfhydryl group although the importance of a possible tyrosine group with a p*K* of about 9.4 cannot be eliminated. Further evidence and information about the reactivity of a sulfhydryl group in this enzyme comes from reaction with bromoacetate. A 40% inhibition is obtained in the presence of two equivalents of bromoacetate at high dilution. The enzyme is not inhibited, however, by glutaconate, malate, mesaconate, or tartrate but is by deoxycholate.

The isomerase is specific for maleate. No other compound tested underwent cis–trans isomerization. These include: maleamide, *cis*-aconitate, ascorbate, *cis*-butene-1,4-diol, citraconate, crotonate, diethylmaleate, *trans*-epoxysuccinate, glutaconate, isocrotonate, linolenate, maleurate, oxaloglycolate, tiglate, and ferulate. This wide array suggests not only that the active site might be crowded but also that two carboxylate groups in the *cis*-1,2 position are required. The K_m for sodium or potassium maleate is 3×10^{-4} *M*. Maleate concentrations 100-fold higher than this produce no inhibitory effects.

Another maleate isomerase has been extracted from *Alcaligenes faecalis* IB-14 by Takamura and his co-workers (*7*). The purified enzyme

6. W. Scher and W. B. Jakoby, *JBC* **244,** 1878 (1969).

7. Y. Takamura, I. Kitamura, M. Iikura, K. Kono, and A. Ozaki, *Agr. Biol. Chem.* (*Tokyo*) **30,** 346 (1966); Y. Takamura, M. Soejima, and T. Aoyama, *ibid.* **31,** 207 (1967); Y. Takamura, T. Nakatani, M. Soejima, and T. Aoyama, *ibid.* **32,** 88 (1968); Y. Takamura, T. Takamura, M. Soejima, and T. Vemura, *ibid.* **33,** 718 (1969); **34,** 1501 (1970).

strongly resembles that from *P. fluorescens*. Here, too, the bacteria synthesizing the enzyme can be stimulated to produce the isomerase by the presence of maleate. What is more interesting is that malonate, hydroxymalonate, and ketomalonate are more effective than maleate itself in inducing isomerase synthesis; malonate is ten times more effective than maleate. The induction can be brought about by growing the bacteria in a basal medium and then transferring them to the induction medium. It is interesting that although the analogs of malonate induce enzyme synthesis they do not promote growth and thus do not appear to be metabolized. Maleate, on the other hand, when used as an inducer, does promote bacterial growth. Experiments were also performed to see what compounds could inhibit malonate induction. Oxalacetate, malate, fumarate, and succinate prevent induction by malonate. Maleate itself does not inhibit malonate induction. Takamura has pointed out that malonate inhibits the action of succinic dehydrogenase and the net effect is to reduce the supply of fumarate in the tricarboxylic acid cycle. It appears that the organism can overcome the deficiency of fumarate by manufacturing large quantities of maleate isomerase when it is presented with a supply of maleate. This induction is overridden when fumarate itself or its precursor, succinate, or subsequent products, malate and oxalacetate, are supplied; consequently, the requirement for fumarate synthesis is diminished.

As a further indication that malonate represses the tricarboxylic acid cycle is the fact that malonate induction is greatest when the bacteria are confined to anerobic conditions. Other studies indicate that the tricarboxylic acid cycle is further repressed when oxygen is omitted. Furthermore, ammonium ion, which is also known to inhibit the tricarboxylic acid cycle, also stimulates production of the isomerase.

The isomerase, obtained through malonate induction, was purified about 100-fold at which point it is monodisperse during ultracentrifugation. Sedimentation velocity studies and rates of gel filtration on Sephadex G-100 suggest that the molecular weight is less than 100,000.

Similar pH-rate data are observed for the isomerase from this source; the rate maximum occurs at pH 8.3, and at pH 6.8 and about 9.3 half-maximal rate is observed. The enzyme is highly unstable. Although highest activity is found at pH 8.3, it is rapidly lost at 30° at pH values more alkaline than 7 or more acidic than 5. Between 0° and 25°, acetate buffers near pH 6 provide the highest stabilizing effect if they contain small quantities of dithiothreitol.

The effect of various metal ions was examined. Cupric and mercuric ions at 10^{-4} M inhibit reaction 89 and 100%, respectively. Inhibition can be reversed rapidly with dithiothreitol and mercaptoethanol but only slowly with sodium thioglycolate. Zinc, lithium, nickelous, magnesium, and ferric ions have little effect while cobaltous, manganous, and ferrous ions appear to increase activity by 20–30%. That these metals are probably not specifically required is shown by the negligible decrease in rate of the purified isomerase in the presence of EDTA. Of course, this observation does not rule out the possibility that the enzyme binds a metal much more tightly than EDTA. It is conceivable that those metal ions enhancing activity are providing a mild reducing medium.

Here as with the maleate cis–trans isomerase studied by Otsuka, complete loss of activity is achieved with low concentrations of *p*-mercuribenzoate and can be reactivated with dithiothreitol, mercaptoethanol, and cysteine. Similarly, iodoacetate inactivates the isomerase but is less efficient than *p*-mercuribenzoate. The enzyme is completely inhibited with low concentrations of sodium periodate but moderate to small inhibition is achieved with *N*-bromosuccinimide, cyanide, and semicarbazide at 10^{-3} *M*. It is difficult to pinpoint the effects of the moderate inhibitors at this time, but excellent evidence for a functional sulfhydryl group has emerged for all three maleate cis–trans isomerases.

The isomerase from *A. faecalis* is also specific for maleate; neither fumarate, citraconate, mesaconate, nor malate undergoes change in the presence of the enzyme. The Michaelis constant for maleate is reported to be 2.8×10^{-3} *M*. This is about a factor of ten higher than the K_m reported for the *P. fluorescens* enzyme. It is difficult to compare the maximal rates of the two enzymes, but a crude calculation from the data reported for the *A. faecalis* isomerase suggests a turnover number one-fifth to one-tenth as large as that reported for the *P. fluorescens* isomerase.

Additions to the assay mixture of fumarate, malate, succinate, malonate, citrate, and aspartate at 10^{-3} *M* produce no inhibition. Thus, malonate, which itself induces the synthesis of the isomerase, has no apparent effect on the catalytic efficiency of the enzyme.

Aspects of the general mechanism and relevant chemistry of possible models for this and subsequent isomerases discussed in this section are considered later in this chapter.

2. *Maleyl Isomerases*

The metabolism of aromatic substrates is widespread. During this process, bacterial and mammalian enzymes catalyze the oxidation of the aromatic ring by successive introduction of hydroxy groups (*8*).

8. G. Guroff, J. W. Daly, D. M. Jerina, J. Renson, B. Witkop, and S. Undenfriend, *Science* **157**, 1524 (1967).

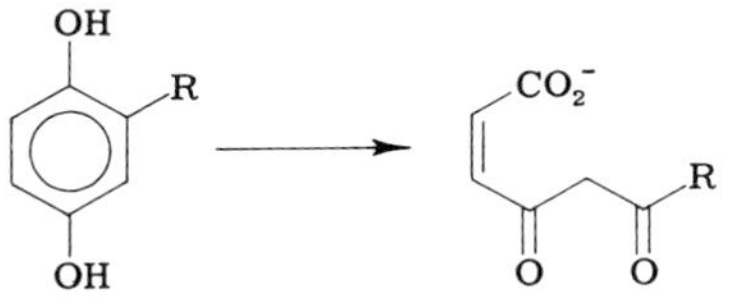

(Ia) R = CO_2H (IIa, b)
(Ib) R = CH_2CO_2H

(1)

After side chain degradation, a common type of intermediate is formed [(Ia) gentistic acid, (Ib) homogentistic acid] which upon further enzymic oxidation undergoes ring opening [Eq. (1)] to a new intermediate, maleylpyruvate (IIa) or maleylacetoacetate (IIb). These compounds are rapidly isomerized to fumaryl derivatives (IIIa,b) [Eq. (2)] which

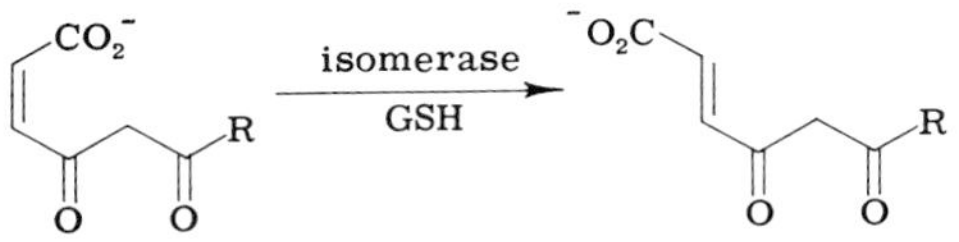

isomerase, GSH

(IIa, b) (IIIa, b)

(2)

are now suscepitble to enzymic hydrolysis to fumarate and acetyl derivatives. An extra dimension of difficulty is encountered in the study of these isomerases for the substrates are not commercially available, and in only one case at present has the substrate been chemically synthesized. They are very labile but they can be obtained through enzymic oxidation of gentisic and homogentisic acids in the absence of isomerase.

a. Maleylacetoacetate Isomerase. This isomerase from mammalian liver was first studied by Knox and Edwards (*9*) and has been reviewed by Knox (*1*) in the previous edition of this treatise. A brief recapitulation follows. Knox and Edwards partially purified the isomerase and demonstrated that sulfhydryl-containing compounds activated the enzyme. Two, however, thioglycolate and hydrogen sulfide, inhibited. Mercuric and cupric ions, *p*-mercuribenzoate, *N,N*-diethyldithiocarbamate, and *N*-ethylmaleimide inhibit reaction. All but the last was demonstrated to be reversible in the presence of glutathione. Complexing agents as bipyridyl, 8-hydroxyquinoline, and phenylthiourea have little effect. Consequently, it appears at first glance that a metal is not required. Glutathione is specifically required for activity and half-maximal activity occurs at 10^{-5} *M* GSH. Maximum rate occurs at pH 8.5. In an assay for isomerase activity the system contains a hydrolase that rapidly

9. W. E. Knox and S. W. Edwards, "Methods in Enzymology," Vol. 2, p. 295, 1955.

cleaves the fumarylacetoacetate that forms. The kinetics that are observed, however, are characteristically first order (*vide infra*). Continued turnover of the partially purified isomerase appears to inactivate the enzyme.

Maleylacetoacetate isomerase has also been obtained from a bacterial source by Chapman and Dagley (*10*). *Vibrio* 01, originally isolated by Happold and Key (*11*) from sewerage, can be grown on phenylacetic acid as the sole carbon source whereupon the organism furnishes maleylacetoacetate isomerase in addition to iron-containing oxidases and hydrolases required for metabolism. The enzyme was partially purified and like the mammalian enzyme requires glutathione for activity.

b. *Maleylpyruvate Isomerase.* Roof *et al.* (*12*) isolated an aerobic *Bacillus* bacterium capable of growth in a medium of sodium gentisate as the sole carbon source and noted that the aromatic system probably suffers oxidation and ring opening as with homogentistic acid. Similar results were obtained by Tanaka and co-workers (*13*) with *Pseudomonas ovalis.* In this study fumarylpyruvate was characterized as an oxidation product. At about the same time, Lack (*14*) independently isolated a strain of *Pseudomonas* from rat feces that could also metabolize gentisate. Extracts of the bacteria were shown to catalyze the oxidation of gentisate to maleylpyruvate when glutathione is absent. In the presence of GSH the extract converted gentisate to a product capable of oxidizing NADH in the presence of lactic acid dehydrogenase. Oxidase, isomerase, and hydrolase activities were separated resulting in a fivefold increase in the specific activity of the isomerase. The properties of the partially purified isomerase were studied.

The activity appears to be constant between pH 7.2 and 8.0; at pH 6.0, however, it drops to 60%. The isomerase is very specific; it catalyzes the isomerization of maleylacetoacetate but by only one fortieth the rate that it exhibits with maleylpyruvate. Liver maleylacetoacetate isomerase, on the other hand, is relatively nonspecific; it catalyzes maleylpyruvate isomerization at a slightly faster rate than it isomerizes maleylacetoacetate. Here, too, glutathione is required; cysteine, thioglycolate, or 2-mercaptoethanol cannot replace GSH. Some indications about the

10. P. J. Chapman and S. Dagley, *J. Gen. Microbiol.* **28**, 251 (1962).

11. F. C. Happold and A. Key, *J. Hyg.* **32**, 573 (1932); G. H. G. Davis and R. W. A. Park, *J. Gen. Microbiol.* **27**, 101 (1962).

12. B. S. Roof, T. J. Lannon, and J. C. Turner, *Proc. Soc. Exptl. Biol. Med.* **84**, 38 (1953).

13. H. Tanaka, S. Sugiyama, K. Yano, and K. Arima. *Bull. Agr. Chem. Soc. Japan* **21**, 67 (1957).

14. L. Lack, *BBA* **34**, 117 (1959); *JBC* **236**, 2835 (1961).

specificity for GSH come from an addition product of GSH and fumarylpyruvate. Increasing amounts gradually inhibit the isomerization. Fumarylpyruvate itself does not inhibit. One way to interpret these results is to suggest that the GSH-fumarylpyruvate adduct associates with the enzyme at a GSH binding site and that the adduct functions as a competitive inhibitor. To test the notion of a specific site for glutathione, it would be interesting to look at the effect of compounds which more closely mimic the structure of glutathione but contain no sulfhydryl group.

The isomerization carried out in D_2O proceeds with no incorporation of deuterium in the fumarate isolated from reaction. Nonenzymic isomerization of maleylpyruvate in D_2O containing hydrolase and catalyzed by GSH also proceeds with no deuterium incorporation in the isolated fumarate. A mechanism consistent with these results is shown in Eq. (3) (*14*). Rotation in the intermediate and rapid loss of GSH leads to

HO_2C CO_2H O O + GSH $\xrightarrow{-H^+}$ [GS HO_2C CO_2H O^- O] $\xrightarrow[-GSH]{+H^+}$ HO_2C CO_2H O O (3)

isomerization. Evidence for this kind of intermediate in a model reaction has been presented by Seltzer (*15*) and is discussed in Section II,B.

c. Maleylacetone Isomerase. Maleylacetone was first prepared by Kisker and Crandall (*16*) in impure form from aniline-catalyzed decarboxylation of maleylacetoacetate which in turn was derived from enzymic oxidation of homogentisic acid. They demonstrated rapid nonenzymic spontaneous and metal-catalyzed isomerization of the substrate (*vide infra*). More recently, maleylacetone has been chemically synthesized by Fowler and Seltzer (*17*). This analog undergoes catalyzed cis–trans isomerization in the presence of extracts of *Vibrio* 01 (*18*), a bacterium

15. S. Seltzer, *Chem. & Ind. (London)* p. 1313 (1959); *JACS* **83**, 1861 (1961).
16. C. T. Kisker and D. I. Crandall, *Tetrahedron* **19**, 701 (1963).
17. J. Fowler and S. Seltzer, *J. Org. Chem.* **35**, 3529 (1970).
18. S. Seltzer, unpublished work (1969–1970).

already noted to contain maleylacetoacetate isomerase (*10*). Whether this isomerase is specific for maleylacetone or is the maleylacetoacetate isomerase acting on a substrate analog remains to be determined. For the present it is classified here as a different isomerase.

Maleylacetone isomerase has essentially been purified to homogeneity as evidenced by its single band in acrylamide disc gel electrophoresis (*18*). Studies with sodium dodecyl sulfate solutions in acrylamide gel electrophoresis (*19*) and the elution pattern of the isomerase from Sephadex G-100 tentatively place the molecular weight between 30,000 and 45,000. Its density appears to be abnormal for when it is precipitated by a salt solution with a density of 1.19 g/cm^3, it floats.

The enzyme is relatively unstable. The presence of mercaptoethanol (10–15 mM), however, prevents rapid loss of activity. This enzyme, like the others, appears to have a sulfhydryl or other easily oxidizable functional group. Attempts to revitalize inactive enzyme with dithiothreitol, however, have failed. The enzyme requires glutathione for activity. Ethylenediaminetetraacetate appears to have a small stabilizing effect which probably results from its quenching of the known metal ion-catalyzed oxidation of sulfhydryl groups (*20*). The turnover number for maleylacetone is about 40 μmoles/min-μmole of enzyme assuming a molecular weight of 35,000 for the enzyme.

In the case of the three enzymes requiring glutathione for activity first-order kinetics are observed for the disappearance of the maleyl derivative (*1, 14, 18*). In the first two systems discussed, the enzyme is assayed by measuring the rate of loss of absorption at 330 nm. Since both reactants and products of isomerization absorb strongly at this wavelength, it would be difficult to monitor the progress of the reaction in this way. To overcome this problem, a hydrolase is added which is capable of rapidly cleaving the fumaryl derivatives to products which have no absorption at 330 nm. The hydrolase exhibits zero-order kinetics and has no action on maleyl derivatives.

As noted above, both isomerases are inhibited by the presence of substrate. It does not appear likely that this inhibition may be responsible for the nonzero-order kinetics that are observed. If, for example, substrate inhibition resulted from an allosteric effect, it would be likely that V_{max} would change. The concentrations of substrate used, however, could still give zero-order kinetics.

A possible explanation for the first-order kinetics comes from the fact that the rate of reaction of sulfhydryl compounds with this type of

19. K. Weber and M. Osborn, *JBC* **244,** 4406 (1969).
20. D. Povoledo, C. De Marco, and D. Cavallini, *Giorn. Biochim.* **7,** 78 (1958).

substrate is very fast. For example, near neutrality, glutathione adds to fumarylacetone with a rate constant of about $5 \times 10^2 M^{-1} \min^{-1}$ (*18*). The addition is accompanied by a loss of the highly conjugated chromophore. In assays for isomerase activity, glutathione is generally in large excess of the substrate and competes with hydrolase for the fumaryl derivative that is formed. First-order kinetics could be observed if the rate of cis–trans isomerization is faster than addition and if glutathione is more effective than hydrolase in capturing fumaryl derivatives.

B. Geometrical Isomerization with Bond Migration

1. *Cis–Trans Isomerization in Unsaturated Fatty Acids*

In the first stomachs of ruminant animals there exist bacteria which reduce unsaturated fatty acids to their saturated analogs (*21*). In the process some of the highly unsaturated acids undergo geometrical isomerization accompanied by migration of the double bond (*22*). In recent studies Kepler *et al.* (*23*) studied the reduction and isomerization of fatty acids by *Butyrivibrio fibrisolvens*. These anerobic bacteria are capable of reducing linoleic acid but in the process the original *cis*-9,*cis*-12-octadecadienoic acid is first isomerized to *cis*-9,*trans*-11-octadecadienoic acid. If the medium is rich in D_2O, deuterium is introduced at C-13 [Eq. (4)]. Subsequent *in vivo* reduction of the isomerized product

$$CH_3(CH_2)_4-CH{=}CH-CH_2-CH{=}CH-(CH_2)_7CO_2H \xrightarrow[D_2O]{\text{isomerase}} CH_3(CH_2)_4-CHD-CH{=}CH-CH{=}CH-(CH_2)_7CO_2H \quad (4)$$

leads to monoenoic acids: *trans*-11-octadecenoic and *trans*-9-octadecenoic acids. It appears that isomerization at the C-9 double bond might also be taking place after or during reduction of the C-11 double bond.

The isomerase has been partially purified by differential centrifugation (*23*). The majority of the activity sediments on centrifugation for 3 hr at 133,000*g*. Attempts to solubilize the particulate fraction with salts, solvents, detergents, ethylenediaminetetraacetate, or ultrasonic vibration have been unsuccessful. The isomerase has been examined with sucrose gradient ultracentrifugation. It concentrates in about the middle of a

21. R. Reiser, *Federation Proc.* **10,** 236 (1951).
22. F. B. Shoreland, R. O. Weenink, A. T. Johns, and I. R. C. McDonald, *Biochem. J.* **67,** 328 (1957).
23. C. R. Kepler, K. P. Hirons, J. J. McNeill, and S. B. Tove, *JBC* **241,** 1350 (1966); C. R. Kepler and S. B. Tove, *ibid.* **242,** 5686 (1967); C. R. Kepler, W. P. Tucker, and S. B. Tove, *ibid.* **245,** 3612 (1970), **246,** 2765 (1971).

0–70% gradient when centrifuged for 2 hr at 82,000*g*. This fraction, however, contains about 50% protein and 45% carbohydrate which suggests that the isomerase may be bound to cell wall constituents.

The demonstration that enzymic activity is not lost on gel filtration through Sephadex G-25 or G-200 indicates that a cofactor is not required for activity. Moreover, additions of ATP, ADP, AMP, Mg^{2+}, CoA, or NAD^+ do not increase enzymic activity. The pH optimum lies between 7.0 and 7.2 with half-maximal rates at about 6.0 and 8.2. The rate of linoleic isomerization is maximal within a very narrow range of concentration. Increasing quantities of linoleic acid show the typical rate-substrate concentration curve up to about 5×10^{-5} *M*, but a further increase of substrate concentration reduces the maximum rate. The K_m for linoleic acid is variable depending on the enzyme preparation, but an average of many determinations place it at about 1.2×10^{-5} *M*.

Linolenic acid [*cis*-9,*cis*-12,*cis*-15-octadecatrienoic acid (IV)] is also isomerized by this enzyme. The cis double bond at C-12 is isomerized here, too, and migrates to C-11 to yield *cis*-9,*trans*-11,*cis*-15-octadecatrienoic acid (V) [Eq. (5)]. The maximal rate is about twice that for linoleic acid and has a K_m of about 2.3×10^{-5} *M*. It too inhibits isomerase activity at high substrate concentrations.

CH_3CH_2 CH_2 CH_2 CH_2 $(CH_2)_6CO_2H$

(IV)

↓ isomerase (5)

$(CH_2)_6CO_2H$ CH_2 CH_3CH_2 $(CH_2)_2$

(V)

Extensive studies have been carried out to determine substrate specificity and functionality of the enzyme through the use of inhibitors (*23*). *cis*-Octadecenoic acids are not isomerized; *cis*-9,*cis*-11, and *cis*-12 acids were shown not to be converted. The effect of configuration and position of the double bond on the rate of isomerization appears to be critical as shown in Table I. Only those substrates that have both *cis*-9 and *cis*-12 double bonds undergo isomerization. The positioning of an additional double bond between *cis*-12 and the terminal methyl increases the rate, while if it is positioned between the carboxyl and the *cis*-9 double bond the rate is reduced. This could mean that the distance between binding sites for carboxyl and olefin groups is quite rigid.

TABLE I
EFFECT OF DOUBLE BOND POSITION ON LINOLEATE ISOMERASE[a]

Cis acid substrate	Rate of isomerization (nmoles/min/mg protein)
$\Delta^{9,12,15}$-C_{18}[b]	35.8
$\Delta^{6,9,12}$-C_{18}	11.6
$\Delta^{6,9}$-C_{18}	0
$\Delta^{8,11}$-C_{17}	0
$\Delta^{9,12}$-C_{18}	32.1
$\Delta^{10,13}$-C_{19}	0
$\Delta^{11,14}$-C_{20}	0
$\Delta^{9,12}$-C_{17}	19.0
$\Delta^{5,8,11,14}$-C_{20}	0
$\Delta^{5,8,11,14,17}$-C_{20}	0
$\Delta^{4;7,10,13,16,19}$-C_{22}	0

[a] From Kepler *et al.* (*23*).
[b] As an example the notation for the first entry refers to *cis*-9,*cis*-12,*cis*-15-octadecatrienoic acid.

Different acyl derivatives of linoleic acid were tested. No isomerization could be detected with the amide, hydrazide, methyl ester, hydroxamate, and triglyceride. Moreover, linoleyl alcohol, its methyl ether, linoleyl amine, linoleyl aldehyde and its oxime, and linoleyl methyl ketone all exhibited no isomerization in the presence of isomerase. Some of these, however, act as competitive inhibitors. Those derivatives that could supply a hydrogen bond (e.g., linoleamide and linoleyl alcohol) increase the apparent K_m for linoleic acid and have little effect on V_{max}.

Saturated C_{10}, C_{14}, C_{15}, C_{16}, C_{17}, and C_{18} fatty acids show little inhibitory effect. Unsaturated acids, on the other hand, inhibit enzymic isomerization. Comparison of *cis*- and *trans*-monenoic acids indicates equal competition for the enzyme. The effect of double bond position and chain length in *cis*-monoenoic acids is shown in Fig. 1. Little or no inhibition is observed for acids shorter than C_{16}. The major effect appears to be in chain length rather than position of the double bond, but a double bond is definitely needed for inhibition. If the position of the double bond is varied in octadecenoic acid it is found that when the double bond is at or near C-4 and when it is at or near C-12 the most potent inhibitors result. It is tempting to point out that if a model of the *cis*-4- and *cis*-12-octadecenoic acids were laid side by side with carboxyl groups in opposite directions, the double bonds would almost overlap with each other. This is to suggest that there is only one region on the enzyme for olefin binding. It might be broad enough, however, to bind a homoconjugated system. If there were two sites, however,

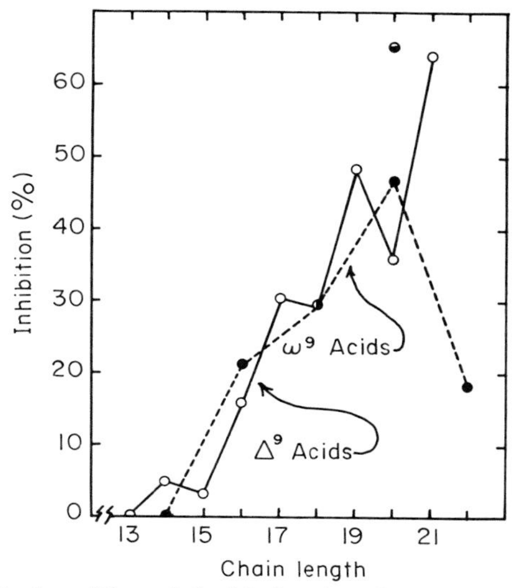

FIG. 1. The effect of position of double bond and chain length of *cis*-monoenoic acids on the enzymic rate of linoleic acid isomerization (*23*). Open circles refer to Δ^9 acids, closed circles to ω^9 acids, and the half-circle refers to *cis*-5-eicosenic acid.

cis-4,*cis*-12-octadecadienoic acid would be a more potent inhibitor than either of the monoenoic acids.

The importance of a sulfhydryl group is demonstrated by the complete loss of activity with 10^{-6} *M* *p*-mercuribenzoate. *N*-Ethylmaleimide at 5×10^{-5} *M* and iodoacetamide at 10^{-3} *M* inhibit 75 and 37%, respectively. A significant observation is that *o*-phenanthroline and EDTA are both noncompetitive inhibitors. Inhibition to half-maximal rate occurs with *o*-phenanthroline and EDTA at 1.5×10^{-4} and 5×10^{-3} *M*, respectively.

The inhibition studies support the presence of three regions involved in binding (*23*). The fact that only the underivatized acid undergoes isomerization and derivatives which can possibly hydrogen bond act as inhibitors suggest a site for binding of the carboxylate group through hydrogen bonding. That the more potent inhibitors are long chain unsaturated acids indicates a hydrophobic region but that long chain fatty acids are without effect suggests a region of olefin binding. Because *o*-phenanthroline is so effective it has been suggested that olefin binding might be through a metal π-complex (*23*). Since *o*-phenanthroline is a noncompetitive inhibitor, however, it would appear that in the inhibited enzyme there would still be room for substrate binding (*24*).

24. W. W. Cleland, "The Enzymes," 3rd ed., Vol. II, p. 1, 1970.

The function of one or more sulfhydryl groups remain obscure. It may be needed for metal binding. Relevant nonenzymic systems will be discussed in another section.

2. *Cis–Trans Isomerization of Aconitate*

cis-Aconitate, a product of citrate dehydration, undergoes catalyzed reversible transformation to *trans*-aconitate [Eq. (6)] in the presence of

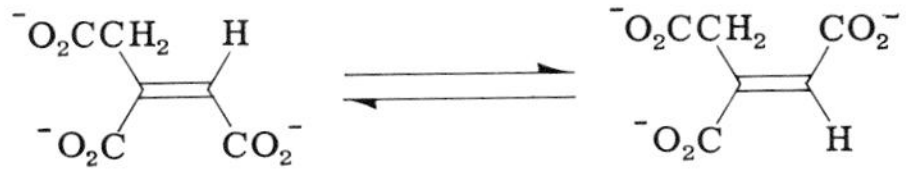

$$\begin{matrix} ^{-}O_2CCH_2 & & H \\ & C{=}C & \\ ^{-}O_2C & & CO_2^- \end{matrix} \rightleftharpoons \begin{matrix} ^{-}O_2CCH_2 & & CO_2^- \\ & C{=}C & \\ ^{-}O_2C & & H \end{matrix} \qquad (6)$$

aconitate isomerase. The enzyme has been partially purified recently by Klinman and Rose (*25*). Having a molecular weight of 78,000, the enzyme exhibits a turnover number in the neighborhood of $4\text{–}50 \times 10^2$ sec^{-1}. This is about a factor of 10^8 or more faster than the rate for specific base catalyzed isomerization.

As with linoleate, aconitate isomerization is accompanied by the incorporation of one atom of hydrogen label from the solvent. Solvent hydrogen can be introduced into the substrate in at least two possible ways: (1) by addition of a proton to give a carbonium ion intermediate (VI), or (2) by proton abstraction to give a carbanion (VII).

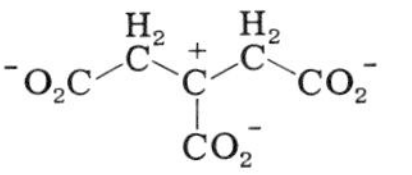

$$^{-}O_2C{-}CH_2{-}\overset{+}{C}(CO_2^-){-}CH_2{-}CO_2^- \qquad ^{-}O_2C{-}CH{\cdots}\overset{-}{C}(CO_2^-){\cdots}CH{-}CO_2^-$$

(VI) 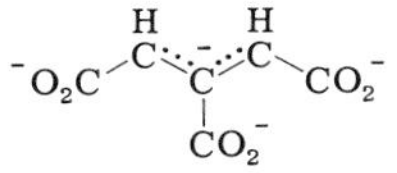 (VII)

Isomerization could be accomplished by internal rotation in either intermediate followed by reversible loss or gain or a proton, respectively. It should be noted, however, that both intermediates are symmetrical and thus the possibility exists that cis–trans isomerization is accompanied by double bond migration. By a set of elegant experiments, Klinman and Rose (*25*) demonstrated that cis to trans conversion is indeed accomplished through a 1,3-proton shift. Identification of the stereochemistry of the malate and glyoxylate from careful degradation of enzymically tritiated *cis*- and *trans*-aconitate indicates that in cis–trans interconversion 1,3-proton transfer is suprafacial. A small but apparently real amount of direct transfer of tritium from the 4-pro-S position of *cis*-aconitate to the 4-pro-S position of *trans*-aconitate can be detected

25. J. P. Klinman and I. A. Rose, *Biochemistry* **10,** 2259, 2267 (1971).

and suggests a carbanion mechanism with a single enzymic base responsible for removal and transfer of the allylic proton. The alternative carbonium ion intermediate appears less likely. For such a mechanism, a proton would first be transferred from the enzymic acid to the substrate followed by removal of tritium by the enzymic conjugate base. In order to introduce this bound tritium into a new substrate molecule, exchange of the enzymic acid would have to be slow during the time it takes for product to dissociate and substrate to be adsorbed.

There are other examples of isomerases in the literature that appear to catalyze 1,3-proton transfers. Some of these include the reaction of *trans*-α,β-hexadecenoyl-CoA to *cis*- and *trans*-β,γ-hexadecenoyl-CoA studied by Davidoff and Korn (*26*), the Δ^5 to Δ^4-3-ketosteroid isomerase investigated by Wang *et al.* (*27*), and the conversion of *trans*-2-decenoyl-*N*-acetylcysteamine to *cis*-3-decenoyl-*N*-acetylcysteamine of Rando and Bloch (*28*).

II. Nonenzymic Cis–Trans Isomerization about Carbon–Carbon Double Bonds

This section presents a few types of reactions that may have some relevance to enzymic isomerizations. The geometry about isolated carbon–carbon double bonds is quite stable. In the absence of catalysts it generally requires about 60 kcal/mole to thermally bring about geometrical isomerization. By now many different catalysts are available to effect this change and some are discussed below.

A. Photoisomerization and Related Reactions

Cis–trans photoisomerization is basic to the study of the visual process as discussed more fully by Heller, Chapter 17, this volume. Brief mechanistic consideration is given here, however, because of its general importance to geometrical isomerization.

In the interaction of light with olefins, the absorption of a photon of light of sufficient energy can lead to the excitation of an electron from the highest occupied bonding π-orbital to the lowest unoccupied π^*-orbital (*29*). In such an electronic configuration, theoretical calcu-

26. F. Davidoff and E. D. Korn, *JBC* **240,** 1549 (1965).
27. S-F. Wang, F. S. Kawahara, and P. Talalay, *JBC* **238,** 576 (1963).
28. R. R. Rando and K. Bloch, *JBC* **243,** 5627 (1968).
29. N. J. Turro, "Molecular Photochemistry," p. 176. Benjamin, New York, 1965.

lations indicate that such an excited triplet state could be most stable if the planes containing the groups around the central carbon–carbon bond were twisted 90° as in (VIII). In the absence of other effects the

$$\text{A(B)C=C(D)E} \underset{}{\overset{h\nu}{\rightleftharpoons}} \text{(VIII)} \underset{h\nu}{\rightleftharpoons} \text{A(B)C=C(E)D} \tag{7}$$

(VIII)

excited intermediate can then decay with equal probability to the initial reactant or isomerized product [Eq. (7)]. If equal amounts of cis and trans isomers are initially present no net geometrical isomerization would take place on photolysis were it not for the fact that isomers generally absorb at somewhat different wavelengths with different extinction coefficients. This being the case one isomer, generally the trans, can be selectively excited at longer wavelength. Isomerization can also be carried out with efficient triplet photosensitizers thereby avoiding the possibility of olefin-excited singlets causing other reactions. In at least one case, however, excited singlets can cause isomerization.

Oxygen exists as a ground state triplet ($^3\Sigma_g^-$) but can be excited chemically or by photosensitizers to singlet states $^1\Delta_g$ or $^1\Sigma_g^+$, which lie about 23 and 37 kcal/mole, respectively, above the ground state. Efficient geometrical isomerization of 15,15′-*cis*-β-carotene to all *trans*-β-carotene by singlet oxygen [Eq. (8)] has been reported recently (*30*).

$$O_2(^1\Delta_g) + \beta\text{-carotene} \rightarrow {}^3\beta\text{-carotene} + O_2(^3\Sigma_g^-) \tag{8}$$

In the reaction, singlet oxygen is produced from interaction with triplet methylene blue and therefore must be in the lower excited singlet state ($^1\Delta_g$) (*31*). Since 23 kcal are available, only those conjugated olefins which absorb at relatively high wavelength ($E_{\pi\rightarrow\pi^*} < 23$ kcal) might be expected to undergo isomerization by singlet oxygen. Aside from the fact that benzene is oxidatively cleaved by singlet oxygen to *trans,trans*-2,4-hexadienedial [Eq. (9)] (*32*) and that the production of maleyl

$$\text{C}_6\text{H}_6 + {}^1O_2 \longrightarrow \text{OHC–CH=CH–CH=CH–CHO} \tag{9}$$

derivatives is dependent on a similar remarkable enzymic ring cleav-

30. C. S. Foote, Y. C. Chang, and R. W. Denny, *JACS* **92,** 5218 (1970).
31. A. U. Khan and D. R. Kearns, *Advan. Chem. Ser.* **77,** 143 (1968).
32. K. Wei, J-C. Mani, and J. N. Pitts, Jr., *JACS* **89,** 4225 (1967).

age [Eq. (1)], there appears to be no direct evidence at the present time for significant concentrations of *free* singlet oxygen or singlet-like oxygen capable of causing geometrical isomerization in the enzymic systems discussed earlier.

B. Nucleophilic Catalysis

Geometrical isomerization catalyzed by nucleophiles is most efficient for α,β-unsaturated carbonyl compounds. In an early kinetic study, Nozaki and Ogg (*33*) investigated maleic to fumaric acid isomerization catalyzed by inorganic acids and salts. Their data show a correlation of rate with anion nucleophilicity and also demonstrate rate enhancement with increasing hydronium ion concentration. Nucleophilic attack on the carbon–carbon double bond can be aided by protonation of carbonyl oxygen [Eq. (10)]. That protonation is on oxygen rather than on

$$H_3O^+ + \mathrm{HC(CO_2H)}{=}\mathrm{CH(CO_2H)} \rightleftharpoons \mathrm{HC(C(OH)_2^+)}{=}\mathrm{CH(CO_2H)} + H_2O$$

$$X^- + \mathrm{HC(C(OH)_2^+)}{=}\mathrm{CH(CO_2H)} \rightleftharpoons \left[\mathrm{HC(C(OH)_2)}{-}\mathrm{CHX(CO_2H)}\right] \rightleftharpoons X^- + \mathrm{H_2\overset{+}{O}_2C(H)C}{=}\mathrm{C(H)CO_2H} \qquad (10)$$

olefinic carbon during hydrogen chloride [X = Cl, Eq. (10)] catalyzed isomerization of maleic acid was first shown by Horrex (*34*). No deuterium is incorporated into the vinyl hydrogens of fumaric acid when the isomerization is carried out in D_2O. Similarly, potassium thiocyanate-catalyzed isomerization of maleic-2,3-d_2 acid in H_2O leads to no loss of deuterium (*15*).

Further support for the direct attack of a nucleophile on the carbon–carbon double bond comes from the studies of Seltzer (*15*) on kinetic secondary α-deuterium isotope effects in the thiocyanate ion-catalyzed cis–trans isomerization of maleic-2,3-d_2 acid. An inverse effect, $k_H/k_D = 0.86$ at 25°, is in harmony with the mechanism proposed by Nozaki and Ogg [Eq. (10)]. In the rate-controlling step a tricoordinated vinyl car-

33. K. Nozaki and R. Ogg, Jr., *JACS* **63,** 2583 (1941).
34. C. Horrex, *Trans. Faraday Soc.* **33,** 570 (1937).

bon is being converted to a tetracoordinated carbon and the major effect is the replacement of the C–D out-of-plane bending force constant in the reactant with the larger D–C–X bending force constant in the transition state of nucleophilic attack on the carbon–carbon double bond (*35*).

Nucleophilic catalysis by amines has also been studied extensively. Nozaki (*36*) reported many years ago that primary and secondary amines actively catalyze cis to trans isomerization of diethyl maleate in ether. Piperidine is particularly efficient but tertiary amines are without activity. In this and a subsequent study, the reaction is reported to be second order in amine (*37*). In recent work, however, the reaction has been shown to be first order or less in amine when the medium is 95% alcohol (*38*). Consistent with the molecularity of reaction in ether is the mechanism shown in Eq. (11), i.e., one amine molecule functions as a general base and the other as a nucleophile. Cilento and Ferreira do Amaral have attempted to incorporate both functions in one molecule

$$2\,R_2NH + EtO_2C\text{–}HC{=}CH\text{–}CO_2Et \longrightarrow \left[\begin{array}{c} EtO_2C\text{–}HC{=}CH\text{–}CO_2Et \\ \vdots \\ R_2N\cdots H \\ \vdots \\ R_2NH \end{array}\right]^{\ddagger} \longrightarrow \begin{array}{c} CO_2Et \\ | \\ EtO_2C\text{–}HC(NR_2)\text{–}\overset{-}{C}H\text{–}C{\cdots}O \end{array} \qquad (11)$$

(*39*). Of several 1,ω-diaminoalkanes studied, 1,3-diaminopropane is the best catalyst for isomerization of dimethyl maleate in ether. Comparison of its rate constant with that for *n*-butylamine suggests that there may be enhancement but it still falls far short of the piperidine-catalyzed rate.

The previous studies have dealt with maleic acid and its esters. In an attempt to study catalyzed isomerization of another molecule resembling maleylacetoacetic or maleylpyruvic acids, *cis*-β-acetylacrylic acid was synthesized (*40*). Nuclear magnetic resonance, ultraviolet, and infrared spectra are in accord with the cyclic pseudo-acid structure (IX) for the compound but the anion (X) is open. The mechanism of spontaneous and catalyzed isomerizations has been studied (*41*).

35. See, e.g., A. Streitwieser, Jr., R. H. Jagow, R. C. Fahey, and S. Suzuki, *JACS* **80,** 2326 (1958).
36. K. Nozaki, *JACS* **63,** 2681 (1941).
37. M. Davies and F. P. Evans, *Trans. Faraday Soc.* **51,** 1506 (1955).
38. Z. Grünbaum, S. Patai, and Z. Rappoport, *J. Chem. Soc., B* p. 1133 (1966).
39. G. Cilento and C. Ferreira do Amaral, *J. Chim. Phys.* **64,** 1547 (1967).
40. S. Seltzer and K. D. Stevens, *J. Org. Chem.* **33,** 2708 (1968).
41. K. D. Stevens and S. Seltzer, *J. Org. Chem.* **33,** 3922 (1968).

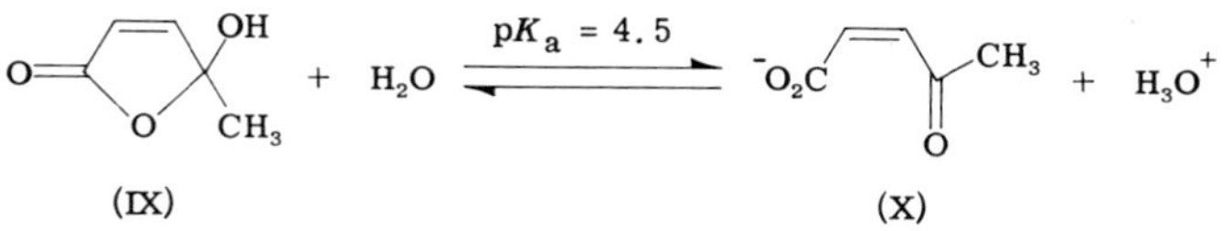

(IX) + H_2O ⇌ ($pK_a = 4.5$) (X) + H_3O^+ (12)

In 10^{-3} *M* hydrochloric acid and in the absence of any other nucleophile, the rate of isomerization is relatively slow. It is noteworthy that during the reaction there is a loss and then a reappearance of the conjugated chromophore; these observations are in accord with a hydrated intermediate. The major path of isomerization in this medium appears to be addition of water across the double bond followed by elimination to give the trans acid.

In the presence of a good nucleophile, thiocyanate ion, the model substrate is estimated to undergo isomerization about 3000 times faster than maleic acid at comparable conditions at 25° (*41*). In contrast to the isomerization in 10^{-3} *M* hydrochloric acid, the isomerization catalyzed by thiocyanate is not accompanied by the loss of the conjugated chromophore. In the case of the latter nucleophile, internal rotation and loss of X is faster than protonation. When the nucleophile is water, however,

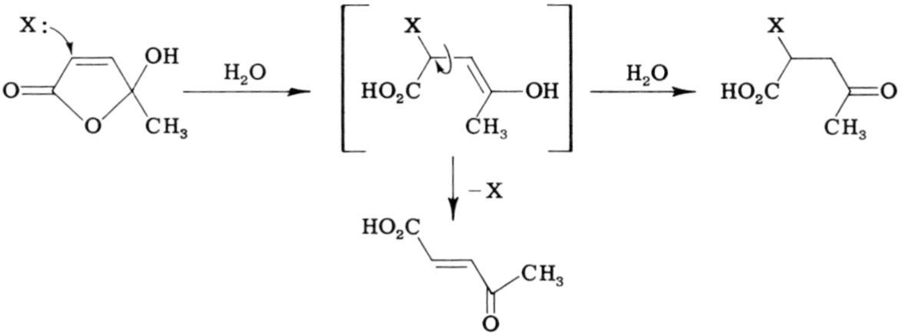

(13)

loss of X appears to be too slow to compete with protonation of carbon, and a sizable concentration of hydrated substrate forms.

The studies presented lend support to the thought that enzymic cis–trans isomerization of maleate and maleyl derivatives is brought about by the attack of a nucleophile on the substrate. Since inhibition studies on maleate isomerases have identified one or more sulfhydryl groups as necessary for activity it seems quite plausible at the present to suggest it as the enzymic nucleophile. Although amines have also been shown to be effective, there appears to be no indication that a functional one is present. Similarly, in the isomerization of maleyl derivatives, glutathione is required as a cofactor while there appears to be other sulfhydryl or other easily oxidizable groups on the enzyme. As suggested previously (*14, 41*), glutathione may be the important nucleophile in this case. A possible role of other enzymic sulfhydryls in this type of isomerase is mentioned below.

In maleate isomerase the enzyme probably provides electrophilic sites for binding of the carboxylate groups. This would be especially important in helping to stabilize the increased negative charge of the carboxyl groups that results from the formation of a sulfur–carbon covalent bond. Yet that site must be mobile enough after sulfhydryl addition to allow release of the carboxylate in order to permit internal rotation in the substrate.

This kind of electrostatic role for the enzyme can also be suggested for maleylacetoacetate, maleylpyruvate, and maleylacetone isomerases but one can speculate on another mechanism. Maleylacetone and most likely the other two maleyl substrates exist predominantly as the ketoenol system (XI) near neutral pH (*17*). These substrates could easily

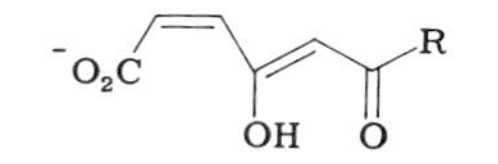

$^{-}O_2C$ … OH O … R

$R = CH_3,\ CH_2CO_2^-,\ CO_2^-$

(XI)

form Schiff bases (XII) with the isomerase if a lysyl ε-amino group were available. The Schiff base being more basic allows a greater fraction of

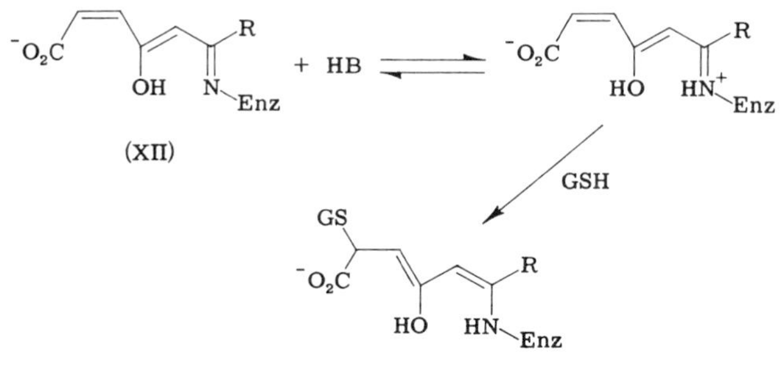

$^{-}O_2C$ … OH N–Enz … R (XII) + HB ⇌ $^{-}O_2C$ … HO HN^{+}–Enz … R

GSH ↓

GS, $^{-}O_2C$ … HO HN–Enz … R

(XIII)

(14)

the substrate to be protonated and thus could promote and stabilize attack by glutathione. Internal rotation in the intermediate (XIII), loss of glutathione, and hydrolysis could lead to fumaryl derivative products.

Accepting such a mechanism one can speculate further why glutathione is required for some isomerases while maleate isomerase uses an enzyme-bound sulfhydryl. If a Schiff base formed and instead of glutathione in (XIII) an enzyme-bound sulfhydryl were to form a covalent bond with the substrate, it would appear that internal rotation would be almost impossible without accompanying severe conformation changes in the enzyme. In the intermediate (XIII), however, rotation can take place by relaxation of any electrostatic binding that may exist between enzyme and glutathione and carboxylate groups. Extension of these thoughts to maleate isomerase leads to the conclusion that an enzyme-bound sulfhydryl is operationally sufficient if all other binding is electrostatic.

The possibility of Schiff base catalysis has been studied with *cis*-β-acetylacrylate (X) (*42*). With several amines the predominant reaction is addition to the carbon–carbon double bond of the substrate. A Schiff base could be formed, however, with semicarbazide. Because nitrogen is less electronegative than oxygen, the unprotonated Schiff base (XIV)

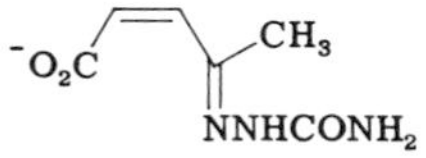

$^{-}O_2C$ CH_3 $NNHCONH_2$

(XIV)

reacts about 25 times more slowly than (X) with thiocyanate to give the *trans* product. The protonated form which is a mixture of (XV) and (XVI) reacts at about only one-half the rate of (IX). Since one might expect a factor of 25 for (IX) vs. (XV), the catalytic effect of (XVI)

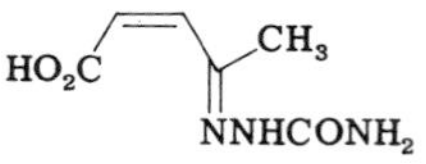

HO_2C CH_3 $NNHCONH_2$ (XV)

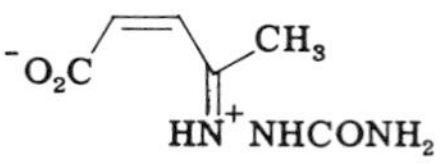

$^{-}O_2C$ CH_3 $HN^{+}NHCONH_2$ (XVI)

might be substantial in order that the combined rate [i.e., for (XV) and (XVI)] is only one-half that of (IX). It has not been possible to detect the presence of (XVI), but these kinetic experiments suggest a catalytic effect for the protonated Schiff base. Further studies to place this on a firmer basis are required.

C. Catalysis by Metals and Metal Ions

It has been known for some time that metallic films that catalyze the addition of hydrogen to olefins may, at the same time, cause cis–

42. C. Santiago and S. Seltzer, *JACS* **93**, 4546 (1971).

trans isomerization and double bond migration (*43, 44*). This can be thought of as an addition of a single hydrogen atom to an olefin adsorbed on the catalytic surface followed by the loss of a geminal or γ-hydrogen atom [Eq. (15)]. As expected, hydrogen exchange is observed. A similar type of reaction has been demonstrated in homogeneous solution with

H_3C CH_3
$HC{=}CH$ + Pt $\xrightleftharpoons{H_2}$ [H_3C CH_3 / $H_2C{-}CH$ / Pt] $\rightleftharpoons$ [CH_3 / $H_2C{-}CH$ / H_3C Pt]
cis

1-Butene $\qquad$ CH_3 / $HC{=}CH$ / H_3C $\qquad$ (15)

trans

complexes of rhodium(I) (*45*), iron(0) (*46*), and palladium(II) (*47, 48*). Evidence from deuterium exchange studies suggest that rhodium(I), for example, complexed to olefin might undergo oxidative addition (*49*) of a molecule of hydroxylic solvent or acid to yield a rhodium(III)–hydride complex as shown for (XVII). Transfer of hydrogen

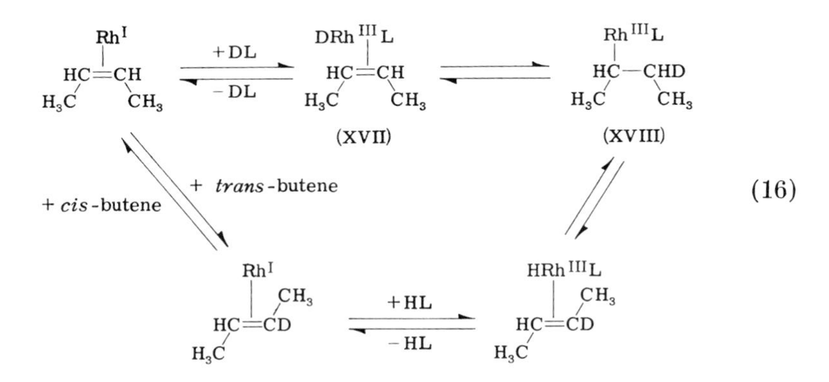

43. G. C. Bond, J. J. Phillipson, P. B. Wells, and J. M. Winterbottom, *Trans. Faraday Soc.* **60,** 1847 (1964).
44. J. Basset, F. Figueras, M. V. Mathieu, and M. Prettre, *J. Catal.* **16,** 53 (1970).
45. For information on catalysis by rhodium complexes and for a leading reference to other metal catalysis of isomerization, see R. Cramer, *JACS* **88,** 2272 (1966).
46. T. A. Manuel, *J. Org. Chem.* **27,** 3941 (1962).
47. M. A. Sparkle, L. Turner, and A. J. Wenham, *J. Catal.* **4,** 332 (1965).
48. N. R. Davies, *Aust. J. Chem.* **17,** 212 (1964).
49. J. P. Coleman, *Accounts Chem. Res.* **1,** 136 (1968).

from rhodium to olefin converts the complex to an alkyl rhodium compound (XVIII). Rotation about the central carbon–carbon single bond and reversal of the preceding steps lead to cis–trans isomerization with expected exchange of hydrogen with solvent. In some cases, however, the reversible transfer of hydrogen between carbon and metal is strictly intramolecular and exchange does not result (*48*).

A particular interesting set of reactions occurs with iron(0) complexes and nonconjugated dienes (*50*). 1,4-Pentadiene and triiron dodecacarbonyl react to form the iron tricarbonyl complex of *trans*-1,3-pentadiene (*51*) [Eq. (17)]. Hydrogen transfer appears to be intramolecular (*52*). If metallic nickel is used instead, conjugation with skeletal rearrangement results (*53*). It appears that stronger binding between

$$\text{1,4-pentadiene} + Fe_3(CO)_{12} \longrightarrow (\text{CH}_3\text{-diene})Fe(CO)_3 \qquad (17)$$

the iron *d*-orbitals and the diene π-orbitals is obtained when the olefin is conjugated, and this appears to be the driving force for the rearrangement. In the simplest model, 1,3-butadiene-iron(0) tricarbonyl, the iron atom resides outside the plane containing the four carbons of butadiene and is equally distant from each. It is interesting that in iron(0) complexes of this type the net effect is the donation of electron density from metal to olefin.

It should be recalled that during enzymic cis–trans isomerization of linoleic acid the Δ^9,Δ^{12}-diene system is similarly converted to a conjugated diene (*vide supra*). The presence of a protein-bound metal is suggested by the strong inhibition observed with *o*-phenanthroline. The isomerization shown in Eq. (17) may serve as a model. Iron(II) complexed to two enzyme-bound thiolate groups might have similar properties as iron(0) tricarbonyl and provide the proper electronic environment around the metal to effect the isomerization.

Recently, vanadous ion-catalyzed isomerization of maleate has been reported (*54*). Isomerization accompanies reduction of maleate to succinate and the oxidation of vanadous ion. Moreover, the isomerization

50. R. Petit and G. F. Emerson, *Advan. Organometal. Chem.* **1,** 1 (1964); H. Alper, P. C. Le Port, and S. Wolfe, *JACS* **91,** 7553 (1969).

51. R. B. King, T. A. Manuel, and F. G. A. Stone, *J. Inorg. & Nucl. Chem.* **16,** 233 (1961).

52. H. Alper, P. C. Le Port, and S. Wolfe, *JACS* **91,** 7553 (1969).

53. R. G. Miller, P. A. Pinke, and D. J. Baker, *JACS* **92,** 4491 (1970).

54. E. Vrachnou-Astra and D. Katakis, *JACS* **89,** 6772 (1967); E. Vrachnou-Astra, P. Sakellaridis, and D. Katakis, *ibid.* **92,** 3936 (1970).

when carried out in 50% deuterated water leads to the introduction of about 0.2 vinyl deuterium atoms. A mechanism which fits these observations is shown in Eq. (18). The lifetime of the intermediate radical ion (XIX) is apparently long enough to permit rotation about the central

$$\begin{matrix} CHCO_2V^+ \\ \| \\ CHCO_2H \end{matrix} + H^+ \longrightarrow \underset{(XIX)}{\left[\begin{matrix} HCHCO_2V^{2+} \\ | \\ \cdot CHCO_2H \end{matrix}\right]} \longrightarrow \begin{matrix} CHCO_2V^+ \\ \| \\ HO_2CCH \end{matrix} + H^+ \qquad (18)$$

carbon–carbon bond. The somewhat higher than statistical amount of deuterium incorporation suggests that the addition–elimination of hydrogen ions cannot be stereospecific because of the rapid inversion of the radical center. The nonstatistical distribution of vinyl deuterium can be attributed to a primary deuterium isotope effect in the conversion of (XIX) to fumarate. The mechanism is further supported by the observation that the monomethyl ester undergoes reduction but the diethyl ester is not isomerized. Vanadous ion also catalyzes the isomerization, and at the same time, the reduction of chloromaleic and citraconic acids. It, however, does not reduce itaconic acid. It appears then that it is necessary to have the carbon–carbon double bond in the center of a highly conjugated system for reduction or isomerization to take place.

The small amount of information pertaining to exchange between solvent and vinyl hydrogen in maleyl isomerases indicates no deuterium incorporation and would tend to rule out such an oxidation–reduction mechanism for the enzyme.

Both maleylacetone and maleylacetoacetate undergo silver ion–catalyzed cis–trans isomerization (*55*). The catalyzed rate with maleylacetone is maximal at pH 5 or greater and drops with decreasing pH until about pH 1 there is no catalysis. The pH of half-maximal rate is about 3.5 which corresponds nicely to the pK_1 of 4.0 for maleylacetone (*17*, *55*). In aqueous solution the structure of maleylacetone in the acid form can best be represented as the cyclic pseudo-acid (XX) but on loss of a proton is converted mainly to the enol carboxylate (XXI) with smaller amounts of the diketo carboxylate (XXII) (*17*). Analogous properties are expected for maleylacetoacetic acid. It is interesting that silver nitrate was the only one of several salts tested (i.e., aluminum nitrate, cadmium sulfate, cobaltous nitrate, cuprous chloride, cupric sulfate, chromic chloride, ferrous sulfate, ferric nitrate, mercurous nitrate, mercuric acetate, magnesium sulfate, manganous sulfate sodium nitrate,

55. See ref. *16*.

(XX)

$\underset{-H^+}{\overset{+H^+}{\rightleftharpoons}}$ (19)

(XXI) + (XXII)

nickelous sulfate, lead hydroxyacetate, platinic chloride dihydrochloride, and zinc sulfate) that catalyzed the reaction. It is well known that silver ion forms stable π-complexes with carbon–carbon double bonds. Extensive orbital reorganization accompanies complexation. The olefin carbons of cyclopentene and cyclohexene complexed to silver ion have their C-13 NMR signals shifted upfield by 4.4 ppm while adjacent carbons are shifted downfield (*56*). These results are consistent with greater silver *d*-orbital donation to carbon than carbon π-orbital donation to silver (*57*). In aqueous solution the stability of silver(I) and copper(I) π-complexes appear to increase if a C–O group is substituted on the double bond as in allyl alcohol and α,β-unsaturated acids (*58*). It should be noted that two such C–O groups are present in (XXI) and suggests that the active intermediate in the silver ion-catalyzed isomerization of maleylacetone is the silver-π-complex. The detailed mechanism remains to be elucidated.

Silver(I) can complex with two, three, or four ligands. In a π-complex with maleylacetone its function may be to bring a nucleophile (ligand) and the substrate into close proximity and to help stabilize an addition product [Eq. (20), where N indicates nucleophile] long enough for internal rotation to take place. That copper(I) does not also catalyze

(20)

56. R. G. Parker and J. D. Roberts, *JACS* **92,** 743 (1970).
57. M. J. S. Dewar, *Bull. Soc. Chim. France* **18,** C79 (1951).
58. R. M. Keefer and L. J. Andrews, *JACS* **71,** 1723, 2379, 2381, and 3906 (1949).

isomerization, however, casts doubts upon such a mechanism. Moreover, one might expect that addition of a nucleophile might be better stabilized by chelation with the ketoenol group. Cupric and ferric ions bind strongly to acetylacetone and would thus be expected to act similarly toward maleylacetone. Yet these metal ions appear to show no catalytic effect.

The possibility also exists, however, that silver(I) is undergoing reversible oxidation of the type already discussed above for vanadous ion [Eq. (18)]. If this is the mechanism, exchange of vinyl hydrogen with the solvent should take place.

D. Catalysis by Reversible Addition of Radicals

This kind of catalysis has been touched upon in the previous edition *(1)* and will not be considered further here.

Acknowledgments

The author wishes to thank Dr. Irwin A. Rose for providing copies of manuscripts in advance of publication. He would also like to acknowledge support for research in his own laboratory by the U. S. Atomic Energy Commission and to the National Institutes of Health for a Special Fellowship, 1F03 GM06392-01, at the Biochemistry Department, Brandeis University, where some of these thoughts were formulated. The author would also like to express his appreciation to Dr. Robert H. Abeles for his help in the purification of maleylacetone isomerase.

12

Phosphomutases

W. J. RAY, JR. • E. J. PECK, JR.

I. Introduction

The existence of enzymes which catalyze an apparent intramolecular phosphate transfer, e.g., between the 1 and 6 positions of glucose or the 2 and 3 positions of glyceric acid, was established in the mid-1930's by the Coris and Colowick (*1–3*) and by Meyerhof and Kiessling (*4*). In metabolic pathways these phosphomutases generally prepare phosphorylated carbon skeletons for subsequent energy-producing or important biosynthetic reactions. Thus, phosphoglucomutase provides a link in the pathway for glycogen metabolism by means of the glucose-1-P:glucose-6-P interconversion, while phosphoglycerate mutase prepares the three-carbon acid for subsequent energy-producing steps in glycolysis. In addition, phosphopentomutases, which catalyze the transfer of phosphate between the 1 and 5 positions of ribose or deoxyribose, are involved in the maintenance of the ribose-1-P and deoxyribose-1-P pools in those tissues which utilize the "salvage" pathway for the synthesis of nucleotides (*5*), and also appear to be important in the catabolism of deoxyribose-1-P (*6*). Brief histories recounting the early development of some of our present ideas about phosphoglucomutase and phosphoglycerate mutase appear in the second edition of this treatise (*7, 8*).

II. Phosphoglucomutase

Because most of the work on phosphoglucomutase has been conducted with the rabbit muscle enzyme this form of the enzyme is implied in this section unless otherwise indicated.

A. Preparation

1. *Isolation*

Phosphoglucomutase is isolated from rabbit muscle extracts by a series of precipitations involving pH, heat, and addition of ammonium sulfate;

1. C. F. Cori, S. P. Colowick, and G. T. Cori, *JBC* **121**, 465 (1937).
2. G. T. Cori, S. P. Colowick, and C. F. Cori, *JBC* **124**, 543 (1938).
3. G. T. Cori and C. F. Cori, *Proc. Soc. Exptl. Biol. Med.* **36**, 119 (1937).
4. O. Meyerhof and W. Kiessling, *Biochem. Z.* **276**, 239 (1935).
5. H. O. Kammen and R. Koo, *JBC* **244**, 4888 (1969).
6. C. E. Hoffmann and J. O. Lampen, *JBC* **198**, 885 (1952).
7. V. A. Najjar, "The Enzymes," 2nd ed., Vol. 6, p. 161, 1962.
8. L. Pizer, "The Enzymes," 2nd ed., Vol. 6, p. 179, 1962.

a final purification on modified cellulose or Sephadex resins is usually performed (see below). Several isolation procedures differing somewhat in detail have been published (*7, 9–11*) and possible variations in the product resulting from such differences have not been examined. This causes difficulty in comparing reports of multiple forms of the enzyme (see Section II,B,5); moreover, because of differences in assay procedures (see Section II,G) the reported specific activity of a given preparation cannot always be related to its purity. The rabbit muscle enzyme is also available commercially; however, such enzyme may be largely in the dephospho form (*12, 13*), whereas the freshly isolated product is predominately the phospho-enzyme (*7, 9, 11, 14, 15*).

Procedures for isolation of phosphoglucomutase from a variety of other sources have been published: flounder and shark muscle (*14*), human muscle (*9*), baker's yeast (*16*), *Escherichia coli* (*17*), potato tubers (*18*), *Bacillus cereus* and *Micrococcus lysodeikticus* (*19*). The enzyme has also been obtained from rabbit liver, rat muscle, and rat liver (*9, 20*), and partial purification from a number of other sources has been reported: pea extracts (*21*), tomato leaf plastids (*22*), broad bean extracts (*23*), green gram seeds (*24*), and jack beans (*25, 26*).

2. *Chromatography*

In addition to variations in isolation procedure, substantial differences in chromatographic techniques have been employed in final purification steps (*5, 7, 9, 27, 28*).

9. J. G. Joshi, J. Hooper, T. Kuwaki, T. Sakurada, J. R. Swanson, and P. Handler, *Proc. Natl. Acad. Sci. U. S.* **57,** 1482 (1967).

10. V. A. Najjar, "Methods in Enzymology," Vol. 1, p. 294, 1955.

11. J. A. Yankeelov, Jr., H. R. Horton, and D. E. Koshland, Jr., *Biochemistry* **3,** 349 (1964)

12. J. B. Alpers and G. K. H. Lam, *JBC* **244,** 200 (1969).

13. O. H. Lowry and J. V. Passonneau, *JBC* **244,** 910 (1969).

14. T. Hashimoto and P. Handler, *JBC* **241,** 3940 (1966).

15. J. B. Sidbury and V. A. Najjar, *JBC* **227,** 517 (1967).

16. M. Hirose, E. Sugimoto, R. Sasaki, and H. Chiba, *J. Biochem.* (*Tokyo*) **68,** 449 (1970).

17. J. G. Joshi and P. Handler, *JBC* **239,** 2741 (1964).

18. R. Pressey, *J. Food Sci.* **32,** 381 (1957).

19. K. Hanabusa, H. W. Dougherty, C. del Rio, T. Hashimoto, and P. Handler, *JBC* **241,** 3930 (1966).

20. T. Hashimoto, R. Sasaki, and Y. Haruhisa, *BBRC* **27,** 368 (1967).

21. C. S. Hanes, *Proc. Roy. Soc.* **B128,** 421 (1940).

22. N. M. Sisakyan and A. M. Kobyakova, *Dokl. Akad. Nauk SSSR* **67,** 703 (1949).

23. K. Onodera, *J. Agr. Chem. Soc. Japan* **25,** 377 (1951–1952).

24. T. Ramasarma, J. Sri Ram, and K. V. Giri, *ABB* **53,** 167 (1954).

A recurring and apparently little-appreciated problem with procedures employing both CM-cellulose and CM-Sephadex is that phosphoglucomutase binds to these resins only at low ionic strengths where neighboring charge–charge interactions are strong and where column load grossly affects chromatographic behavior. In fact, the commonly observed "breakthrough peak" appears to be produced by an overloading effect. Thus, the electrophoretic pattern of the phosphoglucomutase isozymes in such breakthrough peaks is essentially the same as that of the starting material (*29*). Although there is other evidence that the phosphoglucomutase in this fraction is different from the enzyme that is later eluted from the column (*9, 27*), the evidence does not appear to be conclusive.

Moreover, the elution of the enzyme from these resins does not involve true chromatography; instead, desorption appears to accompany a sharp change in the ionic state of the resin produced by the elution gradient. In such cases, the observation of artifactual peaks may be the rule rather than the exception. Separation problems also have been encountered during DEAE-cellulose chromatography: isozyme fractions that appear to be cleanly separated are badly cross-contaminated, as judged by their electrophoretic patterns (*29*). In fact, no entirely satisfactory chromatographic procedure has as yet been published, although a reproducible procedure has been developed (*30*).

3. ^{32}*P-Phosphate-Labeled Enzyme*

Labeled phosphoglucomutase is usually prepared by equilibrating the phospho-enzyme (E_P) with, for example, 0.1 mole fraction of ^{32}P-glucose-P (*28*); no activation step (see Section II,G) is required for this process (*31*). If the enzyme is partially or completely in the dephospho form, sufficient glucose diphosphate should be added to ensure conversion to the phospho form (*14*); however, excess diphosphate will reduce the fractional incorporation of label. The labeled substrate may be eliminated by either absorption of the enzyme on CM-cellulose followed by thorough washing and subsequent elution of the enzyme (*28*), or by selective absorption of phosphates on short anion-exchange columns (*14*). Neither

25. C. E. Cardini, *Enzymologia* **15,** 44 (1951).
26. S. F. Yang and G. W. Miller, *BJ* **88,** 505 (1963).
27. J. G. Joshi and P. Handler, *JBC* **244,** 3343 (1969).
28. W. J. Ray, Jr. and D. E. Koshland, Jr., *JBC* **237,** 2493 (1962).
29. D. M. Dawson and A. Mitchell, *Biochemistry* **8,** 609 (1969).
30. W. J. Ray, Jr., unpublished results (1971).
31. T. Hashimoto, J. G. Joshi, C. del Rio, and P. Handler, *JBC* **242,** 1671 (1967).

dialysis nor gel filtration is adequate for this separation, apparently because the enzyme binds glucose phosphates quite tenaciously (see Section II,B,1). The dephospho-enzyme (E_D) has also been labeled with ^{32}P-glycerate 1,3-diphosphate (*12*); presumably this procedure could also be used with the phospho-enzyme if a small amount of glucose-P (equilibrium mixture of glucose 1- and 6-phosphates) were present to act as a label carrier.

4. *Dephospho-Enzyme*

Treatment of the phospho form of phosphoglucomutase with excess glucose-1-P at enzyme concentrations much less than 10^{-8} *M* converts the phospho-enzyme almost exclusively to the free dephospho form; however, under conditions likely to be useful for preparative purposes, i.e., at concentrations of several milligram per milliliter (about 10^{-4} *M*), only a very small fraction of the enzyme is present as free E_D in the presence of excess glucose 1- and 6-phosphates (see Section II,D,2). Although a number of workers (*13, 19, 32, 33*) have isolated dephospho-enzyme by multiple treatments with excess glucose phosphate interspersed by extensive dialysis, other workers (*34, 35*) have shown that only small amounts of dephospho-enzyme are formed in a given equilibration step prior to dialysis. Although it is not clear precisely how the equilibrium among E_P + glucose-P, E_D + glucose diphosphate, and the various central complexes (see Section II,D,1) would shift during dialysis, so little of the enzyme is initially present as free E_D that such a procedure is not attractive; it is conceivable that the dialysis procedure is successful only when a phosphatase is present as a contaminant (*35*). When present in a mixture, the phospho and dephospho forms of the enzyme can be separated chromatographically (*11, 12, 27*).

A more reasonable way to prepare dephospho-enzyme would be to force the equilibrium to the free dephospho form prior to removal of glucose phosphates. The success of a procedure employing a short anion exchange column for removal of glucose phosphates may depend on such considerations, although both the mono- and diphosphates of glucose are absorbed under the reported conditions (*14*). However, phosphate transfer from the phospho-enzyme to the 1 position of xylose in the presence of phosphate, or to water in the presence of xylose-1-P (see

32. J. A. Yankeelov, Jr. and D. E. Koshland, Jr., *JBC* **240,** 1593 (1965).
33. V. A. Najjar and M. E. Pullman, *Science* **119,** 631 (1954).
34. E. J. Peck, Jr., D. S. Kirkpatrick, and W. J. Ray, Jr., *Biochemistry* **7,** 152 (1968).
35. A. D. Gounaris, H. R. Horton, and D. E. Koshland, Jr., *BBA* **132,** 41 (1967).

Section II,F) seems to hold the greatest promise for a quick and easy preparation of sizable quantities of the dephospho-enzyme.

5. *Storage*

Purified rabbit muscle phosphoglucomutase (phospho form) can be stored for a week or longer with no detectable loss in activity in 0.15 *M* acetate buffer, pH 5, either at 4° (*10*) or at room temperature, or at neutral pH in the presence of saturating concentrations of metal ions (*30*). For longer time periods, storage of frozen solutions in liquid nitrogen (*11*) is recommended for both phospho and dephospho forms; however, repeated freezing and thawing may cause loss of activity (*7*). Lyophilization of relatively concentrated solutions of purified enzyme (e.g., 60 mg/ml) can be conducted with insignificant activity loss (*30*). Storage of the crystalline form as an ammonium sulfate slurry has been used (*10*) but is not recommended since less active polymers are apparently produced thereby (*27*).

B. Physical and Chemical Properties

1. *Purity*

Because of differences in preparation (see Section II,A,1) criteria of purity published by one author may not be applicable to the enzyme isolated by another; moreover, because activation of the enzyme by complexing agents (see Section II,G) was poorly understood at the time when much of the work on purity was done, comparison of reported activities as an index of purity cannot be made with certainty. However, homogeneity by electrophoresis (*11, 36*), equilibrium centrifugation (*11, 37*), and sedimentation (*27, 38, 39*) has been reported for the rabbit muscle enzyme. Enzymic activities of a variety of other enzymes are absent from phosphoglucomutase preparations prior to chromatography (*7, 10*), although phosphoglucomutase obtained commercially may contain substantial transaminase activity (*40*). In addition, several workers have noted that the isolated enzyme frequently contains bound glucose phosphates, even after treatments such as ammonium sulfate precipitation, gel filtration, or extensive dialysis (*14, 19, 41–43*).

36. V. A. Najjar, *JBC* **175,** 281 (1948).
37. D. L. Filmer and D. E. Koshland, Jr., *BBA* **77,** 334 (1963).
38. S. Harshman and H. Six, *Biochemistry* **8,** 3423 (1969).
39. P. J. Keller, C. Lowry, and J. F. Taylor, *BBA* **20,** 115 (1956).
40. A. Waksman and E. Roberts, *ABB* **109,** 522 (1965).
41. J. G. Joshi, T. Hashimoto, K. Hanabusa, H. W. Dougherty, and P. Handler,

The flounder enzyme is nearly homogeneous by paper electrophoresis and is essentially free of 14 common enzymic activities, as is the shark enzyme (*14*); the human muscle enzyme is homogeneous by sedimentation and density gradient centrifugation (*27*), the yeast enzyme by sedimentation and electrophoresis (*16*), and the *E. coli* enzyme may be homogeneous (*17*). Enzymes from potato tubers (*18*), *M. lysodeikticus*, and *B. cereus* (*19*) have been partially purified but are inhomogeneous by one or more criteria.

2. *Physical Properties*

Molecular weights in the range 62,000–67,000 have been reported most frequently for the rabbit muscle enzyme: 62,000 $\pm$ 2000 (*11, 37*) and 64,900 (*38*) by equilibrium sedimentation, 65,900 and 67,000 based on summation of integral residues (*11, 44*), and 60,000–70,000 based on gel filtration (*38*). Surprisingly, the two determinations involving equilibrium centrifugation (*37, 38*) show rather different concentration dependencies under nearly identical conditions except for identity of the buffer anion. Early work was based on a molecular weight of 74,000 (*39*); hence, those comparisons of the rabbit muscle enzyme made herein have been recalculated on the basis of 65,000, which is an average of the more reliable values reported.

Phosphoglucomutases from other sources have molecular weights similar to the rabbit muscle enzyme: *E. coli*, 62,000–65,000 (*17*); shark and flounder 63,000 (*14*); *B. cereus*, 63,000 (*41*); potato tuber, 63,000 (*18*); human muscle, 60,000 (*27*); *M. lysodeikticus*, 58,000 (*19*); and jack bean, 63,000 (*25*). Most of these values were determined by sedimentation and were not examined for concentration dependency; hence, none should be considered as more than a first approximation of the true molecular weight.

Extrapolated values of 3.82 and 3.69 for $s_{20,w}$ are given for the rabbit muscle enzyme (*38, 39*). The partial specific volume calculated from the amino acid composition is 0.73 (*11, 38*). A $D_{20,w}$ of 4.83×10^{-7} cm^2 sec^{-1} has been obtained from diffusion measurements (*39*); however, a value of 5.45×10^{-7} was calculated from the molecular weight, partial specific volume, and sedimentation constant (*38*). Electrophoretic mobilities (in cm^2 V^{-1} sec^{-1} $\times 10^5$) are: 2.0 at pH 5.0 (*36*), 1.2 at pH 6.5, and

Evolving Genes Proteins, Symp., Rutgers Univ., 1964 p. 207. Academic Press, New York, 1965.

42. C. Milstein and F. Sanger, *BJ* **79,** 456 (1961).
43. C. P. Milstein and C. Milstein, *BJ* **109,** 93 (1968).
44. N. H. Sloane, D. W. Mercer, and M. Danoff, *BBA* **92,** 168 (1964).

—1.9 at pH 7.7 (*45*) (measurements at near 0° and $\mu = 0.05–0.1$). An isoelectric point of near pH 7 has been reported via moving boundary electrophoresis (both phospho and dephospho forms) (*37*), starch gel electrophoresis (*11*), and isoelectric focusing (*38*). However, since phosphoglucomutase binds a variety of anions (Section II,H,1), the observed isoelectric point may be dependent on the type of buffer employed in the measurement.

The optical density of a 1% aqueous solution of phosphoglucomutase (1 cm light path) is 7.7 at 278 nm (*36*); hence ϵ_{max} is about 5.0×10^4 liters mole cm^{-1} based on a molecular weight of 65,000. Values reported for λ_0 and $-\beta_0$, 220 nm and 112° (*46*), and 212 nm and —340° (*47*), are in very poor agreement. Part of the difference lies in the choice of an arbitrary value for λ_0 (*47*) as opposed to one based on curve fitting (*46*); however, other discrepancies exist, e.g., $[\alpha]_{233}$ values of —7200° (*46*) and —5000° (*47*) are reported. Part of this difference may have been caused by the use of a commercial sample of phosphoglucomutase in one study (*46*). A small difference in $[\alpha]_D$ is reported for phospho and dephospho forms of the enzyme (*48*).

Large crystals of the enzyme can be obtained from ammonium sulfate solutions at pH 6.8 and room temperature (*30*).

3. *Chemical Properties*

A number of different amino acid analyses of acid hydrolyzates of rabbit muscle phosphoglucomutase have been published (*7, 11, 27, 28, 38, 44, 49*). Of these, the four most recently published analyses are in reasonable agreement. Average values from these are: Asp, 66; Thr, 32; Ser, 37; Glu, 54; Pro, 28; Gly, 55; Ala, 53; half-Cys, 6; Val, 38; Met, 12; Ile, 46; Leu, 45; Tyr, 17; Phe, 32; Lys, 41; His, 12; Arg, 26; and NH_3, 57; *p*-mercuribenzoate analysis gives 3–7 cysteine residues (*28, 50, 51*). Four tryptophan residues have been reported by ion exchange chromatography of basic hydrolysates (*28*) and by spectrophotometric procedures (*11, 52*), but 8–9 residues by *N*-bromosuccinimide titration (*44*). A similar composition has been reported for the human muscle enzyme (*27*), while that of the *E. coli* enzyme is quite different (*17*).

45. V. Jaganathan and J. M. Luck, *JBC* **179,** 561 (1949).
46. B. Jirgensons, *JBC* **240,** 1064 (1965).
47. N. H. Sloane, *BBA* **92,** 171 (1964).
48. J. P. Robinson and V. N. Najjar, *Federation Proc.* **20,** Suppl. 10, 90 (1961).
49. H. Boser, *Z. Physiol. Chem.* **300,** 1 (1955).
50. V. Bocchini, M. R. Alioto, and V. A. Najjar, *Biochemistry* **6,** 313 (1967).
51. C. Milstein, *BJ* **79,** 591 (1961).
52. E. J. Peck, Jr. and W. J. Ray, Jr., *JBC* **244,** 3748 (1969).

In the rabbit muscle enzyme exposure of tryptophan and tyrosine residues to added perturbants is extensive when assessed with D_2O but is quite limited when assessed with dimethyl sulfoxide or ethylene glycol. Evidently, aromatic residues tend to occur in crevices that are generally accessible only to molecules the size of water (*52*). Although 5 accessible and 3–4 buried residues of tryptophan have been reported by *N*-bromosuccinimide titration (*44*), it is questionable whether the "buried" residues thus observed represent tryptophans or other slowly reacting residues [e.g., see Spande and Witkop (*53*)].

Phosphoglucomutase is unusually large for a single polypeptide chain; however, no evidence for multiple chains has been obtained. Thus, there is only one active site phosphate per 65,000 (*11, 15, 27, 35*), and only one site that binds metal ions tightly (*31, 54, 55*). In addition, the reduced, carboxymethylated protein sediments in 5 *M* guanidine hydrochloride at a rate similar to native protein in the absence of guanidine (*38*). Also, 44–46 of the theoretical 68 spots in a "fingerprint" of a tryptic digest can be identified (*9*).

Although no N-terminal groups have been detected for the rabbit muscle and *E. coli* enzymes by means of the dinitrofluorobenzene reaction conducted by the standard procedure (*17*), N-terminal lysine is found for the former when the reaction is conducted in the presence of 6 *M* guanidine hydrochloride (*30*).

Rabbit muscle phosphoglucomutase contains an equivalent of covalently bound phosphate when freshly isolated (*11, 15*), as does the enzyme from yeast (*56*). Chromatographic data (*27*) suggest that a small amount of dephospho-enzyme may remain in the human muscle enzyme even after partial purification although the enzyme is primarily in the phospho form. The phosphate group of the enzyme from flounder and shark is more readily transferred to water (*14*) than that of the rabbit muscle enzyme; hence, when isolated, these enzymes are only partly phosphorylated and the *E. coli* enzyme (*17*) is devoid of phosphate. However, a phospho-enzyme can be formed with the *E. coli* enzyme (*17*), and the fraction of phospho-enzyme in the flounder and shark enzymes can be increased by treatment with glucose diphosphate (*14*). By contrast, the *B. cereus* and *M. lysodeikticus* enzymes do not contain a covalently attached phosphate group either on isolation or after treatment with glucose diphosphate (*19*).

53. T. F. Spande and B. Witkop, "Methods in Enzymology," Vol. 11, p. 498, 1967.
54. E. J. Peck, Jr. and W. J. Ray, Jr., *JBC* **244,** 3754 (1969).
55. W. J. Ray, Jr. and A. S. Mildvan, *Biochemistry* **9,** 3886 (1970).
56. E. E. McCoy and V. A. Najjar, *JBC* **234,** 3017 (1959).

4. *Stability*

Solutions of purified rabbit muscle phosphoglucomutase, phospho form, are stable for days at pH 5–8 and room temperature and at least for short periods of time at 30° between pH values of 4.5 and 10 in the enzymic assay. At pH 5 and high ionic strength, a temperature of 63° causes little activity loss in 3 min and this "heat treatment" constitutes one of the steps in the purification procedure (*7, 10, 11*). The enzyme is also stable at ionic strengths as low as 0.003 (neutral pH) at 0° (*9, 11*) as well as room temperature. The enzyme withstands exposure to 5 *M* but not 6.6 *M* urea (*31*) as well as limited freezing and thawing cycles (see Section II,A,4). The dephospho-enzyme is less stable toward some if not all of these conditions (*7, 12, 56*). Also, enzyme from other sources can be less stable than the rabbit muscle enzyme toward extremes of pH and low ionic strength (*14, 19*). However, the *E. coli* enzyme is stable for 2 hr at pH 11 (at 0°); moreover, cysteine markedly increases the stability of this enzyme toward heat (*17*) although this is not the case for other phosphoglucomutases.

5. *Polymorphism*

Polymorphism of phosphoglucomutase has been attributed to the presence of (a) phospho and dephospho forms of the enzyme [which can be separated by column chromatography (*11, 28*)], (b) genetic variants [which can be separated by starch gel electrophoresis (*29, 57*) and column chromatography (*9*)], and (c) to conformational isomers (*29*). However, the possibility of artifacts (see Section II,A,2) makes it difficult to evaluate reports of genetic variants that were separated only by column chromatography (*9, 29*).

What was probably the first demonstration of genetic polymorphism of phosphoglucomutase was reported by Roberts and Tsuyuki for muscle extracts of rainbow trout (*58*), although Tsoi and Douglas first suggested that this polymorphism was genetic in origin and was an expression of separate genes (*59*). The latter workers showed that mutants of *Saccharomyces* unable to ferment galactose were deficient in a major phosphoglucomutase component, while a minor component with mutase activity was unaffected by the mutation. Polymorphism of human red cell phosphoglucomutase was first described by Spencer *et al.* (*60*); later studies

57. D. A. Hopkinson and H. Harris, *Ann. Human Genet.* **31,** 359 (1968).
58. E. Roberts and H. Tsuyuki, *BBA* **73,** 673 (1963).
59. A. Tsoi and H. C. Douglas, *BBA* **92,** 513 (1964).
60. N. A. Spencer, D. A. Hopkinson, and H. Harris, *Nature* **204,** 742 (1964).

showed that the enzyme can be separated into at least 20 different phenotypic patterns by starch gel electrophoresis. These phenotypes originate in the alleles of three distinct gene loci, PGM_1, PGM_2, and PGM_3. Locus PGM_1 is not linked to either PGM_2 or PGM_3, but the linkage relation between PGM_2 and PGM_3 has not been tested. Evaluating the polymorphism of phosphoglucomutase appears to be a valuable procedure for examining genetic variation among populations and for linkage studies as indicated by recent reviews (*61, 62*). Polymorphism has also been described in rabbits and rats (*9, 29*), mice (*63*), chimpanzees (*64*), chickens, (*29*), herring (*65*), flounder and potato tubers (*9*). Polymorphism of the enzyme appears to be characteristic of evolutionarily "advanced" forms since no polymorphic patterns have been found in any of the lower forms which have been examined, including *Neurospora crassa* (*66*), *Entamoeba* (*67*), and *E. coli* (*9*).

C. Assay

Phosphoglucomutase is usually assayed by measuring the efficiency with which it converts glucose-1-P to glucose-6-P. Although Najjar originally expressed activity in terms of milligrams of phosphorus converted per 5 min (*10*), International Units (micromoles per minute) are now used; one unit according to Najjar is equal to 6.4 IU. All acceptable assay procedures involve a preassay activation step (see Section II,G), although this step can be conducted during preparation of the enzyme dilution for assay (see procedures referred to below). If no preassay activation is used, in addition to lower activities, nonlinear product time plots may be obtained (*68*).

Unfortunately, many workers depend on the glucose-1,6-di-P that is usually present as a contaminant in glucose-1-P isolated from natural sources to provide an adequate level of the diphosphate in the enzymic assay. Although such a procedure may prove adequate for routine assays, it is inadequate for many investigations, e.g., the effect of

61. H. Harris, *Proc. 3rd Int. Congr. Human Genet., Univ. Chicago, Chicago, 1966* p. 337. Johns Hopkins Press, Baltimore, Maryland, 1967.
62. D. A. Hopkinson and H. Harris, *in* "Biochemical Methods in Red Cell Genetics" (J. J. Yunis, ed.), p. 337. Academic Press, New York, 1969.
63. T. B. Shows, F. H. Ruddle, and T. H. Roderick, *Biochem. Genet.* **3,** 25 (1969).
64. M. Goodman and R. E. Tashian, *Human Biol.* **41,** 237 (1969).
65. I. E. Lush, *Comp. Biochem. Physiol.* **30,** 391 (1969).
66. S. Brody and E. L. Tatum, *Proc. Natl. Acad. Sci. U. S.* **58,** 923 (1967).
67. R. E. Reeves and J. M. Bischoff, *J. Parisitol.* **54,** 594 (1968).
68. W. J. Ray, Jr. and G. A. Roscelli, *JBC* **241,** 2596 (1966).

inhibitors on the enzyme (see Section II,H,1). The commercial availability of the diphosphate and its facile synthesis via the recently published modification (*69*) of the McDonald procedure (the direct phosphorylation of tetraacetylglucose-6-P) will hopefully lead to the disappearance of this practice.

1. *Colorimetric Assay*

The rate of conversion of glucose-1-P into glucose-6-P can be conveniently determined by following the decrease with time in the acid-labile phosphate remaining in the assay, since glucose-1-P is completely hydrolyzed by heating in N acid at 100° for 10 min, while glucose-6-P is essentially unaffected. Early procedures [e.g., Najjar (*36*)] utilized the Fiske-Subbarow determination of inorganic phosphate (*70*); later, a modification of the Bartlett procedure (*71*) was introduced in which hydrolysis and color development occurs at the same time (*72, 73*). The Bartlett procedure is some 7-fold more sensitive than the Fiske-Subbarow procedure and the color produced is stable; the latter characteristic allows the assay to be automated (*74*). A typical assay procedure is described by Peck and Ray (*75*). A modified assay has been described (*73*) which can be used at substrate concentrations as low as 6 μM; however, the procedure is awkward and extreme cleanliness of assay tubes is required.

The main disadvantage of the colorimetric assay is the necessity for quantitating disappearance of substrate rather than appearance of product. Fortunately, product-time plots are linear for an appreciable fraction of the reaction at saturating substrate concentrations and a 30–40% decrease in substrate can be utilized for relatively precise work and a 50–60% conversion for routine activity measurements (see Section II,D,3). A variable dilution technique can be used to considerably extend the range of the assay for following activity loss on chemical modification (*28*).

The Nelson-Somogyi determination of reducing sugars (*76*) has been used to follow glucose-6-P production. However, activators such as

69. R. Hanna and J. Mendicino, *JBC* **245,** 4031 (1970).
70. C. H. Fiske and Y. Subbarow, *JBC* **66,** 375 (1925).
71. G. R. Bartlett, *JBC* **234,** 466 (1959).
72. C. Milstein, *BJ* **79,** 574 (1961).
73. W. J. Ray, Jr. and G. A. Roscelli, *JBC* **239,** 1228 (1964).
74. D. A. Hopkinson and W. H. P. Lewis, *3rd Automat. Anal. Chem., Technicon Symp. 1967* Vol. 2, p. 227 (1967).
75. E. J. Peck, Jr. and W. J. Ray, Jr., "Specifications and Criteria for Biochemical Compounds," 2nd ed. Natl. Res. Council, Washington, D. C., 1971.
76. M. Somogyi, *JBC* **195,** 19 (1952).

cysteine, histidine, or imidazole, which are normally present in the phosphoglucomutase assay, cause high blanks in this procedure unless they are eliminated prior to color development [e.g., see Reissig (*77*)].

Appearance of glucose-1-P, rather than its disappearance, can of course be measured if the reaction is initiated with glucose-6-P, although the equilibrium is unfavorable and product-time plots are markedly nonlinear (see Sections II,E,1 and II,D,3). Moreover, although glucose-6-P is relatively stable to the conditions of color development (*72*), slight hydrolysis produces an appreciable and somewhat variable blank when conversions to glucose-1-P are kept very low (*78*).

2. *Coupled Assay*

Glucose-6-P dehydrogenase can be used to follow production of glucose-6-P by phosphoglucomutase (*13, 14, 19, 27, 29, 30, 56, 73, 79*). All procedures except that of Lowry and Passonneau (*13*) involve following the optical density increase at 340 nm; Lowry and Passonneau's procedure involves following fluorescence increases and is thus much more sensitive than the others. Since glucose-6-P dehydrogenase is specific for the β isomer of glucose-6-P (*80*) while phosphoglucomutase produces the α isomer from glucose-1-P (see Section II,F) the apparent rate of the phosphoglucomutase reaction can be limited by the rate of anomerization (*13, 80*). However, if phosphoglucomutase concentrations are properly limited, the rate of the optical density change in the coupled assay (no added isomerase) is linear in phosphoglucomutase concentration and is in good agreement with rates measured with the colorimetric assay (*30*). The coupled assay is limited by sensitivity of the dehydrogenase to pH and temperature and is essentially useless below pH 7 and 20°. However, aliquots of assays run at low pH and/or temperature may be quenched and analyzed for glucose-6-P content under conditions more amenable to the dehydrogenase reaction, with substantial saving on the amount of dehydrogenase required—especially if the dehydrogenase reaction is conducted with a spectrophotometer equipped with a 0 to 0.1 OD scale (*30*).

3. *Radiometric Assay*

The colorimetric assay procedure can be used only with great difficulty to evaluate $K_{m(\text{Glc-1-P})}$ because of the low value of this constant (see Section II,D,3), and the coupled assay is useless in the required concen-

77. J. L. Reissig, *JBC* **219,** 753 (1956).
78. W. J. Ray, Jr. and G. A. Roscelli, *JBC* **239,** 3935 (1964).
79. O. Bodansky, *JBC* **236,** 328 (1961).
80. M. Salas, E. Viñuela, and A. Sols, *JBC* **240,** 561 (1965).

tration range. However, by using ^{32}P-labeled glucose-1-P [which is readily available from the sucrose phosphorylase reaction (*81*)] restraints on attainable concentrations are effectively eliminated. The ^{32}P-glucose-6-P produced can be determined by hydrolyzing the remaining glucose-1-P with acid, precipitating the inorganic phosphate thus produced with the Sugino-Miyoshi reagent (*82*), and measuring the radioactivity of the soluble glucose-6-P (*68*). The primary limitation on such an assay is the purity of the ^{32}P-glucose-1-P used; if no contaminating, ^{32}P-acid-stable phosphates are present and if appropriate precautions are taken (*34*), blanks (no enzyme) as low as 0.3% of the total radioactivity can be obtained. In such a case a single assay can be used to measure substrate conversions of as little as 3% with reasonable accuracy. Low conversions (actually 20% or less) must be used for initial velocity measurements if substrate concentrations that are below the K_m of glucose-1-P are employed (see Section II,D,3).

4. *All-or-None Assays*

All-or-none assays measure total enzyme species in solution that can react in a particular manner, irrespective of the catalytic efficiency in the reaction, and they are useful in assessing the effect on enzymes of chemical modifications (*83*). One type of all-or-none assay for phosphoglucomutase involves determining the fraction of ^{32}P-labeled phosphoenzyme that can transfer its label to substrate during an extended assay interval (*83*). A second assay of this type has been used to distinguish between the enzyme·Mg complex and the free enzyme; the former is catalytically active for a short period of time and the latter completely inactive when added to a substrate mixture in the presence of excess Zn^{2+} (*84*) or, better, excess EDTA (*85*). In fact, since Mg^{2+} completely dissociates from the enzyme within 3 min under these conditions, the total product formed, which is proportional to the amount of enzyme·Mg complex originally present, can be assessed at any convenient time after a minimum 3-min reaction interval. In the presence of saturating substrate the total product formed under these conditions, P_∞, is given by

$$P_\infty = V_{\max}[\text{Enz·Mg}]/k_d \text{E}_\text{T} \tag{1}$$

where k_d is the rate of dissociation of Mg^{2+} from the catalytically active

81. M. Douderoff, "The Enzymes," 2nd ed., Vol. 5, p. 229, 1961.
82. Y. Sugino and Y. Miyoshi, *JBC* **239**, 2360 (1964).
83. W. J. Ray, Jr. and D. E. Koshland, Jr., *JACS* **85**, 1977 (1963).
84. W. J. Ray, Jr. and G. A. Roscelli, *JBC* **241**, 1012 (1966).
85. W. J. Ray, Jr., *JBC* **244**, 3740 (1969).

enzyme·Mg·substrate complex and [Enz·Mg]/E_T is the fraction of total enzyme in the Mg^{2+} form prior to the assay (*84*).

D. Catalytic Reaction

1. *Reaction Sequence*

Leloir *et al.* (*86–88*) showed that under normal assay conditions glucose-1-P is efficiently converted to glucose-6-P by rabbit muscle phosphoglucomutase only in the presence of a catalytic amount of glucose 1,6-diphosphate (*89*). Later, Sutherland *et al.* (*92*) demonstrated the exchange both of phosphate groups and glucose residues between glucose monophosphate and glucose diphosphate. Two basically different reaction sequences are consistent with these results:

Reaction sequence *a*:

$$E + {}^*\text{glucose-1-P} + \text{glucose-1,6-di-P} \rightleftharpoons {}^*\text{glucose-1,6-di-P} + \text{glucose-6-P} + E$$

Reaction sequence *b*:

$$E_P + \text{glucose-1-P} \rightleftharpoons E_D + \text{glucose-1,6-di-P}$$
$$E_D + \text{glucose-1,6-di-P} \rightleftharpoons E_P + \text{glucose-6-P}$$

Najjar and Pullman (*33*) showed that the rabbit muscle enzyme exists in both a phospho form and dephospho form, as required by reaction sequence *b*, and that the phospho form reacts with glucose monophos-

86. R. Caputto, L. F. Leloir, R. E. Trucco, C. E. Cardini, and A. Paladini, *Arch. Biochem.* **18**, 201 (1948).

87. C. E. Cardini, A. C. Paladini, R. Caputto, L. F. Leloir, and R. E. Trucco, *Arch. Biochem.* **22**, 87 (1949).

88. L. F. Leloir, R. E. Trucco, C. E. Cardini, A. C. Paladini, and R. Caputto, *ABB* **19**, 339 (1948).

89. It is interesting to note that many years earlier, Kendal and Stickland (*90*) claimed that fructose 1,6-diphosphate acted as the coenzyme for the phosphoglucomutase reaction. After their discovery of glucose diphosphate, Caputto *et al.* (*86*) suggested that the earlier results were caused by glucose diphosphate present as a contaminant in the fructose diphosphate used by Kendal and Stickland. However, recent specificity studies (*91*) show that fructose diphosphate can act in a capacity analogous to that of glucose diphosphate (see below), albeit in a much less efficient manner. Hence, the original conclusions of Kendal and Stickland were correct, although possibly for the wrong reasons.

90. L. P. Kendal and L. H. Stickland, *BJ* **32**, 572 (1938).

91. J. Passonneau, O. H. Lowry, D. W. Schulz, and J. G. Brown, *JBC* **244**, 902 (1969).

92. E. W. Sutherland, M. Cohn, T. Posternak, and C. F. Cori, *JBC* **180**, 1285 (1949).

phates but not glucose diphosphate while the dephospho form reacts with glucose diphosphate but not glucose monophosphates. The results of later studies involving steady state kinetics and isotope exchange experiments (see Sections II,D,3 and II,D,4) are also consistent with this reaction sequence and exclude sequence *a*.

Subsequently, Ray and Roscelli (*73*) showed that free dephospho-enzyme is not formed during each catalytic cycle (at pH 7.4 and 30°) since the phosphate label of ^{32}P-phospho-enzyme appears predominantly in the glucose-6-P fraction in a rapid exchange experiment in which the labeled enzyme is treated with a very large excess of glucose-1-P and glucose-1,6-di-P for a short time interval (less than 2% of the equilibrium concentration of glucose-6-P was produced). This result is consistent with the following modification of the Najjar-Pullman reaction sequence in which free E_D and free glucose-di-P are not obligatory intermediates:

Reaction sequence *b1*:

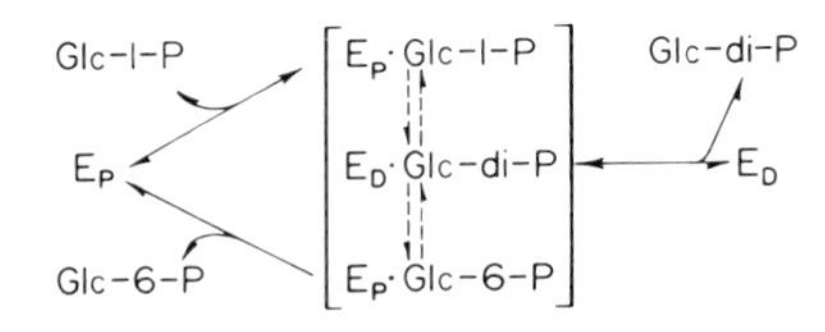

This formulation, which was later confirmed by other workers (*35*, *93*), is also consistent with the kinetic results described in Section II,D,3.

By assuming that E_D·glucose-di-P is an obligatory intermediate in the reaction, the results of the preceding experiment can be rationalized if the conversion of E_D·glucose-di-P to E_P·glucose-6-P and thence to E_P + glucose-6-P is more rapid than the dissociation to free E_D + glucose-di-P. In view of the high affinity of dephospho-enzyme for glucose-di-P (see Section II,D,3), a dissociation rate sufficiently slow to produce the observed results is not unreasonable. However, reaction conditions may have a marked influence on the partitioning of E_D·glucose-di-P, and under different conditions the probability of forming free E_D and glucose-di-P in a given catalytic cycle might increase or decrease markedly.

In any case, the overall reaction with glucose phosphates at 30° and neutral pH is best represented as a one substrate–one product process which involves an enzyme intermediate (E_D·glucose-di-P) capable of dead-end dissociation. Hence, the official name for phosphoglucomutase, "α-D-glucose-1-phosphate: α-D-glucose-1,6-diphosphate phosphotransfer-

93. H. G. Britton and J. B. Clark, *BJ* **110**, 161 (1968).

ase" (*94*), is mechanistically incorrect for the rabbit muscle enzyme although it is correct for the *M. lysodeikticus* and *B. cereus* enzymes (see below). Since the bracketed complexes in reaction sequence *b1* represent three different but isomeric forms of the central complexes (CC), they cannot be distinguished; nor can the rates of the processes by which they are interconverted (dashed arrows) be measured under steady state conditions. Hence, no distinction among them will be made in the following discussion and (CC) will represent the collection of complexes with the stoichiometry enzyme·glucose·(phosphate)$_2$, regardless of the position to which the phosphate group is attached (*95*). Note that in the presence of bound substrate (and metal ion activator) most techniques do not allow a distinction between the phosphate group "belonging to the enzyme" and that "belonging to the substrate."

Phosphoglucomutase also catalyzes 1-phosphate to 5- or 6-phosphate interconversion with a variety of other sugar phosphates, although its efficiency is much lower than with glucose phosphates (see Section II,F). Although other explanations are possible it seems probable that the actual catalytic steps are much slower in such reactions; in such a case the E_D·sugar-di-P intermediate would dissociate during each cycle. Thus, the reaction of fructose-6-P in the presence of glucose-1,6-di-P gives glucose-6-P plus fructose-1,6-di-P (*13, 91*) in a truly Ping Pong (*96*) reaction (*97*). Hence, in reality, phosphoglucomutase action is subject to kinetic control; at one extreme it is truly Ping Pong, at the other it is uni–uni with a dissociable intermediate.

The reaction sequence for phosphoglucomutases from *E. coli* (*17*), shark and flounder (*14*), and human muscle (*27*), is probably analogous to that for the rabbit muscle enzyme. However, the extent to which the E_D·glucose-di-P is partitioned between dissociation and reaction to give E_P + glucose-6-P may be much different, and it would not be surprising if dissociation is sufficiently rapid in the case of at least one of these enzymes so that the reaction sequence must be considered as truly Ping Pong, even with glucose-P as the substrate; in such a case, glucose-di-P acts as both the "first product" and the "second substrate."

94. M. Florkin and E. H. Stotz, *Comp. Biochem.* **13,** 92 (1964).

95. The composition of the central complex is actually enzyme·glucose·(phosphate)$_2$·M, where M is a metal ion, usually Mg^{2+}. Initially, the metal ion will be ignored (since studies were conducted at saturating Mg^{2+}); later, the more complete formulation for the central complex will be used and will sometimes be referred to as the ternary central complex.

96. W. W. Cleland, *BBA* **67,** 104 (1963).

97. Ribose 1,5-diphosphate was first isolated by virtue of the analogous reaction with ribose-1-P (*98*).

98. H. Klenow, *ABB* **46,** 186 (1953).

The generally increased values of $K_{m(\text{Glc-di-P})}$ observed for these enzymes *vis-à-vis* that of the rabbit muscle enzyme supports this suggestion (see Section II,D,3).

By contrast, the *B. cereus* and *M. lysodeikticus* enzymes are isolated devoid of phosphate, cannot be phosphorylated by reaction with glucose-di-P [and do not serve as glucose-di-P phosphatases (*14*)], and are completely inactive in the absence of added glucose-di-P (*19*) (as opposed to the rabbit muscle enzyme, see following section). These observations plus substrate-velocity profiles suggest that *B. cereus* and *M. lysodeikticus* phosphoglucomutases utilize reaction sequence *a*, or a modification thereof. Reaction sequence *a* has also been proposed for the yeast enzyme (*16*); however, a coupled assay was used for this study and difficulties have been encountered with such assays in other phosphoglucomutase systems (*73*).

In a sequential reaction, addition of glucose monophosphate and glucose diphosphate might be noncompulsory, noncompulsory-rapid equilibrium, or ordered; in the latter case the reaction might be ordered because of structural considerations, i.e., if the presence of one substrate is required for binding of the other or because of kinetic behavior, i.e., if glucose-di-P dissociates from the enzyme much more slowly than substrate to product conversion. A combination of these possibilities is shown below.

Reaction sequence *a1*:

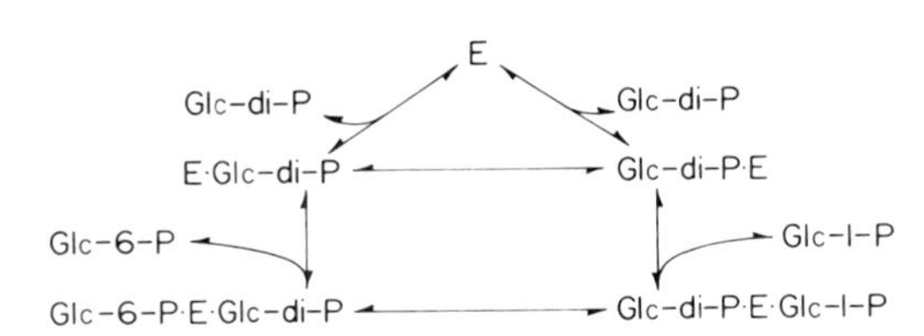

In the *a1* sequence, bound glucose-di-P essentially replaces the phosphoenzyme of sequence *b1* (see above) and the E·Glc-di-P complex may dissociate into its component parts only rarely. Sequences *a* and *a1* can be distinguished by isotope exchange studies (see Section II,D,4) and both can be distinguished from sequences *b* and *b1* by substrate-velocity profiles (Section II,D,3), although circumstances may make the distinction difficult (*93*).

2. *The Roles of Glucose Diphosphate*

In terms of reaction sequence *b1*, added glucose-di-P would not be required in the enzymic assay initiated with the phospho form of the enzyme if (a) the assay interval were short with respect to the rate of dissociation of E_D·glucose-di-P or (b) an enzyme concentration much larger than the Michaelis constant of glucose-di-P were used. Although both conditions are impractical for enzymic assays, both illustrate the involvement of glucose-di-P in the enzymic reaction. Since in the steady state ν_0 is proportional to (CC) and becomes equal to V_{max} when (CC) = E_T, where $E_T = E_P + (CC) + E_D$, in brief assay intervals (a few milliseconds) in the presence of saturating glucose-1-P, no appreciable fraction of the enzyme would be converted to the dephospho form via dissociation of E_D·glucose-di-P; hence, (CC) $\approx E_T$, and $\nu_0 = V_{max}$ without added diphosphate. Alternatively, in the steady state in the presence of saturating substrate *(99)* $(E_D)/(CC) = K_{m(Glc\text{-}di\text{-}P)}/(Glc\text{-}di\text{-}P)$ (see Section II,D,3), and if a sufficiently high concentration of enzyme (phospho form) is used, $(E_D)/(CC)$ will be small and ν_0 will again approach V_{max} without added diphosphate *(100)*. Under these conditions an $(E_T)/K_m$ value of about 400 is required to give $\nu_0 = 0.95\ V_{max}$ *(101)*; this is a much higher concentration of enzyme than usually used in the enzymic assay. However, by using an E_T value of about $6 \times 10^{-8}\ M$ together with suitably short assay intervals, a ν_0/V_{max} of about 0.25 has been achieved in the absence of added diphosphate *(73)*. Thus, when glucose-di-P concentration is severely limited, the apparent specific activity of the enzyme can be concentration dependent *(73)*, and some nonlinear assay results *(102)* may perhaps be rationalized in this manner. Indeed, a diagnostic test for the presence of the phospho form of phosphoglucomutase reacting according to sequence *b* or *b1* is the observation of enzymic activity in the absence of added diphosphate such that the specific activity under these conditions increases with increasing enzyme concentration. (This test of course presupposes the enzyme is not contaminated with glucose-di-P.)

In contrast with the conditions noted above, the normal enzymic assay interval is several minutes and the enzyme concentration is more than an order of magnitude smaller than $K_{m(Glc\text{-}di\text{-}P)}$. Hence, in the normal assay added diphosphate is required to prevent a significant accumulation of free E_D and thus to elicit maximal activity by ensuring that $(CC)/E_T$

99. Saturating glucose-1-P ensures that $E_P/E_T \approx 0$.

100. Although the system involves a steady state rather than a true equilibrium, its behavior is analogous to the dissociation of a weak acid in aqueous solution; only a small fraction dissociates when the total concentration of acid is much larger than its dissociation constant.

101. Under these conditions the quantitative expression relating ν_0 to V_{max} is $\nu_0/V_{max} = 1 - [(1 + 4f)^{1/2} - 1]/2f$ where f is $(E_T)/K^{app}_{m(Glc\text{-}di\text{-}P)}$ *(73)*.

102. A. G. Pasynskii and V. N. Tvertinov, *Zh. Fiz. Khim.* **40,** 1634 (1966).

approaches 1.0. The levels of glucose-di-P found in mammalian tissues (*91*) are more than adequate to ensure that none of the phosphoglucomutase present in such tissues is in the free dephospho form.

For reactions proceeding according to sequence *a* or *a1* no activity should be observed in the absence of added diphosphate; the reaction of the *B. cereus* and *M. lysodeikticus* enzymes conforms to this criterion, as is expected from their substrate-velocity patterns (*19*).

3. *Kinetics*

Product-time plots for the rabbit muscle enzyme obtained by use of standard assay procedures with glucose-1-P as the substrate (see Section II,C) are linear within experimental error to about 40% substrate conversion and are only slightly curved at 60% conversion. As a measure of the linearity of the assay, the fractional error if V_{max} is taken as equal to P/t is given in Table I as a function of the fractional approach to equilibrium P/P_e at time t (substrate is assumed to be saturating). The relatively high degree of linearity with glucose-1-P as the substrate is the result of a combination of a favorable equilibrium constant, about

TABLE I
FRACTIONAL ERROR IN VELOCITY (SATURATING SUBSTRATE) AS A FUNCTION OF EXTENT OF REACTION IF $(P)/t$ IS TAKEN AS EQUAL TO V_{max}

Reaction	Extent[a]	Fractional error[b] × 100
Forward	0.1	1
	0.2	2.5
	0.3	4
	0.4	5.5
	0.5	8
	0.6	10
	0.7	13
	0.8	18
	0.9	25
Reverse	0.1	6.5
	0.2	13
	0.3	19
	0.4	26
	0.5	33

[a] Expressed as (glucose-6-P)/(glucose-6-P)$_e$ or (glucose-1-P)/(glucose-1-P)$_e$ for the forward and reverse reactions, respectively, where the subscript denotes the equilibrium value.

[b] Calculated with an integrated rate equation (*78*, Eq. 3) by using $\theta = 0.18$ and $K_e = 17.2$, and expressed as $1 - (P/t)/V_{max}$.

17 (Section II,E,1), and a favorable ratio of Michaelis constants for substrate and product $[K_{m(\text{Glc-1-P})}/K_{m(\text{Glc-6-P})} \approx 0.18$ *(78)*].

Initial velocities at subsaturating glucose-1-P can be approximated as P/t if conversions are limited to 20% or less and if the average substrate concentration during the assay interval, i.e., $(S_0) - (P)/2$, is taken as the initial substrate concentration. This approximation, together with the error involved at a given conversion, can also be verified with an integrated rate equation (see below).

By contrast, when glucose-6-P is used as the substrate, product-time plots are markedly curved, even when the fractional conversion to product is low, i.e., when $P/P_e < 0.2$ *(13, 78)*. (Because of the unfavorable equilibrium, a P/P_e value of 0.2 is equivalent to about 1% conversion of glucose-6-P to glucose-1-P.) In such a case V_{max} can be approximated by extrapolating P/t vs. t plots to $t = 0$ *(103)*; however, even these plots are somewhat curved. The error produced by equating V_{max} to $(P)/t$ for conversion of glucose-6-P to glucose-1-P is also given in Table I. Analogous data for other phosphoglucomutases are not available, and the data for the rabbit muscle enzyme should not be used for other enzyme systems since $K_{m(\text{Glc-1-P})}/K_{m(\text{Glc-6-P})}$ is probably not invariant among phosphoglucomutases.

For reactions that proceed according to sequence *b* or *b1*, the integrated rate equation for a uni–uni reaction *(104)* gives the time course of the reaction *(78)* if the reaction is conducted at saturating concentrations of glucose-1,6-di-P *(105)*.

Substrate-velocity studies in the concentration range where glucose-1-P $\approx K_{m(\text{Glc-1-P})}$ are difficult to conduct with the rabbit muscle enzyme because of the small value of this constant, about 5 μM *(106)*. However, by using techniques noted in Sections II,C,1 and 3 and by using the initial velocity approximation noted above, reasonable v_0 values can be obtained. Double reciprocal plots of v_0 and glucose-1-P concentration at several levels of glucose-di-P in the region of its Michaelis constant, and similar plots of velocity and glucose-di-P concentration at several levels of glucose-1-P form a series of parallel lines characteristic of Ping Pong reactions *(96)* in accord with reaction sequences *b* or *b1* *(73)*. This means that the expressions for both $K^{\text{app}}_{m(\text{Glc-1-P})}$ and $K^{\text{app}}_{m(\text{Glc-di-P})}$ are of the form, $K^{\text{app}}_{m(\text{A})} = K_{m(\text{A})}\ (1 + K_{m(\text{B})}/\text{B})$; moreover, physically these apparent Michaelis constants are equal to (E_P) (Glc-1-P)/(CC) and

103. R. A. Alberty and B. M. Koerber, *JACS* **79,** 6379 (1957).
104. R. A. Alberty, "The Enzymes," 2nd ed., Vol. 1, p. 143, 1959.
105. In Ray and Roscelli *(78)*, ϕ is incorrectly defined; it should be equal to $[\text{Glc-6-P}]_t/[\text{Glc-6-P}]_e$.
106. W. J. Ray, Jr., G. A. Roscelli, and D. S. Kirkpatrick, *JBC* **241,** 2603 (1966).

(E_D) (Glc-di-P)/(CC) where (CC) is the initial steady state concentration of the central complexes in the forward and reverse reactions, respectively. The overall velocity equation, including the Mg^{2+} dependence, is given in Section II,I,2. Analogous equations apply to the *E. coli* (*19*), shark and flounder (*14*), and human muscle enzymes (*27*).

In all cases examined, substrate inhibition by glucose monophosphate and glucose diphosphate has been detected (see Section II,H,1), and in some cases such inhibition becomes quite important at concentrations not greatly exceeding the respective K_m values [e.g., the *E. coli* enzyme (*19*)]. In such cases the above equation for K_m^{app} represents a limiting relationship that can be verified only by extrapolation; however, this is not the case for the rabbit muscle enzyme (*73*). *All substrate inhibition effects are ignored in the present section,* although they are later discussed in Section II,H,1.

Values of the various kinetic constants for the rabbit muscle enzyme at pH 7.4 and 30° are as follows: $V_{max}^f/E_T = 1000$ IU/mg, $V_{max}^f/V_{max}^r = 3.1$ (*78*), $K_{m(Glc-1-P)} = 5 \times 10^{-6}\,M$ (*106*), $K_{m(Glc-6-P)} = 3 \times 10^{-5}\,M$ (*107*). The apparent Michaelis constant for glucose-di-P is consistently smaller in the reverse than the forward reaction under the same conditions. When the effects of all inhibitors are taken into account [$K_{m(Glc-id-P)}$ is strongly influenced by the concentration of Mg^{2+} and the presence of anionic inhibitors, including the substrate itself (see Section II,H,1)], the true Michaelis constant is about $10^{-8}\,M$ for the forward reaction and $3 \times 10^{-9}\,M$ for the reverse reaction (*34*). This difference indicates that (E_D)/(CC) is not the same for the forward and reverse reactions; this is reasonable because the relative amounts of the isomeric complexes which make up the central complex are not expected to be the same under different steady state conditions. It can be shown that the true equilibrium constant for (E_D) (Glc-di-P)/(CC) must lie between the values of $K_{m(Glc-di-P)}$ for the forward and reverse reactions; thus, the dissociation constant must be equal to or less than $10^{-8}\,M$ at 30° and pH 7.4.

All anions tested inhibit the rabbit muscle enzyme competitively with both glucose mono- and diphosphates (see Section II,H,1). Although the above Michaelis constants were measured at sufficiently low concentrations of chloride that competition was not excessive, the possibility of such effects hinders comparisons with constants measured under somewhat different conditions—especially where the sulfate instead of the chloride of Mg^{2+} was used in the assay. Thus, the true $K_{m(Glc-1-P)}$ for

107. The latter constant was estimated by the Haldane relationship [see Ray and Roscelli (*78*)] from V_{max}^f/V_{max}^r and $K_{m(Glc-1-P)}$ with a value of 17.2 for the equilibrium constant.

the human muscle enzyme is probably essentially the same as that of the rabbit muscle enzyme (*27*) if approximate corrections are made for sulfate inhibition. The true Michaelis constants for both mono- and diphosphates in reactions with the shark and flounder enzymes (*14*) are certainly much larger than those for the rabbit muscle enzyme, even if a generous allowance for anionic inhibition is made, while the constants for the *E. coli* enzyme are apparently still larger (*19*). Values of V^{f}_{max}/E_T are similar for the rabbit muscle, shark, flounder, human muscle, and *M. lysodeikticus* enzymes; they are somewhat smaller for the yeast, *B. cereus*, and *E. coli* enzymes (see above references).

Although the reaction catalyzed by the *B. cereus* and *M. lysodeikticus* enzymes has not been examined as thoroughly as that of the rabbit muscle enzyme, kinetics studies (*19*) suggest that the velocity equation at saturating metal ion is

$$\frac{V_{max}}{\nu_0} = 1 + \frac{K_{m(\text{Glc-di-P})}}{[\text{Glc-di-P}]} + \frac{K_{m(\text{Glc-1-P})}}{(\text{Glc-1-P})}\left(1 + \frac{K_{i(\text{Glc-di-P})}}{[\text{Glc-di-P}]}\right) \tag{2}$$

Such an equation would account for the intersecting double reciprocal plots observed if, for glucose-di-P, K_i is approximately equal to K_m. Of the two enzymes whose kinetics is described by Eq. (2) the *B. cereus* enzyme exhibits much lower Michaelis constants for both glucose mono- and diphosphates than does the *M. lysodeikticus* enzyme. [The Michaelis constants quoted for the latter enzyme (*19*) are actually much smaller than they should be according to replots of the primary data.]

4. *Isotopic Exchange Reactions*

Isotopic exchange at equilibrium has been used in the enzyme system from rabbit muscle to distinguish between reaction sequences *a1* and *b1*, which may be difficult to distinguish by means of steady state kinetics (*93*); sequence *a1* has been suggested for the shark and flounder enzymes (*14*). The ratio of rates of glucose transfer to phosphate transfer between substrate and product pools should be 2:1 for sequence *b1* but 3:2 for sequence *a1* (*108*). The results indicate, in fact, that ^{14}C-glucose-1-P is converted to ^{14}C-glucose-6-P twice as fast as ^{32}P-glucose-1-P is converted to ^{32}P-glucose-6-P, in accord with sequence *b1* (*93*). Transport

108. In reaction sequence *b1* a glucose residue is transported from the substrate to product pool by each catalytic cycle whereas two cycles are required for phosphate transfer (one cycle produces only half-transfer, i.e., transfer to E_P); hence, the 2:1 ratio. In mechanism *a1*, two and three cycles, respectively, are required; thus, a 3:2 exchange rate ratio is expected for this sequence (*93*).

exchange experiments (*109*) can also differentiate between sequences *a1* and *b1*. If the label is in the glucose moiety sequence *b1* requires independent transport while sequence *a1* requires co-transport, whereas if the label is in the phosphate group, both mechanisms require co-transport. Independent transport of glucose residues and co-transport of phosphate moieties, which is again consistent with sequence *b1*, have been observed for the rabbit muscle enzyme (*93*). Since the rabbit, flounder, and shark enzymes appear to be similar in a number of respects (*14*) it seems doubtful that sequence *a1* applies to any of these enzymes.

Isotopic exchange also has been used to examine the role of Mg^{2+} in the phosphoglucomutase reaction (*106*) (see Section II,I,2). As in other systems (*110*) the expression for equilibrium exchange rate is the same as the initial velocity equation [Eq. (15)], except that (a) isotopic exchange constants, K_x, are substituted for Michaelis constants, K_m; (b) r_0, the initial isotopic exchange rate is substituted for v_0; and (c) R^f, the maximum exchange rate in the forward direction, is substituted for V^f_{max} (*34*). Physical relationships defining $K_{x(Glc-1-P)}$ and $K_{x(Glc-di-P)}$ are $(E_P \cdot Mg)$ (Glc-1-P)/(CC) and $(E_D \cdot Mg)$ (Glc-di-P)/(CC), respectively, where (CC) is the concentration of the ternary enzyme·Mg·substrate complex at equilibrium; relationships defining K_x constants for Mg^{2+} are given in Section II,I,2. Differences between values of the exchange constants (*34*) and those of the analogous steady state constants again point to the fact that the complexes comprising (CC) are not in true equilibrium in the steady state (see Section II,D,3).

5. *Possible Isomeric Forms of the Phospho-Enzyme*

The possible existence of two forms of the phospho-enzyme was recognized by Najjar and Pullman (*33*). If the two, E_P and E_P', exchange their phosphate group with the 6 position of glucose-1-P and the 1 position of glucose-6-P, respectively, a phospho-enzyme isomerization

109. Transport exchange experiments are conducted by adding excess unlabeled glucose-1-P to an equilibrium mixture of labeled glucose monophosphates and determining whether the total radioactivity in the glucose-1-P fraction remains the same (independent transport) or undergoes a transient increase (countertransport) or decrease (co-transport). Independent transport occurs when both substrate and product compete for the same form of the enzyme and when the label is transferred from the substrate to product pool by a single catalytic cycle; co-transport occurs when the label requires two or more catalytic cycles to move from the substrate to the product pool and increases in extent with the number of cycles required; counter-transport occurs when substrate and product compete for different, interconvertible forms of the enzyme if interconversion is partially rate limiting (*93*).

110. P. D. Boyer, *ABB* **82,** 387 (1959).

step is required to complete a catalytic cycle, as indicated below (*111*).

Reaction sequence *b2*:

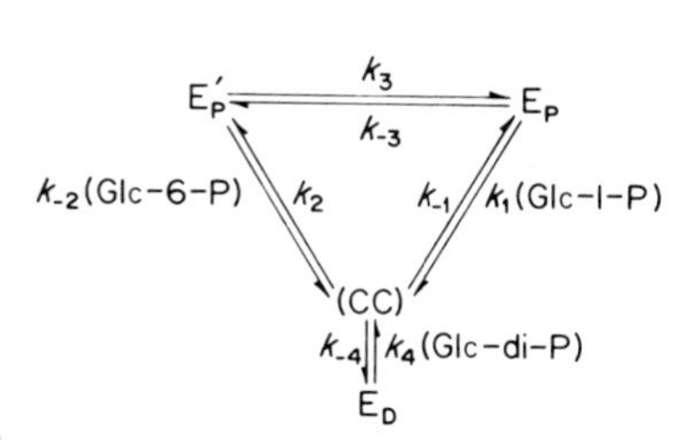

Two alternative kinetic procedures have been used to investigate this possibility. The basic rationale for both procedures is as follows: in the above sequence, if k_3 for the process $E_P' \rightarrow E_P$ is small with respect to the rate at which glucose-6-P binds to E_P', i.e., the rate (*112*) of

$$E_P' \xrightarrow{k_a(\text{Glc-6-P})} E_P'(\text{Glc-P}) \qquad (3)$$

then the reverse reaction (glucose-6-P → glucose-1-P) will have a selective advantage over the forward reaction that will increase with increasing glucose-6-P concentration, even when a constant ratio of glucose-6-P to glucose-1-P is maintained. In product inhibition studies this effect will show up as noncompetitive inhibition, as opposed to competitive inhibition in the absence of a phospho-enzyme isomerization step (*78*); in isotope exchange studies, countertransport rather than independent transport (*109*) will be observed for glucose residues (*93*).

Product inhibition studies in both the forward and reverse reactions fail to show any trace of noncompetitive inhibition, even at high concentrations of product. The results thus impose an unusually stringent requirement on the reaction: If sequence *b2* is valid, the rate constant for either $E_P' \rightarrow E_P$ or a catalytic step involving the exchange of a phosphate group with the substrate must be larger than 10^7 sec^{-1}, i.e., larger than the apparent rate constant for Eq. (3), k_a(Glc-6-P) (*78*). Actually, an E_P to E_P' conversion can be detected by quantitating the extent of countertransport at high concentrations of glucose phosphates; the apparent rate constant is about 4.5×10^7 sec (*93*). Since such a large rate constant is exceedingly improbable for a *covalent* change at the phosphorus nucleus of a phosphate group, even in enzymic systems, it is correspondingly improbable that the phosphate group of the phospho-

111. Here E_P actually refers to the $E_P \cdot Mg$ complex.

112. The rate constant k_{-2} in sequence *b2* is a complex constant which is a function of both k_a [Eq. (3)] and other constants.

enzyme undergoes an intramolecular migration during the catalytic cycle. Instead, the results are more reasonably rationalized if a minor, non-covalent change in the phospho-enzyme is required to complete a catalytic cycle, *viz.*, if sequence *b2* is ruled out for cases where E_P and E_P' involve different attachments for the active site phosphate group. The results do not exclude the existence of two phospho-enzymes with different phosphate attachments; only their obligatory interconversion during each catalytic step is excluded. However, there is no evidence for the existence of isomeric phospho-enzymes in the rabbit muscle system.

6. *Possible Structural Differences among the Central Complexes*

Since intramolecular migration of the phosphate group of E_P is not an obligatory step in the phosphoglucomutase reaction a problem arises as to how such a group can be transferred from a single position on the enzyme to either the 1 or 6 positions of bound glucose phosphate, or alternatively, how glucose-1,6-di-P can transfer either of its phosphate groups to a single serine residue on the enzyme. Three different possibilities are noted in terms of the latter question:

(a) A static model in which phosphate groups of glucose-di-P are bound in such a way as to simultaneously bring both to within potential bonding distance of the hydroxyl group of serine. Molecular models indicate that when the distance between the 1- and 6-phosphate groups of glucose-di-P is minimized, an alkyl hydroxyl group can be placed snugly between the phosphorus nuclei in an approximately equivalent position with respect to both, so that either could be transferred to the hydroxyl group with minimal structural changes.

(b) A "seesaw" model in which only one phosphate group at a time of glucose-di-P is within potential bonding distance of the hydroxyl group of serine and the other is brought into an analogous position by structural alterations which produce a seesaw motion of the phosphate groups.

(c) A dual-site model in which the glucose residue of glucose-di-P can be bound at either of two discrete positions on either side of the reactive serine residue and can change positions without completely dissociating from the enzyme.

At present there is no experimental evidence to favor one over the other of these possibilities. However, all three require a pivoting of the serylphosphate group of E_P about the –O–C bond, depending on whether glucose-1-P or glucose-6-P is bound; moreover, this motion would be an obligatory step in the catalytic cycle. Perhaps this pivoting constitutes the E_P' to E_P conversion detected by isotope transport experiments (see Section II,D,5).

E. Thermodynamics

1. *Overall Reaction*

The equilibrium constant K_e for the reaction α-D-glucose-1-P $\rightleftharpoons$ $(\alpha + \beta)$-D-glucose-6-P is 17.2 at 30° and neutral pH (*113, 114*). From this value and the value of K_e for α-D-glucose-6-P $\rightleftharpoons$ β-D-glucose-6-P, 1.4 (*13*), a K_e of about 7 can be calculated for α-D-glucose-1-P $\rightleftharpoons$ α-D-glucose-6-P. Since the pK_a values of the 1- and 6-isomers are similar (*1*), no effect of pH on the overall equilibrium constant is expected or observed between 6.2 and 7.5 (*114*); similarly, since Mg^{2+} is bound approximately equally by both isomers (*72*) no effect on K_e is expected or observed from 8 to 25 mM (*113*). K_e has a negative temperature coefficient (*114*) such that $\Delta S°$ for the reaction must approach zero, i.e., $\Delta G° \approx \Delta H°$ (*7*).

The corresponding K_e values for the phosphates of glucosamine, *N*-acetylglucosamine, galactose, and mannose, respectively, are as follows: 4.2 at pH 7.5 and 30° (*115*), 6.1 at pH 7.7 and 37° (*116*), about 25 at pH 7.5 and 30° (*117*), and 10 at pH 7 and 25° (*13*). A K_e of about 1.5 for the phosphates of fructose can be approximated from the equilibrium constant for fructose-1,6-di-P + glucose-6-P $\rightleftharpoons$ glucose-1,6-di-P + fructose-6-P, about 12 (*91*), if the free energies of hydrolysis of the phosphate group in the 6 position of glucose-1,6-di-P and fructose-1,6-di-P are taken as the same (*118*).

2. *Phosphate Transfer to Glucose Phosphate*

Direct attempts (*15, 17, 35*) to measure the equilibrium constant for the half-reaction

$$E_P + \text{glucose-6-P} \rightleftharpoons E_D + \text{glucose-1,6-di-P} \tag{4}$$

113. M. R. Atkinson, E. Johnson, and R. K. Morton, *BJ* **79,** 12 (1961).
114. S. P. Colowick and E. W. Sutherland, *JBC* **144,** 423 (1942).
115. D. H. Brown, *JBC* **204,** 877 (1953).
116. T. Posternak and J. P. Rosselet, *Helv. Chim. Acta* **36,** 1614 (1953).
117. T. Posternak and J. P. Rosselet, *Helv. Chim. Acta* **37,** 246 (1954).
118. Combining data (*34*) for the difference in $\Delta G°$ for hydrolysis of the 1-phosphate in glucose-1-P and glucose-di-P (which should be about the same as the difference in $\Delta G°$ for hydrolysis of the 6-phosphate in glucose-6-P and glucose-di-P) with the $\Delta G°$ for hydrolysis of glucose-6-P (*113*) gives a $\Delta G°$ of −3.9 kcal for hydrolysis of the 6-phosphate of glucose-di-P; this value is quite close to that of the corresponding group in fructose-di-P, −3.7 kcal (*119*).
119. O. Meyerhof and H. Green, *JBC* **178,** 655 (1949).

have produced widely divergent values (*120*). This lack of agreement was at least partially caused by failure in some studies to distinguish between free and bound forms of both enzyme and substrates. However, two *kinetic* procedures for evaluating the equilibrium constants for half-reactions in Ping Pong sequences were later used for such measurements; these avoided the problem of bound vs. free forms of the enzyme and substrates by utilizing the substrates in large excess. A third procedure was also used which took advantage of the much less tenacious binding of the pseudo-substrate, glucose 6-sulfate, and involved the direct measurement of the analogous equilibrium,

$$\text{Glc-6-S} + E_P \rightleftharpoons \text{Glc-1-P-6-S} + E_D \quad (5)$$

From this measurement K_e for Eq. (4) was calculated by use of the constant, 1.2–1.5 (*34*) for

$$\text{Glc-6-S} + \text{Glc-1,6-di-P} \rightleftharpoons \text{Glc-1-P-6-S} + \text{Glc-6-P} \quad (6)$$

All three of the latter procedures gave similar equilibrium constants for Eq. (4) (30°): $1\text{—}3 \times 10^{-3}$ at pH 8.5 and about 2×10^{-4} at pH 7.4 (*34*).

The value of K_e [Eq. (4)] actually is not as informative about the phosphate transfer process as assumed by early workers since (E_D + glucose-di-P) does not represent an intermediate state in the catalytic process but rather the products of a dead-end branch in the pathway (Section II,D,1); however, this constant can be used in calculations of the free energy change accompanying hydrolysis of the phospho-enzyme (see below).

3. *Hydrolysis of the Phospho-Enzyme*

In addition to transferring phosphate between the 1-hydroxyl and the 5- or 6-hydroxyl groups of various sugars (see Section II,F), the phospho-enzyme can transfer its phosphate group to water, although with samples of carefully purified enzyme the reaction is very slow—less than 10^{-9} (*121*) to 10^{-10} (*122*) as fast as transfer to glucose-1-P—under normal assay conditions but in the absence of substrate. The transfer reaction to water appears to be unimolecular over a wide concentration range and probably represents a *bona fide* phosphate transfer by phosphogluco-

120. The value reported by Hashimoto and Handler (*14*) is actually the reciprocal of the value that was measured because of a calculation error [P. Handler, personal communication (1967)].

121. D. S. Kirkpatrick, Ph.D. Thesis, Purdue University (1967); *Dissertation Abstr.* **28,** 4414-B (1968).

122. J. W. Long and W. J. Ray, Jr., unpublished results, 1971.

mutase rather than a hydrolysis caused by a contaminating phosphatase. In such a case the half-time for hydrolytic phosphate transfer would be about a month under the above conditions (*122*).

Although the equilibrium

$$E_P \rightleftharpoons E_D + P_i \tag{7}$$

cannot be measured directly, the standard free energy change associated with this reaction, ΔG°_{hyd}, can be determined as the sum of ΔG° for Eq. (4), about 3.7 kcal, and ΔG° for

$$\text{Glc-1,6-di-P} \rightleftharpoons P_i + \text{Glc-6-P} \tag{8}$$

about —5.6 kcal (*34*); thus ΔG°_{hyd} is about —1.9 kcal (*123*). Hence, in the classic sense phosphoglucomutase has a "low energy" phosphate bond.

However, ΔG°_{hyd} is a poor parameter for comparing phosphate bond cleavage of phosphoproteins and simple phosphate esters because of the possibility that alterations in noncovalent interactions occur during or subsequent to bond cleavage in phosphoproteins (*34*) [see also Levine *et al.* (*125*)]. Thus, ΔG° for phosphoprotein hydrolysis includes, in addition to ΔG° for bond breaking [Eq. (9)], a ΔG° for separation of products [Eq. (10)].

$$E_P + H_2O \rightleftharpoons E_D{\cdot}P_i \tag{9}$$

$$E_D{\cdot}P_i \rightleftharpoons E_D + P_i \tag{10}$$

Here, $E_D \cdot P_i$ is the dephospho-enzyme·orthophosphate complex. Since for phosphoglucomutase, ΔG°_{hyd} is —1.9 kcal and ΔG° for phosphate dissociation [Eq. (10)] is +4.2 kcal, ΔG° for bond breaking [Eq. (9)] must be —5.9 kcal (*34*). This value is considerably more negative than the ΔG° deduced for the analogous bond-breaking process involving phosphoserine, Ser-P $\rightleftharpoons$ Ser·P, —0.9 kcal; hence, the phosphate bond-breaking step of Eq. (9) must be accompanied by changes in noncovalent interactions involving the enzyme moiety which strongly favor this process (*34*). If analogous alterations occur during transfer of phosphate to the 1 position of glucose-6-P, a facile rationale is provided for the apparent anomaly of the efficient conversion of a "low energy" serine phosphate to a "high energy" glucosyl 1-phosphate.

123. A value of —4.8 kcal is given for the ΔG° value of Eq. (8) (*34*) based on an early estimate of ΔG° for hydrolysis of glucose-6-P (*124*); by using a more recent value for the latter constant (*113*), ΔG°_{hyd} [Eq. (7)] becomes —1.9 instead of —1.1 kcal.

124. L. M. Ginodman, *Biokhimiya* **19**, 666 (1954).

125. D. Levine, T. W. Reid, and I. B. Wilson, *Biochemistry* **8**, 2374 (1969).

No comparable thermodynamic data is available for other phosphoglucomutases. However, the *rate* of transfer to water of the phosphate group of the shark and flounder muscle enzymes is much more rapid than that observed for the rabbit muscle enzyme. Indeed, the former enzymes act as rather specific glucose diphosphate phosphatases (*14*). Hence, equilibrium between E_P and $E_D + P_i$ should be attained much more readily than with the rabbit muscle enzyme, and the possibility of labeling the dephospho forms of these enzymes with ^{32}P-phosphate, as in the alkaline phosphatase system at low pH (*126*), might be considered. However, if in these systems $E_D \cdot P_i$ is much more stable than E_P, as it is for the rabbit muscle enzyme, an insignificant fraction of label would be incorporated covalently, *viz.*, the system would resemble the alkaline phosphatase system at high pH where phosphate incorporation becomes vanishingly small because of increased stability of $E_D \cdot P_i$ relative to E_P (*127*).

F. Specificity

Phosphoglucomutase catalyzes the interconversion of the 1- and 6-phosphate isomers of many α-D-hexoses other than glucose: mannose (*13, 116, 128*), galactose (*13, 117, 128*), allose and altrose (*128*), fructose (*13*), 2- and 3-deoxyglucose (*128*), 2- and 3-*O*-methylglucose (*128*), glucosamine (*115*), and *N*-acetylglucosamine (*77*); the enzyme also catalyzes the interconversion of the 1- and 5-phosphates of α-D-ribose (*13, 98*) and probably of α-D-deoxyribose. In addition, the diphosphates of the α-D-sugars, galactose (*129*), mannose (*116, 129*), *N*-acetylglucosamine (*129*), and α-D-ribose (*98, 129*) may serve as activators of the enzymic reaction.

Most, if not all, of the above sugar phosphates are much less reactive with phosphoglucomutase than glucose phosphates. For example, the bimolecular rate constant, $V_{max}/E_T K_m$, for glucose-1-P is about 6×10^6 that for fructose-1-P (*13*). Of those sugar phosphates whose bimolecular rate constants are known, mannose-1-P is the best substrate other than glucose 1- and 6-phosphates; however, the value of its constant is only 1/500 that of glucose-1-P (*13*).

All β-phosphates are assumed to be inactive since (a) the rabbit muscle enzyme does not appear to act on β-D-glucose-1-P (*92, 130*); (b)

126. J. H. Schwartz, A. M. Crestfield, and F. Lipmann, *Proc. Natl. Acad. Sci. U. S.* **49**, 722 (1963).
127. T. W. Reid, M. Pavlic, D. J. Sullivan, and I. B. Wilson, *Biochemistry* **8**, 3184 (1969).
128. L. G. Együd and W. J. Whelan, *BJ* **86**, 11P (1963).
129. H. Mulhausen and J. Mendicino, *JBC* **245**, 4038 (1970).
130. M. E. Krahl and C. F. Cori, *Biochem. Prep.* **1**, 33 (1949).

β-D-glucose-1,6-di-P does not serve as an activator (*73, 92*); and (c) α-D-glucose-6-P, rather than the β-isomer, is produced from α-D-glucose-1-P (*80*). Thus, isomerization of α-D-glucose-6-P at the 2, 3, or 4 position does not eliminate activity (see references above) while isomerization at the 1 position apparently does. However, the distinction between complete elimination of activity and reduction by many fold is obviously difficult to make. The best test for low-level activity is treatment of ^{32}P-labeled phospho-enzyme with the substrate in question, followed by measurement and identification of the TCA-soluble radioactivity thus produced [e.g., see Alpers and Lam (*12*)]; however, such a test has not been widely used in specificity studies.

For all sugar phosphates studied, except fructose phosphate, the forward reaction, conversion of the 1-isomer to the 6-isomer, is faster than the reverse reaction by factors ranging from 3- to 200-fold; for fructose phosphate the "forward" reaction is only 0.07 as fast as the "reverse" process (*13*). This inversion raises the possibility that the binding of fructose phosphate by phosphoglucomutase is such that the phosphate of fructose-6-P occupies the position normally taken by the phosphate group of bound glucose-1-P, and the phosphate of fructose-1-P takes the position normally occupied by the phosphate group of glucose-6-P. Although it seems probable that the phosphate groups of glucose 1- and 6-phosphates always occupy distinct and separate sites one cannot be certain that this is the case for all sugar phosphates, and decreased V_{max}/E_T values with substrates other than glucose phosphate may result either from reduced rate constants or from the existence of nonproductive complexes produced by alternative binding modes ["wrong-way" binding, e.g., see Rapp *et al.* (*131*)].

Glucose 6-sulfate accepts a phosphate group from the phospho-enzyme in a reversible reaction. However, transfer of the sulfate group from the glucose-1-P-6-S thus formed to the dephospho-enzyme to give a sulfo-enzyme does not occur to any appreciable extent. Moreover, the mixed ester is bound to dephospho-enzyme much less tenaciously than is glucose-di-P (*34*).

The only nonsugar phosphate known to transfer its phosphate to phosphoglucomutase is glycerate-1,3-di-P; in the absence of added glucose-di-P, glycerate-1,3-di-P acts as an activator of phosphoglucomutase (*132*) by phosphorylating the dephospho form of the enzyme (*12*). The activating efficiency of this diphosphate is much less than that of glucose-di-P so that inhibition by glycerate-1,3-di-P is easily observed in reac-

131. J. R. Rapp, C. Niemann, and G. E. Hein, *Biochemistry* **5**, 4100 (1966).
132. J. B. Alpers, *JBC* **243**, 1698 (1968).

tions conducted in the presence of limiting glucose-di-P (*132*). Since the reaction of glycerate-1,3-di-P involves only the 1-phosphate group and is reversible to only a slight degree, if at all (see below), phosphoglucomutase has the potential of converting mixtures of glucose-6-P and glycerate-1,3-di-P to glucose-1,6-di-P and glycerate-3-P. (Hydrolysis of the glycerate-1,3-di-P and slowing of the reaction because of product inhibition would make this process unattractive for preparative purposes, however.)

In an analogous manner, a mixture of glucose-1-P and glycerate-1,3-di-P can be converted to glucose-6-P, glucose-1,6-di-P, and glycerate-3-P. Surprisingly, at saturating glucose-1-P, changes in glycerate-1,3-di-P concentration alter the ratio of glucose-6-P to glucose-1,6-di-P thus produced (*132*). Such a result is inconsistent with a simple reaction based on sequence *b1* in which E_D·glucose-di-P is partitioned between dissociation to give glucose-di-P and further reaction to give glucose-6-P, unless partitioning of E_D·glucose-di-P is altered by either glycerate-1,3-di-P or glycerate-3-P; however, experiments to prove this point are lacking. Glycerate-1,3-di-P also acts as a cofactor for the *M. lysodeikticus* enzyme (*133*), although the reaction mechanism for this enzyme is thought to be different from that of the rabbit muscle enzyme.

Since glycerate-1,3-di-P transfers a phosphate group to the dephospho-enzyme, glycerate-3-P must also be a phosphate acceptor relative to the phospho-enzyme, although it would seem to be an immeasurably poor one because of problems of reversibility. [The $\Delta G°$ for hydrolysis of glycerate-1,3-di-P to glycerate-3-P and P_i is about —14 kcal (*134*).] However, the product, glycerate-1,3-di-P, is readily hydrolyzed, and it is possible that such hydrolysis could displace the equilibrium in the direction of the dephospho-enzyme; tentative evidence for a limited phosphate transfer of this type has been suggested (*12*).

Transfer of the phosphate group of phosphoglucomutase to hydroxyl groups of materials other than sugar phosphates usually occurs either very slowly or not at all. Thus the bimolecular rate constant for phosphate transfer to glucose appears to be less than 10^{-10} that of the similar constant for transfer to glucose-1-P (*13*) although the transfer can be measured with ^{32}P-labeled enzyme (see also Section II,E,2). However, the phosphate group of the enzyme can be transferred readily to the 1 position of xylose in the presence of inorganic phosphate (but very slowly in its absence) or to water in the presence of xylose-1-P. In

133. J. B. Alpers, *FEBS Symp.* **19,** 241 (1969).

134. H. R. Mahler and E. H. Cordes, "Biological Chemistry," p. 426. Harper, New York, 1966.

the latter case the enzyme becomes a very specific and relatively efficient glucose-di-P phosphatase (*135*).

The specificity of other phosphoglucomutases has not been examined extensively. However, the enzyme from *E. coli* catalyzes isomerization of the phosphates of galactose, glucosamine, and *N*-acetylglucosamine (*19*) and the enzyme from human muscle interconverts ribose phosphates (*27*).

G. Activation

Both glucose-di-P and Mg^{2+} may be considered as activators of phosphoglucomutase (see Sections II,D,2 and II,I,1); however, only activation by chelating agents is discussed in this section.

Phosphoglucomutase was originally assayed in the presence of cysteine (*36*) or sulfite plus albumin (*45*). Subsequently, histidine (*33*) and later imidazole (*48*) were substituted for cysteine; a variety of less effective agents such as 8-hydroxyquinoline and dithizon also have been used (*136*). Sutherland (*136*) first suggested and Milstein (*131*) later demonstrated that such materials activate by forming complexes with "inhibiting" metal ions. In fact, Milstein showed that in reality such activators were unnecessary for enzymic activity if both reagents and enzyme were carefully freed of bivalent metals (other than Mg^{2+}) prior to the assay (*72*). He also pointed out that the efficiency of cysteine and histidine, relative to chelators such as EDTA, could be rationalized in terms of their high affinity for "heavy metals" in contrast with their inappreciable interaction with the normal activator, Mg^{2+} (*68, 137*).

Milstein, who introduced the use of metal buffers (*138*) for studying metal ion binding to phosphoglucomutase, also showed that Cu^{2+} and Zn^{2+}, which are ubiquitous trace contaminants of aqueous solutions, bind very tenaciously to the enzyme and act as inhibitors in the normal enzymic assay (*51*) (see Section II,H,3); for example, at pH 7.4 the dissociation constant for Zn^{2+} is about 10^{-11} *M* (*85, 139*) and that for Cu^{2+} (*51*) is appreciably smaller (*140*). Since in the absence of complex-

135. J. W. Long, J. Owens, and W. J. Ray, Jr., unpublished results (1971).
136. E. W. Sutherland, *JBC* **180**, 1279 (1949).
137. C. Milstein, *BJ* **79**, 584 (1961).
138. C. N. Reilley, *Federation Proc.* **20**, Suppl. 10, 22 (1961).
139. W. J. Ray, Jr., *JBC* **242**, 3737 (1967).
140. Milstein's data (*51*) are inconsistent with the values he gives for Zn^{2+} and Cu^{2+} binding; presumably an error was made in calculating free Cu^{2+} and free Zn^{2+} concentrations from the metal buffer used (*138*); however, his data do show that Cu^{2+} is bound more tenaciously than Zn^{2+}.

ing agents one scarcely expects the concentrations of free Zn^{2+} in aqueous solutions in contact with glass to be less than 10^{-8} to 10^{-9} *M*, and since the assay concentration of enzyme is routinely 10^{-9} to 10^{-10} *M* (see Section II,C,1) only a small fraction of phosphoglucomutase may be in a metal-free form in solutions sufficiently dilute for the enzymic assay unless chelating agents are present or other precautions are taken to eliminate such metal ions (*141*). Moreover, the disparity between the binding constants for Mg^{2+} and Zn^{2+} is such that the level of Zn^{2+} in even m*M* solutions of Mg^{2+} may be sufficiently high to produce inhibition in the absence of the appropriate chelator (*72*).

In addition to their binding efficiency the rate at which complexing agents remove inhibiting metal ions from phosphoglucomutase is also a factor in determining their relative activating efficiencies. Thus EDTA-type chelators act as very efficient scavangers for free metal ions, but remove metal ions from the enzyme–metal complex in a concentration-independent manner which depends only on the enzyme–metal dissociation rate (*139*); for Zn^{2+} this rate is relatively slow (*142*). By contrast, ligands such as histidine and imidazole can greatly accelerate the enzyme–metal dissociation rate, although they do not bind the free metal ion as tenaciously as EDTA; presumably, a metal–ligand complex is actually the dissociating species in such cases. Thus, 0.1 *M* imidazole, 1 m*M* EDTA, pH 7.5, is highly efficient for removing bound metals from phosphoglucomutase with respect to both rate and extent of removal (*139*). Cysteine probably has a similar effect on metal ion dissociation rates since treating the enzyme with cysteine at 0° for only 15 sec produces maximal activation (*143*).

A third factor that is important in chelate-induced activation is the decreased dissociation rate of Zn^{2+} and other metal ions in the presence of saturating concentrations of substrate. Thus, unless metal ion con-

141. The literature on activation of phosphoglucomutase is both contradictory and confusing. For example, while some workers were able to mimic the activating effect of several preassay treatments with enzyme deliberately converted to the Zn^{2+} form (*68*), others found essentially no Zn^{2+} in their purified enzyme—and, in fact, much less than one equivalent of bound metal (*27*). Moreover, while the initial activity of some samples of untreated enzyme is reported as essentially zero (*68*), the activity of untreated enzyme studied by other workers (*19, 48, 72*) was far from zero. One possible explanation of these differences is related to the treatment of the enzyme prior to the assay: the higher the concentration of enzyme used to initiate the assay, the less chance for adventitious metal ions to bind to an appreciable fraction of the enzyme. Differences in the metal content of the undiluted enzyme and of the diluting solution would of course compound the problem.

142. The rate constant for dissociation of Zn^{2+} is incorrectly quoted by Ray (*139*) as about 0.019 sec^{-1}; the actual value is about 0.0019 sec^{-1}.

143. W. N. Aldridge and M. Thomas, *BJ* **98,** 100 (1966).

taminants are removed from the enzyme prior to the assay the enzyme may not exhibit maximal activity, even if the assay contains concentrations of complexing agents that would be sufficient for rapid removal of metals in the absence of substrate. When the assay is conducted at very low substrate concentrations, prior removal of contaminating metal ions affects activity only slightly (because efficient removal can occur during the assay); however, at high substrate concentrations a large effect is produced by a preassay activation step (because efficient metal removal does not occur during the assay) (*68*). Thus, there is no evidence that a "second" or "further" activation (*48, 137, 144*) is a separate and distinct phenomenon from activation induced by complexing agents, except that several types of inhibiting metals may be bound to the enzyme (see Section II,H,3) and the relative ease of removing these varies markedly.

Of the variety of preassay treatments that have been used (*30, 48, 68, 72, 137, 144*), the most efficient utilizes imidazole, EDTA, and excess Mg^{2+} (*30*). Other less attractive procedures involve passage of the enzyme solution through a column of cation exchange resin, Tris form (*72*), or brief exposure to urea (*27, 31, 145*), guanidine, low pH, or elevated temperatures at pH 5 (*145*). Apparently the latter procedures decrease metal binding and increase dissociation rates of bound metals by altering the structure of the enzyme (*145*). In at least the case of the beryllium complex, metal dissociation on treatment with 4 *M* urea has been verified (*31*); however, since initial activation by some of these treatments is accompanied by subsequent inactivation the results are difficult to interpret quantitatively.

Flounder and shark muscle phosphoglucomutases are also activated by treatment with metal-complexing agents prior to the enzymic assay, although to a smaller extent than the rabbit muscle enzyme (*14*), while such treatment is without effect on the human muscle enzyme (*27*). Whether these differences are the result of more rapid metal ion exchange during the assay, less tenacious metal ion binding per se, or higher selectivity for Mg^{2+} binding *vis-à-vis* other metal ions is not known. Mg^{2+} plus EDTA also activates the yeast enzyme; however, the mechanism for this effect appears to be different from that of the rabbit muscle enzyme (*16*). The enzymes from *E. coli* (*17*), *M. lysodeikticus*, and *B. cereus* (*19*) also fail to respond to treatment with chelating agents prior to the assay. The first of these requires relatively high concentrations of cysteine in the assay for activity, the second requires cysteine

144. S. Harshman, J. P. Robinson, V. Bocchini, and V. A. Najjar, *Biochemistry* **4**, 396 (1965).

145. V. Bocchini, M. R. Alioto, and V. A. Najjar, *Biochemistry* **6**, 3242 (1967).

for maximal activity, and the last has no requirement for cysteine at all. The potato tuber enzyme is also activated prior to the assay by treatment with cysteine and Mg^{2+} but not by EDTA and Mg^{2+} (*18*). In contrast with the mammalian muscle enzymes where the activator requirement is related to protection against inhibiting metals, the cysteine requirement for the *M. lysodeikticus* and *E. coli* enzymes and the activating effect of cysteine on the potato tuber enzyme is probably related to the sensitivity of their sulfhydryl groups toward chemical modification (*17, 19*). The lack of inhibition of the *E. coli* enzyme by m*M* Zn^{2+} (*17*) shows that cysteine activation of this enzyme is not related to protection against Zn^{2+} inhibition.

H. Inhibition

Because of the complexity of the phosphoglucomutase system, many authors have inadvertently studied effects of enzyme inhibitors under conditions where one or more of the normal reactants was limiting; hence, it is frequently impossible to deduce the true nature of the inhibition from the published data. In most cases an effect on V_{max} is implied, while the true effect, especially for anionic inhibitors (see below), is almost certainly on K_m^{app}—usually $K_{m(Glc\text{-}di\text{-}P)}^{app}$. Inhibitors whose mode of action is not well documented are not described herein (see, however, references *18, 25–27, 45, 98, 146–150*).

1. *Anions*

Inhibition of phosphoglucomutase by added salts was recognized quite early (*2*) and all salts tested have subsequently proved to be inhibitory, frequently because of anionic effects. All inhibitions by anions that have been carefully examined are of the competitive type (*68, 151*), and even though such inhibition has been examined primarily under conditions where competition is with glucose-di-P (i.e., is caused by binding to the dephospho-enzyme) inhibition competitive with glucose-1-P (binding to phospho-enzyme) also is to be expected, although sulfate and chloride are the only simple anions that have been reported to act in this manner (*68, 106*). In general, polyvalent anions are better inhibitors than monovalent anions; thus, sulfate is a better inhibitor than chloride

146. F. A. M. Alberghina, *Life Sci.* **3**, 49 (1964).
147. N. Bargoni, *Giorn. Biochim.* **13**, 68 (1964).
148. B. C. Loughman, *Nature* **191**, 1399 (1961).
149. K. I. Rubchins'ka, *Ukr. Biokhim. Zh.* **37**, 553 (1965).
150. L. H. Stickland, *BJ* **44**, 190 (1949).

and pyrophosphate than phosphate (*68, 151*). Other anions that inhibit competitively with glucose diphosphate are 6-phosphogluconic acid, ATP, ADP, AMP, UTP, UDP (*151*), 3-phosphoglycerate, 2,3-diphosphoglycerate (*132*), acetate, bromide, nitrate, iodide, and perchlorate (*68*). Glucose monophosphates and other sugar phosphate substrates also may inhibit competitively with glucose diphosphate, as discussed below.

Inhibition by fluoride was originally interpreted in terms of a "magnesium-fluoro-glucose phosphate" complex (*36*) because enhancement of fluoride inhibition by glucose-1-P and Mg^{2+} was observed. However, in light of later studies (*68, 151*) such an enhancement is expected for simple anionic inhibition when glucose-di-P is limiting (see below) and reversal of fluoride inhibition by increased glucose-di-P concentration has been observed (*151*); hence, there is no reason to postulate the existence of the above complex.

As in the case of many enzymes that exhibit Ping Pong kinetics (*152*), competitive substrate inhibition by both substrates can be demonstrated in the phosphoglucomutase reaction. Thus, inhibition by glucose-1-P is competitive with glucose-di-P (*68, 73, 79, 106*) because glucose-1-P binds in a dead-end manner to the dephospho-enzyme; however, this binding is much less tenacious, $K_I \approx 0.5$ mM (*106*), than its equilibrium binding to the phospho-enzyme, $K_x \approx 1$ μM (*106*). Conversely, glucose-di-P inhibits competitively with glucose-1-P by dead-end binding to the phospho-enzyme, but again the abortive binding is much less tenacious, $K_I \approx 0.7$ mM (*73*), than its equilibrium binding to dephospho-enzyme, $K_d \leqq 10^{-8}$ M (Section II,D,3). Earlier claims of glucose-di-P inhibition in the micromolar range (*7, 115, 153, 154*) can probably be explained by impurities in the diphosphate preparation used (*73*).

Substrate inhibition by both glucose mono- and diphosphates has also been detected in the *E. coli*, *M. lysodeikticus*, *B. cereus* (*19*), shark, and flounder enzyme systems (*14*). Inhibition by glucose-di-P is unusually pronounced in the *E. coli* system and inhibition by glucose-1-P is quite apparent in the *M. lysodeikticus* system; however, the analogous substrate inhibitions for the shark and flounder enzymes are quite similar to the weak inhibitions noted above for the rabbit muscle enzyme.

Anions which bind Mg^{2+} can inhibit both as anions per se as well as by reducing the concentration of free Mg^{2+}, e.g., citrate and ATP (*155,*

151. E. F. Kovács and G. Bot, *Acta Physiol. Acad. Sci. Hung.* **27,** 328 (1965).
152. W. W. Cleland, *BBA* **67,** 188 (1963).
153. G. R. Bartlett, *JBC* **234,** 449 (1959).
154. V. A. Najjar and E. E. McCoy, *Federation Proc.* **17,** 1141 (1958).
155. G. Bot and E. F. Kovács, *Acta Physiol. Acad. Sci. Hung.* **21,** 44 (1962).

156). However, the simple reversal of anionic inhibition by added Mg^{2+} does not in itself demonstrate that the anion inhibits by complexing Mg^{2+}; the effect may instead be anionic inhibition competitive with glucose diphosphate which is relieved by formation of the anion·Mg complex. Alternatively, inhibition (by glucose-1-P) that is partially relieved by increasing the Mg^{2+} concentration, even though no inhibitor·Mg complex actually forms, has also been described (*106*).

2. *Poor Substrates and Substrate Analogs*

Sugar phosphates which are poor substrates for phosphoglucomutase (see Section II,F) can, of course, act as inhibitors of the normal enzymic reaction; however, in such cases the observed inhibition may be difficult to rationalize in terms of molecular interactions since the inhibiting sugar phosphate may exert its effect by binding to either the phospho or dephospho forms of the enzyme or by depleting the glucose diphosphate level via a Ping Pong type of reaction (see Section II,D,1). Diphosphates such as galactose diphosphate (*157*), fructose diphosphate (*13*), and glycerate 1,3-diphosphate (*132*), which are poor activators (see Section II,F), can also inhibit. The substrate analog, glucuronic acid-1-P, which binds quite tenaciously to the enzyme·Mn complex (*55*), also should act as a good competitive inhibitor.

3. *Cations*

Inhibition by Mg^{2+}, which is also an activator (see Section II,I,1), has been recognized by several workers (*2*, *36*, *68*, *79*, *144*, *155*). Mg^{2+} inhibition is exceedingly complex, although at pH 7.5 such inhibition becomes important only in the concentration range above that required for maximum activation (*158*). Mg^{2+} inhibition is competitive with both glucose monophosphate (*106*) and glucose diphosphate (*68*), but it is usually seen as competitive with the diphosphate (*159*) because the latter is more likely to be limiting than the monophosphate. Since the uncomplexed sugar phosphates are the normal substrates (see Section II,I,2), Mg^{2+} inhibition is caused by formation of inactive sugar phosphate·Mg complexes. Moreover, because glucose-di-P can bind two equivalents of Mg^{2+} (*68*), inhibition competitive with the diphosphate is of the parabolic

156. H. Zwarenstein and V. van der Schyff, *BBRC* **26**, 372 (1967).
157. Y. L. Huang and K. E. Ebner, *Experientia* **25**, 917 (1969).
158. However, at limiting glucose diphosphate, 2 mM $MgCl_2$ causes a 50% inhibition (*68*).
159. G. Bot and E. Polyik, *Acta Biochim. Biophys.* (*Budapest*) **2**, 349 (1967).

type (*160*). In addition, inhibition by Mg^{2+} and glucose-1-P together is equal to the product of these inhibitions taken separately at limiting glucose-di-P (*68*). Plots of $K^{app}_{Glc\text{-}di\text{-}P}$ vs. glucose-1-P concentration at several levels of magnesium chloride have been published for the rabbit muscle enzyme (*68*).

The observation that free sugar phosphates are the normal substrates does not indicate that, for example, the glucose-1-P·Mg complex could not bind to free E_P. In reality, the dissociation constant for the E_P·Mg complex (see Section II,I,2) is sufficiently small, 0.025 m*M*, that the fraction of free enzyme becomes inappreciable at a Mg^{2+} concentration too low to convert a significant fraction of glucose-1-P into glucose-1-P·Mg, *viz.*, at a concentration much less than 10 to 20 m*M*, the K_d for glucose-1-P·Mg (*68, 72, 161*). Hence, by using steady state kinetics one can only show that glucose-1-P·Mg does not bind efficiently to E_P·Mg.

Numerous metal ions such as Mn^{2+}, Co^{2+}, Ni^{2+}, Zn^{2+}, Cd^{2+}, and possibly Ca^{2+} apparently act as inhibitors of phosphoglucomutase; however, these metals are actually activators that are simply less efficient than is Mg^{2+} (see Section II,I,1), although most bind more tenaciously than Mg^{2+} (*30, 85*).

On the other hand, Cu^{2+} is not an activator (*162*), and inhibits (*137*) by binding at sites other than the activator site; thus, for example, the optical properties of E_P·Cu are almost identical with the difference in the properties of E_P·Zn·Cu and E_P·Zn (*162*). By contrast beryllium inhibition involves binding at what is probably the activator site, since its binding and the binding of Mg^{2+} are mutually exclusive and since other metal ions must be removed before beryllium inhibition is observed (*31, 143*). Moreover, beryllium binding reduces activity many orders of magnitude and probably to zero since, after exposure to beryllium, ^{32}P-labeled enzyme will not exchange its label with glucose-P during an extended reaction period (*31*). Beryllium inhibition has been referred to as "generally irreversible" (*31*), and as such its effect on phosphoglucomutase might at least partially account for the cumulative effect of beryllium poisoning (*31, 163*). Nevertheless, a degree of reversibility is possible under some conditions, since 4 *M* urea causes sufficient dissociation so that beryllium can be removed from the enzyme via gel filtration; however, this treatment elicits only partial return of enzymic activity (*31*). Some workers have reported "mitigation of beryllium inhibition"

160. W. W. Cleland, *BBA* **67,** 173 (1963).
161. H. B. Clarke, D. C. Cusworth, and S. P. Datta, *BJ* **58,** 146 (1954).
162. J. W. Long and W. J. Ray, Jr., *BBA* **221,** 522 (1970).
163. K. W. Cochran, M. M. Zerwic, and K. P. DuBois, *J. Pharmacol. Exptl. Therap.* **102,** 165 (1951).

by Mg^{2+}, Co^{2+}, and Mn^{2+} (*164, 165*) although the work of others (*31, 143*) indicates that Mg^{2+} is ineffective in reversing inhibition. The extent to which reactivation of beryllium-inhibited enzyme is possible under physiological conditions is obscure.

It should be pointed out that aqueous solutions of divalent beryllium contain a mixture of hydrolysis products, especially at higher pH values, and in fact Be^{2+} is a minor component of the mixture at pH 7 (*166*). Whether or not Be^{2+} or some other form such as $BeOH^{+}$, binds to phosphoglucomutase is not known; however, if a minor component of the system is involved the rather large temperature coefficient for inactivation of phosphoglucomutase by beryllium could be at least partially attributed to processes such as $Be(OH)^{+} \rightleftharpoons Be^{2+} + OH^{-}$, as opposed to possible structural changes in the enzyme or to covalent bond formation with the enzyme as has been suggested (*31, 143*). Differential effects of beryllium on phosphoglucomutase from various sources have also been reported (*31*).

Some workers (*45*) have observed only slight inhibition of rabbit muscle phosphoglucomutase by Hg^{2+} whereas others (*136*) reported extensive inhibition by both Hg^{2+} and Ag^{2+}; Hg^{2+} does not inhibit the *E. coli* enzyme (*17*), but 10^{-5} *M* concentrations cause 50% inhibition of the *B. cereus* enzyme (*19*).

4. *Chemical Modification*

One of the first covalent alterations of phosphoglucomutase that was reported involved reaction with diisopropylphosphofluoridate (*167*); however, subsequent workers were unable to obtain a reaction (*42, 168*) or even to observe inhibition (of the *E. coli* enzyme) by this reagent (*17*). Perhaps the original observations were the result of impurities that are found in some commercial samples of the fluoridate (*169*).

Phosphoglucomutase is slowly inactivated by iodoacetamide; the rate of the reaction is altered by the presence of substrate (see Section II,K,3), and the initial reaction is with cysteine residues (*32*). Iodoacetic acid also produces inhibition (*45*). Although iodination apparently produces no ac-

164. Y. T. Liu, *Vopr. Med. Khim.* **7,** 605–8 (1961).

165. J. V. Pavlova, Y. T. Liu, and V. I. Fedorova, *Proc. Toksikol. Klin. Prof. Zabolevanii Khim. Etiol.* **56,** 213 (1962).

166. L. G. Sillen and A. E. Martell, "Stability Constants of Metal-Ion Complexes," pp. 40–41. Chem. Soc., London, 1964.

167. E. P. Kennedy and D. E. Koshland, Jr., *JBC* **228,** 419 (1957).

168. S. Harshman and V. A. Najjar, *Biochemistry* **4,** 2526 (1965).

169. N. Gould, R. C. Wong, and I. E. Liener, *BBRC* **12,** 469 (1963).

tivity loss (*29*), exposure to iodosobenzene, *N*-ethylmaleimide, or chloroacetophenone produces slow inactivation (*137*). Activity is reduced 20–30% by reaction with two equivalents of mercuribenzoate, and by 40–60% by reaction with four equivalents (*14, 50*); the partial inhibition after reaction with two equivalents is reversed by treatment with cysteine (*137*).

By contrast, reaction of one equivalent of mercuribenzoate with the *M. lysodeikticus* (*19*) or flounder enzymes (*14*) increases activity significantly, although further additions either substantially reduce activity (*19*) or return it to its original level (*14*). The reaction of the *E. coli* enzyme with two equivalents of this reagent completely abolished activity (*17*), while the shark enzyme is unaffected (*14*).

Photooxidation of rabbit muscle phosphoglucomutase in the presence of methylene blue destroys activity primarily by modification of a unique and critical methionine and a unique histidine residue within groups of several of these residues that are rapidly oxidized. Oxidation of the histidine residue reduces activity by about 12-fold (*28*), while oxidation of the critical methionine (to the sulfoxide) decreases activity by several orders of magnitude (*83*); however, although changes in V_{max} are implied, this was not verified and the partial activity loss accompanying histidine oxidation actually might have been produced, for example, by a large increase in $K_{m(Mg)}$. Since the histidine residue adjacent to the phosphoseryl group in the active site peptide (see Section II,K,2) is relatively unreactive toward photooxidation (*170*), it is not this histidine whose oxidation effects enzymic activity.

5. *Miscellaneous Effects*

Rabbit muscle phosphoglucomutase can be reversibly denatured by urea (*167*); if the urea concentration is 5 *M* or less the process is fully reversible after 40 min. Moreover with some samples an apparent activation actually may be observed (*27, 31*), apparently because of dissociation of inhibiting metal ions. Although urea concentrations of 6 *M* produce irreversible inactivation (*27*), at concentrations of 0.1 mg/ml the enzyme apparently retains residual activity in 8 *M* urea in the presence of 0.1 *M* mercaptoethanol (*29*). At low concentrations of urea, inhibition of the *E. coli* enzyme is overcome by increasing the glucose diphosphate concentration (*17*).

Because the enzyme requires a metal ion for activity (see Section II, I,1), chelating agents also can act as inhibitors (*42, 85*).

170. C. Milstein, *Anales Asoc. Quim. Arg.* **51,** 272 (1963).

I. Metal Ion Effects

1. *Metal Ion Activation*

Activation of the phosphoglucomutase reaction by Mg^{2+} was first observed in washed, minced, frog muscle (*1*). However, an unequivocal evaluation of the effect of Mg^{2+} and other metal ions on enzymic activity was not possible until purified samples of "metal-free" enzyme (*171*) were available (*85*). Metal-free enzyme can be prepared by treatment with an imidazole–EDTA mixture, dialysis against EDTA for several hours, and subsequent removal of EDTA by further dialysis against a suspension of solid phase chelating agent (*85*). The activity of metal-free enzyme is less than 10^{-5} that of the Mg^{2+} enzyme; however, complete activity regain is observed (*85*) [see also Milstein (*72*)] on addition of a slight excess of metal ion.

A number of different metal ions activate the metal-free enzyme although Mg^{2+} is the traditional activator (*2, 10*) and the most efficient one. The relative efficiencies at pH 7.4 and 30° for the Mg^{2+}, Ni^{2+}, Mn^{2+}, Co^{2+}, Cd^{2+}, Ca^{2+}, and Zn^{2+} complexes of the enzyme are 100, 70, 15, 5, 0.8 < 0.5, and 0.3, respectively (*30, 85*). Since Mg^{2+} is both the most efficient activator *in vitro* as well as the important activator *in vivo* (see Section II,J,6), Mg^{2+} activation will be described in the greatest detail (see following section). However, some aspects of metal ion activation are more easily studied with other metals. For example, metal-binding stoichiometry is easily determined for the more tightly bound Mn^{2+}, Co^{2+}, and Zn^{2+}; the maximal activity attainable with each of these metals is observed at a 1:1 stoichiometry relative to the enzyme. The assumption that binding of a single equivalent of Mg^{2+} maximally activates phosphoglucomutase seems reasonable by analogy with the binding of other metals, especially since the tight binding of Mg^{2+}, Mn^{2+}, and Zn^{2+} is mutually exclusive (*85*). Because the dissociation constant of Mg^{2+} is greater than $10^{-5}\,M$ it would be technically difficult to measure the number of Mg^{2+} equivalents required for maximal activity. Moreover, additional metal ions may be bound at high concentrations, as indicated by studies of Mn^{2+} and Co^{2+} binding (*30, 55*); however, such binding probably involves nonspecific sites since activity is not affected (*52, 85*).

Actually the activity of all phosphoglucomutases probably depends on the presence of a bivalent metal ion. Usually Mg^{2+} is the most efficient

171. An elemental analysis of "metal-free" phosphoglucomutase has not been conducted; there is thus a possibility that such enzyme retains a tightly bound metal ion essential to either its structure or the activity elicited by subsequent addition of a single metal ion. However, this possibility seems remote.

activator reported, although thorough studies of other metal ions have not been made in most cases. Activation of yeast phosphoglucomutase by Zn^{2+} is reported to be nearly as efficient as activation by Mg^{2+}, with Mn^{2+} and Co^{2+} activating to a smaller degree (*56*); however, other workers were unable to show any dependence on added metal ion (*16*). Activation of the *E. coli* enzyme by Mn^{2+} and Ni^{2+} has also been reported (*17*), while enzyme from jack beans functions more efficiently with Mn^{2+} than Mg^{2+} (*26*).

2. *Addition and Release of* Mg^{2+}

The results of steady state kinetics studies at (a) saturating glucose-di-P with varying Mg^{2+} and glucose-1-P, and (b) saturating glucose-1-P with varying Mg^{2+} and glucose-di-P rule out mechanisms in which Mg^{2+} complexes of either glucose mono- or diphosphates are substrates for the enzymic reaction. In addition, isotope exchange experiments at equilibrium show that Mg^{2+} cannot suppress dissociation of glucose phosphates from the central complex, nor can glucose phosphates suppress the dissociation of Mg^{2+} from the same complex. These results rule out the ordered addition of Mg^{2+} and glucose phosphates in the enzymic reaction (*106*); they also provide values for several kinetic constants associated with Mg^{2+}. At 30° and pH 7.4 the Michaelis constant for Mg^{2+} (saturating glucose-1-P and glucose diphosphate), $K_{m(\mathrm{Mg})}$, is about $2.4 \times 10^{-5}\,M$ [a value of $5 \times 10^{-5}\,M$ has also been reported (*72*)]; also the Mg^{2+} concentration producing half-maximal isotope exchange at equilibrium, $K_{x(\mathrm{Mg})}$, is about $4 \times 10^{-5}\,M$. The limiting Michaelis constant for Mg^{2+} at saturating glucose diphosphate and glucose-1-P concentration approaching zero, $K^{1}_{i(\mathrm{Mg})}$, is $2.5 \times 10^{-5}\,M$, while the limiting Michaelis constant at saturating glucose-1-P as glucose diphosphate approaches zero, $K^{1,6}_{i(\mathrm{Mg})}$, is $1.7 \times 10^{-4}\,M$. According to the mechanism discussed below, these constants may be represented by the following relationships; although the first of these refers to a steady state equilibrium (subscript "s") the remaining three describe true equilibria (subscript "e"):

$$K_{m(\mathrm{Mg})}/(\mathrm{Mg}^{2+}) = ([\mathrm{E_P{\cdot}Glc\text{-}1\text{-}P}] + [\mathrm{E_D{\cdot}Glc\text{-}di\text{-}P}] + [\mathrm{E_P{\cdot}Glc\text{-}6\text{-}P}])_s/(\mathrm{CC})_{ss} \quad (11)$$

$$K_{x(\mathrm{Mg})}/(\mathrm{Mg}^{2+}) = ([\mathrm{E_P{\cdot}Glc\text{-}1\text{-}P}] + [\mathrm{E_D{\cdot}Glc\text{-}di\text{-}P}] + [\mathrm{E_P{\cdot}Glc\text{-}6\text{-}P}])_e/(\mathrm{CC})_e \quad (12)$$

$$K^{1}_{i(\mathrm{Mg})}/(\mathrm{Mg}^{2+}) = (\mathrm{E_P})_e/(\mathrm{E_P{\cdot}Mg})_e \quad (13)$$

$$K^{1,6}_{i(\mathrm{Mg})}/(\mathrm{Mg}^{2+}) = (\mathrm{E_D})_e/(\mathrm{E_D{\cdot}Mg})_e \quad (14)$$

A comparison of $K^{1}_{i(\mathrm{Mg})}$ with $K^{1,6}_{i(\mathrm{Mg})}$ indicates that under the above conditions, Mg^{2+} binds to E_P about 7-fold more tenaciously than to E_D (*172*).

172. A study of binding of Mg^{2+} to E_P via equilibrium dialysis (*72*) yielded a dissociation constant much closer to that of $E_D \cdot Mg$ than $E_P \cdot Mg$; possibly the enzyme was converted to the dephospho form during the extended dialysis period.

Similarly, a comparison of $K_{x(\mathrm{Mg})}$ with $K^1_{i(\mathrm{Mg})}$ indicates that the additional phosphate group provided by bound substrate alters Mg^{2+} binding to only a minor degree.

Later studies showed that the Mg^{2+} enzyme is active for a finite time in the presence of a large excess of either EDTA (*85*) or Zn^{2+} (*139, 173*). Because of the divergent nature of these inhibitors and because the rate of inactivation by both is not only identical but independent of inhibitor concentration, the slow step in both inactivations is undoubtedly the dissociation of Mg^{2+} from the enzyme·Mg·substrate complex. At saturating substrate concentrations the time-course of the enzymic reaction under conditions such that Mg^{2+} dissociation is essentially irreversible is given by Eq. (1). Hence, k_d can be evaluated as 0.03 sec^{-1} (*85, 173*) from the appropriate semi-log plot. Additional studies indicate that the rate constant for Mg^{2+} dissociation from E_P·Mg is only slightly larger than the above k_d (*173*). Since $V_{\max}/[E_T] \approx 1000\ sec^{-1}$, during steady state conversion of glucose-1-P to glucose-6-P a single E_P·Mg molecule will, on the average, participate in many thousands of catalytic events before dissociating—even if the substrate concentration is well below its Michaelis constant (*173*).

Moreover, the ratio, $V_{\max}/[E_T]K_{m(\mathrm{Glc\text{-}1\text{-}P})}$, about $10^8\ M^{-1}\ sec^{-1}$, is a minimal estimate of the rate constant for binding of glucose-1-P to E_P·Mg (*78*) while $k_{d(\mathrm{E_P \cdot Mg})}/K^{1,6}_{i(\mathrm{Mg})}$, about $10^3\ M^{-1}\ sec^{-1}$, is the rate constant for binding of Mg^{2+} to E_P·glucose-1-P (*173*). Thus, not only is the addition of Mg^{2+} and glucose-P kinetically ordered to an extreme degree in favor of glucose-P binding but the release of Mg^{2+} and glucose-P from ternary E_P·Mg·glucose-P complex is also kinetically ordered in favor of prior release of glucose-P (see above). However, in 10 min assays such as those used for initial velocity and isotope exchange studies (*106*), adequate time is available for complete equilibration of Mg^{2+} with the various enzyme forms so that the noncompulsory nature of the metal- and substrate-binding process is observed.

The velocity equation for the reaction involving noncompulsory addition of both Mg^{2+} and glucose phosphates to phosphoglucomutase, as derived by the King-Altman procedure (*174*), is exceedingly complex and contains numerous terms involving the square of the concentrations of glucose-1-P and Mg^{2+}. However, by taking advantage of the above inequalities in rate constants the equation can be simplified to eliminate all squared terms. The final equation (*106*) is identical with that derived

$$\frac{V_{\max}}{\nu_0} = 1 + \frac{K_{m(\mathrm{Mg})}}{(\mathrm{Mg})} + \frac{K_{m(\mathrm{Glc\text{-}1\text{-}P})}}{(\mathrm{Glc\text{-}1\text{-}P})}\left(1 + \frac{K^1_{i(\mathrm{Mg})}}{(\mathrm{Mg})}\right) + \frac{K_{m(\mathrm{Glc\text{-}di\text{-}P})}}{(\mathrm{Glc\text{-}di\text{-}P})}\left(1 + \frac{K^{1,6}_{i(\mathrm{Mg})}}{(\mathrm{Mg})}\right) \quad (15)$$

173. W. J. Ray, Jr. and G. A. Roscelli, *JBC* **241**, 3499 (1966).
174. E. L. King and C. Altman, *J. Phys. Chem.* **60**, 1375 (1956).

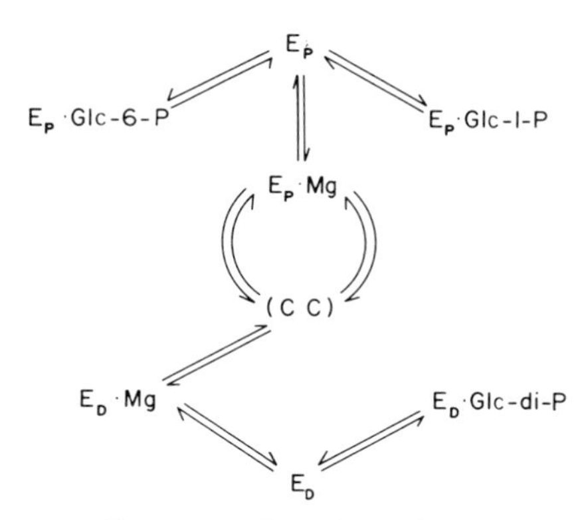

SCHEME 1. Representation of steady state reaction.

for the reaction sequence in Scheme 1, which is a good representation of the steady state reaction. In this scheme all slow steps (those involving addition and release of Mg^{2+}) are omitted except those that are required to maintain a continuous pathway interconnecting all enzyme species. Thus, it appears that the properties of the Mg^{2+} form of phosphoglucomutase are borderline between those of "metal-activated" enzymes and "metallo-enzymes" (*175*).

3. *The Addition and Release of Other Metal Ions*

Because of relatively rapid dissociation of Mg^{2+}, linear product-time plots cannot be obtained with the Mg^{2+}-activated enzyme in the presence of excess EDTA without using inconveniently short assay intervals (*85*); however, the Mn^{2+}, Co^{2+}, Ni^{2+}, and Zn^{2+} forms of the enzyme can be assayed in this manner (*30, 85*). The actual time allowable for the assay varies with substrate concentration, pH, and temperature but can be as long as 40 min with the Zn^{2+} enzyme (*85*) and as short as 20 sec for the Ni^{2+} enzyme (*30*). In all of these cases the metal dissociates appreciably faster from the $E_P \cdot M$ complex than from the $E_P \cdot M \cdot$glucose-P complex. In fact, in the presence of excess substrate, the dissociation rate for a given metal ion appears to vary with the efficiency of the enzymic reaction elicited by that ion: The greater the efficiency, the faster the rate of dissociation (*85*). The reason for this relationship is not known, although a steric effect might be involved; however, the presence of bound substrate does not appreciably alter the binding constant of at least one metal, Zn^{2+}, whose dissociation rate is markedly retarded by bound substrate (*139*). Apparently some type of structural alteration is required for metal association and dissociation, and bound

175. B. G. Malmström and A. Rosenberg, *Advan. Enzymol.* **21**, 131 (1959).

substrate makes this alteration more difficult for metals other than Mg^{2+}.

In addition to assaying in the presence of excess EDTA, the Co^{2+} and Mn^{2+} enzymes may also be assayed in the presence of NTA metal buffers, e.g., Co·NTA. In these cases, linear product-time plots can be obtained over extended assay intervals. Presumably, this is also true of the Zn^{2+} enzyme; however, metal buffers have not been successfully employed in the assay of the Ni^{2+} enzyme (*30*).

4. *Structural Changes Produced by Metal Ion Binding*

Ultraviolet difference spectra are produced by the binding of bivalent metal ions to "metal-free" phosphoglucomutase (*52*); the changes in extinction at 286 and 293 nm involve a 1:1 stoichiometry, and the relative intensities of these peaks indicate that the environment of both tyrosine and tryptophan residues is altered. Since maximal activity is also elicited by binding of a single equivalent of metal, the changes in ultraviolet absorbance are produced by metal binding at the "active site." The difference spectra generated by Mg^{2+}, Mn^{2+}, Co^{2+}, Ni^{2+}, Zn^{2+}, or Cd^{2+} are all identical; hence, binding of any of these must produce a similar structural effect. However, solvent perturbation studies indicate that any such structural alteration must be relatively minor in magnitude since, of several different perturbants such as D_2O, dimethyl sulfoxide, ethylene glycol, and 1,2-propanediol, only D_2O perturbs more aromatic residues in the free enzyme than in the binary enzyme–metal complex. Thus the metal-induced difference spectrum probably arises from minor changes in solvation of aromatic residues which are accessible to water but not to other small solutes (*52*).

By contrast, metal-specific structural differences can be detected by both ultraviolet difference spectroscopy and solvent perturbation techniques in the various ternary enzyme–metal–substrate complexes involving the above metals. Also, in contrast with metal binding in the absence of substrate, metal binding in the presence of glucose-P produces essentially a pure tyrosine difference spectrum. Moreover, the absorbance of tryrosine residues is shifted either to the blue or to the red, depending on the identity of the bound metal ion: Binding of Co^{2+} or Zn^{2+} produces red shifts, while binding of Mg^{2+}, Ni^{2+}, Mn^{2+}, Ca^{2+}, or Cd^{2+} produces blue shifts. In both series the extent of the shift increases in the order given. In no case is the spectral shift accompanied by either significant exposure or shielding of tyrosine residue from small perturbants such as dimethyl sulfoxide or ethylene glycol; however, with larger perturbants such as glycerol or even better, sucrose, the red or blue spectral shift is related, respectively, to shielding or exposure of tyrosines. The binding of various metal ions to E_P·glucose-P thus produces effects which suggest a change in the size of a structural crevice so that the solvation of several partially exposed tyrosine residues is either increased or decreased, depending on the identity of the metal ion. The extent of structural change is related to the activity of the ternary complex thus produced—the smaller the structural change, the higher the catalytic activity (*54*).

5. *Structure of the Metal-Binding Site*

The electronic absorption spectrum of the E_P·Ni complex has bands at 410, ~730, and 1300 nm, consistent with Ni^{2+} surrounded by an octahedral-like arrangement of coordinating ligands. Although substrate binding produces slight increases in both wavelength and intensity of the 410 nm band, the coordination geometry appears to remain the same. Since the Ni^{2+} enzyme is about 70% as active as the Mg^{2+} enzyme, and since temperature effects from 0° to 40° are parallel, it is likely that bound Mg^{2+} also is in an octahedral-like coordination sphere (*30*); Mn^{2+} may be bound similarly (*55*).

The position of the electronic absorption bands of the Ni^{2+} enzyme suggest that the metal-binding ligands are primarily carboxylate groups rather than nitrogenous groups. The relative binding affinities for Mg^{2+}, Mn^{2+}, Co^{2+} and Zn^{2+}, about 10^5, 10^7, 10^9, and 10^{11} *M*, are in accord with this suggestion, since a much larger increase in binding usually occurs from Mg^{2+} to Co^{2+} with nitrogenous ligands (such as ethylenediamine), followed by either a decrease, or at least no further increase, from Co^{2+} to Zn^{2+} (*176*). Of course, several different factors may operate to invalidate generalizations such as these.

The question of whether the phosphate groups of E_P·M·glucose-P act as ligands for the bound metal ion has not been established, although the present indications are negative. For example, both Mg^{2+} and Zn^{2+} bind nearly equally to E_D, E_P and E_P·glucose-P (*85*, *173*), and Mg^{2+} dissociates from both E_P·Mg and E_P·Mg·glucose-P at essentially the same rate (*173*). Moreover, inorganic phosphate binds to the E_P·Mn complex competitively with glucose-P, but it does so without displacing water ligands remaining in the hydration sphere of bond Mn^{2+}. If bound inorganic phosphate occupies the same site as the phosphate group of either glucose 1- or 6-phosphates, the latter phosphate groups do not themselves displace water from bound Mn^{2+} and thus are not ligands (*55*). However, identical binding may be a poor assumption since binding of glucose-P produces a marked structural alteration of the E_P·Mg

176. A. E. Dennard and R. J. P. Williams, *Transition Metal Chem.* **2**, 115 (1966).

complex while binding of inorganic phosphate produces a comparatively small effect (*32*).

Apparent displacement of water from bound Mn^{2+} on sugar phosphate binding has been observed; displacement is maximal with good substrates and inhibitors (glucose-P and glucuronic acid-1-P, respectively) but does not occur with poor substrates (fructose-P and mannose-P). Since 2-deoxyglucose-P and ribose-P apparently displace less water from bound Mn^{2+} than does glucose-P, the 2-hydroxyl and possibly the 3-hydroxyl groups of glucose-P may act as ligands for bound metal ions. In such a case the metal would be too distant from the phosphate groups to participate directly in bond-breaking and bond-making process (*55*).

6. *Metal Complexes in Vivo*

Analyses of purified samples of phosphoglucomutase (*27, 137*) cannot be utilized to identify the "physiological" metal ion or ions, since the most probable candidate, Mg^{2+}, dissociates readily and other metal ions, e.g., Zn^{2+}, are bound much more tenaciously than Mg^{2+} and may replace it during purification. To identify the physiologically important metal, rapid analyses were conducted on muscle extracts in which as nearly physiological conditions as possible were maintained. By taking advantage of the observations that (a) different metal ions dissociate from the enzyme–metal–substrate complex at different rates in the presence of excess EDTA and (b) 30-sec initial velocity studies can be conducted in the presence of excess Mg^{2+} without significant replacement of any of the common bivalent metals in the enzyme–metal–substrate complex by Mg^{2+}, Peck and Ray (*177*) were able to show that in the muscle of normal rabbits only Zn^{2+} and Mg^{2+} are significantly bound to phosphoglucomutase *in vivo*. No evidence was found for metal-free enzyme. The results further indicate that the fraction of phosphoglucomutase in the Mg^{2+} form *in vivo* is about 35% and that the remainder is present as the essentially inactive Zn^{2+} complex.

A sizable increase in the fraction of phosphoglucomutase in the Mg^{2+} form *in vivo, viz.*, from 35 ± 10 to $83 \pm 8\%$, is produced in starved rabbits by insulin shock. Because of the relative activities of the Zn^{2+} and Mg^{2+} enzymes, this increase produces an increased average catalytic efficiency of the endogenous phosphoglucomutase, which is undoubtedly a reflection of an increase in the ratio of free Mg^{2+} to free Zn^{2+} in the muscle tissue (*177*). Effects of insulin and adrenalin on phosphoglucomutase activity in extracts of rat liver and muscle also have been reported by other workers (*9, 20*).

177. E. J. Peck, Jr. and W. J. Ray, Jr., *JBC* **246**, 1160 (1971).

J. pH AND TEMPERATURE EFFECTS

A number of authors have published bell-shaped pH-velocity profiles for phosphoglucomutase; however, if V_{max}—as opposed to v_0—is plotted, the profile is flat in the neutral pH range and V_{max} is essentially invariant from pH 6.5 to 8.0 (*85, 106, 159*). There is some indication that decreased activity along the high pH limb of plots involving v_0 is caused by increases in $K_{m(Glc\text{-}di\text{-}P)}$ (*106*); from the effect of pH on $K_{m(Mg)}$ it is certain that the low pH limb is strongly influenced by a greatly decreased affinity for Mg^{2+} (*106*).

The V_{max}-pH profile for the Ni^{2+}-activated enzyme is quite similar to that for the Mg^{2+} enzyme, although it is slightly more curved in the high pH range (*30*). The V_{max}-pH profiles are anomalous for the Mn^{2+}, Co^{2+}, and Zn^{2+} forms of the enzyme and appear to be sigmoidal in the neutral pH range at 30°. It has been suggested that this sigmoidicity is caused by the presence of two active forms of the enzyme when Mn^{2+}, Co^{2+}, or Zn^{2+} are activators (*85*).

The pH optimum for bacterial phosphoglucomutases (Mg^{2+} form) appears to occur at higher pH values than for other phosphoglucomutases (*19*); however, since there is no evidence that V_{max} was actually measured in this study, such a conclusion is questionable. Actually, except for the rabbit muscle enzyme, no pH studies on phosphoglucomutases have been conducted in such a way as to guarantee that V_{max} as opposed to v_0 was measured.

Early data on the variation in velocity with temperature for the phosphoglucomutase reaction (pH 7.5) suggested an activation energy of 19 kcal/mole (*144*); however, later data indicate that the Arrhenius plot in question is, in reality, curved (*30*). The actual value of the apparent activation energy seems to be of little importance because of the complexity of the overall reaction. However, the Arrhenius plot for the Ni^{2+}-activated enzyme is essentially congruent with that for the Mg^{2+} enzyme, and this congruency suggests a similarity between the binding of Ni^{2+} and Mg^{2+} to the enzyme (*30*).

K. STRUCTURAL STUDIES

1. *The Active Site Phosphate Group*

Jaganathan and Luck (*45*) first labeled phosphoglucomutase by an exchange reaction with ^{32}P-glucose phosphate. Anderson and Jolles (*178*)

178. L. Anderson and G. R. Jolles, *ABB* **70,** 121 (1957).

later showed that the labeled enzyme could be hydrolyzed with acid to give serine phosphate and Kennedy and Koshland (*167*) demonstrated that the phospho-enzyme was stable toward a variety of agents, as would be expected if the phosphate group were esterified with the hydroxyl group of serine. Kennedy and Koshland also observed the then rather unexpected phosphate lability of the phospho-enzyme in the presence of 0.5 *M* base; subsequently, this lability was identified as a β elimination that is characteristic of phosphoserine peptides (*179*).

No evidence to date indicates that the active site phosphate group becomes covalently bonded to any group in the enzyme other than the serine hydroxyl group at any stage of the phosphoglucomutase reaction; moreover, the early concern about the unreactivity of phosphoserine toward nucleophilic displacement, in contrast to the rapid rate of phosphate transfer in the enzymic reaction, has been largely dispelled by the proven reactivity of the active site serine in chymotrypsin in an analogous transfer reaction. However, no one has as yet demonstrated that the *initial* transfer of the active site phosphate to the 6 position of glucose-1-P, for example, is as rapid as the steady state reaction.

2. *The Amino Acid Sequence of the Active Site Phosphopeptide*

Koshland and Erwin (*180*) suggested that the amino acid sequence containing the active site phopshate group was Asp–SerP–Gly–Glu–Ala–Val–Thr–Leu. A sequence involving some of these amino acids was subsequently reported by Harshman and Najjar (*181*) who found Asp–Gly–Glu–SerP–Ala–Gly. However, by using a beautifully elegant technique which involved quantitation of changes in chromatographic behavior and electrophoretic mobility of the phosphopeptide on chemical modification, Milstein and Sanger (*42*) were able to avoid the main problem with previous sequence work, *viz.*, purification of the active site phosphopeptide, and they proposed the presently accepted sequence Thr–Ala–SerP–His–Asp. Subsequently, Harshman and Najjar (*168*) found evidence for both the Milstein-Sanger sequence and their earlier sequence and later proposed that the active site peptide contained 2 serine residues, either one or the other but not both of which might be phosphorylated in the intact enzyme: Glu(Asp, Leu)–Gly–Val–Thr–Ala–Ser(P)–His–Asp–Gly–Glu–Ser(P)–Ala–Gly–Leu–Asp–Leu. Sloane and Mercer (*182*) also provided evidence for two phosphoserine sequences

179. L. Anderson and J. J. Kelley, *JACS* **81**, 2275 (1959).
180. D. E. Koshland, Jr. and M. J. Erwin, *JACS* **79**, 2657 (1957).
181. S. Harshman and V. A. Najjar, *Federation Proc.* **21**, 233 (1962).
182. N. H. Sloane and D. W. Mercer, *BBA* **89**, 563 (1964).

but claimed that the isolated phosphopeptides from these two sequences had unequal specific activities and thus could not be part of a contiguous sequence. Additional doubt about a dual sequence was provided by the isolation of a 13-residue phosphopeptide from the flounder enzyme (*183*) in which the 5 amino terminal residues were identical with those in the Milstein–Sanger rabbit sequence, but contained only a single serine residue: Thr–Ala–SerP–His–Asp–Pro–Gly–Gly–Pro–Asp–Asp–Gly–Phe. Later, Hooper *et al.* (*184*) claimed evidence for a generalized phosphoglucomutase active site sequence similar to that above: Ala–Ile–[Gly–Glu–Thr–Ala–SerP–His–Asp–Pro–Gly–Pro–Gly–Asp–Gly–Asp–Phe]–(Ile,Leu)–Lys. The sequence is given for the rabbit muscle enzyme and the sequence from the flounder enzyme contains substitutions only outside the brackets; the active site peptide from human muscle also resembles this sequence (*27*). Milstein and Milstein (*43*) later published an expansion of the Milstein–Sanger sequence which, to the right of the threonine residue, is the same as the sequence of Hooper *et al.*, except for a single inversion at residues 15 and 16 (in the sequence below), but differs substantially to the left of the threonine residue: Ala–Ile–Gly–Gly–Ile–Ile–Leu–Thr–Ala–SerP–His–Asx–Pro–Gly–Gly–Pro–(Asx_2, Gly)–Phe–Gly–Ile–Lys. Subsequently, Harshman *et al.* (*185*) confirmed the Milstein and Milstein sequence to the right of threonine, resolved the ambiguity in that sequence, and specified the side chain amide groups. Combining the latter two sequences gives Ala–Ile–Gly–Gly–Ile–Ile–Leu–Thr–Ala–SerP–His–Asp–Pro–Gly–Gly–Pro–Asn–Gly–Asn–Phe–Gly–Ile–Lys. There is thus no structural evidence for the attachment of a phosphate group at two different sites in phosphoglucomutase.

An active site pentapeptide has been isolated from the *E. coli* enzyme which is identical with the 5 residues in the above sequence beginning with Thr (*17*).

3. *Conformational Studies*

The high reactivity of the phosphate group of phosphoglucomutase toward the 6-hydroxyl group of glucose-1-P, coupled with its very low reactivity toward the hydroxyl group of the much smaller water molecule, was considered by Koshland in formulating the "induced-fit" theory of enzymic action (*186*). Subsequent studies (*32, 187*) have produced

183. T. Hashimoto, C. del Rio, and P. Handler, *Federation Proc.* **25**, 408 (1966).
184. J. Hooper, J. G. Joshi, T. Sakurada, T. Kuwaki, and J. R. Swanson, *Federation Proc.* **27**, 639 (1968).
185. S. Harshman, H. Six, and V. A. Najjar, *Biochemistry* **8**, 3417 (1969).
186. D. E. Koshland, Jr., *Proc. Natl. Acad. Sci. U. S.* **44**, 98 (1958).
187. D. E. Koshland, Jr., *Cold Spring Harbor Symp. Quant. Biol.* **28**, 473 (1963).

several lines of evidence that substrate binding does indeed induce a conformational change in the structure of the protein. Thus binding of substrate increases the reactivity of cysteinyl residues toward iodoacetamide (*32*) and increases the heat sensitivity of the enzyme (*187*); however, binding of α-glycerophosphate is without effect on these properties. Moreover, substrate binding also produces a sizable ultraviolet difference spectrum involving both tryptophan and tyrosine residues and a substantially enhanced fluorescence of tryptophan residues. By contrast, anions such as sulfate or glycerophosphate, both of which are competitive inhibitors, produce very much smaller optical changes (*32*). Although it has been argued (*50*) that glucose-P produces these effects by converting the phospho-enzyme to the dephospho form, this contention was based on an incorrect assessment of the equilibrium constant for the dephosphorylation reaction [Eq. (4)]; later data (*34, 35*) do not support this contention (see Section II,E,2).

Minor conformational changes which accompany metal ion binding are discussed in Section II,I,4.

III. Other Sugar Phosphate Mutases

A number of preparations with phosphomutase activity toward sugar phosphates other than α-D-glucose phosphate have been reported; however, the demonstration of a rather broad substrate specificity for muscle phosphoglucomutase (see Section II,F) suggests that some of these preparations will require further characterization before their individuality can be considered established. At present phosphoacetylglucosamine mutase from *Neurospora* (*188*) and porcine submaxillary gland (*189*), phosphomannosamine mutase (*190*), and phosphopentomutase, as well as a phosphoglucomutase specific for β-D-glucose-1-P (*191*) have been sufficiently separated from the normal phosphoglucomutase to allow a separate classification for each. Of these, the latter two have been the most extensively characterized and will be described briefly.

Phosphoglucomutase specific for β-D-glucose-1-P has been partially purified from *N. perflava*. Several properties such as metal ion requirement and anionic inhibition appear to resemble those of phosphoglucomutase with α-specificity; however, the possible requirement for a co-

188. J. L. Reissig and L. F. Leloir, "Methods in Enzymology," Vol. 8, p. 175, 1966.
189. D. M. Carlson, "Methods in Enzymology," Vol. 8, p. 179, 1966.
190. L. Glaser, "Methods in Enzymology," Vol. 8, p. 183, 1966.
191. R. Ben-Zvi and M. Schramm, *JBC* **236,** 2186 (1961).

factor has not been studied. Presumably the enzyme acts to provide a nonhydrolytic pathway for metabolism of maltose (*191*).

A phosphodeoxyribomutase activity was first demonstrated by Manson and Lampen (*192*) in *E. coli* and later shown to be quite sensitive to phosphate and sulfate inhibition (*6*). The enzyme is involved in the first step in the catabolism of nucleosides to formate, acetate, and ethanol (*6*); it has been partially purified from *Salmonella typhimurium* (*193*), *E. coli* (*194*), and rabbit tissue (*5*). The *S. typhimurium* enzyme has a molecular weight of 32,000 ± 3,000 [by gel filtration (*194*)] and is classed as phosphopentomutase because both ribose and deoxyribose 1-phosphate serve as substrates (*5, 194*); the K_m value for deoxyribose-1-P and ribose-1-P are in the range of 10^{-5} M. The enzymes from *S. typhimurium* and *E. coli* appear to require Mn^{2+} as the metal ion activator; by contrast with phosphoglucomutase, Mg^{2+} appears to be a poor activator of either system (*194*). Binding of metal ion to the enzyme is required for catalytic activity as demonstrated by the effect of a preassay treatment with EDTA (*195*). Ribose-1,5-di-P, deoxyribose-1,5-di-P, and glucose-1,6-di-P are all cofactors in the enzymic reaction with K_m values in the 0.1 μM range (*194*). None of the results obtained thus far with phosphopentomutase are inconsistent with a reaction sequence similar to that found for rabbit muscle phosphoglucomutase (see Section II,D,1); however, no experimental proof for such a sequence has been reported.

IV. Phosphoglycerate Mutase

A. Preparation and Purity

Phosphoglycerate mutase has been detected in over 70 different plant and animal tissues (*196, 197*) and has been purified from several of these, primarily by salt and solvent fractionation and by batchwise absorption steps: chicken breast (*198*), rabbit muscle (*199, 200*), pork

192. L. A. Manson and J. O. Lampen, *JBC* **191,** 95 (1951).
193. P. A. Hoffee and B. C. Robertson, *J. Bacteriol* **97,** 1386 (1969).
194. K. Hammen-Jepersen and A. Munch-Petersen, *European J. Biochem.* **17,** 397 (1970).
195. A. Munch-Peterson, personal communication (1970).
196. S. Grisolia, "Homologous Enzymes in Biochemical Evolution, Colloquium" (N. Van Thoai, ed.), p. 167. Gordon & Breach, New York, 1968.
197. S. Grisolia and B. K. Joyce, *JBC* **234,** 1335 (1959).
198. A. Torralba and S. Grisolia, *JBC* **241,** 1713 (1966).
199. V. W. Rodwell, J. C. Towne, and S. Grisolia, *JBC* **228,** 875 (1957); *BBA* **20,** 394 (1956).
200. S. Grisolia, "Methods in Enzymology," Vol. 5, p. 236, 1962.

kidney (*201*), blood cells (*202*), yeast (*203, 204*), wheat germ (*205*), and rice germ (*206*). Unfortunately because of problems related to the enzymic assay (see Section IV,D), specific activities reported by different workers usually cannot be used as an index of relative purity for enzyme preparations from a given source.

The most nearly homogeneous PGA-mutase preparation is probably the yeast enzyme, "component I," which is purified electrophoretically (*203*); however, a simpler isolation procedure has been published which gives higher yields of this enzyme, apparently because proteolytic degradation that occurs during the usual autolysis, and that gives rise to four modified (*207*) and less active mutases (*209*), is curtailed by use of a more basic medium (*204*). Unfortunately, the purity of the enzyme obtained by the latter procedure has not been extensively investigated; presumably a procedure which combined basic autolysis and electrophoretic purification would optimize both yields and purity. Preparations of the pork kidney enzyme, which is the only mutase that has been purified by column chromatography, and of the chicken breast enzyme, which is one of the simplest preparations, appear to contain isozymes (*201, 210*) and assessment of purity is thus difficult. Either no evidence for homogeneity or definite evidence for inhomogeneity has been reported for other preparations noted above. The rabbit muscle (*211*), chicken breast (*198*), and yeast enzymes (*203, 204*) are reported to be crystalline.

B. Physical and Chemical Properties

The molecular weights reported for PGA-mutases from mammalian tissue are in the range of 54,000–65,000; the higher values were determined by sedimentation (*199*) or gel filtration (*198*) and are probably less accurate than the values of 57,000 [equilibrium centrifugation (*211*)]

201. D. Diederich, A. Khan, I. Santos, and S. Grisolia, *BBA* **212,** 441 (1971).
202. D. R. Harkness and J. Ponce, *ABB* **134,** 113 (1969).
203. H. Chiba, E. Sugimoto, and M. Kito, *Bull. Agr. Chem. Soc. Japan* **24,** 428 (1960).
204. E. De le Morena, I. Santos, and S. Grisolia, *BBA* **151,** 526 (1968).
205. S. Grisolia, B. K. Joyce, and M. Fernandez, *BBA* **50,** 81 (1961).
206. M. Fernandez and S. Grisolia, *JBC* **235,** 2188 (1960).
207. The enzyme responsible for proetolysis of the mutase during autolysis has been partially purified and characterized and its mode of action clarified (*208*).
208. R. Sasaki, E. Sugimoto, and H. Chiba, *ABB* **115,** 53 (1966).
209. H. Chiba, E. Sugimoto, and M. Kito, *Bull. Agr. Chem. Soc. Japan* **24,** 555 and 558 (1960).
210. E. James, T. G. Flynn, and R. O. Hurst, *Federation Proc.* **29,** 891 (1970).
211. L. I. Pizer, *JBC* **235,** 895 (1960).

and 54,000 [acrylamide gel electrophoresis in the presence of dodecyl sulfate (*212*)] for the rabbit muscle enzyme. By contrast, the yeast enzyme has a molecular weight of about 111,000 [by sedimentation plus diffusion measurements and by equilibrium centrifugation (*199, 213, 214*)], while the wheat and rice germ enzymes have molecular weights in the 30,000–35,000 range [by sedimentation (*200, 206, 215*)].

A structure composed of dimeric and tetrameric subunits has been reported for the rabbit muscle and yeast enzymes, respectively; the electrophoretic mobility of the former enzyme in dodecyl sulfate (*212*) and the sedimentation of the latter enzyme at high pH (*214*) correspond to that of monomers with a molecular weight of about 27,000. In the case of the yeast enzyme the monomers may not be identical since only one sulfhydryl and five methionyl groups (*216*) have been reported per tetramer (see below), and in 8 M urea evidence for dimers only was found (*213*). However, both subunits of the rabbit muscle enzyme are probably identical since this enzyme has two active sites (*212*).

For the yeast and rabbit muscle enzymes, extinction coefficients at 280 nm (1% solution) are 14.2 [component I (*216*)] and about 10.5 (*200*); the diffusion coefficients ($D_{20,w}$) are 5.6×10^{-7} [at zero concentration (*213*)] and 6.6×10^{-7} cm^2 sec^{-1} [provisional estimate (*217*)], respectively, while both isoelectric points are about pH 5.3 (*217*). For the yeast enzyme the optical rotatory parameters a_0 and b_0 are —95° and —38°, respectively, if λ_0 is taken as 212°, while $\alpha_{233} = -2900$ (*213*); extrapolated values for $[\eta]$ and $s_{20,w}$ are 3.93 ml/g and 6.4, respectively (*213*).

Both chicken breast (*198*) and rabbit muscle PGA-mutase (*211*) contain four sulfhydryl groups (by reaction with mercuribenzoate) while the yeast enzyme contains either one such group (*216*) or none (*196*). The complete amino acid analysis of the yeast enzyme (component I) has been published; it contains an unusually large amount of proline (*216*).

Early work suggested the presence of small but significant amounts of phosphate in the yeast and rabbit muscle enzymes (*199*); however, later preparations of the yeast enzyme were essentially free of phosphate

212. Z. B. Rose, *ABB* **140,** 508 (1970).
213. E. Sugimoto, R. Sasaki, and H. Chiba, *ABB* **113,** 444 (1966).
214. R. Sasaki, E. Sugimoto, and H. Chiba, *Agr. Biol. Chem.* **34,** 135 (1970).
215. N. Ito and S. Grisolia, *JBC* **234,** 242 (1959).
216. E. Sugimoto, R. Sasaki, and H. Chiba, *Agr. Biol. Chem.* (*Tokyo*) **27,** 222 (1963).
217. H. Edelhoch, V. W. Rodwell, and S. Grisolia, *JBC* **228,** 891 (1957).

as well as bound sugars (*213*); the wheat germ enzyme has no bound phosphate (*200*).

C. Stability and Storage

Both the rabbit muscle and yeast enzymes are denatured by high concentrations of urea. In the latter case dissociation into subunits (dimers) occurs in 8 *M* urea, pH 7, accompanied by a change in the rotatory dispersion parameter a_0 from $-95°$ to $-604°$; however, both the chemical and physical properties of the native enzyme reappear on removal of urea (*213*). The yeast enzyme can be dissociated into tetrameters at pH 11.5 and extensive activity regain is obtained on neutralization (*214*). Since the muscle enzyme consists of dimers (*212*), it may also dissociate into subunits during inactivation in 6 *M* urea; about 70% of the original activity can be recovered on removal of the denaturant (*211*).

At neutral pH in the presence of ammonium sulfate or glycerate-di-P, the rabbit muscle, chicken breast, pork kidney, and yeast enzymes are stable at 55°–60° for several minutes (*196, 205*) and such treatment constitutes one step of the preparation for the mammalian enzymes; however, one report indicates that the rabbit muscle enzyme is not stable in very dilute solutions at room temperature (*199*). The yeast enzyme can be stored either under ammonium sulfate (*218*) or in the frozen state (*219*); freezing is recommended for the rabbit muscle enzyme (*200, 219*); however, repeated freeze–thaw cycles produce inactivation (*219*).

The wheat and rice germ enzymes are less stable toward heat at neutral pH than the yeast and mammalian enzymes and are not markedly stabilized by added ammonium sulfate or glycerate-di-P; both are more stable at pH 9 than 5 (*206, 215*).

D. Assay

No entirely satisfactory assay for PGA-mutase has been described; thus, all reported assays give activities that are sensitive to small changes in the assay conditions and none allows direct measurement of V_{max}/E_T. As in the case of the phosphoglucomutase reaction (see Section II,C), it no longer seems tenable to depend on the diphosphate impurity present in glycerate monophosphates from natural sources to saturate the enzyme. Indeed, because of the use of suboptimal conditions for activity

218. H. Chiba and E. Sugimoto, *Bull. Agr. Chem. Soc. Japan* **23**, 207 (1959).
219. R. W. Cowgill and L. I. Pizer, *JBC* **223**, 885 (1956).

measurements, much of the published data on PGA-mutase activity is difficult or impossible to interpret quantitatively, and specific activities obtained by different workers usually cannot be compared. The lack of a satisfactory assay is undoubtedly one of the reasons why studies on PGA-mutase have lagged behind those on phosphoglucomutase.

The large difference in optical rotation at neutral pH between the molybdate complexes of glycerate 2- and 3-phosphates serves as the basis for one assay method: $\alpha_D = +5°$ and $-745°$ for the two isomers, respectively (*220*). Because of the relative sizes of the rotations and because the forward reaction is not linear for greater than 2% conversions to product, this assay is useful only for following the reverse reaction (2-isomer → 3-isomer), where reasonable linearity is maintained for conversions of up to 30% (*199*). As usually performed, the assay is limited to studies involving relatively high concentrations of glycerate-2-P; however, the sensitivity of the method might be improved by conducting the rotation measurements at shorter wavelengths. A typical polarimetric assay procedure is given in Ballou and Fischer (*220*) and Chiba *et al.* (*221*).

A second type of assay involves measurement of the product, glycerate-2-P, by conversion, with enolase, to an equilibrium mixture of glycerate-2-P and phosphoenolpyruvate. In most such assays glycerate-3-P has been employed as the substrate and enolase used as a coupling enzyme; the reaction then can be followed by appearance of color at 240 nm (*199, 200*). Alternatively, the disappearance of color at 240 nm on addition of PGA-mutase to an equilibrium mixture of glycerate-2-P and phosphoenolpyruvate also could be measured. Unfortunately, in such assays, the absorbance change per micromole of glycerate-2-P formed or converted is a function of pH, salt and Mg^{2+} concentration, and temperature (*222*); hence, in assays conducted under different conditions it is difficult to correlate "ΔOD per unit time" with activity. In some, if not all, cases the enolase reaction should be conducted as a separate step in the assay, under a standard set of conditions, e.g., as described by Rodwell *et al.* (*199*). The glycerate-2-P produced in the PGA-mutase reaction also can be converted into lactate, which increases the sensitivity of the assay (*223*).

For a further increase in assay sensitivity, hypoiodite (*224*) or mercuric

220. C. E. Ballou and H. O. L. Fischer, *JACS* **76,** 3188 (1954).
221. H. Chiba, E. Sugimoto, and M. Kito, *Bull. Agr. Chem. Soc. Japan* **24,** 418 (1960).
222. F. Wold and C. E. Ballou, *JBC* **227,** 301 (1957).
223. E. W. Sutherland, T. Posternak, and C. F. Cori, *JBC* **181,** 153 (1949).
224. H. Chiba, E. Sugimoto, R. Sasaki, and M. Hirose, *Agr. Biol. Chem.* (*Tokyo*) **34,** 498 (1970).

chloride (*225*) can be used to convert the phosphoenolpyruvate formed in the enolase reaction to inorganic phosphate, which is subsequently measured. Again, however, the enolase reaction probably should be conducted as a separate step (*225*). In such an assay the 2-isomer would be the substrate of choice because of the limitation on substrate conversion when the 3-isomer is used (see above); however, the use of the 2-isomer requires measurement of the disapparance rather than appearance of color.

That glycerate-2-P can be hydrolyzed under conditions where the 3-isomer is stable (*226*) could serve as the basis for both a routine as well as a sensitive and accurate enzyme assay, since the Bartlett procedure for determination of inorganic phosphate (*153*) is both accurate and convenient at the 0.2 μmole level, and by taking special precautions, can be used with amounts as small as 0.04 μmole (*73*). Alternatively, ^{32}P-labeled glycerate phosphates could be employed and the increase or decrease in molybdate-labile phosphate measured according to the procedure of Sugino and Miyoshi (*82*) [see, for example, Peck *et al.* (*34*)]; such an assay could be used at exceedingly low substrate concentrations.

Recent studies with the muscle and yeast enzymes show that assay concentrations of substrates previously used are not adequate to saturate the enzyme because of substrate inhibition effects. For a routine assay of the muscle enzyme at pH 7.0, 30 mM glycerate-3-P and 1 mM glycerate-di-P should be adequate to allow a direct measurement of $V_{\max}$, as deduced from kinetic data (*227*); a similar set of concentrations should saturate the yeast enzyme under these conditions. However, at substrate concentrations such as these a sizable lag phase occurs in the reaction when it is coupled with enolase (*227*). This lag is probably caused by an increase in the K_m^{app} of enolase and should be eliminated by increasing the concentration of enolase used; alternatively, the enolase reaction could be conducted in a separate step, as was suggested above. Insufficient evidence is available to define analogous concentrations for the reverse reaction; however, the 2-isomer may have a somewhat lower K_m than the 3-isomer (*199*).

E. Reaction Sequence

A preliminary study of the reaction sequence for PGA-mutase from mammalian muscle was conducted only a few years after the enzyme

225. J. C. Lee, Master's Thesis, Purdue University (1964).
226. Z. B. Rose, *JBC* **243**, 4806 (1968).
227. S. Grisolia and W. W. Cleland, *Biochemistry* **7**, 1115 (1968).

was first recognized; in 1938; Meyerhof *et al.* (*228*) showed that ^{32}P-inorganic phosphate did not equilibrate with glycerate-P during the enzymic reaction. A hydrolysis-esterification sequence involving free inorganic phosphate was thus ruled out. Later, Sutherland *et al.* (*223*) demonstrated the activating effect of glycerate-2,3-di-P and observed phosphate exchange between ^{32}P-glycerate-3-P and glycerate-di-P, similar to the exchange demonstrated for the phosphoglucomutase reaction (see Section II,D,1). The scheme used by these workers in presenting their results represents a sequential reaction analogous to sequence *a* for the phosphoglucomutase reaction; however, it does not appear that this scheme was offered as an exclusive reaction sequence, i.e., as a sequential as opposed to Ping Pong reaction.

Subsequent descriptions of the muscle PGA-mutase reaction usually have been either vague or contradictory. This was caused by a variety of problems including (a) the unexpected lability of the phospho-enzyme, (b) the intrinsic phosphatase activity of the enzyme (see Section IV,I), (c) failure to consider the possibility that the enzyme–substrate complex was a mixture of E_P·glycerate-P and E_D·glycerate-2,3-di-P, (d) failure to quench the enzymic reaction before initiating experiments to determine the fate of the phosphate label, and (e) failure to distinguish between the fate of a phosphate label in short-interval experiments as opposed to the fate of the same label at equilibrium. However, with recent developments a semblance of order is becoming apparent—at least in retrospect.

Pizer (*211*) was undoubtedly the first to label the enzyme with ^{32}P-phosphate; because PGA-mutase is not isolated in the phospho form (see Section IV,B) he utilized ^{32}P-phosphate labeled glycerate-di-P. However, Pizer apparently labeled a phosphoglucomutase contaminant at the same time (*229, 230*) and inadvertently hydrolyzed the labile phospho form of PGA-mutase; hence, his subsequent experiments were conducted with a labeled artifact.

Grisolia *et al.* (*205*) obtained labeled enzyme, also by treatment with glycerate-di-P, but removed the excess glycerate-di-P with a short column of anion exchange resin and thereby avoided the lengthy dialysis previously used for this purpose. An apparent ^{32}P-phosphate incorporation of 2.2 moles/mole of enzyme was claimed in this study. However, as noted by the authors, the calculation of stoichiometry involved questionable assumptions; in fact, the published data do not support this stoichiometry,

228. O. Meyerhof, P. Ohlmeyer, W. Gentner, and H. Maier-Leibnitz, *Biochem. Z.* **298**, 396 (1938).
229. N. Zwaig and C. Milstein, *BJ* **98**, 360 (1966).
230. N. Zwaig and C. Milstein, *BBA* **73**, 676 (1963).

even if the stated assumptions are made (*231*). However, the lability of the enzyme–phosphate bond toward ammonium sulfate fractionation was established and the exchangeability of the radioactive label with cold substrate verified.

Zwaig and Milstein (*229*) also labeled the enzyme with glycerate-di-P and verified the acid lability of the phospho-enzyme thus produced. In addition they studied the PGA-mutase·glycerate-di-P complex obtained by treating the enzyme with less than stoichiometric amounts of glycerate-di-P. This complex behaved like E_P·glycerate-P under some conditions if E_P is assumed to be acid labile (see below); a 1:1 combining stoichiometry was also reported. However, no entirely satisfactory explanation is yet available for their observation that more than one mole of P_i could be produced from PGA-mutase·glycerate-P under mild conditions, *viz.*, at pH 3.5; however, the absence of a phosphatase that could hydrolyze glycerate-P at this pH was not verified.

Jacobs and Grisolia (*232*) also studied the PGA-mutase·glycerate-di-P complex. When isolated by means of gel filtration at pH 7.0 and $\mu = 0.01$, the enzyme contained 0.85 equivalent of glycerate-di-P. These authors also reported that nearly all of the label present in the complex could be converted to P_i by treatment with acid or base. However, from the results of Rose (*212*) it seems probable that in cases where more than 50% of the label was converted to P_i, part of the glycerate-P had dissociated from the enzyme and had been separated from it during isolation, *viz.*, the enzyme studied was a mixture of E_P and [E_P·glycerate-P + E_D·glycerate-di-P].

Thus, when Torralba and Grisolia (*198*) later studied the binding of glycerate-di-P by gel filtration in the presence of the mono- and diphosphates, the separated complex was only partially converted to P_i on acid treatment (*198*); moreover, these authors indicated that the apparent binding of ^{32}P-phosphate by the enzyme was more extensive than binding of ^{14}C-glycerate in an isotopic mixture of glycerate-di-P, as would be expected if the above mixture had been isolated. Torralba and Grisolia further stated that PGA-mutase binds 2 moles of glycerate-di-P; however, the assumptions made in the calculation are questionable.

Rose (*212*) brought a great deal of order to a confused field by isolating a phospho-enzyme free of bound glycerate that contained two phosphate groups per 54,000 molecular weight. Her successful elimination of glycerate-P and glycerate-di-P from the enzyme preparation may have

231. Apparently an error was made in designating the specific activity of labeled substrate [S. Grisolia, personal communication (1970)].

232. R. J. Jacobs and S. Grisolia, *JBC* **241,** 5926 (1966).

been made possible by the use of substantially higher ionic strengths ($\mu = 0.13$) during gel filtration than were used by previous workers. Since the ^{32}P-label was attached to histidine side chains of the protein, the lability of the phospho-enzyme under mildly acidic conditions, that caused so much confusion in the earlier work, is now explained. Rose also provided excellent evidence for a combining stoichiometry of 2 moles of diphosphate per mole of PGA-mutase.

During this time a variety of suggestions were made as to the nature of the reaction sequence of muscle PGA-mutase. Pizer suggested a sequential as opposed to Ping Pong reaction (*8*); this designation was accepted by Grisolia *et al.* (*205*) in spite of their apparent success in labeling the muscle enzyme. Although Zwaig and Milstein did not suggest a specific mechanism, their interpretation of their data supports a sequential reaction (*229*, *230*). Jacobs and Grisolia (*232*) called attention to similarities between the reaction of PGA-mutase and that of phosphoglucomutase, which was then known to be Ping Pong. Because of incorrect interpretations of isotope exchange data, subsequent claims were made for a reaction sequence which did not involve glycerate-di-P at all (*233*) or which was sequential (*234*) or which changed its identity with a change in salt concentration (*196*). Later, Grisolia and Cleland published kinetic data and reinterpreted the isotope exchange data in favor of a Ping Pong reaction analogous to sequence *b1* for the phosphoglucomutase reaction (*227*); Mantel and Garfinkel (*235*) claimed that these data could be fit equally well to a very complicated sequential mechanism. However, the labeling studies of Rose (*212*) and the observation by Britton and Clarke (*236*) of co-transport of phosphate and independent transport of glycerate, analogous to the case of muscle phosphoglucomutase (see Section II,D,4), strongly suggest that the Grisolia–Cleland proposal is correct and that the reaction sequence for muscle PGA-mutase is uni–uni with a dissociable intermediate (see Section II,D,1). Thus the role of the diphosphate in the muscle PGA-mutase reaction appears to be analogous to that of the diphosphate in the phosphoglucomutase reaction (although the latter enzyme is isolated predominately as the phospho form), and dissociation of central complex to E_D plus glycerate-di-P occurs much more slowly than the substrate to product conversion—possibly as little as 0.01 as fast as the catalytic reaction (*227*). In such a case the official name for the enzyme, "2,3-diphospho-D-glycerate:2-

233. S. Grisolia and M. Cascales, *BBRC* **2,** 200 (1966).
234. S. Grisolia, A. Torralba, and J. Tecson, *BBA* **151,** 367 (1968).
235. J. Mantel and D. Garfinkel, *JBC* **244,** 3884 (1969).
236. H. G. Britton and J. B. Clarke, *BJ* **112,** 10P (1969).

phospho-D-glycerate phosphotransferase" (*94*), is mechanistically incorrect for the muscle enzyme. Moreover, the effect of salt on the isotope exchange processes (*237*) may easily be rationalized in terms of this model: increasing ionic strength increases the rate of dissociation for E_D·glycerate-di-P; thus, high salt concentrations can drastically increase $K^{app}_{m(GA\text{-}di\text{-}P)}$ while producing only minor effects on V_{max} (*227*).

By contrast, initial velocity studies with the yeast enzyme suggest a sequential reaction similar to sequence *a* or *a1* (*224*), which has been proposed as the reaction sequence for *E. coli* and *B. cereus* phosphoglucomutases (Section II,D,1). However, the similarities between the muscle and yeast enzymes are too striking to ignore: both have similar molecular weights (*212, 214*); both exhibit similar intrinsic phosphatase activity (*196*); the phosphatase activity of both is similarly affected by binding of glycolate-2-P (which represents part of the substrate) and whose mode of action is readily rationalized in terms of a Ping Pong reaction sequence (see Section IV,I); both show similar substrate inhibition patterns (*224, 227, 238*); and both have similar Michaelis constants (see Section II,F). Moreover, both show residual activity in the complete absence of glycerate-di-P (*209, 218, 223, 225*); as indicated in Section II,D,2, the latter is characteristic of phosphomutases whose reaction sequence is Ping Pong. However, in the case of PGA-mutase only a small activity can be generated in this manner, apparently because the enzyme is isolated predominately in the dephospho form. In addition, although earlier labeling experiments failed (*205*), Rose has recently labeled the yeast enzyme and demonstrated the electrophoretic similarity in dodecyl sulfate between labeled monomers of the yeast and muscle enzymes (*239*). Since the reaction of the muscle enzyme is almost certainly Ping Pong, it would be surprising if the reaction of the yeast enzyme were not also Ping Pong, in spite of the reported kinetic patterns.

In contrast with the muscle and yeast enzymes, the wheat germ and rice germ enzymes apparently function with maximal efficiency in the complete absence of glycerate disphosphate. As yet no definitive studies have been reported on the reaction sequence for the enzymes. However, a process involving direct isomerization via a cyclic phosphate intermediate, as in the acid-catalyzed isomerization of glycerate phosphate (*221*), is conceivable and would fit the known facts; a synchronous transfer of two phosphate groups involving two bound glycerate phosphate molecules is also feasible, but without precedent.

237. M. Cascales and S. Grisolia, *Biochemistry* **5**, 3116 (1966).
238. H. Chiba and E. Sugimoto, *Bull. Agr. Chem. Soc. Japan* **23**, 213 (1959).
239. Z. B. Rose, personal communication (1970).

F. Kinetics

A study of the kinetics of diphosphate-dependent PGA-mutases has been hampered by the lack of a facile assay sufficiently sensitive to measure initial velocities at low substrate concentrations, by the presence of substrate inhibition, and by very large salt effects (at least in the case of the muscle enzymes). However, the substrate-velocity pattern for the chicken breast enzyme is that expected for a Ping Pong reaction with competitive inhibition by both substrates (*227*), although the experimental error was rather large. By contrast, the substrate-velocity patterns observed for the yeast enzyme are those expected of a sequential reaction with substrate inhibition by glycerate-3-P (see, however, the previous section) (*224*).

A critical evaluation of the various values reported for the kinetic parameters of PGA-mutases is not possible because in general insufficient attention has been devoted to important details in such studies, e.g., substrate inhibition has not been handled adequately, the coupled assay has been conducted with suboptimal concentrations of enolase, the contribution of the glycerate-2-P $\rightleftharpoons$ phosphoenolpyruvate equilibrium to the reaction has been overlooked, the concentration of diphosphate in standard assays has been suboptimal, and defined activity units have been expressed in such a way that conversion to International Units involves a complex correction (see Section IV,D).

The best set of constants for muscle enzyme (chicken breast) is $V^{f}_{max}/E_T \sim 2000$ IU/mg; $K^{app}_{m(GA\text{-}3\text{-}P)}$ and $K^{app}_{m(GA\text{-}di\text{-}P)}$, 0.6 m$M$ and 1.4 μM, respectively (*227*). The K_m values were calculated to eliminate the inhibitory effect of the alternative substrate, but not buffer anions; moreover, no correction was made for Mg^{2+} binding to the mono- and diphosphates [5 mM Mg^{2+} were used to optimize the reaction of the coupling enzyme, enolase; since PGA-mutase does not require metal ions for activity (Section IV,J), the Mg^{2+} complexes of the substrate are probably inactive]. Since the competitive inhibition constant for glycerate-3-P, $K_{I(GA\text{-}3\text{-}P)}$ is about 3 mM, i.e., only 5-fold larger than $K_{m(GA\text{-}3\text{-}P)}$, it is not possible to saturate the enzyme with glycerate-3-P without substantially increasing $K^{app}_{m(GA\text{-}di\text{-}P)}$ (see Section IV,D) ; by contrast $K_{I(GA\text{-}di\text{-}P)}$, about 2.5 m$M$, is much larger than the corresponding K_m, and substrate inhibition by the diphosphate is much less important than by the monophosphate. The primary difference in the substrate velocity patterns for muscle PGA-mutase and muscle phosphoglucomutase is that the K_I/K_m ratio for the monophosphate is much larger for the latter reaction. It is also interesting to note that the K_I values for the mono- and diphosphate are approximately the same, as is the case in the phosphoglucomutase system; the possible significance in the PGA-mutase system of the similarity in these constants has been noted (*227*).

The K_m, K_I, and V^f_{max}/E_T values (at pH 5.9 and 24°) reported for yeast enzyme (*224*) are within a factor of about two of those given for the muscle enzyme at 30° and pH 7.4 (*227*); probably $K_{I(GA\text{-}di\text{-}P)}$ (value not reported) is also similar to that for the muscle enzyme; at pH 7.4 the K_m values are somewhat smaller than at pH 5.9 (after correction for Mg^{2+} binding effects, see above) while the K_I values are about 10-fold smaller (*240*); hence, it should be more difficult to obtain maximum velocities at the latter pH, although the largest reported activity for the yeast enzyme was obtained at pH 7 (and 30°), about 1800 IU/mg (*204*); the largest value reported for the reverse reaction is 530 IU/mg (*224*) at pH 5.9 and 25°.

Values of $K_{m(GA\text{-}3\text{-}P)}$ within a factor of two of the values reported for the chicken breast enzyme (see above) have also been found for the diphosphate-independent wheat (*215*) and rice germ (*206*) mutases. However, neither of these enzymes shows the marked substrate inhibition effects prominent with the diphosphate-dependent muscle and yeast enzymes.

G. Thermodynamics

A number of different values have been reported for the equilibrium constant [glycerate-2-P]/[glycerate-3-P] (*8*); of those values for which raw data are provided, the most reliable at near neutral pH appear to be 5.8 ± 0.2 [at 25° and 5 m*M* Mg^{2+}, as recalculated (*8*) from earlier data (*218*)], and 6.3 ± 0.3 [at 30° but with no added Mg^{2+} (*198*)]. The temperature dependence of K_e appears to be small (*4*) and consistent with a $\Delta S°$ of near zero, both by analogy with the phosphoglucomutase reaction (*7*) and from calculations (*8*) which involve data from several different studies. In addition, the value of K_e is reported to be independent of pH between about 5 and 7 (*199, 218*) [K_e is about 4.2 in acid (*8*)]. Such a pH invariance is surprising since pK_{a3} values for glycerate-2- and 3-phosphates differ by about 0.7 (cf. *222, 241*) and, as has been pointed out (*8*), this difference should produce a change in K_e of several fold over the pH range 5–7 [e.g., see Wold and Ballou (*222*)]. Thus, a redetermination of K_e as a function of pH seems de-

240. R. Sasaki, E. Sugimoto, and H. Chiba, *BBA* **227,** 584 (1971).
241. W. Kiessling, *Biochem. Z.* **273,** 103 (1934).

sirable; the assay procedure of choice would be the recently reported molybdate-catalyzed hydrolysis of glycerate-2-P (*226*) rather than assays involving the use of enolase or polarimetry of the molybdate complexes of the equilibrium products (see Section IV,D). No systematic study of the effect of Mg^{2+} on K_e has been conducted; however, the correspondence between K_e values in the presence of 5 m*M* Mg^{2+} (added to optimize conditions for the coupling enzyme, enolase) (*218*) and in the absence of Mg^{2+} (*198*) suggests that the effect would be small, although the dissociation constant of the Mg·glycerate-2-P is about 3.6 m*M* (*222*); presumably the corresponding constant for the 3-isomer is similar.

H. SPECIFICITY

Rabbit muscle PGA-mutase has a high (*4*) but not absolute specificity (*242*) for the D-isomer of glyceric acid (and its homologs), and DL-glycerate-2-P has frequently been used as a substrate for the enzyme instead of the optically pure D-isomer [e.g., Chiba and Sugimoto (*218*)]. Other substrates (phosphate acceptors) are D-*erythro*-dihydroxybutyrate monophosphates, DL-1,2-dihydroxyethanesulfonate-2-P (presumably only the D-isomer reacts), and hydroxypyruvate-3-P (*242*). The fact that water is also an acceptor, albeit a very inefficient one (see Section IV,I), suggests that the other compounds containing hydroxyl groups would probably accept phosphate also, if phosphate transfer was assessed with ^{32}P-phosphate-labeled phospho-enzyme. However, efficient transfer probably requires the presence of a negative charge in the 1 position, an *erythro*-glycol structure in the 2 and 3 positions, and a phosphate (or possibly sulfate) group esterified with one of the hydroxyl groups (*242*). Whether inorganic phosphate plus glycolate would produce glycolate-2-P in a reaction similar to that of xylose and inorganic phosphate in the phosphoglucomutase system (Section II,F) is not known.

Theoretically, the phosphorylated form of any compound which accepts phosphate from E_P should in turn be able to donate phosphate to E_D. However, the capacity to act as a cofactor has been demonstrated for only D-glycerate-2,3-di-P, D-*erythro*-dihydroxybuterate-di-P (*242*), and D-glycerate-1,3-di-P (*133*).

No analogous specificity studies have been conducted with PGA-mutase from other sources.

242. L. I. Pizer and C. E. Ballou, *JBC* **234,** 1138 (1959).

I. Phosphate Transfer to Water

A specific glycerate 2,3-diphosphate phosphatase activity is associated with most PGA-mutase preparations (*211*). Whether or not such an activity represents an impurity or an intrinsic activity of PGA-mutase is a complex problem. Thus, several glycerate-di-P phosphatases have been described (*243–246*) and in several cases (*244*) the phosphatase preparation also had PGA-mutase activity. Moreover, the mutase activity may co-purify with the phosphatase activity or *vice versa* (*202*, *244*). However, the relative activities and various properties of phosphatase and mutase preparations suggest that the phosphatase is not simply an altered mutase and vice versa; however, it is conceivable that both might have evolved from a common precursor. Immunological studies of these possibilities thus far have been inconclusive (*247*).

In some partially purified mutase preparations at least part of the phosphatase activity is caused by a contaminating phosphatase, e.g., PGA-mutase from yeast (*199*) and erythrocytes (*202*) and probably from wheat and rice germ (*206*, *215*). However, the diphosphate-specific phosphatase activity in purified samples is best rationalized as an abortive phosphate transfer to water by the mutase itself. Thus, parallel losses of phosphatase and mutase activities on treatment with trinitrobenzenesulfonate have been observed; moreover, glycerate-3-P protects against the inactivation of both types of activities efficiently and to about the same extent (*248*). In addition, the diphosphate phosphatase reaction of the mutase is very inefficient; under normal conditions (but in the absence of the glycerate monophosphates) it is typically only 10^{-4} to 10^{-5} as rapid as the mutase reaction (in the presence of glycerate monophosphates). Moreover, both glycerate 2- and 3-phosphates inhibit the phosphatase action of both the muscle and yeast enzymes (*211*, *244*, *248*) and neither is hydrolyzed by purified mutase preparations from muscle that also exhibits diphosphate phosphatase activity (*211*, *229*).

In terms of a Ping Pong reaction sequence, the observation that the K_m value for glycerate-di-P in the phosphatase reaction is similar to that of K_I and not K_m for glycerate-di-P in the mutase reaction (*248*) tends

243. D. R. Harkness and S. Roth, *BBRC* **34**, 849 (1969).
244. B. K. Joyce and S. Grisolia, *JBC* **233**, 350 (1958); G. T. Zancan, C. R. Krisman, J. Mordoh, and L. F. Leloir, *BBA* **110**, 348 (1965).
245. S. Rapoport and J. Luebering, *JBC* **189**, 683 (1951).
246. Z. B. Rose and J. Liebowitz, *JBC* **245**, 3232 (1970).
247. S. Grisolia and J. C. Detter, *Biochem. Z.* **342**, 239 (1965).
248. R. Sasaki, M. Hirose, E. Sugimoto, and H. Chiba, *BBA* **227**, 595 (1971).

to suggest that E_P·glycerate-di-P is the species responsible for phosphate transfer to water since this complex is probably responsible for inhibition by the diphosphate in the mutase reaction. In addition, the free phospho-enzyme is undoubtedly capable of hydrolysis, also (*212*). A rationale for the above correspondence of kinetic parameters has also been prosposed in terms of a sequential mechanism for the mutase reaction (*248*).

The phosphatase activity of some PGA-mutases can be stimulated (*201*) by 30–80% by treatment with m*M* Hg^{2+} (*249*) [but not by Cd^{2+} or Pb^{2+} (*250*)], although stimulation is the exception rather than the rule (*251*). A mechanistically more interesting observation is that glycolate-2-P enhances phosphatase activity of the yeast and muscle enzymes by several fold (*201, 248*). Since the 3-hydroxymethyl group of glycerate-2-P is missing in glycolate-2-P, the enhanced phosphatase activity of PGA-mutase in the presence of this compound bears at least a formal analogy to the enhanced phosphatase activity of phosphoglucomutase produced by xylose-1-P (see Section II,F); the observation that ^{32}P-labeled enzyme produces ^{32}P-inorganic phosphate on treatment with glycolate-2-P (*212*) substantiates this suggestion. Indeed, it seems possible that all PGA-mutases in which phosphatase activity is increased by glycolate-2-P exist in a phospho-enzyme form and have a Ping Pong reaction sequence. However, the observation that glycolate-2-P enhances the activity of erythrocyte glycerate-di-P phosphatase by 1600-fold (*246*) may make the effect on PGA-mutase difficult to study unless samples of the mutase free of this phosphatase are used. Also, the unusually large degree of stimulation of phosphatase activity of erythrocyte mutase (*201*) and of commercial samples of mutase (*246*) by glycolate-2-P may be simply an indication of the lack of purity of these enzymes. As might be expected, glycolate-2-P is also an inhibitor of the mutase reaction (*248*).

In addition to glycolate-2-P, hydroxypyruvate-3-P also stimulates phosphatase activity of both yeast and muscle enzymes (*242, 248*) but via a different reaction sequence. Thus, phosphate is apparently transferred to the hydrated form of hydroxypyruvate-3-P to give

249. Hg^{2+} also stimulates glycerate-di-P phosphatase activity (*245*).

250. S. Grisolia and J. Tecson, *BBA* **132**, 56 (1967).

251. Treatment by Hg^{2+} and other sulfhydryl agents may simultaneously reduce mutase activity by many fold in a variety of PGA-mutase preparations (*201*), but it affects phosphatase activity to only a small extent. However, because of the widely disparate phosphatase-to-mutase activities, initially, there is no compelling reason to accept the suggestion (*250*) that this change in activity ratio represents a conversion of a mutase to a phosphatase.

$$\begin{array}{c} \quad\quad\quad\quad OH \\ \quad\quad\quad\quad | \\ CH_2\text{—}C\text{—}CO_2^- \\ | \quad\quad\quad | \\ {}^{2-}O_3PO \quad OPO_3^{2-} \end{array}$$

Such enolphosphates are readily hydrolyzed nonenzymically (*242*). In this case the phosphatase action produced is an indirect result of a transfer that is more nearly normal than that discussed above.

Other compounds which stimulate glycerate-di-P phosphatase activity in mammalian PGA-mutase preparations are pyrophosphate and bisulfite (*201*); however, the observed effects could be caused by stimulation (*202, 243*) of the activity of a phosphatase impurity.

J. Effects of Inorganic and Simple Organic Compounds

By analogy with muscle phosphoglucomutase (Section II,H,1) it may be presumed that most anions can inhibit PGA-mutase competitively with both the mono- and diphosphates; however, because of the high levels of substrate usually used in the assay, such inhibition has been detected only with inhibitors such as pyrophosphate, oxalate, or citrate (*218, 248*) that bind efficiently to the enzyme, and only inhibition competitive with the diphosphate (*252*) has been detected. The substrate analog lactate-2-P (*242, 248*) is also an inhibitor as might be expected; however, no inhibition by the isomeric 3-hydroxypropionate-3-P was found under the same conditions. This difference may indicate that glycerate-2-P binds to the enzyme more tenaciously than glycerate-3-P. The substrate-like analogs, phosphoenolpyruvate, glycolate-2-P, and hydroxypyruvate-3-P, inhibit competitively with the diphosphate and have inhibition constants in the 0.1 m*M* range (*240*). Since *erythro*-dihydroxybutyrate 2- and 3-P are both poor substrates (*242*), both should also act as inhibitors.

Fluoride inhibition is puzzling. Thus, both the increasing inhibition by fluoride with increasing assay time and the protection against fluoride inhibition provided by EDTA and other chelating agents (such as citrate) (*252*) are inconsistent with a rationale of anionic competitive inhibition; unfortunately, the available data do not indicate whether pretreatment of the enzyme with fluoride in the absence of substrate produces inhibition, or prove whether fluoride inhibition in the presence of substrate is actually irreversible as suggested. Strangely, an EDTA concentration in the same range as the fluoride concentrations is required to relieve

252. H. Chiba, E. Sugimoto, and M. Kito, *Bull. Agr. Chem. Soc. Japan* **24**, 424 (1960).

inhibition (*218, 252*), as though a noninhibitory fluoride–EDTA complex were being formed.

Muscle PGA-mutase is not inhibited if dialyzed against EDTA prior to the assay or if EDTA is included in the assay; moreover, added metal ions fail to activate the enzymic reaction (*219*). Hence, if PGA-mutase activity requires a metal ion the metal must be firmly bound to the enzyme. In addition, of 16 different metal ions tested with the yeast enzyme, including Cu^{2+}, Hg^{2+}, Ag^{2+}, and Pb^{2+}, only Zn^{2+} produced as much as 50% inhibition at millimolar concentrations under the conditions used, and none produced activation (*218*). Since Zn^{2+} inhibition increases with increasing substrate concentration (*199*), and since the diphosphate is limiting in the assay (see Section IV,D), Zn^{2+} inhibition is caused by competition with the enzyme for glycerate-di-P (*252*). The muscle enzyme is much more sensitive to those metal ions which can react with sulfhydryl groups, Hg^{2+}, Ag^{2+}, and Cu^{2+} (*219*); in the case of Hg^{2+} the inhibition is largely reversed by treatment with dithioerythritol (*250*). However, the extent of Hg^{2+} inhibition appears to vary markedly with enzyme from embryonic to adult tissue, possibly as a result of the presence of different isozymes during development (*253*). Iodoacetate, *p*-mercuribenzoate, and *N*-ethylmaleimide also inhibit both enzymes when treatment is conducted prior to the assay (*8, 219, 250*); substrate protection is observed for these processes (*8, 219*).

Trinitrobenzenesulfonate reacts preferentially with 4 (*240*) of a possible 92 lysine residues (*224*), while binding of substrate and substrate-like analogs dramatically protects all 4 of the reactive residues. The kinetics of the reaction indicates that one reactive residue per monomer is essential for activity (*240*) if it is assumed that each monomer is equally and independently active.

K. Effects of pH and Temperature

The effect of pH and temperature on the enzyme has been measured by several workers; however, v_0 as opposed to V_{max} values have been used and the results are therefore of questionable significance. This is graphically illustrated by the report of "pH optima" differing by 1 pH unit when the enzymic assay was conducted by two different procedures (*199*). In spite of this, it seems quite probable that muscle and yeast enzymes do indeed have pH optima at significantly lower pH than the wheat and rice germ enzymes (*200*).

253. J. Grisolia, D. Diederich, and S. Grisolia, *BBRC* **41,** 1238 (1971).

V. Diphosphoglycerate Mutase

Diphosphoglycerate mutase was first studied by Rapoport and Luebering (*254, 255*) in rabbit red cell preparations and has since been found in a number of plant and animal tissues (*197*). The enzyme has been partially purified from chicken breast muscle (*256*) and extensively purified from erythrocytes of rabbits and humans (*257*). The erythrocyte enzyme can be stored in the frozen state for several months (*257*).

The radiometric assay of Rose (*257*), based on the acid and heat stability of glycerate-2,3-di-P, represents a marked improvement over coupled assays originally used with the enzyme (*255*), and it has been used to study the reaction sequence for the erythrocyte enzyme. Although an ordered bi–bi sequence was suggested (*257*), the results are actually ambiguous because of the identity of one of the substrates with one of the products:

$$\text{GA-1,3-di-P} + \text{GA-3-P} \rightarrow \text{GA-2,3-di-P} + \text{GA-3-P}$$

This identity precludes the exclusion of a random order of substrate addition and product release on the basis of the available date. Either glycerate 2- or 3-phosphates can act as the phosphate acceptor but the same products are produced in both cases. The values of K_m for glycerate-1,3-di-P and glycerate-3-P are in the micromolar range; the product, glycerate-2,3-di-P, is a potent competitive inhibitor of the reaction with a K_I also in the micromolar range. An examination of the distribution of label from ^{32}P-glycerate-1,3-di-P between the 2 and 3 positions of glycerate-2,3-di-P in the presence of various ratios of glycerate-2- and 3-phosphates suggests that a second active site on the enzyme may become available in the presence of glycerate-2-P (*257*).

A number of potential physiological effectors (inhibitors) have been examined both for primary inhibition and for the relief of product inhibition, but no significant effects were observed (*257*) despite previous reports that ADP regulates enzymic activity (*258*). Glycerate-2,3-di-P appears to be the most likely candidate for control of diphosphoglycerate mutase in red cells; such control would be exerted via product inhibition. Thus, the binding of glycerate-2,3-di-P to deoxyhemoglobin but not

254. S. Rapoport and J. Luebering, *JBC* **183,** 507 (1950).
255. S. Rapoport and J. Luebering, *JBC* **196,** 583 (1952).
256. B. K. Joyce and S. Grisolia, *JBC* **234,** 1330 (1959).
257. Z. B. Rose, *JBC* **243,** 4810 (1968); *Federation Proc.* **29,** 1105 (1970).
258. W. Schröter and H. von Heyden, *Biochem. Z.* **341,** 387 (1965).

to oxyhemoglobin (*259*) suggests that changes in diphosphate concentration may occur within the red cell during transitions from low to high oxygen tension; these changes might in turn regulate diphosphoglycerate mutase activity (*257*) as well as activities of hexokinase, transaldolase, and transketolase (*260*).

259. R. Benesch, R. E. Benesch, and C. I. Yu, *Proc. Natl. Acad. Sci. U. S.* **59,** 526 (1968).

260. R. C. Darling, C. A. Smith, E. Asmusser, and F. M. Cohen, *J. Clin. Invest.* **20,** 739 (1941).

13

Amino Acid Racemases and Epimerases

ELIJAH ADAMS

I. Introduction

Amino acid racemases catalyze formation of a racemic mixture from either the D or L form of a free amino acid by equilibrating configuration at the α-carbon. Amino acid epimerases (those presently known are enzymes acting on isomers of hydroxyproline and of 2,6-diaminopimelic acid, respectively) also catalyze equilibration of configuration at the α-carbon; since epimerases act on compounds possessing an additional asymmetric carbon, a diastereomer or epimer of the substrate is formed rather than its antipode. Both groups of enzymes are frequently referred to as racemases since they share a common reaction at the carbon directly involved in the catalytic change. While, in principle, amino acid epimerization could also occur by configurational change at an asymmetric carbon other than the α-carbon, enzymes that directly catalyze such reactions have not been described. Other types of amino acid isomerization, such as the conversion of L-lysine to β-lysine (*1*) or L-glutamate to β-methylaspartate (*2*) are considered elsewhere in this volume (*3*).

Type reactions involving the functional groups common to many amino acids such as enzymic decarboxylation, deamination, or transamination are commonly referred to as general reactions of amino acids. Of the various classes of enzymes catalyzing such reactions, the racemases were the most recent to be demonstrated and are least well understood in terms of structure and mechanism. Amino acid racemases have been the subject of brief surveys in several recent monographs devoted to amino acid metabolism in general (*4, 5*) or to molecular asymmetry in biology (*6*).

Unlike the other major classes of enzymes catalyzing general reactions of amino acids, racemases have thus far been clearly identified only in bacteria. D-Amino acids occur widely, and perhaps universally, in the cell walls of bacteria (*7, 8*), as well as in the peptide antibiotics produced by specialized groups of microorganisms (*9*). One major metabolic function of racemases appears to be the formation of free D-amino acids for cell wall synthesis. In contrast, in the synthesis of at least certain of the peptide antibiotics, free L-amino acids may not be substrates for direct conversion to the D antipodes; rather, racemization reactions may pro-

1. L. Tsai and T. C. Stadtman, *ABB* **125**, 210 (1968).
2. H. A. Barker, V. Rooze, F. Suzuki, and A. A. Iodice, *JBC* **239**, 3260 (1964).
3. T. C. Stadtman, "The Enzymes," 3rd ed., Vol. VI, p. 539, 1972.
4. A. Meister, "Biochemistry of the Amino Acids," 2nd ed., Vol. 1, p. 369. Academic Press, New York, 1965.
5. H. J. Sallach and L. A. Fahien, *Metab. Pathways* **3**, 60 (1969).
6. R. Bentley, "Molecular Asymmetry in Biology," Vol. 2, p. 196. Academic Press, New York, 1970.
7. J. M. Ghuysen, *Bacteriol. Rev.* **32**, 425 (1968).
8. H. J. Rogers, *Bacteriol. Rev.* **34**, 194 (1970).
9. S. G. Waley, *Advan. Protein Chem.* **21**, 2 (1966).

ceed with amino acids already bound or activated as the antibiotic precursors (*10, 11*).

II. History and Survey

Historically, the existence of pyridoxal phosphate-dependent racemases was predicted by nutritional observations on lactic acid bacteria made by Snell and his associates in the 1940's. These studies indicated an alternative growth requirement for vitamin B_6 or D-alanine (*12*) by *Streptococcus faecalis* and *Lactobacillus casei*. While D-amino acids had earlier been shown to occur in bacteria, Snell's findings were the first to indicate bacterial growth dependence on a D-amino acid and the replacement of this requirement by a form of vitamin B_6. The explicit postulation of Holden and Snell (*12*) that D-alanine was formed from L-alanine and that pyridoxal phosphate acted as a "coracemase" was confirmed two years later, in 1951, by Wood and Gunsalus in a report (*13*) which constitutes the first description of an amino acid racemase as an enzyme. D-Alanine was formed from L-alanine in a reaction catalyzed both by dried cells of *S. faecalis* and by slightly purified extracts. Extracts were shown to catalyze racemization of either D- or L-alanine at similar rates. The possibility that coupled reactions involving transamination of both L- and D-alanine might be the route of racemization was excluded by an unsuccessful effort to trap any postulated pyruvate so formed with lactate dehydrogenase and NADH; furthermore, addition of keto acids had no effect on the reaction. The racemase preparations from *S. faecalis* were easily resolved from pyridoxal phosphate; dried cells were already 2-fold stimulated by added coenzyme, while ammonium sulfate fractionation yielding only minor purification produced 90% resolution of the enzyme from pyridoxal phosphate. This initial study not only introduced the racemases as enzymes but also utilized a widely followed assay (for D-alanine with D-amino acid oxidase) and reported features of the racemase reaction that were not qualitatively extended by later studies for a number of years. Thus, Wood and Gunsalus concluded that alanine racemase was a pyridoxal phosphate enzyme, that

10. A. Hurst, *in* "The Bacterial Spore" (G. W. Gould and A. Hurst, eds.), p. 167. Academic Press, New York, 1969.

11. Y. Saito, S. Otani, and S. Otani, *Advan. Enzymol.* **33**, 337 (1970).

12. J. T. Holden and E. E. Snell, *JBC* **178**, 799 (1949), and references therein.

13. W. A. Wood and I. C. Gunsalus, *JBC* **190**, 403 (1951).

the enzyme was specific for alanine among a number of other amino acids, and that this activity could be demonstrated in a variety of other bacterial species, but not in the yeasts, fungi, or animal tissues examined.

Subsequent studies during the 1950's confirmed and extended the conclusion that alanine racemase activity was demonstrable in a variety of bacterial species (*14–17*) and in spores (*15*). Racemization of other amino acids was also reported, including lysine (*18*), methionine (*19, 20*), glutamic acid (*21, 22*), threonine (*23*), and proline (*24*). During this period also, epimerases for 2,6-diaminopimelic acid (*25*) and hydroxyproline (*26*) were described. Many of these studies, however, were carried out with dried cells or relatively unpurified extracts and did not in themselves offer convincing evidence for direct racemization.

During the subsequent decade (1960–1970) more extensive purification of several racemases permitted fuller understanding of the nature of the reactions. Early examples of partially purified racemases were the alanine racemase of *Bacillus brevis* (*27*) and the glutamic acid racemase of *Lactobacillus arabinosus* (*28*). The latter enzyme, after more than 300-fold purification, was apparently free of transaminase activity, judged by its failure to catalyze exchange of ^{14}C between glutamate and α-ketoglutarate; this was the clearest demonstration to that date that a racemization reaction did not involve concurrent action of a D- and L-specific transaminase. Another first demonstration with this preparation was the labilization of hydrogen from the substrate, demonstrated by incorporation of tritium from water into glutamate. Only about 0.1 μatom of tritium was exchanged per micromole of substrate, but the experimental design was such that even a modest isotope effect could have prevented stoichiometric labeling.

The first amino acid racemizing enzyme to be characterized in an

14. A. G. Marr and P. W. Wilson, *ABB* **49,** 424 (1954).
15. B. T. Stewart and H. O. Halvorson, *J. Bacteriol.* **65,** 160 (1953).
16. D. C. Jordan, *Can. J. Microbiol.* **1,** 743 (1955).
17. C. B. Thorne, C. G. Gomez, and R. D. Housewright, *J. Bacteriol.* **69,** 357 (1955).
18. H. T. Huang and J. W. Davisson, *J. Bacteriol.* **76,** 495 (1958).
19. G. D. Schockman and G. Toennies, *ABB* **50,** 9 (1954).
20. R. E. Kallio and A. D. Larson, *in* "Amino Acid Metabolism" (W. D. McElroy and B. Glass, eds.), p. 616. Johns Hopkins Press, Baltimore, Maryland, 1955.
21. S. A. Narrod and W. A. Wood, *ABB* **35,** 462 (1952).
22. P. Ayenger and E. Roberts, *JBC* **197,** 453 (1952).
23. H. Amos, *JACS* **76,** 3858 (1954).
24. T. C. Stadtman and P. Elliott, *JBC* **228,** 983 (1957).
25. M. Antia, D. S. Hoare, and E. Work, *BJ* **65,** 448 (1957).
26. E. Adams, *JBC* **234,** 2073 (1959).
27. T. Hameda, *J. Osaka City Med. Center* **9,** 1927 (1960).
28. L. Glaser, *JBC* **235,** 2095 (1969).

essentially homogeneous state, permitting direct investigation of cofactor status, was hydroxyproline epimerase (*29*), which had earlier been briefly described as an inducible enzyme in hydroxyproline-grown *Pseudomonas* (*26*). Because of the earlier kinetic indications that racemases in general might be pyridoxal phosphate enzymes, it was an unexpected finding that hydroxyproline epimerase appeared neither to contain nor to require pyridoxal phosphate or other possible prosthetic groups (*29, 30*).

Other amino acid racemizing enzymes which have been studied after extensive purification are the proline racemase of *Clostridium* (*31*), alanine racemase of *Pseudomonas* (*32, 33*) and of *Lactobacillus* (*34*), glutamate racemase of *Lactobacillus* (*35*), phenylalanine racemase of *Bacillus brevis* (*36, 37*), arginine racemase of *Pseudomonas* (*38*), and a nonspecific enzyme from *Pseudomonas*, initially recognized as racemizing leucine and α-aminobutyrate (*39*). Of these, some were obtained as crystalline proteins (*38, 39*) or as proteins of demonstrated high absolute purity (*33, 37*). Table I summarizes certain properties of those racemases which have been characterized as enzymes. At the close of the second decade since the discovery of this group of enzymes, it can be said that there are a number of individually specific enzymes more or less widely distributed in bacteria, that several have been purified to homogeneity, that some are clearly pyridoxal phosphate enzymes while others are clearly not, and that this group of enzymes is just beginning to be studied in structural and mechanistic detail (*1, 13–15, 17–25, 27–63*).

29. E. Adams and I. L. Norton, *JBC* **239,** 1525 (1964).
30. E. Adams, *BBRC* **10,** 327 (1963).
31. G. J. Cardinale and R. H. Abeles, *Biochemistry* **7,** 3970 (1968).
32. C. A. Free, M. Julius, P. Arnow, and G. T. Barry, *BBA* **146,** 608 (1967).
33. G. Rosso, K. Takashima, and E. Adams, *BBRC* **34,** 134 (1969).
34. M. M. Johnston and W. F. Diven, *JBC* **244,** 5414 (1969).
35. W. F. Diven, *BBA* **191,** 702 (1969).
36. M. Yamada and K. Kurahashi, *J. Biochem.* (*Tokyo*) **63,** 59 (1968).
37. M. Yamada and K. Kurahashi, *J. Biochem.* (*Tokyo*) **66,** 529 (1969).
38. T. Yorifuji, K. Ogata, and K. Soda, *BBRC* **34,** 760 (1969).
39. K. Soda and T. Osumi, *BBRC* **35,** 363 (1969).
39a. J. R. Martin and N. N. Durham, *BBRC* **14,** 388 (1964).
40. B. T. Stewart and H. O. Halvorson, *ABB* **49,** 168 (1954).
41. B. D. Church, H. Halvorson, and H. O. Halvorson, *J. Bacteriol.* **68,** 393 (1954).
42. H. K. Kuramitsu and J. E. Snoke, *BBA* **62,** 114 (1962).
43. W. F. Diven, R. B. Johnston, and J. J. Scholz, *BBA* **67,** 161 (1962).
44. W. F. Diven, J. J. Scholz, and R. B. Johnston, *BBA* **85,** 322 (1964).
45. A. Gaffer, D. R. Terry, and R. D. Sagers, *J. Bacteriol.* **91,** 1618 (1966).
46. J. L. Strominger, E. Ito, and R. H. Threnn, *JACS* **82,** 998 (1960).
47. U. Roze and J. L. Strominger, *Mol. Pharmacol.* **2,** 92 (1966).
48. R. Y. Stanier, N. J. Palleroni, and M. Doudoroff, *J. Gen. Microbiol.* **43,** 159 (1966).

TABLE I
AMINO ACID RACEMASES AND EPIMERASES[a]

Amino acid	Source of enzyme	Comments	Ref.
Alanine	*Streptococcus faecalis*	First demonstration of an amino acid racemase. Marked stimulation by pyridoxal phosphate after slight purification	(*13*)
	Bacillus terminalis	Comparison of enzyme in extracts of spores and vegetative cells: Spore enzyme heat resistant and pepsin resistant until solubilized. After dialysis, enzyme in spore extracts stimulated 2- to 3-fold by pyridoxal phosphate	(*15, 40, 41*)
	Brucella abortus	No labeling of alanine from $^{15}NH_3$ during enzymic racemization. Crude sonic extracts about 10-fold stimulated by pyridoxal phosphate	(*14*)
	Bacillus subtilis	Postulated as the source of D-glutamic acid for capsular polyglutamate synthesis, through D-specific transamination	(*17*)
	Bacillus licheniformis	Alanine racemase in crude sonic extracts; other amino acid (glutamic acid and phenylalanine) racemases not demonstrable, but a variety of D-specific transminase reactions present	(*42*)
	Bacillus brevis	Purified 40-fold, specific for alanine, inhibited by sulfhydryl reagents	(*27*)
	Bacillus subtilis	Up to 140-fold purification; no indication of absolute purity. Partly purified enzyme (25-fold) completely resolved from pyridoxal phosphate; FAD participation inferred from association of fluorescence with enzymic activity on zone electrophoresis, partial inactivation, and restoration on acid ammonium sulfate precipitation	(*43, 44*)
	Clostridium acidi-urici	Alanine racemase activity of extracts measured at 37° and 44°: no correlation with difference in D-alanine content of cell walls after growth at these two temperatures	(*45*)
	Staphylococcus aureus	100-fold purified enzyme not stimulated by pyridoxal phosphate; inhibited by hydroxylamine and hydra-	(*46, 47*)

TABLE I (*Continued*)

Amino acid	Source of enzyme	Comments	Ref.
		zine and by D-cycloserine but not L-cycloserine	
	Pseudomonas sp.[b] (Squibb culture 3550)	150-fold purified enzyme not stimulated by pyridoxal phosphate. Inhibited by hydroxylamine and aminooxyacetate	(*32*)
	Pseudomonas putida	Enzyme induced by growth on alanine and purified 1000-fold to homogeneity. Homogeneous enzyme contains approximately 1 molar equivalent of pyridoxal phosphate. No evidence for flavins; pyridoxal phosphate not removed on purification	(*33*)
	Lactobacillus fermenti	1500-fold purified enzyme estimated 30–60% pure. Pyridoxal phosphate required after 4-fold purification	(*34*)
	Escherichia intermedia	Alanine racemization catalyzed by crystalline tyrosine phenollyase. Racemase activity a small fraction of β-elimination activity converting tyrosine to phenol, pyruvate, and NH_3. Spectral evidence for enzyme–pyridoxal phosphate–alanine Schiff-base complex	(*49*)
Arginine	*Pseudomonas graveolens*	Enzyme purified 2300-fold and crystallized.[c] Enzyme contains 4 molar equivalents of pyridoxal phosphate by spectral data on phenylhydrazine derivative	(*38, 51*)
Aspartic acid	*Pseudomonas* sp.	Presence of aspartate racemase in crude extracts inferred from formation of alanine from L- or D-aspartate, and cycloserine inhibition of alanine formation from D-aspartate but not from L-aspartate	(*52*)
	Lactobacillus fermenti	Aspartate racemase separated from alanine and glutamate racemase by gel filtration	(*34*)
Diaminopimelic acid	*Escherichia coli*	Enzyme studied in crude extracts; specific for interconversion of L ⇄ *meso* epimers. Inhibited by hydroxylamine, semicarbazide, isoniazid, and mercurials. Enzyme present in a number of bacterial strains	(*25*)

TABLE I (*Continued*)

Amino acid	Source of enzyme	Comments	Ref.
	Bacillus megaterium	Enzyme 23-fold purified. No pyridoxal phosphate requirement after partial inactivation by acid precipitation, dialysis, Dowex-1, or charcoal treatment. No inhibition (qualitative test) by cycloserine, hydroxylamine, semicarbazide, or deoxypyridoxine	(*53*)
Glutamic acid	*Lactobacillus arabinosus*	Over 300-fold purified.[d] No stimulation by pyridoxal phosphate after dialysis, salt precipitation, charcoal, hydroxylamine, or cyanide treatment. Inhibition by hydroxylamine. Substoichiometric hydrogen exchange	(*28*)
	Lactobacillus fermenti	500-fold purified. No inhibition by hydroxylamine, semicarbazide, or phenylhydrazine. FAD participation inferred from slight inhibition by riboflavin, restoration by FAD. No stimulation by pyridoxal phosphate	(*54*)
	Lactobacillus fermenti	2000-fold purified. Molecular weight 23,000 by gel filtration. Enzyme inhibited by hydroxylamine (competitive with substrate) and by FMN (noncompetitively); inhibited also by FAD, riboflavin, and FAD analogs	(*35*)
Hydroxyproline	*Pseudomonas putida*	Enzyme induced by hydroxyproline and purified to homogeneity. Acts as racemase for α-carbon of all epimers of 4-hydroxyproline and of 3-hydroxyproline. Molecular weight 64,000; neither contains nor requires pyridoxal phosphate, flavins, or pyridine nucleotides. Evidence for active site sulfhydryls. α-Hydrogen exchanged at rate concomitant with racemization	(*29, 30, 55*)
Lysine	*Pseudomonas* sp.	Lysine racemization in cell free preparation[e] detected by oxidation of D-lysine in presence of *Neurospora* L-amino acid oxidase. Direct racemization inferred from racemization activity in salt-precipitated fraction without added cosubstrates	(*57*)

TABLE I (*Continued*)

Amino acid	Source of enzyme	Comments	Ref.
	Pseudomonas sp.	About 10-fold purified, not increased by growth on lysine. No stimulation by pyridoxal phosphate. Inhibited by hydroxylamine	(*58*)
	Clostridium sticklandii	Presence in extracts noted without details, in connection with a detailed study of lysine mutase	(*1*)
Methionine	*Pseudomonas* sp.	Enzyme[f] required pyridoxal phosphate after limited fractionation	(*20*)
Phenylalanine	*Bacillus brevis*	Phenylalanine racemized by almost-homogeneous enzyme that activates L- or D-phenylalanine. Racemization requires ATP and yields stoichiometric AMP. Racemase activity only a small fraction of activating activity: enzyme not inhibited by hydroxylamine, cyanide, or isoniazid	(*36, 37*)
Proline	*Clostridium sticklandii*	In first description (*24*), enzyme purified about 9-fold; no pyridoxal phosphate stimulation after dialysis, ultraviolet treatment; hydroxylamine not inhibitory. In later study (*31*), enzyme purified about 300-fold to at least 90% purity. No evidence for pyridoxal phosphate by inhibitor data. Exchange of α-H at a rate concomitant with initial racemization rate	(*24, 31*)
Threonine	*Escherichia coli*	In crude extracts, formation of L-threonine from D-threonine detected by NH_3 formation in presence of L-threonine specific deaminase. Activity stimulated by ATP or AMP. No subsequent report following this preliminary communication	(*23*)
Tryptophan	*Pseudomonas* sp.	Anaerobic formation of racemate from L or D antipode catalyzed by crude extract. Fourfold stimulation by pyridoxal phosphate after limited fractionation. Inhibited by hydroxylamine and semicarbazide	(*59*)

TABLE I (*Continued*)

Amino acid	Source of enzyme	Comments	Ref.
General (Alanine, α-aminobutyric acid, arginine, ethionine, leucine,[g] lysine, methionine, norleucine, and norvaline)	*Pseudomonas striata*	Purified about 250-fold to homogenity and crystallized. Molecular weight about 110,000. Contains pyridoxal phosphate by spectral data and requires pyridoxal phosphate after hydroxylamine treatment	(*39, 60–62*)

[a] This table includes racemases which have been demonstrated by rather direct assays in cellfree preparations. Particularly where little purification is indicated, indirect racemization has not always been excluded. An example of the latter is the conversion of D-tryptophan to L-tryptophan by D-specific oxidation to indole pyruvate followed by transamination, described in *Flavobacterium* extracts (*39a*).

[b] This designation is used for a laboratory strain that has not been identified as one of the recognized taxonomic groups (*48*).

[c] This enzyme is induced by growth in arginine-containing media (*50*).

[d] Prior reports of glutamic acid racemization were based on whole cell preparations (*21, 22*).

[e] Prior reports of lysine racemization were based on whole cell preparations (*18, 56*).

[f] Prior report of methionine racemization based on whole cells (*19*).

[g] Leucine racemization in extracts of *Bacillus brevis* has also been inferred from the apparent equivalent incorporation of both L-leucine and D-leucine into gramicidin S (*63*).

49. H. Kumagai, N. Kashima, and H. Yamada, *BBRC* **39,** 796 (1970).
50. K. Soda, personal communication (1969).
51. K. Soda, T. Yorifuji, and K. Ogata, *BBA* **146,** 606 (1967).
52. A. J. Markovetz, W. J. Cook, and A. D. Larson, *Can. J. Microbiol.* **12,** 745 (1966).
53. P. J. White, B. Lejeune, and E. Work, *BJ* **113,** 589 (1969).
54. M. Tanaka, Y. Kato, and S. Kinoshita, *BBRC* **4,** 114 (1961).
55. T. H. Finlay and E. Adams, *JBC* **245,** 5248 (1970).
56. H. T. Huang, D. A. Kita, and J. W. Davisson, *JACS* **80,** 1006 (1958).
57. P. S. Thayer, *J. Bacteriol.* **78,** 150 (1959).
58. A. Ichihara, S. Furiya, and M. Suda, *J. Biochem.* (*Tokyo*) **48,** 277 (1960).
59. E. J. Behrman, *Nature* **196,** 150 (1962).
60. K. Soda and T. Osumi, *Agr. Biol. Chem.* (*Tokyo*) **31,** 1097 (1967).
61. K. Soda, T. Osumi, T. Yorifuji, and K. Ogata, *Agr. Biol. Chem.* (*Tokyo*) **33,** 424 (1969).
62. T. Osumi, T. Yamamoto, and K. Soda, *Agr. Biol. Chem.* (*Tokyo*) **33,** 430 (1969).
63. K. J. Figenschou, L. O. Frøholm, and S. G. Laland, *BJ* **105,** 451 (1967).

III. Assay Methods

A. Coupling to L- or D-Specific Enzymes

The most general assay principle applied is enzymic detection of a specific antipode formed as product. Most commonly, this has involved incubating the L antipode as substrate, and measuring the D antipode with D-amino acid oxidase. The latter step has been carried out manometrically in studies of alanine racemase (*13–17, 42, 49*), of phenylalanine racemization (*36*) or of the racemization of other amino acids (*61*). Alternatively, colorimetric methods have been used for the ketoacid resulting from D-amino acid oxidase action (*32, 44, 45, 61*), or the pyruvate formed from D-alanine has been measured with lactate dehydrogenase (*33*). An obvious variation is the use of a D antipode as substrate and detection of the L product with L-amino acid oxidase (*20, 57, 59*). Where the above methods have not been applicable, as in the racemization of amino acids (e.g., glutamate) which are not good substrates for commonly available amino acid oxidases, other specific enzymes have been used to measure the formation of a D or L product: arginase (*38*), *meso*-diaminopimelate decarboxylase (*53*), L-glutamate decarboxylase (*60*), D-proline reductase (*24*), allohydroxy-D-proline oxidase (*29*), L-lysine oxygenase (*58*), or L-threonine deaminase (*23*).

B. Polarimetric

A direct assay method, which has the advantage of permitting continuous monitoring of the reaction as well as an identical type of rate measurement in either reaction-direction, is that of polarimetry. This has been applied as a secondary assay in some instances (*24, 34*), and as the principal assay in others (*29, 31, 55*), used either as a continuous assay (*29, 55*) or in single-point samples (*31*). While polarimetry would seem to be a method of choice when possible, it may be limited by the sensitivity of available instruments, particularly when the difference in specific rotation of substrate and product is small. The sensitive and relatively inexpensive Bendix polarimeter (model ETL-NPL, type 143a) has proved very useful for the assay of hydroxyproline-2-epimerase (*55*), in part because of the large difference in specific rotation (134°) between hydroxy-L-proline and allohydroxy-D-proline (*29*). The same instrument, however, has not been useful for the assay of alanine racemase (specific rotation difference, 3–4° at neutral pH) (*64*); alanine racemization, however, was measurable at 240 nm, using a Cary model 60 spectropolarimeter (*34*).

IV. Specificity, Equilibria, and Kinetics

A. Substrate Specificity

With a single exception—the broad substrate-specificity racemase crystallized from extracts of *Pseudomonas* (*39*)—the racemases and epimerases show rather high specificity for the amino acid substrate in instances when this feature has been carefully examined. A striking example is the specificity of diaminopimelate epimerase for L- or *meso*-2,6-diaminopimelate (*25*): Only these two epimers are substrates, indicating that the configuration at one asymmetric carbon must be L to permit racemization of configuration at the other asymmetric carbon. This is not a generalizable finding for epimerases since hydroxyproline-2-epimerase acts on the α-carbon of hydroxyproline with approximately equal efficiency regardless of the configuration at the hydroxyl-carbon (*55*); in the case of the latter enzyme, even the 3-hydroxyprolines are substrates, although with much reduced V_{max} values (*55*). However, a number of common amino acids other than hydroxyproline were not racemized by purified hydroxyproline-2-epimerase (*29*), and purified proline racemase had little action on hydroxyproline and none on valine or alanine (*31*). Other purified racemases which have been explicitly examined for substrate specificity are the glutamic racemase of *Lactobacillus fermenti*, reported inactive with aspartic acid, alanine, α-aminobutyric acid, or α-methylglutamic acid (*35*); an ATP-dependent phenylalanine racemase of *Bacillus brevis* which had no action on tryptophan, tyrosine, valine, or leucine (*37*); and the arginine racemase of *Pseudomonas graveolens* reported inactive for a number of other amino acids (*38*). The most widely distributed enzyme, alanine racemase, does not appear to have been examined critically for substrate specificity with highly purified preparations.

B. Equilibrium Position

The expected equilibrium constant of unity for the racemization of antipodes has been demonstrated for a number of reactions, for example, the racemization of alanine (*13, 14, 17*), of arginine (*51*), of lysine (*58*),

64. E. Adams, unpublished data (1969).

of proline (*24, 31*), of glutamic acid (*28*), of leucine and α-aminobutyrate (*60*). The equilibrium constant of epimerization reactions is not necessarily unity since substrate and product are not energetically equivalent. It was found, however, that the equilibrium constant for epimerization of the 4-hydroxyprolines and 3-hydroxyprolines was close to unity (*29, 55*); the calculated values of K_{eq} from the values of K_m and V_{max} for various substrates were in agreement with observed values (*29, 55*). The observed equilibrium position of diaminopimelate epimerization favored the *meso* epimer over the L epimer by approximately a 2-fold ratio (*53*). Phenylalanine racemization by the ATP-dependent enzyme of *Bacillus brevis* (*36, 37*) has an apparently anomolous equilibrium position favoring the D-isomer; however, this reaction is probably *not* represented by

$$\text{L-Phenylalanine} \rightleftharpoons \text{D-phenylalanine} \quad (1)$$

but rather by a more complex net equation involving the hydrolysis of ATP, so that an equilibrium constant of unity is not necessarily to be expected (see Section VI).

C. Kinetic Features

To the extent that this has been examined, the general kinetic features of racemase reactions fit a reversible one substrate–one product reaction; thus, the appropriate Haldane expression has agreed with the experimentally determined K_{eq} (*29–31, 55*), or, when calculated from data given [Table IV in Johnson and Diven (*34*)], the Haldane expression for such a reaction type agrees with the expected K_{eq}. The most detailed kinetic studies have been carried out on proline racemase (*31*) and hydroxyproline epimerase (*55*), primarily in connection with deuterium effects or with substrate analogs; some of these findings are discussed in more detail in Section VI.

Table II summarizes data concerning the pH optima and substrate K_m's for essentially all the enzymes for which these data are available. In addition, for the few enzymes which have been purified to or close to homogeneity, the specific activity of the purest fraction is given. In several instances, where it is possible to do so, turnover numbers are also calculated (see footnotes to Table II) using data for the highest specific activity available, substrate K_m, and molecular weight.

This survey permits the generalization that the racemases as a group have neutral-to-alkaline pH optima and relatively high substrate K_m's, within the range of 2–50 mM. A single exception to the latter statement is the phenylalanine racemase activity associated with a phenylalanine-activating enzyme in *Bacillus brevis*. There is reason to believe, how-

TABLE II
KINETIC FEATURES AND MOLECULAR WEIGHT OF SOME PURIFIED RACEMASES

Enzyme and source	pH optimum	Substrate: K_m (mM)	Specific activity[a]	Molecular weight	Ref.
Alanine racemase					
Streptococcus faecalis	>8.5	L-Alanine: 8.5	—	—	(*13*)
Bacillus brevis	—	L- or D-alanine: 31–38	—	—	(*27*)
Lactobacillus fermenti	9–9.5	L- or D-alanine: 7–16[b]	634	—	(*34*)
Pseudomonas sp.[c]	8.9–9.5	L-Alanine: 30	2500	—	(*32*)
Pseudomonas putida	7.8	L-Alanine: 30	3680[d]	60,000	(*33*)
Escherichia intermedia	8.3[e]	L-Alanine: 26	0.051[f]	170,000	(*49*)
Arginine racemase					
Pseudomonas graveolens	8.0[e]	—	1250	167,000	(*51*)
Diaminopimelate epimerase					
Bacillus megaterium	6.5–8.0	L-Diaminopimelate: 6.7 *meso*-Diaminopimelate: 100	—	—	(*53*)
Glutamate racemase					
Lactobacillus arabinosus	7.5	D-Glutamate: 3.6	—	—	(*28*)
Lactobacillus fermenti	7.5[e]	D-Glutamate: 2.2	—	23,000	(*35*)
Hydroxyproline epimerase					
Pseudomonas putida	7.5–8.0	Epimers of 4-hydroxyproline, 3-hydroxyproline: 11–49	298[g]	64,000	(*29, 55*)

Phenylalanine racemase					
Bacillus brevis	7.8–8.5 (depends on direction)	L-Phenylalanine: 0.02	0.015[h]	100,000	(*37*)
Proline racemase					
Clostridium sticklandii	8.0[e]	L-Proline: 2.3 D-Proline: 3.8	600	—	(*31*)
Nonspecific racemase					
Pseudomonas striata	8.0–9.0	L-Methionine: 38[i]	708[j]	110,000	(*39*)

[a] Cited for most purified fraction, homogeneous or highly purified enzymes; μmoles formed per mg of enzyme per min under the specific assay conditions used in each case.

[b] Depends on other additions such as acetate.

[c] See footnote *b*, Table I.

[d] Turnover number for V_{max} activity is 2.3×10^5; the turnover value cited in (*33*) was erroneously cited there as 2.3×10^4.

[e] pH of assay.

[f] Specific activity for the tyrosine phenol-lyase reaction itself is 1.9 (*65*).

[g] Turnover number for V_{max} activity is about 2×10^4.

[h] Markedly influenced by AMP and PP concentrations.

[i] K_m values for other substrates were also in the range of 30–40 m*M*.

[j] Specific activity for L-methionine. Turnover for V_{max} activity (methionine) is about 1.5×10^5, calculated from data in (*39*).

ever, that this enzyme is fundamentally different in mechanism from most other racemases (see Section VI). For most of the highly purified enzymes cited, catalytic activity is quite high, represented by specific activities ranging from several hundred to several thousand micromoles per minute per milligram of protein (Table II). These rates are consistent with the fact that at least some of these (*29, 33, 51*) represent inducible enzymes whose rates may need to support growth of the cells on a single substrate. Two exceptions appear in Table II: the very low specific activity of alanine racemization by crystalline tyrosine phenol-lyase (*49*) and the low specific activity of phenylalanine racemization by the phenylalanine-activating enzyme (*37*). In each case it should be noted that racemization may well represent a reaction incidental to that primarily catalyzed by the enzyme involved.

V. Cofactors

A. Pyridoxal Phosphate

Much of the information on the coenzyme status of various racemases is summarized in Table I. A relationship between amino acid racemases and pyridoxal phosphate was first inferred by Snell from nutritional experiments (*12*) and was seemingly confirmed by the pyridoxal phosphate stimulation of the first preparations of alanine racemase studied (*13*). As experience with other racemases accumulated it appeared that in a number of cases pyridoxal phosphate was similarly loosely bound; thus, dialysis or only modest purification sufficed to demonstrate relative or absolute dependence on added pyridoxal phosphate (Table I). With other enzymes, however, a pyridoxal phosphate requirement was not demonstrable even after high purification; examples of these are the alanine racemases purified from strains of *Pseudomonas* (*32, 33*) or from *Staphylococcus aureus* (*47*), the arginine racemase of *Pseudomonas graveolens* (*38*), the wide specificity racemase of *Pseudomonas striata* (*39*), and the glutamate racemase of *Lactobacillus arabinosus* (*28*). Of this latter group, from which pyridoxal phosphate is not readily dissociable, three enzymes have been purified to homogeneity; in the case of alanine racemase of *Pseudomonas putida* (*33*), arginine racemase of *Pseudomonas graveolens* (*51*), and the nonspecific racemase of *Pseudomonas striata* (*39*), stoichiometrically significant pyridoxal phosphate

65. H. Kumagai, H. Yamada, H. Matsui, H. Ohkishi, and K. Ogata, *JBC* **245**, 1767 (1970).

was then demonstrable by the enzyme spectrum (*33, 39*), by the phenylhydrazine-derivative method (*31*), by microbiological assay (*33*), or by fluorometry (*33*). The crystalline wide specificity racemase could be resolved from pyridoxal phosphate by hydroxylamine treatment and then became dependent on addition of the coenzyme (*39*). The action of typical pyridoxal phosphate inhibitors such as aminoxyacetate (*32, 33*) or hydroxylamine (*29, 32*) on the alanine racemase of *Pseudomonas* further supports the functional role of pyridoxal phosphate in these cases. From the combined information constituting kinetics, inhibitor, or direct analytical evidence for the participation and/or presence of pyridoxal phosphate, it seems safe to conclude that in general the racemases of primary amino acids contain and require pyridoxal phosphate.

B. No Pyridoxal Phosphate

Not all amino acid racemases are pyridoxal phosphate enzymes. The absence of pyridoxal phosphate, pyridine or flavin nucleotides, or of any phosphate-containing prosthetic group, was demonstrated for homogeneous hydroxyproline-2-epimerase by a variety of methods including spectral and fluorometric investigation, content of ^{32}P after purification from cultures containing this isotope, and ineffectiveness of a variety of inhibitors including excess borohydride (*29, 30*). It is of particular interest that the same cells that make hydroxyproline-2-epimerase also produce an inducible alanine racemase which does contain and require pyridoxal phosphate (*29, 33*). While a less complete effort has been made to investigate the coenzyme status of proline racemase, its activity was unaffected by charcoal, dialysis, gel filtration, or inhibitors such as borohydride, cyanide, and hydroxylamine (*31*), and it may be concluded that pyridoxal phosphate does not participate in the racemization of proline or epimerization of hydroxyproline.

C. Pyridoxal Status Uncertain

A small number of enzymes do not clearly fit either of the above two categories on the basis that purification has been insufficient to exclude pyridoxal phosphate, but, on the other hand, kinetic and inhibitor data do not support pyridoxal phosphate participation. The best example of this group is the diaminopimelate epimerase of *Bacillus megaterium* (*53*). Although the enzyme was not extensively purified, residual activity after acid ammonium sulfate precipitation could not be increased by pyridoxal phosphate, nor did tests of 1 mM cycloserine, hydroxylamine, isoniazid, cyanide, or semicarbazide indicate inhibition by a semiquantitative assay. While failure to demonstrate a requirement is not strong evidence against the involvement of pyridoxal phosphate, most racemases so tested have been sensitive to one or more of the inhibitors noted above, and the possibility remains that this enzyme, like those racemizing the secondary amino acids, does not utilize pyridoxal phosphate. It is worth noting that the crude diaminopimelate epimerase of *Escherichia coli* was inhibited by hydroxylamine, semicarbazide, and isoniazid (*25*).

D. Possible Flavin Prosthetic Groups

A more nebulous question concerns the possible participation of flavin coenzymes in the action of certain racemases. This possibility was first raised in a preliminary study of glutamate racemase extensively purified from extracts of *Lactobacillus fermenti* (*54*) by the reported slight inhibition (up to 26%) of the reaction by riboflavin, FMN, tetracycline, or acriflavin, partly overcome by FAD. A similar type of observation was described for the alanine racemase of *Bacillus subtilis* by Diven and associates (*43, 44*); the role attributed to a postulated flavin nucleotide was that of a reversible acceptor of hydrogen lost from the substrate during ketimine formation with pyridoxal phosphate, or during α,β unsaturation of the substrate, as a step in racemization. Additional findings have been the inhibition by FMN, riboflavin, or FAD of glutamate racemase purified from *Lactobacillus fermenti* (*35*). A final decision concerning flavin content in the specific racemases cited above will require their availability as pure proteins. However, the present evidence for flavin participation in racemase reactions is weak in that it rests on the presence of flavins in enzyme preparations which have not been characterized as single proteins and on inhibition–reactivation data that are not quantitatively impressive. It should be noted that homogeneous hydroxyproline-2-epimerase, which was demonstrably free of flavins, also showed inhibition by atabrine and acriflavin, with occasional partial reactivation by FMN or FAD (*29*). Furthermore, no evidence for flavins by spectral or fluorescence data could be obtained with homogeneous alanine racemase of *Pseudomonas putida* (*33*).

E. Metal Ion Status

The possibility of metal ion participation in the action of racemases has not yet been considered decisively in any case. Only a few enzymes have been purified sufficiently to permit meaningful direct analysis, and such data have not been reported. On speculative grounds, metal participation has been proposed in the proline racemase reaction since a metal might provide a mechanism for activating the α-hydrogen (*31*). Hydroxyproline epimerase, however, was not inhibited by a number of metal chelating reagents (*29*).

VI. Mechanism of Action

A. Pyridoxal Racemases

The best model available for racemization of amino acids by pyridoxal phosphate enzymes is the nonenzymic racemization of alanine studied by Olivard *et al.* (*66*) in 1952. The basic postulation in this model involves labilization of the α-hydrogen through aldimine-ketimine tautomerization of the amino acid–pyridoxal phosphate complex. Racemization would result from the sterically random return of hydrogen to the α-carbon before amino acid release from the complex. Secondary questions concern the involvement of other forms of the bound amino acid such as an α,β-unsaturated intermediate, the possibility that a separate mechanism promotes dissociation of the α-hydrogen, and the ever-present questions concerning the manner in which the protein contributes to the high specificity and high catalytic efficiency of specific enzymes. These questions have not yet been seriously approached with the racemases, although considerable information is now available concerning pyridoxal phosphate mechanisms in other enzymic reactions, most recently reviewed by Snell and Di Mari (*67*).

Only fragmentary experimental information has been reported relative to the above scheme. The partial labilization observed for a hydrogen of glutamate in the course of the glutamate racemase reaction (*28*) has been referred to earlier; proof that this was indeed the α-hydrogen, and the α-hydrogen exclusively, has not been supplied; nor is it known whether or not hydrogen labilization proceeded at the same rate as the overall reaction. A similar study of alanine racemase reported only 2–3% of a molar equivalent of tritium-labeled hydrogen incorporated from water into alanine during 24% racemization (*68*); the likelihood of a prominent tritium isotope effect, however, makes it difficult to interpret this as an observation excluding significant hydrogen exchange. It is worth noting

66. J. Olivard, D. E. Metzler, and E. E. Snell, *JBC* **199,** 669 (1952).

67. E. E. Snell and S. J. Di Mari, "The Enzymes," 3rd ed., Vol. II, p. 335, 1970.

68. R. B. Johnston, J. J. Scholz, W. F. Diven, and S. Shepherd, *in* "Pyridoxal Catalysis: Enzymes and Model Systems" (E. E. Snell *et al.*, eds.), p. 537. Wiley (Interscience), New York, 1968.

that a number of racemases, presumably pyridoxal phosphate enzymes, also require reduced cysteines for activity or stability (*27, 34, 35, 54*), but there is not yet sufficient further information to accept sulfhydryl participation as universal, or to relate the sulfhydryl function to the action of the coenzyme. Speculatively, it is conceivable that cysteines act as reversible acceptors for the substrate hydrogens labilized during racemization, as has been suggested for hydroxyproline-2-epimerase, an enzyme independent of pyridoxal phosphate (*55*).

The existence of a known pyridoxal enzyme, tyrosine phenol-lyase, which racemizes alanine at a small fraction of the rate of its primary catalyzed reaction (*49*), is reminiscent of other pyridoxal enzymes that catalyze reactions of several types (*69–72*). It is of particular interest that L- or D-alanine converted this enzyme's absorption spectrum (*49*) to one including a peak at 500 nm. A similar peak was reported for the complex formed between L-alanine and tryptophanase and interpreted as a quinonoid complex of enzyme-bound pyridoxal phosphate and deprotonated alanine (*73*). In the case of tryptophanase it is notable that only L-alanine, and not D-alanine, was bound to the enzyme and that the enzyme did not act as a racemase even though the α-hydrogen of alanine was labilized (*73*). It would therefore appear that the precise nature of the complex formed, rather than merely the fact of ketimine formation, is important in determining whether racemization will occur. In support of this postulation is the observation that when aspartate β-decarboxylase acts with an *N*-methyl pyridoxal phosphate analog, D-aspartate is both decarboxylated and deaminated, the decarboxylation product being L-alanine (*74*).

B. Nonpyridoxal Enzymes

The indication that hydroxyproline-2-epimerase does not utilize pyridoxal phosphate, and the likelihood that this is true also for proline racemase, raise the question of an alternative mechanism for these enzymes. Because a carbonyl group of another type might replace pyridoxal phosphate without necessarily introducing a basically different reaction mechanism [as may be the case with bacterial histidine decarboxylase and similar enzymes (*67*)] it is important to note that

69. H. C. Dunathan, *Proc. Natl. Acad. Sci. U. S.* **55,** 712 (1966).
70. W. A. Newton, Y. Morino, and E. E. Snell, *JBC* **240,** 1211 (1965).
71. E. W. Miles, M. Hatanaka, and I. P. Crawford, *Biochemistry* **7,** 2742 (1968).
72. S. S. Tate and A. Meister, *Biochemistry* **7,** 3240 (1968).
73. Y. Morino and E. E. Snell, *JBC* **242,** 2800 (1967).
74. S. S. Tate and A. Meister, *Biochemistry* **8,** 1056 (1969).

TABLE III[a]

COMPARISON OF THE RATE OF ISOMER INTERCONVERSION AND DEUTERIUM INCORPORATION DURING INITIAL ACTION OF PROLINE RACEMASE

Substrate[b]	% Product isomer in reaction mixture		% proline-α-^{2}H in total proline[e]
	A[c]	B[d]	
L-Proline	5.0	5.0	5.2
L-Proline	9.8	9.6	9.6
D-Proline	7.2	6.6	7.4
D-Proline	12.2	13.5	13.4

[a] Modified from Table IV of Cardinale and Abeles (*31*).

[b] Refers to isomer initially added to enzyme; samples were taken when the reaction had proceeded to approximately 5 or 10% of racemization.

[c] Determined by optical rotation of reaction mixture.

[d] Determined by optical rotation of 2,4-dinitrophenylproline from reaction mixture.

[e] Calculated from deuterium content of 2,4-dinitrophenylproline.

both hydroxyproline-2-epimerase (*29*) and proline racemase (*31*) are not inhibited by sodium borohydride, making unlikely the participation of prosthetic groups such as pyruvate. A kinetic finding common to both enzymes was the evidence that ^{2}H uptake from ^2H_2O paralleled the initial rate of racemization. These data for proline racemase and hydroxyproline-2-epimerase are shown respectively in Table III and Fig. 1. The

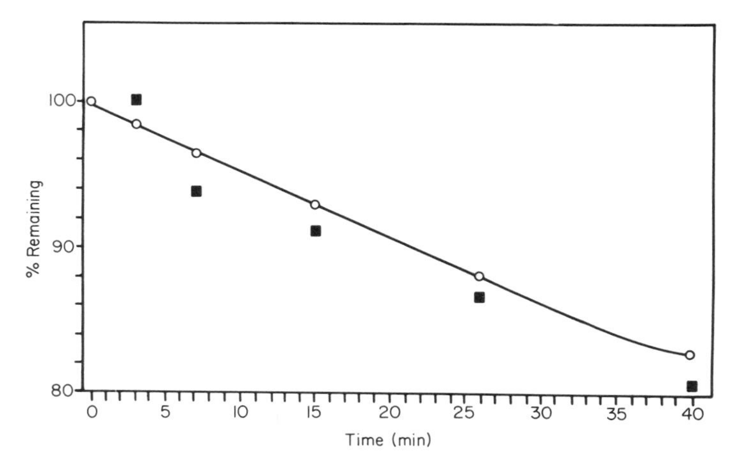

FIG. 1. Enzymic epimerization of allohydroxy-D-proline in D_2O [from Finlay and Adams (*55*)]. Percent of initial allohydroxy-D-proline remaining was calculated from optical rotation; replacement of α-hydrogen ("2-hydrogen") by deuterium was determined by nuclear magnetic resonance spectra of hydroxyproline in samples of the reaction mixture: (■) % 2-hydrogen and (○) % allohydroxy-D-proline.

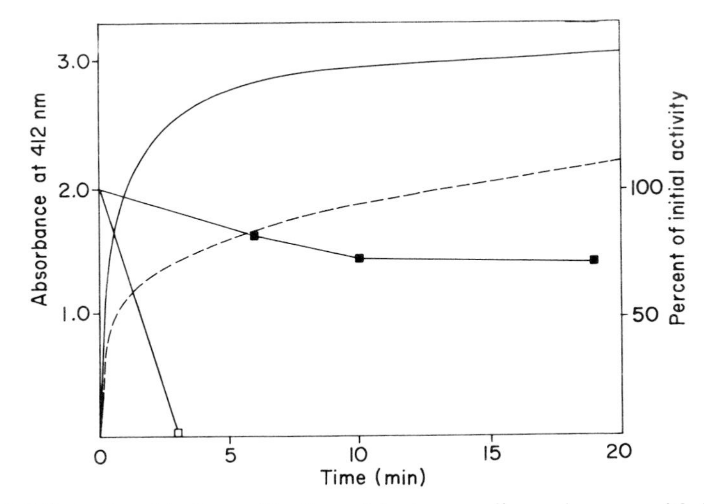

FIG. 2. Effect of susbtrate on titration of hydroxyproline epimerase with Ellman's reagent (*55*). The top curve (——) shows increasing reaction with enzyme thiols (absorbance at 412 nm) in absence of substrate; the lower curve (- - -) shows a similar reaction after addition of allohydroxy-D-proline. Enzymic activity was measured in aliquots removed and diluted at the times shown: (□) substrate absent and (■) substrate present.

interpretation of such data has been discussed by Rose (*75, 76*); the findings cited are consistent with a two-site model in which two groups on the enzyme act as reversible acceptor and donor sites. In the case of hydroxyproline-2-epimerase, 2 cysteines, of the 12 per mole of enzyme, appeared to be required for activity; the presence of substrate prevented reaction of Ellman's reagent with two thiol groups on the enzyme, and at the same time protected the enzyme from the expected inactivation during thiol titration with Ellman's reagent (Fig. 2).

The similarity of hydroxyproline-2-epimerase and proline racemase in apparent dependence on a nonpyridoxal mechanism and in kinetic evidence suggesting a two-site mechanism invites the generalization that racemases for primary amino acids are pyridoxal phosphate enzymes, while racemases for secondary amino acids are not. This postulation is in accord with the lack of known transaminases or decarboxylase for proline and hydroxyproline (*77*) and with the failure to demonstrate nonenzymic transamination of proline with pyridoxal (*78*) or nonenzymic racemization of hydroxyproline with pyridoxal (*29*).

75. I. A. Rose, *Ann. Rev. Biochem.* **35,** 23 (1966).
76. I. A. Rose, "The Enzymes," 3rd ed., Vol. II, p. 281, 1970.
77. E. Adams, *Intern. Rev. Connective Tissue Res.* **5,** 1 (1970).
78. E. E. Snell, *JACS* **67,** 194 (1945).

C. Racemization Associated with Activation as Aminoacyl Complex

A third mechanism of racemization may be represented by the ATP-dependent phenylalanine racemase of *Bacillus brevis* (*36, 37*), an enzyme which appears to be involved in the synthesis of gramicidin S (*36*). While the precise partial reactions involved are not entirely clear, from the recent data of Yamada and Kurahashi (*37*) it seems clear that a single enzyme catalyzes phenylalanine activation and racemization. They demonstrated parallel heat inactivation of the L-phenylalanine-dependent ATP-PP_i exchange, the D-phenylalanine-dependent ATP-PP_i exchange, and phenylalanine racemase activity. Also, there is a strict dependence of phenylalanine racemase activity on ATP and phenylalanine racemization is accompanied by stoichiometric formation of AMP (and presumably PP_i). These findings led to a formulation of the conversion of L-phenylalanine to D-phenylalanine as

$$\text{L-Phenylalanine} + \text{ATP} + \text{enzyme} \leftrightarrows \text{L-phenylalanine} \cdot \text{AMP} \cdot \text{enzyme} + PP_i \quad (2)$$

$$\text{L-Phenylalanine} \cdot \text{AMP} \cdot \text{enzyme} \leftrightarrows \text{D-phenylalanine} \cdot \text{AMP} \cdot \text{enzyme} \quad (3)$$

$$\text{D-Phenylalanine} \cdot \text{AMP} \cdot \text{enzyme} + H_2O \rightarrow \text{D-phenylalanine} + \text{AMP} + \text{enzyme} \quad (4)$$

$$\text{Sum: L-Phenylalanine} + \text{ATP} + H_2O \rightarrow \text{D-phenylalanine} + \text{AMP} + PP_i \quad (5)$$

A similar set of reactions with D-phenylalanine as the initial substrate was implied by the finding that L-phenylalanine was formed from D-phenylalanine, although, as noted in Section IV,B, with an equilibrium constant far from unity. Since the formulation [Eqs. (2)–(4)] could not explain the observed L-phenylalanine-dependent incorporation of ^{14}C-AMP into ATP, a separate reaction was postulated:

$$\text{L-Phenylalanine} \cdot \text{AMP} \cdot \text{enzyme–X} \rightleftharpoons \text{L-phenylalanine–X–enzyme} + \text{AMP} \quad (6)$$

where X could represent a thiol group on the enzyme as suggested by Gevers *et al.* (*79*). The properties of this enzyme may represent an example of general application to D-amino acid synthesis in peptide antibiotics, in which racemization of an enzyme-bound amino acid might proceed without the subsequent release of free D-amino acid. The D-amino acid so formed in normal biosynthesis could be incorporated directly into the growing peptide chain, and this would account for the many

79. W. Gevers, H. Kleinkauf, and F. Lipmann, *Proc. Natl. Acad. Sci. U. S.* **63**, 1335 (1969).

instances in which a free L-amino acid is a better source of the D-amino acid residue in a peptide antibiotic than is the corresponding free D-amino acid [briefly reviewed by Saito *et al.* (*11*)].

The low specific activity of phenylalanine racemization by the *Bacillus brevis* enzyme favors the conception that release of *free* D-phenylalanine may occur at a slow rate because the natural acceptors for the D-phenylalanyl residues in gramicidin synthesis (i.e., the appropriate fragments of the growing peptide chain) are not present. The mechanism of the racemization step itself, however, is not known, nor is it clear whether it occurs while phenylalanine is bound as an enzyme-AMP-aminoacyl residue or to another group on the enzyme, possibly as a thioester (*79*). No evidence for pyridoxal phosphate has been adduced by the use of inhibitors such as phenylhydrazine, isoniazide, hydroxylamine, cysteine, or cyanide (*36*).

VII. Physiological Aspects

As with the other general reactions of amino acids, given racemases serve diverse physiological functions from a step in a biosynthetic pathway or an obligatory preparation for oxidative utilization of a D-amino acid to the synthesis of universal bacterial components such as cell wall peptides or of specialized products such as the peptide antibiotics.

A. Cell Wall Biosynthesis

The synthesis of the bacterial cell wall peptide, or peptidoglycan, is believed to result from a series of steps beginning with UDP-*N*-acetylmuramic acid, in the course of which free amino acids are activated and added as single residues, with the exception that D-alanine is added as a dipeptide unit [reviewed by Perkins (*80*)]. The basic peptide unit common to many bacteria contains a residue of D-glutamate and one of D-alanine (*7*) while *meso*-diaminopimelate is a commonly found residue (*7*). Free D-glutamate appears to be a substrate for direct addition to the cell wall peptide (*81*), while free D-alanine is first condensed to form the D-alanine dipeptide (*82, 83*). The only known source of D-alanine as a substrate for the action of D-alanyl-D-alanine synthetase is

80. H. R. Perkins, *Advan. Pharmacol. Chemother.* **7**, 283 (1969).
81. E. Ito and J. L. Strominger, *JBC* **237**, 2689 (1962).
82. F. C. Neuhaus, *JBC* **237**, 778 (1962).
83. E. Ito and J. L. Strominger, *JBC* **237**, 2696 (1962).

alanine racemase; the distribution of alanine racemase appears to be extremely wide [Table I, and (*13, 61*)] and perhaps universal in bacteria. Glutamate racemase in contrast has not been reported in as many strains, and it appears to be absent from some of the bacilli (*17, 42*) even though D-glutamate is also a component of the basic peptidoglycan subunit. This may be in part the consequence of a less diligent survey of many strains for glutamate racemase and the fact that the assay for this enzyme is not as convenient as the widely employed D-amino acid oxidase-coupled assay for alanine racemase (see Section III,A). It is also the case, however, that an alternative pathway for formation of D-glutamate is known through transamination reactions of α-ketoglutarate with D-alanine or D-aspartate as amino donors (*17, 84*). Data do not appear to be available that would permit a clear decision as to the relative importance of racemases or of D-specific transaminases coupled to a racemase in supplying free D-amino acids for cell wall synthesis. It would be of interest to know how general is the finding of Kuramitsu and Snoke (*42*) for *Bacillus licheniformis* that in the absence of demonstrable racemases for glutamate, aspartate, or phenylalanine their D-antipodes could be made through the combined action of alanine racemase and D-specific transamination.

B. Peptide Antibiotic Biosynthesis

At present there does not appear to be conclusive information on the source of the D-amino acid residues of numerous peptides made by bacteria or of capsular poly-γ-D-glutamate or poly-γ-DL-glutamate in certain bacilli (*9–11*). In a number of studies, both with whole cells and cellfree preparations, L-amino acids served as better precursors of the corresponding D-amino acid residues in peptide antibiotics than did the free D antipodes themselves [summarized by Saito *et al.* (*11*)]. In other cases, there is support for the direct incorporation of free D-amino acids (*11, 63, 85, 86*) with as great, or greater, efficiency than the L antipodes. The ATP-dependent phenylalanine racemase of *Bacillus brevis* (*36, 37*), discussed in Section VI, provides a possible explanation for some of the quantitative differences in various systems between L- and D-amino acid as peptide precursors. In the case of gramicidin S synthesis, the incorporation of

84. R. D. Housewright, *Bacteria* **3**, 389 (1962).
85. A. B. Banerjee and S. K. Bose, *Indian J. Biochem.* **5**, 35 (1968), and references therein.
86. S. Tomino, M. Yamada, H. Itoh, and K. Kurahashi, *Biochemistry* **6**, 2552 (1967).

either L-phenylalanine or D-phenylalanine into the D-phenylalanine residues of the antibiotic (*86*) could be explained by the observed activation of either L- or D-phenylalanine by the enzyme; the preferential or exclusive incorporation of an L-amino acid into a D residue in other cases could be explained simply by the relative affinities of L or D antipodes for an activating enzyme which, in analogy with the *Bacillus brevis* enzyme, acts also as a racemase for its bound amino acid. The specificity of insertion of a D residue in the proper position of the growing peptide is presumably guaranteed by absolute stereospecificity of the transfer reaction to the appropriate acceptor fragment.

C. Biosynthesis and Degradation of Free Amino Acids

1. *Lysine Biosynthesis*

The single example of amino acid racemization as a step in a biosynthetic pathway is the reaction catalyzed by 2,6-diaminopimelate epimerase, a presumably obligatory step in lysine biosynthesis (*87*). While the relationship of this enzme to lysine biosynthesis does not appear to have been studied directly in terms of epimerase-deficient mutants or end product regulatory phenomena, there is little reason to doubt its role; thus, the action of *N*-succinyl-2,6-diaminopimelate deacylase would yield the L isomer (*88*), while 2,6-diaminopimelate decarboxylase [whose biosynthetic relationship to lysine has been shown both by mutant (*89, 90*) and repression (*91*) studies] is specific for the *meso* isomer (*92, 93*). The epimerase reaction would seem to represent an essential bridge between these two epimers. In addition to its status as an intermediate in L-lysine synthesis, *meso*-2,6-diaminopimelate is directly incorporated into the cell wall peptide in many bacteria (*7*), a further example of activation and insertion of a D-amino acid into these structures.

2. *Amino Acid Degradation*

Two enzymes, proline racemase and hydroxyproline-2-epimerase, have the unusual metabolic role of converting an L-amino acid to its D antipode

87. C. Gilvarg, *Federation Proc.* **19,** 948 (1960).
88. S. H. Kindler and C. Gilvarg, *JBC* **235,** 3532 (1960).
89. D. L. Dewey and E. Work, *Nature* **169,** 533 (1952).
90. B. D. Davis, *Nature* **169,** 534 (1952).
91. J. C. Patte, T. Loviny, and G. N. Cohen, *BBA* **58,** 359 (1962).
92. D. S. Hoare and E. Work, *BJ* **61,** 562 (1955).
93. P. J. White and B. Kelly, *BJ* **96,** 75 (1965).

or D epimer so that the latter can be utilized for further energy-yielding degradative steps. In the case of proline utilization by *Clostridium sticklandii*, D-proline formed from L-proline is the selective substrate for reduction to δ-aminovalerate (*24*); presumably, absence of the racemase would prevent these cells from utilizing L-proline as an energy source. In the case of the induced pathway for hydroxyproline oxidation by *Pseudomonas putida*, a similar conclusion was drawn on the basis that allohydroxy-D-proline, but not hydroxy-L-proline, could be oxidized by an induced particulate enzyme (*94*). A more conclusive demonstration of the dependence of cells on the epimerase for utilization of hydroxy-L-proline (by far the most common epimer in nature) came from observations with a mutant lacking this enzyme. In the absence of hydroxyproline-2-epimerase, cells could grow normally on allohydroxy-D-proline, while hydroxy-L-proline not only failed to support growth but also prevented growth on the D epimer by inhibiting entry of the latter into the cell (*95*).

In the case of the methionine racemase and lysine racemase of *Pseudomonas* strains (Table I) the further reactions outlined in these strains (*20, 58*) indicate oxidation or other degradative reactions selective for the L isomer; thus, racemization can be viewed as an additional mechanism for utilizing the D isomer through these pathways. The lysine racemase was reported not to be induced by growth on lysine (*58*), while the methionine racemase was detected in cell extracts only after growth on D- or DL-methionine (*20*). The arginine racemase of *Pseudomonas graveolens* is apparently induced by arginine (Table I, footnote c), but it is not clear whether L- or D-arginine, or both, are substrates for further reactions.

While the role of alanine racemase is generally understood to be the provision of D-alanine for cell wall peptide synthesis (see Section VII,A), its induced level in *Pseudomonas* grown on alanine as a sole carbon-nitrogen source (*33*) is obviously related to energy-producing reactions. The high level of alanine racemase induced by growth on alanine suggested the possibility of racemization as an obligatory step for alanine utilization (*33*). Consistent with this possibility was the effect of 10^{-4} *M* aminooxyacetate on alanine-induced cells or homogenates obtained from them: While uninhibited preparations consumed oxygen at equal rates with L- or with D-alanine as substrate, aminooxyacetate markedly inhibited oxidation of L-alanine but not D-alanine. A possible interpretation is based on the action of aminooxyacetate as an inhibitor of alanine

94. T. Yoneya and E. Adams, *JBC* **236**, 3272 (1961).
95. R. M. Gryder and E. Adams, *J. Bacteriol.* **97**, 292 (1969).

racemase (*32*). Induced oxidation of D-amino acids, including alanine, by preparations from various *Pseudomonas strains* has been reported previously (*94, 96–98*). It therefore seems possible, although not yet proven, that in this case also the induced racemase serves to permit utilization of L-alanine through a D-specific oxidative pathway.

D. Alanine Racemase and Spore Germination

That alanine racemase may participate in the regulation of spore germination was first suggested by reports that L-alanine was a good initiator and D-alanine a good inhibitor of this process in a number of bacterial spore types [reviewed by Gould (*99*)]. The presence of alanine racemase both in spores and vegetative cells of various *Bacillus* species was reported in one of the earlier studies of this enzyme (*15, 40, 41*). An investigation of the possibility that alanine racemase was itself required for germination (*41*) indicated no apparent relationship between the activity of the enzyme and the germination process; the enzyme was not demonstrable in spores of *Bacillus globigii* although L-alanine stimulated germination. Furthermore, in spores of *Bacillus terminalis,* extensive spore germination occurred at pH 11.3, at which the racemase was inactive. Alanine racemase, however, while not essential for germination, is influential in this process, as indicated by the studies of Jones and Gould (*100*). A number of inhibitors of the alanine racemase of *Bacillus cereus* stimulated germination induced by L-alanine but inhibited germination induced by D-alanine plus adenosine.

E. Source of D-Amino Acids in Animal Tissues

In recent years several D-amino acids, either free or found as portions of other structures such as lombricine or octopine, have been identified in a number of invertebrates; this subject has been reviewed recently (*101*). While evidence for direct enzymic racemization of an amino acid in animals is yet lacking, D-serine ethanolamine phosphate, a precursor of lombricine, has been shown to arise in earthworm tissue minces from

96. J. E. Norton, G. S. Bulmer, and J. R. Sokatch, *BBA* **78,** 136 (1963).
97. K. Tsukada, *JBC* **241,** 4522 (1966).
98. V. P. Marshall and J. R. Sokatch, *J. Bacteriol.* **95,** 1419 (1968).
99. G. W. Gould, *in* "The Bacterial Spore" (G. W. Gould and A. Hurst, eds.), p. 397. Academic Press, New York, 1969.
100. A. Jones and G. W. Gould, *J. Gen. Microbiol.* **53,** 383 (1968).
101. J. J. Corrigan, *Science* **164,** 142 (1969).

free D-serine (*102*) leaving open the possibility that the latter arises by racemization of L-serine. Homogenates of silkworm prepupae or larvae were also shown to catalyze slow formation (up to 5 nmoles/mg protein/hr) of D-serine from L-serine and somewhat slower formation of L-serine from D-serine (*103*).

102. A. K. Allen and H. Rosenberg, *BBA* **152,** 208 (1968).
103. N. G. Srinivasan, J. J. Corrigan, and A. Meister, *JBC* **240,** 796 (1965).

14

Coenzyme B_{12}-Dependent Mutases Causing Carbon Chain Rearrangements

H. A. BARKER

I. Introduction

Three B_{12} coenzyme-dependent enzymes catalyzing carbon skeleton rearrangements are presently known: glutamate mutase (methylaspartate mutase; *threo*-3-methyl-L-aspartate carboxy-aminomethylmutase, EC 5.4.99.1), methylmalonyl-CoA mutase (2-methylmalonyl-CoA CoA-carbonylmutase; 5.4.99.2), and methyleneglutarate mutase (2-methylene glutarate carboxy-methylene methyl mutase; 5.4.99.x). The reactions catalyzed by these enzymes are similar in that they all involve rearrangements [Eq. (1)] in which a substituent group is moved between the α and β positions of a propionate residue while a hydrogen atom is moved in the opposite direction.

$$\underset{\displaystyle R}{\underset{|}{CH_2}}CH_2COO^{\ominus} \rightleftharpoons CH_3\underset{\displaystyle R}{\underset{|}{CH}}\cdot COO^{\ominus} \quad (1)$$

Methylmalonyl-CoA mutase, the first enzyme of this group to be recognized, was discovered during studies of the oxidation of propionate by extracts from animal tissues. Flavin *et al.* (*1*) and Katz and Chaikoff (*2*) independently observed the formation of both succinate and methylmalonate during the utilization of propionate in the presence of bicarbonate. The conversion of propionate to succinate was soon shown (*3–5*) to involve propionyl-CoA, methylmalonyl-CoA, and succinyl-CoA as intermediates. The enzyme catalyzing the interconversion of methylmalonyl-CoA and succinyl-CoA was partially purified from sheep kidney cortex and called methylmalonyl-CoA isomerase (later changed to mutase). No cofactor requirement for the enzyme was detected at that time.

The glutamate mutase reaction was discovered during studies of the path of glutamate fermentation by *Clostridium tetanomorphum* (*6–8*). Extracts of this bacterium were found to convert glutamate reversibly to ammonia and mesaconate (2-methylfumaric acid), a 5-carbon, branched chain compound. When charcoal-treated extracts were used the formation of mesaconate from glutamate was prevented, whereas mesaconate was still readily aminated to form the corresponding branched chain amino acid, β-methylaspartate. The reversible conversion of this compound to glutamate, catalyzed by glutamate mutase, was found to require a charcoal adsorbable coenzyme that was isolated and found to be a derivative of pseudovitamin B_{12} (*9*). The corresponding derivative of

1. M. Flavin, P. J. Ortiz, and S. Ochoa, *Nature* **176**, 823 (1955).
2. J. Katz and I. L. Chaikoff, *JACS* **77**, 2659 (1955).
3. M. Flavin and S. Ochoa, *JBC* **229**, 965 (1957).
4. W. S. Beck, M. Flavin, and S. Ochoa, *JBC* **229**, 997 (1957).
5. W. S. Beck and S. Ochoa, *JBC* **232**, 931 (1958).
6. H. A. Barker, "Bacterial Fermentations," Chapter 3. Wiley, New York, 1956.
7. H. A. Barker, *Bacteria* **2**, 151 (1961).
8. H. A. Barker, *Federation Proc.* **20**, 956 (1961).
9. H. A. Barker, H. Weissbach, and R. D. Smyth, *Proc. Natl. Acad. Sci. U. S.* **44**, 1093 (1958).

vitamin B_{12} was prepared and found to be active both with glutamate mutase (*10*) and with methylmalonyl mutase from animal (*11–13*) and bacterial sources (*14*). The coenzyme was later shown to be the Co-5′-deoxyadenosyl derivative of the vitamin (*15*).

The third enzyme of this group, α-methyleneglutarate mutase, was recently detected (*16*) in extracts of a *Clostridium* that ferments nicotinic acid (*17*). It catalyzes the reversible conversion of methyleneglutarate to methylitaconate.

II. Methylmalonyl-CoA Mutase

This enzyme catalyzes the reversible conversion of (*R*)-methylmalonyl-CoA (*18–20a*) to succinyl-CoA [reaction (2)].

$$CH_3\text{--}C(COSCoA)(COO^{\ominus})\text{--}H \rightleftharpoons {}^{\ominus}OOCCH_2CH_2COSCoA \qquad (2)$$

(*R*)-Methylmalonyl-CoA   Succinyl-CoA

The enzyme is known to participate in the oxidation of propionate, methylmalonate, and compounds such as isoleucine and valine that are converted to propionyl-CoA or methylmalonyl-CoA during degradation by animal tissues or microorganisms. The oxidation of propionate involves its conversion to propionyl-CoA and the carboxylation of the latter in a ATP-dependent reaction to form (*S*)-methylmalonyl-CoA

10. H. Weissbach, J. Toohey, and H. A. Barker, *Proc. Natl. Acad. Sci. U. S.* **45,** 521 (1959).
11. R. M. Smith and K. J. Monty, *BBRC* **1,** 105 (1959).
12. J. R. Stern and D. L. Friedman, *BBRC* **2,** 82 (1960).
13. S. Gurnani, S. P. Mistry, and B. C. Johnson, *BBA* **38,** 187 (1960).
14. E. R. Stadtman, P. Overath, H. Eggerer, and F. Lynen, *BBRC* **2,** 1 (1960).
15. P. G. Lenhert and D. C. Hodgkin, *Nature* **192,** 937 (1961).
16. H. F. Kung, S. Cederbaum, L. Tsai, and T. C. Stadtman, *Proc. Natl. Acad. Sci. U. S.* **65,** 978 (1970).
17. E. R. Stadtman, T. C. Stadtman, I. Pastan, and L. DS. Smith, *J. Bacteriol.* (1972) (in press).
18. J. Retey and F. Lynen, *BBRC* **16,** 358 (1964).
19. M. Sprecher, M. J. Clark, and D. B. Sprinson, *JBC* **241,** 872 (1966).
20. M. Sprecher, M. J. Clark, and D. D. Sprinson, *BBRC* **15,** 581 (1964).

20a. Before the absolute configurations of isomeric forms of methylmalonyl-CoA were known, the (*S*) and (*R*) forms were referred to as the *a* and *b* forms.

(*21*). This compound, which is not a substrate for the mutase, must be converted to the (*R*) isomer before it can be further utilized. This conversion is catalyzed by a specific methylmalonyl-CoA racemase which appears to be present in all tissues and organisms containing the mutase (*22*). In the degradation of valine, methylmalonyl-CoA is formed by the DPN- and CoA-dependent oxidation of methylmalonic semi-aldehyde, but the configuration of the initial product has not been established.

In the synthesis of propionate by propionibacteria the mutase reaction functions in the reverse direction from succinyl-CoA to (*R*)-methyl-malonyl CoA. A methylmalonyl-CoA racemase is also required in propionate synthesis to form (*S*)-methylmalonyl-CoA which then reacts with pyruvate in a transcarboxylation reaction to give propionyl-CoA and oxalacetate.

A. Assays

The two most convenient and commonly used assays will be outlined. Mazumder *et al.* (*22, 23*) described an assay applicable to crude extracts of animal tissues and other preparations containing holoenzyme of relatively low specific activity. The method uses propionyl-CoA, ATP, and ^{14}C-labeled bicarbonate as substrates, and involves the following reactions catalyzed by propionyl-CoA carboxylase, methylmalonyl-CoA racemase, and the mutase:

$$\text{Propionyl-CoA} + HCO_3^{\ominus} + \text{ATP} \rightleftharpoons (S)\text{-methylmalonyl-CoA} + \text{ADP} + P_i \quad (3)$$

$$(S)\text{-Methylmalonyl-CoA} \rightleftharpoons (R)\text{-methylmalonyl-CoA} \quad (4)$$

$$(R)\text{-Methylmalonyl-CoA} \rightleftharpoons \text{succinyl-CoA} \quad (5)$$

When purified carboxylase and racemase are added in excess, the amount of succinyl-CoA formed is proportional to the concentration of mutase. Succinyl-CoA and succinate formed from succinyl-CoA are estimated by measuring the radioactivity remaining after the deproteinized reaction mixture is heated with acid permanganate to destroy methylmalonate. Succinate is stable to this treatment. In this method, B_{12} coenzyme is not added since the mutase in animal tissues is normally present as holoenzyme that does not readily dissociate.

21. J. Retey and F. Lynen, *Biochem. Z.* **342,** 256 (1965).
22. R. Mazumder, T. Sasakawa, Y. Kaziro, and S. Ochoa, *JBC* **237,** 3065 (1962).
23. R. Mazumder and S. Ochoa, "Methods in Enzymology," Vol. 13, p. 198, 1969.

Another assay method (*24, 25*) applicable to mutase preparations of higher specific activity involves the following reactions catalyzed by mutase, methylmalonyl-CoA racemase, oxalacetate transcarboxylase, and malate dehydrogenase, respectively.

$$\text{Succinyl-CoA} \xrightleftharpoons{B_{12}\text{ coenzyme}} (R)\text{-methylmalonyl-CoA} \quad (6)$$

$$(R)\text{-Methylmalonyl-CoA} \rightleftharpoons (S)\text{-methylmalonyl-CoA} \quad (7)$$

$$(S)\text{-Methylmalonyl-CoA} + \text{pyruvate} \rightleftharpoons \text{propionyl-CoA} + \text{oxalacetate} \quad (8)$$

$$\text{Oxalacetate} + \text{DPNH} \rightleftharpoons \text{malate} + \text{DPN} \quad (9)$$

The rate of DPNH oxidation, estimated spectrophotometrically, provides a measure of the rate of the mutase reaction when the other enzymes and B_{12} coenzyme are present in excess.

Other assay methods for the mutase have been described (*4, 14*).

B. Distribution

The mutase has been found in several animals. Acetone powders from various tissues of rat, ox, rabbit, and sheep contain appreciable activity, whereas pig heart shows a scarcely detectable level (*4*). In sheep, the kidney cortex and liver have higher levels of mutase than brain, skeletal muscle, heart, pancreas, and intestinal mucosa. In rats, the level in the heart is twice that in the liver. However, insufficient systematic data are available to justify firm conclusions concerning the relative levels in different tissues. No data are available on methylmalonyl-CoA mutase levels in human tissues.

In microorganisms, the mutase has been demonstrated in *Propionibacterium shermanii* (*26*), *Micrococcus lactilyticus* (*27*), and several *Rhizobium* species (*28*). Less direct evidence indicates the enzyme is present in propionate-adapted cells of *Ochromonas malhamensis* (*29*), *Micrococcus denitrificans* (*30*), *Rhodospirillum rubrum* (*31*), and in valine-grown cells of *Pseudomonas aeruginosa* (*32*).

24. R. W. Kellermeyer, S. H. G. Allen, R. Stjernholm, and H. G. Wood, *JBC* **239,** 2562 (1964).
25. R. W. Kellermeyer and H. G. Wood, "Methods in Enzymology," Vol. 13, p. 207, 1969.
26. R. Stjernholm and H. G. Wood, *Proc. Natl. Acad. Sci. U. S.* **47,** 303 (1961).
27. J. H. Galivan and S. H. G. Allen, *JBC* **243,** 1253 (1968).
28. A. A. De Hertogh, P. A. Mayeux, and H. J. Evans, *JBC* **239,** 2446 (1964).
29. H. R. V. Arnstein and A. M. White, *BJ* **83,** 264 (1962).
30. J. Smith and H. L. Kornberg, *J. Gen. Microbiol.* **47,** 175 (1967).
31. M. Knight, *BJ* **84,** 170 (1962).
32. J. R. Sokatch, L. E. Sanders, and V. P. Marshall, *JBC* **243,** 2500 (1968).

C. Purification and Molecular Properties

1. *Mutase from Propionibacterium shermanii*

The only bacterial methylmalonyl-CoA mutase that has been extensively purified and investigated in some detail is that obtained from *P. shermanii*. The specific activity of the enzyme in crude extracts of this organism (0.01–0.1 unit/mg of protein) is 10–100 times higher than in sheep liver extracts (0.001 unit/mg), the best animal source of the mutase.

The enzyme has been highly purified by Kellermeyer *et al.* (*24*) and by Overath *et al.* (*33, 34*). The most active preparations, which contain little or no racemase activity, have specific activities between 7.3 and 14.4 units/mg. The variation in specific activity appears to be caused at least in part by differences in the amount of inactive corrinoid compound bound to the enzyme. The highest specific activity preparation reported was virtually colorless, and it was estimated that less than 2% of the enzyme molecules contained inactive corrinoids. In a less active but relatively homogeneous pink preparation of the mutase about 38% of the molecules contained inactive corrinoids. The absorption spectrum of the pink preparation in the visible region shows a peak at about 352 nm and some absorption at 490 and 530 nm which is characteristic of hydroxocobalamin. This compound is readily formed by photolytic decomposition of B_{12} coenzyme in bacterial extracts and binds strongly to protein because of its positive charge in neutral solution. The binding of hydroxocobalamin to the mutase probably blocks the site normally occupied by the coenzyme and consequently inactivates the enzyme. Some purified mutase preparations also contain a very small amount of tightly bound B_{12} coenzyme (*33*) which is not removed by treatment with charcoal and which is relatively stable to light. In these respects, a small amount of bound coenzyme in bacterial mutase behaves like the bulk of the bound coenzyme in animal mutase.

The molecular weight of the mutase, determined by the sedimentation equilibrium method, was estimated to be 56,000. The sedimentation coefficient $s_{20,w}$ was reported to be 7.0 (*24*) and 7.8 (*33*). No information on the subunit structure of bacterial mutase is available.

The bacterial mutase is relatively stable at low temperatures. The specific activity of a colorless mutase preparation, after precipitation

33. P. Overath, E. R. Stadtman, G. M. Kellerman, and F. Lynen, *Biochem. Z.* **336,** 77 (1962).

34. P. Overath, G. M. Kellerman, F. Lynen, H. P. Fritz, and H. J. Keller, *Biochem. Z.* **335,** 500 (1962).

with ammonium sulfate, declined from 14.8 to 8.8 during storage for 1 year at —10°.

Dilute neutral solutions of the enzyme lost 27% of the activity after 10 min at 42° and became completely inactive after 10 min at 55°.

The activity of the mutase is not increased by exposure to reduced glutathione and is only partially inhibited by incubation with relatively high levels of *p*-hydroxymercuribenzoate or *N*-ethylmaleimide (*24*). Consequently, sulfhydryl groups appear not to be essential for activity. In contrast the animal apoenzyme is strongly inhibited by –SH binding reagents (see below).

2. *Animal Mutase*

a. Holoenzyme Isolation and Properties. Unlike the bacterial mutase, the enzyme from sheep liver or kidney binds cobamide coenzyme very tightly and most of the coenzyme remains attached to the enzyme during isolation. Consequently, a pink holoenzyme is the main product of isolation.

The holoenzyme from sheep liver has been most extensively purified (*35, 36*). The best preparations have a specific activity of 5.5–7.2 units/mg of protein which is similar to that of the bacterial mutase. These preparations appear to be relatively homogeneous as judged by sedimentation velocity measurements ($s_{20,w} = 7.7$ S). The molecular weight of the holoenzyme, estimated by sedimentation equilibrium measurements, is about 165,000. Such holoenzyme preparations always contain some apoenzyme as judged by a small increase (about 12%) in specific activity when assayed in the presence of a saturating level of B_{12} coenzyme.

The pink holoenzyme absorbs light in the visible region up to about 600 nm with a small absorbance maximum at 520 nm (*35*). The coenzyme content of holoenzyme can be estimated approximately from the absorbance at 520 nm on the assumption that this is caused entirely by the presence of DBC coenzyme, and no change in its absorbance occurs on binding to protein. On this basis pure holoenzyme was estimated to contain 1 mole of coenzyme per 75,000 g of protein or about 2 moles of coenzymes per mole of holoenzyme (*36*). Coenzyme is not removed from holoenzyme by treatment with charcoal, and illumination of the enzyme solution does not reduce its specific activity. Since free coenzyme is readily inactivated by light, it is apparent that the coenzyme

35. R. Mazumder, T. Sasakawa, and S. Ochoa, *JBC* **238,** 50 (1963).

36. J. J. B. Cannata, A. Focesi, R. Mazumder, R. C. Warner, and S. Ochoa, *JBC* **240,** 3249 (1965).

is bound to the protein in such a way as to prevent the photolytic reaction.

The identity of the corrinoid compound in the holoenzyme has not been established. The presence of an absorbance peak at about 520 nm and the absence of a peak at 350–355 suggests that the prosthetic group is a benzimidazole-containing cobamide coenzyme, probably cobalamin coenzyme. However, the spectrum of the holoenzyme is not sufficiently detailed to permit identification of the corrinoid compound and, because of the strong binding of the prosthetic group to the mutase, it has not yet been possible to isolate and characterize the compound. All attempts to separate the prosthetic group have been unsuccessful and resulted in its modification as indicated by a change in the absorption spectrum of the enzyme (see below).

b. Apoenzyme. Apoenzyme has been prepared from holoenzyme by brief exposure of the latter to pH 3.5 in the presence of ammonium sulfate (*36, 37*). The enzyme is partially inactivated (21%) by this procedure, but the inactive protein can be separated from the enzyme and the final product has been found to have a higher specific activity (7.4 units/mg) in the presence of excess coenzyme than the starting material (5.7 units/mg) although the yield of activity is only about 37% of that of the starting holoenzyme. Apoenzyme prepared in this way still contains a considerable amount of holoenzyme. In one preparation, the activity in the absence of added coenzyme was 16% of that with excess coenzyme.

The conversion of holoenzyme to apoenzyme by acid ammonium sulfate treatment does not remove bound coenzyme but converts it to an inactive form. This is evident from the fact that the apoenzyme retains the pink color of the holoenzyme but the absorption spectrum is altered to resemble photoinactivated cobalamin coenzyme with absorbance maxima at 354, 405, 505, and 535 nm (*36*). The reason for this transformation is not clear since it occurs under conditions in which free corrinoid coenzyme is stable. It is noteworthy that the presence of inactivated coenzyme apparently does not prevent access of added coenzyme to its normal binding site; specific activity of the apoenzyme when assayed in the presence of excess coenzyme is as high as that of the holoenzyme. In this respect the sheep liver apoenzyme differs from the *P. shermanii* apoenzyme which appears to be partially inactivated by tightly bound hydroxocorrinoid compounds.

The apoenzyme is considerably less stable than the holoenzyme par-

37. P. Lengyel, R. Mazumder, and S. Ochoa, *Proc. Natl. Acad. Sci. U. S.* **46**, 1312 (1960).

ticularly in solutions of low ionic strength. It can be partially stabilized by addition of glutathione or DBC coenzyme. When the holoenzyme is reconstituted from apoenzyme and DBC coenzyme, the bound coenzyme becomes relatively resistant to inactivation by light. This indicates that coenzyme is bound in reconstituted holoenzyme in the same way as in native holoenzyme.

The apoenzyme is much more sensitive than the holoenzyme to –SH binding reagents such as *p*-hydroxymercuribenzoate, iodoacetamide, and *N*-ethylmaleimide (*36*). However, prior addition of DBC coenzyme markedly protects the apoenzyme against these reagents so that it is no more inhibited than the native holoenzyme. These results indicate that resolution of the holoenzyme exposes essential sulfhydryl groups and suggests that they may be involved in coenzyme binding. There is some evidence that the coenzyme is also attached to one or more additional sites on the apoenzyme.

The molecular weight of the apoenzyme is similar to that of the holoenzyme. Although a precise value is not available, it is clear that conversion of holoenzyme to apoenzyme does not cause a large change in molecular weight.

D. Catalytic Properties

1. *Substrates*

(*R*)-Methylmalonyl-CoA and succinyl-CoA are the only known substrates of the sheep liver and *P. shermanii* mutase (*24, 34, 36*). (*S*)-Methylmalonyl-CoA is not used by the bacterial enzyme and is inhibitory to the animal enzyme. Normal Michaelis–Menten kinetics are observed with the substrates. Both methylmalonyl-CoA and succinyl-CoA have relatively low apparent K_m values (Table I) characteristic of most thioesters of coenzyme A. The bacterial mutase has a higher affinity for its substrates than the animal mutase.

2. *Coenzymes*

The *P. shermanii* mutase is isolated mainly as the apoenzyme which has only slight activity in the absence of added coenzyme. The bulk of the coenzyme is readily removed from the enzyme during isolation by adsorption on charcoal although some coenzyme often remains with the enzyme even after several treatments with charcoal.

The most abundant corrinoid coenzyme in *P. shermanii* is the Co-5′-deoxyadenosyl derivative of cobalamin (DBC coenzyme and B_{12} co-

TABLE I
SUBSTRATE AND COENZYME AFFINITIES OF SHEEP LIVER AND *P. shermanii* METHYLMALONYL MUTASES

Substrate or coenzyme[a]	Sheep liver enzyme: Apparent K_m (μM)	Sheep liver enzyme: Ref.	Bacterial enzyme: Apparent K_m (μM)	Bacterial enzyme: Ref.
(R)-Methylmalonyl-CoA	240	(*36*)	80	(*24*)
Succinyl-CoA	62	(*36*)	34.5	(*24*)
DBC coenzyme	0.021	(*37*)	0.021–0.035	(*26, 33, 37*)
BC coenzyme	0.20	(*37*)	0.13–0.34	(*24, 33, 37*)
AC coenzyme	Inactive		0.10–0.11	(*33, 37*)
PC coenzyme			1.4	(*33*)

[a] Abbreviations for B_{12} coenzymes: DBC, dimethylbenzimidazolyl cobamide; BC, benzimidazolyl cobamide; AC, adenyl cobamide; PC, purine cobamide.

enzyme) which contains 5,6-dimethylbenzimidazole in the nucleotide side chain, and this coenzyme has the highest affinity for the mutase (Table I). The benzimidazolyl cobamide (BC) coenzyme which differs from the DBC coenzyme by the absence of two methyl groups has a somewhat lower affinity for the enzyme as does the adenyl cobamide (AC) coenzyme which contains adenine in place of benzimidazole in the side chain. The purine cobinamide (PC) coenzyme, which differs from the AC coenzyme by the absence of the amino group of adenine, has a still lower affinity for the mutase. Relative V_{max} values for these coenzymes have not been determined with this enzyme. The cobinamide coenzyme ("conjugate"), which differs from DBC coenzyme by lacking the entire nucleotide moiety, was reported (*33*) to be active for the *P. shermanii* mutase with a K_m of 1.7×10^{-6} M. However, this result could not be confirmed (*24*), and it is therefore probable that the apparent activity with this "incomplete" corrinoid coenzyme was caused by contamination with DBC coenzyme which has a much higher affinity for the enzyme (2.4×10^{-8} M). Corrinoid compounds lacking a Co-5′-deoxyadenosyl group in the sixth coordination position, such as cyanocobalamin and other B_{12} "vitamins," show no coenzyme activity (*33*). The DBC coenzyme analog containing 5′-deoxyinosine in place of 5′-deoxyadenosine is also inactive.

The conclusions derived from these observations are that (1) only "complete" corrinoid coenzymes, i.e., those containing a nucleotide side chain, are active; (2) either a purine or benzimidazole base on the nucleotide side chain is permissible and various substituent groups are tolerated; and (3) the deoxyadenosyl group attached to cobalt is essential for activity and a comparatively minor modification in this structure, such as substitution of an amino by a hydroxyl group, destroys the activity of the coenzyme.

The cobamide coenzymes containing a purine moiety in the nucleotide side chain differ from those containing a benzimidazole moiety in the affinity of the base for the cobalt atom. In the purine cobamide coenzymes the base is weakly bound to cobalt whereas in the benzimidazole cobamide coenzymes the base is tightly bound (*38*). Since both types are active and do not differ greatly in their affinities for the enzyme, it appears that this structural feature of the coenzymes is not important with the bacterial mutase.

In contrast to the bacterial mutase, the sheep liver enzyme requires the more highly structured benzimidazole-containing cobamide coenzymes; the AC coenzyme is entirely inactive (Table I). The animal mutase, like the bacterial mutase, binds the DBC coenzyme more strongly than the BC coenzyme.

3. *Effect of pH on Activity*

A typical bell-shaped pH-activity curve is obtained with the *P. shermanii* mutase (*24*). Detectable activity extends between about pH 4.0 and 9.0 with a relatively flat maximum between pH 6.0 and 8.0 (*33*).

The sheep liver mutase is active between pH 5.0 and 10.0 with a rather sharp maximum at pH 7.0 (*36*).

4. *Equilibrium of the Methylmalonyl Mutase Reaction*

With racemase-free mutase preparations, the equilibrium constant, K = (succ-CoA)/(MM-CoA), for the conversion of (R)-methylmalonyl-CoA to succinyl-CoA was estimated to be 23.1 at pH 7.0 and 25° (*24*). This corresponds to a $\Delta F_0'$ of −1.86 kcal. A somewhat lower value for the equilibrium constant of 18.6 was obtained by Cannata *et al.* (*36*). Under *in vivo* conditions in the presence of racemase, the apparent equilibrium constant based upon the total concentration of methylmalonyl-CoA would be about 11.5.

5. *Mechanism of the Reaction*

The chemistry of the methylmalonyl mutase reaction has been investigated extensively. Early experiments with relatively crude mutase preparations containing both propionyl-CoA carboxylase and methylmalonyl-CoA racemase activities suggested that the reaction involves

38. J. N. Ladd, H. P. C. Hogenkamp, and H. A. Barker, *JBC* **236**, 2114 (1961).

an intermolecular transcarboxylation (*5*). This hypothesis was disproved by Eggerer *et al.* (*39*) and Swick (*40*) who showed that the carbonyl thioester group is actually moved between the α- and β-carbon atoms of the propionate moiety [Eq. (1)]. Eggerer *et al.* used 2-^{14}C-methylmalonyl-CoA as a substrate for mutase and determined the position of the isotope in the resulting succinyl-CoA by suitable degradation reactions. About 80% of the ^{14}C was found in the carbon atom adjacent to the carboxyl group [Eq. (10)].

$$\begin{array}{c} HOOC-{}^{*}CH-CoSCoA \\ | \\ CH_3 \end{array} \longrightarrow \begin{array}{c} HOOC-{}^{*}CH_2 \\ | \\ CH_2COSCoA \end{array} \qquad (10)$$

This result could be accounted for most readily by transfer of the carbonyl thioester to the methyl group of methylmalonyl-CoA. The relatively small fraction (20%) of the isotope in the other methylene carbon was attributed to deacylation of part of the succinyl-CoA and CoA transfer to the resulting succinate. This would result in randomization of the isotope between the two methylene carbon atoms of succinyl-CoA. Coenzyme A migration between the carboxyl groups of succinate or methylmalonate has been excluded as a significant component of the mutase reaction by Hegre *et al.* (*41*). By the use of methylmalonyl-CoA labeled with ^{14}C in either the carbonyl or the thioester group, they showed that the coenzyme A moiety in the succinyl-CoA formed in the mutase reaction is attached to the same carbon atom as in the substrate.

The mutase reaction could theoretically involve either an intramolecular or an intermolecular transfer of the carboxyl thioester (*42*). The occurrence of an intermolecular transfer has been excluded by experiments with ^{13}C-methylmalonyl-CoA (*43–45*). In one experiment a mixture of carboxyl-labeled and carbonyl thioester-labeled methylmalonyl-CoA was used. An intramolecular transfer would yield a product still labeled in only one position. An intermolecular transfer would yield both unlabeled and doubly labeled product. Mass analysis of the succinate derived from the resulting succinyl-CoA showed no change in the distribu-

39. H. Eggerer, E. R. Stadtman, P. Overath, and F. Lynen, *Biochem. Z.* **333,** 1 (1960).

40. R. W. Swick, *Proc. Natl. Acad. Sci. U. S.* **48,** 288 (1962).

41. C. S. Hegre, S. J. Miller, and M. D. Lane, *BBA* **56,** 538 (1962).

42. H. G. Wood, R. W. Kellermeyer, R. Stjernholm, and S. H. G. Allen, *Ann. N. Y. Acad. Sci.* **112,** 661 (1964).

43. R. W. Kellermeyer and H. G. Wood, *Biochemistry* **1,** 1124 (1962).

44. E. F. Phares, M. V. Long, and S. F. Carson, *BBRC* **8,** 142 (1962).

45. E. F. Phares, M. V. Long, and S. F. Carson, *Ann. N. Y. Acad. Sci.* **112,** 680 (1964).

tion of ^{13}C during the reaction, thus proving that the mutase reaction involves an intramolecular transfer of the carbonyl thioester group.

The mutase reaction involves a transfer of a hydrogen atom between the α- and β-carbon atoms of the propionate moiety as well as a transfer of the carbonyl thioester group in the opposite direction. The hydrogen might be transferred as a proton, a hydrogen atom, or a hydride ion. The participation of a proton has been investigated by carrying out the mutase reaction in the presence of ^3H_2O. With racemase-containing mutase, tritium is rapidly incorporated into the product (*46*). However, when racemase-free mutase is used, the product (succinyl-CoA) contains very little tritium, less than 0.01 atom/mole (*34*). The small incorporation of solvent hydrogen is caused by a slow nonenzymic exchange of hydrogen between methylmalonyl-CoA and water. It must be concluded that the methylmalonyl mutase reaction does not involve a free proton. Further confirmation of this conclusion was provided by showing that when D_3-methyl-methylmalonyl-CoA is converted to succinyl-CoA virtually all of the deuterium is retained in the product (*47*).

The conversion of (*R*)-methylmalonyl-CoA-2-D to succinyl-CoA-2-D has been shown to occur with retention of configuration at C-2 (*19, 20*). In this respect the methylmalonyl mutase reaction differs from the glutamate mutase reaction which involves an inversion of configuration (*48, 49*).

Following the demonstration by Frey and Abeles (*50*) that the B_{12} coenzyme-dependent dioldehydrase catalyzes an intermolecular hydrogen transfer from substrate to the 5′-carbon of the deoxyadenosyl moiety of the coenzyme and back to product, a similar hydrogen transfer was shown to occur in the methylmalonyl mutase reaction (*51, 52*). The specific reaction demonstrated is the transfer of tritium from coenzyme to product during the conversion of methylmalonyl-CoA to succinyl-CoA. The tritiated coenzyme, prepared either synthetically or enzymically by means of the dioldehydrase reaction, was labeled exclusively on the 5′-carbon. Tritium transfer from coenzyme to product is catalyzed by both *P. shermanii* and sheep liver methylmalonyl mutase.

The rate-limiting step in the conversion of methylmalonyl-CoA to

46. P. Overath, *Vitamin B_{12} Intrinsic Factor, 2. Europ. Symp., Hamburg, 1961* p. 155. Enke, Stuttgart, 1962.
47. J. D. Erfle, J. M. Clark, Jr., and B. C. Johnson, *Ann. N. Y. Acad. Sci.* **112,** 684 (1964).
48. M. Sprecher and D. B. Sprinson, *Ann. N. Y. Acad. Sci.* **112,** 655 (1964).
49. M. Sprecher, R. L. Switzer, and D. B. Sprinson, *JBC* **241,** 864 (1966).
50. P. A. Frey and R. H. Abeles, *JBC* **241,** 2732 (1966).
51. J. Retey and D. Arigoni, *Experientia* **22,** 783 (1966).
52. G. J. Cardinale and R. H. Abeles, *BBA* **152,** 517 (1967).

succinyl-CoA appears to be the breaking of a carbon–hydrogen bond in the substrate (*53*). This is indicated by the fact that the reaction rate is decreased when the hydrogen on the methyl group of methylmalonyl-CoA is replaced by deuterium. The kinetic isotope effect, measured by the k_H/k_D ratio, has a value of at least 3.5. After hydrogen has been abstracted from methylmalonyl-CoA the resulting substrate intermediate is estimated to have a 79% chance of going to a succinyl-CoA and a 21% chance of being reconverted to methylmalonyl-CoA under conditions in which the reaction is far from equilibrium.

The transfer of hydrogen between substrate and the 5′-carbon of the coenzyme has been investigated by determining the distribution of deuterium in the products obtained from the partial enzymic conversion of a mixture of methylmalonyl-CoA containing no deuterium and methylmalonyl-CoA containing three deuterium atoms in the methyl group. By comparing the observed distribution of deuterium with the theoretical distribution based upon the presence of one, two, or three equivalent hydrogen atoms in the coenzyme intermediate, Miller and Richards (*53*) concluded that the intermediate contains three equivalent hydrogen atoms, two derived from the coenzyme and one from the substrate. One of these three hydrogen atoms is returned to the substrate intermediate to complete the reaction either with the formation of product or regeneration of substrate. These results are consistent with a reaction mechanism involving a cleavage of the cobalt–carbon bond of the coenzyme with formation of a methyl group, presumably the C-5′ methyl group of deoxyadenosine. The formation of deoxyadenosine has not yet been demonstrated in the methylmalonyl-CoA reaction, but a small amount of a substance tentatively identified as this compound has been detected as a product of a noncatalytic reaction between glycolaldehyde and the dioldehydrase-B_{12} coenzyme complex (*54*).

Although the role of B_{12} coenzyme in the hydrogen transfer component of the methylmalonyl mutase reaction is clearly established, its function in the carbon chain rearrangement is still a matter of speculation. No substrate intermediate, either free or bound to the holoenzyme, has been detected (*40*). However, several investigators have postulated that formation of a bond between the reactive carbon atoms of the substrate and the cobalt atom of the coenzyme is involved in the rearrangement, and several types of substrate–coenzyme compounds have been suggested. One of the postulated intermediates would be formed by transfer of the methylmalonyl group from coenzyme A to the cobalt atom of B_{12}

53. W. W. Miller and J. H. Richards, *JACS* **91**, 1498 (1969).
54. O. W. Wagner, H. A. Lee, Jr., P. A. Frey, and R. H. Abeles, *JBC* **241**, 1751 (1966).

coenzyme (*55*). Such an acyl transfer reaction has been virtually excluded by showing that no exchange occurs between ^{14}C-labeled coenzyme A and succinyl-CoA during the mutase reaction (*56*). Another possible mechanism for the mutase reaction (*53*), which involves the displacement of the 5′-methylene carbon of the deoxyadenosyl moiety of the coenzyme by the methyl group of methylmalonyl-CoA with the formation of deoxadenosine as an intermediate, is shown in Fig. 1.

The rearrangement of the carbon chain is postulated to occur while the substrate is attached to cobalt. One of the hydrogen atoms on the

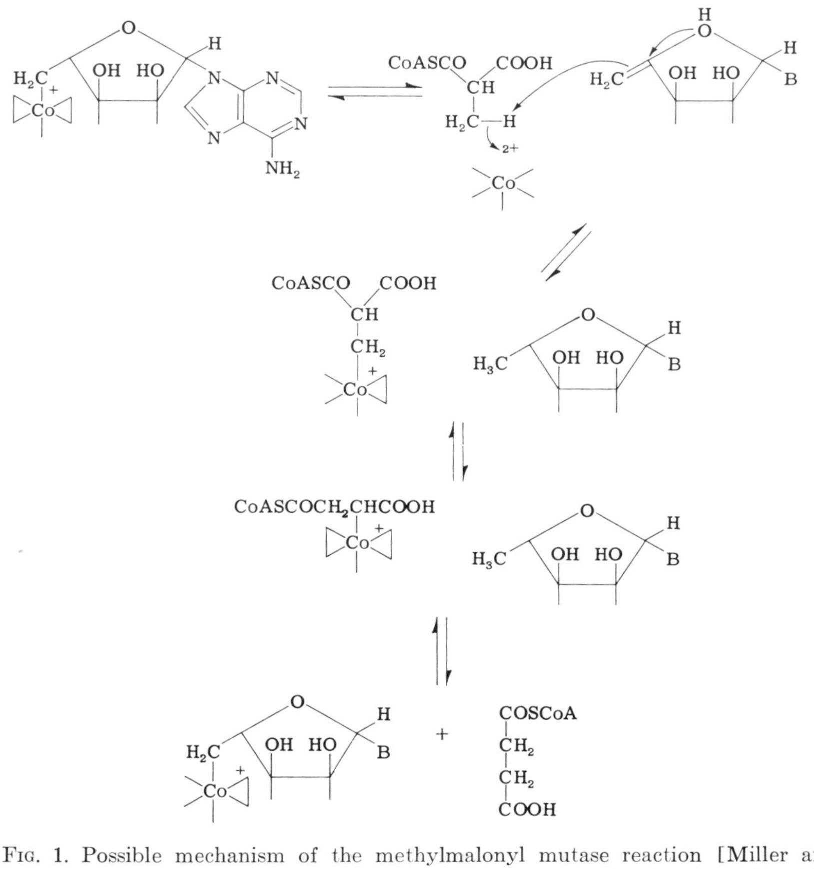

FIG. 1. Possible mechanism of the methylmalonyl mutase reaction [Miller and Richards (*53*)]. Reprinted with permission of the *Journal of the American Chemical Society*.

55. H. W. Whitlock, *Ann. N. Y. Acad. Sci.* **112,** 721 (1964).
56. J. Retey, U. Coy, and F. Lynen, *BBRC* **22,** 274 (1966).

methyl group of deoxyadenosine is then returned to the substrate with simultaneous cleavage of the carbon–cobalt bond. It should be emphasized that this plausible scheme is only one of several possible mechanisms that are consistent with the known facts. No information is available concerning the specific role of the enzyme in the reaction.

III. Glutamate Mutase

Glutamate mutase catalyzes the reversible conversion of L-glutamate to *threo*-β-methyl-L-aspartate [(2 *S*,3 *S*)-3-methylaspartate] [reaction (11)].

$$\begin{array}{c} COO^{\ominus} \\ | \\ H_3N \cdot CH \\ | \\ CH_2 \\ | \\ CH_2 \\ | \\ COO^{\ominus} \end{array} \rightleftharpoons \begin{array}{c} COO^{\ominus} \\ | \\ H_3\overset{+}{N} \cdot CH \\ | \\ HC \cdot CH_3 \\ | \\ COO^{\ominus} \end{array} \qquad (11)$$

L-Glutamate　　*threo*-β-Methyl-L-aspartate

The enzyme has been reported to be present in *Clostridium tetanomorphum* grown in a medium containing glutamate as a major energy source and in two photosynthetic bacteria, *Rhodospirillum rubrum* and *Rhodopseudomonas spheroides* (*57*, *58*). No systematic study of the occurrence of the mutase in other glutamate-fermenting anaerobes has been made. In *C. tetanomorphum*, the enzyme catalyzes the first of a sequence of energy yielding reactions resulting in the conversion of glutamate to acetate, butyrate, carbon dioxide, ammonia, and hydrogen. Although the glutamate mutase reaction is readily reversible and the equilibrium favors glutamate formation, the function of the enzyme in *C. tetanomorphum* is the conversion of glutamate to β-methylaspartate.

A. Assay

The mutase is most conveniently assayed by using L-glutamate as substrate and coupling the mutase reaction [reaction (11)] with the deamination of β-methylaspartate [reaction (12)] catalyzed by β-

57. S. Fukui, K. Sato, H. Ishitani, and S. Shimizu, *Vitamins* (*Kyoto*) **41,** 442 (1970).
58. H. Omori, K. Sato, S. Shimizu, and S. Fukui, *Vitamins* (*Kyoto*) **42,** 334 (1970).

methylaspartase (*59*). The formation of mesaconate can be followed

$$\beta\text{-Methylaspartate} \rightleftharpoons \text{mesaconate} + NH_4^+ \quad (12)$$

spectrophotometrically at 240 nm. When a saturating level of a corrinoid coenzyme, preferably benzimidazolyl cobamide coenzyme, and an excess of β-methylaspartase are present, the rate of mesaconate formation provides a measure of mutase activity.

Deaminase-free glutamate mutase can also be assayed less conveniently by using *threo*-β-methyl-L-aspartate as substrate and following its disappearance as it is converted to glutamate (*59*). The reaction is stopped by acidifying the solution after a suitable period of incubation, and the residual β-methylaspartate is estimated spectrophotometrically by converting it enzymically to mesaconate.

The protein components (E and S) of the mutase (see below) can be estimated by either of the above methods by adding an excess of one component and making the other rate limiting. The accuracy of these assays depends in some degree upon the purity of the component added in excess. The accuracy is decreased when the latter contains either an inhibitor or the rate-limiting component.

B. Purification and Molecular Properties of the Mutase Components

As already mentioned, glutamate mutase has been shown to consist of two dissimilar proteins, referred to as the E and S components (*60*). Since neither component alone possesses appreciable catalytic activity, whereas glutamate mutase activity is restored when the two components are combined under appropriate conditions, they are considered to be subunits of a single enzyme. Crude extracts of *C. tetanomorphum* after charcoal and protamine treatment contain approximately equal activities (0.02 to 0.09 unit/mg) of the two components.

1. *Component E*

This protein has been purified 120- to 180-fold over the crude extract in 18–25% yield. The best preparations have a specific activity of 3.5–4.0 units/mg of protein. They are not homogeneous but usually contain 25–50% contaminating proteins that are detected by ionophoresis or

59. H. A. Barker, V. Rooze, F. Suzuki, and A. A. Iodice, *JBC* **239,** 3260 (1964).
60. F. Suzuki and H. A. Barker, *JBC* **241,** 878 (1966).

ultracentrifugation. The sedimentation coefficient ($s_{20,w}$) of component E is 6.9 S. The molecular weight is estimated to be 128,000.

Component E slowly loses activity during storage in a neutral buffer at 0°; the half-life under these conditions is about 40 days. The activity can be maintained for long periods by storage at —196°.

Component E is relatively unstable in the presence of various cationic buffers, particularly above pH 7. The half-lives of the activity at 0° and pH 8.0 in the presence of 50 m*M* imidazole chloride or tris chloride, for example, are 0.5 and 0.6 days, respectively. Other amino buffers are somewhat less damaging. The harmful effect of tris chloride is significant at levels as low as 3 m*M* and increases with concentration up to at least 50 m*M*. These effects are irreversible. Inactivation by tris buffer is completely prevented by the presence of 0.3 *M* sodium L-glutamate and greatly diminished by 50 m*M* glutamate. β-Methyl-L-aspartate (50 m*M*) is about 90% as effective and L-aspartate (50 m*M*) about 70% as effective as glutamate against inactivation by tris, whereas potassium phosphate is entirely ineffective at the same concentration. These observations are of interest because they indicate that glutamate, β-methylaspartate, the substrates of glutamate mutase, and the closely related L-aspartate bind to component E. This is the only evidence presently available for substrate binding to a specific mutase component. It is not known whether the protection against inactivation by tris is the result of binding at the catalytic site or elsewhere.

Most preparations of component E are pink because of the presence of a tightly bound corrinoid, probably a photolysis product of a corrinoid coenzyme. This bound corrinoid compound appears to be an impurity rather than a prosthetic group. This is indicated by the observation that such component E preparations, in combination with component S, are virtually inactive unless a corrinoid coenzyme is added. The absorption spectrum of component E shows a maximum at 280 nm and a minimum at 251 nm; the $A_{280\ nm}/A_{260\ nm}$ ratio is about 1.6. There is a small and progressively decreasing adsorption in the region between 300 and 550 nm attributable to the contaminating corrinoid compound. Component S does not contain significant amounts of any vitamin B_6 derivatives.

2. *Component S*

This component has been purified about 350-fold over the crude extract in 25–30% yield (*61*). The specific activity of most preparations is in the range of 28–35 units/mg. The best preparations appeared to

61. R. L. Switzer and H. A. Barker, *JBC* **242,** 2658 (1967).

contain less than 10% contaminating protein as judged by polyacrylamide gel electrophoresis and gel filtration. The sedimentation coefficient ($s_{20,w}$) of reduced component S is 1.69 S. The molecular weight determined by sedimentation equilibrium, sedimentation velocity, and gel filtration techniques is 17,000 ± 1,000. On exposure to air the reduced protein is readily oxidized with formation of a dimer (molecular weight 34,000) presumably held together by one or more disulfide bonds. Other oxidation products are also formed (see below).

Component S is a relatively stable protein. No activity is lost during exposure to an acetate buffer, pH 3.5, for 20 min at 30° and less than 10% activity is lost in 24 hr at 0° in solutions of pH 4.5–7.5. In the absence of thiols, no activity is lost during storage for a month at —10°.

The absorption spectrum of component S is similar to that of other unconjugated proteins. It shows an absorbance maximum at 280 nm, a minimum at 254 nm and a $A_{280\ nm}/A_{260\ nm}$ ratio of 1.44. Although some preparations show a slight pink color, the very low absorption in the visible region indicates the absence of a corrinoid prosthetic group. Component S does not contain significant amounts of any vitamin B_6 compound.

The amino acid composition of component S has been determined (*61*). The protein contains one tryptophan residue and probably five half-cystine residues. Native component S, reduced by treatment with borohydride, appears to contain five sulfhydryl groups per molecule (Fig. 2). Somewhat less than four of these sulfhydryl groups react with DTNB in the native state, and the fifth is partially exposed to reaction with DTNB after treatment with sodium dodecyl sulfate, but not with 8 *M* urea. Only one sulfhydryl group in the reduced native protein reacts with iodoacetate (VI); this causes complete inactivation. In contrast, component E is not sensitive to iodoacetate inhibition. Component S apparently also reacts with arsenite which inhibits the mutase reaction. Arsenite presumably reacts with the two adjacent sulfhydryl groups that can also be oxidized to form an intramolecular disulfide bond [(II) and (V)].

When reduced component S is exposed to air, it is oxidized and simultaneously inactivated. The oxidized protein retains two sulfhydryl groups which can react with sulfhydryl reagents and do not participate in disulfide bond formation. The remaining two sulfhydryl groups are readily oxidized by air with formation of intra- or intermolecular disulfide bonds [Fig. 2, (II) and (IV)]. The relative amounts of these two types of oxidation products vary with the experimental conditions, but considerable dimer is always found.

The oxidized forms of component S have little, if any, activity in

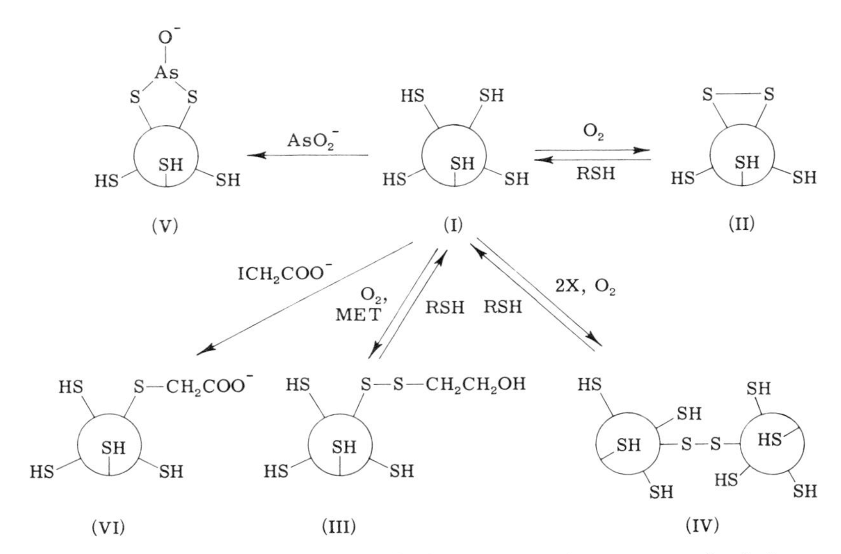

FIG. 2. Possible reactions of the sulfhydryl groups of component S of glutamate mutase [Switzer and Barker (*61*)]. Reprinted with permission of the *Journal of Biological Chemistry*. MET, 2-mercaptoethanol.

the complete glutamate mutase system. However, the oxidation and the associated inactivation are completely reversible processes. The oxidized form can be fully reactivated by incubation with thiols in the absence of oxygen. This is a relatively slow reaction. At 0° in the presence of 20 m*M* 2-mercaptoethanol, for example, about 45 min are required to reach full activity and the activation is slower at lower thiol concentrations.

C. CATALYTIC PROPERTIES

1. *Interaction of the Components and Coenzyme*

The two mutase components combine rapidly in the presence of coenzyme and substrate to form the active enzyme (*59–61*). The maximal rate of mesaconate formation from glutamate in the coupled assay with excess β-methylaspartase is reached within 20–40 sec after adding the last component. This short lag phase must be caused at least in part by the necessity of accumulating β-methylaspartate in the steady state concentration required for the coupled assay.

To obtain maximal mutase activity component S must be fully reduced (see above) before it is added to the assay solution, whereas component E need not be reduced. A low level (1 or 2 m*M*) of 2-mercaptoethanol is normally added to the assay solution. This produces a small apparent increase in activity by retarding the gradual inactivation of the system by oxygen.

The level of activity observed in the mutase assay is dependent upon the absolute and relative amounts of the two components (*60*). When one component is held constant and the second is increased, a typical saturation curve is obtained. With highly purified components, the rate becomes virtually constant at high levels of the variable component. The inhibition sometimes observed at high levels of the variable component (*60*) is apparently caused by impurities in the preparations. With the best preparations of mutase components and a saturating level of coenzyme, half-maximal activity is obtained when the molar ratio of E:S is about 1.5. The absolute concentrations of the reacting components are illustrated by an experiment in which 8×10^{-6} *M* BC coenzyme and 1.3×10^{-7} *M* component E were used; a component S concentration of 0.9×10^{-7} *M* was required for half-maximal activity and 5×10^{-7} *M* to reach almost full activity. The low concentrations required to give half-maximal activity show that the components have a strong affinity for each other. Nevertheless, the reconstituted active complex must dissociate rather easily since attempts to separate it from the two components by gel filtration have been unsuccessful.

The affinity of the mutase components for each other, measured by their ability to form a catalytically active complex, increases markedly with coenzyme concentration. This is shown by the generally inverse relation between coenzyme concentration and the level of component S required to reach half-maximal activity with a constant amount of component E. In one experiment, for example, with 0.80×10^{-7} *M* component E, 4.4×10^{-7} *M* component S was required to give half-maximal activity with 8.0×10^{-7} *M* BC coenzyme, whereas only 0.59×10^{-7} *M* was required with 1.6×10^{-5} *M* coenzyme. Conversely, the apparent K_m of the coenzyme decreases markedly as the S:E ratio increases. An increase in the weight ratio from 0.043 to 2.1, corresponding to an increase of component S from 0.46 to 23 μg, decreased the apparent K_m from 1.29 to 0.031 μM. The apparent K_m continues to decrease at higher S:E ratios that cause no further change in catalytic activity. This indicates that the components interact at more than one site.

Direct measurements of coenzyme binding have shown that at a subsaturating concentration (64 μM) of cobalamin coenzyme, component E alone binds 0.7 mole of coenzyme per mole, component S alone binds 0.024 mole/mole, and the two components together in an S:E molar ratio of about 7 bind 1.7 moles/mole of component E. Since the purity of the component E preparation used in these experiments was not precisely known, these values for coenzyme binding may be somewhat low. Furthermore, estimates of coenzyme binding by an indirect but apparently reliable method have given much higher values averaging 6.8 moles of cobalamin coenzyme per mole of component E in the presence of excess component S (*62*). An explanation for this discrepancy is not available.

No spectral change is associated with binding of coenzyme to the E–S complex.

2. *Substrate and Coenzyme Specificity*

The only known substrates for glutamate mutase are L-glutamate and *threo*-β-methyl-L-aspartate (*59*). D-Glutamate, *threo*-β-methyl-D-aspartate, *erythro*-β-methyl-DL-aspartate, and *threo*-β-ethyl-DL-aspartate do not react at a significant rate. The apparent K_m values for L-glutamate and *threo*-β-methyl-L-aspartate are approximately 1.5 and 0.5 mM, respectively.

The mutase shows no detectable activity in the absence of a suitable B_{12} coenzyme, i.e., Co-5′-deoxyadenosyl derivative of a complete corrinoid compound. The heterocyclic base on the nucleotide side chain of the corrinoid may be either a purine or a benzimidazole (*63, 64*). Incomplete corrinoid coenzymes, lacking a heterocyclic base, are inactive as are corrinoid compounds containing a complete nucleotide side chain but lacking the Co-deoxyadenosyl group. Other modifications of the structure of corrinoid enzymes also result in complete loss of activity. The inactive compounds include the 5′-deoxyuridyl, 5′deoxythymidyl, 2′,5′-dideoxyadenosyl, methyl, carboxymethyl, and carboxyethyl derivatives of cobalamin and the 5′-deoxyadenosyl derivatives of dehydrocobalamin, 10-chlorocobalamin, and *N*-methylcobalamin (a methyl group on N-3 of the dimethylbenzimidazolyl moiety) (*65, 66*). Evidently minor modifications in any one of several parts of the coenzyme interfere with catalytic activity. Some of the inactive coenzyme analogs function as inhibitors and therefore are able to bind to the enzyme.

The various corrinoid coenzymes known to be active with glutamate mutase are listed in Table II (*64, 67*) along with their apparent K_m

62. R. L. Switzer, B. G. Baltimore, and H. A. Barker, *JBC* **244,** 5263 (1969).
63. H. A. Barker, *Vitamin B_{12} Intrinsic Factor, 2. Europ. Symp., Hamburg, 1961* p. 82. Enke, Stuttgart, 1962.
64. J. I. Toohey, D. Perlman, and H. A. Barker, *JBC* **236,** 2119 (1961).
65. H. P. C. Hogenkamp and T. G. Oikawa, *JBC* **239,** 1911 (1964).
66. H. A. Barker, *in* "The Vitamins" (W. H. Sebrell, Jr. and R. S. Harris, ed.), 2nd ed., Vol. 2, p. 184. Academic Press, New York, 1968.
67. A. G. Lezius and H. A. Barker, *Biochemistry* **4,** 510 (1965).

TABLE II
KINETIC PROPERTIES OF COBAMIDE COENZYMES IN THE GLUTAMATE MUTASE ASSAY[a]

Nucleotide base of coenzyme	K_m ($M \times 10^{-7}$)	Relative V_{max}
5,6-Dimethylbenzimidazole	128	1.0
5-Hydroxybenzimidazole	18	
2,6-Diaminopurine	16.4	1.0
Adenine	12.4	1.0
5(6)-Nitrobenimidazole	5.3	1.2
5(6)-Aminobenzimidazole	4.6	1.4
5(6)-Methylbenzimidazole	4.3	1.5
5(6)-Trifluoromethylbenzimidazole	4.0	1.4
Benzimidazole	2.3	1.3

[a] Data from Toohey *et al.* (*64*) and Lezius and Barker (*67*).

and relative V_{max} values. All of the coenzyme analogs have similar V_{max} values, but they differ markedly in apparent K_m values. The coenzyme containing benzimidazole has the highest affinity for the enzyme and that containing dimethylbenzimidazole (cobalamin coenzyme) has the lowest affinity. The adenine-containing coenzyme, which is the main naturally occurring form in *Clostridium tetanomorphum*, has an intermediate affinity for the enzyme. The K_m values in Table II give a measure of the relative affinities of the coenzyme analogs for the mutase since they were all obtained with the same type of enzyme preparation. However, it should be noted (see above) that the affinity of a coenzyme for mutase reconstituted from its components varies markedly with the S:E ratio in the assay system. Consequently, the absolute value of the apparent K_m depends upon the experiment conditions. Also, it should be noted that the K_m values in Table II are not a measure of a readily reversible binding process. Although coenzyme can be more or less completely removed from the active complex by treating the enzyme with charcoal, experiments with labeled coenzyme (see below) indicate that little if any exchange occurs between free coenzyme and enzyme-bound coenzyme during the mutase reaction over periods as long as 80 min (*62*). Therefore, the physical meaning of the observed K_m values is not clear.

3. *Conditions Affecting Activity*

The pH-activity curve is asymmetrically bell-shaped with a range from pH 5.0 to 9.7 and an optimum at pH 8.5.

The reaction rate increases about 3-fold when the temperature is increased from 25° to 37°. A maximal rate is obtained at 40° when a

30-min incubation time is used; with a shorter incubation the optimal temperature is undoubtedly higher. The enzyme is rapidly inactivated at 55°.

4. *Equilibrium of the Reaction*

The equilibrium constant K = (L-glutamate)/(*threo*-β-methyl-L-aspartate) at pH 8.2 and 30° is 10.7 ± 0.4. This corresponds to a $\Delta F_0'$ of -1.43 kcal for the conversion of methylaspartate to glutamate (*59*).

5. *Mechanism of Reaction*

The carbon chain rearrangement in the conversion of glutamate to β-methylaspartate could occur theoretically either by a cleavage of the bond between carbon atoms 2 and 3 of glutamate and formation of a new bond between carbon atoms 2 and 4 or by migration of the carbon-5 carboxyl group from carbon atom 4 to carbon atom 3. In the first type of rearrangement the methyl group of β-methylaspartate would be derived from carbon-3 of glutamate; in the second rearrangement, the methyl group would contain carbon-4 of glutamate. A rearrangement of the first type, involving a migration of the "glycine" residue (carbon atoms 1 and 2 of glutamate plus substituent atoms) from the β- to the α-carbon of the "propionate" residue (carbon atoms 3, 4, and 5) of glutamate, was established by converting 4-^{14}C-glutamate to ^{14}C-mesaconate by means of glutamate mutase and β-methylaspartate (*68*). By suitable degradation procedures the mesaconate was shown to be labeled exclusively in the carbon atom adjacent to the methyl group. Therefore, the methyl group of mesaconate must have been derived from carbon-3 and the adjacent carbon from carbon-4 of glutamate. Since no carbon chain rearrangement is involved in the conversion of β-methylaspartate to mesaconate, the same conclusion applies to β-methylaspartate.

When the "glycine" residue migrates between the α- and β-carbons of the "propionate" residue of glutamate, a hydrogen atom migrates in the opposite direction. By carrying out the mutase reaction in D_2O or ^3H_2O, Iodice and Barker (*69*) demonstrated that the hydrogen migration does not involve a free proton; no significant amount of solvent hydrogen was incorporated into the product.

The stereochemistry of the glutamate mutase reaction was investigated by Sprecher *et al.* (*49*) by using *threo*-β-methyl-L-aspartate-3-*d*

68. A. Munch-Petersen and H. A. Barker, *JBC* **230**, 649 (1958).
69. A. A. Iodice and H. A. Barker, *JBC* **238**, 2094 (1963).

as a substrate and determining the absolute configuration of the resulting 4-deuterioglutamate. They found that the deuterioglutamate has the 4*R* configuration and concluded that the intramolecular rearrangement of *threo*-β-methyl-L-aspartate to L-glutamate proceeds by net inversion of configuration of carbon atom 3. The hydrogen from the methyl group must have approached C-3 of β-methylaspartate from the side opposite to that of the leaving "glycine" moiety. In the reverse direction, the "glycine" moiety approached C-4 of glutamate from the side opposite to that of the leaving hydrogen resulting in the *threo* configuration of the product.

Since other B_{12} coenzyme-dependent reactions involve hydrogen transfer between substrate and the 5′-carbon of the deoxyadenosyl group (see above), the occurrence of a similar hydrogen transfer in the glutamate mutase reaction was investigated (*62*). Tritium from 5′-tritiated cobalamin coenzyme was found to be readily transferred to product and substrate during the conversion of β-methylaspartate to glutamate. Under suitable conditions, virtually all of the tritium in the coenzyme was transferred to the amino acids. Since the two hydrogens on C-5′ were randomly labeled, this result proves that both hydrogen atoms on C-5′ of the coenzyme participated in the mutase reaction.

The fraction of the added coenzyme tritium transferred to amino acids is greatly dependent on the experimental conditions, particularly the relative amounts of coenzyme and enzyme. The results of several experiments indicate that only enzyme-bound coenzyme participates in tritium transfer and that essentially no exchange occurs between enzyme-bound and free coenzyme. On the assumption that all the tritium in the enzyme-bound coenzyme is transferred to amino acids during an 80-min incubation and that none of the free coenzyme participates in the reaction, it was estimated that approximately 6.8 moles of coenzyme were bound per mole of component E in the E–S complex. As previously mentioned, this value is considerably higher than the value obtained by other methods.

The number of equivalent hydrogen atoms on the enzyme–coenzyme complex during the glutamate mutase reaction has been investigated (*70*) by the method (see above) previously applied to methylmalonyl mutase (*53*). The conclusions are the same, namely, that hydrogen which moves from the methyl carbon of β-methylaspartate to C-4 of glutamate becomes equivalent with two other hydrogens, presumably those on the C-5′ of the B_{12} coenzyme. This implies that C-5′ methylene group is

70. R. G. Eagar, Jr., B. G. Baltimore, M. M. Herbst, H. A. Barker, and J. H. Richards, *Biochemistry* **10** (1972) (in press).

converted to a methyl group in the intermediate state. These experiments also provide an estimate of the kinetic isotope effects with protium and deuterium. For abstraction of the hydrogen from the methyl group of β-methylaspartate and transfer to C-5′ of the coenzyme, k_H/k_D is about 5, whereas for transfer from the C-5′ carbon of the coenzyme back to C-4 of the glutamate skeleton, k_H/k_D is approximately 1.5.

The possible occurrence of free intermediates in the glutamate mutase reaction has been investigated by tracer experiments (*8*, *60*, *71*). For example, the possible cleavage of glutamate to glycine and acrylate which might recombine to form β-methylaspartate was investigated by carrying out the mutase reaction in the presence of ^{14}C-labeled glycine. Since no ^{14}C was incorporated into the products of the mutase reaction, free glycine was excluded as an intermediate. By similar experiments, acrylate, propionate, α-ketoglutarate, and ammonia have been excluded as free intermediates. No enzyme-bound intermediate has been detected.

Several mechanistic hypotheses have been proposed for rearrangement of the carbon skeleton in the glutamate mutase reaction (*19*, *39*, *55*, *72*, *73*), but no hypothesis is yet supported by specific evidence. It seems possible that the mechanism of the glutamate mutase reaction is similar to that of the methylmalonyl mutase reaction which has already been discussed.

IV. α-Methyleneglutarate Mutase

This inducible enzyme catalyzes the reversible, B_{12} coenzyme-dependent conversion of α-methyleneglutarate to β-methylitaconate (*74*) [reaction (13)].

$$^{\ominus}OOC \cdot CH_2CH_2-\underset{\underset{CH_2}{\|}}{C}-COO^{\ominus} \qquad ^{\ominus}OOC \cdot \overset{\overset{CH_3}{|}}{CH}-\underset{\underset{CH_2}{\|}}{C}-COO^{\ominus} \tag{13}$$

α-Methyleneglutarate **β-Methylitaconate**

The enzyme participates in the anaerobic conversion by *Clostridium barkeri* (*17*) of nicotinic acid to equimolar amounts of propionate,

71. H. A. Barker, F. Suzuki, A. A. Iodice, and V. Rooze, *Ann. N. Y. Acad. Sci.* **112**, 644 (1964).
72. L. L. Ingraham, *Ann. N. Y. Acad. Sci.* **112**, 713 (1964).
73. L. L. Ingraham, "Biochemical Mechanisms," p. 100. Wiley, New York, 1962.
74. H. F. Kung, L. Tsai, and T. C. Stadtman, *JBC* **246**, 6444 (1971).

acetate, carbon dioxide, and ammonia (*75*). α-Methyleneglutarate is formed from nicotinic acid via the following known or probable intermediates: 6-hydroxynicotinic acid, 1,4,5,6-tetrahydro-6-oxonicotinic acid, and 2-formylglutarate (*76–78*). Methylitaconate is readily converted to 2,3-dimethylmaleate by the action of methylitaconate isomerase (*74*). The double bond of dimethylmaleate is thought to be hydrated to form dimethylmalate which is then cleaved to propionate and pyruvate. The latter can be oxidized to acetate and carbon dioxide.

A. Assay

Mutase activity is determined by using α-methyleneglutarate as substrate in an assay solution containing cobalamin coenzyme, glutathione, and an excess of methylitaconate isomerase. The resulting dimethylmaleate is estimated by reaction with 2,4-dinitrophenylhydrazine in acid solution to give *N*-2′,4′-dinitrophenylanilino-3,4-dimethylmaleimide. The characteristic red color of the latter compound in alkaline solution provides a sensitive quantitative measure of dimethylmaleate. Under suitable conditions, mutase activity is directly proportional to the quantity of this product. Since the enzyme is sensitive to O_2, the assay solution is kept under an argon atmosphere.

B. Purification and Molecular Properties

The mutase is an inducible enzyme which is present in cells grown in a medium containing nicotinic acid. The enzyme has not yet been obtained in a homogeneous state, but it has been purified 40-fold from crude extracts of *C. barkeri* by a multistep procedure which gives activity yields ranging from 16 to 27% (*74*). Such preparations have a specific activity of about 0.51 μmole/min/mg of protein; they are heavily contaminated with methylitaconate isomerase. Low yields of isomerase-free mutase have been obtained by chromatography on DEAE-cellulose. Determination of the molecular weight of the mutase by sucrose density gradient centrifugation, using a preparation containing both mutase and isomerase, gave a value of about 170,000.

Purified mutase preparations are virtually free of corrinoid coenzyme and show little activity unless coenzyme is added. Any residual co-

75. I. Harary, *JBC* **227,** 815 (1957).
76. L. Tsai, I. Pastan, and E. R. Stadtman, *JBC* **241,** 1807 (1966).
77. J. S. Holcenberg and L. Tsai, *JBC* **244,** 1204 (1969).
78. L. Tsai and E. R. Stadtman, "Methods in Enzymology," Vol. 18B, p. 233, 1971.

enzyme can be readily removed by treating the enzyme with charcoal or exposing it to light.

The mutase is relatively unstable, particuarly when the protein concentration is low (1 mg/ml). More concentrated solutions (10 mg/ml) are much more stable. They retain up to 90% of their activity during storage for 6 months at —80°C.

The mutase appears to be activated by glutathione and other thiol compounds. It is partially inhibited by iodoacetamide but not by 10^{-3} M arsenite. The enzyme is not inhibited by EDTA nor dipyridyl and also is not activated by either mono- or divalent cations.

C. Catalytic Properties

1. *Substrates and Inhibitors*

α-Methyleneglutarate and β-methylitaconate are the only known substrates for the mutase. The apparent K_m value for α-methyleneglutarate is 7.1 mM; the value for β-methylitaconate has not been determined (*74*).

Several dicarboxylic acids inhibit the activity of the mutase. Itaconate, mesaconate, succinate, 1-methyl-1,2-*trans*-cyclopropanedicarboxylate and L-malate are competitive inhibitors with K_I values of 1.2, 1.2, 1.8, 3.3, and 7.0 mM, respectively. Glutaconate and 1-methyl-1,2-*cis*-cyclopropanedicarboxylate behave as noncompetitive inhibitors with K_I values of 4.5 and 38 mM, respectively.

2. *Coenzymes*

5,6-Dimethylbenzimidazolyl, benzimidazolyl, and adenyl cobamide coenzymes activate the mutase; the apparent K_m values are 7.3×10^{-8}, 3.0×10^{-7}, and 1.2×10^{-6} M, respectively. The V_{max} values of the first two coenzymes are nearly the same, whereas that of adenyl cobamide coenzyme appears to be about 40% lower. Hydroxocobalamin inhibits the mutase, probably by competing with coenzyme for the catalytic site (*16*).

3. *Equilibrium of the Reaction*

The mutase reaction is readily reversible, the equilibrium favoring the formation of α-methyleneglutarate. The equilibrium constant [K_m = (MI/MG)], estimated by the use of enzyme preparations containing both mutase and methylitaconate isomerase, was found to be approximately 0.23 at pH 7.9 and 34°. This corresponds to a $\Delta F_0'$ of +0.89

kcal. As in other B_{12}-dependent mutase reactions, the more branched-chain acid is less stable thermodynamically than the straighter chain acid.

4. *Mechanism of Reaction*

The mechanism of the α-methyleneglutarate mutase reaction appears to be similar to that of other B_{12} coenzyme-dependent mutase reactions. This is indicated by several experiments on possible free intermediates and hydrogen transfer reactions. Acrylate and 1-methyl-1,2-cyclopropanedicarboxylate (both *cis* and *trans* forms) have been excluded as free intermediates by showing that ^{14}C-acrylate does not exchange with the propionate residue of the products and 1-methyl-1,2-cyclopropanedicarboxylate does not serve as a substrate for the enzyme (*74*). When the mutase reaction is carried out in the presence of $^{3}H_2O$, no tritium is incorporated into substrate or product unless methylitaconate isomerase is also present (*79*). This indicates that the hydrogen transfer does not involve a free proton. As in other mutase reactions, hydrogen is transferred between the substrates and the 5′-carbon of the deoxyadenosyl moiety of the coenzyme. This was established by using 5′-^{3}H-cobalamin coenzyme and showing that tritium is transferred to both α-methyleneglutarate and α-methylitaconate.

79. H. F. Kung and L. Tsai, *JBC* **246,** 6436 (1971).

15

B_{12} *Coenzyme-Dependent Amino Group Migrations*

THRESSA C. STADTMAN

I. Introduction

There are two types of amino group migration reactions presently recognized that are B_{12} coenzyme mediated (*1*). In one type the amino group, after its migration, is liberated as ammonia (e.g., ethanolamine

1. In this review, the terms B_{12} coenzyme and cobamide coenzyme are used interchangeably in a generic sense.

deaminase) whereas in reactions involving certain diamino acid substrates (e.g., β-lysine conversion to 3,5-diaminohexanoate) the ω-amino group is shifted to the adjacent carbon atom and retained in the isomeric amino acid. In both types of reactions B_{12} coenzyme is the intermediate carrier of the hydrogen that migrates in the reverse direction to replace the amino group. This hydrogen-carrying function of B_{12} coenzyme is the common feature of several seemingly diverse types of reactions in which the coenzyme participates (see Abeles, Chapter 17, Volume V, and Barker, Chapter 14, this volume).

II. Deaminases

A. Ethanolamine Deaminase (Ethanolamine Ammonia Lyase)

Only one example of a B_{12} coenzyme-dependent deamination reaction has been reported. This reaction, the deamination of ethanolamine, was discovered by Bradbeer (*1a*). It is catalyzed by ethanolamine deaminase which converts ethanolamine to acetaldehyde and ammonia [reaction (1)]. The aldehyde carbon of the product is derived from the carbinol carbon of the substrate (*2*).

$$CH_2(NH_3)^+CH_2OH \rightarrow CH_3CHO + NH_4^+ \quad (1)$$

1. *Occurrence*

Ethanolamine deaminase is formed by an anaerobic microorganism which originally was isolated by Hayward (*3*) from a large batch of contaminated culture medium containing choline as substrate. A pure culture of the contaminant (a *Clostridium* not identified as to species) was shown by Hayward to ferment choline to trimethylamine, ethanol, and acetate. Bradbeer (*4*), in a study of the ability of the organism to use various analogs of choline, found that ethanolamine could replace choline as a substrate for growth. With the latter substrate the overall fermentation products are ammonia, ethanol, and acetate. In both of these fermentations the primary attack on the substrate yields equimolar amounts of acetaldehyde plus trimethylamine from choline or acetaldehyde plus ammonia from ethanolamine (*3, 4*). The acetaldehyde then

1a. C. Bradbeer, *JBC* **240**, 4675 (1965).
2. B. H. Kaplan and E. R. Stadtman, *JBC* **243**, 1794 (1968).
3. H. R. Hayward, *Bacteriol. Proc.* p. 68 (1961).
4. C. Bradbeer, *JBC* **240**, 4669 (1965).

is dismuted to ethanol and acetate. The primary cleavage reactions are catalyzed by two different inducible enzymes (*4*); each one is formed only when the growth medium contains its specific substrate. The enzyme which deaminates ethanolamine [reaction (1)] requires B_{12} coenzyme for activity (*1a*) and, as will be discussed later, the coenzyme participates directly in the reaction as a hydrogen carrier. Little is known about the enzyme which cleaves choline except that it appears not to be B_{12} coenzyme dependent. Although the same choline fermentation (*5–7*) is catalyzed by *Vibrio cholinicus* (*5, 8*), this organism did not grow in a medium containing ethanolamine instead of choline as substrate and, therefore, may not have the capacity to synthesize ethanolamine deaminase. Other microorganisms appear not to have been surveyed as possible sources of the enzyme.

2. *Isolation*

Clostridial ethanolamine deaminase was isolated (*9*) in homogeneous form by two different procedures that resulted in 40- to 50-fold enrichment of the enzyme. The enzyme therefore comprises 2–2.5% of the total protein extracted from the cell when the bacteria are cultured on ethanolamine as the fermentable substrate (*1a, 9, 10*). Both purification procedures involved preliminary treatment of crude bacterial extracts with streptomycin sulfate to remove nucleic acids followed by fractionation with ammonium sulfate. In the original procedure the pure enzyme then was obtained by two successive chromatographic steps using DEAE-cellulose followed by DEAE-Sephadex. When it was observed that the purified enzyme is relatively insoluble in solutions of ionic strength greater than 0.11 a modified procedure was developed that exploited this property of the protein and eliminated the tedious chromatographic steps. The orange-colored protein precipitate from the ammonium sulfate step was washed with several portions of buffer containing 0.4 *M* NaCl to remove contaminating proteins; the washed pellet, which was intensely orange in color, was suspended in dilute buffer, dialyzed, and then sub-

5. H. R. Hayward and T. C. Stadtman, *J. Bacteriol.* **78,** 557 (1959).
6. H. R. Hayward and T. C. Stadtman, *JBC* **235,** 538 (1960).
7. H. R. Hayward, *JBC* **235,** 3592 (1960).
8. This organism was later determined to be a strain of *Desulfovibrio desulfuricans* [J. R. Postgate and L. L. Campbell, *Bacteriol. Rev.* **30,** 732 (1966)].
9. B. H. Kaplan and E. R. Stadtman, *JBC* **243,** 1787 (1968).
10. In contrast, extracts of cells of the same organism cultured and processed at the Enzyme Center in Boston failed to yield active enzyme preparations even though procedures apparently identical to those used in Bethesda were followed [B. M. Babior, personal communication (1970)].

jected to a gel filtration step on Bio-Gel P-200 to achieve final purification. The yield of pure enzyme obtained by either procedure was about 10%.

3. *Enzyme Assays*

The activity of ethanolamine deaminase is measured spectrophotometrically at 340 nm in a coupled assay wherein the acetaldehyde formed from ethanolamine is reduced to ethanol by DPNH and an excess of ethanol dehydrogenase (*9*). In this system a change in optical density at 340 nm of 0.125 per microgram of ethanolamine deaminase per minute was reported. A unit of ethanolamine deaminase activity is defined as that amount of enzyme required to produce a decrease in optical density at 340 nm of 6.2 in 1 min. This is equivalent to the formation of 1 μmole of acetaldehyde per minute. Impure preparations of the deaminase that are contaminated with DPNH oxidase are assayed by an alternative procedure in which ^{14}C-labeled acetaldehyde formation from ^{14}C-ethanolamine is estimated (*4, 11*). In the radioactive assay the residual substrate, cationic at pH 2, is removed on Dowex-50 H^+ and the labeled neutral product in the supernatant solution is estimated by scintillation spectrometry.

4. *Physical Properties*

A summary of some physical properties of ethanolamine deaminase is given in Table I (*2, 11–16*). The native enzyme is a large protein molecule (molecular weight about 500,000) which, when treated with 5 *M* guanidine hydrochloride in the presence of an alkylating agent, dissociates into subunits (about 50,000 MW) that sediment in the ultracentrifuge as a single component (*2*). Estimation of the minimum molecular weight from amino acid analysis (28,700) indicates a slightly larger subunit of about 57,000 MW. Thus, on the basis of available data the native protein probably consists of eight or ten such subunits which may be identical (*2*).

11. B. M. Babior and T. K. Li, *Biochemistry* **8**, 154 (1969).
12. Computed from amino acid composition by method of E. J. Cohn and J. T. Edsall, *in* "Proteins, Amino Acids and Peptides as Ions and Dipolar Ions," p. 375. Reinhold, New York, 1943.
13. Determined by miniscus depletion method of D. A. Yphantis [*Biochemistry* 3, 297 (1964)].
14. Subunits formed by treatment of native enzyme with 5 *M* guanidine·HCl and 10 m*M* iodoacetamide.
15. Reactive with 5,5′-dithiobis(2-nitrobenzoic acid); Ellman's reagent.
16. Determined by kinetic studies and circular dichroism spectroscopy (*11*).

TABLE I
SOME PHYSICAL PROPERTIES OF ETHANOLAMINE DEAMINASE

Form of enzyme	Property	Average value	Reference
Native	Sedimentation coefficient ($s_{20,w}$)	14.5	*2*
Native	Partial specific volume	0.74	*2, 12*
Native	Weight average molecular weight (M_w)	520,000 g	*2, 13*
Subunits	Weight average molecular weight (M_w)	51,000 g	*2, 13, 14*
Native	Minimum molecular weight from amino analysis	28,700 g	*2*
Native	Number of half-cystine residues per 28,700 MW subunit	3.1	*2*
Native	Titratable sulfhydryl groups per mole of holoenzyme (520,000 g)	1.19	*2, 15*
Unfolded	Titratable sulfhydryl groups (per 520,000 g) after treatment of holoenzyme with 2% sodium dodecyl sulfate	≥14.9	*2, 15*
Native apoenzyme	Active sites (cobamide binding) per mole	2	*11, 16*

The native holoenzyme (containing bound cobamide; see later) possesses one sulfhydryl group per 520,000 g that is accessible to titration with Ellman's reagent (*2*). In the presence of 5 *M* guanidine hydrochloride additional sulfhydryl groups are exposed, and a total of at least 15 per 520,000 MW can be titrated. Thus, a considerable number (perhaps 25%) of the half-cystine residues detected by amino acid analysis are in the reduced form in the native protein. It is not known whether additional titratable sulfhydryl groups are exposed when the apoprotein is prepared by removal of bound cobamide as is the case with mammalian methylmalonyl-coenzyme A mutase (*17*).

5. *Cobamide Binding Sites*

The amount of cobamide bound to the native protein, isolated by either of the purification procedures, varied from 1.35 to 3.1 moles/mole of enzyme (*2*). Removal of the bound cobamide by treatment with acid ammonium sulfate yielded an apoprotein that was capable of binding 7 moles of hydroxycobalamin per mole of enzyme. Binding of only 2–3 equivalents of hydroxycobalamin was sufficient to inhibit completely the catalytic activity of the enzyme. None of this hydroxycobalamin, whether specifically or nonspecifically bound, was released by extensive

17. J. J. B. Cannata, A. Focesi, R. Mazumder, R. C. Warner, and S. Ochoa, *JBC* **240**, 3249 (1965).

dialysis (2 days at neutral pH). However, treatment with acid ammonium sulfate removed some of the bound inhibitory cobamide and a partially active holoenzyme was reconstituted upon the addition of 5′-deoxyadenosylcobalamin (*2*). From other studies it has been concluded that there are two catalytically important binding sites for cobamides per mole of the native apoprotein (*11*). The data of kinetic studies (*11*) indicate that these are catalytically active sites and that they are separate and independent. The formation of the enzyme–coenzyme complex at one site imparts activity to the enzyme whether or not the second site is also filled. Binding of either the active coenzyme (5′-deoxyadenosylcobalamin) or the inhibitor (hydroxycobalamin) to the apoenzyme caused distinctive changes in the circular dichroism spectra of the two cobamides (*11*). The change in the specific region of the spectrum where the signal is the result of complex formation between enzyme and the particular cobamide added showed that the increase in intensity was maximal when 2 moles of cobamide per mole of enzyme had been added. The stoichiometry was the same whether the enzyme was titrated with coenzyme or with hydroxycobalamin. It is presumed that binding of additional equivalents of hydroxycobalamin to the enzyme at high cobamide concentrations (*2*) is nonspecific in nature.

6. *Specificity of Cobamide Coenzyme Requirement*

The form of B_{12} coenzyme synthesized by the ethanolamine-fermenting *Clostridium* is α-(adenyl)-Co-5′-deoxyadenosylcobamide, and it is this derivative or degradation products thereof that remains bound to the native form of ethanolamine deaminase which is isolated from the bacterial extracts (*2*). After removal of these cobamides by treatment with acid ammonium sulfate, the apoenzyme is reactivated by the following naturally occurring B_{12} coenzymes with the apparent K_m values as indicated: α-(adenyl)cobamide coenzyme, $7.7 \times 10^{-6}\ M$; α-(dimethylbenzimidazolyl)cobamide coenzyme, $1.5 \times 10^{-6}\ M$; and α-(benzimidazolyl)cobamide coenzyme, $1.9 \times 10^{-7}\ M$ (*9*). Among a number of synthetic coenzyme analogs, in which the 5′-deoxyadenosyl moiety covalently attached to the cobalt atom is replaced by other alkyl groups, only the 2′,5′-dideoxyadenosyl analog was equally active as a coenzyme (*18*).

7. *Cobamides That Are Inhibitors of Ethanolamine Deaminase*

Inhibition of ethanolamine deaminase by hydroxycobalamin, methylcobalamin, and cyanocobalamin (*1a*) in a concentration-dependent and

18. B. M. Babior, *JBC* **244,** 2917 (1969).

time-dependent fashion is a complex process which may appear competitive in nature with respect to the normal coenzyme if initial reaction rates are considered (*18*) or irreversible if reaction rates are measured after a few minutes of exposure of enzyme to inhibitor (*2, 18*). Babior (*18*) studied a number of coenzyme analogs and found the following to be inhibitors in addition to those listed above: β-hydroxyethylcobalamin, β-(2-tetrahydropyryloxy)ethylcobalamin, 1-*O*-methyl-5-deoxyribosylcobalamin, δ-(9-adenyl)butylcobalamin, 5′-deoxyinosylcobalamin, and 5′-deoxyuridylcobalamin. Among all of these compounds, only 5′-deoxyinosylcobalamin behaved as a simple competitive inhibitor with respect to α-(dimethylbenzimidazolyl)cobamide coenzyme with a K_i of 2.3×10^{-6} *M*. The other inhibitors caused a progressive decay in enzymic activity which, although usually not complete, nevertheless often was as little as 0.1% the uninhibited rate. The fate of the inhibitor during the inactivation process was examined using methylcobalamin labeled with tritium in the methyl group. No detectable cleavage of the methyl–cobalt bond occurred suggesting that a chemical change of this type is not responsible for the enzyme inactivation. Also, in light scattering experiments no association or dissociation of oligmers of the enzyme molecule was observed to occur during the inactivation process. The inhibitory compounds were suggested to induce a change in the enzyme from the native form to a form with a lower V_m and a much greater affinity for the inhibitor (*18*).

8. *Activation of Ethanolamine Deaminase by Monovalent Cations*

The activity of ethanolamine deaminase is dependent on the presence of a monovalent cation (*9*) such as potassium ion (apparent K_m of 4.7×10^{-4} *M*) or ammonium ion (same K_m). Rubidium ion is somewhat less effective (apparent K_m of 7.6×10^{-4} *M*), whereas lithium ion and sodium ion act as competitive inhibitors for potassium ion (apparent K_i value for both is 3.2×10^{-2} *M*). At nonsaturating concentrations of potassium ion the addition of cesium ion (10^{-2} *M*) had no effect on the activity of ethanolamine deaminase.

9. *Substrate Specificity and Binding*

Ethanolamine ($K_m = 2.2 \times 10^{-5}$ *M*) is the only compound known to serve as an effective substrate for the clostridial ethanolamine deaminase (*9*). Several analogs of ethanolamine serve as competitive inhibitors in the reaction. Among these, with their respective K_i values, are DL-2-amino-1-propanol, 7.1×10^{-6} *M*; L-2-amino-1-propanol, 6.3×10^{-6} *M*;

1-amino-2-propanol, 2×10^{-5} M; 2-amino-1-butanol, 2.7×10^{-5} M; and 2-(methylamino)ethanol, 3.9×10^{-3} M. Other weakly competitive inhibitors are 3-aminopropanol and 2-(dimethylamino)ethanol. 2-Amino-2-methyl-1-propanol behaves as a mixed inhibitor and several other analogs, namely, glycerol, ethylene glycol, propanediol, 2-amino-2-methylpropanediol, glycine, sarcosine, DL-serine, choline, ethylenediamine, and ethylamine, are not inhibitory or only slightly inhibitory (*9*). Studies by Babior (*19*) show that ethylene glycol serves as a quasi-substrate for ethanolamine deaminase in an abortive type of reaction which results in cleavage of coenzyme at the carbon–cobalt bond, generation of acetaldehyde, and decreased activity of the enzyme (see Section IV,C). Binding of substrate to the enzyme requires the presence of a cobamide at the active sites, and when an inhibitory cobamide such as methylcobalamin is present on the enzyme a relatively stable enzyme–substrate complex is formed which can be isolated by gel filtration (*20*). Maximum binding of substrate is observed when the ratio of methylcobalamin to apoenzyme is 2 to 1, and strongly competitive substrate analogs inhibit the binding of substrate under these conditions (*20*).

Ethanolamine (10 mM) protects the enzyme against loss of activity during isolation and storage (*9*).

10. *Other Properties*

The optimum range of pH for ethanolamine deaminase activity is between 6.8 and 8.2. The same activity of the enzyme was observed whether the buffer was potassium phosphate, potassium citrate, or tris hydrochloride (*9*).

The activity of ethanolamine deaminase in crude enzyme preparations was inhibited by the addition of intrinsic factor, and this inhibition was overcome by the addition of B_{12} coenzyme (*1a*). Treatment with small amounts of charcoal also yielded enzyme preparations that were completely dependent on added B_{12} coenzyme for activity (*1a*).

Native ethanolamine deaminase shows no requirement for an added mercaptan for maximal catalytic activity. Treatment with the alkylating agents iodoacetamide (10 mM) or N-ethylmaleimide (10 mM) was not inhibitory (*9*). However, the activity of the enzyme was inhibited by p-chloromercuriphenylsulfonate (10^{-4} to 10^{-3} M); approximately 50% inhibition was caused by a 3×10^{-4} M concentration of the organic mercurial. Whether sulfhydryl groups on the apoenzyme (deaminase resolved of bound cobamides) would be more accessible to these reagents is not known.

19. B. M. Babior, *JBC* **245,** 1755 (1970).
20. B. M. Babior, *JBC* **244,** 2927 (1969).

Stability of enzyme bound B_{12} coenzymes and various bound coenzyme analogs to irradiation with visible light was investigated by Babior and associates (*21*). A comparison of the ultraviolet and visible adsorption spectra of enzyme-bound 5′-deoxyadenosylcobalamine, 5′-deoxyinosylcobalamin, and β-hydroxyethylcobalamin, before and after aerobic photolysis, showed that they changed to one resembling that of enzyme-bound hydroxycobalamin whereas that of enzyme-bound methylcobalamin shifted to one resembling the spectrum of B_{12r}. Peaks in electron paramagnetic resonance spectra of the irradiated complex between methylcobalamin and enzyme were also attributed to the formation of B_{12r}. Babior suggested that this apparent ability of the enzyme to stabilize the photolysis product of methylcobalamin, which normally under aerobic conditions is immediately oxidized to hydroxycobalamin, is related to a unique property of the methyl radical.

III. Enzymes That Catalyze Migrations of ω-Amino Groups of Diamino Acids

Three mutases (B_{12} coenzyme dependent) that catalyze the same type of amino group migration reaction but act upon different diamino acids occur in *Clostridium sticklandii*. These are β-lysine mutase (*22*, *23*), D-α-lysine mutase (*24*, *25*), and ornithine mutase (*26*, *27*) which catalyze reactions (2), (3), and (4), respectively.

$$\text{L-}\beta\text{-Lysine}\quad \underset{NH_2}{CH_2}CH_2CH_2\underset{NH_2}{CH}-CH_2COOH \rightleftharpoons CH_3\underset{NH_2}{CH}-CH_2\underset{NH_2}{CH}-CH_2COOH \tag{2}$$

$$\text{D-}\alpha\text{-Lysine}\quad \underset{NH_2}{CH_2}CH_2CH_2CH_2\underset{NH_2}{CH}-COOH \rightleftharpoons CH_3\underset{NH_2}{CH}-CH_2CH_2\underset{NH_2}{CH}-COOH \tag{3}$$

$$\text{Ornithine}\quad \underset{NH_2}{CH_2}CH_2CH_2\underset{NH_2}{CH}-COOH \rightleftharpoons CH_3\underset{NH_2}{CH}-CH_2\underset{NH_2}{CH}-COOH \tag{4}$$

The two lysine mutases have been obtained in highly purified form and a number of their cofactor requirements and properties are known, but

21. B. M. Babior, H. Kon, and H. Lecar, *Biochemistry* **8**, 2662 (1969).
22. L. Tsai and T. C. Stadtman, *ABB* **125**, 210 (1968).
23. T. C. Stadtman and P. Renz, *Federation Proc.* **26**, 343 (1967); *ABB* **125**, 226 (1968).
24. T. C. Stadtman and L. Tsai, *BBRC* **28**, 920 (1967).
25. C. G. D. Morley and T. C. Stadtman, *Biochemistry* **9**, 4890 (1970).
26. J. K. Dyer and R. N. Costilow, *J. Bacteriol.* **101**, 77 (1970).
27. Y. Tsuda and H. C. Friedman, *JBC* **245**, 5914 (1970).

the ornithine mutase has been studied only in crude enzyme preparations. Therefore, although the superficial similarities of ornithine mutase and the lysine mutases will be discussed here, detailed comparisons of the three enzymes cannot be made.

A. L-β-Lysine Mutase (L-3,6-Diaminohexanoate Mutase)

The second enzymic step in the overall fermentation of L-α-lysine to fatty acids and ammonia is the conversion of L-β-lysine (3,6-diaminohexanoate) to 3,5-diaminohexanoate [reaction (2)]. The discovery that this new amino acid is formed by a B_{12} coenzyme-dependent enzyme, β-lysine mutase (*22, 23*), explained the earlier observation that B_{12} coenzyme is one of the essential cofactors for the fermentation process (*28, 29*) and identified the specific reaction in which the coenzyme participated.

1. *Occurrence*

The reaction catalyzed by β-lysine mutase has been demonstrated in three different lysine fermenting anaerobic bacteria, *Clostridium sticklandii* (*22, 23*), *Clostridium* strain M-E (*22*), and *Clostridium* strain SB4 (*30*).

Another possible source of the enzyme is suggested by the report several years ago (*31*) that a mixed culture of two strains of *Escherichia coli* fermented lysine to 1 mole each of acetate and butyrate and 2 moles of ammonia (the same products formed by the clostridia). Although β-lysine is a constituent of several polypeptide antibiotics produced by *Steptomyces lavendulae* and related organisms (*32*) the isomeric 3,5-diamino acid has not been found in these microbial products (*33*).

2. *Enzyme Assay*

Conversion of L-β-lysine to 3,5-diaminohexanoate by β-lysine mutase can be estimated in crude bacterial extracts by thin-layer chromato-

28. T. C. Stadtman, *JBC* **237,** PC2409 (1962); *JBC* **238,** 2766 (1963).
29. T. C. Stadtman, *Ann. N. Y. Acad. Sci.* **112,** 728 (1964).
30. E. E. Dekker and H. A. Barker, *JBC* **243,** 3232 (1968).
31. P. M. Dohner and B. P. Cardon, *J. Bacteriol.* **67,** 619 (1954).
32. E. E. Smissman, R. W. Sharpe, B. F. Aycock, E. E. van Tamelen, and W. H. Peterson, *JACS* **75,** 2029 (1953); T. H. Haskell, S. A. Fusari, R. P. Frohardt, and Q. R. Bartz, *ibid.* **74,** 599 (1952); H. E. Carter, W. R. Hearn, E. M. Lansford, Jr., A. C. Page, Jr., N. P. Salzman, D. Shapiro, and W. R. Taylor, *ibid.* p. 3704.
33. T. C. Stadtman, unpublished data (1970).

graphic analysis of the amino acids in the reaction mixture (*22*). The 3,5-diaminohexanoate product is easily separated and recognized by its characteristic orange color following reaction with ninhydrin. In partially purified enzyme preparations which lack the ability to form α-lysine from β-lysine, a direct colorimetric assay using an acid ninhydrin reagent (*34*) can be used for estimation of 3,5-diaminohexanoate formation (*22*). Alternatively, a spectrophotometric assay (*35*), which couples β-lysine mutase and the next enzyme in the L-lysine fermentation pathway, namely, the DPN^+-linked 3,5-diaminohexanoate dehydrogenase (*36–38*), can be used once enzyme preparations are relatively free of a powerful DPNH-oxidase. The final products in the coupled assay system are 3-keto-5-aminohexanoate, ammonia, and DPNH. The optical assay has the additional advantage of consuming less β-lysine, a substrate that is expensive and difficult to prepare in large amounts.

3. *Purification and Physical Properties*

β-Lysine mutase consists of two dissimilar protein moieties: One is an acidic orange protein containing tightly bound cobamide coenzyme as its chromophore and the other is a colorless sulfhydryl protein. The approximate molecular weights, estimated by Sephadex gel filtration, are 150,000 for the cobamide protein and 60,000 for the sulfhydryl protein (*23*). In crude extracts the mutase exists chiefly as a complex of the two proteins. This complex is readily separated from the crude extracts by adsorption on DEAE-cellulose followed by a batch type of elution step which yields preparations that are enriched about 5-fold in mutase activity. During subsequent purification steps which include precipitation with ammonium sulfate, dialysis, gradient elution from microgranular DEAE-cellulose, chromatography on Sephadex G-150, and preparative disc gel electrophoresis, the complex from *C. sticklandii* is separated into its two component proteins and highly purified preparations of each are obtained. In contrast the β-lysine mutase complex from *Clostridium* M-E remains tightly associated throughout the same purification steps, and only upon acidification to pH 3–4 has partial dissociation of the two dissimilar protein components been achieved (*39*). The molecular basis of this difference in stability of the β-lysine mutase complexes from the two microorganisms is not known. As judged

34. F. P. Chinard, *JBC* **199**, 91 (1952).
35. J. J. Baker, unpublished experiments (1971).
36. E. A. Rimerman and H. A. Barker, *JBC* **243**, 6151 (1968).
37. J. J. Baker and H. A. Barker, *Federation Proc.* **29**, 343 (1970).
38. J. J. Baker, Ph.D. Thesis, University of California (1970).
39. C. G. D. Morley and T. C. Stadtman, unpublished experiments (1969).

from the relative affinities of the complex and its two component protein moieties for the anion exchange cellulose (DEAE), it appears that the complex is more acidic than the sulfhydryl protein but considerably less acidic than the cobamide protein (*40*). Whereas the sulfhydryl protein moiety is eluted from DEAE-cellulose with low ionic strength buffers (0.05–0.1 *M* NaCl; pH 6–7) more readily than is the complex, the cobamide protein moiety remains tightly bound and is eluted only with higher ionic strength buffers (0.3–0.5 *M* NaCl; pH 6–7).

The smaller of the two protein moieties of β-lysine mutase is inactivated at neutral pH by treatment with iodocetamide under conditions that normally favor the alkylation of sulfhydryl groups. Also, the protein tends to aggregate unless kept reduced with mercaptans in the buffer solutions. On the basis of these observations the smaller protein is considered to contain essential sulfhydryl groups. The number of sulfhydryl groups has not been determined.

The larger acidic protein moiety contains tightly bound cobamide which is not removed by any of a number of procedures that are effective in resolving several other enzymes of bound B_{12} compounds. These include prolonged treatment at 0° with ammonium sulfate at pH 3, precipitation with perchloric acid, charcoal treatment, dialysis for 2–3 days against 1 *M* KBr (*41*), and treatment with Na_2SO_3 (*42*). This failure to resolve the protein of its chromophore has prevented investigation of properties of the apoprotein. Treatment with iodoacetamide at neutral pH had no effect on the catalytic activity of the cobamide protein indicating that in its holo form it has no essential sulfhydryl groups accessible to the neutral alkylating agent.

4. *Cofactor Requirements*

An unusually large number of cofactors are required for maximal activity of β-lysine mutase. These are B_{12} coenzyme, a mercaptan, a monovalent cation, a divalent cation, ATP, pyruvate, and an unidentified charcoal adsorbable factor. The only other B_{12} coenzyme-dependent process of similar complexity yet described is that catalyzed by D-α-lysine mutase. The specificity and possible roles of the cofactors required by these two mutases are discussed in Section III,C,1.

40. T. C. Stadtman and M. A. Grant, "Methods in Enzymology," Vol. 17B, Section [168], p. 206, 1971.

41. V. Massey and B. Curti, *JBC* **241,** 3417 (1966).

42. Z. Schneider, E. G. Larsen, G. Jacobson, B. C. Johnson, and J. Pawelkiewicz, *JBC* **245,** 3388 (1970).

5. *Inhibitors*

The activity of β-lysine mutase can be inhibited completely by the addition of intrinsic factor in amounts sufficient to bind the cobamide compounds present on the enzyme. This inhibition is specifically reversed by the addition of a slight excess of cobalamin coenzyme. Coenzyme degradation products such as hydroxycobalamin also are markedly inhibitory.

Various amine buffers such as tris and its analogs are inhibitory, whereas 2-methylimidazole (pH about 8) is a satisfactory buffer. Ammonium salts at concentrations greater than 20 mM inhibit β-lysine mutase activity significantly (*43*) if the reaction mixtures already contain potassium ion (30–60 mM). Under the same conditions sodium ion is only slightly inhibitory. In the absence of potassium ion and at low concentrations, ammonium ion may activate the enzyme partially (*33*). The ammonium ion appears to fulfill the requirement for a monovalent cation activator to some extent, but at higher concentrations some specific inhibitory effect obscures its role as an activator.

6. *Reversibility of β-Lysine Mutase Reaction*

As indicated in reaction (2), β-lysine mutase catalyzes the interconversion of the two diaminohexanotes (*23*). The yield of β-lysine from 3,5-diaminohexanoate is considerably greater in the back reaction if pyruvate is omitted from the incubation mixtures. All of the other cofactors required for the forward reaction (see Section III,C) appear to be needed. Addition products between pyruvate and 3,5-diaminohexanoate, which undoubtedly tend to favor the reaction in the forward direction, probably inhibit the mutase reaction in the reverse direction. In complete reaction mixtures lacking added pyruvate (10–40 mM), 3,5-diaminohexanoate conversion to β-lysine occurs to the extent of 33%. In the forward direction substrate can be converted to product to the extent of about 50%. The precise equilibrium position of the reaction is not known.

B. D-α-LYSINE MUTASE

A second lysine fermentation pathway catalyzed by *C. sticklandii* results in the conversion of carbons 5 and 6 of α-lysine to carbons 1

43. R. C. Bray and T. C. Stadtman, *JBC* **243**, 381 (1968).

and 2, respectively, of acetic acid and conversion of the remainder of the carbon chain to butyric acid (*44*). Although the ability of various lots of cells to carry out this overall fermentation sequence is exceedingly variable, most lots of cells and enzyme preparations thereof do catalyze the B_{12} coenzyme-dependent conversion of D-α-lysine to 2,5-diaminohexanoate [reaction (3)]. This new amino acid is presumed to be the first intermediate in the series of reactions that yields acetic acid from carbons 5 and 6 of α-lysine (*24, 25*). The reaction that forms 2,5-diaminohexanoate has many superficial similarities to that catalyzed by β-lysine mutase. In both reactions an ω-amino group migrates to an adjacent carbon atom to replace a hydrogen that migrated in the opposite direction. The enzyme that forms 2,5-diaminohexanoate, D-α-lysine mutase, has been detected so far only in *C. sticklandii* and *Clostridium* M-E.

1. *Purification of* D-α-*Lysine Mutase Complex*

A purification procedure that yields essentially homogeneous preparations of the D-α-lysine mutase complex has been devised (*25*). In this procedure the two dissimilar protein moieties that make up the complex are maintained in the associated form throughout the purification and then, if desired, are dissociated from the pure complex by an acidification step. Once separated the two D-α-lysine mutase components, an acidic cobamide protein of about 150,000 MW and a sulfhydryl protein of about 60,000 MW, are indistinguishable by a variety of physical means from the β-lysine mutase components described above. Although the sulfhydryl protein or E_2 components of both mutases appear to be interchangeable, the cobamide protein components differ in catalytic activity and are specific for their respective substrates. The activity of the reconstituted D-α-lysine mutase complex is usually somewhat lower than that of the intact isolated complex. Whether this is the result of partial inactivation of the more labile sulfhydryl protein moiety (E_2) or the loss of some, as yet, unrecognized activator is not known.

2. *Cofactor Requirements*

Like β-lysine mutase, D-α-lysine mutase is unusual in its requirement for several cofactors. These are B_{12} coenzyme, a mercaptan, a monovalent cation, a divalent cation, ATP, and pyridoxal phosphate. Evidence concerning the roles of some of these factors is discussed later (Section III,C,1).

44. T. C. Stadtman, *in* "Amino Acid Metabolism" (W. D. McElroy and B. Glass, eds.), p. 493. Johns Hopkins Press, Baltimore, Maryland, 1955.

3. *Inhibitors*

Several amino acids structurally related to α-lysine are potent inhibitors of D-α-lysine mutase activity (*25*). The substrates for the two related amino acid mutases, namely, L-β-lysine and ornithine, are especially inhibitory even in the presence of a 10-fold molar excess of D-α-lysine. Also effective are *S*-aminoethylcysteine and DL-ε-*N*-acetyllysine.

Compounds known to be antagonists of pyridoxal phosphate, e.g., 1-amino-D-proline and isonicotinic acid hydrazide, are inhibitors of D-α-lysine mutase (*45*). The mutase is inactivated by treatment with hydroxylamine (1–10 m*M*), by acetylation with *N*-acetylimidazole, by reduction with sodium borohydride, and by nitration with tetranitromethane (*45*). The hydroxylamine-treated enzyme, after dialysis, is reactivated by the addition of pyridoxal phosphate. Treatment of the acetylated enzyme with hydroxylamine followed by dialysis and addition of pyridoxal phosphate restores its catalytic activity. Inactivation of D-α-lysine mutase by nitration occurs only when the apoprotein freed of pyridoxal phosphate is treated. The enzyme is protected from tetranitromethane inactivation by the addition of pyridoxal phosphate. It is known that treatment of the apoenzyme form of aspartate aminotransferase with tetranitromethane nitrates one specific tyrosine residue on the protein, and the resulting enzyme is inactive by virtue of the fact that it can no longer bind its essential cofactor, pyridoxal phosphate (*46*). It will be interesting if a tyrosine of D-α-lysine mutase also is the group that is protected from nitration by pyridoxal phosphate.

Inhibitors or treatments which inactivate B_{12} coenzyme also affect the activity of D-α-lysine mutase. Inhibition by intrinsic factor, which is complete if the glycoprotein is added in sufficient amounts, is reversed by the addition of B_{12} coenzyme. Hydroxycobalamin, either added or formed by photolysis of B_{12} coenzyme bound to the protein, is markedly inhibitory.

4. *Partial Reactions Catalyzed by Cobamide Protein Moiety*

The cobamide protein moiety of D-α-lysine mutase catalyzes a slow exchange of a hydrogen on carbon-6 of lysine with water (*45*). Both release of tritium from 6-^{3}H-α-lysine and incorporation of tritium from ^3H_2O into lysine were demonstrated. The only cofactors that appear

45. C. G. D. Morley and T. C. Stadtman, *Biochemistry* Feb. 15 issue (1972).

46. C. Turano and co-workers; quoted by P. Christen and J. F. Riordan, *Biochemistry* **9**, 3025 (1970).

to be required for this exchange reaction are pyridoxal phosphate and magnesium ion. Other components essential for catalysis of the overall mutase reaction, namely, the sulfhydryl protein moiety, B_{12} coenzyme, a mercaptan, and ATP, are either without effect or somewhat inhibitory. Neither the hydrogen exchange activity nor the mutase activity of the cobamide protein moiety is sensitive to alkylation with iodoacetamide. This treatment, which effectively inactivates residual contaminating lysine racemase, yields cobamide protein preparations that are specific for either D-α-lysine or 2,5-diaminohexanoate as substrates and are inactive on L-α-lysine. In addition to the identical substrate requirements for mutase and hydrogen exchange activities of the D-α-lysine mutase cobamide protein moiety, both activities are inhibited by hydroxylamine, by the vitamin B_6 antagonists, 1-amino-D-proline and isonicotinic acid hydrazide, and are completely dependent on pyridoxal phosphate. Although the rate of the hydrogen exchange reaction is only about one-tenth that of the overall mutase reaction under comparable experimental conditions, it nevertheless suggests that binding of the amino acid substrates to the cobamide protein moiety by Schiff base formation between the ω-amino groups and pyridoxal phosphate may be an integral part of the amino group migration reaction.

5. *Formation of α-Lysine from 2,5-Diaminohexanoate*

The D-α-lysine mutase reaction is reversible and the formation of lysine from 2,5-diaminohexanoate is readily demonstrable by thin-layer chromatographic analysis (*47*). Each of the partial reactions involving transfer of a hydrogen from carbon-5 of lysine to B_{12} coenzyme and a hydrogen from carbon-6 of 2,5-diaminohexanoate to B_{12} coenzyme is also reversible as will be discussed later. Although the equilibrium position of the overall reaction has not been determined, under similar reaction conditions (i.e., about 2.5 mM amino acid substrate) lysine was converted to 2,5-diaminohexanoate to the extent of 40–45%, and in the back reaction lysine yields of 25% were obtained. The absolute configuration at carbon-5 of the 2,5-diaminohexanoate formed in the enzyme reaction has not been determined.

C. Comparison of Some Properties of L-β-Lysine and D-α-Lysine Mutases

The complex cofactor requirements, the wide variability in catalytic activity of enzyme preparations of similar degrees of purity, and the

47. C. G. D. Morley and T. C. Stadtman, *Biochemistry* **10**, 2325 (1971).

heat-stable activator of β-lysine mutase is tightly bound to the complex in such a way that it is not removed by dialysis. However, the material is gradually separated from the proteins, particularly during chromatographic steps involving prolonged exposure to DEAE-cellulose and high ionic strength buffers. The resulting large apparent loss of β-lysine mutase activity can be restored by the addition of a heated extract of the crude β-lysine mutase preparation (*48*). The active component, which is dialyzable after removal from the protein by boiling, is adsorbed by charcoal and eluted with ammoniacal ethanol. It is amphoteric and is not inactivated by heating at 100° for 10 min in either 1 N HCl or 1 N NaOH. Treatment for 16 hr with 7% H_2O_2 at room temperature destroyed the factor. No decrease in activity of the factor was detected after incubation with snake venom phosphodiesterase suggesting that the substance is not a dinucleotide. A large number of known cofactors of similar properties failed to replace the active material. No enzyme preparations have been studied that exhibit an absolute requirement for the factor; rather its effect is to stimulate the reaction about 40–50%. This suggests that either the unknown factor merely modulates the activity of β-lysine mutase or the enzyme preparations examined were incompletely resolved of the material. No effect of the purified factor on D-α-lysine mutase activity has been detected.

D. Ornithine Mutase

The ornithine mutase of *C. sticklandii* (*25–27*) catalyzes the migration of the δ-amino group of ornithine to carbon-4 to form a new amino acid, 2,4-diaminopentanoate [reaction (4)]. Although not yet studied in detail, this enzyme appears to resemble D-α-lysine mutase in many of its properties. It requires B_{12} coenzyme, pyridoxal phosphate, and a mercaptan for activity. Pyridoxamine phosphate does not replace pyridoxal phosphate as cofactor, and hydroxylamine (0.5 mM) inhibits the reaction (*27*). Inhibition of the mutase by intrinsic factor is reversed by the addition of B_{12} coenzyme (*26, 27*). The substrate presumably is D-ornithine (*27*).

IV. Reaction Mechanisms

A. General Considerations

In the ethanolamine deaminase reaction [reaction (1)] a hydrogen that was abstracted from the carbinol carbon of the substrate replaces the amino group that is subsequently lost and becomes one of the hydrogens of the methyl group of acetaldehyde (*50*). Similarly, in the α- and β-lysine mutase reactions [reactions (2) and (3)] a hydrogen abstrated from the carbon-5 position of the amino acid replaces the amino group which migrates from carbon-6 to carbon-5 and in the back reactions the process is reversed (*47*, *51*). In a formal sense the ethanolamine deaminase and amino acid mutase reactions may be considered to involve a similar mechanism. However, migration of the amino group of ethanolamine to the carbinol carbon after abstraction of the hydrogen would result in the formation of an unstable 1-amino-1-ol derivative which would spontaneously decompose to acetaldehyde and ammonia. In the analogous reaction catalyzed by diol dehydrase, it was deduced that a 1,1-diol is an intermediate and this eliminates water to form the aldehyde product (*52*). Whereas one-half of the oxygen in the aldehyde product of the diol dehydrase reaction is derived from the migrating hydroxyl group (*52*), in the ethanolamine deaminase reaction the aldehyde oxygen comes exclusively from the 1-carbinol oxygen (*50*) and the migrating group ($-NH_2$) is lost quantitatively. In the amino acid mutase reactions the amino group migrates to a carbon atom whose other substituent is a hydrogen and the resulting derivative is stable.

In the diol dehydrase reaction it was postulated that the unstable 1,1-diol was preceded by a half-acetal intermediate (*52*). Formation of a hydroxy aziridine type of intermediate prior to the shift of the amino group would be the analogous mechanism in the ethanolamine deaminase reaction. If an aziridine derivative is involved in the amino acid mutase reactions (*53*), then in the transformation of α-lysine to 2,5-diaminohexanoate the intermediate might be a pyridoxal phosphate-substituted aziridine (*25*).

It has been established in the β-lysine mutase reaction that the nitrogen of the ε-amino group is transferred quantitatively to the C-5 amino group of 3,5-diaminohexanoate and that no exchange with ammonia nitrogen of the medium occurs (*43*).

B. Cobamide Coenzyme as Hydrogen Carrier

The hydrogen-carrying role of B_{12} enzyme has been established for several enzymes that utilize it for catalysis of carbon–carbon bond

50. B. M. Babior, *JBC* **244,** 449 (1969).
51. J. Rétey, F. Kunz, T. C. Stadtman, and D. Arigoni, *Experientia* **25,** 801 (1969).
52. J. Rétey, A. Umani-Ronchi, J. Seibl, and D. Arigoni, *Experientia* **22,** 502 (1966).
53. J. Rétey, personal communication (1969).

cleavages or carbon–oxygen bond cleavages (see Abeles, Chapter 17, Volume V, and Barker, Chapter 14, this volume). The coenzyme plays the same role in the reactions involving carbon–nitrogen bond cleavages discussed in this chapter. The migrating hydrogen in each case is transferred to and from the 5′-methylene position of the 5′-deoxyadenosyl moiety covalently linked to the cobalt atom of the coenzyme. Transfer of tritium from B_{12} coenzyme specifically labeled in the 5′-methylene position of the 5′-deoxyadenosyl moiety (*54*) to acetaldehyde by ethanolamine deaminase (*55*), to 3,5-diaminohexanoate by β-lysine mutase (*51*) and to 2,5-diaminohexanoate by α-lysine mutase (*47*) was demonstrated. In each instance the tritium was located in the terminal methyl group of the product. In the case of the two amino acid mutases the residual substrate also was labeled with tritium indicating that the half-reaction involving transfer of hydrogen from substrate to coenzyme is appreciably reversible. Moreover, in the β-lysine mutase reaction it was demonstrated that tritium was introduced from the coenzyme into the 5-carbon position of β-lysine stereospecifically, and in a subsequent reaction with unlabeled coenzyme the same labeled atom was removed from this position and transferred to the 6 position of 3,5-diaminohexanoate (*51, 56*). Hence, both the hydrogen abstraction and transfer processes are stereospecific in the β-lysine mutase reaction. There is a pronounced isotope effect of tritium on the overall transfer reaction from C-5 of β-lysine to C-6 of 3,5-diaminohexanoate.

In the ethanolamine deaminase reaction there is also discrimination between the hydrogens on the carbinol carbon and there is a pronounced isotope effect with either deuterium or tritium (*50*).

Formation of 5′-^{3}H-deoxyadenosyl-B_{12} coenzyme from 4,5-^{3}H-α-lysine and from 6-^{3}H-2,5-diaminohexanoate by D-α-lysine mutase (*47*) and from 1,1-^{3}H-ethanolamine by ethanolamine deaminase (*50*) has been demonstrated.

C. Mechanism of Hydrogen Transfer Reaction

Although there is still some uncertainty as to the species of hydrogen that is transferred in the B_{12} coenzyme-dependent reactions, there is increasing evidence that it is a hydrogen radical that is abstracted from the substrate and that a transient homolytic cleavage of the carbon–cobalt bond of the coenzyme enables this radical to be accepted by the

54. P. A. Frey, S. S. Kerwar, and R. H. Abeles, *BBRC* **29**, 873 (1967).
55. B. Babior, *BBA* **167**, 456 (1968).
56. F. Kunz, Dissertation No. 4519, Edg. Technische Hochschule, Zürich, Verlag Zürich, 1970.

5′-deoxyadenosyl moiety. In the studies of Babior and associates with substrate levels of ethanolamine deaminase the appearance of unpaired electrons during the course of the enzyme-catalyzed reaction was demonstrated spectroscopically (*57*) and also the cleavage of the coenzyme to a fragment that appears to be 5′-deoxyadenosine was detected (*58*). The deoxyadenosine moiety is tightly bound to the protein and does not dissociate in the normal course of the reaction. Instead, after giving up a hydrogen from the 5′-methyl group it is presumed to re-form its bond with the cobalt atom. This alternate breaking and remaking of the carbon–cobalt bond would enable the nonequivalent hydrogens of the 5′-methylene group to become equivalent in the methyl group and also could explain how inter- as well as intramolecular hydrogen transfers occur (*59, 60*).

Another poorly understood aspect of the hydrogen transfer process catalyzed by B_{12} coenzyme-dependent enzymes is the exchange of hydrogen between two different enzyme systems in a mixture. For example, in a reaction mixture containing β-lysine mutase plus substrate, B_{12} coenzyme and other cofactors together with tritiated propanediol and diol dehydrase, there was significant incorporation of tritium into both residual β-lysine and the reaction product 3,5-diaminohexanoate (*60*). It is not known whether the entire coenzyme, a part of the coenzyme, or only the hydrogen was transferred from one enzyme to the other.

V. Role of Amino Group Migration Reactions in Bacterial Fermentations

In the overall fermentation of lysine to fatty acids and ammonia the 6-carbon atom chain eventually is cleaved between carbon atoms two and three on one pathway and between carbon atoms four and five on the other. Preliminary migration of amino groups, two of which are mediated by B_{12} coenzyme, prepare the amino acid substrate for conversion to ketoacid derivatives that can then undergo normal thiolytic cleavage and conversion to the final fatty acid products. This enables the same enzymes that form fatty acid products from diverse substrates to catalyze the energy-yielding portion of the fermentations once the suitable isomers of the diamino acid have been formed. In the ethanolamine

57. B. Babior and D. G. Gould, *BBRC* **34**, 441 (1969).
58. B. M. Babior, *JBC* **245**, 6125 (1970).
59. P. A. Frey, M. K. Essenberg, and R. H. Abeles, *JBC* **242**, 5369 (1967).
60. J. Rétey and T. C. Stadtman, unpublished experiments (1970).

fermentation the B_{12} coenzyme-mediated process serves the same function, i.e., a substrate which is not a common intermediate is converted to acetaldehyde which then is readily utilized by enzymes shared by many metabolic pathways of the cell. In this respect the B_{12} coenzyme-catalyzed amino group migration reactions resemble several of the reactions that involve carbon–carbon bond cleavages or carbon–oxygen bond cleavages (see Abeles, Chapter 17, Volume V, and Barker, Chapter 14, this volume).

16

Isopentenylpyrophosphate Isomerase

P. W. HOLLOWAY

I. Introduction

A. Occurrence

The enzyme isopentenylpyrophosphate isomerase (EC 5.3.3.2) catalyzes the reaction shown in Eq. (1).

$$\text{Isopentenyl pyrophosphate} \rightleftharpoons \text{dimethylallyl pyrophosphate} \quad (1)$$

The enzyme was first recognized in extracts of baker's yeast by Agranoff *et al.* (*1*) who purified the enzyme 6-fold from yeast autolyzates. It has since been partially purified from pig liver (*2, 3*) and pumpkin fruit (*4*)

1. B. W. Agranoff, H. Eggerer, U. Henning, and F. Lynen, *JBC* **235,** 326 (1960).
2. D. H. Shah, W. W. Cleland, and J. W. Porter, *JBC* **240,** 1946 (1965).
3. P. W. Holloway and G. Popják, *BJ* **106,** 835 (1968).
4. K. Ogura, T. Nishino, and S. Seto, *J. Biochem.* (*Tokyo*) **64,** 197 (1968).

and has been detected in orange juice vesicles (*5*), *Mycoplasma laidlawii* (*6*), and *Pinus radiata* (*7*).

B. Function

The realization that polyisoprenoids were synthesized from a C_5 precursor, isopentenyl pyrophosphate, prompted speculation as to the mechanism of the condensation reactions. Both Lynen *et al.* (*8*) and Rilling and Bloch (*9*) suggested that before the initial condensation of two isopentenyl pyrophosphates could take place one of the isopentenyl pyrophosphate molecules had to be isomerized to dimethylallyl pyrophosphate. The subsequent condensation reaction would involve the electrophilic attack of the allylic pyrophosphate, dimethylallyl pyrophosphate, on the terminal methylene group of the isopentenyl pyrophosphate with concerted elimination of a proton and the allylic pyrophosphate group. This reaction, catalyzed by the prenyltransferase, results in the formation of geranyl pyrophosphate, which is also an allylic pyrophosphate that can react with another molecule of isopentenyl pyrophosphate to yield farnesyl pyrophosphate. In animal systems the prenyltransferase can be terminated at farnesyl pyrophosphate; two of these C_{15} pyrophosphates can react in the presence of squalene synthase to yield squalene. Higher prenyl pyrophosphates can be formed in some systems, especially plants, by further addition of isopentenyl pyrophosphate to the allylic pyrophosphate. Since the isomerase forms the dimethylallyl pyrophosphate required to initiate the condensation, the isomerase is presumed to be present in any system synthesizing polyisoprenoid compounds, although difficulty has been experienced in demonstrating a requirement for the enzyme in rubber biosynthesis (*10*).

The isomerase may also be involved in the synthesis of nonpolyisoprenoids. Dimethylallyl pyrophosphate may be the immediate precursor of the dimethylallyl group found attached to nitrogen or carbon in various natural products. This suggestion has been supported by the incorporation of radioactivity from radioactive dimethylallyl pyrophosphate into

5. V. H. Potty and J. H. Bruemmer, *Phytochemistry* **9,** 1229 (1970).
6. C. V. Henrikson and P. F. Smith, *J. Bacteriol.* **92,** 701 (1966).
7. E. Beytia, P. Valenzuela, and O. Cori, *ABB* **129,** 346 (1969).
8. F. Lynen, B. W. Agranoff, H. Eggerer, U. Henning, and E. M. Moslein, *Angew. Chem.* **71,** 657 (1959).
9. H. C. Rilling and K. Bloch, *JBC* **234,** 1424 (1959).
10. B. L. Archer and B. G. Audley, *Advan. Enzymol.* **29,** 221 (1967).

N[6]-dimethylallyladenosine in tRNA (*11*) by a partially purified enzyme from yeast; isopentenyl pyrophosphate could not substitute.

II. The Catalytic Reaction

A. Synthesis of Substrate and Product

All assays of isopentenylpyrophosphate isomerase depend upon the availability of radioactive isopentenyl pyrophosphate. This material can be conveniently synthesized on a small scale from 2-^{14}C-mevalonic acid by a crude enzyme system from pig liver (*12, 13*). Larger quantities of radioactive substrate can be made from 1-^{14}C-isopentenol (*14*). Dimethylallyl pyrophosphate can be made chemically (*15, 16*) or, in small quantities, by the isomerase (*11*).

B. Evaluation of Assays

Several assays have been used to measure the isomerization of radioactive isopentenyl pyrophosphate. Most assays are based on the acid lability of allylic pyrophosphates. At the end of the enzymic reaction, the reaction is terminated by addition of acid which hydrolyzes the product, dimethylallyl pyrophosphate, to dimethylallyl alcohol and dimethylvinyl carbinol. In contrast, the substrate, isopentenyl pyrophosphate, is stable to mild acid treatment. The radioactive alcohols can be isolated by extraction with ether or light petroleum and the extract assayed for radioactivity (*2, 3*). Alternatively, the amount of isomerization can be deduced by estimating the amount of radioactive isopentenyl pyrophosphate remaining after acidification and removal of the dimethylallyl alcohol and dimethylvinyl carbinol. These alcohols may be removed either by extraction as described above (*1*) or by evaporation (*17*). The assays which determine the amount of product formed

11. L. K. Kline, F. Fittler, and R. H. Hall, *Biochemistry* **8**, 4361 (1969).
12. J. K. Dorsey, J. A. Dorsey, and J. W. Porter, *JBC* **241**, 5353 (1966).
13. P. W. Holloway and G. Popják, *BJ* **104**, 57 (1967).
14. C. Donninger and G. Popják, *BJ* **105**, 545 (1967).
15. H. Plieninger and H. Immel, *Chem. Ber.* **98**, 414 (1965).
16. R. H. Cornforth and G. Popják, "Methods in Enzymology," Vol. 15, p. 359, 1969.
17. T. T. Tchen, "Methods in Enzymology," Vol. 6, p. 505, 1963.

are most sensitive, especially when low levels of enzymic activity are present.

The product of the isomerase is, as stated previously, the substrate of the prenyltransferase; hence, in crude enzyme systems synthesizing polyprenyls, where both enzymes will be present, the ultimate product will be higher prenyl pyrophosphates. When the incubation is terminated with acid these too will be hydrolyzed and extracted and this will lead to erroneously high values for "isomerase" activity. This observation was exploited in a coupled assay (*3*). The isomerization was allowed to occur in the presence of endogenous, or added, prenyltransferase and the ultimate product, in this case (*3*) farnesyl pyrophosphate, was hydrolyzed and extracted with light petroleum. The specific activity of the isomerase calculated from assay of the extract was three times the true isomerase activity as three radioactive C_5 units appear in the final product. This coupled assay also circumvents difficulties resulting from the presence of phosphatase activity in the crude enzyme system. Phosphatase activity would liberate isopentenol, which would give erroneously high values for isomerase activity with the simple assay but would not affect the coupled assay (*3*).

None of the above assays is unequivocal for assay of enzymic activity in crude enzyme preparations. The direct assay can only be valid if prenyltransferase and phosphatase are absent; this has only been rigorously shown in two enzyme preparations to date (*3, 4*).

C. Properties of the Enzyme

The enzyme has not been obtained pure from any source; the highest specific activities obtained, in millimicromoles of substrate formed per minute per milligram of protein, are yeast enzyme 100 (*1*) and pig liver enzyme 22 (*3, 18*). The molecular weight has been estimated to be 60,000 by gel filtration and ultracentrifugation. The enzyme is stable at —20° for several weeks.

1. *pH Optimum*

The yeast enzyme had a broad pH optimum from 5.5 to 9.3 (*1*). The same pH profile was observed with one preparation from pig liver (*2*) whereas a second preparation from pig liver had a sharp optimum at pH 6 (*3*).

18. G. Popják, "Methods in Enzymology," Vol. 15, p. 393, 1969.

2. *Metal Ion Requirement*

The yeast enzyme had an absolute requirement for Mg^{2+} (*1*), and one preparation from pig liver had a partial requirement for Mg^{2+} which could not be met by Mn^{2+} (*2*). The preparation from pig liver made by Holloway and Popják (*3*) had an absolute requirement for divalent metal ion, but in this instance Mn^{2+} was more effective than Mg^{2+}. After dialysis of the enzyme preparation against EDTA there was a specific requirement for Mn^{2+} (*3*). The discrepancy between the data of Shah *et al.* (*2*) and Holloway and Popják (*3*) for the pig liver enzyme may result from an interdependence of pH optimum and metal ion specificity or from different states of purity of the two preparations.

3. *Michaelis Constant*

The K_m for isopentenyl pyrophosphate was established as $3.6 \times 10^{-5}\ M$ for the yeast enzyme (*1*) and $4 \times 10^{-6}\ M$ for the pig liver enzyme (*3*).

4. *Inhibitors*

The enzyme from yeast, pig liver, and pumpkin is completely inhibited by 2 m*M* iodoacetamide. Agranoff *et al.* (*1*) observed that the susceptibility of the isomerase to iodoacetamide inhibition was unique among the enzymes required for conversion of mevalonate to squalene. This observation, made in yeast, has led others to conclude that during polyisoprenoid synthesis in other systems the only enzyme inhibited by iodoacetamide will be the isomerase. This conclusion cannot be universally made since it has been demonstrated in liver that the prenyltransferase is also susceptible to iodoacetamide (*13*). If the addition of iodoacetamide to a crude enzyme system causes inhibition of mevalonate conversion into products or causes the accumulation of isopentenyl pyrophosphate, this is not unambiguous proof of the existence of isopentenylpyrophosphate isomerase.

D. Reversibility and Position of Equilibrium

The reaction catalyzed by the yeast enzyme was shown to be reversible by the incorporation of tritium into isopentenol when dimethylallyl pyrophosphate was incubated with the enzyme in tritiated water (*1*). The same technique was used by Shah *et al.* with pig liver enzyme (*2*). The direct conversion of dimethylallyl pyrophosphate to isopentenyl pyrophosphate was also demonstrated with the pig liver enzyme (*3*). When mevalonate, stereospecifically labeled at C-2 with tritium, was converted by various systems into terpenoids, randomization of label was observed. This randomization was explained by the reversibility of the isomerase (*19–21*).

The equilibrium constant [dimethylallyl pyrophosphate]/[isopentenyl pyrophosphate] is about nine for yeast (*1*) and pig liver enzymes (*2, 3*).

E. Mechanism

During the isomerization of isopentenyl pyrophosphate to dimethylallyl pyrophosphate a "new" methyl group is created from the original methylene group. That the new methyl group ($^*CH_3$) (Fig. 1) was trans to the $-CH_2OPP$ group was indicated by experiments on the incorporation of 2-^{14}C-mevalonate into soyasapogenol A (*22*) and mycelianamide (*23*). With the availability of purified isomerase (*3*) it should be possible to perform the isomerization in D_2O and isolate the product dimethylallyl alcohol. The only hydrogens replaced by deuterium should be those on the asterisked carbon and the replacement should be complete (*1*). Nuclear magnetic resonance spectroscopy would then be able to distinguish between the expected product or the alternative compound containing deuterium on the carbon of the un-asterisked methyl group.

The stereochemistry of the eliminated hydrogen was deduced by Cornforth *et al.* (*24*) to be of the pro-*R* configuration (*25*) (Fig. 1). It has been suggested that this stereochemistry is only adhered to in systems synthesizing *trans*-polyprenyls and that the isomerization of isopentenyl pyrophosphate to dimethylallyl pyrophosphate in systems synthesizing

$$H^+ + H_2\overset{*}{C}{=}C(CH_3){-}C(H_R)(H^{\dagger}_S){-}CH_2OPP \rightleftharpoons H_3\overset{*}{C}{-}C(CH_3){=}C(H^{\dagger}){-}CH_2OPP + H^+$$

Fig. 1. Stereochemistry of the reaction catalyzed by isopentenylpyrophosphate isomerase.

19. A. R. H. Smith, L. J. Goad, and T. W. Goodwin, *Chem. Commun.* p. 926 (1968).

20. G. F. Gibbons, L. J. Goad, and T. W. Goodwin, *Chem. Commun.* p. 1212 (1968).

21. C. J. Coscia, L. Botta, and R. Guarnaccia, *ABB* **136**, 498 (1970).

22. D. Arigoni, *Experientia* **14**, 153 (1958).

23. A. J. Birch, M. Kocor, N. Sheppard, and J. Winter, *JCS* p. 1502 (1962).

24. J. W. Cornforth, R. H. Cornforth, G. Popják, and L. Yengoyan, *JBC* **241**, 3970 (1966).

25. G. Popják, "The Enzymes," 3rd ed., Vol. 2, p. 115, 1970.

cis-polyprenyls (e.g., rubber) would involve loss of the pro-*S* hydrogen of isopentenyl pyrophosphate (*3*). It has not been possible to verify this latter suggestion experimentally since rubber biosynthesis appears to proceed via addition of C_5 units onto preexisting rubber particles. Hence, only very little, if any, isomerization would be required (*10*). The stereochemistry of the isomerization which occurs during formation of *C*- or *N*-dimethylallyl residues in natural products is not known.

On the basis of the stereochemistry shown in Fig. 1 and the observed stereochemistry of the prenyltransferase, Cornforth *et al.* (*24*) proposed that the isomerization followed the overall stereochemistry shown in Fig. 2. The nature of the X group was not specified; it could be a second sulfhydryl group on the enzyme. The stereochemistry of the isomerization is still incomplete in that it has not been shown that the proton is definitely added to the side of the double bond shown in Fig. 2. It was suggested (*24*) that this remaining stereochemical problem of squalene biosynthesis could be solved by use of all three isotopes of hydrogen, and the theoretical and practical aspects of interconversions of vinyl and methyl groups have recently been discussed by Popják (*25*).

Agranoff *et al.* (*1*) were able to demonstrate the incorporation of tritium from tritiated water into isopentenol when dimethylallyl pyrophosphate was incubated with the isomerase. Shah *et al.* (*2*) made a thorough analysis of the incorporation of tritium into both isopentenyl and dimethylallyl pyrophosphates and observed that when tritiated water was added to an incubation which had already reached equilibrium then (1) isopentenyl pyrophosphate exchanged a proton at a much faster rate than dimethylallyl pyrophosphate at all times; and (2) initially, almost all of the tritium was found at C-4 of isopentenol, however, as time elapsed C-2 of isopentenol gradually became labeled. These data were consistent with a carbonium ion mechanism (Fig. 3) but would be inconsistent with either a carbanion mechanism, which would require the initial incorporation of tritium to be at C-2 of isopentenol, or a concerted mechanism. This mechanism is similar to one proposed earlier by Agranoff *et al.* (*1*) and is in agreement with the stereochemical data (Fig. 2).

Although the data of Shah *et al.* (*2*) are in agreement with the car-

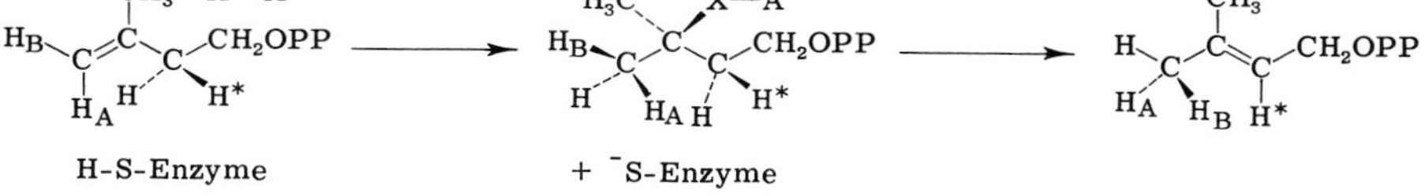

CH₃ ⁻X—A
HB C=C C CH₂OPP
HA H H*
H-S-Enzyme

H₃C X—A
HB C C C CH₂OPP
H HA H H*
+ ⁻S-Enzyme

CH₃
H C C=C CH₂OPP
HA HB H*

FIG. 2. Hypothetical reaction scheme for isopentenylpyrophosphate isomerase.

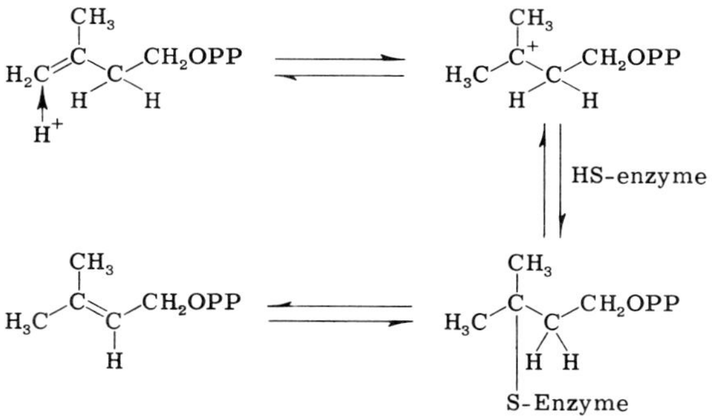

FIG. 3. Carbonium ion mechanism for isomerization of isopentenyl pyrophosphate.

bonium ion mechanism proposed by them they are, as Rose has pointed out (*26*), also consistent with a two-site carbanion mechanism, provided no exchange occurs with the carbanion intermediate. Rose has also pointed out how the carbonium ion or carbanion mechanisms could be distinguished by measuring deuterium isotope effects. Such studies should await a more highly purified preparation, or, at least, a preparation known to be uncontaminated with prenyltransferase and phosphatase.

Investigations of the active site of the isomerase have been limited to demonstrating the susceptibility of the enzyme to sulfhydryl reagents and one preliminary publication on the affect of substrate analogs. Ogura *et al.* (*27*) found similar degrees of inhibition with inorganic pyrophosphate, allyl pyrophosphate, *trans*-2-butenyl pyrophosphate, neryl pyrophosphate, and geranyl pyrophosphate but found little inhibition with the corresponding monophosphates. This would suggest that the major force which binds the substrate to the enzyme is concerned with the pyrophosphate and that the active site must be relatively unhindered. This is in contrast to similar experiments performed with prenyltransferase where the lipophilic forces appear to be more important than the pyrophosphate binding force (*27, 28*).

26. I. A. Rose, "The Enzymes," 3rd ed., Vol. II, p. 301 (1970).

27. K. Ogura, T. Koyama, T. Shibuya, T. Nishino, and S. Seto, *J. Biochem.* (*Tokyo*) **66,** 117 (1969).

28. G. Popják, P. W. Holloway, R. P. M. Bond, and M. Roberts, *BJ* **111,** 333 (1969).

17

Isomerizations in the Visual Cycle

JORAM HELLER

I. Introduction

The aim of this chapter is to describe the properties of visual pigments and their interactions with light and with enzymes. Visual pigments are conjugated proteins with either retinal or 3-dehydroretinal as the prosthetic chromophoric group. The multiple double bonds of the chromophore allow several geometrical isomers, only one of which is present in native visual pigment. The absorption of light results in the *cis* → *trans* isomerization of the chromophore. A series of dark reactions are necessary to reisomerize retinal to its original shape and thus regenerate the native pigment. The known transformations of retinal by light, the changes which then occur in the protein, and the reactions that lead to the regeneration of the native photosensitive pigment make it possible to describe in some detail the initial steps in vision which take place in the photoreceptors.

The biochemistry of visual pigments was recently discussed in several reviews (*1–3*), and the reader is referred to these reviews for information about particular visual pigments.

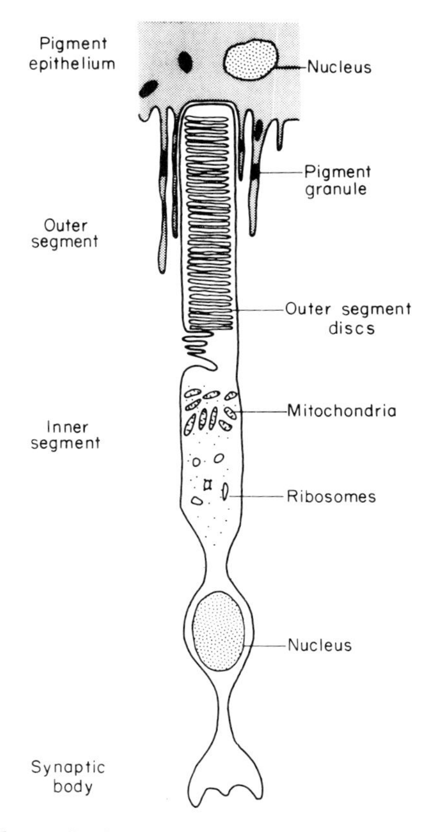

FIG. 1. Rod photoreceptor cell—schematic.

1. G. Wald, *Science* **162**, 230 (1968).
2. C. D. B. Bridges, *Compr. Biochem.* **27**, 31 (1967).
3. R. A. Morton and G. A. J. Pitt, *Advan. Enzymol.* **32**, 97 (1969).

II. Site of Visual Pigments in the Photoreceptor

Visual pigments are found in the outer segments of rod and cone photoreceptors. The outer segment is an extension of the photoreceptor cell body, and it is filled with a closely packed array of membranes which appear in electron micrographs as a neatly arranged stack of discs (*4, 5*) (Fig. 1). Visual pigments are an integral structural component of the photoreceptor disc membrane (*6–8*) and are highly oriented within the membrane (*6, 9*). Although the precise arrangement of the various components in the membrane is not yet known, several tentative models have been proposed (*8, 10, 11*). The model of Vanderkooi and Sundaralingam consists of a double layer of globular protein molecules with a lipid bilayer filling the spaces between the proteins. It is assumed that the protein molecules "float" half-submerged in a lipid field and that this, presumably, serves to orient them with respect to the incident illumination.

III. Molecular Properties of Visual Pigments

A. Preparation

Visual pigments are prepared from purified photoreceptor outer segments by extraction with various detergents. The methods of isolating and purifying outer segments rely on the unique bouyant density of the disc membranes and involve flotation techniques in either sucrose or Ficoll (*12, 13*). Both ionic and nonionic detergents have been used to extract visual pigments from purified disc membranes (*14, 15*). The

4. F. S. Sjöstrand, *in* "The Structure of the Eye" (G. K. Smelser, ed.), p. 1. Academic Press, New York, 1961.
5. A. I. Cohen, *Biol. Rev.* **38,** 427 (1963).
6. G. Wald, P. K. Brown, and I. R. Gibbons, *J. Opt. Soc. Am.* **53,** 20 (1963).
7. J. K. Blasie, C. R. Worthington, and M. M. Dewey, *JMB* **39,** 407 (1969).
8. J. K. Blasie and C. R. Worthington, *JMB* **39,** 417 (1969).
9. W. J. Schmidt, *Kolloid-Z.* **85,** 137 (1938).
10. G. Vanderkooi and M. Sundaralingam, *Proc. Natl. Acad. Sci. U. S.* **67,** 233 (1970).
11. A. E. Blaurock and M. H. F. Wilkins, *Nature* **223,** 906 (1969).
12. R. G. Matthews, R. Hubbard, P. K. Brown, and G. Wald, *J. Gen. Physiol.* **47,** 215 (1963).
13. R. N. Lolley and H. H. Hess, *J. Cell Physiol.* **73,** 9 (1969).
14. C. D. B. Bridges, *BJ* **66,** 375 (1957).
15. F. Crescitelli, *Vision Res.* **7,** 685 (1967).

detergents vary in the extent to which they break and disperse the membrane into its individual components. Thus, digitonin disperses the membrane into relatively large fragments which presumably still contain visual pigment, other proteins, and lipids in close association. The ionic detergent cetyltrimethylammonium bromide (CTAB), on the other hand, seems to break the membrane into its individual building blocks. Visual pigments can then be further purified and separated from other membrane proteins and from all the lipids to yield pure native pigment (*16*, *17*).

B. Criteria of Purity

Visual pigments have no known measurable enzymic activity; consequently, purity is determined by spectral parameters and by physical characterization of the protein. One of the older and best known criteria of purity is the ratio of absorption at about 280 nm (representing mainly absorption of aromatic amino acid in the protein) to that of the main longer wavelength peak in the visible spectrum (resulting from the chromophore). This ratio is, at least in theory, a unique and constant number defining each visual pigment. For bovine visual $pigment_{500}$ this number is about 1.6–1.85 as determined in various laboratories. Unfortunately, this ratio is not free of some uncertainties. Visual pigments are always measured in solution as complexes with detergents. The ratio of visual pigment to detergent is variable and depends, among other things, on the nature and relative concentration of the detergent. The fact that the various detergents sometimes have a considerable absorption of their own in the UV introduces an obvious element of uncertainty. In addition, solutions of proteins in detergents often show considerable light scattering. This, too, is dependent on the nature of the detergent and on the size of the protein-detergent micelles. No agreement has been reached on how to correct these sources of uncertainty in the ratio of UV to visible absorption.

Purity of visual pigments is also determined by rechromatography (*16*) and by disc gel electrophoresis (*16*, *17*).

C. The Protein

The molecular weight of native bovine, rat, and frog visual $pigments_{500}$ is about 28,000–30,000 as determined by gel filtration chromatography

16. J. Heller, *Biochemistry* **7**, 2906 (1968).
17. H. Shichi, M. S. Lewis, F. Irreverre, and A. Stone, *JBC* **244**, 529 (1969).

on a calibrated column (*16, 18*). Amino acid analyses yield a minimal molecular weight for the above proteins of about 26,400 (*16, 18*) and 28,000 in another study of the bovine protein (*17*). No subunits are found in bovine visual pigment and the protein is a single polypeptide chain. Attempts to determine the molecular weight by ultracentrifugal methods have been unsuccessful to date because of the uncertainty introduced by the variable amount of detergent which is bound to the protein.

The amino acid composition of several visual pigments was determined and was found to contain about 50% nonpolar amino acid residues (*16–18*). This high content of nonpolar residues is related, most probably, to the function of visual pigments as structural components in the rod disc membrane.

The amino terminal residue of bovine visual pigment is apparently blocked (*16, 19*), and in addition, no carboxyl terminal residue can be determined by the usual methods (*16, 19*).

Visual pigments are glycoproteins containing an average of three neutral sugar residues and three *N*-acetylglucosamine residues (*18*) bound to the protein as a single side chain. The sugar moiety is linked to the protein by a linkage involving an *N*-acetylglucosamine and an asparagine residue. The structure of the 9-residue glycopeptide from bovine visual pigment is also known (*20*).

D. The Retinal Chromophore

Following the identification of the yellow chromophore extracted from the retina as retinol (vitamin A) (*21*), it was recognized that the chromophoric prosthetic group in native visual pigment is actually the aldehyde form of retinol or retinal (*22*). Retinal is a 20-carbon polyene containing five conjugated double bonds. The four double bonds in its aliphatic chain make a total of 16 theoretically possible *cis-trans* isomers. The all-*trans* (I), 9-*cis* (II), 11-*cis* (III), 13-*cis*, 9,13-*dicis*, and 11,13-*dicis* isomers of retinol, retinal, and retinoic acid have been synthesized and their properties studied (*23*) (Fig. 2).

The relation between structure and absorption spectra for the different

18. J. Heller, *Biochemistry* **8**, 675 (1969).
19. G. Albrecht, *JBC* **229**, 477 (1957).
20. J. Heller and M. A. Lawrence, *Biochemistry* **9**, 864 (1970).
21. G. Wald, *J. Gen. Physiol.* **18**, 905 (1935).
22. S. Ball, T. W. Goodwin, and R. A. Morton, *BJ* **42**, 516 (1948).
23. J. G. Baxter, *Compr. Biochem.* **9**, 169 (1963).

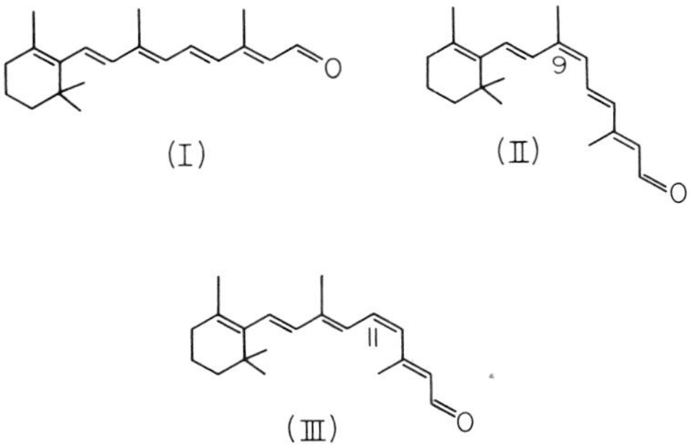

FIG. 2. Isomers of retinal.

isomers is fairly well understood (*24–26*). It was assumed, originally, that the 11-*cis* isomer could not exist because of steric hindrance between the C-10 hydrogen and the C-13 methyl. Therefore, it was quite surprising to find that the hindered 11-*cis* is actually quite stable and is the isomer involved in vision (*27*, *28*). Recently, it was shown that this strain is relieved by a skew geometry around the diene unit C-11=C-12—C-13=C-14 (*29*). The thermodynamic parameters of the isomerization of 11-*cis*- to all-*trans*-retinal have also been measured (*30*).

Only the 11-*cis*- and 9-*cis*-retinal isomers combine *in vitro* with the apoprotein to form photosensitive visual pigments. The combination with the 11-*cis* isomer leads to a pigment that is identical with native visual pigment, while the 9-*cis* isomer produces a photosensitive pigment with a λ_{max} at 487 nm (bovine isorhodopsin) (*27*). All visual pigments which have been investigated to date, mostly by means of absorption spectroscopy and ability of their chromophore to form native pigment with bovine apoprotein, have been found to have the 11-*cis* isomer of either retinal or 3-dehydroretinal as the prosthetic group.

Several analogs of retinal have been synthesized, and their ability to combine with bovine apoprotein to form photosensitive pigments was tested in an attempt to delineate the geometrical constraints of the combining site. It is possible to reduce the 5,6 ring double bond (IV) or to replace either 9 (V) or 13 (VI), or both (VII), side chain methyl groups with hydrogen and obtain a retinal molecule that will combine with the apoprotein. The 4-oxoretinal (VIII), the 5,6-monoepoxy-

24. L. Jurkowitz, J. N. Loeb, P. K. Brown, and G. Wald, *Nature* **184,** 614 (1959).
25. D. E. Balke and R. S. Becker, *JACS* **89,** 5061 (1967).
26. W. Sperling and C. N. Rafferty, *Nature* **224,** 591 (1969).
27. R. Hubbard and G. Wald, *J. Gen. Physiol.* **36,** 269 (1952).
28. P. K. Brown and G. Wald, *JBC* **222,** 865 (1956).
29. D. J. Patel, *Nature* **221,** 825 (1969).
30. R. Hubbard, *JBC* **241,** 1814 (1966).

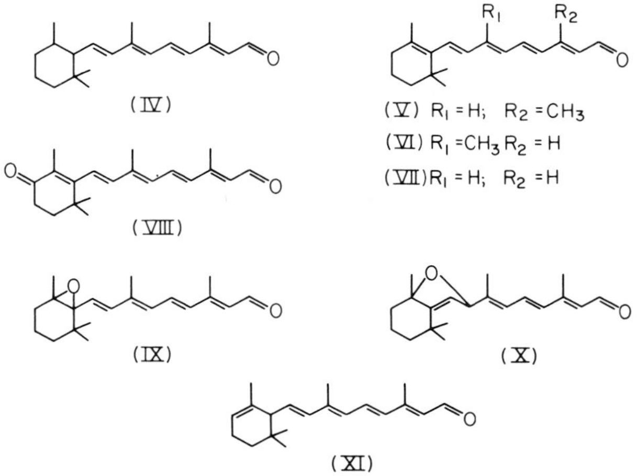

(IV)

(V) R_1 = H; R_2 = CH_3

(VI) R_1 = CH_3 R_2 = H

(VII) R_1 = H; R_2 = H

(VIII)

(IX)

(X)

(XI)

FIG. 3. Derivatives of retinal that will form photosensitive visual pigments.

retinal (IX), the 5,8-monoepoxyretinal (X) and "α-retinal" [where the ring double bond is shifted out of conjugation with the polyene side chain (XI)] will also form photosensitive pigments (Fig. 3). The modified retinals have to be either the 11-*cis* or 9-*cis* isomers. On the other hand, either lengthening or shortening of the polyene side chain or elimination of the ring will destroy the ability of the molecule to combine with the apoprotein to form a photosensitive molecule (*3, 31–33*).

E. LIPIDS

The classic method of preparing visual pigment employs the dissociating agent digitonin to solubilize photoreceptor disc membranes (*34*). The membrane contains approximately 40–50% lipid, most of which is phospholipid (predominantly phosphatidylethanolamine, phosphatidylcholine, and phosphatidylserine) (*35–38*). Digitonin as well as other

31. P. E. Blatz, M. Lin, P. Balasubramaniyan, V. Balasubramaniyan, and P. B. Dewhurst, *JACS* **91,** 5930 (1969).
32. P. E. Blatz, P. B. Dewhurst, V. Balasubramaniyan, P. Balasubramaniyan, and M. Lin, *Photochem. Photobiol.* **11,** 1 (1970).
33. R. Nelson, J. K. deRiel, and A. Kropf, *Proc. Natl. Acad. Sci. U. S.* **66,** 531 (1970).
34. K. Tansley, *J. Physiol. (London)* **71,** 442 (1931).
35. F. D. Collins, R. M. Love, and R. A. Morton, *BJ* **51,** 669 (1952).
36. N. I. Krinsky, *Arch. Ophthalmol.* **60,** 688 (1958).
37. R. P. Poincelot and J. E. Zull, *Vision Res.* **9,** 647 (1969).
38. N. C. Nielsen, S. Fleischer, and D. G. McConnell, *BBA* **211,** 10 (1970).

dissociating agents solubilize the visual pigment at the same time as the membrane lipids, and only fairly drastic conditions such as solvent extraction can remove the lipid from visual pigment. This procedure often leads to some denaturation of the visual pigment. Thus, it was difficult to establish whether lipids are an essential component of visual pigment. Lately, it has been possible to remove all the lipid from native visual pigment by gel filtration chromatography in the presence of the strong dissociating agent CTAB (*16*). Visual pigments purified by this method retain most of the properties that are currently used in defining the native state. They exhibit the characteristic spectrum of visual pigments, are photosensitive, and pass through the same intermediates upon illumination at low temperatures. Unfortunately, visual pigments extracted with CTAB do not regenerate in the dark after illumination and the addition of 11-*cis*-retinal (*39*). It is impossible, therefore, to establish at this time whether the removal of lipid prevents the regeneration of visual pigment from apoprotein and 11-*cis*-retinal *in vitro*. Taken together, the available data indicate that isolated visual pigments do not contain lipid as part of their structure and so are not lipoproteins in the strict sense of the word. As yet, we know little about the interactions of membrane proteins with lipids *in situ*, and it should be borne in mind that the overall function of visual pigments as triggers of visual excitation might involve some such special interactions.

F. Linkage between Retinal and Protein

It has long been known that when a dark adapted retina is extracted with solvents such as petroleum ether only a small amount of retinal (and/or retinol) is obtained. Only after illumination is it possible to extract the full complement of retinal found in photoreceptors (*40*). Moreover, aldehyde reagents such as hydroxylamine do not react with retinal in the dark adapted retina or with native visual pigment in solution. Again the aldehyde group is detected only after illumination (*41*).

In addition, the long wavelength peak of illuminated visual pigment shows a shift from about 365 nm in alkaline solutions to about 440 nm in acid solutions at 20° ("indicator yellow"). On the basis of the above findings, the suggestion was made that retinal is linked to an amino group on the protein through an aldimine bond (Schiff base) (*42*, *43*). More recently it was found that, although the retinal–apoprotein linkage

39. D. M. Snodderly, *Proc. Natl. Acad. Sci. U. S.* **57**, 1356 (1967).
40. G. Wald, *Nature* **134**, 65 (1934).
41. G. Wald and P. K. Brown, *Proc. Natl. Acad. Sci. U. S.* **36**, 84 (1950).
42. F. D. Collins, *Nature* **171**, 469 (1953).
43. R. A. Morton and G. A. J. Pitt, *BJ* **59**, 128 (1955).

cannot be reduced in the native molecule, it was possible to achieve reduction with $NaBH_4$ after illumination. The retinal was then linked through a stable secondary amine linkage to an ϵ-amino group of a lysine residue (*44–47*).

Although it seems well established that at some stage following illumination the linkage between the aldehyde group of retinal and the protein is through an aldimine bond, the properties of the linkage in the native pigment are such as to suggest that some modification of the bond took place. In contrast to the properties of the aldimine linkage in illuminated visual pigment, the linkage in the native pigment cannot be reduced with $NaBH_4$; it is not a pH indicator (the spectrum is independent of pH), and it is not susceptible to the rapid hydrolysis of the chromophore which is observed in the illuminated pigment. Although the exact nature of the modification which alters the properties of the aldimine bond is not known, it is possible that the double bond of the aldimine is substituted by an hydroxyl, amino, or sulfhydryl group (*48*).

In an attempt to find a role for the large amounts of lipid in the photoreceptor discs, it was recently suggested that retinal is covalently linked to the amino group of phosphatidylethanolamine in native visual pigment (*49–53*). In addition, it was proposed that upon illumination the retinal shifts from the phosphatidylethanolamine to an ϵ-amino group of a lysine residue of the protein. The suggestion was made that this is actually the triggering event in visual excitation. This novel theory seems untenable because of experiments that show that native visual pigment purified by gel filtration chromatography does not contain any phospholipids in general nor any phosphatidylethanolamine in particular (*16, 54–57*).

To recapitulate, it seems well established that the retinal chromo-

44. D. Bownds and G. Wald, *Nature* **205**, 254 (1965).
45. D. Bownds, *Nature* **216**, 1178 (1967).
46. A. Akhtar, P. T. Blosse, and P. B. Dewhurst, *Life Sci.* **4**, 1221 (1965).
47. M. Akhtar, P. T. Blosse, and P. B. Dewhurst, *BJ* **110**, 693 (1968).
48. J. Heller, *Biochemistry* **7**, 2914 (1968).
49. R. P. Poincelot, P. G. Millar, R. L. Kimbel, Jr., and E. W. Abrahamson, *Nature* **221**, 256 (1969).
50. F. J. Daemen and S. L. Bonting, *Nature* **222**, 879 (1969).
51. M. Akhtar and M. D. Hirtenstein, *BJ* **115**, 607 (1969).
52. R. P. Poincelot, P. G. Millar, R. L. Kimbel, Jr., and E. W. Abrahamson, *Biochemistry* **9**, 1809 (1970).
53. R. L. Kimbel, Jr., R. P. Poincelot, and E. W. Abrahamson, *Biochemistry* **9**, 1817 (1970).
54. M. O. Hall and A. D. E. Bacharach, *Nature* **225**, 637 (1970).
55. R. E. Anderson, R. T. Hoffman, and M. O. Hall, *Nature* **229**, 249 (1971).
56. R. E. Anderson and M. B. Maude, *Biochemistry* **9**, 3624 (1970).
57. M. D. Hirtenstein and M. Akhtar, *BJ* **119**, 359 (1970).

phore is covalently linked to the apoprotein in native visual pigment. This linkage involves either an aldimine bond with some unique properties or a substituted aldimine linkage. Following illumination, the linkage between the chromophore and the protein changes to a simple aldimine bond with all the usual properties associated with a Schiff base such as spectral dependence on pH, reduction with $NaBH_4$, rapid hydrolysis, and so on.

G. Structure and Color

The chromophore 11-*cis*-retinal has its long wavelength absorption peak around 380 nm (depending on the solvent). On the other hand, native visual pigments which result from the combination of 11-*cis*-retinal with different apoproteins have a long wavelength absorption peak ranging from about 450 nm to about 550 nm. No satisfactory theory has yet been proposed to account for this large red shift. Most of the current proposals are based on the assumption that the chromophore is linked to the protein through an aldimine bond. Some of the theories propose then that the nitrogen of the aldimine linkage is protonated and that this protonation leads to a red shift similar to that observed in many other aldimine bonds (*43, 58, 59*). Other theories propose that the aldimine nitrogen is actually not protonated in native visual pigment and that the cause of the red shift is an interaction of the chromophore polyene chain with the apoprotein (*60*). Other suggestions have combined these two theories proposing that the protonated aldimine linkage and the interaction of the retinyl polyene chain with the unique environment of the protein are both important (*61*). Suggestions of various charge transfer complexes between the retinyl aldimine and the protein also have been made (*47, 62*). Unfortunately, these theories suffer from lack of knowledge as to the precise structure of the carbonyl linkage and retinyl binding site to the apoprotein. It is not clear whether the carbonyl protein linkage in native visual pigment is a simple aldimine bond (see Section III,F); therefore, extrapolation from the behavior of retinyl aldimine model compounds under various conditions to the interpretation of the spectrum of native visual pigment is somewhat questionable. Even more disturbing is the finding that in one of the illuminated inter-

58. A. Kropf and R. Hubbard, *Ann. N. Y. Acad. Sci.* **74,** 266 (1958).
59. R. Hubbard, *Nature* **221,** 432 (1969).
60. H. J. A. Dartnall and J. N. Lythgoe, *Vision Res.* **5,** 81 (1965).
61. C. S. Irving, G. W. Byers, and P. A. Leermakers, *Biochemistry* **9,** 858 (1970).
62. P. E. Blatz, *J. Gen. Physiol.* **48,** 753 (1965).

mediates of visual pigment (metarhodopsin) where the linkage is almost certainly a simple aldimine bond (as revealed by reduction with $NaBH_4$), it is found that increased acidity actually results in a large blue shift (from 478 nm at about pH 8 to 380 nm at about pH 5) (*12*). On the other hand, the theory would predict that increased acidity should result in a red shift.

IV. Interaction of Light with Visual Pigment

A. The Overall Reaction

Light is absorbed by the chromophoric prosthetic group of visual pigment. The absorption of a photon produces the corresponding excited state of the chromophore–apoprotein complex. The excited state in returning to the ground state produces several chemical transformations. Some of these transformations are a direct consequence of the absorption of light such as the *cis* → *trans* isomerization of the chromophore, while others are probably related only in a secondary manner such as conformational changes in the visual pigment apoprotein. Although visible light acts directly only on the chromophore, it is important to realize that the various changes observed in visual pigment after illumination represent reactions and transformations of both the chromophore and the protein. Some of the reactions are the breaking and making of covalent bonds, while others are the simultaneous conformational changes of the apoprotein.

It is generally assumed that the overall action of light on the chromophore of visual pigments is to isomerize it from the 11-*cis* to the all-*trans* configuration. This conclusion is based primarily on the studies of Hubbard and Wald (*27*), showing that illumination of native visual pigment with long wavelength light (orange—which is not absorbed by the reaction products) yields a product that has an absorption spectrum similar to all-*trans*-retinal. Moreover, the effects of illuminating this product with white light mimic the effects of illuminating all-*trans*-retinal with white light. Unfortunately, no direct identification and quantitative determination of the retinal isomer(s) released from visual pigment upon illumination with monochromatic light of various wavelength in the range of the visual pigment absorption peak (about 450–550 nm) has yet been reported. There are hints in the literature that other isomers besides the all-*trans* are produced by illumination (*63*), but until

63. R. Hubbard, P. K. Brown, and A. Kropf, *Nature* **183**, 442 (1959).

the various retinal isomers are chromatographically separated and quantitatively determined, it is not possible to judge the relative importance of other isomers as products of illumination.

Various intermediates, or states of illuminated visual pigment have been recognized, mostly by absorption spectroscopy. Some of the intermediates are extremely unstable at ordinary temperatures and, consequently, can be recognized only by low temperature spectroscopy. In this chapter the various intermediates, or states, obtained by illuminating visual pigments are discussed and an effort is made to relate the spectroscopic states to known or possible structural changes. This whole field can be rather confusing for the uninitiated because several intermediates have been given trivial names, sometimes with mechanistic implications as to the reaction route, while others (or possibly the same) intermediates are designated according to the peak position at the long wavelength end of the spectrum. Both systems of nomenclature are used in this chapter.

B. Early Intermediates

When a solution of bovine visual pigment in glycerol is illuminated at 77°K with blue light (440 nm), its absorption peak shifts from 505 nm (the absorption peak of native pigment at 77°K) to 517 nm. When this solution is next illuminated with red light (600 nm) the absorption peak returns to 505 nm. This cycle of illumination with the accompanying shifts in the absorption spectrum can be repeated a number of times. Yoshizawa and Wald suggested that the mixture produced by illumination with blue light, which has a λ_{max} at 517 nm, includes the earliest intermediate produced by illuminating visual pigment (*64*). They called this intermediate *prelumirhodopsin* and calculated its absorption peak at about 543 nm. They interpreted prelumirhodopsin to be similar to the native pigment in all respects, except that the retinyl prosthetic group is isomerized to the all-*trans* form. It was thought that native visual pigment and prelumirhodopsin were freely interconvertible by light of appropriate wavelength and that this conversion represented only the reversible isomerization of 11-*cis*- to all-*trans*-retinyl. Although prelumirhodopsin is stable at 77°K, it can also be observed transiently at higher temperatures up to —25°C by the technique of flash photolysis (*65*).

Measurements of the circular dichroism spectra of native and illuminated bovine visual pigment at 77°K have shown that illumination with

64. T. Yoshizawa and G. Wald, *Nature* **197,** 1279 (1963).
65. K. H. Grellman, R. Livingston, and D. Pratt, *Nature* **193,** 1258 (1962).

light of specific wavelength produces two new states with unique circular dichroic spectra and that light itself cannot regenerate the native pigment (*66*). These experiments make it difficult to interpret the shifts in absorption spectrum of visual pigment upon illumination at 77°K as resulting from a reversible 11-*cis* ↔ all-*trans* isomerization of the chromophore. As yet, there is no satisfactory explanation for the absorption and circular dichroism spectral shifts upon illumination. In this connection it is interesting to note that illumination of visual pigment at 77°K produces an electron paramagnetic resonance signal. On the basis of this observation it was suggested that excited state free radicals of retinals and visual pigment are produced by illumination in solid glasses at this temperature (*67*). Unfortunately, no information is as yet available on the geometry and absorption properties of the excited state(s) of retinal and retinyl-apoprotein (visual pigment). Research in this field is hampered by the recurrent problem as to what exactly is the structure of the linkage and binding between retinal and the apoprotein and by the related problem of interpreting the visual pigment spectrum in terms of the linkage structure.

C. Later Intermediates

When the temperature of a solution of illuminated visual pigment is raised from 77°K (in the dark), the spectrum shifts to produce several new intermediates. Above 133°K and below 233°K the absorption spectrum of the bovine pigment is characterized by a peak at 497 nm (lumirhodopsin). Visual pigments from other species are characterized by absorption peaks at different wavelengths (*63*). Above 233°K and below 258°K the absorption peak of the illuminated bovine pigment is at 480 nm (metarhodopsin I), while above this temperature the peak shifts to 380 nm (metarhodopsin II) (*12*). The equilibrium between metarhodopsin I (λ_{max} 480 nm) and metarhodopsin II (λ_{max} 380 nm) depends not only on the temperature but also on the nature of the solvent. Lowering the pH from 8 to 5, increasing the ionic strength or the concentration of glycerol or methanol shifts the equilibrium to metarhodopsin II. This equilibrium is reversible, and, for example, lowering the glycerol concentration by dilution shifts the equilibrium back to the species with λ_{max} of 480 nm.

When metarhodopsin II (λ_{max} 380 nm) is kept in the dark at 3°C it changes over a period of hours into a new state with a λ_{max} of about

66. J. Horwitz and J. Heller, *Biochemistry* **10**, 1402 (1971).
67. F. J. Grady and D. C. Borg, *Biochemistry* **7**, 675 (1968).

465 nm (*12*). A state absorbing maximally at 467 nm is also produced by illuminating the λ_{max} 380 state with UV light (365 nm). It is not known whether the two states produced from metarhodopsin II (λ_{max} 380) by illumination, or spontaneously in the dark, are identical.

When illuminated visual pigment in aqueous detergent solutions is kept at room temperature, the retinal chromophore is ultimately released from the apoprotein by hydrolysis.

Several chemical changes have been identified in illuminated visual pigment during and following the above series of transformations. The retinyl-apoprotein, which—as discussed above—is resistant to reduction with $NaBH_4$ in the native pigment, now becomes easily reduceable (*44, 46*). Moreover, while in the native pigment the linkage of the symmetrical optically inactive chromophore with an asymmetric environment results in a circular dichroic signal produced from the 500-nm peak, illumination at room temperature leads to loss of the circular dichroism, i.e., the linkage environment becomes symmetrical (*68*). Whether the disappearance of the circular dichroism signal is the result of a new environment of the retinyl, which is still linked to the apoprotein, or whether it results from a complete hydrolysis and release of the chromophore is not known. The isoelectric point of the protein changes after illumination (*69*) and there is an uptake or release of protons (depending on pH) by the pigment (*70*). New sulfhydryl groups become detectable in illuminated visual pigment (*48, 71, 72*), and a major conformational change in the apoprotein take place (*48*).

It is generally accepted that the only observable action of light is to isomerize 11-*cis*- to all-*trans*-retinyl and that this has already happened in the earliest observable intermediate at 77°K. Consequently, all the later intermediates represent "dark" reaction of the protein or the protein–chromophore complex with retinyl fully in the all-*trans* form. Because of technical difficulties, no direct characterization of the retinyl isomer has been attempted under the various experimental conditions employed in detecting the various intermediates. Until such a characterization is achieved, it will probably remain a distinct possibility that the absorption changes observed at low temperature result from various excited states of retinal as well as from various *cis–trans* isomerizations.

68. F. Crescitelli, W. F. H. M. Mommaerts, and T. I. Shaw, *Proc. Natl. Acad. Sci. U. S.* **56,** 1729 (1966).
69. E. E. Broda and E. Victor, *BJ* **34,** 1501 (1940).
70. D. G. McConnell, C. N. Rafferty, and R. A. Dilley, *JBC* **243,** 5820 (1968).
71. G. Wald and P. K. Brown, *J. Gen. Physiol.* **35,** 797 (1952).
72. C. M. Radding and G. Wald, *J. Gen. Physiol.* **39,** 902 (1956).

V. Visual Pigment Regeneration following Illumination; Retinal Isomerase

As discussed above, the effect of light absorption by visual pigments is to isomerize the chromophore from the 11-*cis* to the all-*trans* configuration. As we have already seen, only the 11-*cis* combines with the apoprotein to form the native photosensitive pigment. To understand fully the visual cycle, it is necessary to study the reactions whereby retinal is reisomerized to the 11-*cis* configuration. Theoretically, this can be achieved in two ways: One way would be for the all-*trans* isomer to absorb light and, by a reversal of the initial photoisomerization, return to the 11-*cis* configuration. The second mechanism would achieve the same result through a "dark" reaction(s).

The reverse photoisomerization seems, for several reasons, to be inherently less probable than a dark reaction. The only way possible to selectively excite the all-*trans* isomer in the presence of many other photosensitive molecules (visual pigments) is to use near UV light of about 380–400 nm, the peak absorption of the all-*trans* isomer. Yet, due to the nature of the spectrum of the sun, little light of these wavelengths is available at the retina. Moreover, it is known that illumination of all-*trans*-retinal produces not only the 11-*cis* but several other isomers. It would be necessary, then, to reisomerize these other isomers back to the all-*trans* or 11-*cis* conformation. In addition, if both the forward and reverse isomerization steps are photoreactions, it is difficult to conceive of this cycle as having any biologically controlled step; the reactions would be purely random and determined only by the level of illumination. Lastly, if the reverse reaction is a photoisomerization, it is impossible to account for the well-known fact that the highest content of native visual pigment in the retina is found after prolonged dark adaptation. These considerations, taken together, strongly suggest that visual pigment regeneration is not achieved by reverse photoisomerization but as a result of a dark reaction(s).

Until very recently, it has been generally accepted that after illumination the chromophore is ultimately released by hydrolysis from the apoprotein as the all-*trans* isomer. Following its reduction to retinol by retinol dehydrogenase in the outer segment (*73*); the chromophore is transported to the pigment epithelium. [The pigment epithelium is a layer of cells adjacent to the photoreceptor outer segment (Fig. 1).

73. S. Futterman, A. Hendrickson, P. E. Bishop, M. H. Rollins, and E. Vacano, *J. Neurochem.* **17**, 149 (1970).

These cells are filled with a dark melanin pigment and hence their name. The melanin pigment plays no direct role in vision.] In the pigment epithelium, all-*trans*-retinol is reoxidized to the aldehyde form and then is reisomerized to 11-*cis*-retinal by the enzyme retinal isomerase (*74*). The chromophore then is transported back to the photoreceptor outer segment, where it recombines spontaneously with the apoprotein to form native visual pigment. The experimental evidence for this scheme is detailed in several papers and was discussed in detail in the previous edition of "The Enzymes" (*75–77*). To the reviewer's knowledge, no new work has been reported on this enzyme since the last edition.

If the above scheme for visual pigment regeneration is correct, it can be predicted that in animals kept in darkness there should be no exchange between the retinal chromophore which is bound in visual pigment and that which is the pigment epithelium. On the other hand, in animals kept under ordinary illumination, where the chromophore constantly shuttles between the photoreceptor outer segment and the pigment epithelium, there should be a rapid exchange of retinal.

Recently, labeled retinal was injected into two groups of rats (*78*) and frogs (*79*). The animals were kept either in total darkness or under normal illumination. Animals were sacrificed at intervals, their visual pigment purified, and the specific activity of the labeled retinal chromophore in the pigment was determined. Contrary to expectation, it was found that there is a constant progressive incorporation of labeled retinal into visual pigment and that this incorporation is not effected by illumination. In other words, animals kept in total darkness or under ordinary illumination show practically the same incorporation of retinal into visual pigment.

The obvious explanation for the lack of exchange between visual pigment retinal and the general pool of retinal in animals kept under normal illumination is that the chromophore stays attached to the apoprotein throughout the visual cycle. Illumination isomerizes the 11-*cis*-retinyl to the all-*trans* configuration, while a series of dark enzymic reactions reisomerize the all-*trans* to the 11-*cis* configuration. During this whole cycle the chromophore most probably either stays covalently attached to the specific lysine residue of the apoprotein or it might move,

74. R. Hubbard, *J. Gen. Physiol.* **39,** 935 (1956).
75. J. E. Dowling, *Nature* **188,** 114 (1960).
76. R. Hubbard and J. E. Dowling, *Nature* **193,** 341 (1962).
77. G. Wald and R. Hubbard, "The Enzymes," 2nd ed., Vol. 3, Part B, p. 369, 1960.
78. C. D. B. Bridges and S. Yoshikami, *Nature* **221,** 275 (1969).
79. M. O. Hall, D. Bok, and R. T. Hoffman (personal communication).

within the membrane, from the visual pigment apoprotein to an immediate neighboring protein (an "isomerase"?) and back again without ever becoming available for exchange with other retinal molecules. In this connection, it is interesting to note the recent findings of visual pigment regeneration in isolated rat and frog retinas that were dissected free of pigment epithelium (*80*, *81*).

VI. Concluding Remarks

Conceived in broadest terms the function of visual pigments is to transduce light into chemical energy. This transduction is achieved by a series of reactions initiated by the retinal chromophore absorbing a photon. The 11-*cis*-retinal chromophore present in native visual pigments is thereby isomerized to the all-*trans* form. By a mechanism still unknown, this *cis–trans* isomerization of the tightly bound prosthetic group induces a conformational change(s) in the protein. Recalling that visual pigments are structural components of the photoreceptor outer segment disc membrane, it can be seen that such a conformational transition would be tantamount to a membrane with some new and different properties. This change in the membrane properties leads, in some yet unknown way, to the initiation of a nervous impulse in the retina which is then carried to the brain. The *cis–trans* isomerization of the retinyl chromophore is thus the first step in the transduction of light into chemical energy.

The nature of the retinal-to-apoprotein linkage in native visual pigment is a constant source of difficulty in deciphering the chemistry of visual pigments. It will be imperative to know the exact structure of the linkage in order to account for several observations of fundamental importance. These include the red shift of the retinyl absorption peak on combining with the apoprotein, the chemical nature of the various spectroscopic intermediates produced by illumination, and the recent finding that retinyl does not leave the immediate environment of the protein after illumination.

The structural map of the linkage, at the very minimum, should describe with assurance the type of bond between the carbonyl carbon of retinal and the protein and, in addition, should detail the interactions between the polyene chromophore and the surrounding protein.

Another related problem is the molecular and structural basis for the various intermediates or states produced in visual pigments upon

80. R. A. Cone and P. K. Brown, *Nature* **221**, 818 (1969).
81. E. B. Goldstein, *Vision Res.* **10**, 1065 (1970).

illumination. This is a wide open question of the present time, and there is little, if any, information regarding the nature of the various states.

The problem of visual pigment regeneration after illumination is much less clear today than it seemed a few years ago. We would like to know with certainty whether the retinal chromophore is circulating to the pigment epithelium *in vivo* or whether it is reisomerized *in situ* after illumination. The enzyme(s) and possible cofactors involved in reisomerizing the all-*trans* to 11-*cis* chromophore remain to be identified and characterized.

Finally, it is important to remember that most of our information about visual pigments is derived from a small number of pigments, mostly the bovine and frog pigments. These pigments are derived from rod photoreceptors and are most probably the pigments responsible for dim-light, colorless vision. There is practically no information about the cone photoreceptor visual pigments which mediate color vision. Are the various steps involved in the visual cycle of dim-light, rod pigments identical to the bright-light, color-sensitive cone pigments? Most probably the answer is yes, but we would like to know this with more assurance.

18

Δ^5-3-Ketosteroid Isomerase

PAUL TALALAY • ANN M. BENSON

I. Introduction

An unusual enzymic isomerization reaction involving an intramolecular proton transfer and migration of a double bond into conjugation with a carbonyl group was discovered in 1955 (*1*) during the course of studies on the mechanism of the enzymic transformation of steroid hormones. The action of this enzyme is exemplified by the conversion of Δ^5-androstene-3,17-dione (I) to Δ^4-androstene-3,17-dione (II).

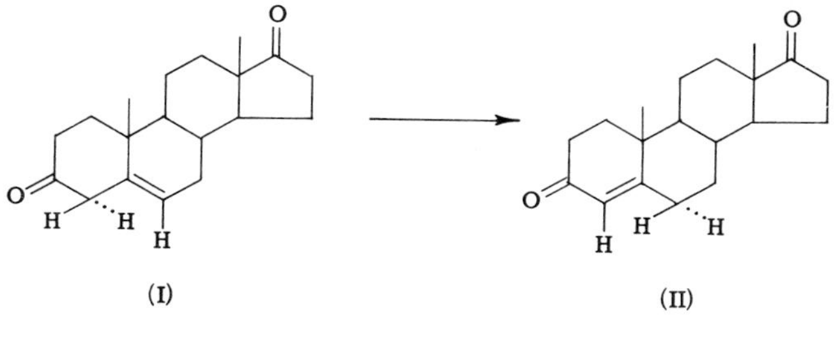

(I) (II)

1. P. Talalay and V. S. Wang, *BBA* **18**, 300 (1955).

The Δ^5-3-ketosteroid isomerase (EC 5.3.3.1) responsible for this reaction promotes the conversion of a variety of $\Delta^{5(6)}$- and $\Delta^{5(10)}$-3-ketosteroids to the corresponding Δ^4-3-ketosteroids. Although the transformation of a variety of Δ^5-3-hydroxysteroids to Δ^4-3-ketosteroids had been described in a number of crude enzymic systems of animal, vegetable, and bacterial origin (*2*), it had not been appreciated that these conversions involved two enzymic steps: a freely reversible nicotinamide-adenine nucleotide-dependent oxidation of the hydroxyl group to the ketone followed by a largely irreversible transposition of the double bond into a position of conjugation. The latter activity was first recognized in soluble extracts of steroid-induced *Pseudomonas testosteroni* (*3*), a soil microorganism capable of growing on testosterone and related steroids as its only source of carbon. A similar enzymic activity has been described in animal tissues (*1*) and is especially prominent in organs concerned with steroid biosynthesis such as the adrenal, testis, and ovary (*4*), and this may be related to the fact that the Δ^5-3-ketosteroid isomerases promote an important reaction in the biosynthesis of steroid hormones of all types. The bacterial enzyme, which has been obtained in crystalline form (*5*), is of unusual interest from the mechanistic viewpoint because of its extraordinarily high catalytic activity. The Δ^5-3-ketosteroid isomerase was perhaps the first enzyme for which direct evidence for base-catalyzed proton transfer was provided and for which the implication was drawn that the enzyme provided a basic group promoting the transfer of a proton between allylic carbon atoms in an enol (*6*).

II. Δ^5-3-Ketosteroid Isomerase of *Pseudomonas testosteroni*

A. Molecular Properties

1. *Induction, Purification, and Crystallization*

The crystallization of Δ^5-3-ketosteroid isomerase from *P. testosteroni* was reported by Kawahara and Talalay in 1960 (*5*). The original preparative procedure has undergone a series of modifications and improvements, and techniques for the preparation of the enzyme on a relatively

2. R. I. Dorfman and F. Ungar, "Metabolism of Steroid Hormones," Academic Press, New York, 1965.
3. P. Talalay, M. M. Dobson, and D. F. Tapley, *Nature* **170,** 620 (1952).
4. F. Kawahara, "Methods in Enzymology," Vol. 5, p. 527, 1962.
5. F. S. Kawahara and P. Talalay, *JBC* **235,** PC1 (1960).
6. I. A. Rose, "The Enzymes," 3rd ed., Vol. II, p. 281, 1970.

large scale have been developed (*7–9*). The enzyme is induced, by the presence of steroids in the growth medium, to levels which are 50- to 100-fold those of the noninduced cells (*10*). There is considerable specificity with respect to the nature of the steroidal inducer. Testosterone, progesterone, deoxycorticosterone, 17β-hydroxy-5α-androstan-17-one, and $\Delta^{1,4}$-androstadiene-3,17-dione are all almost equally efficient inducers of the enzyme, but 19-nortestosterone is considerably less active (*10*). A number of 11-oxygenated steroids which are not metabolized by the organism cannot function as inducers. Progesterone is commonly used for induction purposes for economic reasons. Optimal induction occurs only under carefully controlled conditions of nutrition and environment, and many nutrients if present in excess suppress the induction of the enzyme (*10*). The early preparations of the isomerase were obtained from sonically disrupted cells, but greater ease of isolation of the enzyme was attained when acetone powders were used. For large-scale preparations, lyophilized cells that are treated with lysozyme provide better yields of enzyme and greater ease of preparation (*9*). The purification procedure takes advantage of the fact that the enzyme may be extracted from the cells into buffered 50% ethanol solutions in which many other proteins are insoluble. The enzyme may then be precipitated quantitatively from such solutions by raising the concentration of ethanol to 80% and adding $MgCl_2$ to a final concentration of 1 m*M*. Chromatography on DEAE-cellulose and on calcium phosphate gel provides preparations that may be crystallized by the addition of ammonium sulfate at neutral pH. The enzyme crystallizes in long needles and elongated flat plates (*5*). Conditions for obtaining relatively large crystals (1–2 mm long) of isomerase have been described (*8*).

2. *Ultracentrifuge Studies and Evidence for Purity*

The crystalline enzyme migrates as a single sharp boundary in sedimentation velocity runs. In the Archibald method, the approach to equilibrium was time-independent and agreed at the upper and lower meniscus within experimental precision. On the assumption of a partial specific volume of 0.734 ml/g and a medium density of 1.00 (0.02 *M* sodium phosphate buffer, pH 6.8), an average molecular weight of 40,800 daltons

7. F. S. Kawahara, S.-F. Wang, and P. Talalay, *JBC* **237,** 1500 (1962).
8. P. Talalay and J. Boyer, *BBA* **105,** 389 (1965).
9. R. Jarabak, M. Colvin, S. H. Moolgavkar, and P. Talalay, "Methods in Enzymology," Vol. 15, p. 642, 1969.
10. M. Shikita and P. Talalay, *in* "Biogenesis and Action of Steroid Hormones" (R. I. Dorfman, K. Yamasaki, and M. Dorfman, eds.), p. 41. Geron-X, Los Altos, California, 1968.

was obtained. The sedimentation coefficient, $s_{20,w}$, was 3.3 Svedberg units and the frictional ratio was approximately 1.3. Many recent experiments have shown that the enzyme migrates as a single sharp band upon polyacrylamide gel electrophoresis. Further support for the purity of the crystalline protein is the total absence of cyst(e)ine and tryptophan.

3. *Ultraviolet Absorption and Fluorescence Spectra*

The ultraviolet absorption spectrum (*7, 11*) of crystalline isomerase in dilute sodium phosphate buffer at pH 7.0 shows a principal absorption peak at 277 nm and a well-defined shoulder at 282–284 nm, both of which are characteristic of tyrosine, although slightly displaced toward longer wavelengths than in the case of the free amino acid. In addition, the intact protein displays clearly defined maxima at 253, 259, 266, and 269 nm which are diagnostic of phenylalanine absorptions but are also displaced toward longer wavelengths. Except for these 1–2 nm bathochromic shifts of absorption bands of the component amino acids in peptide linkage (*12*), there is thus a close agreement between the absorption spectrum of the protein and a solution of free tyrosine and phenylalanine of the same molar proportions as exist in the intact protein.

Upon addition of base, the absorption maximum is intensified and shifted to 294 nm, and the minimum is displaced from 250 to 273 nm. These changes are characteristic of tyrosine and, when evaluated quantitatively, indicate quite clearly that only negligible, if any, quantities of tryptophan are present. If the degree of ionization of the phenolic hydroxyl groups of the tyrosine residues of the protein is measured at different pH values, it is seen that almost exactly one-third of the tyrosine residues titrate normally with a pK around 9.7, similar to that of free tyrosine, whereas the remaining two-thirds of the tyrosine phenolic groups become accessible to titration only after exposure of the protein to much higher pH values (above 12), presumably as a consequence of the base-promoted denaturation of the protein.

The fluorescence spectrum of the native isomerase is indistinguishable from that of free tyrosine with respect to excitation and emission maxima (*11*). Isomerase displays relatively high fluorescence efficiency which approaches about 65% of the fluorescence yield from a solution containing the equivalent molar concentration of free tyrosine. It is concluded that some of the buried tyrosine residues that are not directly titratable in the native enzyme by base must contribute quite efficiently to the overall fluorescence of the protein. In solutions containing 2–6 M urea,

11. S.-F. Wang, F. S. Kawahara, and P. Talalay, *JBC* **238**, 576 (1963).
12. D. B. Wetlaufer, *Advan. Protein Chem.* **17**, 303 (1962).

the fluorescence intensity of the isomerase increases and approaches that of free tyrosine when measured under similar conditions. The characteristic tyrosine fluorescence of the isomerase is quite specifically quenched in the presence of substrate analogs (e.g., 19-nortestosterone) that bind to the active site of the enzyme. This quenching of fluorescence is dependent on the concentration of 19-nortestosterone, and under saturating conditions as much as three-quarters of the energy normally emitted as fluorescence is transferred to the steroid (*11*).

4. *Chemical Modification of Isomerase*

The enzyme undergoes photoinactivation in the presence of methylene blue as a first-order rate process up to at least 90% loss of enzymic activity. This destruction of catalytic activity is accompanied by a parallel decline in the histidine content of the protein. Under specified conditions, only minor losses of methionine and tyrosine occur. All of the histidine residues behave in an apparently homogeneous manner. Protection against photoinactivation is provided in the presence of the competitive inhibitor, 19-nortestosterone. The photooxidized enzyme undergoes aggregation (*13–16*).

Reaction of isomerase with diethyl pyrocarbonate (*16*) at pH 6.0 also inactivates the enzyme and protection against inactivation is afforded by 19-nortestosterone. Since diethyl pyrocarbonate (*17*) is a relatively specific reagent for histidine at pH 6.0, these experiments provide further support for the participation of at least one histidyl residue in the active site or center of the enzyme.

Under controlled conditions selective nitration of the isomerase with tetranitromethane has been achieved with conversion of about one-third of the tyrosine residues to 3-nitrotyrosine (*15*). The modified enzyme displays about 50% of the V_{max} and the same K_m for Δ^5-androstene-3,17-dione as native isomerase. The nitro-isomerase migrates as a single homogeneous band with greater mobility on polyacrylamide electrophoresis at pH 8.3 than the unmodified enzyme. These findings point to the role of at least 1 tyrosyl residue in the catalytic and/or substrate binding process.

13. P. Talalay, *Proc. 2nd Intern. Congr. Endocrinol., London, 1964* Intern. Congr. Ser. No. 83, p. 1096. Excerpta Med. Found., Amsterdam.

14. P. Talalay, *Ann. Rev. Biochem.* **34,** 347 (1965).

15. O. M. Colvin and P. Talalay, *Federation Proc.* **27,** 523 (1968) (abstr.).

16. M. Colvin, J. L. Daubek, K. G. Büki, and P. Talalay, in preparation.

17. A. Mühlrád, G. Hegyi, and G. Tóth, *Acta Biochim. Biophys. Acad. Sci. Hung.* **2,** (1), 19 (1967); L.-A. Pradel and R. Kassab, *BBA* **167,** 317 (1968).

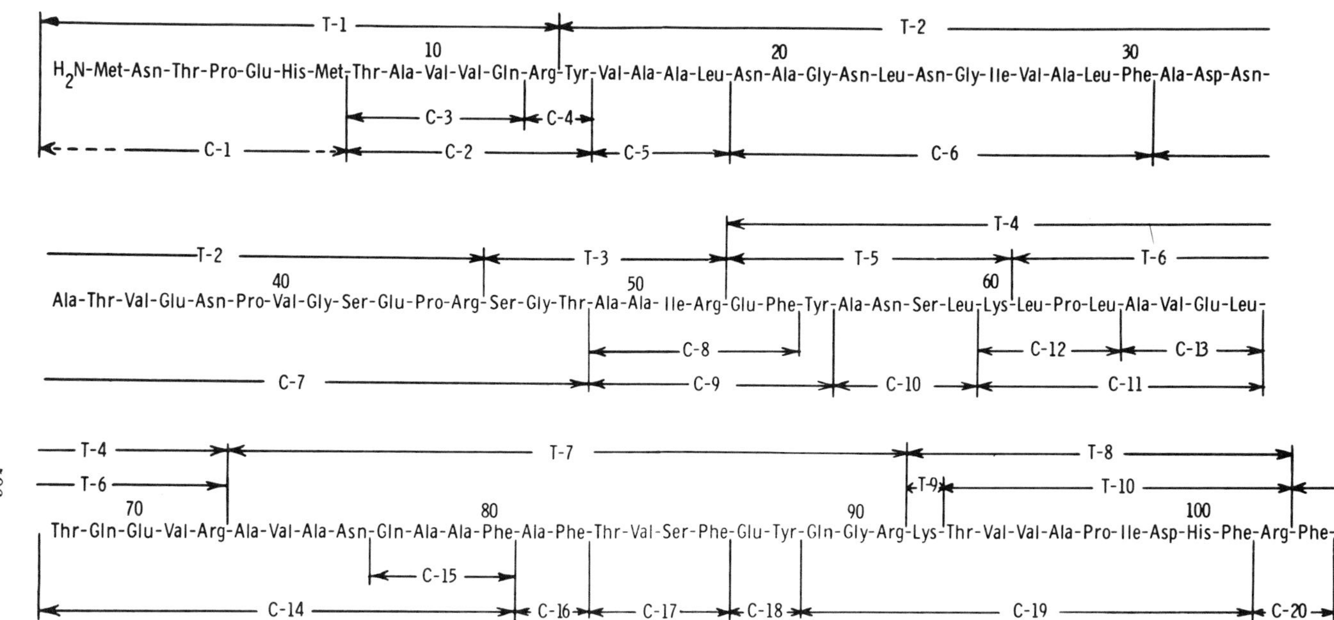

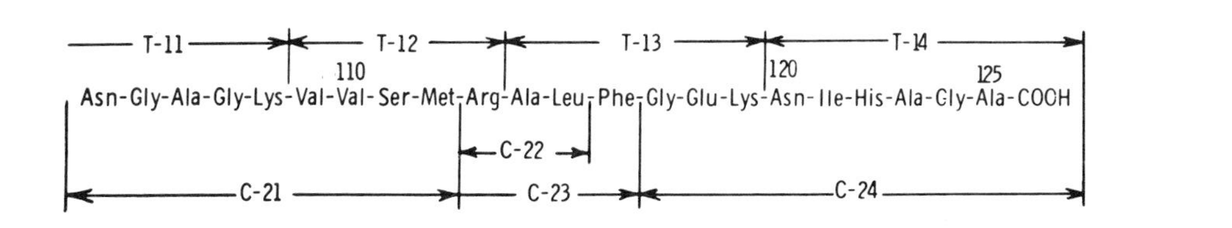

FIG. 1. The complete amino acid sequence of the identical subunits of Δ^5-3-ketosteroid isomerase of *P. testosteroni*. The peptides obtained on hydrolysis with trypsin (T) and with chymotrypsin (C) are shown above and below the sequence, respectively [from Benson *et al.* (*19*)].

5. *Structure*

a. Amino Acid Composition and Primary Structure of Subunits. The composition of the crystalline isomerase from *Pseudomonas testosteroni* (*7, 18*) includes all of the usual amino acids with the exception of tryptophan and cyst(e)ine. The absence of any significant quantities of tryptophan has been demonstrated by spectral methods and is consistent with the ultraviolet absorption spectrum of the protein (*7*). The amino acid composition of the protein determined by hydrolysis and appropriate corrections was reported by Boyer and Talalay (*18*) and is in reasonably good agreement with the amino acid sequence determination (*19*) which gives the following subunit composition: Asp_2, Asn_{10}, Thr_7, Ser_5, Glu_8, Gln_4, Pro_5, Gly_9, Ala_{21}, Val_{14}, Met_3, Ile_4, Leu_8, Tyr_3, Phe_8, Lys_4, His_3, and Arg_7.

Early fingerprints indicated that the number of tryptic peptides was small in relation to the number of lysine and arginine residues and the molecular weight determined by ultracentrifugation. These findings and other evidence pointed to the existence of several identical subunits. The complete amino acid sequence of the 125-amino acid residue subunit has been reported recently by Benson *et al.* (*19*) and is shown in Fig. 1. Hydrolysis of isomerase with trypsin yielded free lysine and 13 peptides which are designated T-1 to T-14 in order of their occurrence in the protein. Eleven of these peptides are distinct portions of the primary structure, whereas T-4 comprises those portions of the sequence represented by T-5 and T-6 and is therefore the product of incomplete tryptic hydrolysis. The peptide T-8 includes T-10 and 1 residue of lysine. Figure 1 designates the fragments obtained by enzymic hydrolysis with trypsin and chymotrypsin.

Since it has been suggested (*13, 14*) that histidine may play a role in the mechanism of action of the enzyme, and since at least 1 histidine residue appears to be essential for catalytic activity, it is of interest to note that 3 histidine residues are to be found in the peptide chain: near the amino terminus (residue 6), near the carboxyl terminus (residue 122), and in the central portion of the chain (residue 100). Tyrosine residues, which may also participate in the active sites, and the fluorescence of which is quenched by binding of steroid substrate analogs, are located in positions 14, 55, and 88. A high incidence of clustering among hydrophobic residues in the chain may also be noted. Such groupings are to

18. J. Boyer and P. Talalay, *JBC* **241,** 180 (1966).
19. A. M. Benson, R. Jarabak, and P. Talalay, *Federation Proc.* **30,** 1242 (1971) (abstr.); *JBC* **246,** 7514 (1971).

be found in positions 7–12, 14–31, 54–59, 61–65, 73–86, 93–98, and 103–107. Whether such concentrations of hydrophobic residues are in any way uniquely related to the physical properties of the enzyme (solubility in high concentrations of ethanol), to the binding of the hydrophobic steroid substrates, or to subunit association is as yet unclear. The presence of hydrophobic regions on the isomerase has been inferred from measurements of the binding of steroidal substrates and inhibitors to the enzyme in mixtures of methanol and water (*20*). The binding of the steroids to the enzyme is a function of their partition between the specific methanol–water mixture and isooctane.

b. Quaternary Structure. The amino acid sequence gives a calculated subunit weight of 13,394 daltons whereas the molecular weight determined by the approach to equilibrium method of Archibald was approximately 40,800 daltons. The chain weight calculated from the sequence is therefore 33% of the ultracentrifugally determined molecular weight. Hydrazinolysis of the intact protein released 3.0 residues of alanine (*19*) and the Edman degradation yielded 1.8 residues of methionine (*18*) as the phenylthiohydantoin per unit of molecular weight 40,800. However, the latter procedure is recognized to give quite low yields. In order to reconcile the molecular weight determined in the ultracentrifuge with the amino acid composition and sequence, it is postulated that there are three (identical) subunits per molecule of weight 40,800 daltons. Efforts (*21*) to determine the chain weight by electrophoresis on polyacrylamide gel in the presence of sodium dodecyl sulfate (*22*) gave a single sharp band indicating a discrete and homogeneous subunit which migrated more rapidly than either ribonuclease (MW 13,700) or cytochrome c (MW 12,270). The observed mobility corresponded to a subunit weight of 11,000 ± 500. Since this value is inconsistent with the primary structure, the reliability of the sodium dodecyl sulfate electrophoresis method in the low molecular weight region is open to question. The precise number of subunits in the oligomeric isomerase molecule remains unresolved. The compilation of the subunit structure of a large number of proteins, assembled by Klotz *et al.* (*23*), reveals no carefully documented examples of proteins containing an odd number of *identical* subunits. In the light of these observations the determination of the precise number of subunits in the isomerase requires careful reexamination. All of the presently

20. F. Falcoz-Kelly, E.-E. Baulieu, and A. Alfsen, *Biochemistry* **7**, 4119 (1968).
21. A. J. Suruda and P. Talalay, unpublished experiments (1971).
22. A. L. Shapiro, E. Vinuela, and J. V. Maizel, Jr., *BBRC* **28**, 815 (1967).
23. I. M. Klotz, N. R. Langerman, and D. W. Darnall, *Ann. Rev. Biochem.* **39**, 25 (1970).

available evidence for the occurrence of three subunits per molecule of isomerase depends upon the reliability of the ultracentrifugally determined molecular weight. There is some *a priori* reluctance to accept an oligomeric structure composed of an odd number of subunits because of the difficulty of formulating a model with properties of symmetry that would not favor further aggregation.

B. Catalytic Properties

1. *Substrate Specificity*

Although Δ^5-androstene-3,17-dione is the commonly employed substrate for the isomerase reaction, a number of other Δ^5-3-ketosteroids may also serve as substrates (*1, 7, 11, 20, 24, 25*). Δ^5-Estrene-3,17-dione and Δ^5-pregnene-3,20-dione are both excellent substrates. A number of $\Delta^{5(10)}$-3-ketosteroids of the 19-nor series are also converted to their corresponding Δ^4-3-ketosteroids. The Michaelis constants for the 19-nor $\Delta^{5(10)}$ compounds are much lower than for the C_{19} steroids, but the rates of isomerization are less than 1% of that of Δ^5-androstene-3,17-dione (Table I). It was originally reported (*1*) that Δ^5-cholesten-3-one is isomerized at only a negligible rate by the bacterial isomerase. Jones and Wigfield (*24*) have pointed out that this observation is probably in error. Because of the very low solubilities of Δ^5-cholesten-3-one and other highly nonpolar steroids, these compounds exist in aqueous solutions in micellar form. By using increasing concentrations of methanol in the reaction system,

TABLE I
Michaelis Constants for Substrates and Competitive Inhibitors of Δ^5-3-Ketosteroid Isomerase of *P. testosteroni*[a]

Compound	pH	Relative V_{max}	K_m (μM)	K_i (μM)	Ref.
Δ^5-Androstene-3,17-dione	7.0	100	320		(*5*)
17β-Hydroxy-$\Delta^{5(10)}$-estren-3-one	7.0	0.25	23		(*5*)
	4.2	0.06	40		(*9*)
	9.4	0.15	38		(*9*)
17β-Estradiol	7.0			10	(*9*)
19-Nortestosterone	7.0			5.2	(*9*)
17β-Dihydroequilenin	7.0			6.3	(*9*)

[a] All determinations were made as described by Kawahara *et al.* (*7*) in 33 mM potassium phosphate, pH 7.0; in 33 mM sodium citrate buffer, pH 4.2; or in 33 mM tris-acetate buffer, pH 9.4.

24. J. B. Jones and D. C. Wigfield, *Can. J. Chem.* **46**, 1459 (1968).

Jones and Wigfield (*24*) have shown that the kinetics of acid and base-catalyzed isomerizations of Δ^5-3-ketosteroids with nonpolar substituents at C-17 can be converted from complex kinetics to strictly first-order processes and that compounds such as Δ^5-androsten-3-one and Δ^5-cholesten-3-one become efficient substrates for the isomerase under these conditions. The differences in apparent substrate efficiency of a series of substituted Δ^5-3-ketosteroids may thus reflect the states of aggregation of the steroids in aqueous systems rather than variations in the intrinsic substrate activity. The presence of substituents (–OH, $-CH_3$, –C≡CH) on the α face of the steroid at C-17 does, however, result in a modest reduction of the rate of isomerization. According to Jones and Wigfield (*24*) the absence of a substituent at C-17 or the presence at C-17β of hydroxyl, methyl, ethyl, 2-propyl, or octyl substituents leads to substrates of comparable activity if measurements are carried out in systems designed to assure the dissociation of steroidal micelles.

2. *Effects of Metal Ions, Chelating Agents, and Urea*

The isomerase is quite insensitive to the action of heavy metal ions and chelating agents (*1, 4*). When added directly to the assay system, ethylenediamine tetraacetate, α,α'-dipyridyl, *o*-phenanthroline, and 8-hydroxyquioline are without effect on the enzymic activity. Urea is a noncompetitive inhibitor, and its action is at least in part reversible (*7*). Guanidine (6 *M*) is also a reversible inhibitor of isomerase activity. Since the isomerase contains no sulfhydryl groups, it is not surprising that sulfhydryl reagents are without effect on enzymic activity.

The enzyme is precipitated from ethanol–water mixtures by 5 m*M* Mg^{2+}, Ca^{2+}, and Zn^{2+}, although these cations are without effect on the enzyme activity in aqueous solutions (*4*). Of the three cations mentioned, Mg^{2+} appears to be the most selective precipitant of the isomerase and is therefore used in the purification procedure.

3. *Competitive Inhibitors*

A number of phenolic and neutral steroids are efficiently bound to the isomerase and behave kinetically as competitive inhibitors (*4, 7, 11, 20*). The most powerful inhibitors (Table I) are those lacking an angular C-19 methyl group, e.g., 19-nortestosterone ($K_i = 5.2\ \mu M$), 17β-estradiol ($K_i = 10\ \mu M$); and 17β-dihydroequilenin ($K_i = 6.3\ \mu M$). Falcoz-Kelly *et al.* (*20*) have pointed out that the K_m and K_i values for substrates and inhibitors, respectively, vary profoundly with the concentration of methanol in the assay cuvette. Under conditions similar to those used by other workers, Falcoz-Kelly *et al.* (*20*) reported the following K_i values (3.3%

methanol): 19-nortestosterone, 13 μM; Δ^4-estrene-3,17-dione, 2 μM; progesterone, 6.4 μM; and 19-norprogesterone, 0.5 μM. Neville and Engel (*26*) and Goldman (*27*) have described the competitive inhibition of the bacterial isomerase by 2α-cyano-4,4,17α-trimethyl-Δ^5-androsten-17β-ol-3-one ($K_i = 0.4$ μM) and 2-hydroxymethyleneandrostan-17β-ol-3-one ($K_i = 1.5$ μM).

4. *Irreversible Active-Site-Directed Inhibitor*

In an effort to delineate the topography of the active site of the isomerase, Büki *et al.* (*28*) have recently discovered that 6β-bromotestosterone acetate is an active-site-directed irreversible inhibitor which appears to alkylate the enzyme. The 6β-bromotestosterone acetate was prepared and carefully characterized with respect to chemical identity and stereochemistry. In experiments of brief duration, this compound inhibited the isomerization of Δ^5-androstene-3,17-dione when the inhibitor was added directly to the assay system. This inhibition, which could be overcome by raising the concentration of substrate, followed strictly competitive kinetics and gave a K_i value of 57 μM for 6β-bromotestosterone acetate (the K_m for Δ^5-androstene-3,17-dione was 330 μM in the same experiment). Upon prolonged incubation with 6β-bromotestosterone acetate, the enzyme became irreversibly inactivated. The rate of this process was dependent upon the concentration of the reagent and markedly accelerated as the pH of the reaction system was raised. At a molar ratio of inhibitor to enzyme of 40:1, at pH 8.0 and 25°, the enzyme was almost totally inactivated in 24 hr, whereas 70% of the activity was retained if the pH was lowered to 5.8. Support for the view that the inhibitor binds to the active site of the enzyme was provided by the finding that 19-nortestosterone protected the enzyme against inactivation. When 6β-bromotestosterone 17-[^{3}H]-acetate was used as an inhibitor, the radioactivity became firmly attached to the protein suggesting covalent binding. The inactivation was accompanied by the formation of at least three radioactive protein species which could be separated by polyacrylamide gel electrophoresis. Some of these components contained more than one mole of steroid per mole of protein. At higher levels of substitution the protein migrated more rapidly than the native enzyme in the electrophoretic system, possibly because of dissociation into subunits.

25. W. R. Nes, E. Loeser, R. Kirdani, and J. Marsh, *Tetrahedron* **19,** 299 (1963).
26. A. M. Neville and L. L. Engel, *J. Clin. Endocrinol. Metab.* **28,** 49 (1968).
27. A. S. Goldman, *J. Clin. Endocrinol. Metab.* **28,** 1539 (1968).
28. K. G. Büki, C. H. Robinson, and P. Talalay, *BBA* **242,** 268 (1971).

5. *Effect of Hydrogen Ions*

The bacterial isomerase is relatively insensitive to changes in pH of the medium, and the pH-activity curve is quite flat between pH 6 and 9 when either Δ^5-androstene-3,17-dione or 17β-hydroxy-$\Delta^{5(10)}$-estren-3-one were used as substrates. Outside of these limits, the spontaneous rate of isomerization of Δ^5-androstene-3,17-dione increases quite rapidly and this complicates measurement of the enzymic reaction (*11*). The $\Delta^{5(10)}$ compounds are much more stable. The pH-activity profile reveals that the activity is reduced to 50% of maximum at pH 4.2–4.5 on the acid side and at pH 9.5–9.7 on the basic side.

6. *Thermodynamic Parameters*

When the bacterial isomerase was first obtained in crystalline form, it was noted that this protein was endowed with remarkably high catalytic activity. Although maximum velocities could not be attained experimentally with Δ^5-androstene-3,17-dione because of the limited solubility of this substrate, extrapolation to V_{max} gave a molecular activity of 1.73×10^7 moles of substrate converted per mole of protein (MW 40,800) per minute at 25° and pH 7.0. Since there are probably three binding sites for steroid per mole of enzyme, but the number of catalytic sites is not known, recalculation of this activity value per catalytic site may be necessary. Nevertheless, even allowing for such corrections, it would appear that the isomerase is the most active catalyst known and promotes its reaction at maximal velocity in perhaps 3 μsec or less. Efforts to explain this remarkable catalytic power have been made on thermodynamic grounds, through a comparison of the rates of isomerization catalyzed by acid, base, and enzyme, as a function of temperature. Jones and Wigfield (*29*) have pointed out that under selected conditions, the product and the rate limiting step are probably the same for all of these reactions, lending validity to this type of experimental approach. The results of these experiments (Table II) suggest that the great facility of the enzymic process may be correlated with its extremely low enthalpy of activation of 5.0 kcal mole^{-1} since the entropy of activation is not greatly different among the acid-, base-, and enzyme-catalyzed processes. Jones and Wigfield (*29*) pointed out that the low enthalpy value indicates that the rate determining step in the reaction, which is the C-4β-H bond cleavage (see Section II,C,1), is accompanied by an energetically favorable compensating bond formation and that the entropy of activa-

29. J. B. Jones and D. C. Wigfield, *JACS* **89**, 5294 (1967).

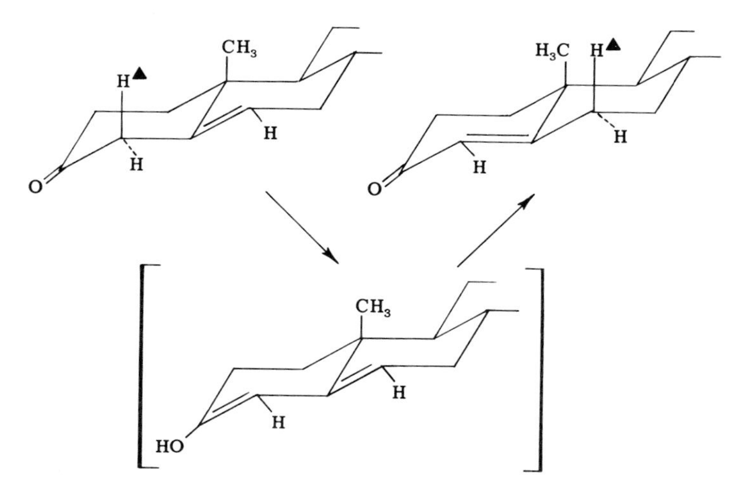

FIG. 2. Partial formulas showing the *cis,cis* diaxial proton transfer (▲) involved in the enzymic isomerization of a Δ^5-3-ketosteroid.

protonation of the anion is favored at C-4 whereas that of the neutral enol is favored at C-6.

Further strong evidence for the occurrence of an enolic intermediate in the isomerization reaction was provided by isotope exchange studies (*11*). The enzymic isomerization of Δ^5-3-ketosteroids in media containing labeled protons occurs essentially without incorporation of isotope from the medium into the product, as already noted above. Yet, when several α,β-unsaturated steroids such as testosterone, 19-nortestosterone, and Δ^4-androstene-3,17-dione were incubated in tritiated water in the presence of relatively large quantities of crystalline isomerase, the reisolated Δ^4-3-ketosteroids were found to have undergone an enzyme-dependent incorporation of isotope (*11*). Although no efforts were made to maximize the exchange reaction, as much as 1–2 atoms of tritium were incorporated into the product under some conditions. The precise locations of the isotope in the steroid molecules have not been established; however, almost the entire quantity of isotope could be removed under enolizing conditions. These findings support the participation of an enolic intermediate in the isomerization reaction and provide indirect evidence for at least partial reversibility of the enzymic reaction.

2. *Ultraviolet Absorption and Fluorescence Spectra of Steroid–Isomerase Complexes*

a. Complexes with 19-Nortestosterone. When the competitive inhibitor, 19-nortestosterone (III), is mixed with stoichiometric quantities of

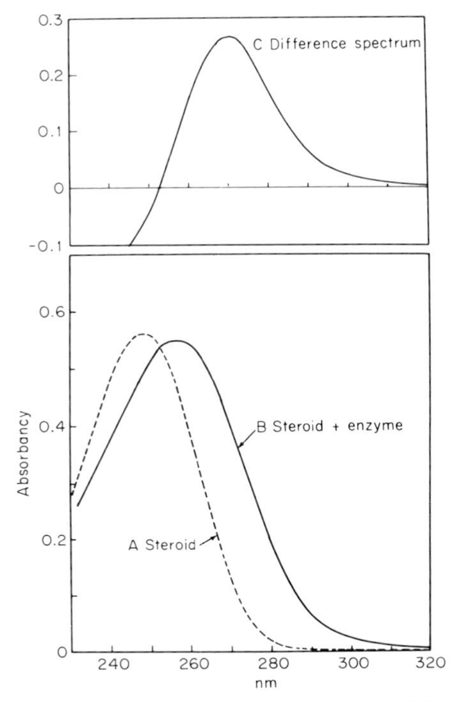

FIG. 3. Ultraviolet absorption spectra showing the combination of 19-nortestosterone with crystalline Δ^5-3-ketosteroid isomerase of *P. testosteroni*. Curve A, 19-nortestosterone (33.0 μM) measured against a buffer blank. Curve B, a mixture of 19-nortestosterone (33.0 μM) and crystalline isomerase (18.3 μM) measured against a buffer blank containing isomerase (18.3 μM). Curve C, the calculated difference spectrum between B and A. All cuvettes contained 0.3 μmole of potassium phosphate, pH 7.0, and 0.01 ml of methanol in a final volume of 0.31 ml [reproduced with permission from Wang *et al.* (*11*)].

crystalline isomerase, the typical ultraviolet absorption spectrum characteristic of the α,β-unsaturated ketone becomes modified. The absorption maximum of the steroid in neutral aqueous solutions is displaced from 248 to 258 nm, and the difference spectrum between the enzyme–steroid mixture and the sums of the individual absorptions of the steroid and the enzyme has a maximum at 270 nm (Fig. 3). A spectral titration of a constant quantity of isomerase with increasing amounts of 19-nortestosterone is shown in Fig. 4. The absorbancy at 270 nm (or 280 nm) rises to a maximum as the enzyme becomes fully saturated with steroid.

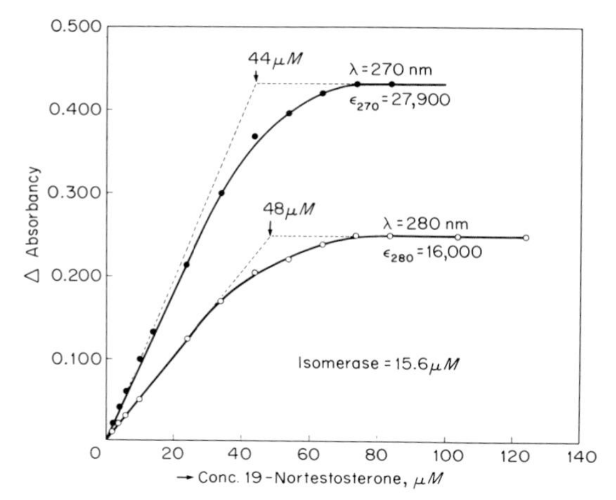

FIG. 4. Spectrophotometric titration of crystalline Δ^5-3-ketosteroid isomerase with 19-nortestosterone. The absorbance changes are shown at 270 and 280 nm as increasing quantities of 19-nortestosterone are added to a constant amount of isomerase and the absorbance of the mixture is measured against a blank containing the equivalent concentration of steroid alone. Corrections have been made for the absorption of the enzyme itself and for volume changes. All the cuvettes contained 1.0 m*M* potassium phosphate buffer, pH 7.0. The isomerase concentration was 15.6 μM [from Wang *et al.* (*11*)].

OH

O

(III)

OH

HO

(IV)

OH

HO

(V)

The molar equivalence point for this interaction, based on a molecular weight of 40,800, is almost exactly 3 moles of steroid per mole of enzyme. From such spectral measurements, the concentration of the complex was calculated at successively increasing concentrations of steroid. If binding sites of equal affinity are postulated, a single value for dissociation constant is obtained only if three binding sites are assumed to be present, and the following relation holds:

$$K = \frac{(E_T - ES_3)(S_T - 3\ ES_3)}{(ES_3)}$$

where E_T = total enzyme concentration, S_T = total steroid concentration, and ES_3 = concentration of complex containing 3 moles of steroid. The average value for K for 19-nortestosterone was found to be 1.64 μM which is close to one-third the value of 5.2 μM obtained for the K_i for 19-nortestosterone (Table I).

The displacement of the absorption maximum of the α,β-unsaturated ketone upon interaction with the isomerase was interpreted on the basis of the formation of an enolic-type structure (*11*). The fact that 19-nortestosterone undergoes isotope exchange with protons of the medium in the presence of the enzyme is consonant with this view.

b. Complexes with Aromatic Steroids. The formation of stoichiometric complexes between isomerase and certain phenolic steroids such as 17β-estradiol (IV) and 17β-dihydroequilenin (V) is associated with profound changes in the absorption spectra which are consistent with enolizations of the steroids (*11*).

The ultraviolet absorption spectrum of 17β-estradiol at pH 7.0 shows the characteristics of a phenol with a high intensity band at 220 nm and a double band of much lower intensity in the 280- to 290-nm region. In basic solution, both bands are intensified and shifted toward longer wavelengths (238 and 295 nm, respectively) (Fig. 5). The spectrum of a mixture of isomerase and 17β-estradiol measured against an equivalent concentration of isomerase resembles that of the ionized form of 17β-estradiol. Both major absorption bands are intensified and shifted toward longer wavelengths (Fig. 5).

An even more striking demonstration of these spectral shifts is obtained with 17β-dihydroequilenin (V) in which rings A and B are both aromatic. The difference spectrum of the complex between isomerase and 17β-dihydroequilenin and the individual components of the mixture shows bands that are closely similar in position to those of the difference spectrum between the ionized and un-ionized forms of the steroid (Fig. 6). These difference spectra have relatively sharp absorption maxima near

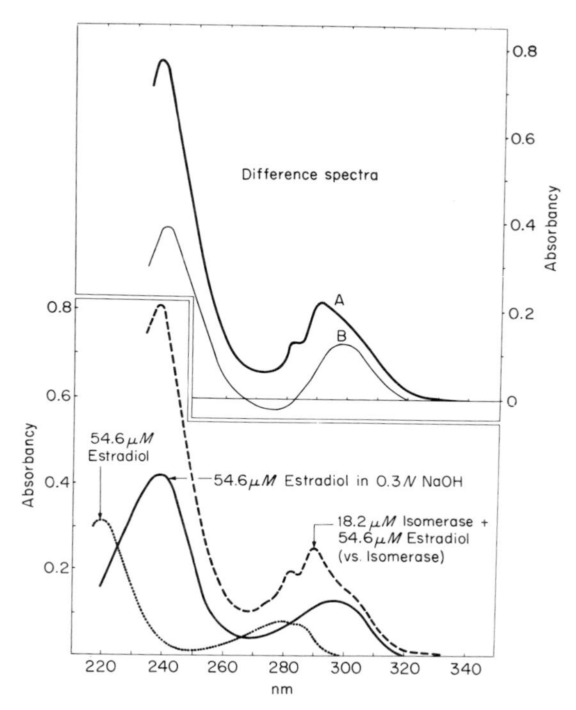

FIG. 5. Ultraviolet absorption spectra showing the interaction of 17β-estradiol with crystalline Δ^5-3-ketosteroid isomerase of *P. testosteroni*. Lower panel: Spectra of 17β-estradiol (54.6 μ*M*) in 1 m*M* potassium phosphate, pH 7.0 (· · ·), and in 0.3 *N* NaOH (——), measured against blanks containing phosphate buffer and NaOH, respectively. The spectrum of a mixture of isomerase (18.2 μ*M*) and 17β-estradiol (54.6 μ*M*) was measured in 1 m*M* potassium phosphate, pH 7.0 (- - -), against a blank containing enzyme only. Upper panel: The calculated difference spectra of: A, the absorption of the enzyme–steroid mixture from which the individual absorptions of enzyme and steroid have been subtracted; and, B, the spectra of 17β-estradiol in basic and neutral solutions [from Wang *et al.* (*11*)].

274, 285, and 297 nm, a broad band in the 355–360-nm region, and a trough near 310 nm.

Further information on the mechanism of the interaction of 17β-dihydroequilenin with the enzyme was obtained by fluorescence spectroscopy. When a neutral aqueous solution of 17β-dihydroequilenin was activated at 292 nm, a sharp fluorescent band was observed at 370 nm, whereas in 0.1 *N* NaOH this band was intensified and displaced to 425 nm. A solution (pH 7.0) that was 2 μ*M* with respect to both isomerase and the steroid revealed a fluorescence spectrum with a principal maximum at 410 nm and an inflection at 270–275 nm. These findings suggest

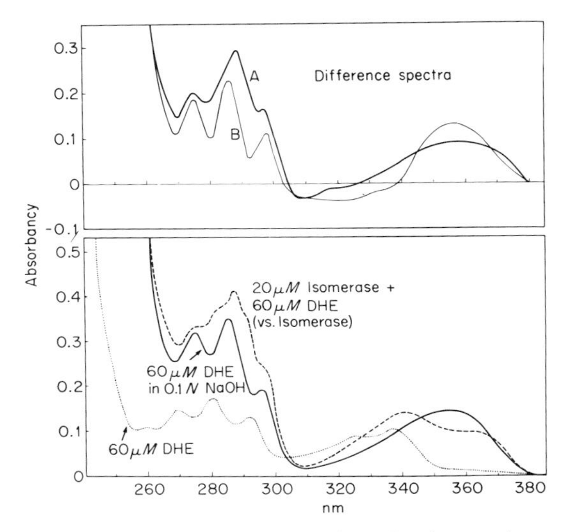

FIG. 6. Ultraviolet absorption spectra showing the interaction of 17β-dihydroequilenin with crystalline Δ^5-3-ketosteroid isomerase of *P. testosteroni*. Lower panel: Spectra of 17β-dihydroequilenin (60 μM) in 1 mM potassium phosphate, pH 7.0 (· · ·) and in 0.1 N NaOH (——). The spectrum of a mixture of 17β-dihydroequilenin, 60 μM, and isomerase, 20 μM (- - -) was measured against a blank containing enzyme only. Upper panel: Calculated difference spectra of: A, the absorption of the enzyme–steroid mixture from which the individual absorptions of enzyme and steroid have been subtracted; and, B, 17β-dihydroequilenin in basic and neutral solutions [from Wang *et al.* (*11*)].

that the enzyme–steroid mixture contains forms of 17β-dihydroequilenin that resemble both the un-ionized and ionized forms of the steroid.

The remarkable changes in the absorption spectra of the two phenolic steroids (17β-estradiol and 17β-dihydroequilenin) on interaction with the enzyme and the alterations in the fluorescence emission spectrum of 17β-dihydroequilenin suggest that in a medium of neutral pH, the enzyme, on combination with these steroids, induces alterations in electron distribution that resemble the ionization of the phenols in basic solution.

3. *Prosposals for Catalytic Mechanism*

An ideal model for the enzymic isomerization of Δ^5-3-ketosteroids by the bacterial Δ^5-3-ketosteroid isomerase should be constructed with the following experimental findings in view:

(a) The reaction involves a stereospecific, direct, intramolecular (*cis, cis* or diaxial) proton transfer of the 4β-hydrogen to the 6β position.

(b) There is a very high primary isotope rate effect suggesting that cleavage of the 4β-hydrogen to carbon bond is the rate limiting step in the reaction.

(c) Direct spectral studies indicate that the interaction of Δ^4-3-ketosteroids or ring A phenolic steroids with the enzyme results in formation of enolic or phenolate anion type of structures.

(d) The enzyme promotes isotope exchange reactions between protons of the medium and Δ^4-3-ketosteroids into positions from which the label may be removed under enolizing conditions.

(e) Kinetic and stereochemical studies in close analogy to the purely chemically promoted isomerizations suggest the conversion of the Δ^5-3-ketone to the enol rather than the enolate anion.

(f) The low enthalpy of activation of the enzymic process in comparison to the acid- and base-catalyzed reactions.

(g) The critical nature of histidyl residues in the enzymic process.

(h) The relatively flat activity curve between the pH values of 6 and 9, with 50% activity occurring at pH 4.2–4.5 and 9.4–9.7.

(i) The enormous catalytic activity of the enzyme.

Malhotra and Ringold (*31, 35*) and Talalay (*13, 14*) proposed the existence on the enzyme of a proton donor (AH) and proton acceptor (B) group. The mechanism of the reaction as visualized by Malhotra and Ringold (*31*) is shown in Fig. 7, and described by these authors as follows:

> Coordination of AH with the steroid carbonyl by hydrogen bonding or full hydrogen transfer concomitant with or followed by abstraction of the C-4β proton by B, yields the enzyme–enol intermediate in a low energy push–pull type mechanism. Protonation at the 6β-position by BH, which can occur with a minimum of bond stretching due to the diaxial nature of the transfer, yields the protonated α,β-unsaturated ketone. Removal of the carbonyl proton then gives the α,β-unsaturated ketone and regenerates the enzyme. The high isotope effect which accompanies the introduction of a deuterium atom at the 4β-position can theoretically arise either in the enolization step or in the reprotonation step since deuterium is transferred in both steps. In common with the chemical mechanism and on the assumption that the enzyme-enol intermediate will be the highest energy complex present, we favor loss of the C-4 proton as the rate-determining step. The assumption that it is the enol rather than the enolate anion which undergoes protonation stems from the strictly chemical finding that C-protonation of the anion is favored at C-4 rather than C-6. The requirement that BH or BD does not exchange with protons or deuterium of the medium is readily explicable on the basis of a very fast reprotonation step and the probability that in the enzyme–enol complex BH is shielded from solvent. It might in fact be suggested that the extremely fast rate of this enolization and reprotonation depends upon reaction occurring in a hydrophobic region.

35. H. J. Ringold and S. K. Malhotra, *Abstr. 6th Intern. Congr. Biochem., New York, 1964* Vol. IV, p. 329.

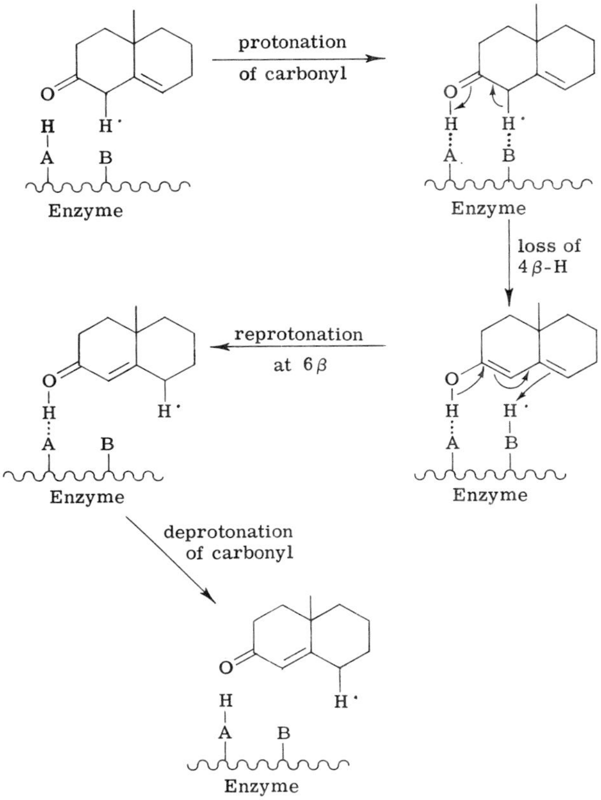

FIG. 7. General mechanistic model proposed for the Δ^5-3-ketosteroid isomerase reaction involving unspecified proton donor (AH) and acceptor (B) groups on the enzyme and an enolic intermediate [from Malhotra and Ringold (*31*)].

The key question as to the chemical nature of the groups AH and B remains speculative at the moment. Talalay (*13, 14*) has proposed that the nitrogen atoms of the tautomeric forms of a single imidazole residue could function alternately as proton donor and acceptor species in the manner shown in Fig. 8. A mechanism of this type would be consistent with the essential nature of a histidyl residue and the pH-activity profile. This proposal is not considered attractive by Jones and Wigfield (*36*) on the grounds that imidazole itself has negligible catalytic activity. These authors further question whether the geometry of the transition state would be favorable for the requisite proton transfer from C-4β to C-6β.

36. J. B. Jones and D. C. Wigfield, *Can. J. Chem.* **47**, 4459 (1969).

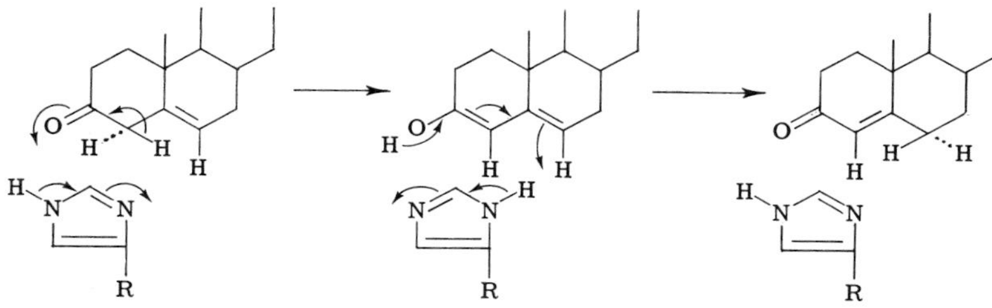

FIG. 8. Mechanistic model proposed for the Δ^5-3-ketosteroid isomerase reaction involving the participation of the two nitrogen atoms of the imidazole ring of histidine, acting alternately as proton donor and acceptor [from Talalay (*13, 14*)].

Jones and Wigfield (*36*) invoke the participation of the phenolic group of tyrosine as a proton donor (AH) responsible for the protonation of the carbonyl group of the substrate and suggest that a nitrogen of imidazole is the proton acceptor (B) and transferring agent. Further speculation on the precise nature of the chemical groups involved in the catalysis is clearly fruitless at this time since the problem must be solved by experimental means for which the tools are at hand now that the primary structure of the enzyme is known.

III. Δ^5-3-Ketosteroid Isomerases of Animal Tissues

A. DISTRIBUTION AND PROPERTIES

In contrast to the detailed state of understanding of the Δ^5-3-ketosteroid isomerase reaction of *P. testosteroni*, comparatively little information is available on the isofunctional enzymes in animal tissues. Two reasons account for this disparity. The mammalian enzymic activities are quite feeble in whole tissue extracts, and most, if not all, of the enzymic activity is firmly bound to the endoplasmic reticulum. Consequently, the majority of studies have been carried out with such crude preparations that little unequivocal information on specificity and mechanism has been elicited.

The widespread presence of Δ^5-3-ketosteroid isomerase activity in animal tissues was recognized many years ago, and it was noted that the highest enzymic activities were present in adrenal, ovary, and testis as well as the liver of the rat (*1, 4, 37, 38*). Most of the more recent

37. P. Talalay, *Physiol. Rev.* **37**, 362 (1957).
38. H. L. Krüskemper, E. Forchielli, and H. J. Ringold, *Steroids* **3**, 295 (1964).

studies have been conducted with preparations of adrenal glands of rat (*38, 39*), beef (*38, 40*), and sheep (*33*), as well as with human placenta (*41*).

Reports from two laboratories suggest that isomerase activity is distributed among the soluble, mitochondrial, and microsomal fractions of beef adrenal cortex (*38, 42*). Murota *et al.* (*43*), however, could find almost no Δ^5-3-ketosteroid isomerase activity (for Δ^5-androstene-3,17-dione) in the soluble fraction. Evidence obtained from velocity measurements with substrate mixtures (*38*) and from fractionation experiments (*44*) points to the existence in bovine adrenal cortex preparations of different enzymes concerned with the isomerization of Δ^5-androstene-3,17-dione and Δ^5-pregnene-3,20-dione. It is claimed that these activities are separable. In addition, Alfsen and her colleagues (*45*) claim to have fractionated beef adrenal cortex to obtain a preparation that isomerizes Δ^5-cholesten-3-one but is inactive with respect to C_{19} and C_{21} Δ^5-3-ketosteroids.

Ward and Engel (*33*) have provided evidence for the reversal of the Δ^5-3-ketosteroid isomerase reaction in acetone powders of sheep adrenal microsomes. Reversibility depended upon the simultaneous presence of a pyridine nucleotide-linked hydroxysteroid dehydrogenase which reduced the ketonic group and thus displaced the normally unfavorable equilibrium. These preparations converted Δ^4-androstene-3,17-dione to 3β-hydroxy-Δ^4-androsten-17-one and 3β-hydroxy-Δ^5-androsten-17-one.

Oleinick and Koritz (*39*) have described an interesting property of the isomerase of the beef adrenal microsome fraction. The enzymic activity of these preparations was greatly augmented in a rather specific manner by low concentrations of NAD and NADH. Both of these nucleotides also protected the enzyme against inactivation by various agents. A careful study of this phenomenon led to the tentative conclusion that the nicotinamide nucleotides are modifiers which bind at a noncatalytic site(s) and that the behavior of these enzyme preparations is compatible with the presence of an allosteric protein.

It is of considerable interest that the Δ^5-3-ketosteroid isomerase of beef

39. N. L. Oleinick and S. B. Koritz, *Biochemistry* **5,** 715 and 3400 (1966).
40. W. Ewald, H. Werbin, and I. L. Chaikoff, *BBA* **81,** 199 (1964); *Steroids* **4,** 759 (1964).
41. G. M. Segal, I. V. Torgov, and T. S. Fradkina, *5th Intern. Sympos. Chem. Nat. Prod., IUPAC, London, 1968* Abstract D14; I. V. Torgov, G. M. Segal, and T. S. Fradkina, personal communication (1968).
42. W. Ewald, H. Werbin, and I. L. Chaikoff, *Steroids* **3,** 505 (1964).
43. S. Murota, C. C. Fenselau, and P. Talalay, *Steroids* **17,** 25 (1971).
44. K. Grėen and B. Samuelsson, *JBC* **239,** 2804 (1964).
45. A. Alfsen, E.-E. Baulieu, and M.-J. Claquin, *BBRC* **20,** 251 (1965).

adrenal is powerfully and competitively inhibited by 2α-cyano-4,4,17α-trimethyl-Δ^5-androsten-17β-ol-3-one ($K_i = 0.6\ \mu M$) and by 2-hydroxymethyleneandrostan-17β-ol-3-one ($K_i = 0.08\ \mu M$) (*27*). Both of these synthetic steroids block the biosynthesis of adrenal corticosteroids and cause the development of pseudohermaphrodite states in experimental animals.

B. Catalytic Mechanism

Several groups of investigators have examined the mechanism of the isomerase reaction in animal tissues in order to determine if the reaction is similar to that established for the isomerase of *P. testosteroni*. These studies have been carried out with isotopes of hydrogen and with systems varying in complexity from whole animals to enzyme fractions.

From studies of the conversion of [3α-^{3}H,4-^{14}C]- and [4β-^{3}H,4-^{14}C]-cholesterol to bile acids in bile fistula rats, Gréen and Samuelsson (*44*) concluded that these reactions involved the stereospecific transfer of hydrogen from the 4β to the 6β position during the isomerization of the olefinic double bond.

Werbin and Chaikoff (*46*) fed a mixture of [4-^{14}C]- and [4β-^{3}H]-cholesterol to guinea pigs and found that the excreted [4-^{14}C]- cortisol and 6β-[4-^{14}C]- hydroxycortisol contained negligible quantities of tritium. Similarly, washed acetone powders of beef adrenal mitochondria converted the above doubly labeled cholesterol mixture to doubly labeled 3β-hydroxy-Δ^5-pregnen-20-one, but the progesterone formed under the same conditions contained negligible quantities of ^{3}H (*46*). These experiments were interpreted as showing that the isomerase reaction did not involve transfer of the 4β proton from C-4 to C-6. Fukushima *et al.* (*47*) carried out the isomerization of [4-^{14}C, 4β-^{3}H]- Δ^5-androstene-3,17-dione to Δ^4-androstene-3,17-dione with beef adrenal microsomes and rat liver 100,000*g* supernatant fluid. Marked losses of the tritium at 4β (73% loss with beef adrenal microsomes and 53% with rat liver supernatant) were obtained in these conversions. Moreover, the residual tritium was located at C-4 rather than C-6, and Fukushima *et al.* (*47*) suggested the possibility that a removal of the 4α proton might be occurring during the isomerization.

Torgov *et al.* (*41*) studied the isomerization of [4β-^{2}H]-Δ^5-androstene-3,17-dione (0.5 atom) to the Δ^4-3-ketone with centrifuged preparations

46. H. Werbin and I. L. Chaikoff, *BBA* **71,** 471 (1963); **82,** 581 (1964).
47. D. K. Fukushima, H. L. Bradlow, T. Yamauchi, A. Yagi, and D. Koerner, *Steroids* **11,** 541 (1968).

of human term placenta homogenates. According to Torgov *et al.* (*41*) about one-half of the deuterium was lost during the course of this conversion by an exchange reaction with the medium, and the authors concluded that the 4α proton was removed during the isomerization. In other experiments conducted with rat adrenal microsome preparations in a medium labeled with D_2O, Oleinick and Koritz (*39*) noted that the conversion of Δ^5-androstene-3,17-dione to Δ^4-androstene-3,17-dione involved the incorporation of deuterium at both C-4 and C-6. Sih and Whitlock (*48*) have pointed out that the experiments of Oleinick and Koritz (*39*) are difficult to interpret since even the nonenzymic isomerization in D_2O at pH 7.0 resulted in isotope incorporation from the medium into both C-4 and C-6 positions, and it cannot be concluded that the isomerase reaction is responsible for the observed isotope incorporation.

In the light of these conflicting results, Murota *et al.* (*43*) reexamined the isomerization of Δ^5-androstene-3,17-dione by fractionated beef adrenal membranes in a medium of D_2O. Under carefully controlled conditions these workers could find no evidence for enzymic incorporation of isotope from the medium into the product, Δ^4-androstene-3,17-dione. Murota *et al.* (*43*) also pointed out the many difficulties involved in these types of experiments with crude enzyme preparations (isotope exchange reactions catalyzed by the isomerase, interfering unrelated reactions, and nonenzymic reactions).

Although definitive information on the mechanism of the adrenal Δ^5-3-ketosteroid isomerase reaction is not available, it would be premature to conclude that a direct proton transfer from C-4 to C-6 does not occur in these systems.

Acknowledgments

The authors wish to express their appreciation to Dr. C. H. Robinson for many helpful and enlightening discussions. Studies from the laboratory of the authors and the preparation of this chapter were supported in part by the National Institutes of Health (Grants AM 07422 and GM 1183) and by the Gustavus and Louise Pfeiffer Foundation of New York.

48. C. J. Sih and H. W. Whitlock, Jr., *Ann. Rev. Biochem.* **37**, 661 (1968).

Author Index

Numbers in parentheses are reference numbers and indicate that an author's work is referred to, although his name is not cited in the text.

A

B

C

D

E

F

G

K

M

O

P

Q

R

S

U

Subject Index

F

H

I

K

L

M

N